Nanostructured Materials for Next-Generation Energy Storage and Conversion

Qiang Zhen • Sajid Bashir
Jingbo Louise Liu
Editors

Nanostructured Materials for Next-Generation Energy Storage and Conversion

Advanced Battery and Supercapacitors

With 186 Figures and 23 Tables

Springer

Editors
Qiang Zhen
Nano-Science and Nano-Technology
Shanghai University
Shanghai, China

Sajid Bashir
Department of Chemistry
Texas A&M University Kingsville
Kingsville, TX, USA

Jingbo Louise Liu
Department of Chemistry
Texas A&M University Kingsville
Kingsville, TX, USA

ISBN 978-3-662-58673-0 ISBN 978-3-662-58675-4 (eBook)
https://doi.org/10.1007/978-3-662-58675-4

This Springer imprint is published by the registered company Springer-Verlag GmbH, DE, part of Springer Nature.
The registered company address is: Heidelberger Platz 3, 14197 Berlin, Germany

Preface to Volume III on Batteries

Benjamin Franklin used a battery to describe an array of charged glass plates in 1748 and within 90 years, the advent of electric doorbells, telegraphs, and telephones required a backup energy source. John Daniell (1836) developed primary battery in the form of a cell which could deliver 1.1 V and specific energy less than 0.1 MJ/kg to power these devices and the demand for greater output never ceased. Almost 115 years later, Lew Urry developed (1949) the Alkaline Manganese Battery with 1.25 V and a specific energy around 0.4 MJ/kg. Almost 20 years later (in 1971) Alexandr Kloss and Boris Tsenter implemented nickel and hydrogen to form the precursor of the Nickel Metal Hydride battery, also able to deliver 1.2 V but at a higher specific energy (~140 MJ/kg). Lastly, the work by Akira Yoshino of Asahi Chemical led to the lithium-ion battery at 1.5 V and a specific energy density of 0.8 MJ/kg. This was rapidly followed by the lithium polymer battery which had a flexible casing and enabled the form factor to be altered. Portable electronics could be used for longer periods, including cell phones, laptops, and multimedia devices, such as tablets, ipads, nanoPCs, and wearable electronics, increasing lithium-ion battery sales past the billion units sold mark.

The advantage of the lithium-ion rechargeable batteries as secondary batteries (LIB) relative to nickel-cadmium or nickel-hydride batteries is the endurance of the battery coupled with high specific energy density and no memory effects. These storage devices make up the top three choices, after which manganese dioxide-zinc and lead acid are the next most common storage device. Unlike the other metals, lithium has distinct advantages. It has a low atomic mass, low ionization enthalpy, and high electrode potential, resulting in high energy density relative to zinc-based batteries.

Lithium-ion batteries are a system where the lithium metal forms the anode with a nonaqueous electrolyte such as propylene carbonate-lithium perchlorate and carbon-based cathode. The lithium at the anode forms whisker-like formations which resemble dendrites, causing the battery to short-circuit, in addition to safety issues related to a strong oxidizer such as perchlorate. Other concerns relate to replacement of lithium aluminum and/or polyacene conducting polymer used for memory backup or powering small devices with solar recharge, which may be less efficient and more costly. The "current" lithium-ion batteries do not have lithium metal but intercalation of lithium ions into the active electrode surfaces, using the Asahi design, which uses the carbon-based anode and lithium cobalt oxide as the cathode. During charging, lithium would be inserted into the carbon anode, and during discharging, lithium would be extracted. The battery design due to conducting polymers could be cylindrical and was 18 mm by 65 mm in length and known as 18650 type with a capacity of 800 mAh/g. During the decades, after World War II, the demands for portable energy have progressively increased, requiring design of a new generation of energy storage devices. Recently, increased battery usage and lifespan have been achieved by the substitution of carbon-based anodes with more flexible graphite-based carbon and changes to electrolyte and modification of lithium cobalt oxides and cell design (polymers, binders, and internal circuitry) relative to lithium only anodes. The early batteries contained mesophase carbon microbead (MCMB with a capacity of 280 mAh/g). By incorporating functional electrolytes that passivate the electrode, higher graphitization of the anode carbon material can be achieved without solvent decomposition, leading to greater catalyticity at the active electrode surface. This graphitization could be accomplished using mesophase carbon fiber, or more active form of MCMBs yielding capacity of 372 $mAh.g^{-1}$. The addition of cyclohexylbenzene would lead to the generation of hydrogen gas at higher voltages and would interrupt the current output, safely shutting down the battery. In the early 2000s, the 18650 type cell had a capacity of 2.4 Ah or an energy density of 200 Wh/kg, which currently have increased to 3.6 Ah and $\sim$ 250 Wh/kg. The passivation of the active surfaces through the introduction of electrolyte additives has enabled the higher energy densities to be reached. Thus, a conductive membrane protects the positive active material in a similar manner to the formation of a solid electrolyte interface to protect the anode; comparison is given below:

Primary cell →	Alkaline	Lithium iron disulfide ($LiFeS_2$)	Lithium manganese dioxide ($LiMnO_2$)
Specific energy (Wh/kg)	200	300	280
Voltage (V)	1.5	1.5	3.3
Continuous voltage output	Low	Intermediate	Intermediate
Passivation layer	No, N/A	Yes, moderate	Yes, moderate
Operating temperature (OT, °C)	$0 < OT < 60$	$0 < OT < 60$	$-30 < OT < 60$
Shelf-life (SL, years)	10	15	$-10 < SL < 20$

Note: The cathode is carbonaceous material and the anode hosts the active material

To achieve even higher capacities, electrode conditioning is required. Here the cell undergoes several charge-discharge cycles at a low rate (1/4 C) to decompose the additives and form the protective film. The cell is held in their charged state for a week (e.g., 0.005 V vs. Li/Li^+ for 10 h, Li^+ possess ultrahigh capacity [3,860 $mAh\ g^{-1}$] and the very low standard negative electrochemical potential [−3.040 V]) to ensure completion of the protective film. In the absence of conditioning, lithium ions can intercalate into graphite (in propylene carbonate-based electrolytes) forming a large number of active sites. These sites are formed because of nonhomogeneous voltage distribution across the anode surface due to underpotential (0 V vs. Li/Li^+) during the Li+ intercalation. Areas of graphite which become exfoliated do not get

Table 1 Summary of the chemistries and configuration of most common secondary battery platforms

Secondary cell →	Lead acid	Ni-Cd	Ni MH	Li cobalt	Li manganese
Specific energy (SE, Wh/kg)	<50	<80	<120	150 < SE < 250	100 < SE < 150
Shelf-life at 80% depth of discharge (cycles)	300	1,000	500	1,000	1,000
Cell voltage (V)	2	1.2	1.2	3.6	3.7
Columbic efficiency (CE, %)	90	90	90	99	99
Decade or year introduced	The 1880s	The 1950s	1970–1990	1991	1996
Time to charge (h)	16	2	4	2	2
Operating charge temperature (OT, °C)	−20 < OT < 50	0 < OT < 45	0 < OT < 45	0 < OT < 45	0 < OT < 45

Note the charge and discharge temperature(s) are different; typically the battery can discharge at lower temperatures but should not be charged if frozen. Cycle life is based on the depth of discharge (DoD), where shallow DoD at moderate to cool temperatures will prolong cycle life.

Li^+ penetration, instead Li metal deposition occurs. This would decrease cell potential to 0.8 V and decrease the cell life cycle. The next stage of development for the 18650 type cells is the use of graphene, planar lithium-nickel-cobalt-aluminum, and nickel-manganese-cobalt cathode-type materials, in addition to new electrolyte additives to promote Li^+ intercalation, suppress Li metal deposition, and promote greater homogeneity across the electrode surface. With these advances, a capacity of 3.0 Ah may be possible by focusing on nano-carbonaceous materials, improved Li^+ intercalation, and use of polyanions other than oxides to increase cell voltages. These different chemistries and formulations are summarized in Table 1.

This book contains an overview of energy policy. Energy policy is critical to long-term planning and stability. While consumer demands related to home and travel account for 4% of the gross domestic product, they account for 28% and 41% of energy usage and CO_2 emissions, respectively. Electrical storage such as LIBs has a CO_2 “tax” associated with them. Only hydroelectric is CO_2-neutral, although fuel cells could fall into this category, LIB, in the long run, could assist in lowering the CO_2 emission profile when compared with electrical generation using coal (baseline comparison), which has been subjected to peak-load pricing and lead to increasing the efficiency of national energy resources, clean-coal development, as well as development of sustainable resources. Lithium-ion batteries have a role to play both as portable power and as storage, where electric vehicles could be used to power homes in emergencies such as earthquakes and hurricanes when the "centralized grid" fails. The role of national energy policy vis-à-vis LIB and energy storage including recycling of lithium-ion batteries are reviewed.

Other chapters are more traditional in offering an in-depth review of electrode materials, electrolytes, and additives for LIB, as well as indicators of the future directions for continued maturation of the LIB. Here, the authors will explicitly discuss advances in electrode design as the anatomy of the battery is the same as in Daniell's time of anode, cathode, electrolyte, and casing/binding. The authors discuss materials composites for forming newer more reactive electrode surfaces without side reactions or dendrite formation, with higher specific capacity, energy densities, and cycling stability. The fabrication using more streamlined procedures where appropriate is discussed, including newer materials, such as antimony trioxide, chalcogens, and titanium/molybdenum disulfide as high potential anodes. Cathode materials based on nickel cobalt manganese oxide have been employed to minimize cation dissolution of the cathode materials. Other approaches include the use of layered oxides, structured materials to encapsulation designed to reduce the degree of material dissociation, particularly after prolonged use. In addition, alternative strategies related to modification of carbon nanotubes, graphite, and graphene with dopants, passivation of electrodes by coating, or alloying to enhance Li-ion diffusivity are explored. The purposes of such strategies and approaches are to minimize lithium metal deposition and to improve electrode conductivity, ionic mobility, and battery capacity during charging and discharging cycles and use.

All-solid-state lithium-ion batteries' solid electrolytes, such as NASICON-type, perovskite-type, anti-perovskite type, garnet-type, sulfide-type, Li_3N-type, and other chemistries, require optimal ionic conductivity and electrochemical stability at room and higher operational temperatures. These performance characteristics are discussed with respect to changes in ion mobility at different temperature regimes. The current challenges related to electrolyte selection, refinement of the interfaces, and interfacial resistance are reviewed beyond LISICON- and LiPON-type of solid electrolytes.

Room temperature molten salts (ionic liquids) have drawn attention for electrochemical energy storage. These systems offer excellent thermal and electrochemical stability, exhibit low volatility, and are enabled as electrolytes in batteries such as LIBs, sodium-ion, lithium-air (O_2), and Li-sulfur batteries for clean energy storage. The issue of low volatility and low flammability is discussed as well as used in all-solid-state batteries to improve the conductivity of the solid electrolyte, in addition to modification of electrode materials to take advantage of the new physical and chemical properties.

The thermochemistry and modeling aspects are also discussed. Lithium-ion batteries' stability is typically evaluated by cycling under a galvanostatic regime ($\sim$1/4 C – C/2-rate) between 2.75 V and 4.2 V for up to 250–1,500 cycles. After each completed 25n (n = 1–4) cycles, the discharge capacity, capacity loss, and average discharge potential are evaluated. Then cells undergo a supplemental charge and discharge cycle at higher C/6 rate followed by thermodynamics measurements to determine the open-circuit potential (OCP), entropy, and enthalpy values. Calculations show that state functions of entropy and enthalpy exhibit higher sensitivity to morphological changes in the electrode surface or crystal structure, such as graphite carbon-based anodes and lithium cobalt oxide cathode through repeated cycling (aging) relative to changes in Gibbs free energy estimated from the state of charge or depth of discharge, and OCP values.

Coupled with ex situ X-ray diffractometry, electron microscopy, and Raman spectroscopy analyses on the electrode materials, the relationship between state functions, electrochemical output, crystal structure composition, and density of states can be assessed at both the anode and cathode during aging and inferences made using computational means of processes to stabilize the cathode and/or anode by increased graphitization during the aging process.

The role of non-lithium electrodes is also discussed as it is apparent that increased electric vehicle usage will deplete lithium and cobalt unless aggressive recycling, reuse, and transition to non-lithium are introduced at an early stage, as the demand for portable energy is likely to increase in the near term.

The general aim has been to learn how to assess the technical impact of new active materials and to increase charge density as well as operational reliability, cost, ease of use, and safety. Lastly, no book would be complete without the stellar contribution of the authors, their research group, and institutions. We also would like to extend our deep appreciation to our institution, colleagues, and students as well as the editorial staff at Springer who have strived to deliver the best possible scholarly product. As always, errors and omission are the responsibility of the editors, for which we as preface authors ask forgiveness.

We also wish to thank the Robert Welch Foundation (AC-0006) Award for assistance with part of the research.

Kingsville, USA
August 2019

Sajid Bashir
Nancy KingSanders
Jingbo Louise Liu

Contents

About the Editors

Qiang Zhen received his doctoral degree in the field of Metallurgical Physical Chemistry at the University of Science and Technology Beijing in 2000. Currently, he is a distinguished Professor of Research Center of Nano-Science and Nano-Technology, Shanghai University. He is the Director of Advanced Functional Materials Laboratory, Engineering Research Center of Material Composition and Advanced Dispersion Technology, Ministry of Education, China. He also serves as the Dean of the Science and Technology Development Academy of Shanghai University, is one of the executive members of the China Instrument Functional Materials Society Council, and is member of China Silicate Society. In 2002, he visited the Department of Mechanical Engineering, Hong Kong University, and then the French National Center for Scientific Research, University of Science and Technology in Lille, France, from 2005 to 2012. His research interests focus on solid electrolyte for fuel-cells and sensors, and design, fabrication, and characterization of micro-/nano-composites. He has directed more than 30 scientific research projects, supported by different funding agencies, such as National Natural Science Foundation of China (NSFC), NSFC International Cooperation Plan, and Excellent Talents Supporting Programs. He has published more than 60 papers as a leading coauthor and has reviewed journal papers and project proposals.

Sajid Bashir received his Ph.D. in Analytical Chemistry from the University of Warwick, England, 2001. He was a postgraduate research associate at Cornell University to conduct research in the field of plant genetics. Currently, he is a Full Professor at Texas A&M University-Kingsville (TAMUK) and a Faculty Fellow at the US Air Force. He has directed and participated in more than 20 projects supported by the Welch Foundation, TAMUK, Texas Workforce Commission, and US National Institutes of Health. He has coauthored more than 80 book chapters and peer reviewed journal articles. He is a Fellow and Chartered Chemist of the Royal Society of Chemistry and a Chartered Scientist. During his service in TAMUK, he trained more than 3,000 students on both undergraduate and graduate levels. He created online courses and established safety training protocols in conjunction with the risk management. Currently, he collaborates with the local law enforcement as a consultant.

Jingbo Louise Liu received her Ph.D. in Materials Science and Engineering from the University of Science and Technology Beijing in 2001. She is a Full Professor at Texas A&M University-Kingsville (TAMUK) and is focused on materials preparation, characterization, and applications. She is the Fellow of Royal Society of Chemistry and DEBI Faculty Fellow at the US Air Force Research Laboratory. She has authored and coauthored books and book chapters and peer reviewed journal articles (more than 90). During her 12-year services in TAMUK, she taught more than 8,000 students and trained about 150 students and scholars to conduct leading edge research. She directed and/or participated in the projects (more than 40) supported by the National Science Foundation (NSF) (USA and China), Natural Sciences and Engineering Research Council of Canada (NSERC), ACS Petroleum Research Funds, and the Department of Education as principal investigator (PI), Co-PI, and senior personnel.

About the Authors

Dr. Nancy KingSanders' background in higher education spans more than 27 years to include a current appointment as Associate Vice President for Student Success at Texas A&M University-Kingsville, a position she has held since 2011. Prior appointments include Associate Dean for Graduate Studies, Administrative Intern for Academic Affairs, Graduate Music Education Coordinator, as well as attaining tenure and the rank of Full Professor in the Music Department at Texas A&M University-Kingsville. Under her administrative leadership at TAMUK, she established the Center for Student Success to include AVID (Advancement Via Individual Determination) for Higher Education, the First-Year Experience, a two-semester First-Year Seminar Course, Centralized Academic Advising, the Graduation Pathway Success Mentoring Center, the Pathway Academic Assistance Tutoring Center, Freshman Interest Groups, Academic Career Literacy, Mapping First-Year Seminar curriculum to General Education Component Area options, and Javelina Common Read Program, resulting in a 15.9% increase in First-Time in College retention at Texas A&M University-Kingsville. She was also selected as a National First-Year Student Advocate by Cengage and the National Resource Center for the First-Year Experience and Students in Transition. Dr. KingSanders' other responsibilities include all academic support for undergraduate students and Co-Principal Investigator for a Title V Individual $2.625 Million Hispanic Serving Institution Grant to embed research into undergraduate courses.

Weixin Song is a Ph.D. student in Materials Science at Imperial College London, United Kingdom. He started his Ph.D. in 2015 funded by the President's Ph.D. Scholarship of Imperial College after he received his B.Sc. in 2012 and M.Sc. in 2015, both from Central South University, China. His research interests are materials electrochemistry and materials characterizations, particularly the applications in energy storage, electrocatalysis, and photoelectrocatalysis. He has 39 publications that have been cited more than 1,700 times and 17 licensed patents. He has been awarded the Armourers and Brasiers Rolls-Royce Scheme in 2018 and the STFC Early Career Award in 2016 from the United Kingdom. He was also awarded the national scholarship and the highest honor when he studied in Central South University. He is a member in Center for Doctoral Training (CDT) in Advanced Characterization of Materials and London Center for Nanotechnology.

Jun Chen is a Ph.D. student in the Department of Materials at the University of Oxford, United Kingdom. She received her Bachelor's degree in Applied Chemistry in 2014 and Master's degree in Physical Chemistry on Metallurgy in 2017, both from Central South University, China. Her Master's research focused on the synthesis, characterization, and electrochemical application of titanium-based nanomaterials for energy storage devices. She was awarded the national scholarship and the honor of "Top Ten Outstanding Students" during her study in Central South University. She then started her Ph.D. in October 2017 in Prof. Jamie Warner's group, fully funded by China Scholarship Council. Her research interest is in two-dimensional (2D) materials, particularly the atomic structure study of transitional metal dichalcogenides and their heterostructures for understanding their properties and behavior. She has 26 publications in peer-reviewed journals, with more than 1,000 citations since 2014, as well as 7 licensed patents.

Carlos M. Costa graduated in Physics in 2005 and obtained his Master's degree in Materials Engineering in 2007. In 2014, he received his Ph.D. in Physics from the Science School of the University of Minho. Currently, he is postdoctoral researcher at the same institution. His work is mainly focused on the development of advanced polymer composites and new materials and formulations for energy storage applications, i.e., lithium-ion batteries and printed batteries.

Renato Gonçalves graduated in Chemistry in 2009, in 2011 received his Master's degree in Characterization Techniques and Chemistry Analysis, and in 2017 obtained his Ph.D. in Materials Engineering, all at the University of Minho, Portugal. Currently, he is working with Electroactive Smart Material group of the Physics Department at the University of Minho, Portugal. His work focuses on polymers composites, inorganic synthesis, and magnetoelectric nanocomposites with advanced applications like sensors and actuators. Recently, his work focuses on energy field at innovative green printed batteries for advanced devices. He also collaborated with the University of Basque Country, Spain; BCMaterials, Spain; and Cambridge University, United Kingdom, among others.

S. Lanceros-Méndez graduated in Physics at the University of the Basque Country, Leioa, Spain. He obtained his Ph.D. at the Institute of Physics of the Julius-Maximilians-Universität Würzburg, Germany. He is Ikerbasque Research Professor and Scientific Director at BCMaterials – Basque Center for Materials, Applications and Nanostructures, Leioa, Spain, and Associate Professor at the Physics Department of the University of Minho, Portugal (on leave). His work is focused in the area of smart and multifunctional materials for sensors and actuators, energy, and biomedical applications.

Bin Liu received his Ph.D. from Huazhong University of Science and Technology in China. He is currently a research scientist in the Electrochemical Materials and Systems Group in the Energy Processes & Materials Division at Pacific Northwest National Laboratory. Dr. Liu has extensive experience in the development of electrodes and electrolytes for next-generation energy storage devices (i.e., Li-metal battery, Li-O_2 battery, flexible Li-ion batteries, fiber supercapacitors, Mg-ion battery, and functional integrated devices). Dr. Liu has published 1 book chapter and more than 45 papers with total citations of over 4,200 and H-index of 30. Seven papers were selected as Thomson Reuters ESI 1% Highly Cited Papers. Eight papers were featured as cover and back cover of peer-reviewed journals. His studies were highlighted by American Association for the Advancement of Science (AAAS), *Nature Nanotechnology*, *Nature Photonics*, Materials Research Society (MRS) *Materials360*, weekly Royal Society of Chemistry (RSC) Publishing press pack, Chemie.DE, *Chemistry World*, and China Central Television Channel 1 (CCTV 1). Dr. Liu holds three issued Chinese patents and one pending US patent as well.

Huilin Pan received her Ph.D. in Condensed Matter Physics from the Institute of Physics Chinese Academy of Sciences. She is currently serving as a research scientist in the field of electrochemical materials and systems at Pacific Northwest National Laboratory. Dr. Pan has 8 years of experience in energy storage materials/chemistry and advanced characterizations in different electrochemistry systems, including lithium, sodium, and zinc batteries. She has published more than 30 papers on battery-related research and had filed 4 US patent applications. She was honored as the "TOP 1% author, highly cited in the Royal Society of Chemistry" in 2014. Her primary research interest includes the development of battery materials and electrolytes for energy storage and the underlying reaction mechanisms in battery systems.

Wei Liu received her Bachelor's in Materials Physics from Beijing Normal University in 2008 and Ph.D. in Materials Science and Engineering from Tsinghua University in 2013. During the year 2010–2011, she visited the University of Tokyo for half a year. From 2013 to 2017, she worked at Stanford University as a postdoctoral scholar. She is currently Assistant Professor at ShanghaiTech University. Her research interests cover the area of materials science, nanotechnology, and solid state ionics, with a focus on studies of lithium batteries and fuel cells. Dr. Liu has been an author or coauthor of 47 papers published in journals such as *Nature Energy*, *Nature Communications*, *Science Advances*, *Advanced Materials*, and *Journal of the American Chemical Society*. She is also peer reviewer of journals including *Joule*, *Nano Letters*, and *Nano Energy*. In 2018, she was awarded as one of the Distinguished Young Scholars by the Thousand Talents Program.

Qiming Tang holds his Bachelor's degree from Hunan University of Technology, China, in 2015, and he received his M.S. in the Department of Material Science and Engineering from Harbin Institute of Technology (Shenzhen) in 2018. Tang is now Research Assistant in the Department of Materials Science and Engineering at HITSZ, in Dr. Junwei Wu's group. His research is mainly focused on the synthesis and application of metal-based chalcogenides anode materials for lithium-ion and sodium-ion batteries, especially in Sn-based and Fe-based chalcogenides.

Qin Jiang received her Bachelor's degree from Nanchang Hangkong University, China, in 2016. She is a graduate student in the Department of Material Science and Engineering, Harbin Institute of Technology (Shenzhen), under the supervision of Prof. Junwei Wu. Her current research mainly focuses on synthesis and applications of iron-based nanomaterials as anode materials for lithium-ion and sodium-ion battery.

Junwei Wu obtained his B.E and M.E from the University of Science and Technology Beijing, China, and a Ph.D. from West Virginia University, USA. He is now an Associate Professor in the Department of Material Science and Engineering at Harbin Institute of Technology (Shenzhen). His research concerns the synthesis and characterizations of nanomaterials and their applications for energy storage and conversion, such as Li/Na-ion batteries, Li-S batteries, and full solid cell. He was awarded R&D 100 Award in 2011. He has published more than 50 papers in international journals.

Xingjun Liu obtained his B.E. and M.E. from Northeastern University, China, in 1988, and Ph.D. from Tohoku University, Japan, 1998. He is a Professor in the Department of Materials Science and Engineering and State Key Laboratory of Advanced Welding and Joining, Harbin Institute of Technology (Shenzhen). He focuses on thermodynamic energy changes related to changes of phase and phase transitions of materials, as well as computational analysis of materials, including analysis of metal materials, and functional materials. The analyses involve electronic packaging materials, composite materials, amorphous materials, high-temperature alloys, rare earth materials, etc. He holds 36 Chinese patents and has published more than 300 papers in national and international journals.

Dr. Yanbin Chen, Professor, candidate of the National Talent Project in China, and candidate of the Beijing Capital Technical Leading Talent Project, is currently the Vice President of Beijing Easpring Material Technology Co., Ltd. He was awarded Ph.D. by the University of Science and Technology Beijing in 2001. He has been engaged in research, development, and industrialization of the key materials for lithium battery for 20 years. Up to now, he has developed more than 50 products of three series cathode materials, including lithium cobalt oxide, lithium manganese oxide, and lithium nickel cobalt manganese oxide, has published more than 50 papers, has obtained 16 licensed patents, and has drafted 9 national and industrial standards. The Easpring is a global top supplier of cathode materials for lithium-ion batteries, providing high-end products to leading battery makers, including LG Chem, Samsung SDI, SK Innovation, Sanyo, Sony, CATL, Farasis, BYD, BAK, etc. The Easpring has been recognized as an Excellent Export Enterprise, National Technical Innovation Demonstration Enterprise, and National Accredited Enterprise Technical Center.

Professor Yafei Liu is currently the Director of Institute of Lithium-ion Battery Materials of Beijing Easpring Material Technology Co., Ltd. He is also an affiliate with the State Chinese Government Council Special Allowance Program as Topic Expert. He was awarded Ph.D. by the University of Science and Technology of China in 2001. He has been engaged in research, development, and industrialization of lithium-ion cathode materials and their precursors, fuel cell materials, and gas-sensitive and functional materials for a long time. So far, he has published more than 50 papers and has applied 52 patents.

Xiujun Qi received her Bachelor's degree in School of Materials Science and Engineering, Liaocheng University, China, in 2015. She attended China University of Mining and Technology in the same year, pursuing her Ph.D. in School of Materials Science and Engineering under the guidance of Prof. Zhicheng Ju. Her current research focuses on the synthesis and electrochemical characterization of heteroatom-doped carbon materials for advanced batteries such as lithium-ion batteries and potassium-ion batteries. She has coauthored six research articles in peer-reviewed journals and has joined several international conferences.

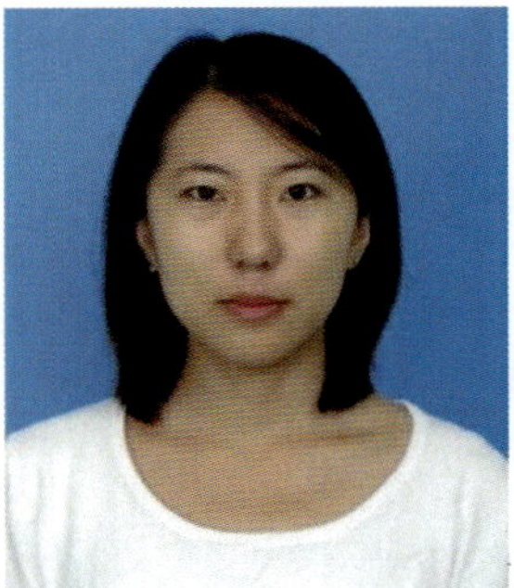

Zheng Xing received her Bachelor and Master degrees in School of Chemistry and Chemical Engineering from Shandong University, China, in 2007 and 2010, following which she began her Ph.D. at the School of Chemistry and Materials Science at the University of Science and Technology of China under the supervision of Prof. Yitai Qian. She received her Ph.D. in Inorganic Chemistry in June 2013. She then joined the School of Materials Science and Engineering at the China University of Mining and Technology as an Assistant Professor. Her research focuses on the study of the transitional metal oxides and new carbon-based functional materials for energy conversion and storage. She has published more than 30 research articles in peer-reviewed journals and has joined many international conferences.

Zhicheng Ju received his Ph.D. in 2011 from the School of Chemistry and Chemical Engineering of Shandong University under Yitai Qian, after which he worked as a postdoctoral researcher in the School of Chemistry and Materials Science at the University of Science and Technology of China. He has been a visiting scholar at the LEE Jim Yang group of National University of Singapore for 1 year. Now he works in the School of Materials Science and Engineering at the China University of Mining and Technology as an Associate Professor. He has coauthored more than 50 peer-reviewed journal publications and his research interests focus mainly on transitional metal oxides and new carbon-based functional materials for energy conversion and storage, especially on new energy materials for alkali-metal ion batteries.

Claudiu B. Bucur obtained his Ph.D. in 2008 from Florida State University under the mentorship of distinguished Leo Mandelkern Professor of Polymer Science, Joseph B. Schlenoff. He studied the manner in which polyelectrolyte multilayers assemble and how doping them with ions changes their mechanical and thermodynamic properties. In 2010, he completed his postdoctoral studies at the USDA Agricultural Research Service Labs, where he investigated corrosion inhibition via biomembranes. Dr. Bucur then joined the Post Lithium

Ion Research Group at the Toyota Research Institute of North America, where he focused on metallic anodes such as magnesium, lithium, sodium, and their electrolytes as well as high capacity conversion cathodes such as the sulfur cathode. He expanded upon his experience with polymers, corrosion, and interfaces and was able to advance many areas in the battery field. Currently, Dr. Bucur is Chief Engineer and Program Director for new battery and solid electrolyte projects at Great Wall Motor, the largest SUV manufacturer in China. He is fascinated by energy storage and dreams of creating the ultimate battery.

Tharamani C. Nagaiah received her Ph.D. in Materials Electrochemistry from Bangalore University, India, in 2005. She joined the Indian Institute of Technology Ropar in 2012 as an Assistant Professor in the Department of Chemistry. Dr. Tharamani has previously worked as a Research Associate at the Indian Institute of Science, Bangalore, and thereafter she moved to the University of Saskatchewan, Canada, as a postdoctoral fellow following which she joined at the Ruhr Universität Bochum, Germany. She has also been a recipient of many prestigious fellowships like Alexander von Humboldt Postdoctoral Fellowship and Ramanujan Fellowship. Dr. Tharamani's research interests include design and development of various carbonaceous materials toward energy conversion and storage, bridging electrochemistry with spectroscopy, and scanning probe techniques, micro-electrochemistry.

Debaprasad Mandal received his Ph.D. in Organometallics Chemistry from the Indian Institute of Technology (IIT) Kanpur, India, in 2006. He joined the Indian Institute of Technology Ropar in 2010 as an Assistant Professor. Prior to join IIT Ropar, Dr. Mandal worked at Erlangen, Germany, as an Alexander von Humboldt fellow for postdoctoral study and then he moved to Texas A&M University as senior scientist. His research work is centered at the interface of organic, organometallics material science, and polymer chemistry. Currently, his group is focused on the development of new ionic liquids, ionic polymers, fluorous materials for catalysis, membrane toward fuel cells, and Li-ion and

Li-S battery. Recently, the group started working on polyoxometalates for electrochemical water oxidation and greener oxidation of organic compound. He is actively involved in research toward energy storage and conversion applications particularly toward advancement of high energy dense batteries by developing highly stable electrolytes and electrode materials.

Aarti Tiwari completed her B.Sc. in Life Sciences (2010) and M.Sc. in Chemistry (2012) from Miranda House, University of Delhi. She later joined the Department of Chemistry, Indian Institute of Technology Ropar in the year 2013 with Dr. Tharamani C. N. Her research interest is toward designing of materials for alkaline fuel cell application followed by in-depth investigation of the processes involved using various physicochemical and electrochemical analysis including the specialized scanning electrochemical microscopy (SECM). She is also involved in the analysis and testing of functionalized electrolytes toward battery applications to enhance the stability and performance of Li-ion cells.

Santosh N. Chavan was born in 1987 and graduated from Balbhim College Beed in 2009. He completed his Master's in Organic Chemistry in the Department of Chemistry, Dr. Babasaheb Ambedkar Marathwada University, Aurangabad. He has completed his Ph.D. under the guidance of Dr. Debaprasad Mandal in March 2018 in the Department of Chemistry, Indian Institute of Technology Ropar. He received Director Fellowship at IIT Ropar after Ph.D. thesis submission. His major research interest is in ionic liquids and polymers with the applications such as battery electrolytes, non-conventional solvents, lubricants, surface coatings, gas absorption, catalysis, membranes, and electrochemical applications.

Michihisa Koyama received his Ph.D. in Chemical System Engineering from the University of Tokyo in 2002. After serving as Assistant Professor at Tohoku University, he moved to INAMORI Frontier Research Center, Kyushu University, as Professor. He is now serving as Unit Director at the National Institute for Materials Science as well as Professor at Shinshu University and Visiting Professor at Hiroshima University. Dr. Koyama has authored and coauthored more than 270 review articles, books and book chapters, and peer-reviewed journal articles. His research activities cover the wide aspects of energy from materials to systems, further to future energy vision. He was awarded the SCEJ Young Investigator Researcher Award, the Society of Chemical Engineers, Japan, in 2009, and the Young Scientists' Prize, and the Commendation for Science and Technology by the Minister of Education, Culture, Sports, Science and Technology in 2014. He directed and participated in more than 60 projects after promotion to Full Professor, which were supported by Japan Science and Technology Agency, Japan Society for the Promotion of Science, New Energy and Industrial Technology Development Organization of Japan, Ministry of the Environment of Japan, leading private companies, etc. He served as committee member of various panels of public organizations for the science policy makings or proposal reviews.

B. Javed received his doctoral degree in the field of Aeronautical and Astronautical Engineering from the University of Tokyo in 2017. He has a multidiscipline research experience in various energy-related technologies, such as thermal-, wave-, and secondary energy devices. He is currently working as postdoctoral fellow in National Institute for Materials Science (NIMS), Japan. His present research focuses on the simulation of the interfacial phenomena for all solid-state Li-ion batteries.

Dajian Li received his Bachelor and Master degrees in Material Science in July 2003 and July 2006, respectively, from Central South University. He obtained his Ph.D. in Material Science from the University of Genoa (Italy) in February 2011. From June 2011, he started his work in KIT as a scientific researcher on the alloy anodes of Li-ion batteries. From May 2017, he started to work on the spinel cathodes on the Li-ion batteries as principal investigator. He is author or coauthor of 25 peer-reviewed publications and a regular speaker in international conferences. His research fields cover experimental and theoretical investigation of phase diagrams. He has been working on lead-free solders, anodes, and cathodes of Li-ion batteries as well as building batteries for various test techniques. He has been awarded for the Highly Cited Research in Intermetallics in December 2016. He was also awarded as recognized reviewer and outstanding reviewer by the Editors of CALPHAD journal.

Weibin Zhang received his Ph.D. in Material Science from Central South University in June 2015. From January to March 2016, he worked as an exchange researcher at the Interdisciplinary Centre for Advanced Materials Simulation (ICAMS), Ruhr-University Bochum. From April 2016 to August 2018, he worked at the Institute for Applied Materials-Applied Materials Physics (IAM-AWP) at Karlsruhe Institute of Technology (KIT) as a scientific researcher. From September 2018, he became a Professor at Shandong University. He is author or coauthor of more than 40 scientific papers in peer-reviewed journals. His research focuses on the integrated computational materials engineering (ICME) and its application for the cemented carbides and Li-ion batteries.

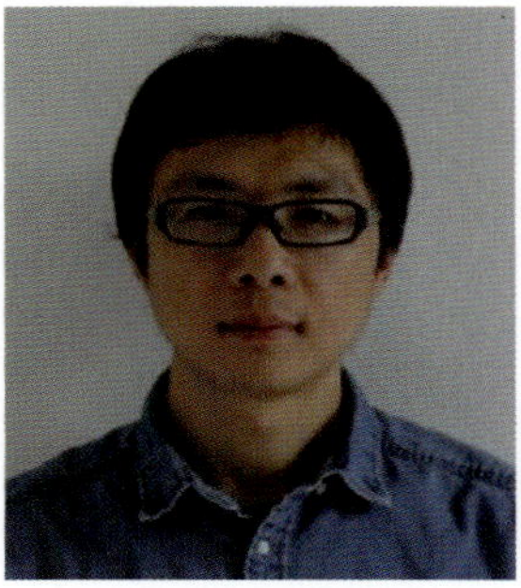

Song-Mao Liang received his Ph.D. from Institute of Metal Research, Chinese Academy of Sciences in 2010. He then worked at the same institute as Assistant Professor. Since July 2017, he works as Research Associate in Clausthal University of Technology. His research interests mainly focus on Calphad modeling and computational thermodynamics applications, such as on light weight alloys, high entropy alloys, Li-ion battery materials and oxides, etc. He has published more than 30 peer-reviewed journal articles and serves as reviewer for more than 10 journals.

Contributors

Irslan Ullah Ashraf Department of Physics, University of Gujrat, Gujrat, Pakistan

Sajid Bashir Department of Chemistry, Texas A&M University-Kingsville, Kingsville, TX, USA

Claudiu B. Bucur Toyota Research Institute of North America, Ann Arbor, MI, USA

Santosh N. Chavan Department of Chemistry, Indian Institute of Technology Ropar, Rupnagar, Punjab, India

Jun Chen Department of Materials, University of Oxford, Oxford, UK

Yanbin Chen Beijing Easpring Material Technology Co., Ltd, Beijing, China

BGRIMM Technology Group, Beijing, China

Carlos M. Costa Centro de Física, Universidade do Minho, Braga, Portugal

Renato Gonçalves Centro de Física, Universidade do Minho, Braga, Portugal

Baber Javed National Institute for Materials Science, Global Research Center for Environment and Energy Based on Nanomaterials Science, Tsukuba, Ibaraki, Japan

Qin Jiang Shenzhen Key Laboratory of Advanced Materials, Department of Materials Science and Engineering, Harbin Institute of Technology (Shenzhen), Shenzhen, China

Zhicheng Ju School of Materials Science and Engineering, China University of Mining and Technology, Xuzhou, People's Republic of China

Nancy KingSanders Texas A&M University-Kingsville, Kingsville, TX, USA

Michihisa Koyama National Institute for Materials Science, Global Research Center for Environment and Energy Based on Nanomaterials Science, Tsukuba, Ibaraki, Japan

Center for Energy and Environmental Science, Shinshu University, Nagano, Japan

Graduate School of Engineering, Hiroshima University, Higashi-Hiroshima, Hiroshima, Japan

S. Lanceros-Méndez BCMaterials, Basque Center for Materials, Applications and Nanostructures, Leioa, Spain

IKERBASQUE, Basque Foundation for Science, Bilbao, Spain

Dajian Li Institute for Applied Materials – Applied Materials Physics (IAM-AWP), Karlsruhe Institute of Technology (KIT), Eggenstein-Leopoldshafen, Germany

Song-Mao Liang University of Wisconsin Madison, Madison, WI, USA

Jingbo Louise Liu Department of Chemistry, Texas A&M University-Kingsville, Kingsville, TX, USA

Department of Chemistry, Texas A&M University (TAMU), College Station, TX, USA

Bin Liu Energy and Environment Directorate, Pacific Northwest National Laboratory, Richland, WA, USA

Wei Liu School of Physical Science and Technology, ShanghaiTech University, Shanghai, China

Xingjun Liu Shenzhen Key Laboratory of Advanced Materials, Department of Materials Science and Engineering, Harbin Institute of Technology (Shenzhen), Shenzhen, China

Yafei Liu Beijing Easpring Material Technology Co., Ltd, Beijing, China

BGRIMM Technology Group, Beijing, China

Abdul Majid Department of Physics, University of Gujrat, Gujrat, Pakistan

Debaprasad Mandal Department of Chemistry, Indian Institute of Technology Ropar, Rupnagar, Punjab, India

Tharamani C. Nagaiah Department of Chemistry, Indian Institute of Technology Ropar, Rupnagar, Punjab, India

Huilin Pan Energy and Environment Directorate, Pacific Northwest National Laboratory, Richland, WA, USA

Xiujun Qi School of Materials Science and Engineering, China University of Mining and Technology, Xuzhou, People's Republic of China

Weixin Song Department of Materials, Imperial College London, London, UK

Qiming Tang Shenzhen Key Laboratory of Advanced Materials, Department of Materials Science and Engineering, Harbin Institute of Technology (Shenzhen), Shenzhen, China

Aarti Tiwari Department of Chemistry, Indian Institute of Technology Ropar, Rupnagar, Punjab, India

Junwei Wu Shenzhen Key Laboratory of Advanced Materials, Department of Materials Science and Engineering, Harbin Institute of Technology (Shenzhen), Shenzhen, China

Zheng Xing School of Materials Science and Engineering, China University of Mining and Technology, Xuzhou, People's Republic of China

Weibin Zhang Key Laboratory for Liquid-Solid Structural Evolution and Processing of Materials (Ministry of Education), Shandong University, Jinan, People's Republic of China

Abbreviations

AC	Activated carbon
AFM	Atomic force microscopy
ASA	Active surface area
^{13}C NMR	Carbon nuclear magnetic resonance spectroscopy
CCCV	Constant current and constant voltage
CCP	Close centered packing
CE	Columbic efficiency
$C_2H_6O_3$	Orthoaceticacid
$C_6H_5CH_3$	Toluene
CID	Current intermitted devices
CMC	Carboxymethyl cellulose
CMD	Chemical manganese dioxide
CPEs	Composite polymer electrolytes
CPR	Current pulse relaxation method
CSTR	Continuous stirred tank reactor
CV	Cyclic voltammetry
DC	Dimethyl carbonate
DFT	Density functional theory
DMSO	Dimethyl sulfoxide
DOD	Depth of discharge
DOT	US Department of Transportation
dQ/dE	Differential capacities
DSC	Differential scanning calorimetry
EC	Ethylene carbonate
EDS	Energy dispersive X-ray spectroscopy
EELS	Electron energy loss spectroscopy
EIS	Electrochemical impedance spectroscopy
EO/PO	Ethylene oxide/propylene oxide
EPDM	Ethylene-propylene-diene methylene
FTIR	Fourier transform infrared spectroscopy
GIC	Graphite intercalation compounds
GO	Graphene oxide
GPC	Gel permeation chromatography

^{1}H NMR	Proton nuclear magnetic resonance spectroscopy
HAADF	High-angle annular dark-field
HCP	Hexagonal close-packing
HEV	Hybrid electric vehicle
HF	Hydrogen fluoride
HFP	Hexafluoropropylene
HOMO	Highest occupied molecular orbital
HOPG	Highly oriented pyrolytic graphite
HR-TEM	High-resolution transition electron microscope
IC	Ion chromatography
IL	Ionic liquid
ITO	Indium tin oxide
LAGP	Lithium aluminum germanium phosphate
LATP	Lithium aluminum titanium phosphate
LCO	Lithium cobalt oxide
LFP	Lithium ferrophosphate
LGPS	Lithium germanium phosphorous sulfide
Li	Lithium
Li^+	Lithium ion
Li_2CO_3	Lithium carbonate
Li_3PO_4	Lithium phosphate
$LiAlO_2$	Lithium aluminate
LiBOB	Lithium bis(oxalato) borate
LIBs	Lithium ion batteries
$LiClO_4$	Lithium perchlorate
$LiPF_6$	Lithium hexafluorophosphate
LiPON	Lithium phosphorus oxynitride
LiTFSI	Lithium bis(trifluoromethylsulfonyl)imide
LLTO	Lithium lanthanum titania
LLZO	Lithium Lanthanum zirconia
LMO	Lithium manganese oxide
LPS	Lithium phosphorous sulfur
LSM	Layer-structured material
LTN	Li-ion transference number
LTO	Lithium titanate
LUMO	Lowest unoccupied molecular orbital
LZP	Lithium zirconium phosphate
MCF	Graphitized mesophase carbon fiber
MCI	Mixed conducting interphase
MEC	Methyl-ethyl carbonate
MO	Molecular orbital
NCA	Lithium nickel cobalt aluminum oxide
NCM	Lithium nickel cobalt manganese oxide
NF	Nanofiber

Ni-Cd	Nickel–cadmium
NiMH	Nickel-metal hydride
NMC	Lithium nickel manganese cobalt oxide
NP	Nanoparticles
OCV	Open circuit voltage
P_2O_5	Phosphorus pentoxide
$P_2O_7^{4-}$	Pyrophosphate
P_2S_5	Phosphorus pentasulfide
PAN	Polyacrylonitrile
PE	Polyethylene
PEA	Polyether-acrylate
PEO	Polyethylene oxide
PLD	Pulsed laser deposition
PP	Polypropylene
PPO	Polypropylene oxide
PS	Polystyrene
PTC	Positive temperature coefficient
PTFE	Poly(tetrafluoroethylene)
PVDF	Polyvinylidene fluoride
rG	Reduced graphene
SEI	Solid electrolyte interphase
SEM	Scanning electron microscope
Sm_2O_3	Samarium(III) oxide
SOC	State of charge
SPEs	Solid polymer electrolyte
SSEs	Solid-state electrolytes
STEM	Scanning transmission electron microscope
TCO	Tin-based composite oxide
TFSILi	(Trifluoromethanesulfonyl)imide lithium
Tg	Glass transition temperature
THF	Tetrahydrofuran
TiO_2	Tantalia
US	United States
UV	Ultraviolet
UV-Vis	Ultraviolet-visible spectroscopy
V_2O_5	Vanadium(V) oxide
XANES	X-ray absorption near edge structure
XPS	X-ray photoelectron spectroscopy
XRD	X-ray diffraction
Zr	Zirconium

Technology Policy and Road Map of Battery 1

Sajid Bashir, Nancy KingSanders, and Jingbo Louise Liu

This chapter is dedicated to Professor Peter J. Derrick who passed away in March 2017 during the writing of this chapter.

Contents

Author Contribution: The first draft was written by LL. The data in the figures and charts was supplied by SB and LL. The final draft was reviewed and edited by NK and LL.

S. Bashir (✉)
Department of Chemistry, Texas A&M University-Kingsville, Kingsville, TX, USA
e-mail: br9@tamuk.edu

N. KingSanders
Texas A&M University-Kingsville, Kingsville, TX, USA
e-mail: Nancy.KingSanders@tamuk.edu

J. L. Liu
Department of Chemistry, Texas A&M University-Kingsville, Kingsville, TX, USA

Department of Chemistry, Texas A&M University (TAMU), College Station, TX, USA
e-mail: kfjll00@tamuk.edu

Q. Zhen et al. (eds.), *Nanostructured Materials for Next-Generation Energy Storage and Conversion*, https://doi.org/10.1007/978-3-662-58675-4_1

Abstract

The increase in global population has resulted in increased demand for stationary and portable energy, as well as increases in carbon dioxide emissions. The increases in greenhouse gases at the present trajectory will increase the global warming potential above the 2° threshold. At this threshold, it is anticipated that climate changes will adversely affect the land and sea masses. To minimize the potential of global warming and meet the increased energy demands, high-energy capacity lithium-ion (Li-ion) batteries in portable electronics and electric vehicles offer one potential solution. Current Li-ion chemistries focus on lithium cobalt oxide (LCO), lithium nickel cobalt manganese oxide (NCM), lithium nickel cobalt aluminum oxide (NCA), lithium iron phosphate (LFP), and lithium titanium oxide (LTO) as cathodes and lithium metal (Li-metal) as an anode. Other semiconductor materials are used as alloying anodes (Si, Sn) that also include chalcogenides (S), metal halides (Br), and polymers. These materials enable the cell to maintain a high charge density through enhanced Li and electron transport, lesser volumetric expansion, material dissolution, and surface reactions between electrode and electrolyte generating heat. In part, battery thermal management is controlled using passive cooling in addition to liquid cooling. However, the widespread usage of Li-ion batteries in electric vehicles also presents a problem in end-of-life disposal and adverse effects on leaching of active materials into the water table. A life cycle assessment of Li material extraction and usage indicates that every watt-hour (Wh) of storage capacity utilizes 328 Wh of energy and generates 110 g carbon dioxide equivalent (CO_2-eq.) emissions. The other components in the cathode, such as Ni and Co, may also pose toxicological adverse effects if released prematurely into the environment. The trend in battery design and evolution appears to indicate size and cost reduction; increased power density; recycling and transition to Li-S, Li-Air, or organic polymers to meet the anticipated energy demands; lower CO_2-eq emissions; and electric vehicle demand by 2030.

US policy alignment on energy, transport, and the environment is rarely aligned, except for the Obama Administration attempt at Advanced Clean Coal projects [1]. US policy via the Energy Policy Act has traditionally focused on energy generation utilities which use coal and natural gas (operators) and separately transmission utilities (independent system operator, ISO). The oil crisis of the mid-1970s focused attention on hydro-, thermal-, and petroleum-ethanol blends derived from corn, biodiesel, wind, photovoltaics, and batteries related to electric vehicles. The driving force for these policy initiatives is broad, relating to air and water quality, job creation, rural development, alternative fuels, and vehicle emission policy to promote American energy independence while adhering to cleaner air initiatives that reduce the carbon dioxide and greenhouse gas footprint to lower the global warming potential below the 2 °C threshold. The policy framework enacted from 1970 to 2006 followed a similar pattern with partisan optics. The members from the Democratic Party initiated more environmental-regulatory policies, while the members of the Republican Party support fossil fuel, tax incentives for grid and operators, and general tax cuts to promote job creation. During a national emergency such as oil embargo and Gulf War, the policy instruments converge, and tax incentives are used to promote research development [2] with the State enacting more regulatory push, particularly from the 1990s. Therefore entities in energy or automakers have volunteerism through federal incentives and compulsion through state regulation. The policies enacted at the state and federal level involved different coalitions with different policies and decision-making priorities but can be broadly classified into four groups.

Innovation and transportation technologies are often misunderstood, and their impact is overstated, leading to cost overruns. This might account for the evolution of the nuclear, clean coal, and solar industries which held promise of a near unlimited energy supply at low cost with low risk, but the actual rollout was quite different since there was no common agreement on adaptation of a specific technology platform, technical ignorance of policymakers, and uncertainty on how quickly early technologies would be adopted by the market [3].

From the 1970s, sustainable policies were added to existing fossil fuel-orientated policy, rather than aligned from the ground up [4]. This interplay is observed with the policy implementation of net metering and solar PV, which many coal and gas utilities failed to challenge because they underestimate the negative impact these policies would have on their fossil fuel business model. Likewise, policymakers did not design these policies to meet a specific challenge, e.g., foster energy independence, but to promote an umbrella of policies that would foster change without changing the current fossil fuel-based businesses. The corn-ethanol blend in petroleum likewise was rolled out as a minor component that was insufficient to meet the policy goals of self-sufficiency and was crafted to allay the fears of the petroleum industry that their profits would not be affected through the introduction of blended and alternative fuels [5]. Once it becomes apparent that some of the policies would lead to revenue loss, the dominant industries resorted to lobbying and lawsuits. The political and financial cost of these policies in terms of tax rebates and payments to the consumer also led to greater pushback from the legislature, in the form of revised legislation. For example, Renewable Fuel Standard (RFS) ethanol from cellulose and biofuel ramp up and California's Zero-Emission Vehicle (ZEV) mandate were scaled back because of increased costs in terms of tax rebates for implementation of the technology to market [6]. These policies, often rolled in as add-ons to mainstream bills or as omnibus bills, are a demonstration that a common federal energy policy has not existed that tied need/want to ability and long-term planning but enabled policymakers to pass legislation that was perceived as difficult or not passing in stand-alone bills. Examples include addition of the Federal Production Tax Credit (PTC) to the 1992 Energy Policy Act or the Investment Tax Credit (ITC) in the 2015 Consolidated Appropriations Act. The main focus of these bills was the use of federal lands for mineral and oil exploitation and the introduction of vehicle fuel economy standards, not electric vehicle (EV) tax exceptions.

In many of the omnibus bills, a cap or sunset provision was added to limit political blowback on "unflavored policies," which enabled renewable energy policy to be advocated as a minor component within an omnibus energy bill that focused on fossil fuels, regulations, and standards rather than a comprehensive energy transition policy to meet the goals of energy independence. Thereby, introduction components placed favorable opportunities piecemeal through Congress rather than as one major component. For this reason, many sustainable policies once introduced underwent incremental increases in either delay of the expiration or increase of the initial state scope or goal. One example was the modest introduction of 50 MW of electricity from solar to be deployed back into the grid that was followed by new standards, ITCs, and growth in solar PV deployment [7] and investment (during the Obama Administration) from beginning at least two decades earlier. Currently, with the Trump Administration, it appears these subsidies may be removed or roll backed. The policy related to batteries in EVs similarly began with California's Zero-Emission Vehicle (ZEV) mandate which included a sunset provision, which was extended through incremental bills, because of the greater lobbying power of these nascent industries at the initial introduction. Fossil fuel tax credits that have been maintained since the Wilson Administration (1916, known as "depletion allowance," which lets oil producers deduct 27.5% of their gross revenues) have been equivalent to more than a $500 billion rebate, while in contrast the tax incentives to the sustainable sector have had more uncertainty because of the provision of sunset clauses [8]. The expansion of renewables such as wind and solar and associated tax incentives has escalated costs and drawn political ire from fiscal conservatives to minimize subsidies. A policy to foster innovation and technological maturity with a pilot to scale up deployment and cost of rebate at specific milestones of deliverables has not been attempted with political and policy costs associated, whereby policymakers are more risk-averse if a particularly favored policy or project had previously failed to meet expectations [9].

An example of the above paradox is net metering, where the consumer receives the same rate for electricity as a provider. The utility companies have objected since the consumer does not have infrastructure or grid costs except the solar panel, whereas solar panel or energy providers support parity since the consumer is providing energy that can be used elsewhere [10]. The RPS mandates have also been criticized by utility operators, who perceive this standard as advocating a specific technology. The tax exemptions were extended when the industry was less than 1% of the entire petroleum industry. However when the biofuels blend became a more significant market force, drawing profits away from petroleum companies, lobbying against further tax exceptions greatly amplified. The policymakers sought to defray criticism from the petroleum industry by converting the gasohol tax exemption into a tax credit, the volumetric ethanol excise tax credit (VEETC) which expired in 2012. A similar trend can be seen with EV tax credits, which when EV sales were less than 200,000 nationally received little criticism from auto- and policymakers, but now EV have sales in the lower 800,000 with project costs of two million by 2020, so many states have allowed the initial rebates to expire. The transition to a sustainable energy mix requires federal and state policy to consider wind, solar, thermal nuclear, and appropriate energy storage technologies in addition to infrastructure and reconstruction to the transmittance grid, cleaner coal-based fuels, examples of ethanol blend past 10%, and risk and environmental factors. Policymakers must take account of the possible failure of some technologies to meet the policy framework of independence from foreign oil, lower greenhouses gas emissions, and creation of American jobs with lower taxation for utility operators and energy providers including mineral and oil exploration and exploitation.

1.1 Possible Battery and Electric Vehicle Road Map from 1970 to 2050

If renewable portfolio standards (RPS)-type mandates are extended including investment tax credits (ITCs), the current market momentum will carry lithium-ion batteries to 70% in portable electronics and mobility by 2030 with the global market for Li-ion batteries doubling to tripling, depending on the market model. The batteries in transportation will range from bicycles, scooters, and mopeds to mini, full-size sedans as well as buses and stationary applications both on-grid and off-grid storage. Market saturation will not occur for some time, because of limitations of electrically propelled vehicle endurance. The miles-per-gallon equivalence (mpg_e) is still lower than internal combustion engine (ICE)-based vehicles. This would be reflected in the popularity of the hybrid automobile, where the ICE powers the automobile at higher speeds and an electric motor coupled with capacitor/generator at lower city speeds. It is estimated by 2030, almost 70% of the automobiles will still be ICE-based.

To ramp up production, battery technology will need to change to greater performance/price and a nearly exponential increase in worldwide battery production to meet the anticipated demand. The trends in the road map (Fig. 1.1) demonstrate using past trends and current US policy over the next decade for battery improvements with respect to anode design, new materials for cathode, energy density, increased life cycle, and increased safety, with the lithium nickel manganese cobalt oxide (NMC) chemistry being dominant by 2020 and silicon-based lithium-ion batteries by 2030, as well as Li-sulfur (Li-S), Li-air, and other Li-alkaline earth metal or organic polymer derivative nano-engineered (Li^{+}_{eng}) batteries in the mix, with the $Li_4Ti_5O_{12}$ (LTO, 175 mAh/g), lead-acid (Pb Acid) being relevant for stationary applications. While nickel-cadmium (NiCd) and nickel-metal hydride (NiMH) have given away to Li-ion batteries in portable electronic applications, they will retain a minor share in low-energy electronics

↓Technology/Policy Time→	1970-1990	1991-2015	2016-2020	2021-2050
Battery Technology	Pb Acid NiCd NiMH	Li-ion III (Li_2CO_3, LiOH) LTO-Anode Li^{+}_{eng}	Li-S (Li_2S- solid electrolyte) Li-Air (Li foil/ Li anode) Li-Organic	
Automobile Technology	LNG Biodiesel E10, E15, E85	Plug-In Hybrid EV FC-EV	PV-EV FC-EV Battery Weight reduction Vehicle Drag Reduction FC-EV Hybrid	Off Grid Storage-EV H_2-FC OnBoard-H_2-FC Engine/Transmission Innovation and redesign AI driven automobiles
Infrastructure	Gasoline/Petroleum /LNG/ E blend refuel stations	Swappable battery station	On-grid recharging station	Off-grid Storage and refuel (H_2) or recharge (ē)
Policy & Regulations	Alternative Fuel Standards	CO_2 emission standards V2G Standards	Efficiency (50 mpg/mpg$_e$), Revised Safety Standards for charging and/or driving Recycling batteries	Refueling/Charging/ Battery standards and Safety Regulations Tax on Charging?

Fig. 1.1 Lithium development road map including application in automobile-based transport. Key: AI, artificial intelligence-driven vehicles. CO_2, carbon dioxide emission standard. ē electrical storage; E10, ethanol-petroleum blends at 10% ethanol; E15, 15% ethanol; and E85, 85% ethanol. EV, electric vehicle. FC, fuel cell; FC-EV, fuel cell-electric vehicle hybrid. H_2, hydrogen gas refilling station, H_2-FC, an electric vehicle with hydrogen as a fuel for the fuel cell. Li^{+}_{eng}, Li-ion with engineered support, Li-ion III, Li-ion generation three battery with Li_2CO_2-based electrode. LNG, liquefied natural gas (mpg); fuel efficiency standards in miles-per-gallon for diesel/gasoline/liquid fuel-based vehicles; (mpg$_e$), miles-per-gallon-equivalence for electric vehicles. NiCd, nickel-cadmium battery; NiMH, nickel-metal hydride battery. Pb Acid, lead-acid battery; PV photovoltaic cell; PV-EV, photovoltaic-powered electric vehicle; and V2G, vehicle-to-grid

and detectors. Material costs are another major hurdle to overcome before mass market flow can be accomplished since about 65% of the final cost is related to the active material (intercalation of lithium). By moving to phosphorous, silicon, and organic-based nanomaterials, these costs can be reduced by 30% per kWh. The $100/kWh should also be reached by 2025, for Si and NMC battery chemistries. It is also likely that ITC will no longer be offered, as the lower battery pack will lead to a lowering of the price of electric vehicles since at present the battery is the largest single unit cost within an electric vehicle.

The policy will revise alternative fuel and car emission standards if ZEV-type mandates are considered, giving rise to liquefied natural gas as a fuel for trucks, to 10%, 15% and 85% ethanol blends (E10, E15, and E85). It is likely that further legislation will be crafted to take account of electric vehicle (EV) emissions, not from the tail pipe but in the generation of electricity to charge these vehicles, as charging becomes more widespread. Other less mature technologies like fuel cell (FC) vehicles are charged using off-grid photovoltaics (PV) or on-board photovoltaics, such as solar transitions from fragile panels to flexible polymers, where the automobile paint or color scheme may have residual PV capacity to charge the car while it is idle or parked. Other hybrid combinations of fuel cells to electric vehicles (FC-EV) include the current internal combustion engine (ICE)-electric stack-based vehicles (ICE-EV). Fuel cells like EV cars require energy, the most ecological of which is hydrogen (H_2); however, hydrogen needs to be generated and requires energy. If photovoltaics/solar is used to electrolyze water to hydrogen and oxygen, these gases can be recombined chemically to generate electricity; therefore the next generation of hybrids could be electric vehicle-fuel cells (EV-FC). If fuel cells become more widespread, refueling with hydrogen, like charging stations, will be required. These charging stations could use water to generate hydrogen and oxygen using grid electricity (such as natural gas, coal) or could use sustainable energy in the form of solar that is stored at the refueling station using batteries which are consumed during hydrogen production. Alternatively, solar power could be used for electrolysis of water into hydrogen and oxygen which are inputs for the fuel cell to generate electricity and water as outputs. This would be a form of onboard hydrogen storage and usage (onboard H_2-FC). Much historic legislation has been passed to preserve the fossil fuel industry via incremental increases as various technological-driven technologies become more widespread. The ability to fabricate all-electric vehicles with the rise of artificial intelligence (AI) for the first time yields the possibility of self-drive, self-park automobiles. Because of the scarcity of these models, there is no legislation to address the technical and legal (liability) issues, but as cars transition to digital, the cockpit will incorporate AI features to replace human error. It is likely laws will be crafted to address AI in transport, as well as battery safety and taxation, as the transport sector transitions to mostly ICE-based to mostly EV-FC-based. It is also plausible to tie increased abundance of EVs to newer CO_2 emissions that incorporated generation of energy materials. Instead of CO_2 per tank, newer rules will be based on CO_2 net weight of the vehicle which in turn takes account of the raw materials and fabrication CO_2 "costs." This, in turn, may lead to more streamlined vehicles that have less CO_2 generated (based on extraction, milling, moldings, fabrication, and energy consumption to manufacture the component) instead of combustion of gasoline. Lastly, like safety standards in gasoline stations, equivalent standards in hydrogen refueling or battery swap station/recharging stations are also anticipated.

1.2 The Electron Economy: Background on Importance of Societal Impact of Energy Storage

Societal needs change, and to understand the role of portable power, such as lithium-ion batteries, it is necessary to understand how and why we have arrived at this juncture.

In the United Kingdom, during the mid-eighteenth century, the mix of technology and innovation led to the first industrial revolution [11].), where for the first time, mass production and consolidation of expertise were housed in specific locations [12]. The cog that connects the import of raw materials, the housing of workers, and the actual operation of machinery and factories was steam and electricity, steam for transportation in addition to the use of inland waterways (canal) and electricity for lighting and operation of machinery [13]. The world population at this time (1780) was approximately 850 million, and today (2018) it is estimated at 7.8 billion with an estimated increase to 11.2 billion by 2100 (Roser and Ritchie 2018) as shown in Fig. 1.2. This rise will increase both economic growths and require additional energy [14]. The International Energy Outlook and other sources estimate the world primary energy demand in 1800 was 0.48 billion ton of oil equivalent (t.o.e), whereas today it is 13.9 billion (t.o.e) and anticipated to rise to 18.5 billion by 2040 under the new policy scenario (Roser and Ritchie 2018; [16]), summarized in Fig. 1.3 [17].

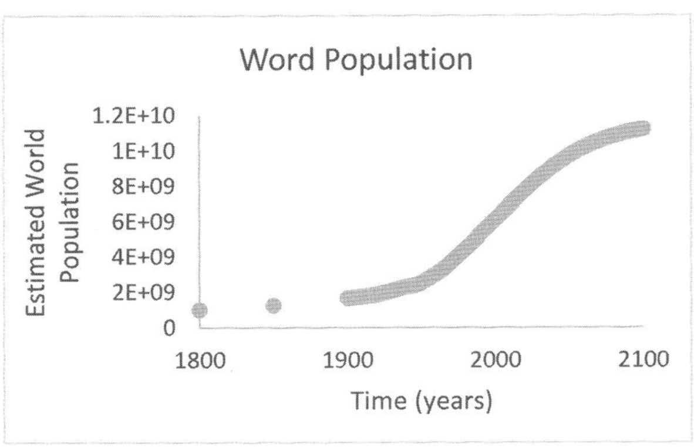

Fig. 1.2 Summary of world population from 1800 to 2100. (Adapted from our world in data [15])

Burning carbon-based fuels to generate electricity or movement, including heat, generates carbon dioxide which contributes toward global temperature fluctuation as a greenhouse gas (GHG) [18]. World energy-related carbon dioxide (CO_2) emissions rose from 33.9 billion metric tons in 2015 to 39.3 billion metric tons in 2040, an increase of 16% over that period [24]. This trend has mirrored the rise in primary energy demand from approximately 6647 million t.o.e (1983 = 100%) to a 39%

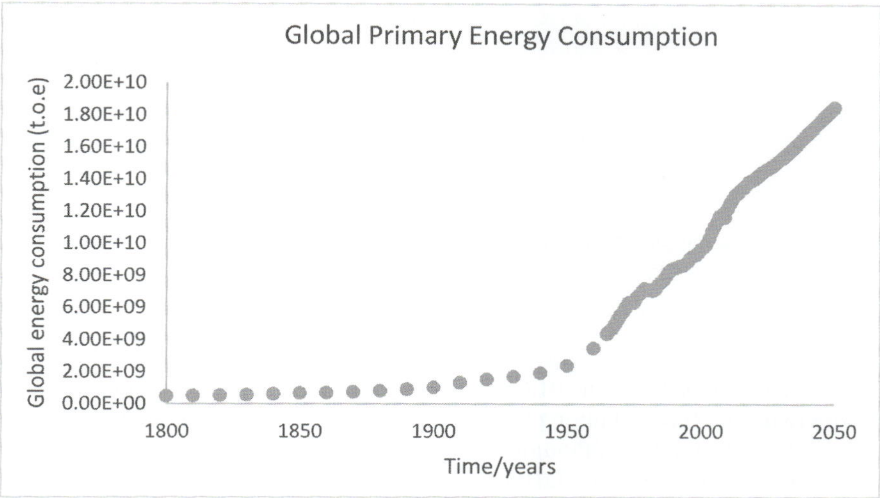

Fig. 1.3 Summary of global energy primary consumption from 1800 to 2100. (Adapted from various sources [19, 20])

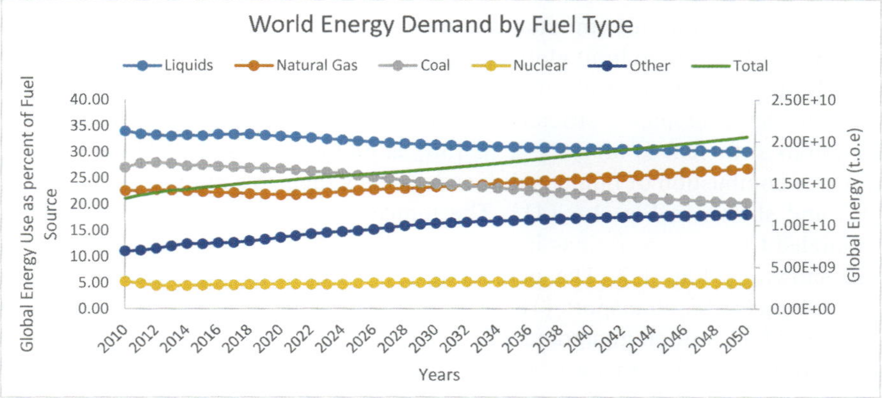

Fig. 1.4 Summary of global energy primary production as a function of fuel source (in percent) from 2010 to 2050. (Adapted from various sources [20–23])

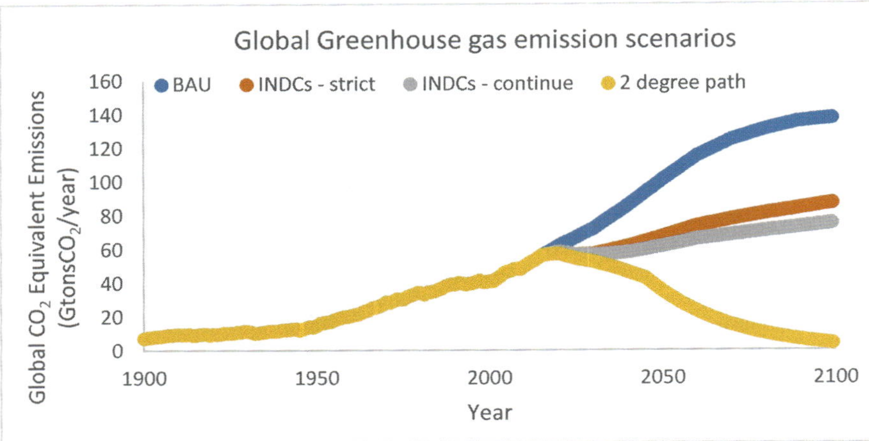

Fig. 1.5 Summary of global greenhouse emission scenarios from 1900 to 2100. (Adapted from various sources [31–34])

increase by 2000 or 100% in 2018, while the sum of non-carbon-emitting sources of energy remained constant during the same time period [21] as illustrated in Fig. 1.4. The continued sum rise (although the percent rate has slowed) has resulted in the more cumulative release of GHGs of which carbon dioxide is by volume the largest component [25]. This emission will increase from approximately 30 gigatons per year to 140 if no action is taken, modeled as business as usual (BAU), intended nationally determined contribution (INDC, [26]). The INDC is modeled with no continued improvement after the target data (strict) [27] or continued improved (continue) after the INDC target date 2025 (United States) or 2030 (most countries) are approaches to tackling GHG emissions [28]. It appears that while the carbon dioxide emission rate and overall CO_2 levels will drop, they will not drop sufficiently to limit global temperature fluctuations by 2 °C, which most models predict will adversely affect global weather patterns [29]. Therefore some countries have adopted a stricter plan above INDC in an attempt to lower CO_2 emissions sufficient to meet the 2 °C [30] as plotted in Fig. 1.5. These are as a result of combustion of liquid, solid, and gaseous fuels in a number of industries [35] and shown in Fig. 1.6 [41]. The tonnage emission of GHG has closely paralleled CO_2 concentrations in the atmosphere (Dlugokencky and Tans 2017) and shown in Fig. 1.7. The percent of CO_2 emissions attributed to energy generation (59%), transport (16%), and commercial and residual (9.8%) in 2010 of 34.42 million gigagrams of CO_2 [45, 46] indicates that transition of the transportation and stationary power systems sector from a carbon-intensive regime to a low-carbon regime, such as battery power (electron economy), can decrease CO_2 emissions, in addition, to extend natural resources, summarized in Fig. 1.8.

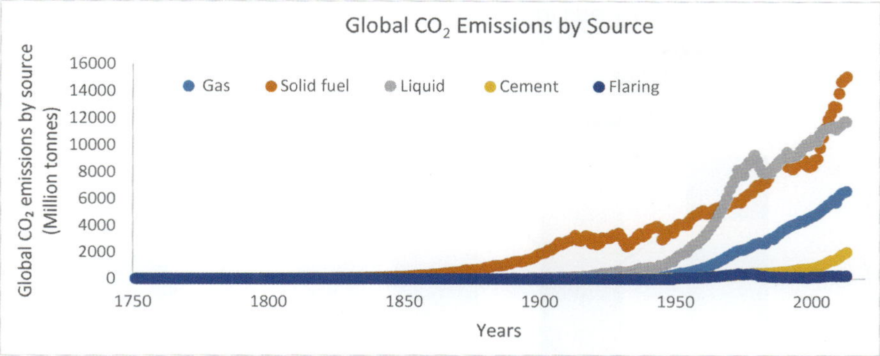

Fig. 1.6 Summary of global carbon dioxide emission as a function of fuel source (in percent) from 1750 to 2020. (Adapted from various sources [36–40])

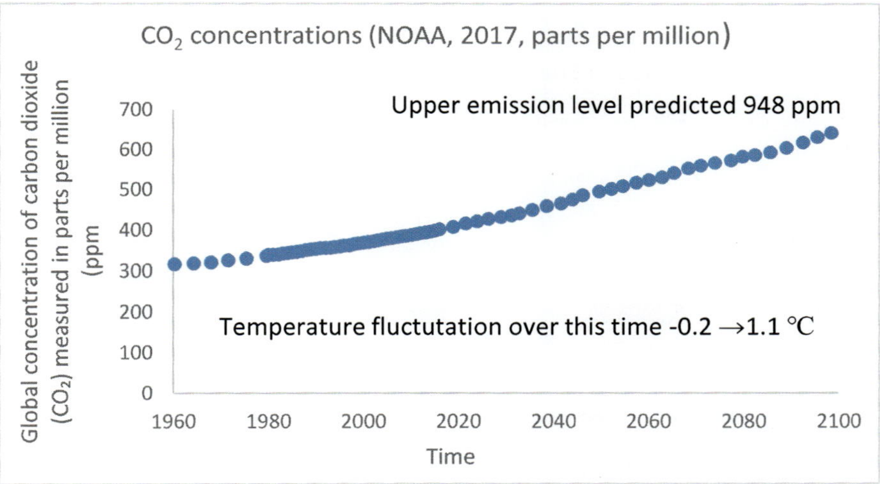

Fig. 1.7 Summary of global carbon dioxide concentration from 1960 to 2100. (Adapted from NOAA [42]. The average temperature fluctuation [43] and modeled CO_2 under various scenarios are also shown [44])

1.3 Environmental Factors Affecting Transportation

While coal and natural gas dominate the source for generation of electricity in transportation, the dominant energy mix is petroleum-derived liquid fuels [49] requiring most industrialized countries to import, since consumption is greater than production. The percent share of oil has averaged 40% (1966–2013), although over the last decade (2006–2016), it has averaged 33% [21] because of improvements in materials (strength and weight) and reduction of friction in

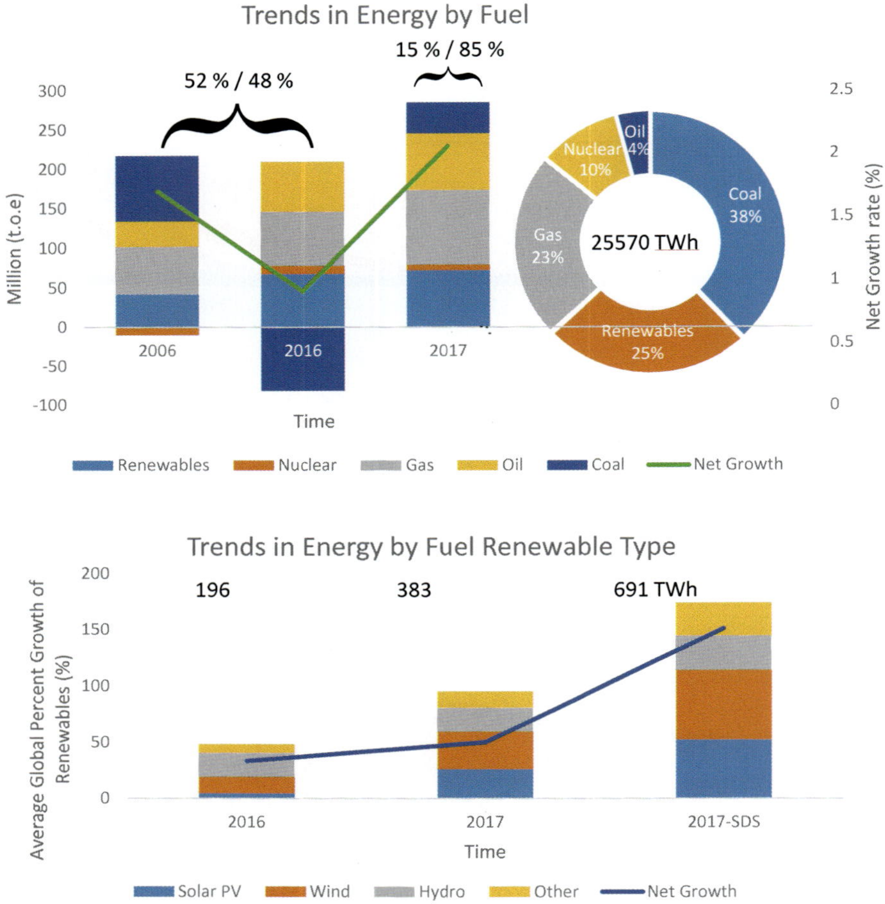

Fig. 1.8 Summary of energy by the source from 2006 to 2015 (2006) to 2017 by millions ton oil equivalent (M toe) and net growth rate (percent, line plot) and percent of the coal used for heating/electricity generation. The pie chart is a breakdown of generation of electricity by fuel type in 2016/2017 which yielded 25,570 TWh globally. The bottom panel represents the breakdown of electricity using renewables from 2000 to 2016 (2016) to 2017 and Sustainable Development Scenario (SDS) modeled to transition to a low-carbon platform using sustainable to sustain global temperatures within the 2 °C band (2017–SDS) including share of net growth (lined plot). (Adapted from various sources [47, 48])

automobiles [50]. With material substitution, streamlined design to generate less friction [51], automobile gasoline consumption is anticipated to drop by 10–20% [52, 53], and rechanneling breaking energy loss because of heat and friction can extend efficiencies and lower dependence on gasoline [54] in automobiles. Heat loss via the exhaust and mechanical losses represent over 70% of the measured losses, and newer developments in the internal combustion engine (ICE) can further extend efficiencies [55] in addition to improvements in recovery of waste heat to work thermoelectric subsystems [56].

1.4 Electric Vehicles and Battery-Based Storage and Electrification

In an electric vehicle, the electric motor replaces the ICE for traction and a flywheel or battery as the energy source. The drivetrain consists of three subsystems, the electric motor (EM), energy pack (B), and auxiliary or secondary power unit (SPU). The motor subsystem is further comprised of a controller unit (CU), the power converter (PU), the electric motor, mechanical transmission (T), and wheels. The energy system contains the battery pack (or flywheel, supercapacitor or fuel cell), energy management unit (EMU), and refueling/recharging assembly. The auxiliary contains the power steering (PS), climate control (CC), and secondary power unit. The control unit channels inputs from the accelerator or braking subsystems to the electric converter to regulate power flow between energy pack and electric motor including reverse power flow because of regenerative braking. This procedure results in energy recovery and efficiency gains. This enables the vehicle to be configured into a number of different conformations with integration and simplification of the drive train. In one configuration the axles point at drive wheels that contain the traction motor, enabling each motor to operate inner wheels at one speed and outer wheels at a different speed when taking corners. Alternatively, the traction motor can be placed inside a wheel (IW), where a gear is used to reduce the motor speed and increase motor torque to provide a responsive reduction ratio [57]. Lastly, an inline configuration of the input and output shaft can also be employed to streamline the mechanical transmission where the out-rotor and in-wheel are connected, thereby eliminating the mechanical gear transmission assembly [58]; therefore the speed control of the motor directs the wheel speed [60] and summarized in Fig. 1.9 [59].

While electric vehicles have the potential to displace ICE-based vehicles in the long term, the performance of battery systems such as energy or power density, charging rate, cycle lifetime, and endurance is not to the same level as with ICEs. The employment of nanomaterials has enabled power profiles to be improved and appear to be within reach at 250 Wh/kg at a full time to charge 3 h are within reach [61]. The cost of a vehicle battery is $ 100 kWh by 2030 [62].

With the US Department of Energy recommendation of a driving range of 500 km, a specific energy of 250–350 Wh/kg would be required [63]. Therefore the selection of the battery is not a critical parameter but does have a bearing on performance and cost. Two parameters used to evaluate the type of battery are mass-energy density (kWh/kg) and volumetric energy density (kWh/m^3) with different batteries exhibiting different profiles.

1.5 Technical Specification for Different Battery Technologies

The US Advanced Battery Consortium initial performance benchmarks for batteries are summarized in Table 1.1 [64] and Fig. 1.10 for the weight-volume energy density [65].

Not all battery technologies can meet all of the goals. The performance criteria of common battery platforms are summarized in Table 1.2 [71].

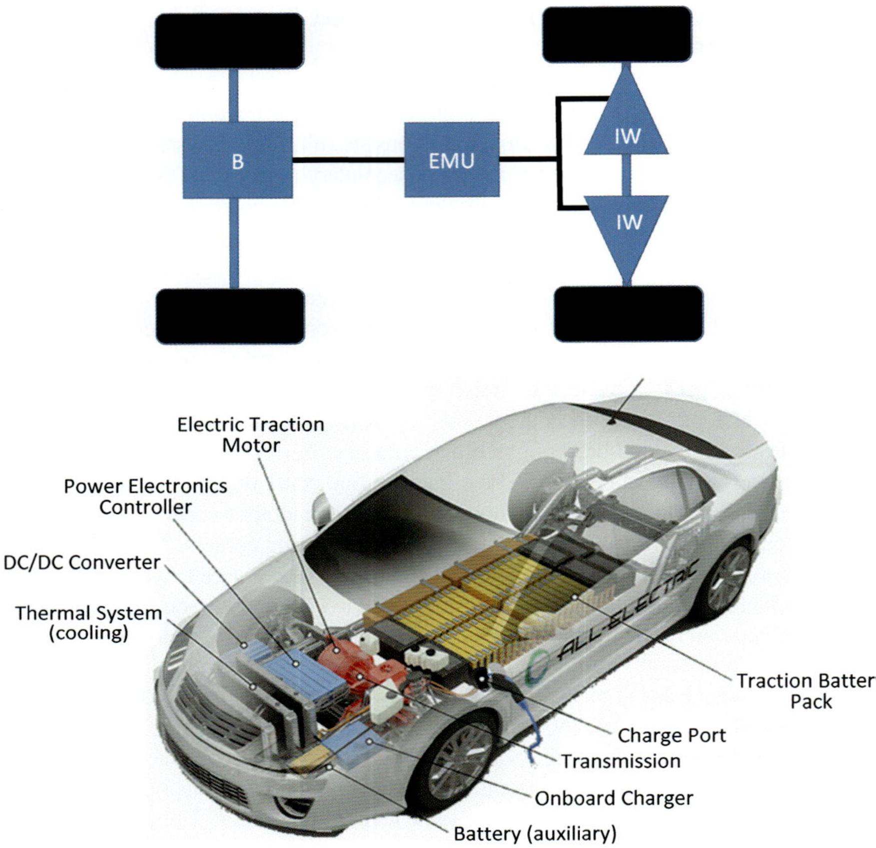

Fig. 1.9 Simplified schematic of the electric vehicle and a 3D cutaway of a vehicle where the battery is connected to the electric control unit which is also connected to the electric motor [57]. The motor, in turn, is connected to the transmission which via the differential is connected to the axle and wheels ([60] with permission, Alternative Fuels Data Center)

While lead-acid (Pb-A) batteries are not ideal, because of their low-energy weight and stored energy/volume ratio, they have had the longest development and are the most inexpensive options and are routinely found in electric golf carts, scooters, wheelchairs, and automotive batteries [74]. The battery supplies power to the starter and ignition system to start the engine. They also supply the extra power necessary when the vehicle's electrical load exceeds the supply from the charging system. A car battery acts as a voltage stabilizer in the electrical system. The battery evens out voltage spikes and prevents them from damaging other components in the electrical system. In smaller powered engines, the battery powers the motor that drives an electric vehicle, such as a forklift truck or wheelchair. These batteries have no memory and can be configured either as

Table 1.1 US Advanced Battery Consortium initial performance benchmarks for batteries

Goal	Benchmark
Specific energy (Wh/kg)	200
Energy density (Wh/L)	300
Specific power (80%, W/kg)	400
Power density (W/L)	600
Life span (years)	10
Cycle life (80% depth of discharge)	1000
Operating temperature (°C)	−40 to +52
Recharge time (h)	3–6
Fast recharge time (40–80% state of charge, h)	0.25
Efficiency (discharge/charge at 6 h, %)	80
Self-discharge (%/time)	< 15 month
Thermal loss (W/kWh)	3.2
Price (US $/kWh)	< 100

Compiled from various sources [66–68]

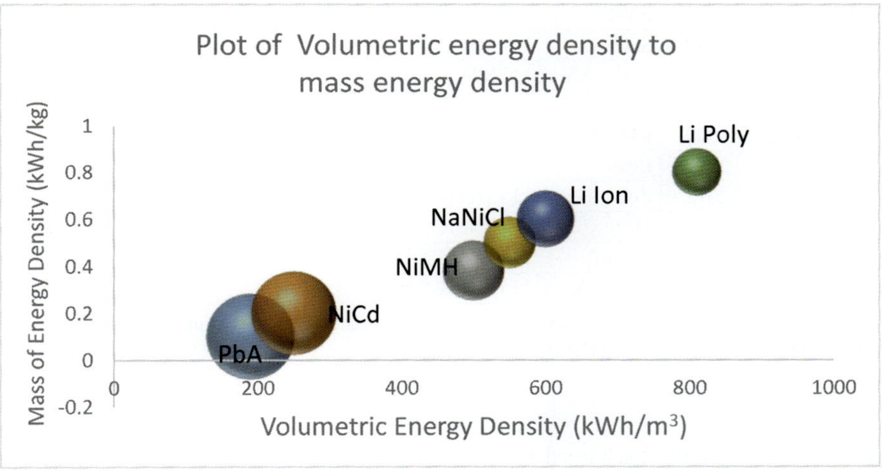

Fig. 1.10 Summary of battery volumetric to mass-energy density ratio for common battery types. (Data from various sources [69, 70])

deep cycle mode for boats or campers, where they are used to power accessories like trolling motors, winches, or lights, or as standby to supply electrical power to critical systems in the event of a power outage or automotive or motive uses. However, they have a short life cycle and lower specific energy and power.

Nickel-cadmium (NiCd) have a low-energy weight ratio but have the highest life span (charge/discharge cycle of 1500); however like Pb, Cd is a heavy metal that is toxic to wildlife and mammals [75]. While they have high specific energy and power, they exhibit a memory effect. Over time, the charge/discharge efficiencies decrease, where

Table 1.2 Summary of key benchmarks of different batteries

Battery type	Specific energy (Wh/kg)	Energy density (Wh/L)	Specific power (W/kg)	Cycle life (cycles)	Self-discharge coefficient (%/24 h)	Cost (US $ kW/h)
PbA	30–45	60–90	200–300	400–600	1	150
NiCd	40–60	80–110	150–350	600–1200	5	300
NiMH	60–70	130–170	150–300	300	2	100–300
NaNiCl	86	149	150	1000	0	230–350
Li^+	90–130	140–200	250–450	800–1200	1	>200
Li^+ Polymer	155	220	315	600	1	>200

Compiled from various sources [72, 73, 75]

Table 1.3 Summary of top EV model sold in any specific year and automobile performance

Company	Year	Model	Battery type	Max range (Km)	Weight (kg)	Battery weight (Kg)	Passengers (#)
GM	1996	EV-1	PbA	152	1348	553	2
GM	1997	S-10	NiMH	128	1909	432	2
Toyota	1998	RAV4-EV	NiMH	202	1564	376	5
Ford	1999	Ranger EV	NiMH	136	1907	568	2
Mitsubishi	2009	i-MiEV	Li^+/NMC	100	1080	200	5
Nissan	2010	Leaf	Li^+/NMC	121	1521	218	5
Renault	2012	Zoe	Li^+/NMC	210	1468	300	5
BMW	2013	I3	Li^+/NMC	130	1195	230	5
Volkswagen	2014	e-Golf	Li^+/NMC	125	1018	344	5
Tesla	2015	Model X	Li^+/NCA	237	2300	540	5

Source: From manufacturers or auto-dealer websites; cobalt aluminum oxide (NCA) and lithium nickel manganese cobalt oxide (NMC)

the batteries partially discharged on a regular basis. This phenomenon should not be confused with voltage depression, which is because of overcharging. Here, the battery is unable to maintain a charge and discharges quickly. This can be partially offset by running the battery through very deep discharge cycles.

Nickel-metal hydride (Ni-MH) have a high load capacity and high discharge coefficient [76]. They also suffer from a slight memory effect and are used in devices that require a high power such as cameras and flashes.

Lithium-ion (Li^+) have a large power storage capacity with intermediate energy density/weight ratio but have a lower life cycle, are prone to overheating, and are expensive. Lithium-ion polymer extends the life cycle of Li^+ [77] but is also not as stable as PbA. Sodium-nickel chloride (NaNiCl) has a molten salt electrolyte but requires high operating temperatures (270–350 °C). They have a high stored energy density but have operational safety and storage concerns [78].

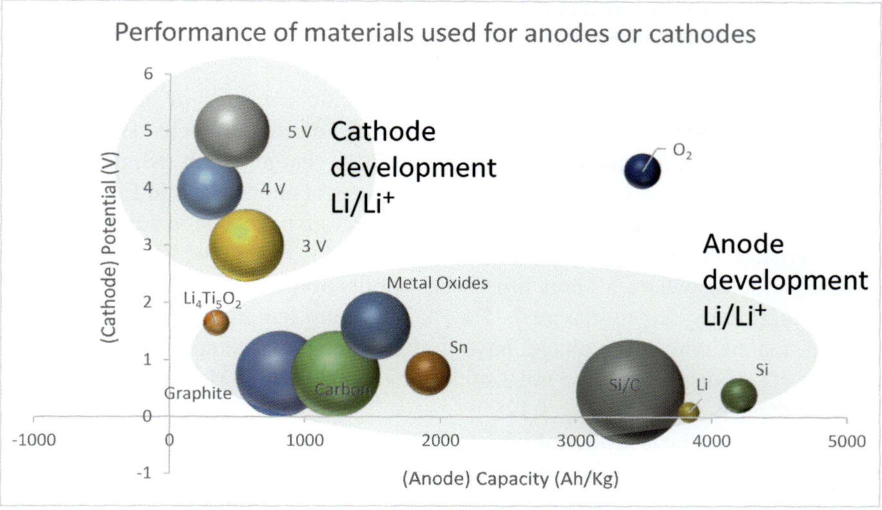

Fig. 1.11 Summary of anode/cathode development related to Li-ion battery catalysts for carbon-based anodic materials and 3 V–5 V cathodic materials [87]

For automobiles with the highest endurance, batteries with the highest coefficient of stored energy density are required. To increase the magnitude of the values in Table 1.2, new anode and cathode designs in modeling, fabrication, and implementation are required [79, 80]. The implementation of different batteries and EV configurations has resulted in extended mileage of commercial EV cars since the early 1990s, summarized in Table 1.3.

1.6 The Extended Range Requires Some Type of Storage Capability

Lithium-ion batteries are common because of the advent of portable electronics. In 2008, two billion cells were produced [81]. Lithium-ion is an electrochemical reaction between lithium cations and cathodic active materials [82]. The cells contain a liquid electrolyte, with the electrodes being separated by a porous membrane composed of polyethylene or polypropylene to transmit the cations [83]. The cathode is not lithium metal but lithium metal oxides, while the anode is graphite and dimethyl carbonates containing dissolved lithium salts [84].

The charging and discharging cycles of the battery occur between the anode and cathode via the electrolyte. During charging of the ions from the cathode (Li metal oxide) to the anode (graphite), the discharge cycle occurs in the reverse direction [85]. The electrochemical performance, therefore, is dependent upon the identity of both the cathode and anode materials as well as the type of electrolyte employed in Fig. 1.11 [86].

1.7 Lithium-Ion Battery Technology Current Development and Road Map

Lithium-ion batteries have high energy and specific densities of 300 Wh/I and 125 Wh/kg, respectively [88, 89], and fast charge/discharge cycles [90]. Adachi reported charging time constant of 200 ms at 90% charge which maintained over 3500 cycles (78% round trip, [91]). These parameters (summarized in Eqs. 1.1, 1.2, 1.3, 1.4, 1.5, 1.6, 1.7, 1.8, 1.9, 1.10, and 1.11) therefore have a good combination of characteristics in terms of response time and battery weight which are ideal for electric vehicles, where a short time scale application is required, because of Li battery high daily self-discharge 1–5% [92]. Lithium batteries are prone to overheating, maintaining safe voltage, have limited operational temperature range, and because of self-discharging are not a suitable backup option [93]. The schematic of a Li-ion battery is shown in Fig. 1.12.

There are four metal oxides of lithium that are traditionally used instead of lithium metal. They are lithium nickel cobalt aluminum oxide (NCA, $LiNiCoAlO_2$, 250 Wh/kg), lithium cobalt oxide (LCO, $LiCoO_2$, 200 Wh/kg), lithium nickel manganese cobalt oxide (NMC, $LiNiCoMnO_2$, 200 Wh/kg), lithium manganese oxide (LMO, $LiMn_2O_4$, 140 Wh/kg), lithium iron phosphate (LFP, $LiFePO_4$, 120 Wh/kg), and lithium titanate (LTO, $Li4Ti5O_{12}$, 80 Wh/kg). These batteries have superior specific energy to traditional batteries such as lead-acid (PbA, $PbO_2|Pb$ 40 Wh/kg), nickel-cadmium (NiCd, NiO(OH), 60 Wh/kg), and even nickel-metal hydride ($Ni(OH)_2$, 85 Wh/kg) [89] (Table 1.4).

$$CoO_2 + Li^+ \rightarrow LiCoO_2 \qquad E^\circ = 1\ V \tag{1.1}$$

$$Li + C_6 + \bar{e}\ LiC_6 \qquad E^\circ = 3.01\ V \tag{1.2}$$

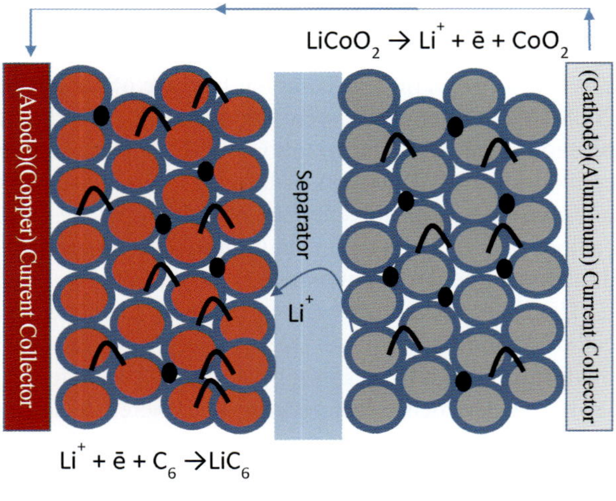

Fig. 1.12 Schematic of lithium-ion battery. The Li-ion battery can be conceptualized as consisting of the cathode, anode, electrolyte, and separator. During charging the lithium cation migrates from the cathode via the electrolyte to the anode and in reverse during discharging

Table 1.4 Summary of the type of Li-ion batteries and general characteristics

Abbreviation	LCO	LMO	LFP	NCA	NMC	LTO
Formulae	$LiCoO_2$	$LiMn_2O_4$	$LiFePO_4$	$LiNiCoAlO_2$	$LiNiMnCoO_2$	Li_2TiO_3
Year introduced	1991	1996	1999	1999	2008	2008
Vmax (V)	4.20	4.20	3.60	4.20	4.20	2.70
Vmin (V)	3.00	2.50	2.00	2.50	2.00	1.50
Cycle life	100	1500	2000	1500	2000	7000
Energy density (max, Wh/kg)	200	140	140	250	200	80

With LTO/graphite as anode and LCO/LMO.LFO/NCA or NMC as potential Li^+ cathode/material and the fine porous film made of polypropylene as a separator and with 1 M $LiPF_6$ dissolved in ethylene carbonate/diethyl carbonate as the electrolyte. Data from EE Times [94] and Battery University [95] websites

$$E_{oc} = E_{cat} - E_{an} \tag{1.3}$$

$$E_{oc} = 1 - [-3.01] = 4V \tag{1.4}$$

$$C\frac{S}{C} = \frac{96485\ As/mol}{72\ g/mol} = 372\,mAh/g_C \tag{1.5}$$

$$LiCoO_2 \leftrightarrow Li_{1-x}CoO_2 + xLi^+ + x\bar{e} \tag{1.6}$$

$$C_n + xLi^+ x\bar{e} \leftrightarrow C_nLi_x \tag{1.7}$$

$$LiCoO_2 + C_n \leftrightarrow Li_{1-x}CoO_2 + C_nLi_x \tag{1.8}$$

$$LiCoO_2 \leftrightarrow Li_{0.45}CoO_2 \tag{1.9}$$

$$0.55\ Li^+ + 0.55\ \bar{e} + Li_{0.45}CoO_2 \leftrightarrow LiCoO_2 \tag{1.10}$$

$$C\frac{S}{LiCoC_2} = \frac{0.55 - 96485\ As/mol}{98\ g/mol} = 150\,mAh/g_{LiCoO_2} \tag{1.11}$$

1.8 Advances in Electrolyte, Membranes, and Energy Density for Electric Vehicle Applications

Compared with lithium metal, the metal oxides yield greater impedance, because of their lower conductivities and diffusion coefficients. This hurdle is overcome by mixing a fine powder of the metal oxide insolvent and binder, such as styrene-butadiene rubber dissolved in N-methyl-2-pyrrolidone. The slurry is shaped into the cathode electrode and pasted onto aluminum foil as the current collector. In a similar manner, the anode is fashioned, using graphite or lithium titanate as the material, and the anode affixed to copper foil as the current collector. The electrolyte are dissolved lithium salts in an organic solvent, such as lithium hexafluorophosphate ($LiPF_6$), in dimethyl carbonate (DMC). For polymer Li-ion a gel precursor is added to the electrolyte in the form of polyvinylidene fluoride. The electrodes are separated using polyethylene [96]. Lithium cobalt oxide (LCO) has high specific energy and can run up to 1000 full cycles, yielding a consumer life span of 2–3 years. Heat is a major disadvantage because of the presence of cobalt oxide which has a low thermal stability. If the battery is overcharged or is left in sunlight and heats up, it short circuits. The excessive heat can decompose the metal casing and metal oxide, because the instability of lithium in air shown in Eqs. 1.12, 1.13, 1.14, and 1.15 is positive and therefore spontaneous.

$$O_2(g) + 4H^+(aq) + 4e \rightarrow 2\ H_2O\ (l)\ [\times 1]\quad E^\circ = +1.229\ V \tag{1.12}$$

$$Li^+ + \bar{e} \rightarrow Li \qquad [\times 4]\quad E^\circ = -3.045\ V \tag{1.13}$$

$$4\ Li + O_2 \rightarrow 2\ Li_2O \tag{1.14}$$

$$E_{cell} = 1.223 - (-3.045) = \sim 4.3V \tag{1.15}$$

The high specific energy of LCO batteries allows them to be the major battery type in portable electronics such as laptops and cell phones, although some laptop manufacturers are using NMC with different ratios of manganese to nickel to cobalt to increase the specific energy of the cell. A major safety concern is overheating [97]. In sensor technology, NMC and NCA cells are utilized because of lower material intensity but with good specific energy. In hybrid electric vehicles, a conventional internal combustion engine with an electric propulsion system is used to drive the vehicle. Here the power recovery from deceleration is stored in the battery for future use during acceleration. The electric system, therefore, is more auxiliary and as a result has a power of 2–3 kWh, insufficient for overnight charging. These types of vehicles do not have a plug-in port for recharging. Cars with 8 kWh or larger batteries require a separate plug-in port to charge the battery. This power is sufficient for an all-electric drive of 30 km, with the load being born by the ICE. In an EV, the ICE is replaced by an electric motor and requires energies of 18 kWh for a drive of 120 km [98]. The EV market is dominated by the NMC (higher lifetime)- and NCA (higher specific energy)-type Li batteries, because of their high specific energy or life cycle. The plug-in vehicle prefers the NMC type, while the EV prefer the NCA. The average ICE car can drive up to 310 miles on a 12-gallon tank, which takes 5 min to refill. The PbA batteries typically have a 3–8-year warranty, usually prorated after the third year. In comparison, the plug-in batteries have a warranty of 100 k miles or 3 years and EV 500–800 cycles or 8 years. However, the plug-in cannot currently drive 310 miles on a full charge or be recharged within 5 min. The current lifetime warranty is less than the possible endurance of these batteries since a 310 mile/cycle at 800 full cycles would travel 248 k miles, well as the warranty period. It, therefore, appears that specific energies around 80 kWh and weight are pressing

Table 1.5 Summary EV and PHEV and their key characteristics

Company	Model	Battery type	Engine power [kW]	Capacity [kWh]	Power-to-capacity ratio (/h)	Range-to-capacity ratio (Km/kWh)
GM	Bolt EV	NMC	149	60	2.48	7
GM	Volt	NMC	111	18.4	6.03	5.5
Toyota	Prius Prime	NMC	20	8.8	2.72	5.5
Ford	Ford Focus EV	NMC	107	33.5	3.19	6.5
Mitsubishi	i-MiEV	NMC	49	16	3.06	7
Nissan	Leaf EV	NMC	80	30	2.67	7
Renault	Zeo	NMC	106	21.4	4.95	7
BMW	BMW i3	NMC	132	36	3.67	6
Volkswagen	e-Golf	NMC	100	35.8	2.79	6.5
Tesla	S 75D	NCA	245	75	3.27	6

Source: various auto manufacturers' websites and industry sales tracker [101]

technical concerns to make automobile performance (drive and refilling) to be on par with gasoline-powered cars [99]. Other related sectors include electric motorcycles, and bicycles, and forklifts that use a mix of LFP-LMO battery chemistries.

The engine power to specific energy (E) capacity of the electric vehicles also needs to be taken into account. The current ratio is between 3 and 5 for EV and 5 and 10 for plug-ins. It is unlikely this ratio may get larger with increases in specific energy and more efficient electric motors, summarized in Table 1.5 [100].

For this reason, global car sales reached a milestone in 2015 (one million units sold) and are the fastest-growing area in the automotive market today (Fig. 1.13) [102]. Bloomberg New Energy Finance predicts that by 2038 the major type of car sold would be the electric vehicle as opposed to the ICE and by 2040 just under 60 million units will be sold, although there will still be over 1.2 billion ICE cars still in operation BNEF [105].

1.9 Stationary Power Supply Systems

The uninterrupted power supply needs to be load-level both for constant power delivery and for frequency and voltage. The electricity demand is not constant but varies by time of day or seasonal or special event such as sports, music, and political events, where audience members at home use electricity for drinks and food during break time of the program or event. In addition, severe weather can affect heating or cooling in offices and households. These variations affect how electricity is generated (coal, nuclear, wind, solar, or gas), and restrictions are placed by grid regulators to ensure no utility operator is near their maximum capacity [106]. Therefore for a steady output, the grid utilizes energy

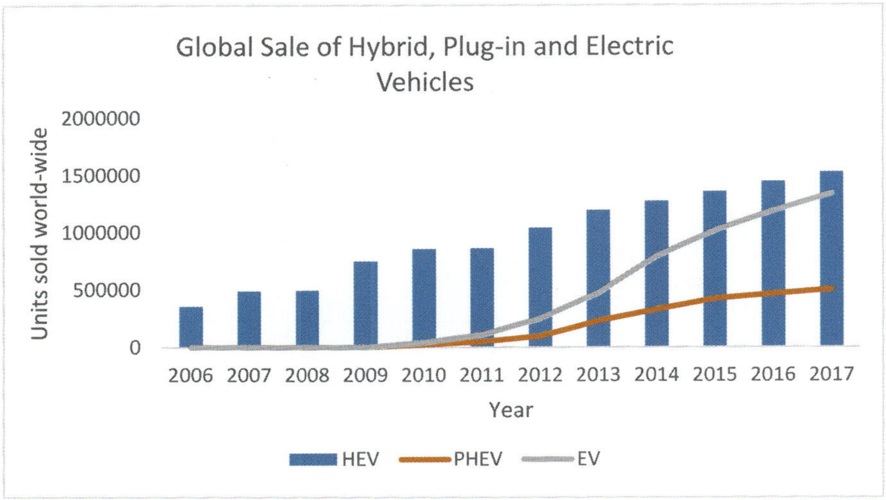

Fig. 1.13 Trend in hybrid, plug-in hybrid, and electric vehicles. (Sources: various [103, 104])

storage units to offset against peak demand via pumped hydro storage. Over the last decade, there has been a slow shift toward more renewable energy mixes such as photovoltaics and wind [107], resulting in decentralization where end-users can input energy into the grid as well as draw energy out [108]. This, in turn, has drawn attention to improved energy efficiency at all levels to lower the total electricity usage and lower carbon dioxide emissions [109]. Electrochemical storage, therefore, can play a critical role in the infrastructure of low carbon buildings [110], where Li-ion batteries have advantages over other electrochemical storage strategies. One disadvantage is the Li battery bank storage is more expensive at approximately \$350/kWh with a lifetime of 2000 cycles. A cost of less than \$175/kWh with a lifetime above 3000 cycles will place storage costs in the c\$5/kWh range and make them competitive toward centralized and distributed power storage units [111]. Here the units could be used to maximize local usage of electricity from distributed renewable power generation units such as windmills that are local to each community of 20 houses [112]. The battery chemistry most favored under these circumstances is related to LFP, vanadium redox, and PbA batteries [113]. A side effect of the millions of units of EVs is the possibility of these vehicles serving as backup distributed units (vehicle-to-grid) in emergencies (hurricanes) where constant power is critical (hospital, care homes, office blocks, police, fire, emergency services) or even as a backup source for home residents who can sell the power during peak hours and recharge at off-peak hours [114].

Lithium-ion batteries can also be employed as electrochemical storage for an off-grid power supply such as with renewables, which is dominated by PbA batteries. Under a renewable-powered home (RPH), the battery can be placed inside the house if it is powered using photovoltaics or a windmill that is close by Jaiswal [115]. The battery packs could power the basic heating/cooling/lighting function of an RPH by 2030 [108]. This has distinct advantages of gasoline or kerosene generators or heating sources since the prices of these fuels fluctuate more than the price of light-emitting diodes (for lighting) or LFP for storage [116]. This is attractive to societies with poor infrastructure since a national grid system would not be necessary. Each community could have their battery bank supplied by a windmill or solar panels centrally located (light harvesting farm) or on each building [117].

1.10 Li Battery Usage Trend

The storage capacity of Li batteries worldwide, either actual or anticipated, is in the lower GWh/y and is doubling every 5 years. In part, this is a consolidation of older portable storage devices, in partial replacement of NiMH and NiCd and in part new demand [118]. The trend is illustrated in Table 1.6.

The Li-ion battery demand for portable electronics is growing at an average annual rate of 10%. The Li-ion battery demand for portable electronics is likely to reach 45 GWh in 2020 and 100 GWh in 2030. The Li-ion battery demand for transport may reach 65 GWh in 2020 mainly attributable to EV because of their higher demand relative to hybrid vehicles. For storage, the demand is coupled with market penetration of wind and solar and is also expected to increase. With increased demand, the likely cost will decrease to $118 $/kWh. One of the largest sub-costs for a Li battery is because of the materials for the cathode (25%), the anode, separator and electrolyte (8% each), and other components (16%) with cell assembly contributing another 35% of the cost. The cell within the battery is approximately 2/3 of the total cost. Most auto manufacturers buy the cell but assemble their own battery packs, the cost of which is rolled into the vehicle, for example, a $3700 Tesla (10 kWh NCA) battery pack, the cell of which is purchased at $470/kWh [118]. As the EV and laptop batteries include, the cost reductions toward a 5 kWh pack at <$1000 are practical in price/performance and market saturation perspectives. As the magnitude of Li batteries increases, not all Li battery chemistries will be increased. The long-term stability of LFP and specific energies of NCA types will increase at the expense of LMO. NiMH and NiCD are expected to decrease significantly, although PbA because of their proven record in stationary and motive applications will remain a strong contender for usage. With Tesla's proposed Gigafactory and the cheaper material costs of LFP batteries, the two dominant platforms are expected to be LFP and

Table 1.6 Summary lithium usage in a portable, automotive, and storage applications

System/GWh	2000	2005	2010	2015	2020
Portable electronics	1.8	6.6	21	31	45
Road transport	n.d	n.d	<4	13	76
Storage	0.5	1.8	5	45	125

Sources: Various [119]

LMO for a market share up to 2030. As the number of Li-ion batteries increases, so does the population of batteries that have reached the end of life (power < 80%) and recycling becomes significant or repacking as batteries for storage [120].

1.11 Life Cycle Assessment for Batteries

The International Organization for Standardization (ISO) has set a number of industry-wide standards that can be utilized in developing a life cycle assessment (LCA) model to understand both energy and material costs. The relevant standards pertain to ISO 14040 (Principles and Framework), ISO 14041 (Goal and Scope definition), ISO 14042 (Inventory analysis), and ISO 14043 (Impact analysis) and summarized in Fig. 1.14.

The LCA has been applied to investigate the gaseous emissions relationship in hybrid vehicles related to battery size [121]. The fuel mix to generate electricity for charging was also investigated in terms of creating a carbon neutral environment [122]. A study on the lifetime driving of an EV (93 k miles) demonstrated that global warming potential could further decrease from 10% to 24% with the substitution of ICE by EV resulting in lowered gasoline consumption [123]. The availability of EVs has led to companies generating more electricity from sustainable resources, and the fewer components resulted in fewer materials in the product chain [124]. When hybrid cars are compared using LCA, the global warming potential was 27% [125]. A simulated model incorporating design elements from LCA based on Pareto set points and a proprietary algorithm demonstrated that NiMH offered optimal fuel economy and Li-ion batteries the lowest operating cost and greatest global warming potential reduction relative to PbA [126]. The impact on the environment in terms of the

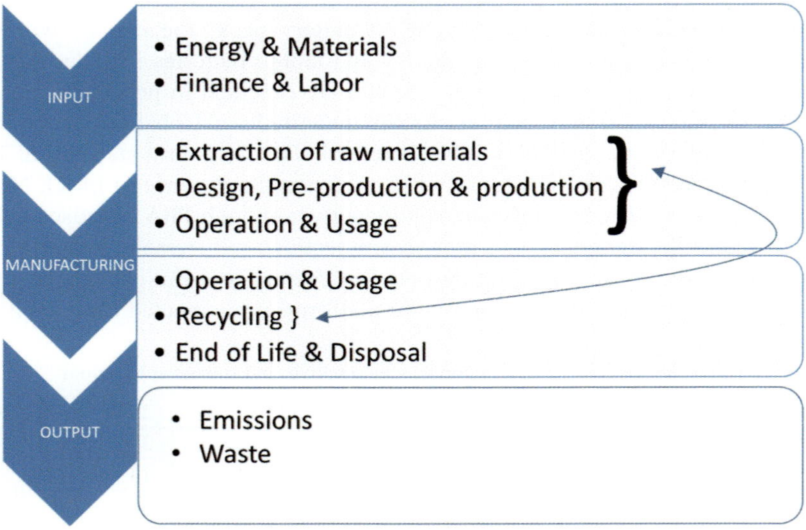

Fig. 1.14 Life cycle assessment for batteries and automobiles

generation of gaseous pollutants and volatile organic compounds was compared between NiMH and LCO cells, and it was concluded the highest environmental impact was with NiMHs.

1.12 Effects of Recycling on Costs and Adaptation in Electric Vehicles: Material Limitations and Potential Toxicity in Disposal at End of Life

Recycling is cost-effective since it minimizes the more energy-intensive extraction of raw materials process. Typically for batteries, the metals are recovered via smelting and refining. From this process transition metals such as cobalt, nickel, and manganese are recovered and can be reutilized in the car. Lithium and rare earth elements reside in the slag and are used as a building material in roadbeds and pavements. An alternative approach to recover rare earth is by leaching, which will become more paramount as the number of Li-ion cells that are at the end of life becomes significant, which is anticipated in the next decade [127].

LCA has also been applied to public and urban transport platforms such as buses, in which battery type, weight, specific energy, and traffic conditions are factors that affect the environment [128].

The LCA model has general ramifications that can apply to all industries, where decarbonization targets the reduction of global warming potential requiring rare earth and transition metals such as dysprosium, neodymium, cobalt, tin, nickel, and manganese. In addition to Li-ion batteries, metal oxides or salts of lithium are required [129]. These resources are not available uniformly and are concerned in a few regions. This requires manufacturers to strike long-term price agreements and researchers to transition to more available materials, in addition to ramping up recycling efforts [130] (Table 1.7).

Lithium is mined from ores of petalite ($LiAl(Si_2O_5)_2$, lepidolite $K(Li, Al)_3(Al, Si, Rb)_4O_{10}(F, OH)_2$, spodumene $LiAl(SiO_3)_2$, and also subsurface brines. Chile and Australia are the world's largest producers of lithium with their combined production being greater than 75% of the world's production [135]. In brine, it is in the form of lithium chloride, which, when treated with soda ash (sodium carbonate, Na_2CO_3), precipitates into lithium carbonate [136]. Almost 80% of extracted lithium is from brine. In 2015, 65,108 tons of lithium carbonate were extracted [137] with 39% being used in the batteries sector, 30% in ceramics and glass, 8% in lubricating greases, 5% in polymer production, 5% in casting mold flux powders, and 13% others. The consumption for lithium in the battery sector has increased from 20,026 tons in 2008 to 77,821 tons in 2016, an almost quadrupling over 8 years [138]. The estimated battery storage capacity between 2020 and 2030 is approximately 2.5 TWh. This capacity translates to approximately 400,000 of lithium carbonate at 0.16 kg/kWh or almost 3% of the global reserve. If recycling, factored in other metals such as cobalt, can be recovered [120], it is now part of the framework incorporating LCA [139]. The average life cycle of Li-ion batteries in portable electronics is 5 years; for automotive applications, it is 12 years; and in power supply and storage, it is 10 years. The lithium recovery percentage is dependent on

Table 1.7 Summary of the major components used in Li cells and battery packs

Crust abundance	Element	Component	Mined (ton/y)	Industry share (%)	Resource as limiting factor (Y/N)
3	Al	Cathode current collector NCA cathode	11,000	< 1%	N
4	Fe	LFP cathode	82,000	< 1%	N
9	Ti	LTO anode	3,960,000	< 1%	N
11	P	LFP cathode	47,000,000	< 1%	N
12	Mn	LMO cathode	16,000,000	< 1%	N
17	C	Anode, electrolyte	1,200,000	< 3%	N
22	Ni	NMC, NCA cathode	2,250,000	< 2%	N
25	Cu	Anode current collector	19,400,000	< 1%	N
30	Co	LCO, NCA, NMC cathode	123,000	30%	Yes
35	Li	Cathode, electrolyte	37,800	39%	Almost

Sources: [131–134]

the type of battery and system and varies from 80% for electronics to 95% for automotive/power systems. At present approximately 60%, ~30%, and <10% of recycled lithium are reused in the newer system, meaning on average, 40% of lithium is recycled into new components. Unfortunately, there is no systematic or uniform back distribution system, where end-users can return the spent batteries [140] and most batteries end up in disposal sites or landfills in the United States at end of life [141].

Lithium usually does not bioaccumulate in humans and its environmental toxicity is low [142]. Lithium is not an essential element for plants although it does stimulate plant growth [143]. Lithium has been used to treat bipolar disorder even though excessive doses can be fatal. For example, 10 mg/L in serum is the recommended treatment regime; however, patients that have taken more exhibit the signs of elemental poisoning (10 mg/L of blood). At higher (15 mg/L) doses, patients exhibit confusion and slurred speech that becomes fatal at 20 mg/L [144]. A provisional recommended daily intake of 14.3 μg/kg body weight lithium for an adult has been suggested [145]. To put these figures in context, 1 g/L alcohol will give the average individual symptoms of slurred speech, and 3–4 g/L (9 μg/kg) is considered fatal [146]. Lithium will undergo price volatility over the next decade as the newer long-term contracts are renewed or older mines close down. In glass manufacturing, other Alkali metal such as potassium and sodium can replace lithium in glass and ceramics, and calcium and aluminum can be used in lubricating greases [147], at present assuming $140/kWh, with lithium material intensity at 0.15 kg/kWh and $13,900 per metric ton (2017). This price is almost double the price from 2016. The percent of lithium in a battery pack and price influence is <1.5%; therefore price fluctuation in the raw materials is unlikely to affect electronics or automakers.

Cobalt is mined in Canada and Morocco but also is recovered as a by-product of copper or nickel production from mines in Democratic Republic of the Congo, Russian Federation, Australia, and Cuba in the form of minerals cobaltite and smaltite (cobalt arsenide). Other cobalt-bearing minerals include erythrite, glaucodot, and linnaeite (cobalt sulfide) [148]. The Congo supply approximately 54% of the global demand (64,000 metric tons). Cobalt is used as a cathode material in Li-ion batteries and also in alloys and cemented carbides [149], in addition to being an essential nutritional element. Approximately 30% of the globally extracted cobalt is utilized by the Li battery sector [148]. Unlike Li, the share of Co to the final cell price is significant. Assuming \$140/kWh, with lithium material intensity at 1.4 kg/kWh \$81,000/ton (from \$30,000 in 2017), the cost share of cobalt in the cell is 81%. For NMC and NCA cells, that share would be <10%. In these cases, price fluctuations could affect battery-pack prices in EVs. This critical need can be addressed by recycling and using cobalt-free cathodes with similar specific energies. Recycling could salvage up to 50% of the cobalt [150], and implementation of cobalt-free chemistries in LFP type, stationary storage, and alternates to cobalt in alloys and structures could also extend its usage [151].

Other less likely shortages are because of nickel and graphite, although these are not as critical as with cobalt [152]. Approximately 1.2 million tons of graphite is mined annually of which 3% is directed to the Li-ion battery sector [148]. The concern is that 65% of the mines are in China. The current mining of nickel is 2.25 million tons over the Philippines (22%), Russian Federation (11%), and Canada (11%) [148] and its use for metal alloys and in NMC, NCA, NiMH, and NiCd batteries, of which 3% is devoted to the battery sector. Li-ion batteries are classified as hazardous because of their lead (6.29 > 5 mg/L), cobalt (0.16 kg > 0.008), copper (0.0987 > 0.0025 kg/kg), and nickel (0.0095 > 0.002) content. The disposal of batteries in landfills often led to leaching of chromium, lead, and thallium leading to eco and human toxicity particularly for Co, Cu, Ni, Th, and Ag [153].

1.13 The Next Evolution of Li Battery Development Technology

The design of batteries has remained constant over the last 100 years with two electrodes, a conducting electrolyte and separator. The electrons move accordingly to the electrochemical potential difference between the electrodes, as a function of their unique chemistries via an external circuit from the more negative to the more positive potential. Ions are transported through the electrolyte/separator to maintain charge balance, and electrical energy is used. In rechargeable batteries, a reverse applied voltage pushes the electrons in the opposite direction. The magnitude of the electrical energy per mass is related to the cells voltage and capacity which in turn is related to cell chemistry.

The chemical to electrical potential can be enhanced by engineering a larger chemical potential difference between the electrodes, decreasing the volume of the reactants per exchanged electron, and maintaining the electrolyte (e.g., for NiMH and Li-ion batteries).

In the lithium-ion battery, Li^+ participates at both electrodes, being either inserted or extracted from the appropriate electrode. Lithium is attractive because of its high specific capacity (3860 mAh/g) and low redox potential (~ 3.0 V SHE). Here Li^+ is exchanged between the anode consisting of graphite (Li_xC_6) and transition metal oxides as a cathode ($Li_{1-x}TMO_2$). The transition metal oxides are cobalt, nickel, or manganese. This system can store up to 180 Wh/kg with an output of 3.8 V, much greater than lead-acid (PbA) batteries. While electric vehicles (EV) do not produce carbon dioxide (CO_2) emissions at the tailpipe, the manufacturer of an EV and/or power source is not carbon-free. Approximately 70 kg COs/kWh is produced [154]. The breakpoint where EV become carbon neutral related to the power cell is approximately 120 charges (with 35% efficiency from the production of electricity from coal and 20% for an ICE), not in itself sufficient to lower the global warming potential by 2°, a suggested target. The next generation of batteries (Fig. 1.15) will need to increase specific energy and lower the carbon footprint.

The initial increase of electrode surface area leads to increased reactions with electrolyte resulting in electrolyte decomposition. However, when the active cathode material is intercalated between graphites, the secondary oxidation or reduction reactions were greatly reduced [155]. By carefully placing the TMO between layers of carbon, the kinetics are optimized, because of lower diffusion path for ions and electrons. Through careful adjustment of the packing plane or lattice, metal insertion or conversion can be achieved [156]. In metal insertion the local covalency and density of states of a semiconductor such as titania (TiO_2) are altered through the careful addition of metal or metal oxide, enabling the band gap to be lowered or increased [157]. In conversion, the TMO is converted to metal (xM^0) nanowire-type morphology embedded within the MO (e.g., Li_2O_2, Fig. 1.16). If the conversion "islands" are not controlled, there will be a large voltage difference between the charge and discharge modes, resulting in poor energy efficiency.

Similarly, nanodesign has been applied to solid electrolytes such as Al_2O_3 or TiO_2 dispersed in polymers for fabrication of polyether-based electrolytes to increase conductivity at higher temperatures (80 °C). Another approach to increasing conductivity is through local ordering using co-block polymeric fibers [159]. Since the hydrophobic and phase separation can be better controlled, this is one approach to optimize the specific energy of the battery. Ionic liquids that have low vapor pressures, which are nonflammable and have high conductivities, will enable batteries without thermal runaway to be fabricated and can potentially assist with overheating, although cost and low-temperature performance have not been demonstrated on par with current technologies.

Since lithium and cobalt are the limiting factors in terms of scale-up and resource allocations, the transition to cobalt-free and recycling of bio-batteries are potential solutions to scarce resources.

The most precise control at the biological level is deoxyribonuclease and protein synthesis that is templated and enzymatically driven to generate precise and well-ordered systems. The enzyme-mediated synthesis of $LiFePO_4$ cathodes would solve

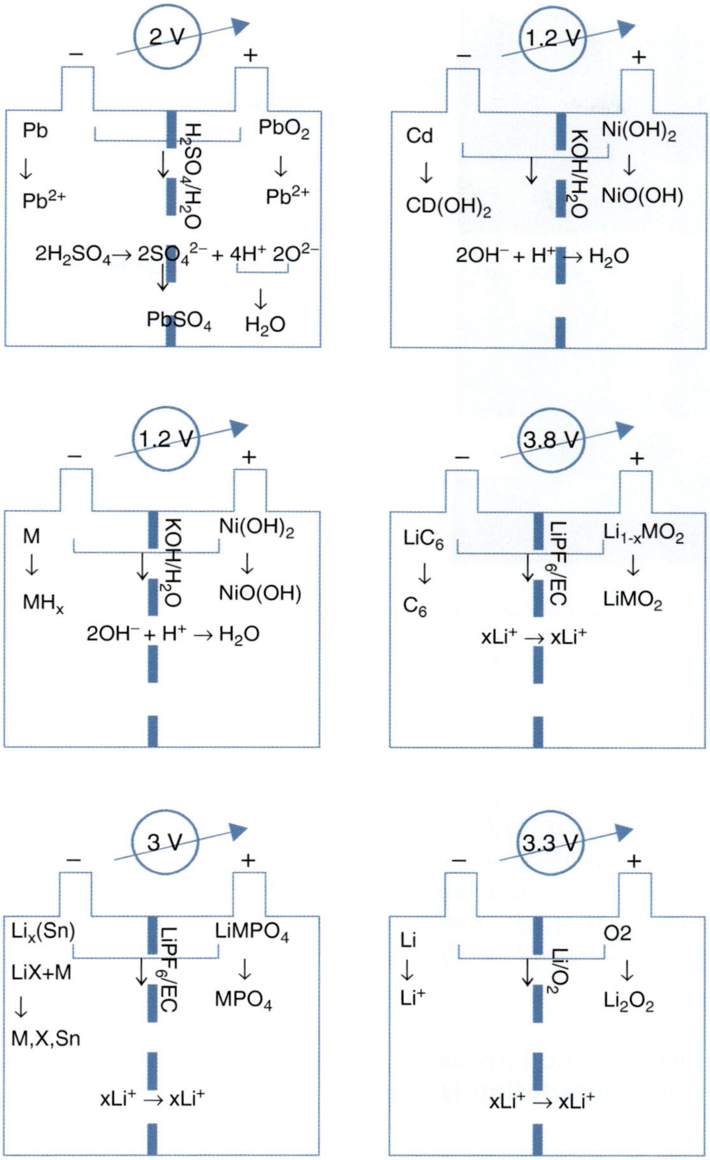

Fig. 1.15 Schematic of batteries showing electrolyte, anode, and cathode materials and generated voltage. (Source: various (including [72]))

many problems related to ordering and scarce resources [160] where virus or bacterium has been used to engineer Co_3O_4- and MnO_2-based electrodes, respectively.

Many systems also employ iron-sulfur or iron storage in the form of apoferritin (protein without iron) and ferritin (with iron) [161]. Mimetic polymers could serve as

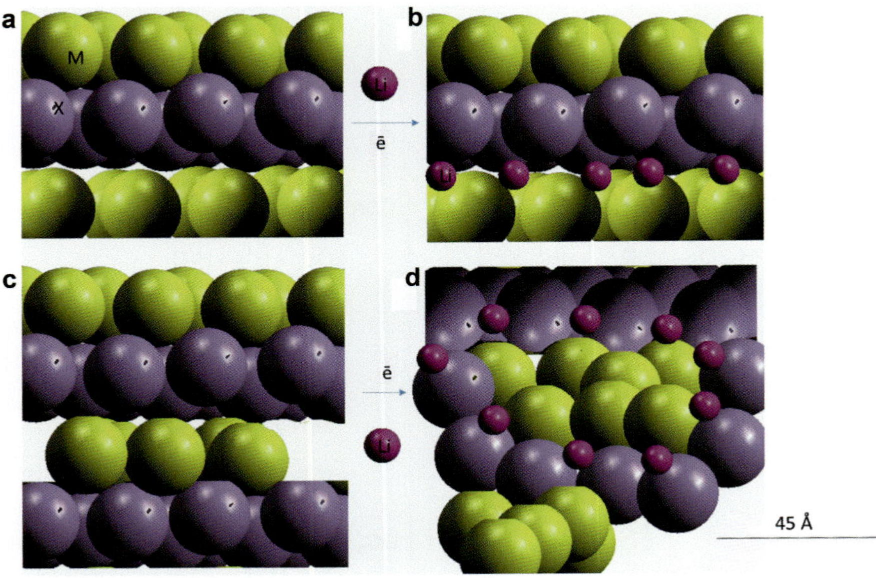

Fig. 1.16 Summary of insertion and conversion mechanism. In insertion, the Li^+ is intercalated between M and X layers, involving 1 electron per metal (MX_2 _discharge→ $LiMX_2$ _charging → MX_2 + ē). In conversion, the M atoms cluster into discrete structures forming a "black-and-white corn-on-the-cob motif." The reaction transfers between 2 and 6 electrons per metal (MX _discharge→ Li_2X + M – _charging → MX_2 + ē). (Source: Various (including [158]))

templates to chelate iron adsorbed on the surface of a conversion electrode to control the deposition/dissolution of TM_xO_y and Li_2O crystals. Other approaches include the use of chemically active conjugative molecules like polyaniline [162] or redox polymers [163, 164] and sugars that can chelate lithium $[C_6H_{12}O_6]n$ + $2nLi^+$ 2nē -> $C_6H_{10}O_5Li_2$ + 2 H^+ [165]. Since carbohydrates and polyketones are structurally related, they could be used for the design of biocompatible electrodes [166].

1.14 Lithium and Lithium-Oxygen Battery Current Performance and Degradation Issues

Air electrodes and metal air are currently used in fuel cells but have the potential to be adopted for Li-ion batteries, where 2Li is oxidized to Li_2O_2 which can, in theory, have a specific energy of 372 mAh/g with the highest being around 350 mAh/g [167]. Because of the large activation energy, in fuel cells, platinum is used to overcome this barrier and facilitate catalysis. Porous 3D electrodes can separate gas (N_2, CO_2) or water away from either Li or Li_2O_2 to increase the energy efficiency, and ionic liquids that are hydrophobic will reduce water entry and contamination of the electrode surface, although at present hydrophobic ionic liquids and hygroscopic lithium salts have low compatibility and solubility.

Lithium as an anode has been deposited unevenly forming dendritic structures that lead to low efficiencies and overheating, because of passivation of the outer layer of lithium metal [168]. The application of dry polymer electrolytes has reduced but not eliminated passivation of lithium. This is for two reasons: (1) high fermi energy resulting in thermodynamic instability promoting reactions with electrolyte and passivation and (2) generation of solid electrolyte interphase (SEI) layers, in addition to consumption of electrolyte resulting in increased internal resistance and lowered cycle life. The high volumetric aspect ratio of Li metal at the anode during periodic plating and stripping processes causes dendritic to be formed and lower efficiency.

1.15 Suppression of Degradation via Electrolytes

Strategies to reduce dendrite growth are via optimization of the electrolytes. If co-additives that promote uniform morphology are added, dendrite formation can be minimized. One example is the use of cesium hexafluorophosphate ($CsPF_6$) in propylene carbonate (PC), resulting in the formation of a smooth Li film with well-ordered nanorods on the copper substrate. The copper surface is passivated by lithium fluoride (LiF) in situ from the reduction of PF_6^- and the shielding effect of Cs^+ [169]. Other approaches include the growth of Li nanorods in 1 M $LiPF_6$/PC electrolyte and trace (50 ppm) water [170]. The hydrolysis of $LiPF_6$ generated HF which was electrochemically reduced to LiF prior to deposition of Li, from the LiF-rich solid electrolyte interphase (SEI) layer on the Cu substratum without an increase in Li Coulombic efficiency. With 2% $LiAsF_6$ and 2%, vinylene carbonate (VC) or fluoroethylene carbonate were added to 1 M $LiPF_6$/PC electrolyte, smooth Li deposition with highly ordered Li nanorods were obtained, and Coulombic efficiency was increased to 96.5% [171] with the electrochemical reduction of $LiAsF_6$ to LiF and Li_xAs. Here, LiF passivated the substrate, and Li_xAs was seed species for the growth of Li. The electrolyte VC or fluoroethylene carbonate was polymerized that bound the inorganic compounds and adhered to the Li metal surface, protecting the Li metal anode and cycle stability of the cell [172]. A similar approach used 4 M lithium bis(fluorosulfonyl)imide (LfSI) in 1,2-dimethoxyethane [173]. The reactions between Li metal and electrolyte were decreased because of the larger particle size, which increased Coulombic efficiency to 99% and the inhibition of dendrite formation in propylene carbonate electrolyte using 4.9 mol/kg of LiFSI ionic liquids [174]. Lithium bis(fluorosulfonyl)imide has also been used with a variety of solvents such as acetonitrile [175], 1,2-dimethoxyethane [173], dimethyl sulfoxide, and lithium nitrate ($LiNO_3$) instead of LiFSI [176]. Lithium bis(trifluoromethanesulfonyl)imide (7 M, LiTFSI) in 1,3-dioxolane and 1,2-dimethoxyethane [178] and LiTFSI in 1,2-dimethoxyethane [177] have all been used to suppress Li dendrite formation in Li-ion and Li-air batteries [174]. In all the above studies, highly concentrated electrolytes were used that promoted salt-solvent coordination, with fewer solvents available for Li metal to react with and the formation of an ionic-liquid interface [179], causing further degradation of the dendritic formation. The lower conductivities of these ionic liquids are offset by their high thermal stability, low volatility, high carrier density, and high reduction/oxidation efficiency [180].

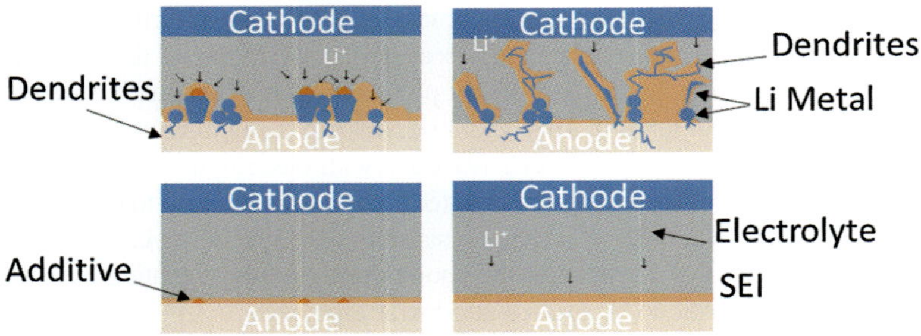

Fig. 1.17 Schematic showing formation of dendrites with co-additive suppressors and without additives. Under cell operation, the Li is stripped and deposited over cycles and forms dendritic formations at the anode. Strategies used to minimize these include high salts, e.g., LiF, polymers and solid electrolytes. The electrolyte with additives deposited more Li during cycling. Upon discharge, the Li stripping is more complete because of larger grain size and more uniform Li deposition, resulting in less dendrite formation and side reactions with electrolyte components. Without any additives, these processes are not inhibited resulting in dendrite formation and lower cycling Coulombic efficiency. (Sources: Various (including [194]))

Li deposition in the form of nanowires can also promote cycling and retard dendritic formation and stability at the anode. The use of dual-salt electrolytes of LiTFSI and lithium bis(oxalato)borate (LiBOB)in ethylene carbonate and ethyl methyl carbonate (E/MC) generates uniform fibrous growth, which protects the metal Li anode. This arrangement leads to long cycling stability and fast charging with high-energy cathode materials ($LiNi_{0.4}Mn_{0.4}Co_{0.2}O_2$/ethyl carbonate, NMC-type battery [181]. The addition of 0.05 M $LiPF_6$ to the LiTFSI-LiBOB/E/MC generates compacted wire-like Li deposits with conductive solid electrolyte interphase (SEI) films on the Li metal surface. In addition, the remaining $LiPF_6$ initiated polymerization of the EC molecules to form polycarbonates on the Li metal surface and also to protect the cathode substrate aluminum from corrosion by TFSI anions. Likewise, the Li metal anode is similarly protected by the inorganic/polymeric nanocomposite SEI layers during periodic Li deposition/stripping cycles, and the uniform Li wires promote high charging capability because of the lower charge-transfer resistance of the SEI films, lowering their charge-transfer resistance [182]. If the polymeric/electrolyte at the SEI layer is conductive, flexible, and compact, protection against dendrite formation and enhanced Coulombic efficiency can occur; however, if the morphology of the wire/fiber/polymeric nanocomposite is highly porous, insulating, or rigid, then protection does not occur, resulting in lower Coulombic efficiency and shorter cycle lifetimes (cf. Figs. 1.16 and 1.17).

1.16 Suppression of Degradation via Membranes

Liquid electrolytes flow and interact with the porous electrolyte resulting in side reactions that degrade Coulombic efficiency [183]. Solid-state electrolytes do not have this limitation; however, their low conductivities at room temperatures ($< 10^{-4}$ Scm^{-1}) require high temperatures (80 °C) as stated earlier in the chapter

[184]. The self-assembled polyanionic BAB triblock copolymers (see Fig. 1.16) finely tune mechanical properties, ionic conductivity, and lithium transport number at the same time. A triblock (ABA)-type polyelectrolyte based on poly(styrene trifluoromethanesulphonylimide of lithium) P(STFSILi) associated with a central A block based on a linear poly(ethylene oxide) (PEO) and terminating with the B-block was evaluated and demonstrated high ionic conductivity (1.3×10^{-5} S cm^{-1} at 60 °C), carrier transport number (> 0.85), and electrochemical stability u to 5 V (relative to Li/Li) [185]. The mechanical strength at these temperatures was reduced, slowing but not eliminating dendrite formation.

Inorganic solid-state electrolytes contain a heteroatom such as sulfur and have been shown to be conducted at room temperature. Other electrolytes based on oxygen, SiO_4, have slightly lower conductivities (10^{-3} Scm^{-1} for LiSICON and 10^{-4} Scm^{-1} for garnet solid-state electrolytes, respectively) [186]. All of these electrolytes are sensitive to moisture or air (sulfur) or are not stable against Li metal (oxygen) and require special insert fabrication environments [187]. While garnet-type solid-state electrolytes are stable with Li metal, the grain boundaries of the garnet promote Li dendrite formation, lowering the cycle life of the cell [188], in addition to the general higher interfacial resistance of solid-state electrolytes relative to the liquid carbonate-based electrolytes [189].

The interfacial resistance between the solid-state electrolyte based on garnet and Li metal anode can be reduced by fabrication of three-dimensional ion-conductive networks deposited by atomic layer deposits. Examples of such as an approach include deposition of $Li_7La_{2.75}Ca_{0.25}Zr_{1.75}Nb_{0.25}O_{12}$ (LLCZNO) electrolyte on Al_2O_3 [190] or $Li_7La_3Zr_2O_{12}$ (LLZO) on Ge [191] or $Li_{6.4}La_3Zr_2Al_{0.2}O_{12}$ (LLZAO, [192]). These composites increase the Li conductivity of the solid-state electrolyte and decrease interfacial resistance. The sandwich of LLCNO-Al_2O_3 has no void space and excellent contact with Li, whereas LLCNO alone has pores and voids with poor interfacial interaction. Thin layers of LLCNO-Al_2O_3 improve the wettability of the garnet surface, similar to $LiZr_5(PO_4)_3$ that was electrochemically stable to 5.5 V with low interfacial resistance to Li-ion transfer [193]. Even after 100 h testing, the electrolyte retained its solid dense morphology without any dendrite formation, suggesting that co-addition of $LiZr_2(PO_4)_3$ solid electrolyte can inhibit Li dendrite formation. Thus, unlike a liquid electrolyte, the solid type can act as a membrane to block dendrite formation and to stabilize the anode (Fig. 1.17).

The Li metal anode can also be stabilized using SEI membranes; examples include use of an elastomer/lithium phosphorous oxynitride (LiPON) SEI membrane by using the mechanical protective properties of the elastomer layer and the chemical protective of the LiPON layer, allowing the anode to be stable up to 100 plating/stripping cycles at 2 $mAcm^{-2}$ [195].

Atomic layered deposition, used to coat Al_2O_3 at the anode, was able to protect the Li surface from corrosion from electrolytes, sulfides, and air, improving cycling stability of lithium-sulfur cells. Through surface structural engineering, a functional space was provided for Li to grow in the forms of columns rather than dendrites on the Cu foil. The arrangement of interconnected hollow carbon shells was able to provide an avenue for the growth of Li-ion flux instead of deposition of Li in the form of treelike dendrites [196]. This two-dimensional arrangement was extended in three dimensions with the fabrication of carbon granules and enable over 95% of Li utilization (as Li-ion instead of Li metal dendrites) [197]. Other approaches include mechanical and chemical treatment of separators, where one face is coated with a conductive material facing the anode. The Li-ion flux toward the conductive face forms a dense dendrite layer without extending toward the cathode. A non-conductive 3D member was recently reported, where Al_2O_3 membranes infused with liquid electrolytes of LiTFSI/1,3-dioxolane/1,2-dimethoxyethane [198] or LiTFSI/1-methyl-3-propylpiperidinium/SiO_2/propylene carbonate. The propensity of dendrite formation was inhibited by using pore size in size-exclusion modality, limiting the size of potential dendrites which could be formed. Other strategies using the conceptualized approach described use porous polyimide 3D separator [199] or polymer nanofiber separators [200].

1.17 Suppression of Degradation via Structural Redesign of the Anode

Most anodes have a tubular or flat 2D morphology which during cell cycling (charging/discharging) undergoes physical change because of Li stripping or deposition and often results in dendrite formation because of uneven Li-ion distribution on the electrode. To alleviate low Coulombic efficiency, the current collectors have been constructed from copper [201], nickel [202], graphene [203], or carbon fibers [204] to minimize morphological changes in the anode structure over prolonged use of the cell. One approach relied on the use of skeletal templating where the current collector was made from a reduction of $CuOH_2$ fiber arrays [205]. The dendrite tips have a large charge density and act as the focus of Li deposition within the electric field. By creating an overlaid Skelton, the ends of the fibers act as nucleation sites for growth because of lower and uniform charge, leading to less sharp contours of deposited Li and higher Li cycling stability. Other potential designs are a fabrication of vertical column geometry, such as nanowires, to decrease Li-ion "hot spots" and to generate a more uniform ion distribution throughout the structure. Since there are fewer charge centers, there are fewer dendrites, relative to a 2D planar electrode as a current collector [206]. Graphene and graphene oxide (GO) have also been evaluated because of their intrinsic properties [207–209], where, often, Li metal is layered between the reduced GOs, generating a Li-rGO-Li-rGO multilayer composite. This arrangement lowers potential volume change at the anode [210]. Alternatively, modified Li anode on stainless steel and titanium formed a "sawtooth"-type arrangement, with Li nucleation being directed by the nano-notches at the expense of dendritic growth [211]. The processes discussed above are summarized in Fig. 1.18 and Table 1.8.

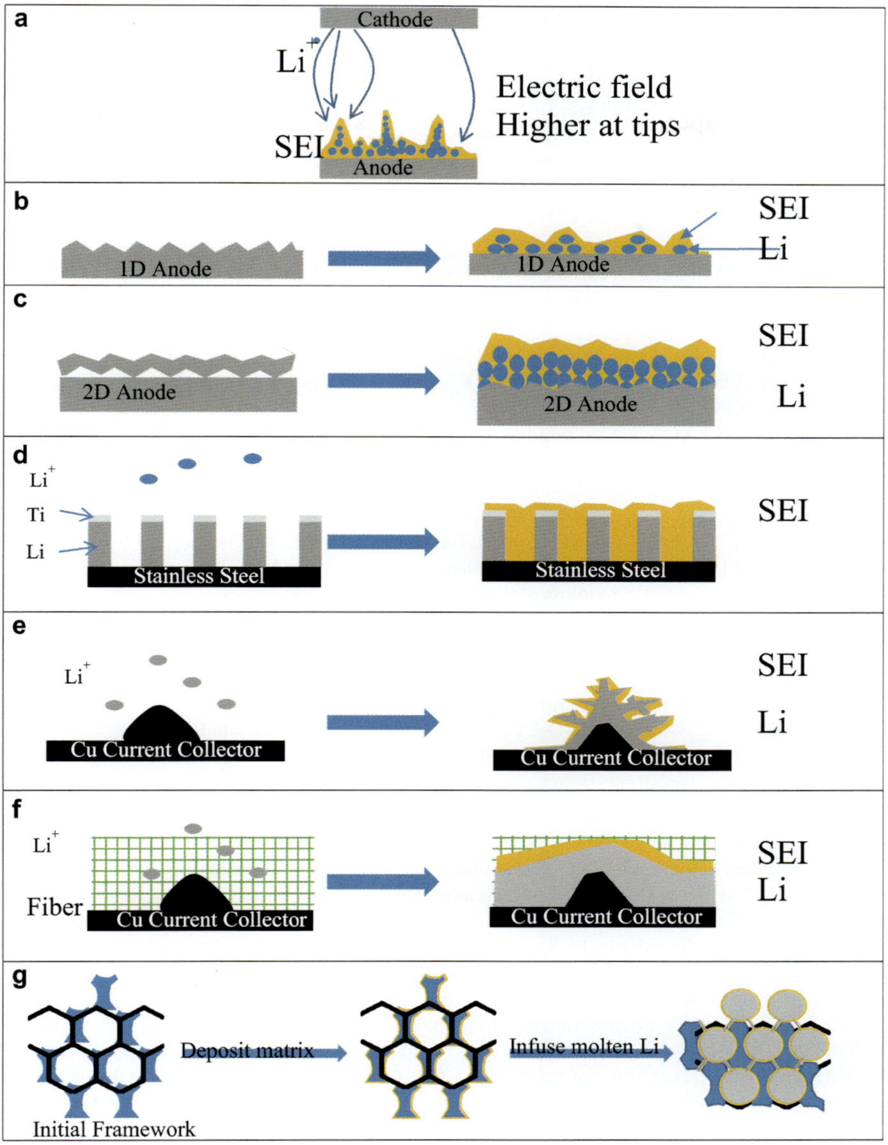

Fig. 1.18 Various schematics of anode geometry/current collector to reduce dendrite formation. Schematic of planar electrode design, where Li-ion flux and Li plating occur between the electrodes (**a** and **e**). Here because of the high charge at the "tips," dendrites are resultant from reactions between Li and electrolyte that generate thick solid electrolyte interphase (SEI) layers on the Li metal surface, adjacent to the current collector. Dendrite formation arising from 1D (**b**) or 2D (**c**) or "sawtooth" (**d**) structured anodes. By preparation of a mesh, nanotubes, nanowires, glass fibers (**f**) modified current collector the electron distribution is more even and the generation of local high current density is diminished resulting in better cell performance and lower Li plating/stripping overpotential

Table 1.8 Summary of use of electrolyte additives used with the anode to enhance SEI stability and cell performance

Electrolyte	SEI	Performance	References
0.18 M Li_2S_8, 5 wt% $LiNO_3$ and 1 M LITFSI in DOL/DME	Li_2S, Li_2S_2, LiF with less $LiCF_3$, $Li_2NSO_2CF_3$ compound	CE: 99%/ 2 mA cm^{-2}/ 300 cycles	[213]
FEC additives (5% v) and 1 M $LiPF_6$ in EC/DEC	LiF-based SEI layer with Li_2CO_3 and $ROCO_2Li$	CE: 95%/ 0.5 mA cm^{-2}/ 100 cycle	[172]
0.05 M $CsPF_6$ additive and 1.0 M $LiPF_6$ in PC	LiF-based SEI layer with $ROCO_2Li$	Formation of nanorods not dendrites of Li	[214]
Trace amounts (25–50 ppm) of H_2O and 1.0 M $LiPF_6$ in PC	LiF-based SEI layer with $C_3H_6(OCO_2Li)_2$	Aligned and high-density films of Li nanorods	[170]
10 mM KPF_6 and 1 M $LiPF_6$ in EC/DEC	Li^+ conductive SEI increased LiF and carbonyl-containing species and reduced insulating	Voltage excursions of 50 mV in symmetrical Li cell for 18 days	[215]
0.5 wt% methyl, 1 wt% $LiNO_3$ and 1 M LiTFSI in DOL/DME	A stable interfacial coating consisting of planar viologen oligomers	CE: 98.2%/ 1 mA cm^{-2}/ 400 cycles	[212]
0.020 M Li_2S_5,5 wt% $LiNO_3$ and 1 M LITFSI in DOL/DME	Mainly inorganic LiF and Li_2S_x	CE: 95%/ 1 mA cm^{-2}/ 233 cycles	[216]
0.2 M Li_2S_6,0.8 M LiTFSI in DOL/DME	Bilayer of $Li_2S_2O_3$, Li_3N, $^-NSO_2CF_3$ (top) and Li_2S, Li_2S_2 (bottom)	Stable overpotential/ 0.4 mA cm^{-2}/ 130 cycles	[217]

Key: $C_3H_6OCO_2Li_2$ = lithium hydroxy({[hydroxy(oxido)methoxy]methoxy})methanolate, CE = Coulombic efficiency, $CsPF_6$ = cesium hexafluorophosphate, DEC = diethyl carbonate, DME = dimethoxy ethane, DOL = 1,3-dioxolane, EC = ethylene carbonate, FEC = fluoroethylene carbonate, H_2O = water Li_2CO_3 = lithium carbonate, $Li_2NSO_2CF_3$ = (trifluoromethoxysulfinyl) azanide lithium, Li_2S = lithium sulfide, Li_2S = dilithium sulfide, Li_2S_2 = (lithiodisulfanyl)lithium, $Li_2S_2O_3$ = lithium thiosulfate, Li_2S_6 = dilithiohexasulfane, $LiCF_3$ = (trifluoromethyl)lithium, LiF = lithium fluoride, LiN_3 = lithium nitride, $LiPF_6$ = lithium hexafluorophosphate, LiTFSI = lithium bis(trifluoromethylsulfonyl)imide, $mAcm^{-2}$ = milliampere per centimeter squared, $^-NSO_2CF_3$ = [Dilithio(sulfamoyl)methyl]lithium, PC = propylene carbonate, ppm = parts per million, $ROCO_2Li$ = lithium(hydroxy) methanebis(olate) if terminating with H; (ROCO2Li = HOCO2Li); if R, use group name, for example, if R = CH_3; then the name is lithium (methoxy)methanebis(olate) ($ROCO_2Li$ = $H3COCO_2Li$), SEI film = solid electrolyte membrane

1.18 Environmental Impact Policy Road Map in Lowering Global Warming Potential

The environmental impact of the disposal of used batteries is a component within the LCA framework. To this analysis, the CO_2 equivalence (eq) must be calculated. This is the energy required to extract, separate, purify, and fabricate the battery cell and is approximately 170 kg CO_2-eq per kWh of battery capacity [218]. Since batteries are electrochemical cells, they discharge and charge in cycles, and the depth of discharge is 85%, with a battery life of 1500 cycles; the approximate travel of an electric vehicle would be 6 km/kWh which is equivalent to 2.2 kg CO_2-eq per 100 km traveled. By increasing the Coulombic efficiency through better anode design, the increase in battery life to 2000 cycles and reduction of CO_2-eq from 2.2 kg to 1 kg CO_2-eq per 100 km traveled, the total carbon emissions can be reduced to 100 kg CO_2-eq per kWh of battery capacity. This corresponds to approximately burning 11.24 gallons of ethanol-free gasoline or 9.8 gallons of diesel [219] or 2.38 kg of CO_2 from burning 1 kg of coal (assuming that 65% by mass is carbon). Other important factors that impact the environment are the use of copper, lithium, and aluminum. While lithium is not toxic, the extraction from seawater by evaporation, ion-exchange cleanup, purification, and carbonation is a high-energy process and will generate more carbon dioxide than is eliminated by savings in gasoline consumption/combustion [220].

A study compared two equivalent auto models (VW Gold and eGolf) and determined the former would utilize 8 l of gasoline to travel 100 km, generating approximately 9 kg of CO_2 through direct combustion, while the latter would utilize 16 kWh of energy, and CO_2-eq was 9 kg CO_2-eq, while in less industrial advanced countries, the output was between 0.9 kfCO_2/kWh (India) and 0.05 kgCO_2/kWh (Paraguay) [221]. To reduce global warming potential, replacement of ICEVs by EVs alone is not sufficient. An integrated approach is required by moving toward sustainable energy sources including nuclear power and carbon capture technologies or decarbonization of generation of power. Over the last 5 years, the fastest-growing sector in energy has been sustainable energies related to wind and solar. There has been a decline in coal and a slight increase in the use of natural gas for the generation of electricity. In the United States, natural gas is the single largest energy carrier used in the generation of electricity, which can also tap into waste in cogeneration systems to heat the plant itself [222].

1.19 Addressing Role of Batteries in Addressing Intermittency: Off-Grid Storage and On-Grid Vehicle-to-Grid Applications

Intermittency of wind and solar is addressed through energy storage with a variety of battery platforms including Li batteries to provide a localized off-grid energy solution [223]. In addition, some communities have adopted a distributed grid system with local wind or solar power within the homes or nearby, depending on generating capacity [224]. The locally generated power is stored and consumed without using the general or national grid. An expansion of this concept would require secondary storage including the use of existing capacity in the form of vehicle-to-grid (V2G) to facilitate the electrification of the road and home (e.g., solar home systems) network [225]. The role of recycling is often overlooked but is critical in reclaiming the Li, Al, Cu, Ni, and Co and reducing the carbon dioxide emitted during the manufacturing processes [226].

The role of energy in the United States was governed by the Second World War [227] followed by the "Space Age" using a mix of regulatory incentives, tax credits, quotas, and caps from foreign competition and privatization of some federal agencies to facilitate innovation [228]. While the utility operators were few, the market price was relatively constant and encouraged private capital investment, overcapacity, and spikes in pricing. The "oil crisis" refocused federal policy on nonpetroleum energy sources such as hydro, nuclear, wind, and battery [229]; however concerns about nuclear plant safety and escalating costs led to less investment and less expansion of nuclear as an energy mix, because of limit profits from such enterprises [230]. During the Reagan Administration, the policy favored privatization and deregulation and restructuring of the electricity marketplace to increase competition and lower consumer pricing. During the last two decades, the utility operators have focused on transmission, rather than a generation to control market prices. This is reflected in the corporate structure of the various utilities with the bulk (prior to the Reagan) administration being private investor-owned utilities which supplied over 80% of the electricity. This decreased to 40% after restructuring, which split energy into generation and transmission [231]. The Federal Energy Regulatory Commission (FERC) regulates interstate natural gas pipelines, wholesale electricity markets, regional transmission authorities, and independent systems operators (ISOs). These changes have a defocused investment on energy generation, adding resilience to the transmission grid and updating aging equipment, in addition to regulating emissions of greenhouses gases, which are regulated by the US Environmental Protection Agency (EPA) [232]. The regulatory framework historically (except for nuclear and biofuels) has been state-level mandates for renewables such as state renewable portfolio standards (RPS), state net energy metering (NEM), and California's Zero-Emission Vehicle (ZEV). The Federal level has focused on rebates and tax concessions such as the Federal Production Tax Credit (PTC), investment tax credit (ITC), and federal tax incentives (FTIs) [233]. The expiration of the business energy tax credits and implementation of the 1992 Energy Policy Act attempted to stabilize the cost of electricity by renewables through a tax credit of 1.5 ¢/kWh for the first 10 years of production from wind [234]; until the Energy Policy Act of 2005 phased out tax credits, the policy was successful in promoting the development of wind power while also meeting state-mandated RPS requirements [235]. The RPS mandates require a fixed percentage of the state 's energy mix be derived from sustainable energy resources, not prescriptive in nature of which technology, and have allowed 37 states [236] to meet the requirement from wind power (68%).

1.20 US Energy Policy (1960–2016) Toward Sustainable Energy Platforms: Photovoltaics and Biofuels

Photovoltaics-based net metering policies have been adopted in 43 states, which guide the level of compensation of distributed power supplied to the grid, first deployed in California [237]. The 2005 Energy Policy Act introduced a 30% ITC for commercial and residential (ceiling of $2000) systems, with residential credits capped at $2000, [238] and are expected to expire in 2021, unless renewed.

Biofuels have had two periods of support, excise tax incentives from the late 1970s to the mid-2000s (1978–2011) and as blending mandates (2005–today), to counter high oil prices and assist the rural farming communities from transition from feed crops to energy crops, as well as address environmental concerns [239]. The 1978 National Energy Act promulgated a 4 ¢/gallon excise tax exemption for "gasohol" (i.e., gasoline with 10% alcohol/v.) which was extended to 1992 by the 1980 Crude Oil Windfall Profit Tax Act, and the 1980 Gasohol Competition Act place a 54 ¢/gallon ethanol import tariff to dissuade foreign imports. The 1992 and 1998 Energy Policy Acts extended the tax credits to other blends up through 2007. The increased use of ethanol fuels after 1998, when methyl tertiary butyl ether (MTBE), a fuel which oxygenates gasoline, was found leaking into groundwater and was a potential source of cancer [240] led to protests. In response, many states banned the use of MTBE as a fuel additive, leading refineries to use blended ethanol as a substitute [241]. The increase in ethanol usage and associated tax exemption leads to the 2004 Volumetric Ethanol Excise Tax Credit (VEETC) which replace a tax exemption with a tax credit to fuel blenders of 51 ¢/gallon of ethanol, which was reduced to 45 ¢/gallon in 2008 and allowed to expire by 2012. This increase in part was fostered by the RFS mandate which required fuel providers to blend a minimum quantity of biofuels into the US petroleum supply each year [242]. The mandate required 2% fuel oxygen content, primarily as a means to lower ozone and smog production because of incomplete fuel combustion. In addition, it promoted US fuel security and rural agro-farming jobs [243]. The oxygenates were utilized to stabilize fuels after lead [244] was phased out during the 1970s [245] because of the toxicity of lead in children and adults [246].

The impact of the initial renewable fuel standard (RFS) mandate was to add four billion gallons of biofuels to the petroleum grid by 2006, raising it by 7.5 million by 2012. The Gulf War during the Bush (W.) Administration led to higher gasoline prices, which in turn fostered a more ambitious expansion of the RFS standard. The revision mandate increased the blending targets to 36 billion gallons by 2022, nearly 1/5 of anticipated motor fuel consumption [247]. The bill conflagrated two distinct policies, the first to increase market uptake and stabilize the marketplace and fuel prices favored by the Republican Party, in addition to promoting energy independence, another Republican Party-favored policy. The House and Senate members of the Democratic Party favored expansion of ethanol biofuels to cellulosic biofuels and submandates for life cycle greenhouse gas reductions. The democratic favored policy toward greener energy and jobs within the framework of the pro-ethanol alliance [248]. Higher startup costs associated with cellulose have slowed cellulosic ethanol, and these submandates have been revised downward. In the current iteration of the RFS mandate, the fuel blend of 10% ethanol (E10) has already been met and poses no technical issues to automobiles or trucks; however additional blends of 1% (E15) and 85% (E85) require hardware changes to automobiles and trucks. While new models (2013 and beyond) can tolerate E15, older models cannot since ethanol at the higher blends is corrosive to rubber gaskets, absorbs more water that dissociates into two layers in the gas tank, and clogs up fuel lines. Only flex-fuel vehicles can tolerate E85 blends; therefore further blending would require additional changes to the ICEs.

1.21 US Energy Policy (1960–2016) Toward Sustainable Energy Platforms: Hybrids, Batteries, and Electric Vehicles

The push for cleaner air centered on a number of initiatives at the state level, principally in California, because of the propensity of smog in the major metropolitan cities. The Zero-Emission Vehicle (ZEV) mandate was designed to meet the Federal Clean Air Act (CAA) ozone standards, a component of smog [249]. The potential to lose Federal funding because of non-compliance with CAA regulations focused on policy in California to examine the role of car emissions. The state legislature set up the California Air Resources Board and tasked it to assess air quality and regulations to reduce vehicle emissions [250]. The board developed a Low Emission Vehicle (LEV) and Clean Fuels Program which included near-zero and Zero-Emission Vehicle standards and mandate [251]. The LEV standard has been periodically revised and currently requires 15% of vehicles sold in California to meet the ZEV standard by 2025, which includes flex fuels and hybrid vehicles. Because of the associated high startup costs, the lawmakers were also able to push for federal tax incentives [252].

The tax policy also coincided with the Bush (W.) Administration goal of reducing oil imports and the 1992 Energy Policy Act of opening up federal lands for oil exploration and the introduction of regulations to make automakers design and build more fuel-efficient automobiles. The bill introduced Production Tax Credit (PTC), extended previous biofuel tax credits including income tax credit for purchasing new EV and an additional income tax deduction for purchasing a clean-fuel vehicle (Clean-Fuel Vehicle Tax Dedication (CFVTD), flex fuel liquid natural gas, E85, EV, Fuel cell) that tapered down from 2002 to 2004 after which the rebates/incentives would expire [253]. The rebate drawdown was revised in further legislation, and in the 2005 Energy Policy Acts, the CFVTD was replaced by the Alternative Motor Vehicle Tax Credit (AMVTC) to promote hybrid vehicle uptake, which gave US automakers tax credits for the first 60,000 hybrids sold. This benefited US makers because of their introduction of hybrids later in the market. By this time, most of the Japanese car makers had sold more than 60,000 models (during the time period when the AMVTC did not exist [254] while allowing the EV tax incentives to expire. The 2005 Energy Policy Act allowed the EV tax credits to expire but was revived via the 2008 Emergency Economic Stabilization Act capped at 250,000 vehicles nationwide. The 2009 American Recovery and Reinvestment Act ("stimulus package") expanded this to 200,000 vehicles per automaker through 2014, which was extended by the 2014 American Taxpayer Relief Act [255] and subsequently rolled back as EV sales rose and the cost of the tax incentive increased.

1.22 Perspectives, Conclusion, and Policy Implications: Li-Ion Battery Pitfalls, Challenges, and Development Road Map

The rollout of Li-ion batteries as the primary platform for high-end electronics, cell phones, and most electric vehicles has indirectly displaced coal as the primary resource for the generation of electricity in the United States. However, Li-ion batteries are not carbon neutral, as the extraction of the metal and metal oxides requires energy that is not offset by usage in electric vehicles, although the emissions are lower than corresponding gasoline internal combustion engine-based vehicles. Li-ion batteries have the potential to become an energy sustainable resource but would need to increase change density through redesign of the anode/current collector, electrolyte, and membrane. The dependence on cobalt will need to be reduced to meet future global demand, and this effort should be integrated with integrated recycling, and more importantly, the carbon footprint should be decreased.

The electrification of the home and road transport has to be integrated with demands for power generation, and transmittance from sustainable resources instead of fossil fuels, because the lower electric vehicle (EV) emissions from vehicles are not sufficient to reduce the global warming potential below 2 °C. This requires a switch to nuclear, hydro, thermal, and sustainable resources with secondary battery storage to offset intermittency issues and load leveling. The fabrication of Li-ion batteries is an energy-intensive process, and this contributes toward greenhouse emissions. Recycling the components would promote more efficient usage and lessen the burden on mining or extraction of Li from seawater, which appear to be the driving force in current pricing and availability. A longer-term potential bottleneck is the availability and price of cobalt, a key component in cathode fabrication, and use of lithium cobalt oxide (LCO)-based cells in portable electronics. Research in cobalt-free electrodes, with high-energy densities, is required, particularly in lithium iron phosphate (LFP) and lithium manganese oxide (LMO) cells. As EVs past one million units sold worldwide (projected 1.9 million by 2018 contrasted with 81 million internal combustion engines (ICEs) globally) newer global standards are required to minimize, waste, greenhouses gases (GHGs,) recycling and battery lifecycle and cost reductions.

With two million units to be sold by 2019, the vehicle-to-grid, secondary life battery storage to extend their utility will become significant because of market penetration and wider availability. These capabilities will ensure that batteries are used to their maximum usage before battery depth of discharge (DoD) (defined as the percentage of the battery that has been discharged relative to the overall capacity of the battery) has been reached before recycling or disposal. Off-grid storage such as that used in solar home systems can provide local infrastructure, a road map for electrification of communities in rural and underdeveloped regions of the world.

The energy density of Li-ion batteries could be accomplished through Li anode redesign, where low Coulombic efficiency (CE) is attributed to the formation of dendrites, as well as overheating and potential chemical fire generation. Some of these disadvantages may be eliminated or reduced through the incorporation of solid-state electrolytes, although the Li metal at anode and anode interface optimization has not been solved. The redox flow battery platform offers potential benefits in high-energy secondary storage scenarios; however, there remain series of technical hurdles before flow batteries become mainstream. This is because of the limited redox potential of species in non-aqueous electrolytes of materials for cathodes and anodes that can generate a stable couple. Newer materials that can push the redox potential of the anode to lower values to maximize the cell voltage are required. Hybrid flow batteries with lithium metal anodes suffer from cycling instability where the lithium metal deteriorates by dendritic formation during the deposition phase, in addition to secondary reactions with the electrolyte leading to low Coulombic efficiency during repeated charging and discharging. If the short-term liquid electrolyte will continue to be used, strategies will need to be employed to reduce Li dendrite growth through restriction of changes to the volume. Current approaches focus on the metal host to limit Li volume changes during deposition. These hosts as 2D or 3D matrices limit dendrite growth resulting in CE degradation and promote nucleation of the Li thorough the matrix because of the porous nature of the template. The anode/electrolyte also needs to form a stable solid electrolyte interphase (SEI) film or a passivated protected layer to promote high CE of the Li metal anode. Generally, this is accomplished through the incorporation of high conductivity oxides or nitrides such as LiF, Li_3N, and Li_3PO_3 compounds.

The cell performance is also related to the uniformity of the electric field on the electrode surface, which in turn is related to morphology changes during deposition/stripping. Incorporation of a porous buffer layer-based transmission channel can promote uniform stripping and plating, where electrostatic shielding because of the presence of cations can facilitate homogeneous Li deposition. Anode metals with lower potential and high-energy capacity are required to replace Li metal. Newer designs focus on electron transfer between active metals at the electrode and current collectors, via a redox shuttle. The shuttle, in turn, is charged/discharged in a secondary storage tank that is filled with the active material, using $LiCoO_2$ and $LiVPO_4F$ as cathode, and lower potential materials, like $Li_4Ti_5O_{12}$ as the anode. To lower the overpotential loss, the potentials of the redox molecules need to be better matched.

The key challenges for efficient and long-term Li battery design should focus on better characterization of the nanomaterials used in the fabrications of the electrodes, using electron microscopy, templating the reaction between anode and electrolyte such that contact is minimized and CE is set at 99% or greater; electrodes should be designed to yield high charge current densities (>1 mA cm^{-2}) and area utilization (4 mAh cm^{-2}), with thermodynamic modeling of lithium intercalation with (N/S/O) heteroatoms than Li||Al or Li||Cu cells. Another redesign should factor in use of non-Li anodes to limit dendrite formation; limit Li metal volume change upon cycling ($<10\%$ volume change) particularly at the Li/Separator face and by using low electrical conductivity 2D/3D structured templates to direct Li deposition on the scaffold. In addition, Li deposition on anode current collector should be better controlled using 1/2/3 D optimized geometry current collectors. Simplifying Li metal stabilization process and overall battery box design for more efficient recycling and second life usage, as well as leach-free disposal, would lower costs and reduce the carbon and environmental burden.

Many of these design limitations require fabrication of specific topologies that rely on the application of nanomaterials to limit dendrite growth but promote Li nucleation and limit gross morphological changes during charging/discharging cycle which will be extended to newer platforms such as Li-sulfur and Li-air technologies. While the focus of the current book chapter is on Li-ion batteries, storage technologies also include lead-acid, zinc bromide, sodium sulfur, sodium-nickel chloride, polysulfide bromine, vanadium redox, and nickel cadmium which offer solid cycling capacity and intermediate power densities and high round-trip efficiencies.

1.23 Perspectives, Conclusion, and Policy Implications: Energy Storage and Recycling

A storage policy should include a mix of battery technologies and wind or photovoltaics (PV) for energy generation. The cost of energy generation using life cycle assessment (LCA) should also include the loss of power probability, intermittency, and environmental impact. As the loss of power probability decreases, the cost of energy and cells in the storage stack increases, requiring some optimization between cost and reliability. As the population of EVs localized sustained energy in the form of wind and solar increases, the environmental impact of lead, cadmium, cobalt, and aluminum should be considered as part of an LCA. Wind/PV systems for battery DoD values ranging from 50% to 80% (stationary power) and 80% to 100% (EVs) would require battery cell replacement at more frequent intervals than the project life replacement of 25 years; therefore increasing battery energy density and lifetime is critical to lower carbon emission and potential toxicity because of end of life. Some of the newer battery technologies address these pitfalls, but currently, no battery platform has the desired lifetime, energy, and cost matrix.

1.24 Perspectives, Conclusion, and Policy Implications: Historic Framework (1970–2006)

The absence of an effective policy regarding battery deployment and recycling is disappointing but not unexpected as these technologies have not reached critical mass but are anticipated before 2025, which would necessitate lawmakers to implement policies similar to the ZEV policy from the State of California in the mid-1990s. The federal government across numerous several administrations (Carter to Bush (W.)) has not adopted a policy that has an integrated environment and energy. In many instances, federal policies have been formulated because of external stress (oil embargo, Gulf War, recession), to decrease dependence on oil import, and protect American jobs in key sectors through tax cuts, rebates, and incentives. The Obama Administration for the first time attempted to align energy and environmental policy through the Clean Coal Act, which limited electricity generation from coal, and promoted natural gas and sustainable resources through a combination of regulative fixes and monetary concessions to end-users, as well as utility operators. Many energy policies were renewed, such as Energy Policy Acts or, extended beyond the initial mandate (e.g., biofuels), or passed as part of large omnibus energy bills. Many policies were incrementally revised upwards or downwards with cost estimates being short of actual targets. Some energy policy was highly partisan and polarizing on renewable energy [256]. US policymaking has not adopted a planned long-term policy but rather a series of policies that have been revised depending on the current stress and political considerations. Sustained advocacy coalitions in many instances had led the states to enact specific mandates related to sustainable energy transitions to avoid policy retrenchment particularly to tax incentives to Li-ion batteries in the EV sector.

1.25 Perspectives, Conclusion, and Policy Implications: Obama Administration Energy Policy (2009–2011) – A Snapshot

Under the Obama Administration to highlight how policies can change during the lifetime of an administration (first term, 2009–2013, House/Senate majorities D + 79/D + 14), the Democratic Party had a supermajority in the Senate for approximately 7 months. From 2009 to 2011, over 50 energy-related policies to clean coal and renewables were started. The general framework of these policies overall was related to clean coal, bioenergy, energy storage, geothermal energy, offshore wind energy, and smart grid:

Federal cost-sharing for projects at the demonstration stage was rigorously maintained, without consideration of risks for specific projects.

The federally funded clean energy project was highly partisan or gained polarized opinion across the political spectrum, partly fueled by limited data sharing from the project to potential end-users and stakeholders.

In some cases, industrial or public pressure accelerated scale-up of demonstration projects before they were mature and ready for deployment, such as carbon capture, utilization, and storage for power generation and lastly risk aversion.

Policymakers wanted experimental technology to behave akin to mature technologies with zero risks. Under the Obama Administration's first term, projects managed by the Department of Energy that did not meet the benchmarks were canceled. This may be because of the passing of the Advanced Research Projects Agency-Energy (ARPA-E) component within the Stimulus Bill [257] or congressional imposed fiscal austerity by the Congress who controlled both Houses of the Congress (Republican Majority, House/Senate R + 49/D + 4) and potential hostility to a president from the Democratic Party. The lessons from the Obama Administration's first term, which from the first midterms until the end of his presidency was controlled by members of the Republican Party, had been that the federal government can effectively shepherd newer technologies past the innovation and testing phase.

1.26 Perspectives, Conclusion, and Policy Implications: Trump Administration (2016–2018) and the Plausible Future (2019–2021)

Before an analysis of the Trump Administration is given, it should be noted that the present administration is only 1.5 years old and the 2018 midterms are still 6 months [258]:

Prioritized economic growth is stimulated through tax cuts and deregulation. It appears that many of the strategies favored under the Obama Administration related to Clean Power Plan (CPP) and its Clean Energy Incentive Program including carbon capture and storage (CCS) are abandoned [259].

Natural gas will continue to be a significant component of US energy portfolio given the magnitudes of natural gas reserves, low natural gas prices, and cleaner combustion relative to coal [16]. It appears that fracking for natural gas, directives to utilize more coal for electricity, and further deregulation [260] of environmental regulations specifically arising from CPP/CCS [261] will continue [262]. Policies towards oil independence will continue through the implementation of the Keystone XL and Dakota Access Pipelines, enabling greater access to and streamlined exploration permits to be issued for federal lands [263]. In addition the markets are being regulated through incentives and tariffs, specifically promoting US steel for pipeline construction tapered with 10–25% tariffs on aluminum and steel [264] for foreign imports which may retard growth and stagnate domestic demand for transportation fuels from US oil refiners, although this may also increase export opportunities to export liquefied natural gas (LNG) to India, China, South America, and Asia [265]. The trade imbalance and potential lifting of international tariffs on US goods may free access to export light crude, LNG, and import inexpensive feedstocks benefiting refiners and petrochemical companies and potentially utility operators that can sell electricity to Canada and Mexico and import fuel supplies using current and proposed pipelines. If US partners (North American Free Trade Agreement (NAFTA consisting of the United States, Mexico, and Canada), European Union (EU)) do not enact trade-free legislation and trade disputes escalate over NAFTA/EU/China, this may negatively impact import of raw materials and export of refined oils and selectivity and is an uncertainty for the energy sectors [266]. If no agreement can be sought and trade tariffs as proposed are implemented, then import and export costs would increase and are not offset by the recent tax bills for US oil, steel, coal, and utility companies [267].

The other environmental and energy policies from the Obama Administration that are likely to be rolled back are exceptions for inadvertently killing birds (operators can inadvertently kill 4200 bald eagles a year without penalty), wind and solar power subsidies, and potential rolling back federal ethanol mandates passed under the Bush (W.) Administration. It is also likely that the Trump Administration will facilitate Yucca Mountain storage facility for spent nuclear fuel to begin operations and potential deregulation and tax concessions for operators to design and build small molten salt reactors powered by thorium. An increase in a hydropower-based generation (reversing policy from the Obama

Administration) and the construction of natural gas export terminals allow American energy companies to export natural gas (a policy that was also blocked under the Obama Administration). It is uncertain how effective this policy will be if countries in the EU/China and NAFTA impose an "energy tax" in response to the current and future tariffs proposed by the Trump Administration [268].

While the present administration is less than 2 years old, a number of general policy frameworks can be discerned. These are:

Focus on policy reversal instead of correction, conservative coalition using fossil fuel industry executives to align environmental policy within the context of energy, and independence of energy policy

Comprehensive reform and expansion of coal and natural gas development, including upstream expansion and downshift deregulation to promote American jobs coupled with import (caps?) and tariffs on select industries

Potential promotion of nuclear power to expand civilian sector nuclear capacity and capability in line with increasing American jobs and a general withdrawal or rollback from multilateral agreements which affect jobs, particularly in the fossil fuel, manufacturing, and infrastructure arena [269]

The post-2018 domestic energy policy is unclear because of the approaching midterm elections. In the case of the Obama Administration, after the Democratic Party lost control of the House in 2011 and Senate (2015), the energy policy was altered from 2011 onwards to take account of the bifurcated nature of governance. Policies that were perceived as partisan or polarizing could not pass both houses (under Republican majority) to be signed by law by the President (from the Democratic Party). The Trump Administration spanned the 114th (House/1320 Senate) R + 57/R + 9 and session of the 115th Congress (R + 42/R + 2), and although currently enjoys a Grand Old Party (GOP) majority in both houses as well as the presidency, the majority in the 116th Congress may not materialize. Like the Obama Administration, a recalibration between energy independence and policies that promote the environment including tax incentives for sustainable energy platforms may be reintroduced including a rollback of favorable coal legislation to reflect the political realities on the ground for the new Congress. Assuming there are no new political realities and the GOP retain small majorities the policies that are likely to be enacted based on current presidential executive orders or memorandum or selection of cabinet secretaries from (conservative states with pro-fossil fuel credentials) are likely to continue.

Acknowledgments The authors wish to thank the College of Arts and Sciences (CoA&S, Dr. Bashir, 160336-00002), ACS-PRF (Liu), SFFP (Bashir) and Welch Departmental Grant (AC-0006, Dr. Hahn), NSF-MRI acquisition (Liu), URA (160315-00015, Liu), and RDF grants (160345-00005, Liu), at Texas A&M University-Kingsville (TAMUK), for funding.

References

1. T. Skodvin, S. Andresen, An agenda for change in US climate policies? Presidential ambitions and congressional powers. Int. Environ. Agreements: Politics, Law Econ. **9**(3), 263–280 (2009)
2. M.F. Sherlock, J.M. Stupak, Energy tax incentives: measuring value across different types of energy resources. Curr. Politics Econ. United States, Canada, Mexico **18**(1), 1–29 (2016)
3. J. Meckling, T. Sterner, G. Wagner, Policy sequencing toward decarbonization. Nat. Energy **2**(12), 918–923 (2017)
4. F.N. Laird, *Avoiding Transitions, Layering Change: The Evolution of American Energy Policy in Germany's Energy Transition* (Palgrave Macmillan, New York, 2016), pp. 111–131
5. S. Mittlefehldt, From appropriate technology to the clean energy economy: renewable energy and environmental politics since the 1970s. J. Environ. Stud. Sci. **8**(2), 212–219 (2018)
6. A.C. Fisher, M.H. Rothkopf, Market failure and energy policy. A rationale for selective conservation. Energy Policy **17**(4), 397–406 (1989)
7. S. Carley, The era of state energy policy innovation: a review of policy instruments. Rev. Policy Res. **28**(3), 265–294 (2011)
8. A. Park, A. Kroll, D. Gilson, B. Hansen-Bundy, Triumph of the Drill: How Big Oil Clings to Billions in Government Giveaways – Mother Jones (2014, Apr), https://www.motherjones.com/politics/2014/04/oil-subsidies-renewable-energy-tax-breaks/
9. M. Graff, S. Carley, D.M. Konisky, Stakeholder perceptions of the United States energy transition: local-level dynamics and community responses to national politics and policy. Energy Res. Soc. Sci. (2018). https://doi.org/10.1016/j.erss.2018.05.017
10. G. Blackburn, C. Magee, V. Rai, Solar valuation and the modern utility's expansion into distributed generation. Electr. J. **27**(1), 18–32 (2014)
11. N.F. Crafts, P. Hudson, The industrial revolution. Econ. Hist. Rev. **47**, 414–414 (1994)
12. J. Mokyr, *The Enlightened Economy an Economic History of Britain 1700–1850* (Yale University Press, New Haven, 2010)
13. S. Pollard, Factory discipline in the industrial revolution. Econ. Hist. Rev. **16**(2), 254–271 (1963)
14. K.K. Goldewijk, Estimating global land use change over the past 300 years: the HYDE database. Glob. Biogeochem. Cycles **15**(2), 417–433 (2001)
15. M. Roser, E. Ortiz-Ospina, World Population Growth. OurWorldInData.Org (2018), https://ourworldindata.org/world-population-growth
16. Energy Information Administration (EIA), Natural Gas-fired Electricity Generation Expected to Reach Record Level in 2016, (2016), https://www.eia.gov/todayinenergy/detail.php?id=27072
17. M. Diesendorf, B. Elliston, The feasibility of 100% renewable electricity systems: a response to critics. Renew. Sust. Energ. Rev. **93**, 318–330 (2018)
18. J. Conti, P. Holtberg, J. Diefenderfer, A. LaRose, J.T. Turnure, L. Westfall, *International Energy Outlook 2016 with Projections to 2040 (No DOE/EIA-0484 (2016)) USDOE Energy Information Administration (EIA)* (Office of Energy Analysis, Washington, DC, 2016)
19. R. Andrews, Emissions reductions and world energy demand growth. Energy Matters: Energy, Environment and Policy (2016, Sept), http://euanmearns.com/emissions-reductions-and-world-energy-demand-growth/
20. H. Ritchie, M. Roser, Energy Production and Changing Energy Sources. OurWorldinData.Org. (2018), https://ourworldindata.org/energy-production-and-changing-energy-sources
21. Global, British Petroleum (BP), BP Statistical Review of World Energy 2017 (2017), https://www.bp.com/content/dam/bp/en/corporate/pdf/energy-economics/statistical-review-2017/bp-statistical-review-of-world-energy-2017-full-report.pdf
22. US Energy Information Administration (EIA), International Energy Outlook Tables (2017), https://www.eia.gov/outlooks/ieo/ieo_tables.php

23. US Energy Information Administration (EIA), International Energy Outlook 2017 (2017), https://www.eia.gov/outlooks/ieo/pdf/0484(2017).pdf
24. M. L. Bacci, The space and strategy of demographic growth. In A Concise History of World Population. (Wiley, 2017), 1–32; ISBN:9781119406822.
25. P.Y. Chen, S.T. Chen, C.S. Hsu, C.C. Chen, Modeling the global relationships among economic growth, energy consumption, and CO_2 emissions. Renew. Sust. Energ. Rev. **65**, 420–431 (2016)
26. T. Qi, Y. Weng, Economic impacts of an international carbon market in achieving the INDC targets. Energy **109**, 886–893 (2016)
27. P. Smith, S.J. Davis, F. Creutzig, S. Fuss, J. Minx, B. Gabrielle, D.P. Van Vuuren, Biophysical and economic limits to negative CO_2 emissions. Nat. Clim. Chang. **6**(1), 42–68 (2016)
28. I. C. Halalay, T. J. Fuller, L. Zou, T. C. Jackson, U.S. Patent 9,077,038 (U.S. Patent and Trademark Office, Washington, DC, 2015).
29. J. Rogelj, M. Den Elzen, N. Höhne, T. Fransen, H. Fekete, H. Winkler, M. Meinshausen, Paris Agreement climate proposals need a boost to keep warming well below 2 °C. Nature **534**(7609), 631–641 (2016)
30. M. den Elzen, N. Höhne, K. Jiang, J. Cantzler, P. Drost, T. Fransen, J. Rogelj, The methodology of national and global studies-The Emissions Gap Report 2017–Chapter 3–Appendix A, in *The Emissions Gap Report 2017: A UN Environment Synthesis Report United Nations Environment Programme* (2017)
31. H.W. Schiffer, T. Kober, E. Panos, (2018). World Energy Council's Global Energy Scenarios to 2060. Zeitschrift für Energiewirtschaft, 42(2), 91–102.
32. United States Environmental Protection Agency (EPA), Global Emissions by Gas (2017), https://www.epa.gov/ghgemissions/global-greenhouse-gas-emissions-data
33. Research Institute of Innovative Technology for the Earth (RITE), Global Scenario for CO_2 and GHG Emissions (2011), https://www.rite.or.jp/English/lab/syslab/research/alps/baselinescenario_co2ghg/E-ScenarioOutline_CO2GHG_20110815.pdf
34. International Energy Agency (IAE), Energy and Climate Change World Energy Outlook Special Report (2015), https://www.iea.org/publications/freepublications/publication/WEO2015SpecialReportonEnergyandClimateChange.pdf
35. M.R. Raupach, G. Marland, P. Ciais, C. Le Quéré, J.G. Canadell, G. Klepper, C.B. Field, Global and regional drivers of accelerating CO_2 emissions. Proc. Natl. Acad. Sci. **104**(24), 10288–10293 (2007)
36. United States Environmental Protection Agency (EPA), Sources of Greenhouse Gas Emissions (2018), https://www.epa.gov/ghgemissions/sources-greenhouse-gas-emissions
37. United States Environmental Protection Agency (EPA), How much is carbon dioxide produced when different fuels are burned? (2018), https://www.eia.gov/tools/faqs/faq.php?id=73andt=11
38. Statista, Energy-related carbon dioxide emissions worldwide from 1990 to 2050, by fuel (in billion metric tons) (2018), https://www.statista.com/statistics/242204/energy-related-global-co2-emissions-by-fuel/
39. Carbon Dioxide Information Analysis Center (CDIAC), CO_2 Emissions Data Browser (2017), http://cdiac.ornl.gov/CO2_Emission/
40. H. Ritchie, M. Roser, CO_2 and Other Greenhouse Gas Emissions. Our World in Data.Org (2018), https://ourworldindata.org/co2-and-other-greenhouse-gas-emissions
41. E.J. Dlugokencky, K.A. Masarie, P.M. Lang, P.P. Tans, Continuing decline in the growth rate of the atmospheric methane burden. Nature **393**(6684), 447–450 (1998)
42. Earth Systems Research Laboratory (ESRL), Trends in Atmospheric Carbon Dioxide (2018), https://www.esrl.noaa.gov/gmd/ccgg/trends/
43. S.A. Marcott, J.D. Shakun, P.U. Clark, A.C. Mix, A reconstruction of regional and global temperature for the past 11,300 years. Science **339**(6124), 1198–1201 (2013)
44. Intergovernmental Panel on Climate Change (IPCC), Working Group I: The Scientific Basis (2001), http://ipcc.ch/ipccreports/tar/wg1/029htm
45. E.G. Hertwich, G.P. Peters, The carbon footprint of nations: a global, trade-linked analysis. Environ. Sci. Technol. **43**(16), 6414–6420 (2009)

46. A. Malik, J. Lan, M. Lenzen, Trends in global greenhouse gas emissions from 1990 to 2010. Environ. Sci. Technol. **50**(9), 4722–4730 (2016)
47. International Energy Agency (IEA), Global Energy and CO_2 Status Report (2017), http://www.iea.org/publications/freepublications/publication/GECO2017.pdf
48. (Dutch: Planbureau voor de Leefomgeving – abbr PBL) Netherlands Environment Assessment Agency, Trends in Global CO_2 Emissions – 2015 Report (2015), http://edgar.jrc.ec.europa.eu/news_docs/jrc-2015-trends-in-global-co2-emissions-2015-report-98184.pdf
49. M. Su, C. Chen, Z. Yang, Urban energy structure optimization at the sector scale: considering environmental impact based on life cycle assessment. J. Clean. Prod. **112**, 1464–1474 (2016)
50. L.W. Cheah, Cars on a diet: the material and energy impacts of passenger vehicle weight reduction in the US, Doctoral dissertation, Massachusetts Institute of Technology, 2010
51. R. Modaresi, S. Pauliuk, A.N. Løvik, D.B. Muller, Global carbon benefits of material substitution in passenger cars until 2050 and the impact on the steel and aluminum industries. Environ. Sci. Technol. **48**(18), 10776–10784 (2014)
52. T.S. Turrentine, K.S. Kurani, Car buyers and fuel economy? Energy Policy **35**(2), 1213–1223 (2007)
53. M. Åhman, Primary energy efficiency of alternative powertrains in vehicles. Energy **26**(11), 973–989 (2001)
54. K. Holmberg, P. Andersson, N.O. Nylund, K. Mäkelä, A. Erdemir, Global energy consumption because of friction in trucks and buses. Tribol. Int. **78**, 94–114 (2014)
55. A. Schäfer, J.B. Heywood, M.A. Weiss, Future fuel cell and internal combustion engine automobile technologies: a 25-year life cycle and fleet impact assessment. Energy **31**(12), 2064–2087 (2006)
56. X. Shi, J. Yang, J.R. Salvador, M. Chi, J.Y. Cho, H. Wang, L. Chen, Multiple-filled skutterudites: a high thermoelectric figure of merit through separately optimizing electrical and thermal transports. J. Am. Chem. Soc. **133**(20), 7837–7846 (2011)
57. C.C. Chan, The state of the art of electric and hybrid vehicles. Proc. IEEE **90**(2), 247–275 (2002)
58. M.W. Verbrugge, C.W. Wampler, On the optimal sizing of batteries for electric vehicles and the influence of fast charge. J. Power Sources **384**, 312–317 (2018)
59. M. Ehsani, Y. Gao, S. Longo, K. Ebrahimi, *Modern Electric, Hybrid Electric, and Fuel Cell Vehicles* (CRC Press, Boca Raton, 2018)
60. Alternative Fuels Data Center (AFDC), How Do All-Electric Cars Work? (2018), https://www.afdc.energy.gov/vehicles/how-do-all-electric-cars-work
61. (Verband Deutscher Maschinen- und Anlagenbau e) VVDMA, Roadmap-Battery Production Equipment by 2030 (2016), https://battprod.vdma.org/documents/7411591/15357859/VDMA+Battery+Production_Roadmap_2016_update_English.pdf/cb2fd37b-348f-467d-a39e-2a3180bb8b5e
62. Mckinsey and Company, Electrifying insights: how automakers can drive electrified vehicle sales and profitability (2017), https://www.mckinsey.com/~/media/McKinsey/Industries/Automotive-and-Assembly/Our-Insights/Electrifying-insights-How-automakers-can-drive-electrified-vehicle-sales-and-profitability_vF.ashx
63. S. Koller, H. Kren, M. Schmuck, B. Fuchsbichler, C. Stangl, C. God, J. Garche, Next-generation materials for electrochemical energy storage–silicon and magnesium. AIP Conf. Proc. **1765**(1), 020007 (2016). AIP Publishing
64. A.A. Pesaran, T. Markel, H.S. Tataria, D. Howell, *Battery Requirements for Plug-in Hybrid Electric Vehicles – Analysis and Rationale (No NREL/CP-540-42240)* (National Renewable Energy Lab (NREL), Golden, 2009)
65. B.D. McCloskey, Attainable gravimetric and volumetric energy density of Li–S, and Li-ion battery cells with solid separator-protected Li metal anodes. J. Phys. Chem. Lett. **6**(22), 4581–4588 (2015)
66. The European Spallation Source (ESS), Battery Technology (2017), https://europeanspallationsource.se/sites/default/files/styles/landscape/public/images/media/2017-08/battery%20technology_440.png?itok=RB1NlkEp

67. US Advanced Battery Consortium (USCAR), News: Energy Storage System Goals (2016), http://www.uscar.org/commands/files_download.php?files_id=472
68. W. Xue, L. Miao, L. Qie, C. Wang, S. Li, J. Wang, J. Li, Gravimetric and volumetric energy densities of lithium-sulfur batteries. Curr. Opin. Electrochem. **6**(1), 92–99 (2017)
69. X. Zhang, H. Xie, C.S. Kim, K. Zaghib, A. Mauger, C.M. Julien, Advances in lithium-sulfur batteries. Mater. Sci. Eng. R. Rep. **121**, 1–29 (2017)
70. J.S. Yeoh, C.F. Armer, A. Lowe, Transition metal oxalates as energy storage materials. A review. Mater. Today Energy **9**, 198–222 (2018)
71. J. Neubauer, A. Pesaran, C. Bae, R. Elder, B. Cunningham, Updating the United States Advanced Battery Consortium and Department of Energy battery technology targets for battery electric vehicles. J. Power Sources **271**, 614–621 (2014)
72. D. Larcher, J.M. Tarascon, Towards greener and more sustainable batteries for electrical energy storage. Nat. Chem. **7**(1), 19 (2015)
73. K. Young, C. Wang, L.Y. Wang, K. Strunz, *Electric Vehicle Battery Technologies in Electric Vehicle Integration into Modern Power Networks* (Springer, New York, 2013), pp. 15–56
74. M. Dimitrijevic, Lithium-Ion Battery Assembly Challenges Electronic Component News Magazine (2011), https://www.ecnmag.com/article/2011/01/lithium-ion-battery-assembly-challenges
75. P.B. Tchounwou, C.G. Yedjou, A.K. Patlolla, D.J. Sutton, Heavy metal toxicity and the environment, in *Molecular, Clinical and Environmental Toxicology*, (Springer, Basel, 2012), pp. 133–164
76. A. Mardani, A. Jusoh, E.K. Zavadskas, F. Cavallaro, Z. Khalifah, Sustainable and renewable energy: an overview of the application of multiple criteria decision making techniques and approaches. Sustainability **7**(10), 13947–13984 (2015)
77. E. Helmers, P. Marx, Electric cars: technical characteristics and environmental impacts. Environ. Sci. Eur. **24**(1), 14–21 (2012)
78. K.B. Hueso, M. Armand, T. Rojo, High temperature sodium batteries: status, challenges and future trends. Energy Environ. Sci. **6**(3), 734–749 (2013)
79. National Petroleum Council (NPC), Advancing Technology for America's Transportation Future: Electric Analysis (NPC, Washington, DC, 2012), http://www.npc.org/reports/trans-future_fuels_summary-2012-lowres.pdf
80. Element Energy, Cost and Performance of EV Batteries (Element Energy, Cambridge, 2012), http://www.element-energy.co.uk/wordpress/wp-content/uploads/2012/06/CCC-battery-cost_-Element-Energy-report_March2012_Finalbis.pdf
81. P.G. Bruce, Energy storage beyond the horizon: rechargeable lithium batteries. Solid State Ionics **179**(21–26), 752–760 (2008)
82. V. Etacheri, R. Marom, R. Elazari, G. Salitra, D. Aurbach, Challenges in the development of advanced Li-ion batteries: a review. Energy Environ. Sci. **4**(9), 3243–3262 (2011)
83. M.V. Reddy, G.V. Subba Rao, B.V.R. Chowdari, Metal oxides and oxysalts as anode materials for Li-ion batteries. Chem. Rev. **113**(7), 5364–5457 (2013)
84. C.S. Kim, K.M. Jeong, K. Kim, C.W. Yi, Effects of capacity ratios between anode and cathode on electrochemical properties for lithium polymer batteries. Electrochim. Acta **155**, 431–436 (2015)
85. M. Majima, S. Ujiie, E. Yagasaki, K. Koyama, S. Inazawa, Development of long life lithium-ion battery for power storage. J. Power Sources **101**(1), 53–59 (2001)
86. L. Ji, Z. Lin, M. Alcoutlabi, X. Zhang, Recent developments in nanostructured anode materials for rechargeable lithium-ion batteries. Energy Environ. Sci. **4**(8), 2682–2699 (2011)
87. Y. Mekonnen, A. Sundararajan, A.I. Sarwat, A review of cathode and anode materials for lithium-ion batteries. IEEE SoutheastCon **2016**, 1–6 (2016, March)
88. J.B. Goodenough, K.S. Park, The Li-ion rechargeable battery: a perspective. J. Am. Chem. Soc. **135**(4), 1167–1176 (2013)
89. N. Nitta, F. Wu, J.T. Lee, G. Yushin, Li-ion battery materials: present and future. Mater. Today **18**(5), 252–264 (2015)
90. K. Zaghib, M. Dontigny, A. Guerfi, P. Charest, I. Rodrigues, A. Mauger, C.M. Julien, Safe and fast-charging Li-ion battery with long shelf life for power applications. J. Power Sources **196**(8), 3949–3954 (2011)

91. K. Adachi, H. Tajima, T. Hashimoto, K. Kobayashi, Development of 16 kWh power storage system applying Li-ion batteries. J. Power Sources **119**, 897–901 (2003)
92. W.F. Pickard, A.Q. Shen, N.J. Hansing, Parking the power: strategies and physical limitations for bulk energy storage in supply-demand matching on a grid whose input power is provided by intermittent sources. Renew. Sust. Energ. Rev. **13**(8), 1934–1945 (2009)
93. C. Liu, Z.G. Neale, G. Cao, Understanding electrochemical potentials of cathode materials in rechargeable batteries. Mater. Today **19**(2), 109–123 (2016)
94. I. Cowie, All About Batteries, Part 12: Lithium Titanate (LTO) Electronic Engineering Times (2015), https://www.eetimes.com/author.asp?section_id=36anddoc_id=1325358
95. Battery University (BU), BU-204: How do Lithium Batteries Work? (2018), http://batteryuniversity.com/learn/article/lithium_based_batteries
96. M. Lowe, S. Tokuoka, T. Trigg, G. Gereffi, Lithium-ion batteries for electric vehicles Duke University Center on Globalization, Governance, and Competitiveness, Tech Rep (2010)
97. B. Molina, This again: lithium-ion batteries behind HP recall USA Today (2017), https://www.usatoday.com/story/tech/news/2017/01/25/hp-expands-recall-laptop-batteries/97042680/
98. A. Opitz, P. Badami, L. Shen, K. Vignarooban, A.M. Kannan, Can Li-Ion batteries be the panacea for automotive applications? Renew. Sust. Energ. Rev. **68**, 685–692 (2017)
99. M.S. Kumar, S.T. Revankar, Development scheme and key technology of an electric vehicle: an overview. Renew. Sust. Energ. Rev. **70**, 1266–1285 (2017)
100. Q. Ren, D.A. Crolla, A. Morris, Effect of transmission design on electric vehicle (EV) performance, in *Vehicle Power and Propulsion Conference, 2009 VPPC'09 IEEE* (IEEE, 2009, Sept), pp. 1260–1265
101. Green Car Congress, BNEF forecasts EVs to hit 54% of new car sales by 2040; decreasing importance of PHEVs (2017), http://www.greencarcongress.com/2017/07/20170706-bnef.html
102. InsideEVs, June Plug-In 2018 EV Sales Scorecard (2018), https://insideevs.com/monthly-plug-in-sales-scorecard/
103. EV-Volumes, Global Plug-in Sales for Q1-2018 (2018), http://www.ev-volumes.com/country/total-world-plug-in-vehicle-volumes/
104. Global, EV (2016). Outlook 2016: Beyond One Million Electric Cars (International Energy Agency, 2016). https://www.ieaorg/publications/freepublications/publication/Global_EV_Outlook_2016.pdf
105. C. McKerracher, Bloomberg New Energy Finance (BNEF) Electric Vehicle Outlook 2018 (2018), https://about.bnef.com/blog/electric-vehicles-accelerate-54-new-car-sales-2040/
106. J. Moradi, H. Shahinzadeh, A. Khandan, M. Moazzami, A profitability investigation into the collaborative operation of wind and underwater compressed air energy storage units in the spot market. Energy **141**, 1779–1794 (2017)
107. I.G. Mason, S.C. Page, A.G. Williamson, A 100% renewable electricity generation system for New Zealand utilizing hydro, wind, geothermal and biomass resources. Energy Policy **38**(8), 3973–3984 (2010)
108. G. Zubi, R. Dufo-López, G. Pasaoglu, N. Pardo, Techno-economic assessment of an off-grid PV system for developing regions to provide electricity for basic domestic needs: a 2020–2040 scenario. Appl. Energy **176**, 309–319 (2016)
109. H. Lund, A.N. Andersen, P.A. Østergaard, B.V. Mathiesen, D. Connolly, From electricity smart grids to smart energy systems–a market operation based approach and understanding. Energy **42**(1), 96–102 (2012)
110. I. Pakere, D. Lauka, D. Blumberga, Solar power and heat production via photovoltaic thermal panels for district heating and industrial plant. Energy **154**, 424–432 (2018)
111. B. Zakeri, S. Syri, Electrical energy storage systems: a comparative life cycle cost analysis. Renew. Sust. Energ. Rev. **42**, 569–596 (2015)
112. J.P. Barton, D.G. Infield, Energy storage and its use with intermittent renewable energy. IEEE Trans. Energy Convers. **19**(2), 441–448 (2004)
113. E. Telaretti, L. Dusonchet, Stationary battery systems in the main world markets: part 2: main trends and prospects, in *Environment and Electrical Engineering and 2017 IEEE Industrial*

and Commercial Power Systems Europe (EEEIC/IandCPS Europe), 2017 IEEE International Conference on (IEEE (2017, June), pp. 1–6
114. A. Foley, B. Tyther, P. Calnan, B.Ó. Gallachóir, Impacts of electric vehicle charging under electricity market operations. Appl. Energy **101**, 93–102 (2013)
115. A. Jaiswal, Lithium-ion battery-based renewable energy solution for off-grid electricity: a techno-economic analysis. Renew. Sust. Energ. Rev. **72**, 922–934 (2017)
116. S.P. Ayeng'o, T. Schirmer, K.P. Kairies, H. Axelsen, D.U. Sauer, Comparison of off-grid power supply systems using lead-acid and lithium-ion batteries. Sol. Energy **162**, 140–152 (2018)
117. G. Bekele, G. Tadesse, Feasibility study of small Hydro/PV/Wind hybrid system for off-grid rural electrification in Ethiopia. Appl. Energy **97**, 5–15 (2012)
118. B. Nykvist, M. Nilsson, Rapidly falling costs of battery packs for electric vehicles. Nat. Clim. Chang. **5**(4), 329–332 (2015)
119. C. Pillot, Battery Market Development for Consumer Electronics, Automotive, and Industrial (2014), http://www.rechargebatteries.org/wp-content/uploads/2015/01/Avicenne-market-review-Nive-2014.pdf
120. X. Zeng, J. Li, N. Singh, Recycling of spent lithium-ion battery: a critical review. Crit. Rev. Environ. Sci. Technol. **44**(10), 1129–1165 (2014)
121. S.B. Peterson, J.J. Michalek, Cost-effectiveness of plug-in hybrid electric vehicle battery capacity and charging infrastructure investment for reducing US gasoline consumption. Energy Policy **52**, 429–438 (2013)
122. R. Faria, P. Marques, P. Moura, F. Freire, J. Delgado, A.T. de Almeida, Impact of the electricity mix and use profile in the life-cycle assessment of electric vehicles. Renew. Sust. Energ. Rev. **24**, 271–287 (2013)
123. T.R. Hawkins, B. Singh, G. Majeau-Bettez, A.H. Strømman, Comparative environmental life cycle assessment of conventional and electric vehicles. J. Ind. Ecol. **17**(1), 53–64 (2013)
124. F.S. Boureima, M. Messagie, J. Matheys, V. Wynen, N. Sergeant, J. Van Mierlo, B. De Caevel, Comparative LCA of electric, hybrid, LPG and gasoline cars in Belgian context. World Electric Veh. J. **3**(3), 469–476 (2009)
125. J. Van Mierlo, F. Boureima, M. Messagie, N. Sergeant, L. Govaerts, T. Denys, C. Macharis, Clean Vehicle Research: LCA and Policy Measures (CLEVER) final report of the project funded by the Belgian science policy (2011), http://etec.vub.ac.be/clever/docs/clever-PUB-003.pdf
126. S.K. Shahi, G.G. Wang, Plug-In Hybrid Electric Vehicle Battery Selection for Optimum Economic and Environmental Benefits Using Pareto Set Points and PSAT™, in *ASME 2010 International Design Engineering Technical Conferences and Computers and Information in Engineering Conference* (American Society of Mechanical Engineers, 2010, Jan), pp. 701–713
127. Q. Song, J. Li, Environmental effects of heavy metals derived from the e-waste recycling activities in China: a systematic review. Waste Manag. **34**(12), 2587–2594 (2014)
128. G. Cooney, T.R. Hawkins, J. Marriott, Life cycle assessment of diesel and electric public transportation buses. J. Ind. Ecol. **17**(5), 689–699 (2013)
129. R.L. Moss, E. Tzimas, H. Kara, P. Willis, J. Kooroshy, *Critical Metals in Strategic Energy Technologies, JRC-Scientific and Strategic Reports* (European Commission Joint Research Centre Institute for Energy and Transport, Luxembourg, 2011)
130. B. Cosic, G. Krajacic, N. Duic, A 100% renewable energy system in the year 2050: the case of Macedonia. Energy **48**(1), 80–87 (2012)
131. W.M. Haynes, D.R. Lide, Geophysics, astronomy, and acoustics; abundance of elements in the Earth's crust and in the sea, in *CRC Handbook of Chemistry and Physics*, ed. by D. R. Lide, 85th edn., (CRC Press, Boca Raton, 2005)
132. C. Grosjean, P.H. Miranda, M. Perrin, P. Poggi, Assessment of world lithium resources and consequences of their geographic distribution on the expected development of the electric vehicle industry. Renew. Sust. Energ. Rev. **16**(3), 1735–1744 (2012)
133. J. Speirs, M. Contestabile, The future of lithium availability for electric vehicle batteries, in *Behaviour of Lithium-Ion Batteries in Electric Vehicles*, (Springer, Cham, 2018), pp. 35–57

134. H. Vikström, S. Davidsson, M. Höök, Lithium availability and future production outlooks. Appl. Energy **110**, 252–266 (2013)
135. P.W. Gruber, P.A. Medina, G.A. Keoleian, S.E. Kesler, M.P. Everson, T.J. Wallington, Global lithium availability. J. Ind. Ecol. **15**(5), 760–775 (2011)
136. J.A. Ober, Mineral commodity summaries 2018 US Geological Survey (2018), https://minerals.usgs.gov/minerals/pubs/mcs/2018/mcs2018.pdf
137. J. Hykawy, T. Chudnovsky, Global consumption of lithium from 2008 to 2016, by battery and non-battery use (in metric tons of lithium carbonate equivalent) (2015), http://static1.squarespace.com/static/535e7e2de4b088f0b623c597/t/55689238e4b09f7c8dacebf1/1432916536166/Stormcrow-Lithium+Industry+Report-May2015-Final.pdf and https://asia.minesandmoney.com/wp-content/uploads/sites/13/2017/04/5Apr17-FEaF-1515-Jon-Hykawy-Stormcrow-1.pdf
138. A. Mancha, A look at some international lithium-ion battery recycling initiatives. J. Undergrad. Res. Univ. Ill. Chic. **9**(1), 1–5 (2016). http://journals.uic.edu/ojs/index.php/JUR/article/download/7546/6038
139. T. Georgi-Maschler, B. Friedrich, R. Weyhe, H. Heegn, M. Rutz, Development of a recycling process for Li-ion batteries. J. Power Sources **207**, 173–182 (2012)
140. M. Asari, S.I. Sakai, Li-ion battery recycling and cobalt flow analysis in Japan. Resour. Conserv. Recycl. **81**, 52–59 (2013)
141. J. Heelan, E. Gratz, Z. Zheng, Q. Wang, M. Chen, D. Apelian, Y. Wang, Current and prospective Li-ion battery recycling and recovery processes. JOM **68**(10), 2632–2638 (2016)
142. G. Leblanc, The mechanism of lithium accumulation in the isolated frog skin epithelium. Pflugers Arch. **337**(1), 1–18 (1972)
143. G.E. Gillaspy, J.S. Keddie, K. Oda, W. Gruissem, Plant inositol monophosphatase is a lithium-sensitive enzyme encoded by a multigene family. Plant Cell **7**(12), 2175–2185 (1995)
144. R.J. Baldessarini, L. Tondo, Recurrence risk in bipolar manic-depressive disorders after discontinuing lithium maintenance treatment: an overview. Clin. Drug Investig. **15**(4), 337–351 (1998)
145. H. Aral, A. Vecchio-Sadus, Toxicity of lithium to humans and the environment-a literature review. Ecotoxicol. Environ. Saf. **70**(3), 349–356 (2008)
146. A.W. Jones, P. Holmgren, Urine/blood ratios of ethanol in deaths attributed to acute alcohol poisoning and chronic alcoholism. Forensic Sci. Int. **135**(3), 206–212 (2003)
147. J. Pessoa, C. Grande, Thermoanalytical and kinetic study of lubricating lithium greases. J. Eng. Appl. Sci. **2**(4), 718–721 (2007)
148. K.B. Shedd, Cobalt Statistics and Information (2018), https://minerals.usgs.gov/minerals/pubs/commodity/cobalt/
149. C. Perks, The role of cobalt in battery supply industrial minerals (2016), http://www.indmin.com/Article/3586106/The-role-of-cobalt-in-battery-supply.html
150. J. Dewulf, G. Van der Vorst, K. Denturck, H. Van Langenhove, W. Ghyoot, J. Tytgat, K. Vandeputte, Recycling rechargeable lithium-ion batteries: a critical analysis of natural resource savings. Resour. Conserv. Recycl. **54**(4), 229–234 (2010)
151. M.J. Deng, F.L. Huang, I.W. Sun, W.T. Tsai, J.K. Chang, An entirely electrochemical preparation of a nano-structured cobalt oxide electrode with superior redox activity. Nanotechnology **20**(17), 175602–175621 (2009)
152. J. Deign, Why lithium isn't the big worry for lithium-ion batteries (2015), https://www.greentechmedia.com/articles/read/Why-Lithium-Isnt-the-Big-Worry-for-Li-ion
153. D.H.P. Kang, M. Chen, O.A. Ogunseitan, Potential environmental and human health impacts of rechargeable lithium batteries in electronic waste. Environ. Sci. Technol. **47**(10), 5495–5503 (2013)
154. K. Ishihara, N. Kihira, N. Terada, T. Iwahori, K. Nishimura, Life cycle analysis of large-size lithium-ion secondary batteries developed in the Japanese national project, in *Proceedings from 5th Ecobalance Conference* (Tsukuba, 2002, Nov), http://www.gbv.de/dms/tib-ub-hannover/494638826.pdf

155. A.S. Arico, P. Bruce, B. Scrosati, J.M. Tarascon, W. Van Schalkwijk, Nanostructured materials for advanced energy conversion and storage devices. Nat. Mater. **4**(5), 366–377 (2005)
156. P.L.S.G. Poizot, S. Laruelle, S. Grugeon, L. Dupont, J.M. Tarascon, Nano-sized transition-metal oxides as negative-electrode materials for lithium-ion batteries. Nature **407**(6803), 496–499 (2000)
157. S. Bashir, J. Liu, H. Zhang, X. Sun, J. Guo, Bandgap evaluations of metal-inserted titania nanomaterials. J. Nanopart. Res. **15**(4), 1572–1582 (2013)
158. G.G. Amatucci, N. Pereira, Fluoride-based electrode materials for advanced energy storage devices. J. Fluor. Chem. **128**(4), 243–262 (2007)
159. D.R. Sadoway, B. Huang, P.E. Trapa, P.P. Soo, P. Bannerjee, A.M. Mayes, Self-doped block copolymer electrolytes for solid-state, rechargeable lithium batteries. J. Power Sources **97**, 621–623 (2001)
160. K.T. Nam, D.W. Kim, P.J. Yoo, C.Y. Chiang, N. Meethong, P.T. Hammond, A.M. Belcher, Virus-enabled synthesis and assembly of nanowires for lithium ion battery electrodes. Science **312**(5775), 885–888 (2006)
161. R.J. Hoare, P.M. Harrison, T.G. Hoy, Structure of horse-spleen apoferritin at 6 Å resolution. Nature **255**(5510), 653–654 (1975)
162. A.G. MacDiarmid, L.S. Yang, W.S. Huang, B.D. Humphrey, Polyaniline: electrochemistry and application to rechargeable batteries. Synth. Met. **18**(1–3), 393–398 (1987)
163. P. Novák, K. Müller, K.S.V. Santhanam, O. Haas, Electrochemically active polymers for rechargeable batteries. Chem. Rev. **97**(1), 207–282 (1997)
164. P. Novák, O. Inganäs, R. Bjorklund Composite polymer positive electrodes in solid - state lithium secondary batteries. J. Electrochem. Soc. **134**(6), 1341–1345 (1987).
165. H. Chen, M. Armand, G. Demailly, F. Dolhem, P. Poizot, J.M. Tarascon, From biomass to a renewable LiXC6O6 organic electrode for sustainable Li-ion batteries. ChemSusChem **1**(4), 348–355 (2008)
166. P. Raveendran, J. Fu, S.L. Wallen, Completely "green" synthesis and stabilization of metal nanoparticles. J. Am. Chem. Soc. **125**(46), 13940–13941 (2003)
167. T. Ogasawara, A. Débart, M. Holzapfel, P. Novák, P.G. Bruce, Rechargeable Li_2O_2 electrode for lithium batteries. J. Am. Chem. Soc. **128**(4), 1390–1393 (2006)
168. M. Rosso, C. Brissot, A. Teyssot, M. Dollé, L. Sannier, J.M. Tarascon, S. Lascaud, Dendrite short-circuits and fuse effect on Li/polymer/Li cells. Electrochim. Acta **51**(25), 5334–5340 (2006)
169. F. Ding, W. Xu, G.L. Graff, J. Zhang, M.L. Sushko, X. Chen, X. Liu, Dendrite-free lithium deposition via self-healing electrostatic shield mechanism. J. Am. Chem. Soc. **135**(11), 4450–4456 (2013)
170. J. Qian, W. Xu, P. Bhattacharya, M. Engelhard, W.A. Henderson, Y. Zhang, J.G. Zhang, Dendrite-free Li deposition using trace-amounts of water as an electrolyte additive. Nano Energy **15**, 135–144 (2015)
171. X. Ren, Y. Zhang, M.H. Engelhard, Q. Li, J.G. Zhang, W. Xu, Guided lithium metal deposition and improved lithium coulombic efficiency through synergistic effects of LiAsF6 and cyclic carbonate additives. ACS Energy Lett. **3**, 14–19 (2017)
172. X.Q. Zhang, X.B. Cheng, X. Chen, C. Yan, Q. Zhang, Fluoroethylene carbonate additives to render uniform Li deposits in lithium metal batteries. Adv. Funct. Mater. **27**(10) (2017). https://doi.org/10.1002/adfm.201605989
173. J. Qian, W.A. Henderson, W. Xu, P. Bhattacharya, M. Engelhard, O. Borodin, J.G. Zhang, High rate and stable cycling of lithium metal anode. Nat. Commun. **6**, 6362–6371 (2015)
174. H. Yoon, P.C. Howlett, A.S. Best, M. Forsyth, D.R. MacFarlane, Fast charge/discharge of Li metal batteries using an ionic liquid electrolyte. J. Electrochem. Soc. **160**(10), A1629–A1637 (2013)
175. Y. Yamada, K. Furukawa, K. Sodeyama, K. Kikuchi, M. Yaegashi, Y. Tateyama, A. Yamada, Unusual stability of acetonitrile-based superconcentrated electrolytes for fast-charging lithium-ion batteries. J. Am. Chem. Soc. **136**(13), 5039–5046 (2014)
176. N. Togasaki, T. Momma, T. Osaka, Enhanced cycling performance of a Li metal anode in a dimethylsulfoxide-based electrolyte using highly concentrated lithium salt for a lithium-oxygen battery. J. Power Sources **307**, 98–104 (2016)

177. B. Liu, W. Xu, P. Yan, S.T. Kim, M.H. Engelhard, X. Sun, J.G. Zhang, Stabilization of Li metal anode in DMSO-based electrolytes via optimization of salt–solvent coordination for Li–O_2 batteries. Adv. Energy Mater. **7**(14) (2017). https://doi.org/10.1002/aenm.201602605
178. B. Liu, W. Xu, P. Yan, X. Sun, M.E. Bowden, J. Read, J.G. Zhang, Enhanced cycling stability of rechargeable Li–O_2 batteries using high-concentration electrolytes. Adv. Funct. Mater. **26**(4), 605–613 (2016)
179. J. Zheng, J.A. Lochala, A. Kwok, Z.D. Deng, J. Xiao, Research progress towards understanding the unique interfaces between concentrated electrolytes and electrodes for energy storage applications. Adv. Sci. (2017). https://doi.org/10.1002/advs.201700032
180. Y. Yamada, A. Yamada, Superconcentrated electrolytes for lithium batteries. J. Electrochem. Soc. **162**(14), A2406–A2423 (2015)
181. J. Zheng, M.H. Engelhard, D. Mei, S. Jiao, B.J. Polzin, J.G. Zhang, W. Xu, Electrolyte additive enabled fast charging and stable cycling lithium metal batteries. Nat. Energy **2**(3), 17012–17015 (2017)
182. S. Jiao, J. Zheng, Q. Li, X. Li, M.H. Engelhard, R. Cao, W. Xu, Behavior of lithium metal anodes under various capacity utilization and high current density in lithium metal batteries. Joule **2**(1), 110–124 (2018)
183. J. Qian, B.D. Adams, J. Zheng, W. Xu, W.A. Henderson, J. Wang, J.G. Zhang, Anode-free rechargeable lithium metal batteries. Adv. Funct. Mater. **26**(39), 7094–7102 (2016)
184. J.W. Fergus, Ceramic and polymeric solid electrolytes for lithium-ion batteries. J. Power Sources **195**(15), 4554–4569 (2010)
185. R. Bouchet, S. Maria, R. Meziane, A. Aboulaich, L. Lienafa, J.P. Bonnet, R. Denoyel, Single-ion BAB triblock copolymers as highly efficient electrolytes for lithium-metal batteries. Nat. Mater. **12**(5), 452–457 (2013)
186. S. Ohta, S. Komagata, J. Seki, T. Saeki, S. Morishita, T. Asaoka, All-solid-state lithium-ion battery using garnet-type oxide and Li_3BO_3 solid electrolytes fabricated by screen-printing. J. Power Sources **238**, 53–56 (2013)
187. A. Manthiram, X. Yu, S. Wang, Lithium battery chemistries enabled by solid-state electrolytes. Nat. Rev. Mater. **2**(4), 16103–16107 (2017)
188. M. Kotobuki, H. Munakata, K. Kanamura, Y. Sato, T. Yoshida, Compatibility of $Li_7La_3Zr_2O_{12}$ solid electrolyte to all-solid-state battery using Li metal anode. J. Electrochem. Soc. **157**(10), A1076–A1079 (2010)
189. L. Cheng, E.J. Crumlin, W. Chen, R. Qiao, H. Hou, S.F. Lux, K. Persson, The origin of high electrolyte–electrode interfacial resistances in lithium cells containing garnet-type solid electrolytes. Phys. Chem. Chem. Phys. **16**(34), 18294–18300 (2014)
190. X. Han, Y. Gong, K.K. Fu, X. He, G.T. Hitz, J. Dai, Y. Mo, Negating interfacial impedance in garnet-based solid-state Li metal batteries. Nat. Mater. **16**(5), 572–579 (2017)
191. W. Luo, Y. Gong, Y. Zhu, Y. Li, Y. Yao, Y. Zhang, E.D. Wachsman, Reducing interfacial resistance between garnet-structured solid-state electrolyte and Li-metal anode by a germanium layer. Adv. Mater. **29**(22) (2017). https://doi.org/10.1002/adma.201606042
192. K.K. Fu, Y. Gong, J. Dai, A. Gong, X. Han, Y. Yao, Y. Li, Flexible, solid-state, ion-conducting membrane with 3D garnet nanofiber networks for lithium batteries. Proc. Natl. Acad. Sci. **113**(26), 7094–7099 (2016)
193. Y. Li, W. Zhou, X. Chen, X. Lü, Z. Cui, S. Xin, J.B. Goodenough, Mastering the interface for advanced all-solid-state lithium rechargeable batteries. Proc. Natl. Acad. Sci. **113**(47), 13313–13317 (2016)
194. Z. Guo, J. Zhu, J. Feng, S. Du, Direct in situ observation and explanation of lithium dendrite of commercial graphite electrodes. RSC Adv. **5**(85), 69514–69521 (2015)
195. A.C. Kozen, C.F. Lin, O. Zhao, S.B. Lee, G.W. Rubloff, M. Noked, Stabilization of Lithium metal anodes by hybrid artificial solid electrolyte interphase. Chem. Mater. **29**(15), 6298–6307 (2017)
196. G. Zheng, S.W. Lee, Z. Liang, H.W. Lee, K. Yan, H. Yao, Y. Cui, Interconnected hollow carbon nanospheres for stable lithium metal anodes. Nat. Nanotechnol. **9**(8), 618–623 (2014)

197. H. Ye, S. Xin, Y.X. Yin, J.Y. Li, Y.G. Guo, L.J. Wan, Stable Li plating/stripping electrochemistry realized by a hybrid Li reservoir in spherical carbon granules with 3D conducting skeletons. J. Am. Chem. Soc. **139**(16), 5916–5922 (2017)
198. Z. Tu, M.J. Zachman, S. Choudhury, S. Wei, L. Ma, Y. Yang, L.A. Archer, Nanoporous hybrid electrolytes for high-energy batteries based on reactive metal anodes. Adv. Energy Mater. **7**(8) (2017). https://doi.org/10.1002/aenm.201602367
199. K. Miyahara, Y. Jin, H. Munakata, K. Kanamura, *3DOM Polyimide Separator for Rechargeable Lithium Batteries with High Rate Performance in Meeting Abstracts (No. 2)* (The Electrochemical Society, 2012, June), pp. 107–107
200. Z. Liang, G. Zheng, C. Liu, N. Liu, W. Li, K. Yan, Y. Cui, Polymer nanofiber-guided uniform lithium deposition for battery electrodes. Nano Lett. **15**(5), 2910–2916 (2015)
201. Q. Yun, Y.B. He, W. Lv, Y. Zhao, B. Li, F. Kang, Q.H. Yang, Chemical dealloying derived 3D porous current collector for Li metal anodes. Adv. Mater. **28**(32), 6932–6939 (2016)
202. L. Yu, N.L. Canfield, S. Chen, H. Lee, X. Ren, M.H. Engelhard, J.G. Zhang, Enhanced stability of lithium metal anode by using a 3D porous nickel substrate. ChemElectroChem **5**(5), 761–769 (2018)
203. A. Zhamu, G. Chen, C. Liu, D. Neff, Q. Fang, Z. Yu, B.Z. Jang, Reviving rechargeable lithium metal batteries: enabling next-generation high-energy and high-power cells. Energy Environ. Sci. **5**(2), 5701–5707 (2012)
204. X. Ji, D.Y. Liu, D.G. Prendiville, Y. Zhang, X. Liu, G.D. Stucky, Spatially heterogeneous carbon-fiber papers as surface dendrite-free current collectors for lithium deposition. Nano Today **7**(1), 10–20 (2012)
205. C.P. Yang, Y.X. Yin, S.F. Zhang, N.W. Li, Y.G. Guo, Accommodating lithium into 3D current collectors with a submicron skeleton towards long-life lithium metal anodes. Nat. Commun. **6**, 8058–8067 (2015)
206. L.L. Lu, J. Ge, J.N. Yang, S.M. Chen, H.B. Yao, F. Zhou, S.H. Yu, Free-standing copper nanowire network current collector for improving lithium anode performance. Nano Lett. **16**(7), 4431–4437 (2016)
207. Y. Zhu, S. Murali, W. Cai, X. Li, J.W. Suk, J.R. Potts, R.S. Ruoff, Graphene and graphene oxide: synthesis, properties, and applications. Adv. Mater. **22**(35), 3906–3924 (2010)
208. E. Yoo, J. Kim, E. Hosono, H.S. Zhou, T. Kudo, I. Honma, Large reversible Li storage of graphene nanosheet families for use in rechargeable lithium ion batteries. Nano Lett. **8**(8), 2277–2282 (2008)
209. X. Zhu, Y. Zhu, S. Murali, M.D. Stoller, R.S. Ruoff, Nanostructured reduced graphene oxide/Fe_2O_3 composite as a high-performance anode material for lithium ion batteries. ACS Nano **5**(4), 3333–3338 (2011)
210. D. Lin, Y. Liu, Z. Liang, H.W. Lee, J. Sun, H. Wang, Y. Cui, Layered reduced graphene oxide with nanoscale interlayer gaps as a stable host for lithium metal anodes. Nat. Nanotechnol. **11**(7), 626–632 (2016)
211. W. Liu, D. Lin, A. Pei, Y. Cui, Stabilizing lithium metal anodes by uniform Li-ion flux distribution in nanochannel confinement. J. Am. Chem. Soc. **138**(47), 15443–15450 (2016)
212. H. Wu, Y. Cao, L. Geng, C. Wang, In situ formation of stable interfacial coating for high performance lithium metal anodes. Chem. Mater. **29**(8), 3572–3579 (2017)
213. W. Li, H. Yao, K. Yan, G. Zheng, Z. Liang, Y.M. Chiang, Y. Cui, The synergetic effect of lithium polysulfide and lithium nitrate to prevent lithium dendrite growth. Nat. Commun. **6**, 7436–7444 (2015)
214. Y. Zhang, J. Qian, W. Xu, S.M. Russell, X. Chen, E. Nasybulin, F. Ding, Dendrite-free lithium deposition with self-aligned nanorod structure. Nano Lett. **14**(12), 6889–6896 (2014)
215. S.M. Wood, C.H. Pham, R. Rodriguez, S.S. Nathan, A.D. Dolocan, H. Celio, C.B. Mullins, K^+ reduces lithium dendrite growth by forming a thin, less-resistive solid electrolyte interphase. ACS Energy Lett. **1**(2), 414–419 (2016)
216. C. Yan, X.B. Cheng, C.Z. Zhao, J.Q. Huang, S.T. Yang, Q. Zhang, Lithium metal protection through in-situ formed solid electrolyte interphase in lithium-sulfur batteries: the role of polysulfides on lithium anode. J. Power Sources **327**, 212–220 (2016)

217. S. Xiong, K. Xie, Y. Diao, X. Hong, On the role of polysulfides for a stable solid electrolyte interphase on the lithium anode cycled in lithium-sulfur batteries. J. Power Sources **236**, 181–187 (2013)
218. L.A.W. Ellingsen, G. Majeau-Bettez, B. Singh, A.K. Srivastava, L.O. Valøen, A.H. Strømman, Life cycle assessment of a lithium-ion battery vehicle pack. J. Ind. Ecol. **18**(1), 113–124 (2014)
219. US Energy Information Administration (EIA), How much electricity is used for lighting in the United States? (2017), https://www.eia.gov/tools/faqs/faq.php?id=307andt=10
220. J.F. Peters, M. Baumann, B. Zimmermann, J. Braun, M. Weil, The environmental impact of Li-Ion batteries and the role of key parameters–a review. Renew. Sust. Energ. Rev. **67**, 491–506 (2017)
221. J. Das, A.P. Abraham, P.C. Ghosh, R. Banerjee, Life cycle energy and carbon footprint analysis of photovoltaic battery microgrid system in India. Clean Techn. Environ. Policy, 1–16 (2017)
222. P. Yodovard, J. Khedari, J. Hirunlabh, The potential of waste heat thermoelectric power generation from diesel cycle and gas turbine cogeneration plants. Energy Sources **23**(3), 213–224 (2001)
223. J.P. Barton, D.G. Infield, Energy storage and its use with wind power, in *Power Engineering Society General Meeting, 2005. IEEE*, (IEEE, Piscataway, 2005), pp. 1934–1938
224. Y.S. Mohammed, M.W. Mustafa, N. Bashir, Hybrid renewable energy systems for off-grid electric power: review of substantial issues. Renew. Sust. Energ. Rev. **35**, 527–539 (2014)
225. W. Kempton, J. Tomic, Vehicle-to-grid power implementation: from stabilizing the grid to supporting large-scale renewable energy. J. Power Sources **144**(1), 280–294 (2005)
226. A.M. Bernardes, D.C.R. Espinosa, J.S. Tenório, Recycling of batteries: a review of current processes and technologies. J. Power Sources **130**(1–2), 291–298 (2004)
227. K. McDermott, Cost of service regulation in the investor-owned electric utility industry Edison Electric Institute June (2012), http://www.ourenergypolicy.org/wp-content/uploads/2012/09/COSR_history_final.pdf
228. J.D. Hamilton, Oil and the macroeconomy since World War II. J. Polit. Econ. **91**(2), 228–248 (1983)
229. M.A. Delucchi, M.Z. Jacobson, Providing all global energy with wind, water, and solar power, part II: reliability, system and transmission costs, and policies. Energy Policy **39**(3), 1170–1190 (2011)
230. J.P. Tomain, The dominant model of United States energy policy. Univ. Colo. Law Rev. **61**, 355–394 (1990). http://www.ourenergypolicy.org/wp-content/uploads/2013/08/The-Dominant-Model-of-United-States-Energy-Policy.pdf
231. J. Kassakian, R. Schmalensee, G. Desgroseilliers, T. Heidel, K. Afridi, A. Farid, H. Michaels, *The Future of the Electric Grid: An Interdisciplinary MIT Study* (Massachusetts Institute of Technology, Cambridge, MA, 2011)
232. M.S. Dorsi, Piedmont environmental council v Ferc. Harv. Environ. Law Rev. **34**, 593 (2010). http://paceeenvironmentalnotes.blogspot.com/2010/11/recent-law-review-articles-november.html
233. P.Z. Grossman, Energy shocks, crises and the policy process: a review of theory and application. Energy Policy **77**, 56–69 (2015)
234. K. Schneider, New York Times "Bush's Energy Plan Emphasizes Gains in Output Over Efficiency", Feb 9 1991 (1991), https://www.nytimes.com/1991/02/09/business/bush-s-energy-plan-emphasizes-gains-in-output-over-efficiency.html
235. R.F. Hirsh, Power loss: the origins of deregulation and restructuring in the American electric utility system. J. Energy Lit. **9**, 77–79 (2003)
236. G. Barbose, US Renewables Portfolio Standards: 2016 Annual Status Report, (2016), https://emp.lbl.gov/sites/all/files/lbnl-1005057.pdf
237. J.W. Stoutenborough, M. Beverlin, Encouraging pollution-free energy: the diffusion of state net metering policies. Soc. Sci. Q. **89**(5), 1230–1251 (2008)
238. The US Internal Revenue Service (IRS), Residential Energy Credits (2017), https://www.irs.gov/pub/irs-pdf/f5695.pdf
239. H. Bernton, W. Kovarik, S. Sklar, B. Griffin, R.J. Woolsey, *The Forbidden Fuel a History of Power Alcohol* (University of Nebraska Press, Lincoln/London, 2010)

240. W.R. Hartley, A.J. Englande, D.J. Harrington, Health risk assessment of groundwater contaminated with methyl tertiary butyl ether (MTBE). Water Sci. Technol. **39**(10–11), 305–310 (1999)
241. K.N. Rask, Clean air and renewable fuels: the market for fuel ethanol in the US from 1984 to 1993. Energy Econ. **20**(3), 325–345 (1998)
242. H. De Gorter, D.R. Just, The economics of a blend mandate for biofuels. Am. J. Agric. Econ. **91**(3), 738–750 (2009)
243. J. Meadowcroft, What about the politics? Sustainable development, transition management, and long-term energy transitions. Policy. Sci. **42**(4), 323–340 (2009)
244. A.J. Pahnke, W.E. Bettoney, Role of Lead Antiknocks in Modern Gasolines (No 710842) SAE Technical Paper (1971), https://www.sae.org/publications/technical-papers/content/710842/
245. V. Thomas, A. Kwong, Ethanol as a lead replacement: phasing out leaded gasoline in Africa. Energy Policy **29**(13), 1133–1143 (2001)
246. R.L. Boeckx, B. Postl, F.J. Coodin, Gasoline sniffing and tetraethyl lead poisoning in children. Pediatrics **60**(2), 140–145 (1977)
247. L. Partzsch, Biofuel research: perceptions of power and transition energy. Sustain. Soc. **7**(1), 14–24 (2017)
248. M. Mondou, G. Skogstad, D. Houle, Policy image resilience, multidimensionality, and policy image management: a study of US biofuel policy. J. Publ. Policy **34**(1), 155–180 (2014)
249. M.E. Kahn, Smog reduction's impact on California county growth. J. Reg. Sci. **40**(3), 565–582 (2000)
250. G. Collantes, D. Sperling, The origin of California's zero-emission vehicle mandate. Transp. Res. A Policy Pract. **42**(10), 1302–1313 (2008)
251. R.J.F. Hoogma, Exploiting technological niches: strategies for experimental introduction of electric vehicles (2000), https://research.utwente.nl/en/publications/exploiting-technological-niches-strategies-for-experimental-intro
252. L.C. Stokes, H.L. Breetz, Politics in the US energy transition: case studies of solar, wind, biofuels and electric vehicles policy. Energy Policy **113**, 76–86 (2018)
253. C. E. Ciocirlan, B. Yandle The political economy of green taxation in OECD countries. Eur. J. Law Econ. **15**(3), 203–218 (2003).
254. M.J. Richards, Regulating automakers for climate change: US reforms in global context. Environ. Policy Gov. **26**(6), 498–509 (2016)
255. The US Internal Revenue Service (IRS), Plug-In Electric Drive Vehicle Credit (IRC 30D) (2014), https://www.irs.gov/businesses/plug-in-electric-vehicle-credit-irc-30-and-irc-30d
256. L.C. Stokes, C. Warshaw, Renewable energy policy design and framing influence public support in the United States. Nat. Energy **2**(8), 17107–17117 (2017)
257. M. Fletcher, Obama leaves DC to sign stimulus bill Washington Post (2009, Feb), http://www.washingtonpost.com/wp-dyn/content/article/2009/02/17/AR2009021700221.html
258. S.L. Sutherland, US Energy Policy in the Trump Administration and 115th Congress (2017, Jan), https://www.ey.com/Publication/vwLUAssets/EY-us-energy-policy-in-the-trump-administration/$File/EY-us-energy-policy-in-the-trump-administration.pdf
259. D. Trump, Executive Order Expediting Environmental Reviews and Approvals for High Priority Infrastructure Projects (2017), https://www.whitehouse.gov/presidential-actions/executive-order-expediting-environmental-reviews-approvals-high-priority-infrastructure-projects/
260. R. Priebus, Memorandum for the Heads of Executive Departments and Agencies (2017), https://www.whitehouse.gov/presidential-actions/memorandum-heads-executive-departments-agencies/
261. D. Trump, President Trump Takes Action to Expedite Priority Energy and Infrastructure Projects (2017), https://www.whitehouse.gov/briefings-statements/president-trump-takes-action-expedite-priority-energy-infrastructure-projects/
262. M. Mulvaney, Presidential Executive Order on Reducing Regulation and Controlling Regulatory Costs (2017), https://www.whitehouse.gov/presidential-actions/presidential-executive-order-reducing-regulation-controlling-regulatory-costs/

263. W. Ross, Presidential Memorandum Regarding Construction of American Pipelines (2017), https://www.whitehouse.gov/presidential-actions/presidential-memorandum-regarding-construction-american-pipelines/
264. W. Ross, Investigating the impact of steel imports (2017), http://thehill.com/policy/finance/329743-trump-administration-investigating-effect-of-steel-imports-on-us
265. S.T. Mnuchin, Reducing tax regulatory burdens (2017), https://www.whitehouse.gov/presidential-actions/presidential-executive-order-identifying-reducing-tax-regulatory-burdens/
266. W. Ross, Investigating the impact of aluminum imports (2017), https://www.whitehouse.gov/the-press-office/2017/04/27/presidential-memorandum-secretary-commerce
267. W. Ross, Reducing the US trade deficit (2017), https://www.federalregister.gov/documents/2017/04/17/2017-07827/public-comments-and-hearing-regarding-administration-report-on-significant-trade-deficits
268. W. Ross, Reviewing domestic manufacturing regulation (2017), https://www.federalregister.gov/documents/2017/03/07/2017-04516/impact-of-federal-regulations-on-domestic-manufacturing
269. D. Trump, Presidential Executive Order on Buy American and Hire American (2017), https://www.whitehouse.gov/presidential-actions/presidential-executive-order-buy-american-hire-american/

Understanding the Energy Storage Principles of Nanomaterials in Lithium-Ion Battery

2

Weixin Song and Jun Chen

Contents

Abstract
Nanostructured materials offering advantageous physicochemical properties over the bulk have received enormous interest in energy storage and conversion. The nanomaterials have greatly enhanced the performance of electrochemical cells through the optimized surface, size and dimension, as well as the synergy between the bulk and interface. The booming nanotechnologies provide effective tools in fabricating nanomaterials at atomic or molecular scale with specific morphological and electronic structure. In this chapter, we will focus on the nanostructured materials used in lithium-ion batteries and supercapacitors by introducing the progress of nanomethodologies and popular nanostructured materials in each application. Novel and powerful functional nanomaterials are being rapidly developed to advance the

W. Song (✉)
Department of Materials, Imperial College London, London, UK
e-mail: weixinsong@hotmail.com

J. Chen
Department of Materials, University of Oxford, Oxford, UK

Q. Zhen et al. (eds.), *Nanostructured Materials for Next-Generation Energy Storage and Conversion*, https://doi.org/10.1007/978-3-662-58675-4_2

technologies of energy storage and conversion. The understanding of the processing mechanisms in material synthesis and the relationship between the material structure and resulted electrochemistry is critical for the rational synthesis of nanomaterials. Unambiguous knowledge in the electrochemistry of nanomaterials benefts the improvement of the synthetic protocols of materials and the advancement of eletrochemical energy-storage devices.

2.1 Introduction

Electrochemical energy storage and conversion have become the unquestionable challenges and critical technologies in the world, in response to the decreasing availability of fossil fuels and environmental pollution. The energy storage and conversion systems that can electrochemically produce energy have been seriously considered as the alternative power sources, as long as these systems are designed economically and environmentally friendly [1]. These systems include batteries, electrochemical capacitors (supercapacitors), and fuel cells, some of which have been widely used in electronic products, electric vehicles, and the smart grid. Although the mechanisms of these three kinds of devices for energy storage and conversion are different, they still share some common features in electrochemistry. Their electrochemical reactions take place at the interface between the electrolyte and electrode where electron and ion transport are separated to storage/release the energy through the charge/discharge process. There are many different kinds of battery technologies on the market or under development, ranging from primary batteries to rechargeable batteries and solar cells. Supercapacitors include electrochemical double-layer capacitors (EDLCs) and pseudocapacitors, distinguished by the processing mechanism whether non-Faradaic or Faradaic. Fuel cells have many categories in terms of the properties of electrolytes. Figure 2.1 displays the operation mechanism of three selected kinds of devices, namely rechargeable lithium-ion batteries, pseudocapacitors, and solid oxide fuel cells (SOFC), and all of them consist of some basic components: cathode, anode, and electrolyte with two electrodes contacting with the electrolyte solution.

Both batteries and fuel cells store and release charges through the redox reaction of the electrode materials and stored fuels, respectively that own chemical energy. These chemical reactions are associated with Faradaic charge transfer between an electrolyte and electrode surface, and attendant phase changes. The discharge process is usually taken as the nominal defining process for an energy storage and conversion device, so the term, negative electrode is anode where the oxidation reaction at this side usually take place at lower electrode potentials than the cathode side (positive electrode). This definition of electrode directions is only suitable for the pseudocapacitors which process the energy via Faradaic reaction but meaningless for the EDLC systems which use physical adsorption/desorption of the ions at the electrolyte/electrode interface, termed as electrical double layer.

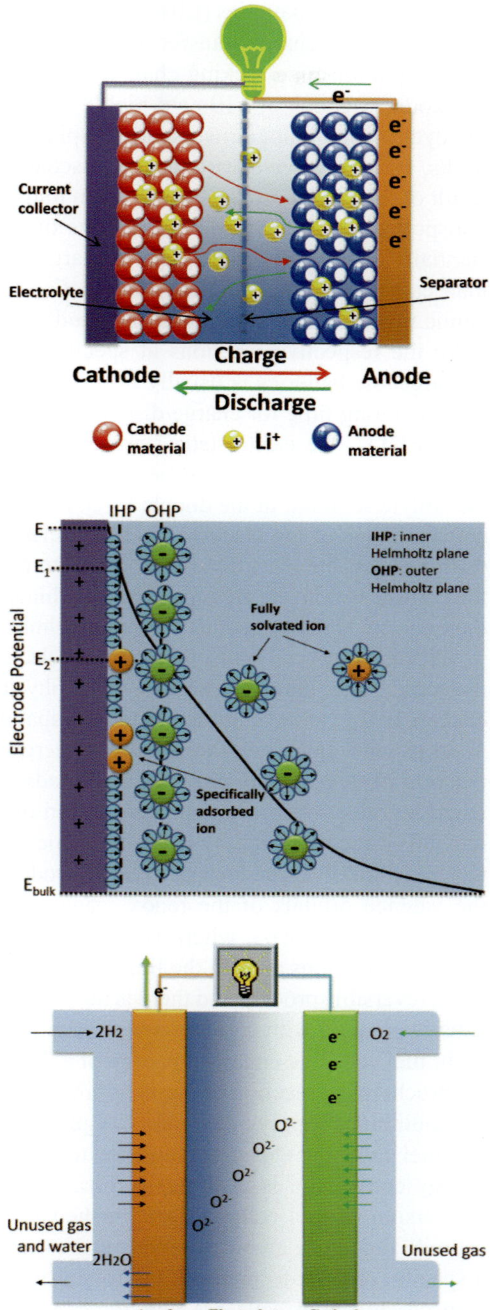

Fig. 2.1 Representation of the electrochemical systems for energy storage and conversion. From left to right are: lithium-ion battery, pseudocapacitor, and solid oxide fuel cell

In lithium-ion batteries (LIBs), the redox reactions of electrodes are accompanied by the Faradaic charge-transfer between the electrolyte and electrode surface, moving lithium ions in a rocking-chair way to realize the discharge/charge process. The electrode reactions of LIBs are highly dependent on the electrode potentials; thus, the cyclic voltammograms usually display typical oxidation and reduction current peaks, and the galvanostatic charge/discharge profiles show voltage plateaus. As a result of the kinetic barriers for electron transfer such as Ohmic resistance, and ionic transport difficulty, the redox reactions of electrodes take place at potentials substantially different in discharge and charge process, with anodic peak more positive than cathodic peak. The electrode reactions are irreversible around the thermodynamic reversible potential; thus, the anodic and cathodic overpotential is required to drive the respective reactions at specific net rates. Accordingly, the cycle life of rechargeable batteries is usually limited and the relatively slow chemical reactions require a long time for charge/discharge.

In EDLC, the electrostatic storage arises by separating charges in a Helmholtz double layer at the interface between the surface of a conductor electrode and an electrolyte solution. In the double layer structure, one electronic layer is located in the surface lattice of the electrode and the other layer, outer Helmholtz plane (OHP), is distributed in the electrolyte with opposite polarity from the dissolved and solvated ions. These two layers are separated by a thin layer of specifically adsorbed ions losing their solvent shell, termed as the inner Helmholtz plane (IHP). The solvent molecules in IHP are physically adsorbed on the electrode surface and act as the molecular dielectric. The intrinsic process of a physical electrostatic adsorption/desorption enables EDLC to have ultrahigh power capability and fast charge/discharge rate.

In pseudocapacitors, the electrical energy is stored by Faradaic charge-transfer reaction of de-solvated and specifically adsorbed ions, achieved through the redox reactions, electrosorption, or intercalation on the electrode surface. This process is inevitably accompanied by a non-Faradaic contribution from static EDLC. In Faradaic reaction of a pseudocapacitive electrode, the electrons are transferred to or from the valence orbitals of the redox reagent clusters near the electrode surface at a distance, $l \ll (2Dt)^{1/2}$, where D is the diffusion coefficient ($cm^2\ s^{-2}$) for charge-compensating ions and t is the time (s). The Faradaic reaction in pseudocapacitors is a fast reversible process and there is no formation or breaking of chemical bonds, and usually coupled with compensating reactions with protons, Li^+ or Na^+ ions [2].

In fuel cells, the chemical energy of a fuel is converted into electricity through an electrochemical reaction between hydrogen fuel and an oxidizing agent, oxygen, and the continuous supply of fuel and oxygen is required to sustain the chemical reaction. For fuel cells, the energy storage means the storage of fuels is in the tank while the energy conversion is in the cell; thus, the local separation in energy storage and conversion of fuel cells is distinguished with other electrochemical systems. The cathode and anode are just charge-transfer media in fuel cells usually assembled with a specific catalyst to promote hydrogen oxidation and oxygen reduction reactions, respectively. Fuel cells are classified by the type of electrolytes, and SOFC, as schematically represented in the figure, uses solid oxide or ceramic electrolyte to allow oxygen ion transport.

Batteries, especially the lithium-ion batteries, have been dominating the market of power sources in comparison with supercapacitors and fuel cells. However, the current performance of batteries is still insufficient to meet the increasing demand in many applications, such as the electric vehicles. In comparison with combustion engines and gas/steam turbines, the electrochemical energy storage and conversion systems still need unremitting development to improve the output capabilities in energy and power densities and lower the cost. To evaluate the energy contents of a system, the terms specific energy, which is expressed in watt-hours per kilogram (Wh kg^{-1}), and energy density, which is expressed in watt-hours per liter (Wh L^{-1}), are used. The rate capability is indicated by specific power in W kg^{-1} and power density in W L^{-1}. In another style of expression, gravimetric and volumetric are used. The as-known Ragone plot which diagrams the energy and power capabilities of the specific energy system is developed in Fig. 2.2. Batteries owning intermediate energy and power characteristics are located in the gap between high-energy fuel cells and high-power supercapacitors. Some new-type electrochemical devices that combine electrodes of different reaction mechanisms and advantageous properties have been developed to improve the whole performance in both energy and power densities. These devices include asymmetric-electrode supercapacitors which consist of EDLC-type and pseudocapacitive materials as the actives of two electrodes and hybrid-electrode supercapacitors which have battery-type and capacitive electrodes. The combination of different types of electrodes in an electrochemical device aims to remarkably improve at least one of the energy or power capability without much sacrifice of another.

The properties of these electrochemical devices in energy storage and conversion are highly dependent on the electrode materials confined by both thermodynamics and kinetics. Nanostructured materials have attracted tremendous interest because of

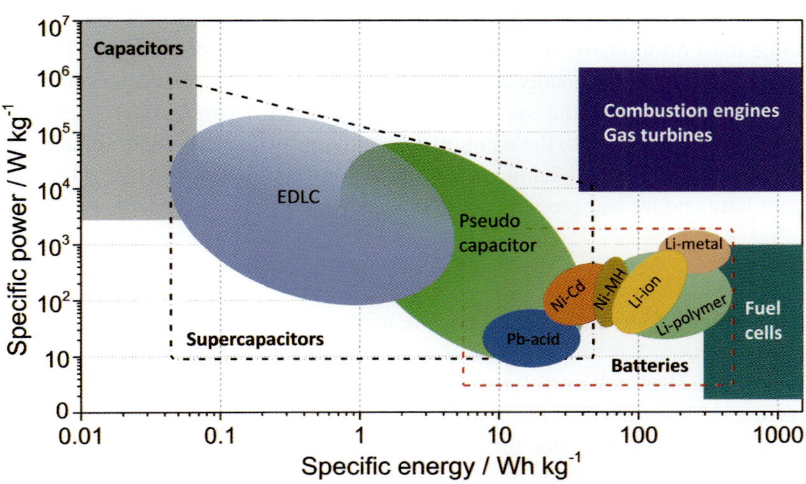

Fig. 2.2 Ragone plot of the current electrochemical energy storage systems with conventional capacitors and internal combustion engines and turbines

their unique mechanical, optical, and electrical properties and the benefits from the synergy of bulk and interfacial properties for efficient and effective energy conversion and storage [3]. Nanostructured materials are playing increasingly important roles in electrochemical energy storage, and especially the booming development of nanotechnologies have afforded emerging but effective tools in designing advanced energy materials [4]. It is a sizeable topic to understand the advantages and disadvantages of the nanomaterials used for energy storage and conversion, as well as the synthesis protocols and the control of the properties [3]. Although there are many novel concepts in fabricating devices and materials, it is beyond the scope of this chapter to present an exhaustive summary of different kinds of electrochemical energy storage and conversion devices and the assembled nanomaterials. Here, we focus on the typical materials and their synthesis strategies in the development of high-performance lithium-ion batteries and supercapacitors. The basic electrochemistry fundamentals associated with the materials are also reviewed to understand the mechanism of electron and ion conduction in the materials.

2.2 Thermodynamics and Kinetics in Electrochemistry

2.2.1 Thermodynamics

The electrochemical reactions in electrochemical energy storage and conversion devices obey the thermodynamic and kinetic formulations. For chemical reactions in electrochemistry, thermodynamics suits the reversible electrochemical reactions and is capable of calculating theoretical cell potentials and electrolytic potentials. An equilibrium in the thermodynamics of an electrochemical cell means the reversibility in both chemical reaction and thermodynamic energy, leading to the reversible reaction in mass and energy. The reversibility in the chemical reaction of an electrochemical cell requires the electrode reaction in charging is the exact reverse reaction in discharging. The reversibility in thermodynamic energy implies that the energy in charging equals to the value in discharging, and both the systems and surroundings return to their original states when the cells work close to the equilibrium conditions where an infinitesimal current is applied. The basic thermodynamic equations for a reversible electrochemical reaction at constant temperature and pressure are given as:

$$G = H - TS \tag{2.1}$$

where G is the Gibbs free energy (J), the useful energy for work from the electrochemical reaction. S is the entropy (J/K), T is the absolute temperature (K), and TS is the heat in an electrochemical reaction associated with the organization and disorganization of materials (J). H is enthalpy, the energy released by the reaction (J). The terms, G, H, and S are state functions that only depend on the properties of the materials and the states of the reactions initially and finally. ΔG, ΔH, and ΔS are the change in the parameters in the reaction. In the definition of H and S,

$$H = U + pV \tag{2.2}$$

$$dQ = TdS \tag{2.3}$$

where U is the internal energy (J) and pV is the volume work arising from volume change (L) V at the specific pressure (atm) P. Q is the heat transfer in a reversible process.

In an electrochemical reaction, the maximum amount of work that can be produced equals to ΔG, representing the net useful energy available from a given reaction. ΔG of an electrochemical cell is given by:

$$\Delta G = -nEF \tag{2.4}$$

$$\Delta G^{\Theta} = -nE^{\Theta}F \tag{2.5}$$

where n is the number of transferred electrons per mole of the reactant (mol), F is the Faraday constant (96,485 C mol^{-1}), and E is the electromotive force (EMF) of a cell in volt (V). nF is the amount of electricity produced by a cell and dependent on the number of materials in the reaction, which is a capacity factor (C mol^{-1}). The cell voltage is considered as an intensity factor. ΔG^{Θ} and E^{Θ} are the standard Gibbs free energy (J/mol) and EMF when all the reactants are of standard states (V). Spontaneous processes have a negative Gibbs free energy and a positive EMF with reactions going forward to a reversible manner. The van't Hoff isotherm gives the relationship between the free energy and bulk chemical reactions:

$$\Delta G = \Delta G^{\Theta} + RT\ \ln(A_P/A_R) \tag{2.6}$$

where R is the gas constant (8.314 J mol^{-1} K^{-1}), T is the absolute temperature in kelvins (K), and A_P and A_R are the activity of reactants and products, respectively (1 mol/L for solutes, 1 atm for gases). The replacement of free energy by nEF derives the Nernst equation, which signifies the relationship between the EMF and activity of the chemicals in the reaction.

$$E = E^{\Theta} + \left(\frac{RT}{nF}\right) \ln(A_P/A_R) \tag{2.7}$$

At the reaction extent of $\xi = 1$ mol, the change in Gibbs free energy is:

$$\Delta G = -\frac{nEF}{\xi} = -zEF \tag{2.8}$$

For a general chemical reaction, the standard free-energy change and the equilibrium constant K_a^{Θ} (dimensionless unit) are related by:

$$-\Delta G^{\Theta} = nE^{\Theta}F = \mathrm{RTln}K_a^{\Theta} \tag{2.9}$$

For T at room temperature, 298 K, the above equation can be transformed to be:

$$E^{\Theta} = \frac{RT}{nF}\ln K_a^{\Theta} = \frac{0.0591V}{n}\lg K_a^{\Theta} \tag{2.10}$$

This equation denotes that the cell voltage E is directly proportional to the logarithm of the equilibrium constant (K_a). The parameters of cell voltage, free-energy change, and equilibrium constant of a chemical reaction are able to calculate from Eqs. (2.9) and (2.10). An electrochemical cell consists of two electrodes and the cell potential is determined by the difference in electrode potentials ρ at two electrodes. Thus, the cell potential $E(V)$:

$$E = \rho_{\text{cathode}} - \rho_{\text{anode}} \tag{2.11}$$

The experimental cell that has only one interesting electrode reaction of a specific material and metallic counter electrode is termed as a half cell. The electrode potentials arise from two electrode reactions, namely anodic and cathodic separately taking place at anode and cathode of a cell. The electrode potential can be obtained from the Nernst equation, pointing out the important relationship between ρ and the activity of the reference chemicals in the electrochemical reaction.

$$\rho = \rho^{\Theta} + (RT/nF)\ln\left(\prod a_O^{\upsilon o} / \prod a_R^{\upsilon r}\right) \tag{2.12}$$

where ρ^{Θ} is the standard electrode potential toward the reference reaction of hydrogen redox; $[H_2(P)|H^+(a)]$, $a_O^{\upsilon o}$, and $a_R^{\upsilon r}$ are the activity of the oxidant and reductant (units are 1 atm or 1 M); and υo and υr are the corresponding coefficient.

Faraday's laws give the quantification relationship between the amount of reaction and the current flowing in the system:

$$g = ItM/nF \tag{2.13}$$

where g is the grams of active material transformed in the reaction (g), I is the current flow in amps (A), t is the flowing time of the current in seconds (s), M is the molecular weight (g), and n is the number of electrons transferred in the reaction (dimensionless). This equation is used to evaluate the theoretical capacity of a material in the electrochemical reaction usually with a unit of mAh g^{-1}.

The differential of H and G gives:

$$dH = dU + pdV + Vdp \tag{2.14}$$

$$dG = dH - TdS - SdT \tag{2.15}$$

According to the firs law of thermodynamics, the reversible process without nonvolume work obeys:

$$dU = dQ - pdV = TdS - pdV \tag{2.16}$$

The combination of Eqs. (2.14), (2.15), and (2.16) derives,

$$dG = -SdT + Vdp \tag{2.17}$$

The differential by T in Eq. (2.17) gives temperature-dependent relation,

$$-S = \left(\frac{dG}{dT}\right)_p \tag{2.18}$$

and the Gibbs–Helmholtz equation is obtained in Eq. (2.19), which is a thermodynamic equation used for calculating changes in the Gibbs energy of a system as a function of temperature.

$$(G - H)/T = \left(\frac{dG}{dT}\right)_p \tag{2.19}$$

Then, we can derive the transformation in Eq. (2.20).

$$H = -zEF + zFT\left(\frac{dE}{dT}\right)_p \tag{2.20}$$

The enthalpy of an electrochemical reaction can be calculated by measuring the cell potential and the cell voltage as a function of the temperature dE/dT, where z is the number of electrons transferred in the cell reaction or half-reaction (C). Reasonably the other thermodynamic parameters of the materials in the electrochemical reaction are clear. A positive dE/dT implies the cells heat on charge and cool on discharge, such as Ni-Cd batteries, while a negative dE/dT means the cells cool on charge and heat on discharge, such as the lead-acid batteries. Any change in the temperature signifies the heat exchange between the cells and the environment.

In the operation of electrochemical cells, the entropic heat is negligible in comparison with the irreversible heat. The heat given off by the system includes the reversible thermodynamic heat and the irreversible joule heat from the working of a cell in an irreversible way during charge and discharge process at finite current densities. As displayed in Eq. (2.21), the irreversible heat manifests as a departure from the equilibrium voltage E_{ocv} which is the open-circuit voltage (ocv) of a cell (V), and the irreversible heat released is dependent on the practical cell terminal voltage E_T(V) and the current (A).

$$q = TS + It(E_{ocv} - E_T) \tag{2.21}$$

Heat is released inside the battery at the reaction site, normally the interface between the electrode and electrolyte. The low rate operation may produce negligible heat, however, high-rate batteries must dissipate the released heat in time to ensure safety.

2.2.2 Kinetics

Electrode kinetics mainly studies the net rate of electrochemical reaction and follows the same general considerations for chemical reactions. However, there is still some difference in the interface reactions. In electrochemical reactions, the potential drop in the double layer along the interface of electrode/electrolyte directly affects the activated species and the interface reaction is in a two-dimensional manner. The potential drop takes place when the current is drawn to the battery in comparison with the open-circuit voltage (E_{ocv}), resulting from the electrode polarization (V). The difference between the irreversible electrode potential and equilibrium potential (E_{ocv}) is termed as overpotential (η) and three types of kinetic limitations lead to the polarization (V).

Ohmic polarization is related to the resistance of the cell components and contact problems between these components (Ω).

Concentration polarization, due to the mass transport limitation in the cell operation, causing the ionic concentration near the electrode is different from the bulk.

Activation polarization, also termed as electrochemical polarization, is associated with the electrochemical redox or charge-transfer reaction at the electrode/electrolyte interfaces where the kinetics affects the charge state of the electrode.

The overpotentials of the anode and cathode in a galvanic cell are expressed as:

$$\eta_{\text{anode}} = E_{\text{anode,ir}} - E_{\text{anode,r}} \tag{2.22}$$

$$\eta_{\text{cathode}} = E_{\text{cathode,r}} - \mathrm{E}_{\text{cathode,ir}} \tag{2.23}$$

where $E_{\text{anode, ir}}$ and $E_{\text{cathode, ir}}$ are the irreversible electrode potentials while $E_{\text{anode, r}}$ and $E_{\text{cathode, r}}$ are the reversible electrode potentials or equilibrium potentials. The practical cell terminal voltage E_T should be smaller than the equilibrium potential resulting from the polarization and internal resistance of a cell.

$$E_T = \left(E_{\text{cathode,r}} - E_{\text{anode,r}}\right) - \left(\eta_{\text{cathode}} + \eta_{\text{anode}} + IR\right) \tag{2.24}$$

where IR is the potential drop arising from the internal resistance of a cell (Ω) which might be negligible at low current densities, and Eq. (2.24) turns to be Eq. (2.25) at the small current rate. However, the large current rates cause large overpotentials and thus small cell terminal potentials.

$$E_T = E_{\text{cathode,ir}} - E_{\text{anode,ir}} \tag{2.25}$$

Concentration polarization arises from the emergence of the concentration gradient in the solution near the electrode surface in the passage of current through the electrode/solution interface. The concentration gradient is determined by the difference in the rate of electrochemical reaction at the electrode and the rate of ion migration in the solution from/to the electrode surface. When a chemical species participating in an electrochemical reaction is depleted and in short supply on the electrode, transport of this matter takes place toward the electrode surface to balance the consumption of the species, termed as diffusion. Fick's first law points out the relationship between the diffusion rate $r_{diffusion}$ and concentration gradient:

$$r_{\text{diffusion}} = D(C_o - C_s)/\delta \tag{2.26}$$

where D is the diffusion coefficient of the actives in electrode reaction (m^2/s), δ is the thickness of the diffusion layer (m), C_s is the concentration of actives near the electrode surface (mol/L), and C_o is the concentration of actives in bulk (mol/L). The parameter of C_s is the result from the competition between the rate of reaction and diffusion and affects the electrode current i, which is called diffusion current (A).

$$i = zFr_{\text{electrode}} = zFr_{\text{diffusion}} = zFD(C_o - C_s)/\delta \tag{2.27}$$

Here, the rate of the electrode reaction and the diffusion current are controlled by the ion diffusion process. The electrode reaction is driven by the electrode potentials and crucial for Cs. As Cs approaches 0, a largest diffusion current is generated, termed as limited diffusion current $i_{limited}$.

$$i_{\text{limited}} = zFDC_o/\delta \tag{2.28}$$

The stable concentration polarization in an electrochemical reaction gives the reversible electrode potential and irreversible electrode potential:

$$E_{ir} = E^{\Theta}\left(M^{Z+}/M\right) - \left(\frac{RT}{ZF}\right)\ln\frac{1}{C_S} \tag{2.29}$$

$$E_r = E^{\Theta}\left(M^{Z+}/M\right) - \left(\frac{RT}{ZF}\right)\ln\frac{1}{C_O} \tag{2.30}$$

and the overpotential from the concentration polarization is:

$$\eta_{\text{concentration}} = E_r - E_{ir} = \frac{RT}{ZF}\ln\frac{Co}{Cs} \tag{2.31}$$

Electrochemical polarization is a collective term for the barriers developed at the electrode/electrolyte interface in an electrochemical process. At equilibrium potential, the rate of the oxidation reaction is equal to the reduction reaction; thus, the related oxidation current ($\overleftarrow{\imath}$) equals the reduction current ($\overrightarrow{\imath}$) and is termed exchange current i_O which is directly related to the reaction rate constant, the activities of reactants and products and the potential drop across the double layer. A large exchange current means a more reversible chemical reaction and low polarization at a given current. The electrode that has an exchange current in the range of 10^{-2} A cm^{-2} at room temperature is favored for use in batteries [1].

$$\overleftarrow{\imath} = \overrightarrow{\imath} = i_O \tag{2.32}$$

The current density of the electrode reaction (A/m^2) can be related with the rate of the chemical reaction.

$$i = \frac{I}{S} = \frac{dQ}{Sdt} = \frac{\mathrm{zFdn}}{Sdt} = zFr = \mathrm{zFkC} \tag{2.33}$$

where k is the constant of the reaction rate, I is partial cathodic (anodic) current (A), S is entropic contribution (J/K), and C is the concentration of the reactants (mol/L). Arrhenius equation gives:

$$k = \mathrm{A}\exp\left(-\frac{\Delta G^{\neq}}{RT}\right) \tag{2.34}$$

where A is a pre-exponential factor and $\Delta G^{\neq}$ is activation energy for a chemical reaction. The changes in the activation energy of the oxidation and reduction reaction in an electrochemical process are in a reverse way when the electrode polarization takes place. The anodic polarization decreases the activation energy of the oxidation reaction while the reduction reaction has increased activation energy, thus resulting in high rate of oxidation reaction and large anodic current $\overleftarrow{i}$. The shift of the electrode potential with η leads to the energy change of a single electron with Fη. If we define the decreased activation energy by using an anodic charge-transfer coefficient $(1 - \alpha)$, the polarized activation energy in anodic reaction ($\Delta G_a^{\neq}$) is smaller than the nonpolarized activation energy ($\Delta G_{oa}^{\neq}$) by an amount of $(1 - \alpha)F\eta$. The cathode polarization increases the cathodic activation energy ($\Delta G_c^{\neq}$) by αFη in comparison with the nonpolarized energy ($\Delta G_{oc}^{\neq}$), where α is termed as the cathodic charge-transfer coefficient (dimensionless). The charge-transfer coefficient signifies the fraction of the over-potential that helps in lowering the free energy barrier for an electrochemical reaction, thus a change in the rate constant for a charge-transfer reaction.

$$\Delta \mathrm{G}_{\mathrm{a}}^{\neq} = \Delta \mathrm{G}_{\mathrm{oa}}^{\neq} - (1 - \alpha)F\eta \tag{2.35}$$

$$\Delta \mathrm{G}_{\mathrm{c}}^{\neq} = \Delta \mathrm{G}_{\mathrm{oc}}^{\neq} + \alpha F\eta \tag{2.36}$$

The substitution of the activation energy expressed in Eqs. (2.35) and (2.36) into Eq. (2.34) derive:

$$\begin{aligned} k_a = A_a^0\exp\left[-\frac{\Delta G_{oa}^{\neq} - (1-\alpha)F\eta}{RT}\right] &= A_a^0\exp\left[\left(-\frac{\Delta G_{oa}^{\neq}}{RT}\right) + \frac{(1-\alpha)F\eta}{RT}\right] \\ &= k_a^o\exp\left[\frac{(1-\alpha)F\eta}{RT}\right] \end{aligned} \tag{2.37}$$

$$\begin{aligned} k_c = A_c^0\exp\left[-\frac{\Delta G_{oc}^{\neq} + \alpha F\eta}{RT}\right] &= A_c^0\exp\left[\left(-\frac{\Delta G_{oc}^{\neq}}{RT}\right) - \frac{\alpha F\eta}{RT}\right] \\ &= k_c^o\exp\left[\frac{-\alpha F\eta}{RT}\right] \end{aligned} \tag{2.38}$$

where k_a^o and k_c^o are the rate constants for the anodic and cathodic reactions. Then, we reconsider the current related with chemical reaction process in Eq. (2.33). The anodic and cathodic current for a single electron transfer are obtained:

$$\overleftarrow{\imath} = FC_o\, k_a^o \exp\left[\frac{(1-\alpha)F\eta}{RT}\right] \tag{2.39}$$

$$\overrightarrow{\imath} = FC_r k_c^o \exp\left[\frac{-\alpha F\eta}{RT}\right] \tag{2.40}$$

Now the exchange current:

$$\imath_O = FC_o\, k_a^o \exp\left[\frac{(1-\alpha)F\eta}{RT}\right] = FC_r k_c^o \exp\left[\frac{-\alpha F\eta}{RT}\right] \tag{2.41}$$

In equilibrium condition of an electrochemical reaction, the reversible currents are the same; thus $C_o k_a^o = C_r k_c^o = Ck_o$. We apply Eqs. (2.39) and (2.40) to calculate the current i arising from the electrochemical polarization.

$$\begin{aligned} \imath = \overrightarrow{\imath} - \overleftarrow{\imath} &= FC_r k_c^o \exp\left(\frac{-\alpha F\eta}{RT}\right) - FC_o\, k_a^o \exp\left[\frac{(1-\alpha)F\eta}{RT}\right] \\ &= FCk_o\left[\exp\left(\frac{-\alpha F\eta}{RT}\right) - \exp\frac{(1-\alpha)F\eta}{RT}\right] \\ &= \imath_O\left[\exp\left(\frac{-\alpha F\eta}{RT}\right) - \exp\frac{(1-\alpha)F\eta}{RT}\right] \end{aligned} \tag{2.42}$$

The electrical current has a dependence on the electrode potential and the relation equation is called the Butler–Volmer equation. There are two limiting cases of the Butler–Volmer equation. The first case is in the small electrochemical polarization region, thus low overpotential in which the current is nearly linear with overpotential and Butler–Volmer equation simplifies to:

$$\imath = \imath_O \frac{F\eta}{RT} \tag{2.42$'$}$$

This equation resembles the Ohm's law by defining a charge-transfer resistance R_{ct}:

$$R_{ct} = \frac{\eta}{\imath} = \frac{RT}{\imath_O F} \tag{2.43}$$

The second limiting case is in large overpotential region, where $E_{anode} \gg E_{eq}$ or $E_{cathode} \ll E_{eq}$. In the case of large polarization on the cathode, $\overleftarrow{\imath}$ is negligible,

$$i \approx \overrightarrow{i} = i_{\mathrm{O}} \exp\left(\frac{-\alpha F \eta_{\mathrm{cathode}}}{RT}\right) \quad (2.44)$$

$$\eta_{\mathrm{cathode}} = \frac{RT}{\alpha F} \ln i_{\mathrm{O}} - \frac{RT}{\alpha F} \ln i \quad (2.45)$$

In another case of large polarization on the node, $\overrightarrow{i}$ is negligible,

$$i \approx \overleftarrow{i} = i_{\mathrm{O}} \exp \frac{(1-\alpha) F \eta_{\mathrm{anode}}}{RT} \quad (2.46)$$

$$\eta_{\mathrm{anode}} = -\frac{RT}{(1-\alpha)F} \ln i_{\mathrm{O}} + \frac{RT}{(1-\alpha)F} \ln i \quad (2.47)$$

In the region of high overpotential for both cathodic and anodic polarization, the overpotential has a mathematical relationship with the current $\eta = a + \mathrm{blg} i$, which is termed as Tafel equation. From the linear relation in Tafel plot, the charge-transfer coefficient from the slope and exchange current by the linear interpolation to the equilibrium potential is then obtained. By using Taylor series expansion to the Butler–Volmer equation, overpotentials presenting slopes smaller than one-third of the Tafel slope display nearly linear relationship between i and η, while overpotentials with slopes larger than half of the Tafel slope have the large net current dominated by the anodic/cathodic reactions.

The fast change in electrode potential or overpotentials leads to a decrease in the thickness of the diffusion layer and as a result, a higher current. In an electrochemically reversible electron transfer process accompanied with freely diffusing redox species, the peak currents i_p (A) increase linearly with the square root of the scan rate υ (V s^{-1}) in a cyclic voltammetry technique according to the Randles–Sevcik equation in Eq. (2.48):

$$i_{\mathrm{p}} = 0.446 \mathrm{nFAC} \left(\frac{\mathrm{nF\upsilon D}}{RT}\right)^{0.5} \quad (2.48)$$

where A is the electrode surface area usually represented with the geometric surface area (m^2), D is the diffusion coefficient (cm^2 s^{-1}) of the analyte, and C is the bulk concentration of the analyte (mol cm^{-3}). For an electrode-adsorbed analyte (electrochemical capacitor), the current response to the scan rate is as follows [5]:

$$i_{\mathrm{p}} = \upsilon A \Gamma \frac{n^2 F^2}{4RT} \quad (2.49)$$

where Γ is the surface coverage of the adsorbed species (mol cm^{-2}). Randles–Sevcik equation is used to describe a diffusion-controlled reaction. For a reaction of the diffusion process, the deviation from linearity in the plots of i_{p} versus $\upsilon^{0.5}$ suggests that either the electrochemical reaction is quasi-reversible, or partial electron transfer takes place via the surface-adsorbed species, or both [5].

2.3 Nanostructured Materials in Lithium-Ion Batteries

Lithium-ion batteries (LIBs) are based on single electron intercalation chemistry [4] and have achieved great success in energy storage used for electronics, smart grid. and electrical vehicles (EVs). LIBs have comparably high voltage and energy density, but their poor power capability resulting from the sluggish ionic diffusion [6] still impedes their wide application in EVs. The performance of LIBs in energy storage has been improved greatly since the first commercialized one in 1991 by Sony, benefiting mainly from the development of advanced electrode materials. The cathode materials have been pointed out to leave less room to further improve the energy density depending on the already existed materials, but the anode part is still promising to achieve higher energy and power densities [7]. This session provides a brief summary on the nanostructured materials for use as the electrode actives in LIBs.

2.3.1 Cathode Materials

The basic requirement for a cathode material in LIBs is high free energy of Li^+ ions and low free energy of electrons in comparison with the anode materials, then the difference in the chemical potentials of both sides establishes a working voltage [8]. The well-known cathode materials mainly process two mechanisms to host the Li^+ ions, namely intercalation and conversion. Basically, the research in cathodic host materials started from the 1970s when a variety of chalcogenides, such as titanium disulfide (TiS_2), work as the cathode and metallic lithium as anode producing a limited cell potential of 2.5 V versus Li^+/Li [9]. Later, manganese and vanadium oxides were investigated. The pioneering work by Goodenough and coworkers invented layered lithium cobalt oxide ($LiCoO_2$) and spinel lithium manganese oxide ($LiMn_2O_4$) as cathode materials for lithium rechargeable batteries in the 1980s and the cell voltage extended above 4 V versus Li^+/Li [10, 11]. After the release of reversible lithium insertion into graphite by Yazami and Touzain [12], the development of lithium-ion batteries using graphite anode and lithium cobalt oxide ($LiCoO_2$) cathode by Sony Inc. led to the successful commercialization of lithium-ion batteries in the 1990s. Thereafter, many other layered transition metal oxides including lithium nickel dioxide ($LiNiO_2$), lithium manganese dioxide ($LiMnO_2$), and vanadium pentoxide (V_2O_5) were developed as potential alternatives to the $LiCoO_2$ cathode. Goodenough and coworkers [13] then reported a polyanion framework material, namely lithium iron phosphate ($LiFePO_4$), as prospective cathode material which has been commercialized by A123 Inc. and Sony Inc. Lithium iron phosphate displays appealing advantages in environmentally benign, low cost, very stable electrochemical properties with structural stability, high thermal stability, and a distinct working plateau [14]. The plots of electrochemical potentials versus specific capacity of the LIBs cathode materials are illustrated in Fig. 2.3a, and some of the commercialized cathode materials with the related properties are listed in Table 2.1. The current

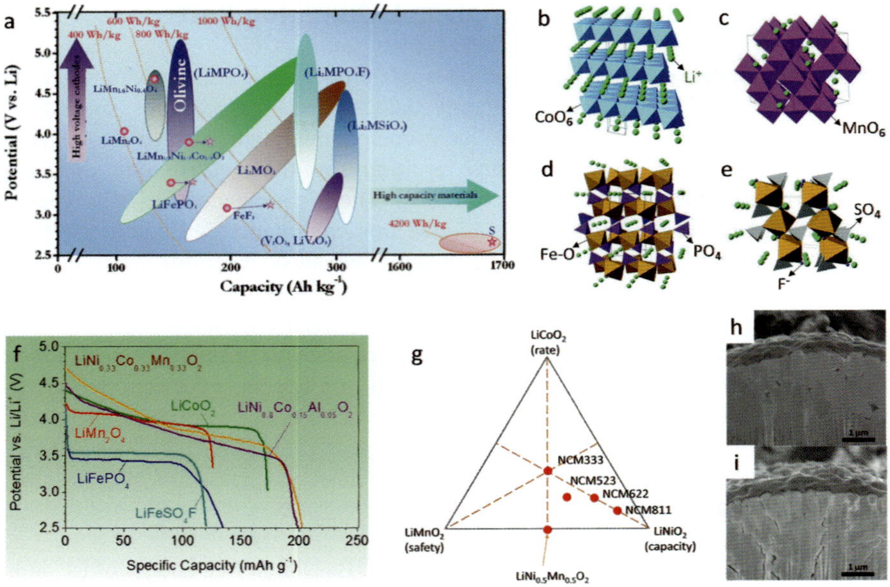

Fig. 2.3 (**a**) Schematic representation of potentials versus specific capacity of the Li-ion battery cathode materials. Lithium metal phosphate ($LiMPO_4$), lithium metal silica (Li_2MSiO_4 [*alpha phase*]), lithium metal fluorophosphates ([*delta phase*] Li_2MPO_4F), where M = Fe, Mn, Co, Ni, etc.; Li_2MO_3, where M = Mn, Mo, Ir, Ru, etc. Adapted with permission from Ref [14], Copyright 2012 Royal Society of Chemistry. Representative crystal structures of intercalation-type cathodes: (**b**) layered $LiCoO_2$, (**c**) [*spinel*] $LiMn_2O_4$, (**d**) *olivine* $LiFePO_4$, and (**e**) [*tavorite*] $LiFeSO_4F$. (**f**) Typical discharge profiles of these intercalation cathodes. Adapted with permission from Ref [15], Copyright 2015 Elsevier. (**g**) Phase diagram of the ternary system between LNO, LCO a, d LMO with representative compositions. The number following NCM demotes the ratio of x:y:z in lithium manganese nickel cobalt oxide ($LiNi_xCo_yMn_zO_2$). Scanning electron microscopy (SEM) images of the cross section of (**h**) a fresh NCM811 electrode and (**i**) after 100 cycles at C/3 at 45 °C. Adapted with permission from Ref [16], Copyright 2016 Electrochemical Society

research focuses on cathode materials are to develop high-voltage and high-capacity materials to increase the energy density (E) of the batteries which is subjected to the reversible capacity (Q) and operating voltage (V_{op}).

$$E = QV_{op} \tag{2.50}$$

These intercalation compounds can be divided into several crystal structures, such as layered, spinel, olivine, and favorite, as shown in Fig. 2.3b–e, and their discharge profiles are displayed in Fig. 2.3f [15]. Among these cathode materials, NCM shows great promise in the target of high specific capacity by changing the ratio of the three transition metals. Figure 2.3g shows the phase diagram of the ternary system between LNO, LCO, and LMO with representative compositions. The number following NCM demotes the ratio of x:y:z in lithium nickel cobalt manganese oxide ($LiNi_xCo_yMn_zO_2$).

Table 2.1 Properties of commercial cathode materials used in lithium-ion batteries and the related advantages, disadvantages, and applications of the full cells. LCO is $LiCoO_2$, LMO is $LiMn_2O_4$, NCA is lithium nickel cobalt oxide ($LiNi_{0.8}Co_{0.15}O_2$), NCM is lithium nickel cobalt manganese oxide ($LiNi_xCo_yMn_zO_2$), and LFP is lithium iron phosphate ($LiFePO_4$). Adapted with permission from Ref [17], Copyright 2016 Electrochemical Society

Materials	Midpoint voltage versus Li	Specific capacity (mAh g^{-1})	Advantages	Disadvantages	Main applications
LCO	3.9	155	High cycle stability and energy density	Moderate charged state thermal stability	Small portable electronics
LMO	4.0	100~200	High thermal stability, low-cost, high power density	Moderate cycle lift, low energy density	High-power tools and electric motive power
NCA	3.7	180	High energy and power density, good cycle stability	Moderate charged state thermal stability, sensitive to moisture	Motive power and premium electronic applications
NCM	3.8	160	High properties in energy and power density, cycle life, thermal stability	Patent issues	Both portable and high power applications including power tools and electric vehicles
LFP	3.4	160	High thermal stability and lifespan, high power density	Low energy density, special preparation conditions	High-power tools and energy storage

Lithium nickel cobalt manganese oxide (NCM523) is a commercial cathode material with high nickel content, but the high reactive Ni^{4+} at the end of charge reaction leads to undesired side reactions with the electrolyte, causing gas evolution and capacity degradation. The temperature stability of the ternary cathode material decreases with an increase in nickel content, magnifying the safety issues. The high nickel content also leads to a Li/Ni cation mixing and formation of inactive phases on the electrode surface. Lithium nickel cobalt manganese oxide (NCM811) in the prolonged cycling leads to cracks along the boundaries of the secondary particles, gradually increasing the surface area and active sites for parasitic reactions. The cracks formed in the cycling are observable from the scanning electron microscopy (SEM) images as shown in Fig. 2.3h, i, the cross section of a fresh NCM811 electrode, and after 100 cycles at C/3 at 45 °C.

Conversion-type electrode materials have a solid-state redox reaction in the process of lithiation/de-lithiation, normally accompanied by the change in the crystal structure and breaking/formation of chemical bonds. These cathode materials include metal halides comprising high valence metal ions and solid chalcogens including sulfur (S), selenium (Se), and tellurium (Te). Most conversion-type materials have very poor electron conductivity and moderate potential plateau as cathode materials, but their specific capacities are relatively high, up to 1675 mAh g^{-1} for sulfur [15].

Nanotechnologies for the synthesis of cathode materials have been widely used to control the morphology and dimension of the products. The synthetic methods mainly include freeze drying, template-direct growth, solvothermal and hydrothermal, supercritical fluid synthesis, ionothermal method, biosynthesis, solid-state reaction, and coprecipitation, but in a practical process, some of these methods are usually combined together. Figure 2.4a shows the SEM images of porous and hollow microspheres of lithium manganese nickel cobalt oxide ($Li_{1.2}Mn_{0.6}Ni_{0.1}Co_{0.1}O_2$) in the synthetic process and Fig. 2.4b schematically explains the route. The precursor solutions coprecipitate the manganese(II) carbonate ($MnCO_3$) nanoparticles and the following calcination forms manganese(IV) oxide (MnO_2) microspheres. As the intermediate, MnO_2 microspheres continue to react with lithium (Li), nickel (Ni), and cobalt (Co) sources at high temperature to obtain the final product. The formation of the hollow structure has been explained by the nanoscale Kirkendall effect and Ostwald ripening [18]. The hollow structured $Li_{1.2}Mn_{0.6}Ni_{0.1}Co_{0.1}O_2$ displays a discharge capacity of 242 mAh g^{-1} at 50 mA g^{-1} with high Coulombic efficiency, as illustrated in Fig. 2.4c. The cobalt(II) hydroxide [$Co(OH)_2$] hexagonal nanoplates via a coprecipitation process was used to synthesize LCO nanoplates by a solid-state reaction as shown in Fig. 2.4d1, d2 [19]. The synthesized nanoplates have a hexagonal structure with exposed (001) planes confirmed by the transmission electron microscopy (TEM) image in Fig. 2.4d3 and selected area electron diffraction (SAED) pattern in Fig. 2.4d4. The synthetic route in Fig. 2.4e keeps the exposed (001) plane of cobalt(II) hydroxide [$Co(OH)_2$] precursor in the final product and the Li^+ ion diffusion is facilitated. The as-synthesized LCO releases a discharge capacity of 140 mAh g^{-1} at 100 mAg^{-1} and 113 mAh g^{-1} at large current rate of 1000 mAg^{-1}, as shown in Fig. 2.4f.

2.3.2 Anode Materials

Anode materials act as a reductant in a reversible battery opposite to the cathode and should be in high energy level but with the low ionic energy of Li^+ ions. The anode materials have been classified into three categories according to the lithiation mechanisms, namely insertion (graphite, titanium-based materials), alloying (metals, semi-metals), and conversion (MaXb, M = metal, X = O, S, F, P, N, etc., and a/b are stoichiometric mole ratios) [7]. All these anode materials have both advantages and disadvantages: insertion-type materials have high cycling stability but low capacity, alloying-type materials have large capacity but suffer the large volume changes, and conversion-type materials display large capacity but unsatisfactory cycling stability and energy efficiency [7].

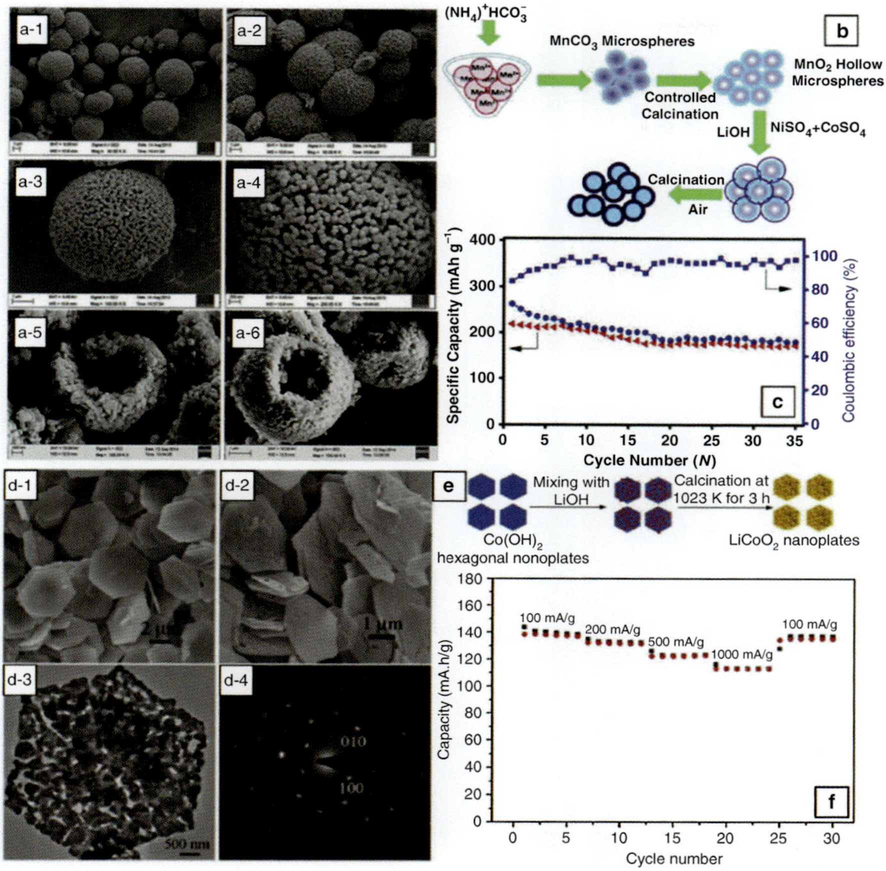

Fig. 2.4 (**a**) Scanning electron microscopic (SEM) images of the lithium manganese nickel cobalt oxide ($Li_{1.2}Mn_{0.6}Ni_{0.1}Co_{0.1}O_2$) hollow microspheres at different resolutions. (**b**) Synthesis protocol diagram of the material. (**c**) Galvanostatic cycling performance with the Coulombic efficiency of the cathode material. Adapted with permission from Ref [18], Copyright 2014 Royal Society of Chemistry. (**d**) SEM image of (d-1) beta cobalt(II) hydroxide [β-$Co(OH)_2$], and (d-2) LCO nanoplates; (d-3) high-resolution transmission electron microscopy (HRTEM) image of LCO nanoplates; (d-4) selected area electron diffraction (SAED) image of LCO nanoplates. (**e**) Two-step synthesis route of LCO nanoplates and (**f**) the rate capability of LCO nanoplates. Adapted with permission from Ref [19], Copyright 2012 Springer Nature

Profiles of voltage versus specific capacity of the studied anode materials are displayed in Fig. 2.5a [20] with a wide range in the specific capacity from 160 to up to 4200 mAh g^{-1} below the voltage range 2 V versus Li^+/Li. The representatives of specific reaction mechanism with the operating voltage range are diagrammed in Fig. 2.5b. Intercalation/de-intercalation mechanism occurs in two-dimensional layer structures or three-dimensional structures that can reversibly store/release lithium ions into/from the lattice without destroying the crystal structure. The alloying/de-alloying

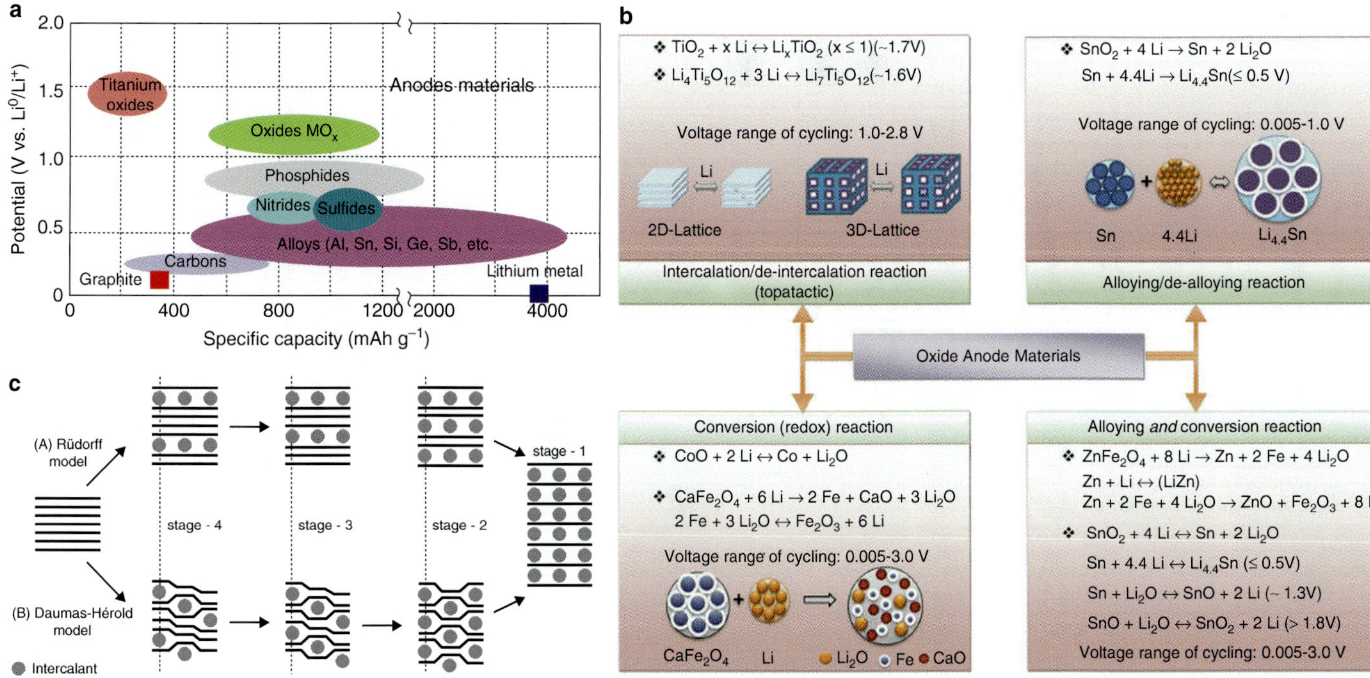

Fig. 2.5 (**a**) Commonly investigated anode materials plotted by voltage versus specific capacity. Reprinted with permission from Ref [20], Copyright 2016 AIMS Press. (**b**) Typical anodic reactions in three types of lithium storage mechanisms, namely intercalation, alloying, and conversion. Adapted with permission from Ref [21], Copyright 2013 American Chemical Society. (**c**) Intercalation reaction of lithium ions in graphite with specific graphite intercalated compounds through two models, namely Rüdorff model and Daumas-Herold model. Reprinted with permission from Ref [22], Copyright 2014 Royal Society of Chemistry

mechanism takes place on metals and nonmetal substances that can form alloys with lithium metal. The alloyed compounds and intermediates have distinct crystal structures and physiochemical properties from the reactants. Conversion mechanism happens in a redox reaction by reducing the materials, e.g. CoO, with lithium and forming electrochemically inactive lithium oxide, Li_2O. Thus, Li_2O leads to large irreversibility in the following reactions. The conversion-type anode materials can deliver high specific capacity by reducing metal ions to zero oxidation and are usually environmentally friendly, attracting great interest for next-generation lithium battery anode materials [23]. Some metal oxides have the synergistic mechanism from alloying and conversion processes and give a large and stable performance in the lithium storage. Nanotechnologies have been widely applied in the fabrication of nanomaterials aiming to improve the electrochemical properties through the control of the size and morphologies.

Graphite, as the most successful anode material in commercialization, has been studied extensively from both theory and experiment. However, there are still contrary understandings to the storage reaction of graphite with lithium in the transformation to graphite intercalated compound (GIC). The staged process takes place under specific potentials and a voltage of 0.2 V versus Li^+/Li is usually the onset of state 4 GIC, and continuous lithiation promotes the staging reaction to stage 1 GIC. In the staging process, there are two developed mechanisms: Rüdorff model gives the sequential filling up mechanism when the lithium ions intercalate alternating graphene interlayer spaces without structure distortions. Daumas–Herold model proposes the intercalation would induce deformation around the intercalated lithium ions in the flexible graphene layers. Many *in situ* techniques, such as Raman spectra, have been developed to study the real-time characteristics of the graphite in the electrochemical reaction and provide insights into the mechanism of lithium storage [22]. Anode materials have been studied extensively within the structures of zero, one, two, and three dimensions, including carbon-based materials, sulfides and nitrides, metals and nonmetals, metal oxides, and polyanions.

Figure 2.6 displays the schematic illustration of a series of hollow carbon nanostructures with high graphitization, and these structures have been used as the sulfur host in lithium-surfer batteries because of the high conductivity and affinity to sulfur [24]. Similarly, these hollow carbon hosts are also able to accommodate other low-dimensional and small-size materials to provide improved performance in the lithium storage because of the suppressed volume change within the space of hollow structures and high conductivity of the carbon matrix. In comparison with the hollow structure, double- and multi-shell structures are proposed to provide more effective confinement to the incorporated materials. Porous shells are developed to contribute large surface area with the actives in the electrolyte and hosted materials, and the porous shell and core components can provide a better electronically connecting matrix. Despite the advantages of the above, the hollow carbon bowls are able to accommodate high-density actives in the hollow space, beneficial for the improvement of the volumetric density of the electrode. When the structures are processed down to one dimension, the high aspect ratio can better act as an electrically conducting network in the electrode leading to improved reaction kinetics and rate capability. The continuous processing within the hollow carbon structures, such as hollow carbon fiber, can derive novel tube-in-tube structures which are rich in the channels and voids for the diffusion of lithium

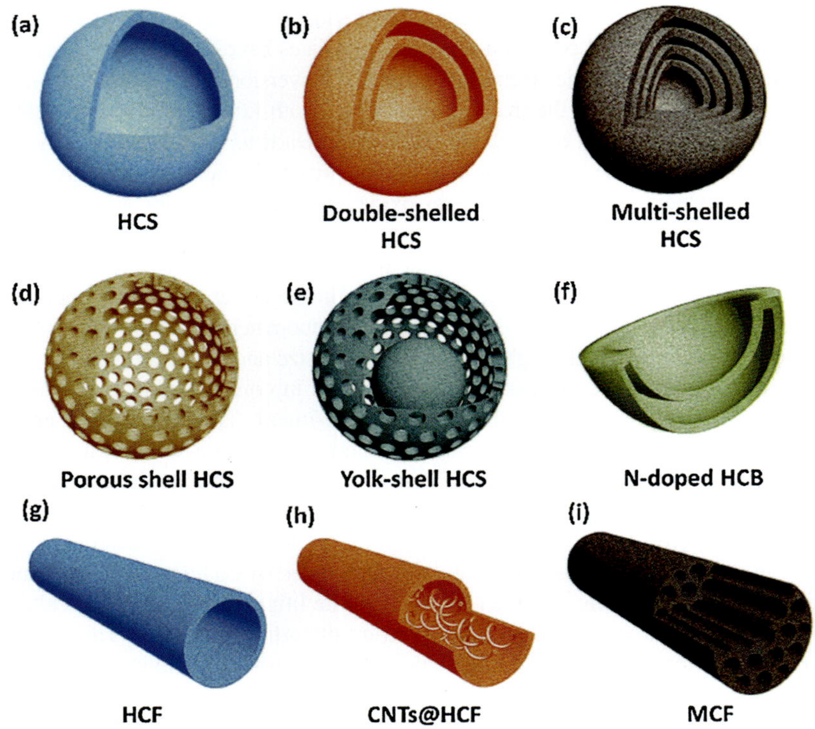

Fig. 2.6 Schematic representation of hollow carbon nanostructures: (**a**) hollow carbon sphere (HCS), (**b**) double-shell (D) HCS, (**c**) multi-shell HCS, (**d**) HCS with tailored porosity, (**e**) yolk-shell structured HCS, (**f**) N-doped hollow carbon bowl (HCB), (**g**) hollow carbon fiber (HCF), (**h**) CNT@HCF, and (**i**) multichannel carbon fiber (MCF). Reprinted with permission from Ref [24], Copyright 2016 Royal Society of Chemistry

ions and the actives accommodation. These structures, shown in Fig. 2.6b–f, h, i, can be considered as the derivatives of hollow spheres (Fig. 2.6a) and hollow tubes (Fig. 2.6g), which can perform better in the energy storage by introducing more channels and conduction matrix.

To understand the benefits of the host structures to the actives in the electrochemical energy storage, we take the easily-dissoluble sulfur as an example of actives and introduce several host structures to accommodate sulfur. In Fig. 2.7a, the schematic protocol shows the steps in forming a composite owning cobalt hydroxide (CH) inner shell and a layered double hydroxide (LDH) out shell loaded with sulfur (CH@LDH/S) [25]. The corresponding TEM images of the product of each stage are displayed in Fig. 2.7b, illustrating the morphology of zeolitic imidazolate framework-67 (ZIF-67), yolk-shell ZIF-67@LDH, double-shell CH@LDH, and CH@LDH/S. The double-shell CH@LDH nanocages have abundant polar surfaces to chemically adsorb lithium polysulfides (LiPSs) and the complex shells are able to suppress their outward diffusion. In addition, the rich hydrophilic/hydroxy groups and the possible electrocatalytic properties of LDH might enhance the adsorption of the soluble LiPSs and promote to convert to the short-chain polysulfides. In the synthesis of the double-shelled CH@LDH nanocages, ZIF-67 polyhedrals are used as a sacrificial template. The reaction between ZIF-67 and nickel(II) nitrate [Ni $(NO_3)_2$] results in the hollow LDH shell-coated ZIF-67 particles (ZIF-67@LDH), which then undergoes a solution reaction with Na_2MoO_4 to convert to CH@LDH. A high content of sulfur (75 wt%) is then loaded into CH@LDH matrix via the melt-diffusion method to form CH@LDH/S and this as-prepared double-shell composite has displayed significantly improved cycling stability in the lithium storage [25].

Figure 2.7c displays a schematic illustration of one-dimensional (1D) hollow carbon fiber (HCF) filled with MnO_2 nanosheets (MnO_2@HCF). This 1D HCF has a high aspect ratio and conductivity, plus the holes enhancing the porosity, together with facilitating both ion and electron conduction in the electrochemical reaction. The inner MnO_2 nanosheets provide strong chemical adsorption for LiPSs and effectively prevent the dissolution [26]. Hollow nanospheres with highly conductive shells composed of titanium monoxide (TiO) nanoparticles and a thin carbon layer (TiO@C-HS) have been prepared as the sulfur host (Fig. 2.7d) [27]. The shell is highly conductive and about 20 nm in thickness (see Fig. 2.7e). After the melt diffusion, the hollow space is filled up with sulfur with a high content of 70 wt%. TiO@C-HS/S thick electrode with a high sulfur loading of 4.0 mg cm^{-2} can deliver high specific capacities at different current densities with a stable cycling performance (Fig. 2.7f). In this protocol, a polar and highly conductive host has been developed for effective accommodating of sulfur.

Hollow nanostructures with multi-level architectures have been attracting great interest in the fabrication of the anode materials in hierarchical structures because of the outstanding buffering to the volume variation in the electrochemical reaction and improved tap density by utilizing the inner cavities in a hollow space, in comparison with the single-shell hollow structures [28]. Multi-level hollow nanostructures including yolk-shell, multi-shell, and multi-chamber structures have been proposed and prepared for the use of energy storage [29]. These complex structures make better use of the inner hollow space and effectively increase the fraction of the electrochemical actives. These multi-level skeletons have strong structural stability [30] to ensure a high capacity retention in the long-term charging-discharging cycles. The widely used synthetic strategies for the hollow nanostructures mainly include Ostwald ripening, nanoscale Kirkendall effect, hard and soft templating method, spray pyrolysis, and heterogeneous contraction.

Figure 2.8a, b displays the scanning electron microscope (SEM) and transmission electron microscope (TEM) images of the yolk-shell nanoprisms of nickel cobalt ($Ni_{0.37}Co$) oxide by a thermally driven heterogeneous contraction reaction. The prism has a pyramid-shaped apex at both ends, and the yolk-shell nanoprisms compose of numerous polycrystalline nanoparticles. The galvanostatic cycling of the as-prepared nanoprisms in Fig. 2.8c indicates that a high reversible capacity of 1028.5 mAh g^{-1} over 30 cycles at 200 mAh g^{-1} is maintained implying high stability of the electrodes [31]. The hard-templating method is common to fabricate the hollow structures and carbonaceous material (CM)-based templates are widely used to form multiple shells

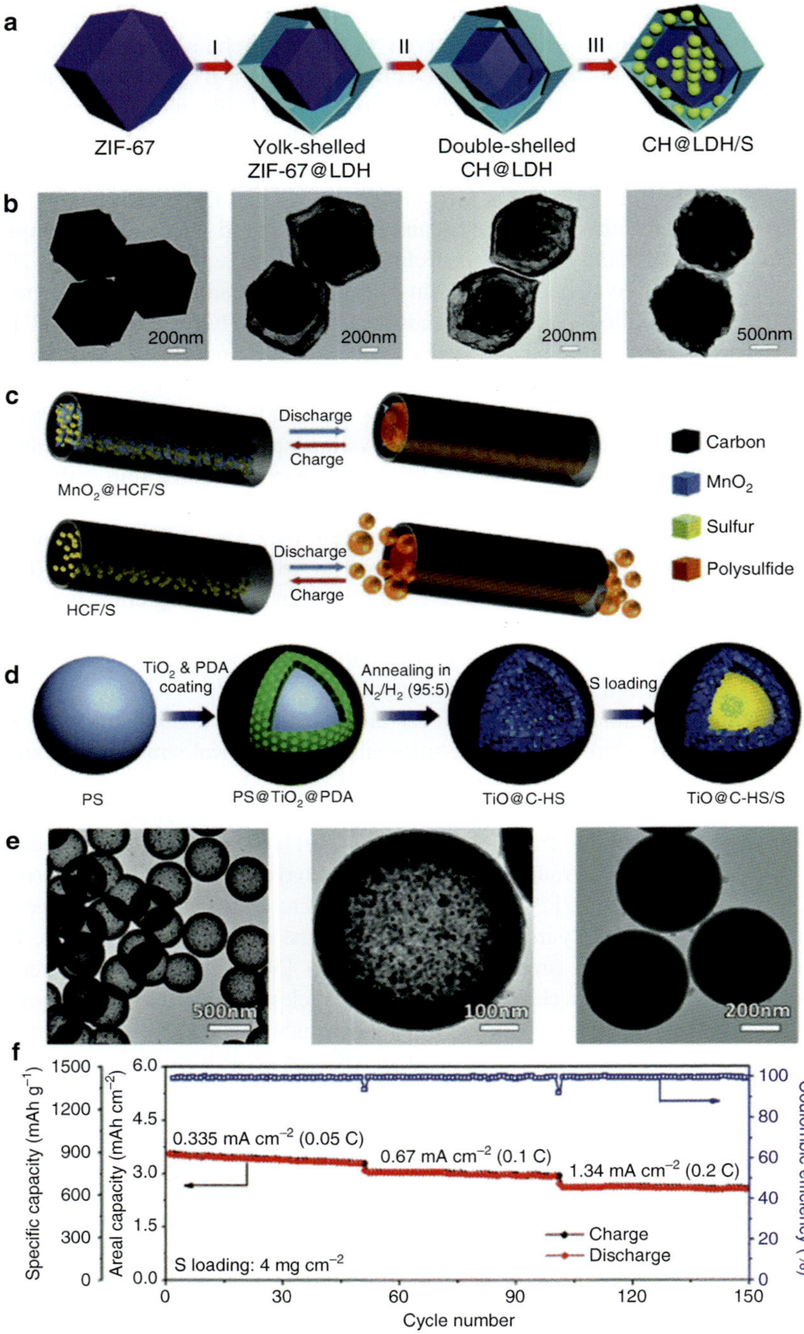

Fig. 2.7 (continued)

and chambers. As shown in Fig. 2.8d, the adsorption and penetration of metal precursors take place in the porous carbon structures usually with rich functional groups to form metal glycolates ($CH_2(OH)COO^- M^+$). The subsequent annealing in the air combusts the carbon template and oxidizes the metal precursors to metal oxides where the temperature gradient along the radial direction in the annealing process results in the gradual removal of carbon and formation of a multi-shell hollow structure. The structure parameters of the products such as the shell number, shell thickness, porosity, and chemical composition can be tuned by the control of metal ion concentration, solvent category, pH value, annealing temperature and rate, reaction atmosphere, etc. TEM images in Fig. 2.8e–g clearly display the identical multi-shell hollow spheres of binary metal oxides ($CoMn_2O_4$, $Co_{1.5}Mn_{1.5}O_4$, and $MnCo_2O_4$) from the penetration-solidification-annealing protocol.

The self-templating method has been extensively used to synthesize complex hollow structures by using the self-engaged templates which are usually the inorganic/organic hybrids composed of metal ions and clusters. Figure 2.8h displays the cobalt sulfide (CoS_4) nanobubble-composed hollow prisms prepared through a two-step diffusion-controlled process in the conversion from cobalt acetate hydroxide nanoprisms to ZIF-67 hollow prisms and subsequent sulfidation reaction to form CoS_4 bubble-like particles [32]. The annealing process preserves the prism morphology and forms interconnected CoS_2 nanobubbles, as shown in Fig. 2.8i. The as-obtained CoS_2 displays extraordinary reversible rate capabilities at current densities ranging from 0.2 to 5 Ag^{-1} (Fig. 2.8j). The chemical compositions of the shells can be manipulated in the synthesis, and the hybridization of two electrochemically active components in one composite can contribute extra benefits to structural stability, conductivity, and electrochemical activity. Multi-shell $Co_3O_4@Co_3V_2O_8$ nanoboxes are fabricated by controlling the amount of vanadium oxytriisopropoxide (VOT, $[OV(OCH(CH_3)_2)_3]$) in the reaction with cubic ZIF-67 precursors [33]. In the annealing process, the intermediates of ZIF-67 nanocubes with the cobalt vanadium oxide ($Co_3V_2O_8$) shell are transformed into double-shell (D) and triple-shell (T) $Co_3O_4@Co_3V_2O_8$ nanoboxes (D-$Co_3O_4@Co_3V_2O_8$ and T-$Co_3O_4@Co_3V_2O_8$), displayed by the TEM images in Fig. 2.8k, l. As evidenced from the elemental mapping results in Fig. 2.8m, the shells have different chemical compositions with a $Co_3V_2O_8$ outermost shell and two Co_3O_4 inner shells in T-$Co_3O_4@Co_3V_2O_8$. The nanoprocessing of structures is as expected, helping to enhance the lithium storage properties resulting from a synergistic effect from the complex configurations and compositions [29].

Fig. 2.7 (**a**) Schematic illustration of the step-by-step synthesis of cobalt hydroxide (CH) inner shell and a layered double hydroxides (LDH) out shell loaded with sulfur (CH@LDH/S) composite from zeolitic imidazolate framework-67 (ZIF-67), to the yolk-shell ZIF-67@LDH, double-shell (D) CH@LDH and CH@LDH/S as the transmission electron microscope (TEM) shown in (**b**). Adapted with permission from Ref [25], Copyright 2016 John Wiley and Sons. (**c**) Schematic representation of MnO_2@HCF as the sulfur host with an advantage over HCF. Adapted with permission from Ref [26], Copyright 2015 John Wiley and Sons. (**d**) Schematics of the synthesis protocol of TiO@C-HS/S composite. (**e**) TEM images of TiO@C-HS (left two) and TiO@C-HS/S (right). (**f**) Cycling performance of the TiO@C-HS/S electrode with a sulfur mass loading of 4.0 mg cmAdapted with permission from Ref [25], Copyright 2016 Springer Nature

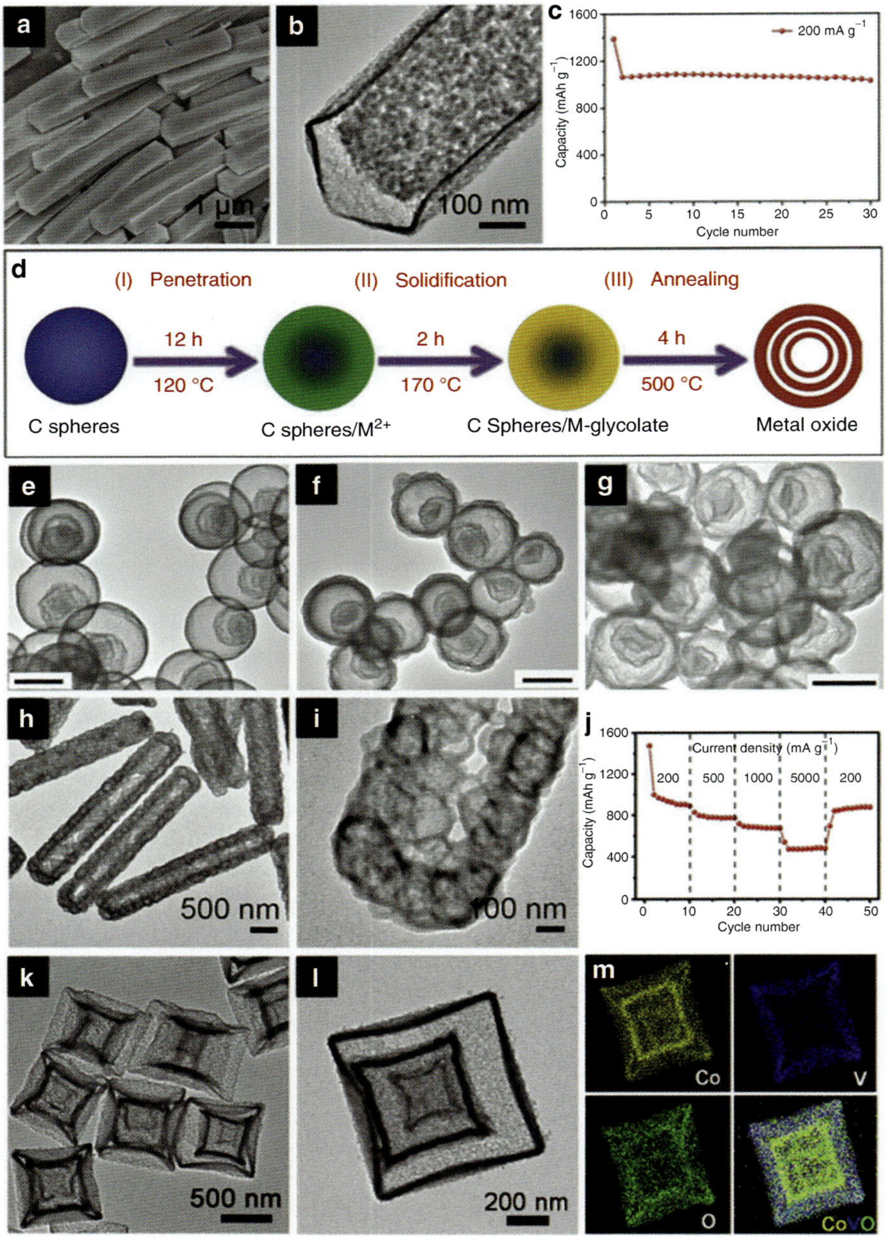

Fig. 2.8 (**a**) Scanning electron microscope (SEM) image, (**b**) transmission electron microscope (TEM) image, and (**c**) the cycling performance of nickel cobalt ($Ni_{0.37}Co$) oxide prisms. (**d**) Schematic illustration of the preparation of multi-shell metal oxide hollow spheres and TEM images of the (**e**) multi-shell cobalt manganese oxides ($CoMn_2O_4$), (**f**) $Co_{1.5}Mn_{1.5}O_4$, and (**g**)

2.4 Nanostructured Materials in Supercapacitors

The development in the physical and chemical properties of nanomaterials and the improved understanding of their synthesis, characterization, and electrochemistry lead to a breakthrough in the field of supercapacitors for energy storage. The principle of supercapacitors is elucidated in terms of the resulting electrochemical characteristics and charge storage mechanisms, i.e., double-layer capacitance or pseudocapacitance. The electrochemical behaviors and charge storage mechanisms are also dependent on the size or thickness; notwithstanding, the materials may be electrically insulating semiconductors. There are increasing numbers of electrode materials, such as transition metal oxides, hydroxides, conducting polymers, etc., that display electrochemical characteristics neither purely capacitive nor purely Faradaic [2]. Before introducing the electrode nanomaterials of the two types of charge storage mechanisms, we use the band theory for semiconductors and electrochemical characteristics in *i-V* and *V-t* curves to understand the origin of the capacitance and the principle.

Electron transfer takes place in electronically interactive or separated redox-active centers, highly depending on the electron conductivity of materials. The electrons in the outer most orbit of an atom experience least attraction force by the nuclei and have high kinetic energy. In Fig. 2.9a, the separated and noninteractive atoms have electrons in the atomic orbits, and these orbits form a discrete set of energy levels. The low-energy levels are filled with electrons with fixed energy level, while the high-energy levels are empty. In a molecule composed of two atoms, the original atomic orbits are split into separate molecular orbits of different energy levels. The amount of molecular orbits is determined by the atomics, p, d, or f orbits, that are occupied by electrons [34]. The molecular orbits of low-energy levels are occupied by electrons and of high-energy levels are empty. When a number of atoms are brought together into a molecule, the electrons of one atom experience forces of other atoms and this effect is the most pronounced on outer most orbits. These atomic orbits split into separate molecular orbitals of different energy. When a large number of atoms (10^{20} or more) are brought together to form a solid, the orbital number is extremely large and the difference in energy between these orbits becomes small. These energy levels form continuous bands rather than the discrete energy levels of the atoms in isolation. The intervals of energy levels containing no orbits form band gaps between the highest level of one band and the lowest level of the next.

The core bands composed of core-level orbits (both the bonding and antibonding levels) are completely populated, so they have small overlapping in molecules and thus little contribution to bonding. The valence orbits of adjacent atoms have strong

Fig. 2.8 (continued) $MnCo_2O_4$ hollow spheres. Scale bars are 500 nm. TEM image of (**h**) CoS_4 nanobubble hollow prisms obtained after sulfidation and (**i**) CoS_2 nanobubble hollow prisms obtained after annealing in nitrogen at 350 °C. (**j**) Corresponding rate capability of the as-prepared cobalt sulfide (CoS_2) nanobubble hollow prisms. TEM image of (**k, l**) [*Tripple Shelled*]-Co_3O_4@$Co_3V_2O_8$ nanoboxes and (**m**) elemental mappings of an individual nano box. Adapted with permission from Ref [29], Copyright 2018 Elsevier

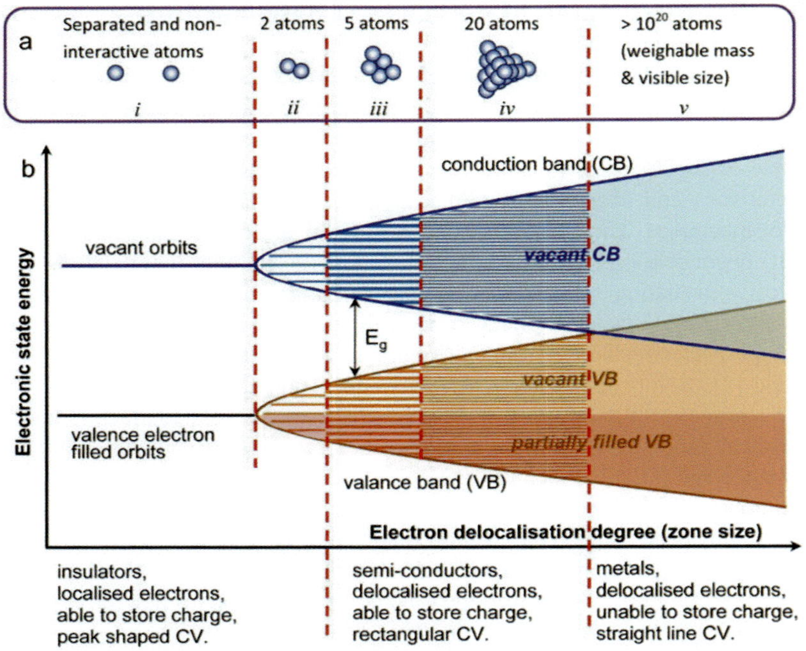

Fig. 2.9 Schematic representation of the band theory for chemical bonding. (**a**) Metal atoms are *i*) separated and noninteractive, forming clusters by (*ii*) 2, (*iii*) 5, (*iv*) 20, and (*v*) 10^{20} atoms. (**b**) The corresponding energy levels of the valence electrons as a function of the delocalization degree of the valence electrons in the cluster or metal atoms. Reprinted with permission from Ref [34], Copyright 2016 Informa UK Limited

interaction and have a large bandwidth. In valence orbits, the bonding orbits have energy lower than the atomic orbits, mostly in-phase, but each of the bonding orbits still has slightly different energies. The antibonding orbits show an increase in energy compared with the atomic orbitals, mostly out-of-phase, and each of the antibonding orbits also has little difference in energy. These energy levels for all the bonding orbitals and antibonding orbits are so close that they form valence band and conduction band, respectively. In most existing solids, the valence band is partially occupied by the valence electrons resulting from the original atomic orbits that are not fully occupied. The valence electrons are able to move through the occupied and unoccupied orbits contributing to semiconductivity, which are delocalized in the valence band [34]. The degree of delocalization of electrons or the zone sizes is the typical difference in metals, semiconductors, and insulators.

The Nernstian reaction, that the Faradaic charge transfer in isolated redox centers, takes place in solid insulators or liquid electrolytes whose valence electrons are localized. All these transferred electrons are from the orbits of the same or very close energy level but different redox centers; thus, the electron transfer occurs at potentials in a very narrow range close to the equilibrium potential. This results in current peaks in the cyclic voltammograms and potential plateaus in galvanostatic charge/discharge profiles. When these redox centers have a strong interaction with each other in the valence band, which might be due to the short separation or high conductivity or both, their energy states are so close to merge together into a broad-band. In this case, the electron transfer at these energy states is continuous over a wide range of potentials, leading to a constant current flow. For pseudocapacitance, each electron in the band model is explained to transfer to a particular energy level shared by many atoms. There is no fixed stoichiometry between the number of electrons and the compensated species over a potential range.

Electrochemical characteristics of electrode materials mainly including *i*-*V* and *V*-*t* curves provide real-time reaction information in the specific electrolyte solution. As explained for the charge-storage mechanisms in electrochemical double-layer capacitors (EDLCs) and pseudocapacitor from surface redox reaction, the capacitor-type responses should have rectangular cyclic voltammograms (Fig. 2.10a, b) and linear voltage response at constant current (Fig. 2.10c). In contrast, battery-type responses have prominent and widely separated peaks in cyclic voltammograms (Fig. 2.10g, h) and nonlinear voltage profiles with characteristic plateaus associated with the redox reactions of the active centers. The redox peaks of a battery reaction can be widened by enlarging the electrochemically active surface area with increased contribution from Faradaic charge-transfer reaction (see Fig. 2.10g). As a result, the voltage-time profile should have a degenerated voltage plateau, which is neither a battery nor a capacitor response. An intermediate behavior between these two extremes indicates the pseudocapacitance from intercalation, as shown in Fig. 2.10d–f. Niobium oxide (Nb_2O_5) in organic electrolyte has displayed a well-defined intercalation/de-intercalation mechanism by lithium ions and MXenes show reversible redox peaks in the same reaction mechanism. The judgment of a material whether belonged to battery-type or capacitor-type can follow this information: (i) electrochemical double-layer capacitors (EDLCs) and surficial pseudocapacitance should have rectangular *i*-*V* curves and triangular-shape charge-discharge profiles; (ii) intercalation pseudocapacitance should display reversible redox peaks although these peaks may be pronounced [6]; (iii) the kinetic parameter, *b* in the equation $i = av^b$ correlating peak currents and sweet rates of potentials, is 1 implying a capacitive material while 0.5 denoting battery material of a semi-infinite diffusion. Any value far from 1 and 0.5 indicates a mixing reaction mechanism from capacitive and diffusion process. These characteristics in electrochemistry are highly dependent on the material type (e.g., metal, metal oxide, etc.), reaction system, and the size and state (e.g., thin film) of materials.

2.4.1 Electrochemical Double-Layer Capacitor

Electrical double-layer capacitors (EDLCs) store charges at an electrode–electrolyte interface when a voltage bias is applied. The maximum capacitance of EDLC is on the order of hundreds of farads per gram, which is numbers of magnitude larger than that of a traditional dielectric capacitor of which the capacitance is only microfarad

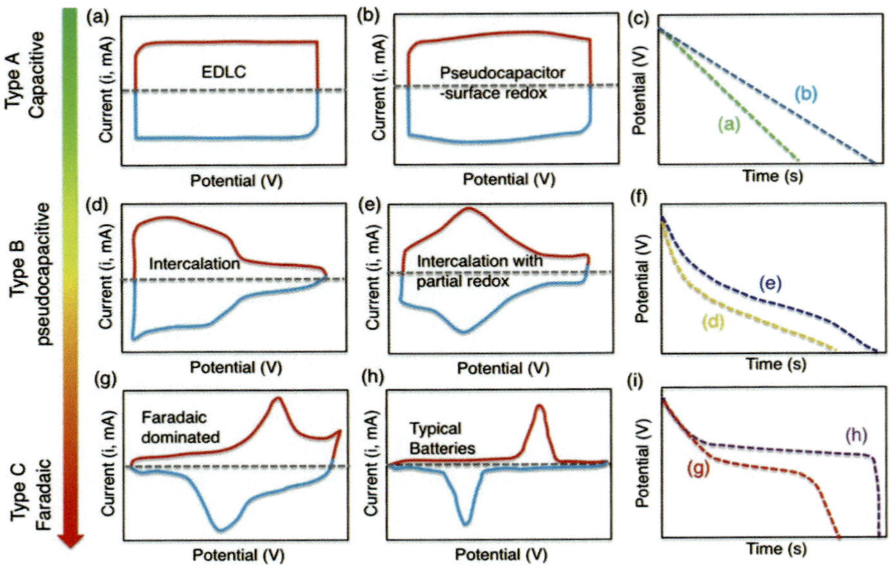

Fig. 2.10 Schematic electrochemical profiles of (**a**, **b**, **d**, **e**, **g**, **h**) cyclic voltammograms and (**c**, **f**, **i**) charge/discharge profiles for specific energy-related materials. (**a**) Electrochemical double-layer capacitors (EDLCs). (**b**) Pseudocapacitance of surface redox reaction (e.g., MnO_2 in neutral, aqueous media). (**d**) Pseudocapacitance of intercalation-type reaction (e.g., lithium insertion in niobium oxide (Nb_2O_5) in organic electrolytes), or (**e**) intercalation-type materials with broad but reversible redox peaks (e.g., titanium carbide (Ti_3C_2) in acidic, aqueous electrolytes). (**g**) Comprehensive reaction from mixing capacitive and diffusion process with Faradaic capacitance-dominated. (**h**) Typical battery-type reaction with diffusion dominated. Adapted with permission from Ref [2], Copyright 2018 American Chemical Society

per gram. A large capacitance is achievable when the charge separation distance is extremely short (<1 nm) at the electrode–electrolyte interface, and the electrode materials have an exceptionally high specific surface area (SSA) which is key to lead to a high gravimetric capacitance and hence high energy density. The highest energy densities of the commercial EDLC using high-SSA activated carbons are nevertheless still below 12 Wh kg^{-1}, much smaller than that of Li-ion batteries. Carbon nanomaterials including carbon nanotubes activated carbon and graphene are typical electrode materials in EDLC and have been widely studied because of their high specific surface area and excellent electrical and mechanical properties. The research now is focusing on increasing the energy density of the carbon nanomaterials in EDLC without sacrificing the power density and cycle life.

Carbon nanotubes (CNTs) are promising electrode materials in supercapacitors as a result of their high aspect ratio, large specific surface area (SSA), and high electrical conductivity. A free-standing mat composed of entangled multiwall nanotubes (MWNTs) has been reported to have a specific capacitance of 102 F g^{-1} with a power density above 8 kW kg^{-1} and an energy density of ~1 Wh kg^{-1} in hydrosulfuric acid (H_2SO_4) electrolyte [35]. A higher specific capacitance of 180 F g^{-1} was achieved on the single-wall nanotube (SWNT) distributed in a random network (Fig. 2.11a) working in potassium hydroxide (KOH) electrolyte [37] and the maximum power density and energy density are 20 kW kg^{-1} and 10 Wh kg^{-1}, respectively. Contrary to the random entanglement of CNTs, the vertically aligned CNT (VA-CNT) arrays have been fabricated with vertically aligned tubes and well-defined tube spacing, as shown in Fig. 2.11b, c [38]. This arrangement allows more contact area between the electrode and electrolyte, thus enhancing the participation in the charging/discharging process of each aligned tube (Fig. 2.11d). Furthermore, the etching processing such as the plasma can open the top caps of these VA-CNTs (see Fig. 2.11e–g), and the access to the inner cavity of these tubes by electrolyte has greatly improved the properties in energy storage [39]. In the opening of the tube tips, plasma etching also introduces defects and oxygenated functionalization to the surface of the nanotubes, enhancing the capacitive behavior by combining the double-layer capacitance and redox pseudocapacitance. As shown in Fig. 2.11h, the aligned carbon nanotubes with excellent contact with the current collector and rich active area reacting with the ionic liquid electrolyte have displayed a high cell voltage (4 V), high specific energy (148 Wh kg^{-1}), and high specific power (315 kW kg^{-1}) [36], which might be the delegate of the high-performance supercapacitor and exceed the current electrochemical capacitor technology.

Graphene, a two-dimensional hexagonal lattice structure of sp^2-hybridized carbon atoms covalently bonded along the single atomic planar sheet [40], has aroused tremendous interest due to its intrinsic characteristics such as physical [41–43], mechanical [44], and electrical properties [40, 45, 46] since the discovery of its freestanding form in 2004. The sp^2 bonds formed by *s*, *px*, and *py* atomic orbitals hybridization can result in 120 °C C-C bond angles and the honeycomb layers. Also the remaining p_z orbital on each carbon could overlap with its neighboring carbons to form π bonds. The valence band and conduction band consist of π orbitals and anti-bonding $\pi*$ orbitals and the cross of bands gives rise to Dirac points K and K′ per Brillouin zone [47]. So far, many protocols including mechanical exfoliation [40], epitaxial growth [48], chemical vapor deposition [49, 50], and oxidation-exfoliation reduction of graphite [51, 52] have been developed for the synthesis of graphene.

Free-hanging graphene is the thinnest conceivable nanocarbon, and, in particular, as a result of the overlarge surface area and mechanical strength [53–55], graphene has been considered promising for energy storage [42, 49, 56–59] including rechargeable lithium-ion batteries [60, 61], supercapacitors [62–64], and fuel cell devices [65]. In comparison with single-walled carbon nanotubes (SWCNTs), graphene exhibits an electrical conductivity ~64 mS cm^{-1}, approximately 60 times higher [45] and owns an extremely high theoretical surface area (2675 m^2 g^{-1}) [66], twice larger than that of CNTs (1315 m^2 g^{-1}) [67]. A fast heterogeneous electron-transfer rate with an order of 0.01 cm s^{-1} has been found at the edges and basal plane defects of graphene [68]. These reported characteristics of graphene imply a high potential of graphene as an electrode material in supercapacitors. The intrinsic specific capacitance of monolayer graphene

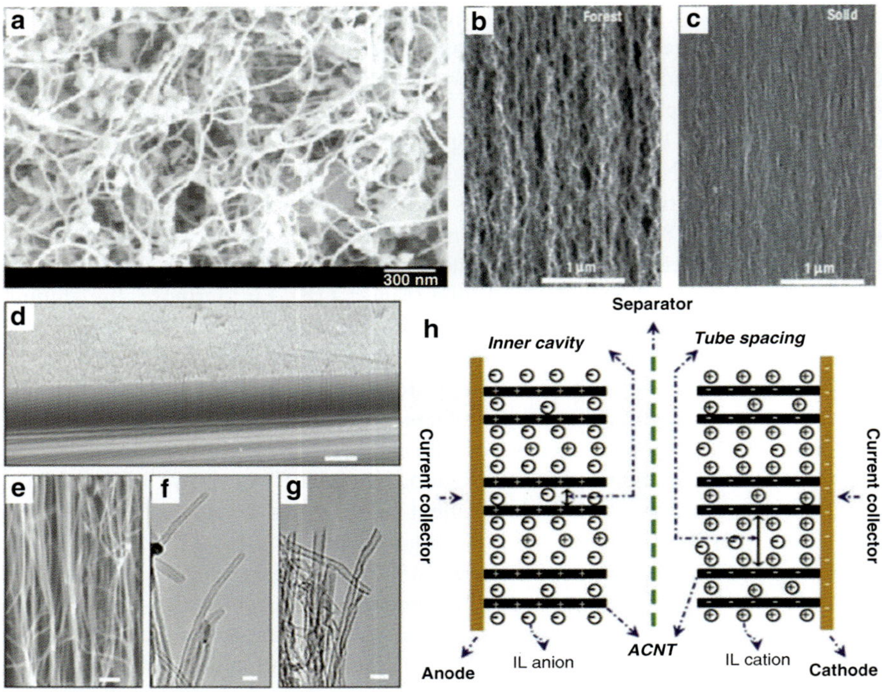

Fig. 2.11 (**a**) Scanning electron microscopy (SEM) image of the randomly entangled SWNTs. Side-view SEM images of (**b**) the single-walled nanotubes (SWNTs) forest and (**c**) solid surface. (**d**) SEM image of a plasma-etched vertically oriented carbon nanotubes (VA-CNT) electrode (scale bar: 100 μm). (**e**) High magnification view of the electrode (scale bar: 100 nm). Transmission electron microscope (TEM) images of the CNTs before (**f**) and after (**g**) plasma etching (scale bar: 20 nm). (**h**) Schematic of an electrochemical capacitor based on plasma-etched VA-CNT electrodes and ionic liquid electrolyte. Adapted with permission from Ref [36], Copyright 2013 Elsevier

was found as 21 μF cm^{-2}, equal to 561 F g^{-1} if the theoretical surface area (2675 m^2 g^{-1}) could be used sufficiently, setting the upper limit for electrical double-layer capacitance for all carbon-based material [66]. Meunier and Ajayan et al. suggest that the specific capacitance of the chemical vapor deposition (CVD) graphene and reduced graphene oxide can increase to 80 and 394 μF cm^{-2} [69]. However, a variety of graphene-based materials derived from graphene oxide (GO) indicated unsatisfied electrochemical behaviors [70], mainly attributed to the restacking of graphene sheets during processing via the sheet-to-sheet van der Waals interaction.

None-restacking curved graphene [71], solvated graphene [72], and laser-scribed graphene [62] have been obtained which show the enhanced performance of 154, 215, and 265 F g^{-1}, respectively. Ajayan et al. use a laser writing technique to directly convert graphene oxide (GO) to reduced graphene oxide (rGO) within a porous structure (Fig. 2.12a) by decomposing the functional groups and adsorbed water molecules [70]. The patterned in-plane and sandwich designs are formed in rGO-GO-rGO configuration by using GO with a substantial amount of water, as the ionic conductor but electrical insulator. The micro-supercapacitors show the highest capacitance of 0.51 mF cm^{-2} for the in-plane structure, nearly twice larger than that of the sandwich structure, resulting from the anisotropic nature of ion mobility and transport distances. Kaner et al. reported a method to reduce GO by the laser using a LightScribe compact disk/digital video disk (CD/DVD) optical drive, and the resulted graphene has an extremely high conductivity of 1738 S m^{-1} and specific surface area (SSA) of 1520 m^2 g^{-1} [62]. Supercapacitor devices with the as-prepared graphene as electrodes exhibit ultrahigh energy density values in different electrolytes of aqueous, organic, and ionic liquid and maintain a high power density in the long-term cycling. The high flexibility of the material and device implies the potential use in high-power, flexible electronics. Ruoff et al. studied the capacitive behavior of graphene in terms of the layer number, from one to five (see Fig. 2.12b). The quantum capacitance effect near the point of zero charge makes the capacitance suppressed near neutrality, and the correlation between the electrode and electrolyte enhances the capacitance only when the thickness is below few layers. They prepared an activated graphene oxide (AGO) with the super-high surface area (3100 m^2 g^{-1}) via the activation with potassium hydroxide (KOH) by microwave [75], and AGO with a three-dimensional porous structure presented high specific capacitance (150~200 F g^{-1}). The enhanced electrochemical performance of AGO is due to its two-dimensional plane and produced edge planes which have much high electron transfer kinetics over the basal [76]. The oxygen-functional groups at the edges of graphene sheets cause a controversial electrochemistry influence [63, 77], not only on the heterogeneous electron transfer rate but also on the adsorption/desorption of molecules [67, 78]. To understand the factors arising from graphene structure that affect the electrochemical results, the electrochemical characteristics of GO, RGO, AGO, and reduced AGO (RAGO) were prepared through a top-down method, as shown in Fig. 2.12c [74]. Graphene oxide was prepared by Hummers method [79] and chemically activated to AGO in situation A (SA), with smaller GO fragments and a larger ratio of edge to basal plane, then going to situation B (SB), with pore defects and higher specific surface area (SSA). The approximate monocrystalline characteristic of GO is transformed to polycrystalline after activation, demonstrating that the activated defects have distorted the crystalline structure of graphene oxides. RGO and RAGO were synthesized via chemical reduction by hydrazine monohydrate using GO and AGO as precursors, respectively. As represented in Fig. 2.12d, the ratio of edge plane to basal plane [63, 76], defects involving edge defects and pore defects, chemical impurities of metal oxides and surfactant [80, 81], oxygen-functional groups [63, 77], and the underlying supporting electrode [76] are key factors to influence the electrochemistry of a graphene supercapacitor. The edge ratio, SSA, edge defects, and metal impurities

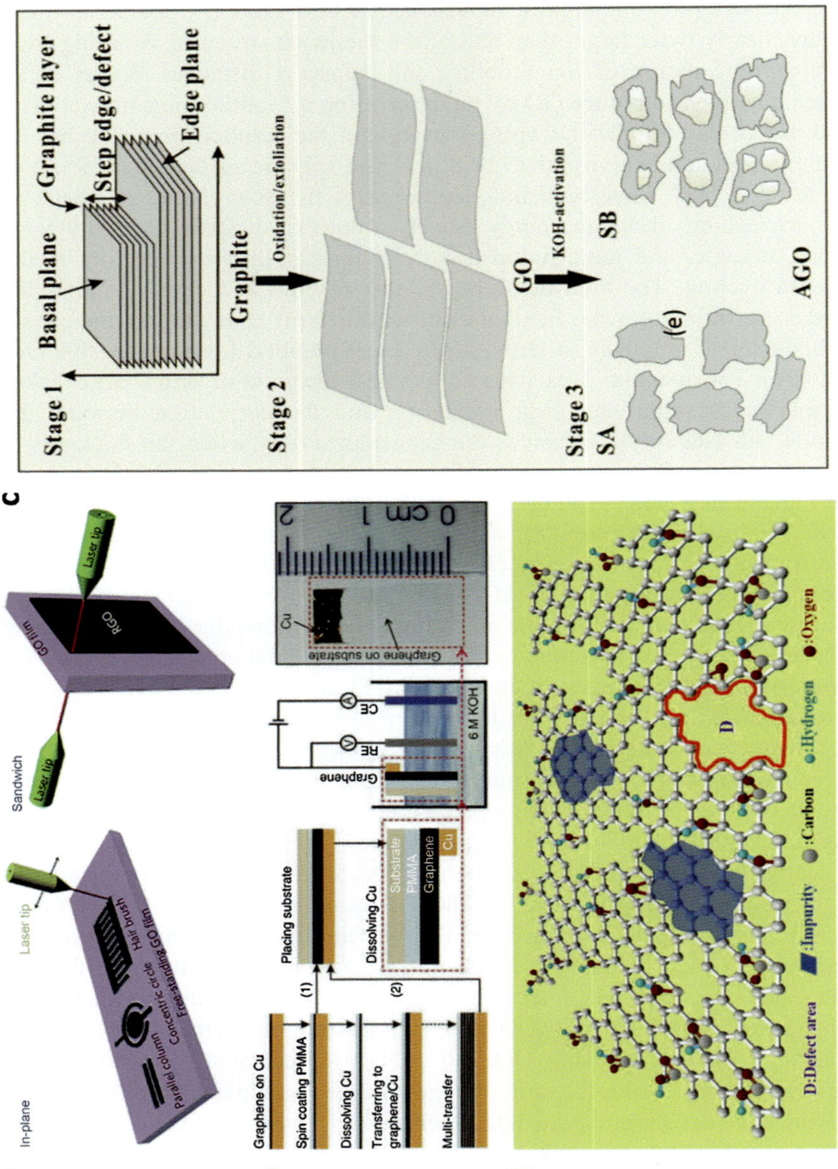

Fig. 2.12 (continued)

could cause a positive impact while the basal/pore defects, oxygen-containing groups, and surfactant impurities could have a negative impact on the electrochemical properties. Optimization of the electrode material is crucial to improving the electrochemical performance of the carbon materials since more than one structure role of the material would cause variation in the electrochemical behaviors.

2.4.2 Pseudocapacitor

In comparison with the carbon materials which mainly contribute double-layer capacitance, the pseudocapacitive electrode materials are highly interesting and the current studied pseudocapacitor materials include metal oxides, metal carbide or MXenes, metal sulfides, metal nitrides, layer hydroxides, and conducting polymers. However, most pseudocapacitive materials suffer from high resistance for electron and/or ion transfer, setting restriction to their capacity, rate capability, and cyclability. Nanotechnologies have offered to enable rational designs to optimize the structure and size of these materials leading to an improvement in the performance of energy storage.

Pseudocapacitors store charges through the transfer of charge between electrode and electrolyte, and the resulted pseudocapacitance (C_{pc}) is proportional to the amount of the charge transferred, which is described in the following equation [82, 83]:

$$C_{\text{pc}} = Q(d\theta/dV) \tag{2.51}$$

where Q is the Faradaic charge required for adsorption/desorption of ions, $d\theta$ is the change in fractional surface coverage, and dV is the change in potential. The value of $d\theta/dV$ is not always linear in a voltage range; the capacitance is thus not always constant and termed as pseudocapacitance [84]. The Faradaic process of pseudocapacitors that can result in pseudocapacitance comprises three mechanisms, as illustrated in Fig. 2.13: (i) underpotential deposition, (ii) redox pseudocapacitance, and (iii) intercalation pseudocapacitance. Underpotential deposition is

Fig. 2.12 (**a**) Patterning reduced graphene oxide-graphene oxide-reduced graphene oxide (rGO-GO-rGO) electrode with in-plane and sandwich configuration by laser. Adapted with permission from Ref [70], Copyright 2011 Springer Nature. (**b**) Graphene electrode with one to five layers in a three-electrode electrochemical cell with 6 M potassium hydroxide (KOH) aqueous electrolyte. Monolayer graphene electrode was prepared by placing a plastic substrate on graphene-coated copper (Cu) foil, and partially dissolving Cu to expose the graphene surface. Multilayer graphene electrodes were prepared by repeating the transfer of poly(methyl methacrylate) (PMMA)-coated graphene films onto a monolayer graphene-coated Cu foil. Adapted with permission from Ref [73], Copyright 2014 Springer Nature. (**c**) Schematic of the top-down method to prepare activated graphene oxide (AGO) from graphite by a chemical activation. (**d**) Representation of the AGO structure with basal and edge planes, functional groups, pores, and impurities. Reproduced from Ref [74] with permission from the PCCP Owner Societies

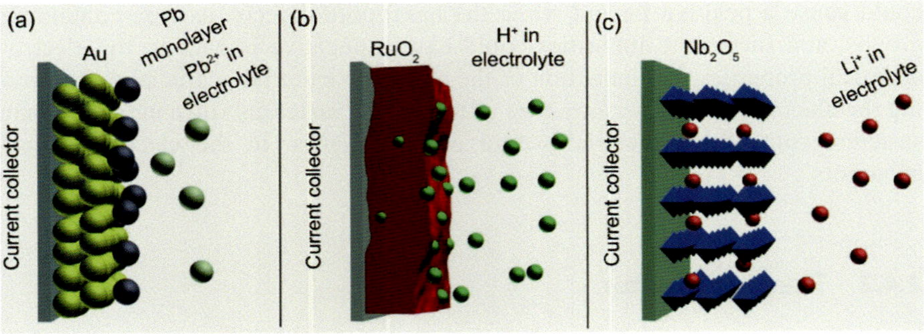

Fig. 2.13 Typical Faradaic reactions of the electrode materials in pseudocapacitors: (**a**) underpotential deposition, (**b**) redox pseudocapacitance, and (**c**) intercalation pseudocapacitance. Adapted with permission from Ref [70], Copyright 2016 Oxford University Press

electrosorption taking place for the metal adatoms to form an absorbing layer on a substrate metal surface. The redox pseudocapacitance occurs in the surface redox reaction by electrochemical adsorbing ions on/near the electrode surface and transferring electrons at the redox active sites. The intercalation pseudocapacitance arises from the electrochemical intercalation of electroactive species in the layered structure of the electrodes without crystallographic phase change. Both the redox and intercalation pseudocapacitance is more popular in the development of pseudocapacitors and their faradaic reactions are very fast without ionic diffusion limitation, which is distinctive from the solid-state diffusion process in batteries. In the following content, we mainly introduce the development of pseudocapacitive materials owning redox and intercalation pseudocapacitance.

Ruthenium oxide (RuO_2) is a typical redox-reaction pseudocapacitive electrode material and has been extensively studied in an acidic electrolyte, giving the reaction mechanism as shown in Eq. (2.52). The charge/discharge process enables the rapid reversible electron transfer which is compensated by the electroabsorption of protons on the surface of RuO_2, and the oxidation states vary from Ru^{+2} to Ru^{+4}:

$$RuO_2 + xH^+ + x\bar{e} \leftrightarrow RuO_{2-x}(OH)_x \quad (2.52)$$

Ruthenium oxide (RuO_2) has high conductivity and this property enables a large utilization of the materials in the electrochemical reaction. The highest capacitance of the hydrated RuO_2 was reported to be 1340 F g^{-1}, close to its theoretical value of 1400 F/g (Fig. 2.14a) [85]. The advantages in high conductivity and reversibility of RuO_2 in the electrochemical reaction allow a high-rate capability in charge storage. As displayed in Fig. 2.14b, the capacitance retains ~75% when the scan rate increases from 2 to 500 mV s^{-1}. Although RuO_2 has a promising performance in the electrochemical energy storage, the high toxicity and cost restrict the large-scale use of RuO_2. Other transition metal oxides have

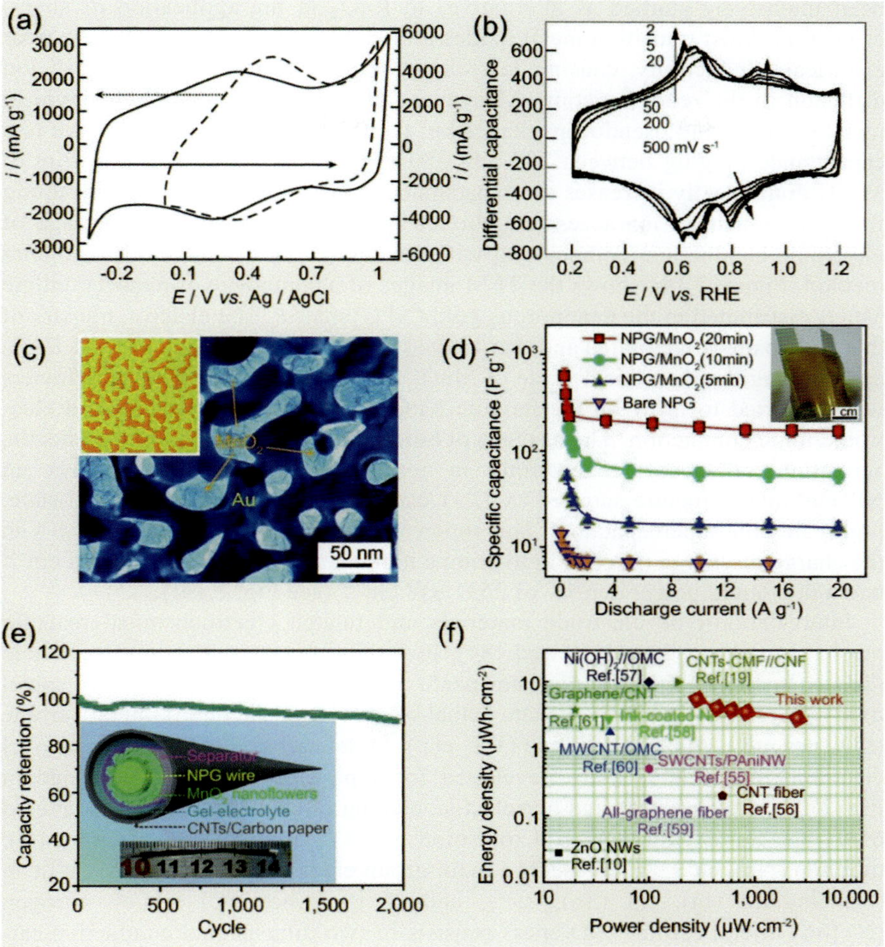

Fig. 2.14 (**a**) Voltammograms of a composite electrode consisting of activated carbon and ruthenium oxide (RuO_x) coated on graphite; the dotted line denotes hydrated ruthenium oxide ($RuO_x•nH_2O$) without annealing and the solid line implies $RuO_x•nH_2O$ with annealing in air at 200 °C for 2 h. (**b**) Cyclic voltammetry profiles of layered $H_{0.2}RuO_{2.1}•nH_2O$ at increasing scan rates from 2 to 500 mV s^{-1}. (**c**) Bright-field TEM image of the nanoporous gold (NPG)/manganese dioxide (MnO_2) composite by plating MnO_2 for 20 min. Inset is a schematic illustration of MnO_2 (orange) grown on NPG. (**d**) Comparison of the specific capacitance at different current densities for bare NPG and NPG/MnO_2 with different plating times. Inset is a photograph of an NPG/MnO_2-based supercapacitor. (**e**) Long-term cycling performance of the device with CNT/carbon-scrolled NPG@MnO_2 electrode at a current density of 0.6 mA cm^{-2}. Inset is a schematic diagram of a coaxial fiber of the electrode. (**f**) Ragone plots of the device in comparison with published results. Adapted with permission from Ref [70], Copyright 2016 Oxford University Press

been intensively studied as alternatives to RuO_2 in the application of supercapacitors. Most transition metal oxides are semiconductors and suffer the poor electrical conductivity, causing low charge-transfer kinetics and sluggish ion diffusion in the redox reaction. Manganese dioxide (MnO_2) is another typical metal oxide giving pseudocapacitance but the thick electrodes usually give low capacitance ranging between 200 and 250 F g^{-1}. The nanosizing thin film of MnO_2 dramatically increases the capacitance to large than 1000 F g^{-1} resulting from the enhanced ion-accessible surface area. To figure out the shortage of electronic conduction, compositing with conductive matrix has been a popular method. Figure 2.14c shows the TEM images of a composite of nanocrystalline MnO_2 distributed in the nanoporous gold (NPG) matrix. The electron transfer of this composite is facilitated and the resulted capacitance is as high as 1145 F g^{-1} approaching the theoretical value of MnO_2 (see Fig. 2.14d). MnO_2 nanoflowers were reported to grow on the flexible NPG wires, which own enhanced electronic/ionic conduction. The absence of binder in this electrode also benefits the activation of electrode materials in reaction. Figure 2.14e displays an NPG@MnO_2 structure scrolled by CNT/carbon paper for use in all-solid-state, highly flexible supercapacitors. This supercapacitor has a high voltage of 1.8 V in the charge/discharge process, delivering a high energy density of 5.4 μWh cm^{-2} and a maximum power density of 2531 μW cm^{-2} (see Fig. 2.14f).

Intercalation-type electrode materials and related electrochemical reaction mechanisms have been reviewed [84], including metal oxides (anatase titania (TiO_2), hydrogen titanate, niobium pentoxide (Nb_2O_5), lithium titanate spinel oxide ($Li_4Ti_5O_{12}$), sodium titanate spinel oxide ($Na_2Ti_3O_7$) etc.), metal carbide (MXenes etc.), and metal sulfide (VS_2 etc.). Metal carbides (MXenes) are novel electrode materials recently developed for supercapacitors. As illustrated in Fig. 2.15a, they are prepared by selective etching of element A from the layered MAX structures, where M is a transition metal (titanium (Ti), vanadium (V), niobium (Nb), etc.), A represents a main-group element (aluminum (Al), silicon (Si), indium (In), tin (Sn), etc.), and X is carbon (C) and/or nitrogen (N) [86]. Metal carbide (MXenes) consists of two-dimensional conductive carbide layers with hydrophilic surface rich in hydroxyl groups, which benefit the electrochemical reaction and electron conduction. Gogotsi et al. [87] demonstrated that a variety of cations including lithium (Li^+), sodium (Na^+), magnesium (Mg^{2+}), potassium (K^+), ammonium (NH_4^+), and aluminum (Al^{3+}) ions can spontaneously intercalate into the $Ti_3C_2T_x$ layers, where T denotes the surface termination (oxygen (O), hydroxide (OH), and/or florine (F)) and x is the number of termination groups. The ionic intercalation is able to be promoted by electrical bias, for large ions, such as Al^{3+}. Figure 2.15b [87] shows a flexible, additive-free, and conductive $Ti_3C_2T_x$ paper electrode, which has been prepared through the filtering of the delaminated titanium carbide (Ti_3C_2). This electrode has nearly doubled the gravimetric capacitance in potassium hydroxide (KOH) electrolyte in comparison with the multilayer $Ti_3C_2T_x$. The volumetric capacitance of this electrode is ultrahigh, 340 F cm^{-3}, as shown in Fig. 2.15c. The in situ XRD study, illustrated in Fig. 2.15d [87], reveals that the intercalation and

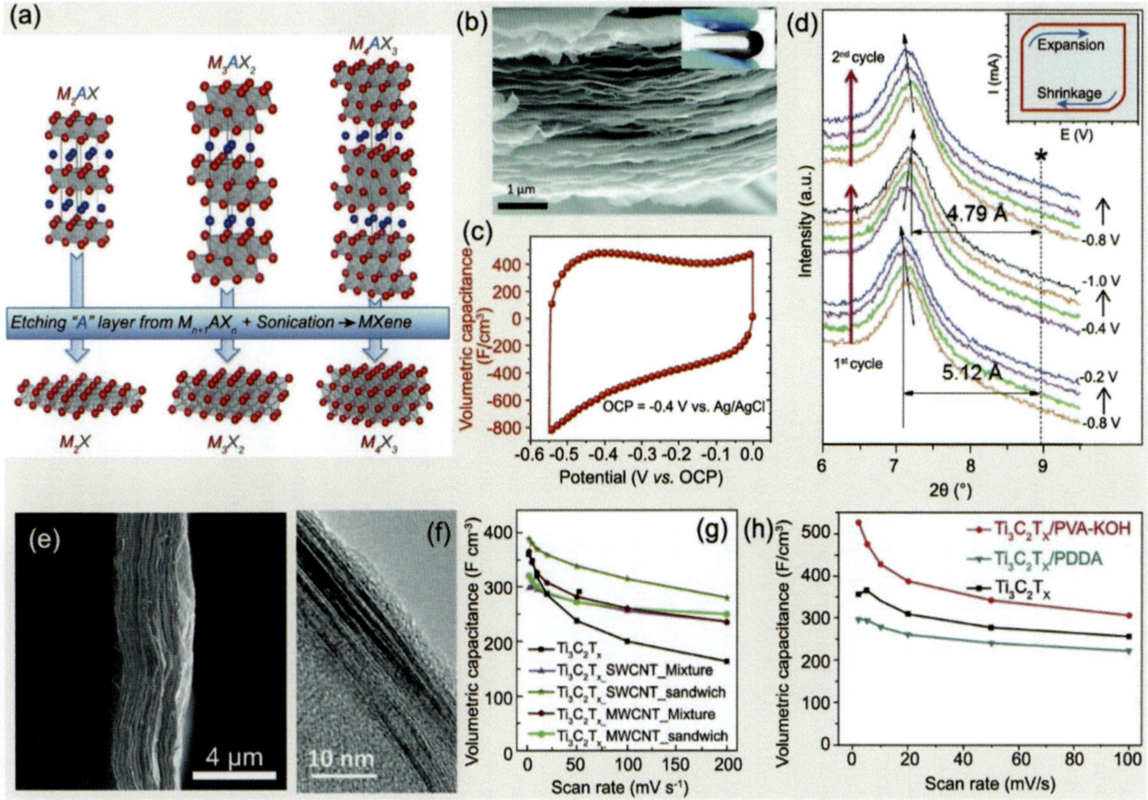

Fig. 2.15 (**a**) Schematic illustration of the structure evolution from MAX to the corresponding metal carbides (MXenes). (**b**) Scanning electron microscope (SEM) image of the metal carbide (Ti_3C_2) with surface terminating groups (T_x) [$Ti_3C_2T_x$] paper electrode and the inset is a photograph displaying the flexibility

de-intercalation in the electrochemical cycling only lead to a small change in the layer distance of $Ti_3C_2T_x$, denoting a strong structure stability. The change in the interlayer distance is only 0.33 Å when the potential shifts from −1 to −0.2 V in a potassium hydroxide (KOH) electrolyte. The capability of hosting larger Na^+, K^+, NH_4^+, Mg^{2+}, Ca^{2+}, and Al^{3+} ions than Li^+ ions would greatly extend the application of MXenes in a range of energy storage devices.

Levi et al. [88] have adopted electrochemical quartz-crystal admittance (EQCA), in situ electronic conductance, and electrochemical impedance spectrum to study the intercalation kinetics of MXenes and conjectured that the multilayers consist of shallow-adsorption sites near the edges and deep-adsorption sites of higher activation energies in the interior. Gogotsi et al. [89] prepared $Ti_3C_2T_x$-based paper electrodes by mixing $Ti_3C_2T_x$ with carbon nanotubes or polymers, which have high mechanical and electrochemical properties (see Fig. 2.15e, f). A sandwich-like MXene/CNT paper electrode was similarly constructed and exhibited a high gravimetric capacitance of 150 F g^{-1} and volumetric capacitance of 390 F cm^{-3} at a scan rate of 2 mV s^{-1} (Fig. 2.15g). KOH was introduced into the $Ti_3C_2T_x$/polyvinyl alcohol (PVA) film [90] and the dramatic improvement in the layered structure has greatly improved the ionic transport, with a high volumetric capacitance of 528 F cm^{-3} at 2 mV s^{-1} and 306 F cm^{-3} at 100 mV s^{-1} (Fig. 2.15h). Metal carbides (MXenes) have displayed both high gravimetric and volumetric capacitance in the supercapacitors and are promising as the electrode materials for high-energy and power devices for energy storage.

Metal carbides (MXenes) have been studied as electrode materials in the nonaqueous devices for energy storage, such as lithium-ion and sodium-ion capacitors. An asymmetric lithium-ion supercapacitor [91] assembled with titanium carbide (Ti_2C) as an anode and activated carbon as cathode delivered a superior specific energy of 239.5 Wh kg^{-1} at the ultrahigh specific power of 10.8 kW kg^{-1}. Yamada et al. [92] demonstrated that in a hybrid sodium-ion supercapacitor, titanium carbide (Ti_2C) nanosheets delivered a reversible capacity of 175 mAh g^{-1} at an operating voltage of 1.3 V (vs. Na/Na^+) and good cycle stability. The hybrid sodium-ion capacitor using sodium iron sulfate ($Na_2Fe_2(SO_4)_3$) as cathode and titanium carbide (Ti_2C) as anode owns a high working potential of 2.4 V and high specific energy of 260 Wh kg^{-1} at the ultrahigh specific power of 14 kW kg^{-1} (based on the weight of Ti_2C anode).

Fig. 2.15 (continued) of the paper. (**c**) Cyclic voltammetry (CV) profiles of the $Ti_3C_2T_x$ paper electrode. (**d**) In situ XRD patterns of two-dimensional metal carbides and carbonitrides ($Ti_3C_2T_x$) in charge/discharge cycle. The electrolyte is 1 M potassium hydroxide (KOH) aqueous solution. (**e**) Cross-section SEM images of the sandwich-like $Ti_3C_2T_x$/single-walled carbon nanotube (SWCNT). (**f**) High-resolution transmission electron microcopy (HRTEM) image of the cross edge of $Ti_3C_2T_x$/poly(vinyl alcohol)-potassium hydroxide (PVA-KOH) film. (**g**) Compared volumetric capacitances of different $Ti_3C_2T_x$-based electrodes. (**h**) Volumetric capacitances of different $Ti_3C_2T_x$-based thin film electrodes. Adapted with permission from Ref [70], Copyright 2016 Oxford University Press

These high properties in both energy and power imply MXenes are promising electrode candidates for use in asymmetric and hybrid-ion supercapacitors of high performance in energy storage.

2.5 Conclusion

Nanomaterials, thanks to the large progress of nanoscience and nanotechnologies, have exhibited great advantages in improving the electrochemical properties of the energy storage and conversion devices. These merits are mainly due to the optimized properties in nanoscale by offering high surface area, large surface-to-volume ratio, high conduction property, altered physical and electronic properties, and confinement effects in nano dimensions. Nanoparticulate electrodes or electrocatalysts are able to lead to high contact area at the interface of electrode/electrolyte, hence the high rate in the electrode reaction. Nanotubes have a high surface-to-volume ratio and display high reactivity in the electrochemical reactions with this particular morphologies. These advantageous nano effects of nanomaterials originate from the methodologies to manipulate and control materials at the atomic or molecular level and the understanding of the synthesis protocol from nanoscale. Nanomaterials have been designed with a specific target in use through a variety of ways, such as morphology fabrication, electronic structure doping, nanosizing, compositing, etc.

A growing number of materials and materials of different structures have been proposed to be used in the electrochemical systems to enhance the performance in energy storage and supply. For example, the carbon nanotubes have been used in lithium-ion batteries as a three-dimensional conductivity network, greatly enhancing the high-rate and cycling performance. However, there are still some issues related to the electrochemical systems by using nanomaterials. The fundamental understanding of the energy chemistry in the process of energy storage and conversion on the nanostructured materials is still far more enough. The scarce knowledge of the relationship between the structure and performance of the material restrains the rational design of specifically targeted materials. There is vague understanding upon the interfacial effects and structure properties of nanomaterials in the form of hybrids or composites, as well as the resulted electrochemical behaviors. Many functional nanomaterials exhibiting high performance in energy storage and conversion have to go through a routine trial-and-error cycle to develop.

In addition, nanostructured materials are not perfect and still need huge development to meet the increasing demand in the energy area. No matter the materials used for lithium-ion batteries or supercapacitors, all the following aspects can be considered for further improvement of nanostructured materials in processing. The high conduction in both electrons and ions is urgent for high-rate electrochemical devices and materials can be developed by heteroatom doping, compositing with conductive networks. The downsizing of materials indeed has brought large improvement in the electrochemical performance of devices but simultaneously caused more side reactions and changed reaction mechanism. Also, effective means are still highly needed to ensure the uniformity of nanomaterials and avoid agglomeration in micro or nanoscale.

References

1. M. Winter, R.J. Brodd, Chem. Rev. **104**, 4245–4270 (2004)
2. Y. Gogotsi, R.M. Penner, ACS Nano **12**, 2081–2083 (2018)
3. A.S. Arico, P. Bruce, B. Scrosati, J.-M. Tarascon, W. van Schalkwijk, Nat. Mater. **4**, 366–377 (2005)
4. X. Zhang, X. Cheng, Q. Zhang, J. Energy Chem. **25**, 967–984 (2016)
5. N. Elgrishi, K.J. Rountree, B.D. McCarthy, E.S. Rountree, T.T. Eisenhart, J.L. Dempsey, J. Chem. Educ. **95**, 197–206 (2018)
6. P. Simon, Y. Gogotsi, B. Dunn, Science **343**, 1210–1211 (2014)
7. N. Mahmood, T.Y. Tang, Y.L. Hou, Adv. Energy Mater. **6**, 1600374 (2016)
8. A. Manthiram, ACS Cent. Sci. **3**, 1063–1069 (2017)
9. M.S. Whittingham, Science **192**, 1126–1127 (1976)
10. K. Mizushima, P.C. Jones, P.J. Wiseman, J.B. Goodenough, Mater. Res. Bull. **15**, 783–789 (1980)
11. M.M. Thackeray, W.I.F. David, P.G. Bruce, J.B. Goodenough, Mater. Res. Bull. **18**, 461–472 (1983)
12. R. Yazami, P. Touzain, J. Power Sources **9**, 365–371 (1983)
13. A.K. Padhi, K.S. Nanjundaswamy, J.B. Goodenough, J. Electrochem. Soc. **144**, 1188–1194 (1997)
14. V. Aravindan, J. Gnanaraj, Y.-S. Lee, S. Madhavi, J. Mater. Chem. A **1**, 3518–3539 (2013)
15. N. Nitta, F. Wu, J.T. Lee, G. Yushin, Mater. Today **18**, 252–264 (2015)
16. F. Schipper, E.M. Erickson, C. Erk, J.-Y. Shin, F.F. Chesneau, D. Auerbach, J. Electrochem. Soc. **164**, A6220–A6228 (2017)
17. G.E. Blomgren, J. Electrochem. Soc. **164**, A5019–A5025 (2017)
18. P. Remith, N. Kalaiselvi, Nanoscale **6**, 14724–14732 (2014)
19. X. Xiao, X. Liu, L. Wang, H. Zhao, Z. Hu, X. He, Y. Li, Nano Res. **5**, 395–401 (2012)
20. A. Mauger, H. Xie, C.M. Julien, AIMS Mater. Sci. **3**, 1054–1106 (2016)
21. M.V. Reddy, G.V. Subba Rao, B.V.R. Chowdari, Chem. Rev. **113**, 5364–5457 (2013)
22. C. Sole, N.E. Drewett, L.J. Hardwick, Faraday Discuss. **172**, 223–237 (2014)
23. J. Yu, S. Chen, W. Hao, S. Zhang, ACS Nano **10**, 2500–2508 (2016)
24. Z. Li, H.B. Wu, X.W. Lou, Energy Environ. Sci. **9**, 3061–3070 (2016)
25. Z. Jintao, H. Han, L. Zhen, L.X. Wen, Angew. Chem. Int. Ed. **55**, 3982–3986 (2016)
26. L. Zhen, Z. Jintao, L.X. Wen, Angew. Chem. Int. Ed. **54**, 12886–12890 (2015)
27. Z. Li, J. Zhang, B. Guan, D. Wang, L.-M. Liu, X.W. Lou, Nat. Commun. **7**, 13065 (2016)
28. X. Wang, J. Feng, Y. Bai, Q. Zhang, Y. Yin, Chem. Rev. **116**, 10983–11060 (2016)
29. Y. Lu, L. Yu, X.W. Lou, Chem **4**, 972–996 (2018)
30. J. Qi, X. Lai, J. Wang, H. Tang, H. Ren, Y. Yang, Q. Jin, L. Zhang, R. Yu, G. Ma, Z. Su, H. Zhao, D. Wang, Chem. Soc. Rev. **44**, 6749–6773 (2015)
31. L. Yu, B. Guan, W. Xiao, W. Lou Xiong, Adv. Energy Mater. **5**, 1500981 (2015)
32. Y. Le, Y.J. Fan, L.X. Wen, Angew. Chem. Int. Ed. **55**, 13422–13426 (2016)
33. L. Yan, Y. Le, W. Minghong, W. Yong, L.X. Wen, Adv. Mater. **30**, 1702875 (2018)
34. G.Z. Chen, Int. Mater. Rev. **62**, 173–202 (2017)
35. C. Niu, E.K. Sichel, R. Hoch, D. Moy, H. Tennent, Appl. Phys. Lett. **70**, 1480–1482 (1997)
36. T. Chen, L. Dai, Mater. Today **16**, 272–280 (2013)
37. K.H. An, W.S. Kim, Y.S. Park, J.-M. Moon, D.J. Bae, S.C. Lim, Y.S. Lee, Y.H. Lee, Adv. Funct. Mater. **11**, 387–392 (2001)
38. D.N. Futaba, K. Hata, T. Yamada, T. Hiraoka, Y. Hayamizu, Y. Kakudate, O. Tanaike, H. Hatori, M. Yumura, S. Iijima, Nat. Mater. **5**, 987 (2006)
39. W. Lu, L. Qu, K. Henry, L. Dai, J. Power Sources **189**, 1270–1277 (2009)
40. K.S. Novoselov, A.K. Geim, S.V. Morozov, D. Jiang, Y. Zhang, S.V. Dubonos, I.V. Grigorieva, A.A. Firsov, Science **306**, 666–669 (2004)

41. K.S. Novoselov, Z. Jiang, Y. Zhang, S.V. Morozov, H.L. Stormer, U. Zeitler, J.C. Maan, G.S. Boebinger, P. Kim, A.K. Geim, Science **315**, 1379 (2007)
42. A.K. Geim, K.S. Novoselov, Nat. Mater. **6**, 183–191 (2007)
43. M.I. Katsnelson, Mater. Today **10**, 20–27 (2007)
44. J.S. Bunch, A.M. van der Zande, S.S. Verbridge, I.W. Frank, D.M. Tanenbaum, J.M. Parpia, H.G. Craighead, P.L. McEuen, Science **315**, 490–493 (2007)
45. D.A.C. Brownson, C.E. Banks, Analyst **135**, 2768 (2010)
46. A. Ambrosi, A. Bonanni, Z. Sofer, J.S. Cross, M. Pumera, Chem. Eur. J. **17**, 10763–10770 (2011)
47. P. Avouris, C. Dimitrakopoulos, Mater. Today **15**, 86–97 (2012)
48. C. Berger, Z. Song, X. Li, X. Wu, N. Brown, C. Naud, D. Mayou, T. Li, J. Hass, A.N. Marchenkov, E.H. Conrad, P.N. First, W.A. de Heer, Science **312**, 1191–1196 (2006)
49. X. Li, W. Cai, J. An, S. Kim, J. Nah, D. Yang, R. Piner, A. Velamakanni, I. Jung, E. Tutuc, S.K. Banerjee, L. Colombo, R.S. Ruoff, Science **324**, 1312–1314 (2009)
50. K.S. Kim, Y. Zhao, H. Jang, S.Y. Lee, J.M. Kim, K.S. Kim, J.-H. Ahn, P. Kim, J.-Y. Choi, B.H. Hong, Nature **457**, 706–710 (2009)
51. S. Park, J. An, I. Jung, R.D. Piner, S.J. An, X. Li, A. Velamakanni, R.S. Ruoff, Nano Lett. **9**, 1593–1597 (2009)
52. D. Li, M.B. Muller, S. Gilje, R.B. Kaner, G.G. Wallace, Nat. Nanotechnol. **3**, 101–105 (2008)
53. A.A. Balandin, S. Ghosh, W. Bao, I. Calizo, D. Teweldebrhan, F. Miao, C.N. Lau, Nano Lett. **8**, 902–907 (2008)
54. C. Lee, X. Wei, J.W. Kysar, J. Hone, Science **321**, 385–388 (2008)
55. S. Park, R.S. Ruoff, Nat. Nanotechnol. **4**, 217–224 (2009)
56. M. Pumera, Energy Environ. Sci. **4**, 668–674 (2011)
57. C. Soldano, A. Mahmood, E. Dujardin, Carbon **48**, 2127–2150 (2010)
58. A. Srivastava, C. Galande, L. Ci, L. Song, C. Rai, D. Jariwala, K.F. Kelly, P.M. Ajayan, Chem. Mater. **22**, 3457–3461 (2010)
59. Y. Sun, Q. Wu, G. Shi, Energy Environ. Sci. **4**, 1113–1132 (2011)
60. D. Pan, S. Wang, B. Zhao, M. Wu, H. Zhang, Y. Wang, Z. Jiao, Chem. Mater. **21**, 3136–3142 (2009)
61. Z. Song, T. Xu, M.L. Gordin, Y.-B. Jiang, I.-T. Bae, Q. Xiao, H. Zhan, J. Liu, D. Wang, Nano Lett. **12**, 2205–2211 (2012)
62. M.F. El-Kady, V. Strong, S. Dubin, R.B. Kaner, Science **335**, 1326–1330 (2012)
63. W. Deng, X. Ji, M. Gomez-Mingot, F. Lu, Q. Chen, C.E. Banks, Chem. Commun. **48**, 2770–2772 (2012)
64. Y. Huang, J. Liang, Y. Chen, Small **8**, 1805–1834 (2012)
65. R. Imran Jafri, N. Rajalakshmi, S. Ramaprabhu, J. Mater. Chem. **20**, 7114–7117 (2010)
66. J. Xia, F. Chen, J. Li, N. Tao, Nat. Nanotechnol. **4**, 505–509 (2009)
67. M. Pumera, Chem. Rec. **9**, 211–223 (2009)
68. D.K. Kampouris, C.E. Banks, Chem. Commun. **46**, 8986–8988 (2010)
69. J.J. Yoo, K. Balakrishnan, J. Huang, V. Meunier, B.G. Sumpter, A. Srivastava, M. Conway, A.L. Mohana Reddy, J. Yu, R. Vajtai, P.M. Ajayan, Nano Lett. **11**, 1423–1427 (2011)
70. W. Gao, N. Singh, L. Song, Z. Liu, A.L.M. Reddy, L. Ci, R. Vajtai, Q. Zhang, B. Wei, P.M. Ajayan, Nat. Nanotechnol. **6**, 496–500 (2011)
71. C. Liu, Z. Yu, D. Neff, A. Zhamu, B.Z. Jang, Nano Lett. **10**, 4863–4868 (2010)
72. X. Yang, J. Zhu, L. Qiu, D. Li, Adv. Mater. **23**, 2833–2838 (2011)
73. H. Ji, X. Zhao, Z. Qiao, J. Jung, Y. Zhu, Y. Lu, L.L. Zhang, A.H. MacDonald, R.S. Ruoff, Nat. Commun. **5**, 3317 (2014)
74. W. Song, X. Ji, W. Deng, Q. Chen, C. Shen, C.E. Banks, Phys. Chem. Chem. Phys. **15**, 4799–4803 (2013)
75. Y. Zhu, S. Murali, M.D. Stoller, K.J. Ganesh, W. Cai, P.J. Ferreira, A. Pirkle, R.M. Wallace, K.A. Cychosz, M. Thommes, D. Su, E.A. Stach, R.S. Ruoff, Science **332**, 1537–1541 (2011)

76. D.A.C. Brownson, L.J. Munro, D.K. Kampouris, C.E. Banks, RSC Adv. **1**, 978–988 (2011)
77. Y. Li, M. van Zijll, S. Chiang, N. Pan, J. Power Sources **196**, 6003–6006 (2011)
78. J. Chmiola, C. Largeot, P.-L. Taberna, P. Simon, Y. Gogotsi, Angew. Chem. Int. Ed. **47**, 3392–3395 (2008)
79. W.S. Hummers, R.E. Offeman, J. Am. Chem. Soc. **80**, 1339 (1958)
80. L. Yan, Y.B. Zheng, F. Zhao, S. Li, X. Gao, B. Xu, P.S. Weiss, Y. Zhao, Chem. Soc. Rev. **41**, 97 (2012)
81. R. Zan, Q.M. Ramasse, U. Bangert, K.S. Novoselov, Nano Lett. **12**, 3936–3940 (2012)
82. B.E. Conway, V. Birss, J. Wojtowicz, J. Power Sources **66**, 1–14 (1997)
83. B.E. Conway, J. Electrochem. Soc. **138**, 1539–1548 (1991)
84. J. Wang, S. Dong, B. Ding, Y. Wang, X. Hao, H. Dou, Y. Xia, X. Zhang, Natl. Sci. Rev. **4**, 71–90 (2017)
85. C.-C. Hu, W.-C. Chen, K.-H. Chang, J. Electrochem. Soc. **151**, A281–A290 (2004)
86. M. Naguib, V.N. Mochalin, M.W. Barsoum, Y. Gogotsi, Adv. Mater. **26**, 992–1005 (2014)
87. M.R. Lukatskaya, O. Mashtalir, C.E. Ren, Y. Dall'Agnese, P. Rozier, P.L. Taberna, M. Naguib, P. Simon, M.W. Barsoum, Y. Gogotsi, Science **341**, 1502–1505 (2013)
88. M.D. Levi, M.R. Lukatskaya, S. Sigalov, M. Beidaghi, N. Shpigel, L. Daikhin, D. Aurbach, M.W. Barsoum, Y. Gogotsi, Adv. Energy Mater. **5**, 1400815 (2015)
89. M.-Q. Zhao, C.E. Ren, Z. Ling, M.R. Lukatskaya, C. Zhang, K.L. Van Aken, M.W. Barsoum, Y. Gogotsi, Adv. Mater. **27**, 339–345 (2015)
90. Z. Ling, C.E. Ren, M.-Q. Zhao, J. Yang, J.M. Giammarco, J. Qiu, M.W. Barsoum, Y. Gogotsi, Proc. Natl. Acad. Sci. **111**, 16676–16681 (2014)
91. J. Luo, W. Zhang, H. Yuan, C. Jin, L. Zhang, H. Huang, C. Liang, Y. Xia, J. Zhang, Y. Gan, X. Tao, ACS Nano **11**, 2459–2469 (2017)
92. X. Wang, S. Kajiyama, H. Iinuma, E. Hosono, S. Oro, I. Moriguchi, M. Okubo, A. Yamada, Nat. Commun. **6**, 6544 (2015)

Advances in Cathode Nanomaterials for Lithium-Ion Batteries

3

Carlos M. Costa, Renato Gonçalves, and S. Lanceros-Méndez

Contents

Abstract

Energy resources, consumption, and management are among the most important issues of this century. Our dependence on fossil fuels should be changed by eco-friendlier renewable energies. Further, the generated electrical energy should be stored for improving mobile applications, being batteries the most used system for this purpose. Lithium-ion batteries are the most promising battery type and are composed of three main elements: cathode, anode, and separator/electrolyte. The active material of the cathode is primarily responsible for the battery capacity, and

Author Contribution: All authors performed the literature revision and critical review of the literature. Further, all authors contributed to the writing of the chapter. The authors further acknowledge that there is no financial relationship with the editors or publisher in this chapter, other than what acknowledged or appropriately cited with copyright permission.

C. M. Costa · R. Gonçalves
Centro de Física, Universidade do Minho, Braga, Portugal
e-mail: cmscosta@fisica.uminho.pt; r.goncalves@fisica.uminho.pt

S. Lanceros-Méndez (✉)
BCMaterials, Basque Center for Materials, Applications and Nanostructures, Leioa, Spain

IKERBASQUE, Basque Foundation for Science, Bilbao, Spain
e-mail: senentxu.lanceros@bcmaterials.net

Q. Zhen et al. (eds.), *Nanostructured Materials for Next-Generation Energy Storage and Conversion*, https://doi.org/10.1007/978-3-662-58675-4_3

for that reason, different cathode materials have been developed and explored for lithium-ion batteries. In the present work, the most relevant cathode materials are presented, their main properties and characteristics described, and their advantage and disadvantage analyzed, allowing to understand to which extent those materials are suitable for specific applications in this field.

3.1 Introduction

In order to reduce dependence on fossil fuels, eco-friendly renewable energies, such as wind, solar, and biomass, must be increasingly used due to global warming caused by increasing energy use and population increase [24, 171].

The energy produced by these systems must be stored, and batteries are the type of energy storage most used for different applications, such as consumer electronics, medical applications, hybrid electric vehicles (HEVs), and pure electric vehicles (PEVs), among others, due to their high power and energy density [50, 68, 209]. In particular, for HEVs and PEVs, it is necessary to increase the energy density of batteries.

The main components of a lithium-ion battery are the two different current collectors, typically copper and aluminum, two electrodes (negative and positive), and the separator/electrolyte [192].

The negative electrode, anode, should present a low potential in order to provide high cell voltage with the positive electrode, cathode, that is responsible for the cell capacity and cycle life [89, 189].

Regardless of the type of electrode, they consist of an active material, a conductive material, and a polymer binder at different ratios [69]. These three components are typically mixed in a solvent where the polymer is dissolved, and the active material and the conductive additive are dispersed to obtain good dispersion without sedimentation.

Considering that the active material of the cathode is primarily responsible for the battery capacity, this chapter focuses on the recent advances in the active cathode materials most commonly used in lithium-ion batteries. Further, new opportunities in this field are discussed.

3.1.1 Electrode Main Components

Figure 3.1 shows a schematic representation of a battery and the main components of the electrodes. In the electrodes, each constituent has a specific function: the polymer binder has the role of binding the active material and conductive additive; the conductive additive increases the electrical conductivity of the electrode; and, in the case of the active material for the cathode, lithium ions must enter and leave repeatedly, its structure containing a readily reducible/oxidable ion [151]. In relation to the conductive additive, the most used materials are carbon black and acetylene black due to higher electrical conductivity and low cost [61].

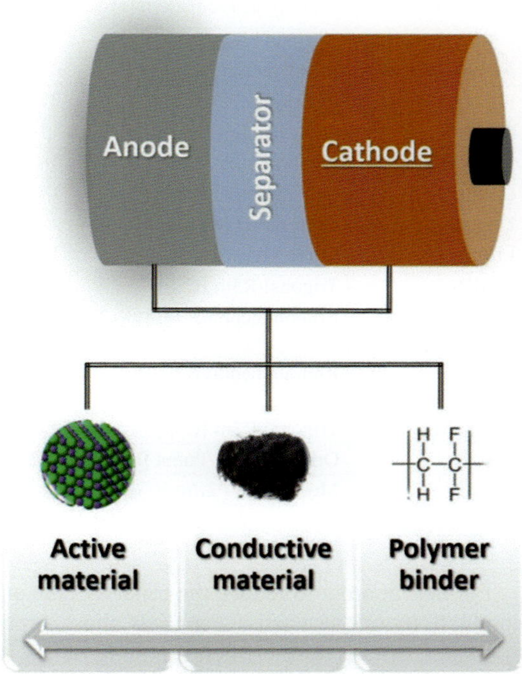

Fig. 3.1 Schematic representation of the battery and the main components of the electrodes

For the polymer binder, poly(vinylidene fluoride), PVDF, and poly(tetrafluoroethylene), PTFE, are the most used binders due to excellent mechanical strength, high chemical resistance, and high melting temperature [66, 239].

The active materials used for cathode development must show high reactivity with lithium, incorporation of large amounts of lithium, reversibility in lithium incorporation with structural stability, high lithium diffusion, good electrical conductivity, and insolubility in the electrolyte. The active materials most used in lithium-ion batteries are presented in Table 3.1 [139]. Table 3.1 shows the crystal system, theoretical capacity, and typical potential for each active material.

These active materials are synthesized by different methods including mechanochemical activation, carbothermal reduction, microwave heating, hydrothermal synthesis, sol-gel synthesis, spray pyrolysis, and coprecipitation, among others, in which their morphology and size are controlled [13, 17, 72, 145, 153, 186, 256].

It is to notice that each synthesis method shows specific advantages and disadvantages. Thus, the hydrothermal method is very used due to its low-energy costs, low cost (relatively cheap reagents), pollution-free (closed system reaction), large dispersion (reaction time), low operating temperature (facilitates industrialization), high reaction rate (higher efficiency), and nucleation control [210].

The control of morphology and particle size is essential for obtaining cathode electrodes with high cycling performance, as particles of small size show shorter Li^+ ion diffusion length, which can further improve Li-ion migration rate [28].

Table 3.1 Most commonly active materials used for lithium-ion battery applications

Active material	Crystal system/space group [point group]	Specific capacity/ $mAh.g^{-1}$	Typical voltage range/V	Ref.
$LiCoO_2$	Orthorhombic/R3m [C_{3V}]	274	2.5–4.25	[87]
$LiFePO_4$	Orthorhombic/Pnma [D_{2h}]	170	2.5–4.2	[269]
$LiCoPO_4$	Orthorhombic/Pnma [D_{2h}]	167	3.0–5.1	[71]
$LiMnPO_4$	Orthorhombic/Pnma [D_{2h}]	171	2.7–4.25	[158]
$LiMnO_2$	Orthorhombic/Pmmn [D_{2h}]	285	1.0–4.6	[91]
$LiMn_2O_4$	Cubic/Fd$\bar{3}$m [O_h]	148	3.4–4.3	[254]
$LiNiO_2$	Trigonal/R3m [C_{3V}]	275	3.0–4.3	[261]
$LiNi_{1-x}Co_xO_2$ ($0.2 \leq x \leq 0.5$)	Rhombohedral/R3m [C_{3V}]	~275	3.5–4.3	[1]
$LiNi_{1/3}Mn_{1/3}Co_{1/3}O_2$	Rhombohedral/R3m [C_{3V}]	278	2.3–4.5	[98]
$LiNi_{0.5}Mn_{0.5}O_2$	Trigonal/R3m [C_{3V}]	280	2.5–4.9	[160]
LiV_2O_5	Orthorhombic/Pnma [D_{2h}]	142	2.50–3.60	[25]
LiV_3O_8	Monoclinic/P2$_1$/m [C_{2h}]	280	2.8	[88]

Further, in order to reduce the capacity fade and improve the electronic and ionic conductivity of these active materials, those structures have been doped with metal, nonmetal, or rare earth elements, among others [143, 259].

3.1.2 Recent Advances in the Most Commonly Used Active Materials

In this chapter, each one of the active cathode materials reported in Table 3.1 will be addressed. For each one, the main properties, synthesis methods, and electrochemical properties will be presented. Some of the active materials are illustrated in Fig. 3.2 by its specific capacity versus cell voltage (Li^+/Li) in order to compare their characteristics.

3.1.2.1 Lithium Cobalt Oxide ($LiCoO_2$)

Lithium cobalt oxide ($LiCoO_2$) has been one of the most widely used cathode materials in commercial Li-ion rechargeable batteries, due to its good capacity retention, high structural reversibility (under 4.2 V vs. Li^+/Li), and good rate capability. This active material was originally suggested by Goodenough et al. [166], and in the early 1990s, Sony Corporation commercialized it [42, 165]. The undoped $LiCoO_2$ cathode material shows a structure along the c-axis with CoO_6 octahedra separated by Li^+ layers, where the LiO_6 octahedra are elongated. The $LiCoO_2$ unit cell in the hexagonal setting is composed of three layers in rhombohedral symmetry [30]. The layered $LiCoO_2$ with the rhombohedral unit cell is characterized by a space group R3m with $a = 2.8166$ Å and $c = 14.045$ Å [211]. The

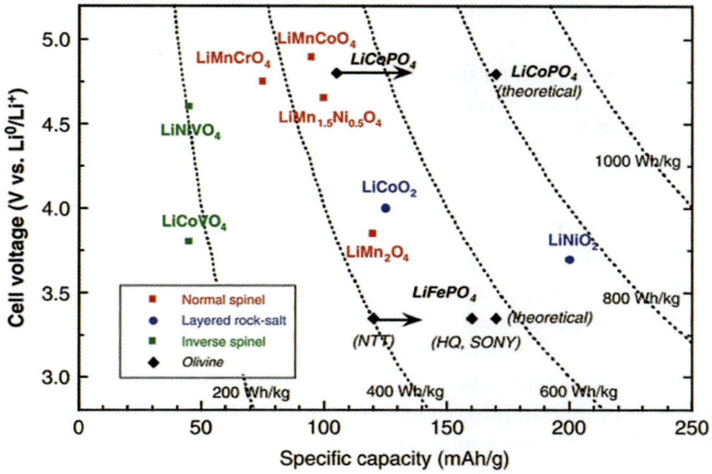

Fig. 3.2 Cell voltage versus capacity for different active materials. (Reprinted by permission from Ref. [97]. Copyright 2016 Springer Nature)

layered crystal structure of $LiCoO_2$ with a density of 5.1 g cm^{-3} is ideal to accommodate lithium ions. The active material also presents magnetic properties [174]. This property led to different applications being its use in cathode material in lithium batteries one of the most promising due to the deintercalation of lithium Li_xCoO_2 ($x < 1$). The experimental band gap (E_g) is about 1.7–2.7 eV [3].

A large anisotropic volume change is verified when $LiCoO_2$ converts to $Li_{0.5}CoO_{1.5}$ during the charging. This volume change of the host lattice is about 7% and occurs in the c-axis direction. Consequently, it is verified a peeling between the conducting material and the active material particles that results in problems in capacity fade and poor rate capability of the cathode. The main reason for this fact is due to the material dispersion into the electrolyte [165]. During the charge, the migration of Li^+ ions from the layered crystal lattice causes a nonstoichiometric $Li_{1-x}CoO_2$ formation, and the oxidation of Co^{3+} changes for charge compensation [42].

The theoretical electrochemical capacity of $LiCoO_2$ is around 274 mAh.g^{-1}, but due to the instability of the structure and the limitation of Li^+ removal from the hexagonal lattice (0.5), this capacity is reduced to around 140 mAh.g^{-1} [42].

Some efforts have been made in order to recycle the active material [182, 275]. Zhang group used ultrasound radiation on $LiCoO_2$ previously calcinated at 500 °C in order to remove all the carbon and binder of the cathode. The treatment consists of the application of hydrothermal ultrasound radiation during 6 h (600 W) at 80 °C in a concentrated solution of lithium hydroxide (LiOH). The renovated active material shows a discharge capacity retention of 99.5 and a first discharge capacity of 133.5 mAh.g^{-1}. It is shown that after 40 cycles the electrochemical properties of commercial batteries were achieved [275].

Different methods and synthesis routes, such as sol-gel [12, 65], flux-coating [165], one-step electrochemical-hydrothermal [8], combustion [200], microwave

[84], hydrothermal [21], thermal decomposition [164], and auto flame [215], were used to produce the $LiCoO_2$.

In Azib's approach [8], thin films of $LiCoO_2$ were synthesized via electrochemical-hydrothermal route using ethanol and water blends as a solvent. The influence of parameters such as ethanol-water ratio, pressure, temperature, current density, and reaction time was evaluated on the R3m layered phase formation and in the electrochemical response of Li batteries. The results show that well-crystalized $LiCoO_2$ with preferred orientation (101) without traces of spinel-like phase and cobalt oxide impurities were achieved and show a capacity of around 120 $mAh.g^{-1}$ [8].

In another study [12], thin films of $LiCoO_2$ were prepared by spin coating deposition via a sol-gel process. The annealing temperature of 275 °C shows to be effective in the crystalline nature with a preferred orientation along the (003) plane of the rhombohedral structure. Above 416 °C, the material degrades. It was found that the conductivity of the samples decreases from 10^{-3} to 10^{-9} Ω^{-1} cm^{-1} and that the band gap energy decreases from 2.51 to 3.4 eV with annealing temperatures from 275 °C to 475 °C, respectively.

Two kinds of crystallographic structures of the $LiCoO_2$ can be achieved according to the synthesis temperature. The cubic spinel structure is formed around 350 °C, and around the 750 °C, the layered trigonal (hexagonal) structure is formed. It is observed that the structure with the layered lattice shows a better electrochemical performance when compared with the structure with a cubic spinel lattice [42].

Nanostructured approaches have been used in order to enable buffering of the stresses caused by the volume variation, improving the electrochemical performance and the structural durability through the increased surface area and decreasing the diffusion length of lithium ions in the extraction/insertion reaction. Higher capacities at high charge/discharge rates are achieved due to the shorter diffusion pathways for ionic transport and electronic conduction [165].

Neutron scattering of a battery with $LiCoO_2$ during the charge/discharge process shows no difference of the active material at 25 °C and 50 °C operation. In the same study, it was verified that the lattice parameter c of the rhombohedral structure and cell volume increases during the discharge process (more noticeable at 25 °C) [30].

Different approaches to overcome the nanosized materials problems, such as particle aggregation that results in capacity fade due to low tap density and the not homogeneous distribution, were evaluated. One of these methods was the growth of an additive-free nanosized $LiCoO_2$ crystal layer directly on the current collector [165].

In order to improve the overcharge of the $LiCoO_2$ (discharge energy of 96%), a coating with $AlPO_4$ was used, and the over-insertion tolerance maintains to over 99% of the discharge energy after ten cycles. Also, the $AlPO_4$ coating prevents crystal structure changes and suppresses irreversible cation Li and Co exchange in the octahedral layers of the $LiCoO_2$ during the over-insertion of Li^+ ions [39].

3.1.2.2 Lithium Manganese Oxide ($LiMnO_2$)

Lithium manganese oxide ($LiMnO_2$) layered structure is a very attractive material (compared to $LiCoO_2$) from the environmental and economic perspective. This material shows two structures: orthorhombic with space group Pmmn and rhombohedral, being the orthorhombic structure thermodynamically stable. In order to obtain the layered material, the ion-exchange technique was used. Contrary, the rhombohedral structure is thermodynamically unstable due to the low spin of the Mn^{3+} [244, 271]. Both of them involve distortion in the oxygen arrays from the ideal cubic close packing. The orthorhombic structure also shows the advantage of being easy to prepare and air stable, showing high capacity (285 $mAh.g^{-1}$) and high potential vs. lithium. Other advantages are also observed once the reduction of manganese oxide (MnO_2) does not release gases and MnO_2 is a low-cost material [211]. Lithium manganese oxide batteries were first used to power calculators, memory backups, and watches. This material is the most common active material used in lithium batteries with 80% of the market share [97].

During the cycling process, it is observed poor electrochemical properties and a drop in the voltage profile, due to both orthorhombic and rhombohedral $LiMnO_2$ transformation in a spinel structure, once this phase shows low crystallinity, Mn dissolution, and large strain (Jahn-Teller activity) [271]. Jin group observed the transformation of orthorhombic lithium manganese oxide ($LiMnO_2$) to spinel $LiMnO_2$ phase during cyclic voltammetry (Fig. 3.3) [95, 194]. The main difficulty in obtaining layered $LiMnO_2$ (compared with $LiCoO_2$ and $LiNiO_2$) is attributed to the fact the ion Mn^{3+} and Li^+ have the closer size [194].

In order to stabilize the layered structure, different approaches with cation doping with aluminum (Al), cobalt (Co), chromium (Cr), and titanium (Ti), among others, were carried out [26].

Different types of synthesis routes were used to obtain this material, including low temperature [288], high temperature [43], hydrothermal [136, 255], flux method [241], solid state [5], and quenching [117].

The work developed by Xiao's group [255] reported that morphologies as rod and nanoparticles show different properties. The rods exhibit higher discharge (200 $mAh.g^{-1}$) capacity and better cyclability. The nanoparticles show a capacity of 170 $mAh.g^{-1}$, but after 30 cycles it decreases to 110 $mAh.g^{-1}$, whereas the rods decrease from 200 to 180 $mAh.g^{-1}$ after 30 cycles. The rod morphology plays an important role in the electrochemical performance once it favors the electron transport along the length pathway and accommodates the volume changes originated by the charge/discharge process.

In order to improve the cycle life of $LiMnO_2$, coatings with cobalt oxide (CoO) and alumina (Al_2O_3) were realized. This improvement is due to the prevention of the Mn dissolution and lattice instability from the Jahn-Teller distortion [31, 32]. Other methods as doping with Boron were also used to improve cycling performance of $LiMnO_2$ batteries, in particular at high temperatures (60 °C), by the reduction of surface reaction and/or transition metal dissolution. The improvement of the B-doping on the Coulombic efficiency ranges from 69.7% in pristine $LiMnO_2$ to 99.0% in ten at % B-doped [288].

3.1.2.3 Lithium Manganese Oxide ($LiMn_2O_4$)

Lithium manganese oxide ($LiMn_2O_4$) was proposed by Thackeray's group [226] in 1986 as the cathode material. This cathode corresponds to the third generation, with $LiCoO_2$ representing the first and $LiFePO_4$ representing the second. Composed by a

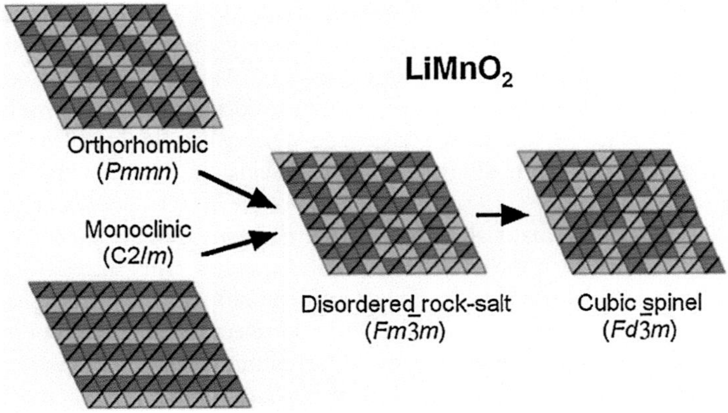

Fig. 3.3 Schematic representation of the structural transformation of orthorhombic $LiMnO_2$ to spinel $LiMnO_2$ phase during the cyclic voltammetry. (Reprinted by permission from Ref. [97]. Copyright 2016 Springer Nature)

transition metal oxide, $LiMn_2O_4$ is characterized by a space group fd3m (a = 8.2476–2), a density of 4.2 g cm^{-3}, the Li and Mn occupying the tetrahedral (8a), and octahedral (16d) sites in the intervening cubic array of oxygen atoms (32e sites). The spinel structure shows a high lithium extraction ratio of 0.6 ($Li_{1-x}Mn_2O_4$, $0 < x < 0.6$) that provides a stable internal spinel structure and exhibits a theoretical capacity of 148 mAh.g^{-1} [74]. The three-dimensional (3D) crystal structure allows also the 3D lithium diffusion [173], the electronic conductivity is about 10^{-4} S cm^{-1} [185], and the $Li_{1-x}Mn_2O_4$ ($0.27 < x < 1$) active material can be cycled at about 4 V vs. lithium.

The main advantages of $LiMn_2O_4$ as a cathode in lithium batteries are the low-cost, low toxicity, and thermal stability. The extra lithiation ($Li_2Mn_2O_4$) ensures a high amount of lithium-ion insertion/extraction in the structure. Therefore, the theoretical capacity of this cathode is doubled as the number of charge carrier (Li^+) is also double, with respect to $LiMn_2O_4$.

MacNeil group reported that $LiMn_2O_4$, when compared with $LiCoO_2$ and $LiNiO_2$, shows higher thermal stability at high temperature, when in contact with the electrolyte. This property shows that spinel $LiMn_2O_4$ suits to high power batteries [140].

On the other hand, the main obstacle to its commercialization is due to the fact that spinel $LiMn_2O_4$ shows significant capacity fading during the cycling and storage processes. This capacity fading at room and high temperatures are due to two main reasons: (1) the manganese (Mn) dissolution into the electrolyte and (2) the Jahn-Teller instability caused by the ion Mn^{3+} at the octahedral sites [140]. Further, at high voltages (>4.3 V) and elevated temperatures (>55 °C), the capacity fade also increases. Capacity degradation of this cathode is thus mainly originated by the dissolution of the Mn ions from the interface between cathode and electrolyte and its subsequent deposition on the anode surface that increases cell impedance [29].

The $LiMn_2O_4$ active material can be synthesized by different synthesis routes, such as spray drying [247], hydrothermal [36], coprecipitation [227], sol-gel [278], and solid state [264], among others. Synthesis routes providing a fine particle size and uniform composition (such as sol-gel) lead to high electrochemical performance [265].

Efforts to overcome/improve the electrochemical problems and high-temperature performance include ion doping [93], reducing particle sizes [265], different morphologies [254], and metal oxide coating [29].

Yi et al. [265] synthesize spinel $LiMn_2O_4$ by adipic acid-assisted sol-gel and studied the influence of the synthesis temperature (350 °C, 800 °C, and 900 °C) on phase formation and electrochemical activity (Fig. 3.4a–d). The results show that the samples calcinated at 800 °C show better ordering of the local structure, higher crystallinity, and lower lattice strain. Thus, it is possible to control the particle size by controlling the sintering temperature, the samples synthesized at 800 °C showing lower lattice defects, being the Mn chemical valence closer to 3.5.

Finally, nanoscale surface modifications have been investigated [29] (Fig. 3.4a). Xiao group [257] evaluated the effect of morphology (porous and hollow) of $LiMn_2O_4$, obtained by precipitation synthesis using manganese cobalt ($MnCo_3$) (Fig. 3.4b) as template and manganese oxide (MnO_2) as intermediate (Fig. 3.4c), in ion the electrochemical response of the cathode. It is concluded that the microspheres with a porous surface (Fig. 3.4e) exhibit a higher reversibility capacity and rate capability than the hollow surface sample (Fig. 3.4d). The current charge/discharge of the porous structure deliver a reversible capacity of 117.2 $mAh.g^{-1}$ at 1 $A.g^{-1}$ and 92.1 $mAh.g^{-1}$ at 10 $A.g^{-1}$.

3.1.2.4 Lithium Ferrous (II) Phosphate ($LiFePO_4$)

Lithium ferrous (II) phosphate ($LiFePO_4$, LFP) was first reported by Padhi's group [187]. It is a member of the olivine family, crystalizes in the orthorhombic system with a Pnma [D_{2h}] structure. Lithium ferrous (II) phosphate exhibits a distorted hexagonal close-packed oxygen framework with lithium and iron located in half the octahedral sites and phosphorous ions located in one-eighth of the tetrahedral sites. The interatomic distances are $a = 10.332(4)$ Å, $b = 6.010(5)$ Å, and $c = 4.692(2)$ Å with a unit cell of 291.4(3) Å3 [81]. The bulk $LiFePO_4$ exhibits paramagnetic (form induced magnetic fields in the direction opposite to that of the applied magnetic field) behavior at room temperature [234].

LFP has received a singular attention as cathode active material due to its ability to reach theoretical capacities of 170 $mAh.g^{-1}$ at moderate current densities and not generating oxygen under conditions such as electric overcharging and shortening [97].

The electrochemical extraction of lithium from the $LiFePO_4$ active material gives (Fe^{2+}/Fe^{3+}) redox potential at ca. 3.5 V vs. lithium, with reversible intercalation/deintercalation of one lithium per each $LiFePO_4$. This active material has excellent cycling performance, the cubic lattice volume and crystal parameters slightly varying in this process, the volume variation during the lithium-ion deintercalation decreasing by 6.81%, and the density increasing by 2.59%.

The advantages of this material are low-cost, iron abundance in nature, no memory effect, non-toxicity, resistance to overcharge, environmental friendliness, stable voltage plateau, and improved safety compared with cobalt oxides materials

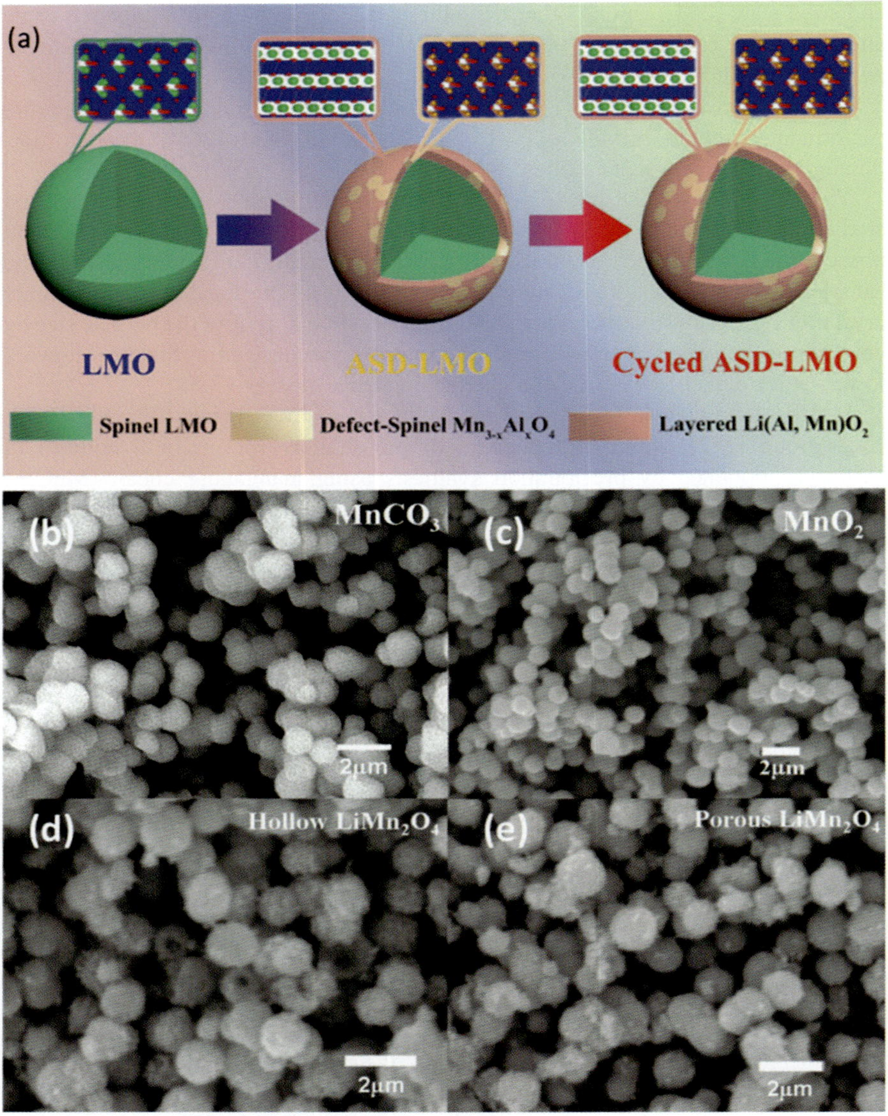

Fig. 3.4 (**a**) Schematic representation of the surface modification of $LiMn_2O_2$. (Reprinted by permission from Ref. [29]. Copyright 2018 American Chemical Society). SEM micrographs of the (**b**) $MnCO_3$ precursor, (**c**) MnO_2 intermediate, (**d**) hollow $LiMn_2O_4$ microspheres, and (**e**) porous $LiMn_2O_4$ microspheres. (Reprinted from Journal of Alloys and Compounds, 738, Haowen Xiao, Yourong Wang, Kai Xie, Siqing Cheng, Xianzhong Cheng, High capacitance $LiMn_2O_4$ microspheres with different microstructures as cathode material for aqueous asymmetric supercapacitors, 7, 2018, with permission from Elsevier [256])

[97]. Consequently, this active material is the first choice to use as a cathode in lithium batteries with medium-large capacity and medium-high power density.

Despite this, the intrinsically low electronic conductivity (10^{-7}–10^{-9} S cm^{-1}) and low diffusion coefficient of lithium ion (10^{-11}–10^{-13} cm^{-2} s^{-1}) of bulk $LiFePO_4$ result in capacity losses during the high-rate discharge. Further, comparing the olivine structure with the spinel and layered ones, the olive structure is less dense, leading to a lower volumetric energy density. The theoretical density of $LiFePO_4$ is just 3.6 g cm^{-3} which is low compared with other materials such as $LiCoO_2$ (5.1 g cm^{-3}) [97], but commercial $LiFePO_4$ exhibits a density around 1.0–1.3 g cm^{-3}, which shows that efforts should be taken to improve this material.

Different types of synthesis techniques were used to produce the $LiFePO_4$ particles, including microwave [266], solid-state reaction [201], template method [137], sol-gel [294], coprecipitation [11], hydrothermal [230], and mechanical activation [236], among others.

In order to enhance the electronic conductivity, different strategies such as the addition of carbon, surface treatment with carbon layers, substitution or doping, and reduction of particle size have been used.

The carbon used on the surface of $LiFePO_4$ should be thin and not exceeding a few wt.%, with a thickness of around 3 nm. With the surface treatment, the electronic conductivity of the active material could increase up to seven orders of magnitude. This carbon layer could be reached by the addition of organic materials as a carbon precursor and is limited to cathodes with low voltage response. In high voltage cathode materials, carbon will be electrochemically oxidized in the coating process and by the highly applied potential in the charging process [271]. The carbon coating (1) provides, after deintercalation, electron channels to balance the charge of active material; (2) increases the surface area of the active material due to the carbon nanometer sized by the production of fine $LiFePO_4$; (3) avoids the production of Fe^{3+}; (4) prevents the polarization during the charge/discharge process; and (5) enhances the electronic conductivity [152, 233, 244].

Nanostructured $LiFePO_4$ particles has been produced in order to overcome some issues such as (1) the decrease to nanosize reduces the distances for the (de)intercalation of lithium ions, decreasing the capacity fading at high current density; (2) the increase of the surface area can increase the theoretical capacity and Coulombic efficiency due to the higher lithium quantity adsorbed on the surface; (3) increases the redox reaction interface; (4) ensures a bigger electrolyte immersion, among others [244].

3.1.2.5 Lithium Cobalt Phosphate ($LiCoPO_4$)

Lithium cobalt phosphate ($LiCoPO_4$) is one of the most stable active material used as cathode due to its intrinsic properties and unique olivine structure with the orthorhombic system (Pnma, $Z = 8$); lattice parameters $a = 10.2048$ Å, $b = 5.9245$ Å, and $c = 4.7030$ Å; and a density of 3.7 g cm^{-3}. The redox potential of this material is 4.8 V vs. lithium and shows a flat voltage profile. With a theoretical capacity of 167 mAh.g^{-1}, this active material also shows a small structural volume change [35].

This material has the advantage of exhibiting a high working potential compared with $LiFePO_4$ and a larger theoretical performance compared with $LiCoO_2$. On the other hand, the active material intrinsically exhibits poor rate capability, low lithium-ion diffusion, and low electronic conductivity ($<10^{-9}$ S cm^{-1}) due to the one-dimensional ion transport channel. The poor cyclability of this material is due to the oxidation of cobalt (Co^{2+} to Co^{3+}) and also by the decomposition of the carbonated electrolyte under charging between the 4.8 and 5.1 V. This reaction contributes to low Coulombic efficiency and capacity fading during cycling.

The Pnma-$LiCoPO_4$ phase is the most explored by the scientific community, but two more metastable $LiCoPO_4$ phases, $Pna2_1$ [C_{2V}] and Cmcm [D_{2h}] $LiCoPO_4$, were discovered. These two phases, in contrast with Pnma [D_{2h}], do not exhibit any significant electrochemical performance, but both exhibit interesting magnetic properties, showing complex magnetic interactions at low temperatures that should be explored [155].

Different methods were used to synthesize this active material and to obtain different electrochemical performances (Fig. 3.5), including coprecipitation, microwave, mechanochemical, hydrothermal, solid state [104], spray drying, and supercritical fluid, among others.

In order to form a stoichiometric $LiCoPO_4$ without impurities, the precursor compounds used during the synthesis should have a strong Co-P bonding, preventing the formation of Co metal, lithium phosphate (Li_3PO_4), and cobalt oxide (Co_3O_4) phases [35]. Further, the synthesis performed in an argon atmosphere and carbon addition improves the conductivity of $LiCoPO_4$ [242, 243].

To enhance the $LiCoPO_4$ performance (increase electrical conductivity and rate capability), efforts in decreasing particle size [250], change morphology [156],

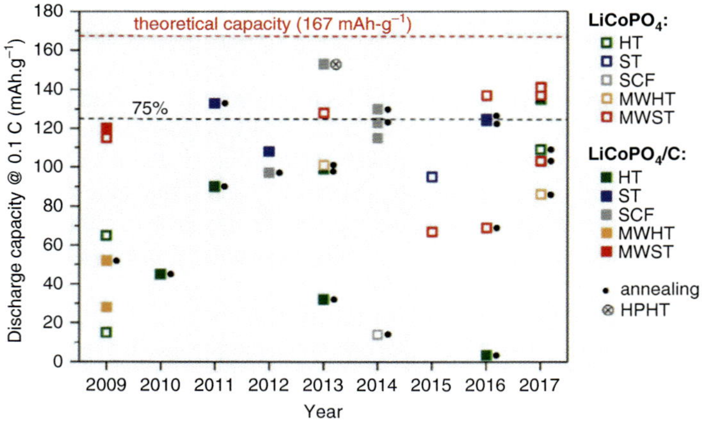

Fig. 3.5 Comparison of the initial discharge capacities at 0.1 C reported for $LiCoPO_4$ (white square) and $LiCoPO_4/C$ (black square) between 2009 and 2017, synthesized by different methods. (Reprinted from Journal of Power Sources, 382, Jennifer Ludwig, Tom Nilges, Recent progress and developments in lithium cobalt phosphate chemistry- Syntheses, polymorphism and properties, 15, 2018, with permission from Elsevier [155])

carbon coating [251], and lattice doping with supervalent metals ions [116] have been made. In the Wang group approach [237], it was shown that the addition of vanadium (V) improves the electrical conductivity of the active material around 5.3 times. Through carbon coating, different factors, such as carbon source, thickness, homogeneity, and carbon content, influence the electrochemical performance.

The nanoflake morphology was investigated by Hou group [83] with the objective of enhancing the electrochemical performance of the $LiCoPO_4$. The nanoflakes were coated with a thin carbon layer that orients the lithium-ion transport channels. With this treatment, the aggregation of particles is avoided improving the stability of the electrode material. The carbon coating was obtained by the carbonization of 5-hydroxymethylfurfural ($C_6H_6O_3$) and levoglucosan ($C_6H_{10}O_5$, (1R,2S,3S,4R,5R)-6,8-dioxabicyclo[3.2.1]octane-2,3,4-triol) that further inhibit the growth of (010) facets that create more exposed lithium-ion transport channels, also creating a conductive layer.

3.1.2.6 Lithium Manganese Phosphates ($LiMnPO_4$)

Lithium manganese phosphates ($LiMnPO_4$) crystallize in an olivine structure with a density of 3.4 g cm^{-3}, and, as all phospho-olivine structures, the oxygens ions are strongly bonded by covalent bonds with the phosphor (+5) to form the PO_4^{3+} tetrahedral polyanion stabilizing the framework. It exhibits an orthorhombic Pnma space group [D_{2h}], and the typically crystallographic parameters are $a = 6.106(1)$ Å, $b = 10.452(1)$ Å, and $c = 4.747(1)$ Å. With this structure stabilization, $LiMnPO_4$ shows improved stability and safety for high operating voltage. Furthermore, manganese (Mn) and lithium (Li) occupy the octahedral 4c and 4a sites, and the P atoms occupy the 4c sites. Comparing with lithium-ion phosphate, $LiMnPO_4$ draws the attention of the scientific community due to its higher theoretical energy density (171 mAh.g^{-1}), higher operating voltage (4.1 V), and working within the stability window of the common nonaqueous electrolytes. Further, this cathode active material has advantages as low cost, environmentally friendlier nature, working ability at high temperatures, and non-toxicity [4].

When compared with Li_xFePO_4, Li_xMnPO_4 presents lower electrochemical performance. Furthermore, the $LiMnPO_4$ active material exhibits low electronic conductivity ($<10^{-10}$ S cm^{-1}), relatively poor cycling stability (Jahn-Teller anisotropic lattice distortion and interface volume strain of 8.9%), and weak lithium diffusion.

In order to overcome these negative properties, different methods and synthesis routes were applied to improve these drawbacks by the manipulation of its morphology/structure [292], cation distribution [252], and electrochemical properties. Hydrothermal synthesis [63], precipitation methods [45], solid-state reaction [187], template methods [134], sol-gel synthesis [195], electrospinning followed by calcination [154], and solvothermal reaction [292], among others, were used to prepare this active material.

To enhance the low intrinsic electronic conductivity, methods such as decrease particle size, carbon coating, and cation doping were explored. The nanoparticle approach is commonly used because with this method it is possible to improve the power rate and rate capability of the $LiMnPO_4$ battery due to the reduction of Li^+ and

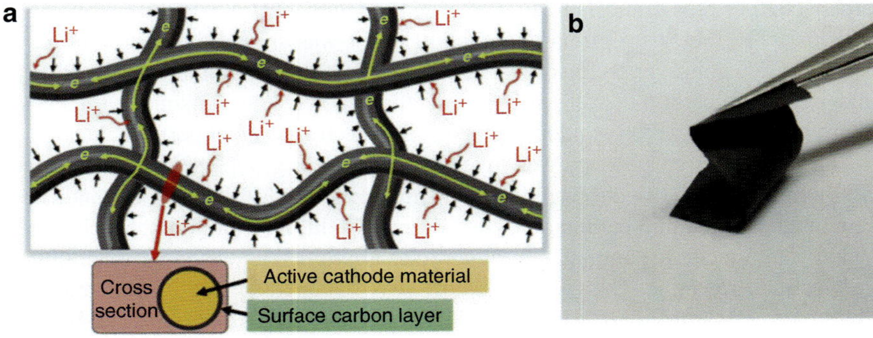

Fig. 3.6 (**a**) Schematic diagram of Mg-$LiMnPO_4$ fibers and (**b**) optical image of the flexibility of the fibers [154]

electron length diffusion. The carbon coating on the $LiMnPO_4$ surface enhances electronic conductivity by the reduction of particle agglomeration and by the formation of an effective conduction network between the layer and particle. Different carbon sources such as graphene and β-cyclodextrin ($C_{42}H_{70}O_{35}$, Heptakis (hydroxymethyl)-tetradecaoxa-octa-cyclonona-tetracontane-tetradecol) were studied. The conductivity can be also improved by the doping with heterovalent and isovalent heteroatoms through the reduction of volume between the lithiation and delithiation phases. $LiMnPO_4$ doping with transition metals is usually used because not only it increases its conductivity but also increases or maintains its capacity. Lu group [154] studied doped Mg-$LiMnPO_4$ with fiber morphology (Fig. 3.6) and obtained a battery performance of 135 $mAh.g^{-1}$ at a C/10 rate (15 $mAh.g^{-1}$) that is superior to the one of $LiMnPO_4$.

This sample shows a good cyclability without capacity fading during 200 cycles and the ability for being applied in flexible batteries. Li's team [134] designed a 3D structure binder-free $LiAlO_2$-$LiMnPO_4$/C cathode material with improvement in the electrochemical performance when compared with $LiMnPO_4$/C. The sample exhibits a capacity of 105 $mAh.g^{-1}$ and 98.4% of capacity retention at a high discharge rate of 10 C, after 100 cycles. This structure can solve drawbacks such as the poor conduction of the active materials.

3.1.2.7 Lithium Nickel Dioxide ($LiNiO_2$)

This active material is very attractive for being used in lithium-ion batteries due to higher discharge capacity (~275 $mAh.g^{-1}$), low cost [261], higher potential [55], and small variation of unit cell volume [183], and it is starting to be used in automotive application [270]. The cell volume varies from 101 to 98.5 Å3 in the composition range $0 < x < 3/4$ for $Li_{1-x}NiO_2$, which corresponds to about 2.5% shrinkage on oxidation [183]. The lithium-ion diffusion of $LiNiO_2$ is higher when the temperature increases, and it is described by simple ion hopping with an activation energy of 0.6 eV [180]. Some disadvantages of this material are the low capacity retention and thermal stability

[289]. As for the other active materials, the morphology and particle size influence the cycling performance of $LiNiO_2$ [213], smaller particles showing higher capacity, but larger particles providing higher thermal stability [121]. At room temperature, the electrical resistivity of this material is 4 $\Omega.cm^{-1}$ [163]. Due to the nickel (Ni) element, this active material shows magnetic properties [15], the antiferromagnetic phase (the magnetic moments related to the spins of electrons align in a regular pattern with neighboring spins pointing in opposite directions) increasing in contact with the ferrimagnetic (whose strength depends on that of the applied magnetizing field and which may persist after removal of the applied field) with nickel/lithium Ni/Li defect $LiNiO_2$ due to increasing Li content [227]. Its electrochemical window, surface energy, and morphology were theoretically studied within the density functional theory (DFT) framework, and it was observed that tetragonal (*t*) Li-terminated surfaces are characterized by lower surface energies than those of the oxygen surfaces irrespective of facet direction. Further, Ni-exposed surfaces are the least stable [33]. Also, this theory was used for studying the effect of non-stoichiometry (extra-Ni-defects) on the electrochemical performance of Ni-rich layered oxides, and it was verified that extra-Ni-defects trigger a charge disproportionation reaction that reduces the Jahn-Teller distortion [114].

One simple synthesis of this structure is by heating a mixture of hydrated lithium hydroxide ($LiOH.H_2O$) and nickel (II) hydroxide, $Ni(OH)_2$ (Li/Ni = 1.1/1 in molar ratio) in the range from 500 °C to 900 °C for 5 h in air or oxygen. It was observed that the structure prepared in an oxygen atmosphere at 700 °C exhibits the highest discharge capacity of 200 $mAh.g^{-1}$ in the voltage range from 3.0 to 4.3 V and the charge transfer resistance is related to the structural transformation of $Li_{1-x}NiO_2$, $0.15 < x < 0.75$ [261]. Different lithium precursors have been used for obtaining this compound, lithium nitrate ($LiNO_3$) [172, 176], $LiOH.H_2O$ [216], and Li_2CO_3 [232], as well as different synthesis methods, such as coprecipitation [205], emulsion [111], combustion [196], solid state [216, 218], and nanoscale $MnCO_3$ treatment [282], among others.

In order to improve its thermal stability and consequently battery performance, different coatings with cobalt (Co) and manganese (Mn) [49], SiO_2 [169], and vanadium (V) oxide (V_2O_5) [258] were carried out. The coating layer of Co, Mn particles improved the electrochemical properties of lithium nickel dioxide ($LiNiO_2$) through decreasing the impedance growth at the interface between the electrodes/electrolyte during battery cycling [49]. It has been shown that silica (SiO_2) coating at 2 wt.% leads to high structural stability and reversible capacity [169] and that a V_2O_5-coating layer works as hydrogen fluoride (HF) inhibitor and/or HF scavenger (from in situ precipitated cathode surface), which contributes to a significant improvement in cycling performance and storage characteristics in electrolyte [258].

Also for improving the electrochemical properties, this compound has been doped with iron [73], cobalt and manganese [178, 283], aluminum [149], and cerium [170]. Further, a novel inorganic nitrate salt synthesis was proposed [73].

The emulsion method has been used to synthesize this compound, and the optimal conditions for this synthesis were 750 °C and 24 h in an oxygen stream, as represented in Fig. 3.7 [111].

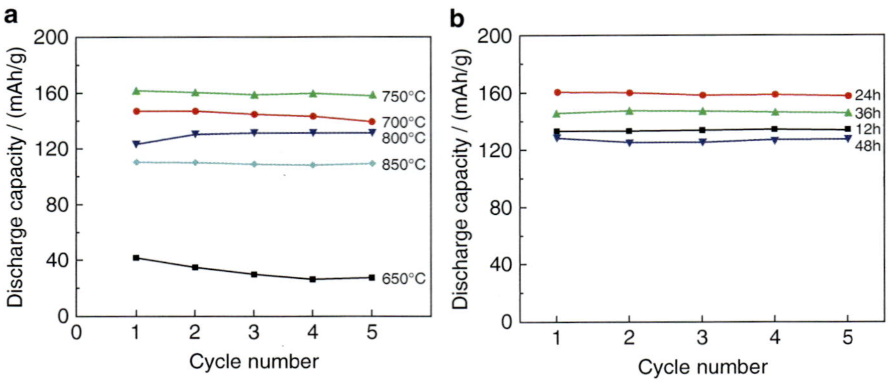

Fig. 3.7 Discharge capacity as a function of cycle number for $LiNiO_2$ powders: (**a**) at various temperatures for 24 h and (b) at 750 °C for various times. (Reprinted from Ceramics International, 33, Bok-Hee Kim, Jong-Hwan Kim, Ik-Hyun Kwon, Myoung-Youp Song, Electrochemical properties of $LiNiO_2$ cathode material synthesized by the emulsion method, 5, 2018, with permission from Elsevier [111])

Aluminum doping in $LiNiO_2$ reduces lithium deficiency, stabilizes the crystal structure, and improves electrochemical cycling [149]. The doping of cerium leads to high Coulombic efficiency, high electrical conductivity, and low irreversible capacity loss [170].

Surface modification of $LiNiO_2$ was also achieved by the addition of titania (TiO_2) or zinc (II) oxide (ZnO), and it was observed a decrease in the discharge capacity but improved cycling performance [216].

3.1.2.8 Lithium Nickel Cobalt Oxide [$LiNi_{1-x}Co_xO_2$ ($0.2 \leq x \leq 0.5$)]

This active material results from the combination of lithium cobalt oxide ($LiNiO_2$) and lithium cobalt oxide ($LiCoO_2$), the addition of cobalt into $LiNiO_2$ improving the thermal stability and cycling performance [46], as well as the delivered capacity, due to the stabilization of the structure with respect to $LiCoO_2$ [119, 202]. Considering the enthalpy of formation of this active material, the cobalt compound is significantly more stable than the nickel analog [232]. The magnetic properties of this material have been studied indicating the existence of randomly oriented magnetic moments even at 2 K, i.e., a disordered state [177].

Different methods have been used for synthesizing these particles, such as sol-gel [60, 175], solid-state reaction [101], heat treatment [112], coprecipitation and rheological phase reaction [124], rheological phase reaction [235], and gel combustion [245], among others.

It was demonstrated that $LiNi_{1-x}Co_xO_2$ ($0.1 < x < 0.3$) shows a low capacity fade rate when compared to that of $LiCoO_2$, being suitable for commercial batteries [119].

Novel synthesis methods – coprecipitation and rheological phase reaction method – were successfully used for this active material in which the calcination temperature effects on the electrochemical performance [124, 235]. This active material was used as a coating for lithium manganese oxide ($LiMn_2O_4$) to improve its thermal and electrochemical properties, which is ascribed to the suppression of electrolyte decomposition and the reduction of Mn dissolution [188].

3.1.2.9 Lithium Nickel Manganese Cobalt Oxide ($LiNi_{1/3}Mn_{1/3}Co_{1/3}O_2$)

Lithium nickel manganese cobalt oxide ($LiNi_{1/3}Co_{1/3}Mn_{1/3}O_2$) is a very promising active material as a cathode electrode, being used in electric vehicles applications [19] and semisolid flow batteries [90]. It shows higher capacity and tap density and better thermal stability, lower cost, less toxicity, and safety in comparison to other active materials [75]. Recently, it was demonstrated that it can be used also for anode electrodes [240]. The magnetic properties of the material were evaluated, and it was verified that both frustration and disorder play important roles in the formation of a spin glass state [18]. In order to investigate the changes in its crystal structure stability, calculations were carried out through the discrete variational (DV)-Xα molecular orbital method based on the self-consistent Hartree-Fock-Slater model compounds. Using known symmetries (C_{3v}) the radial function for molecular orbitals of model cluster atoms method is used to compute the energies in the order of the transition metal layer affecting the charge/discharge cycle characteristics [103]. For this compound, Ni precipitates act a redox species and change from 4^+ to 2^+ during charging. Further, Ni^{4+} becomes Jahn-Teller active Ni^{3+} at intermediate states during charge and Mn [168]. The analysis with in situ Auger electron spectroscopy and X-ray photoelectron spectroscopy of this compound during charging and overcharging shows that Mn is redox active below ~4.3 V and that O becomes redox active above ~4.2 V. Ni and Co also participate in surface redox [225]. Typically, the diffusion coefficient is in the range 10^{-11}–10^{-12} $cm^2.s^{-1}$ [274].

The analysis of overpotentials in this active material shows that the characteristic overpotentials during charge (delithiation) and discharge (lithiation) are state of charge (SOC) and depth of discharge (DOD) dependent, respectively [105]. It was also demonstrated that the specific capacity fades during cycling at constant specific currents can be mainly attributed to the increase of the delithiation (charge) hindrance [106].

This active material presents poor electronic conductivity that affects its electrochemical performance, and, therefore, different approaches have been used to improve its electric and ionic conductivity, including adding graphene, which decreases the charge transfer resistance [231], developing composition between this active material and lithium phosphor sulfur (Li_3PS_4) [6], and co-coating [41] and coating with lithium boron oxide glass [52] and carbon nanotubes/graphene (CNT/G) [132].

In all-solid-state batteries, the interface between the active material and solid electrolyte is essential for obtaining high ionic conductivity, and this interface was studied for argyrodite (Li_6PS_5Cl), demonstrating that the oxidation processes of the solid electrolyte do not hinder the good cyclability of the battery [7].

Several experimental techniques have been used for synthesizing this compound such as redox synthesis [9], coprecipitation [18, 22, 80], solid-state reaction [40, 100], heating [62], pre-sintering [77], convenient rheological phase reaction [75, 191, 197], microwave pyrolysis [76], sol-gel [78], chemical vapor deposition [82], polyol medium [128], spray dry [123], using manganese (II) carbonate ($MnCO_3$) both as a self-template and Mn source [126], acetic acid ($C_2H_4O_2$) leaching and maleic acid ($C_4H_4O_4$, cis-butenedioic acid) leaching [135], Pechini method [204], nitrate-melt decomposition [206], microwave-mediated hydrothermal [212], high-temperature ball milling [228], hydrothermal [249], mechanochemical activation [260], combustion [268], sacrificial template [277], and combining coprecipitation with the hydrothermal method [284].

The preparation of this compound through the Pechini method allows reducing calcination times and minimal lithium evaporation [204].

In the combustion method, it was studied the effect of the addition of triethanolamine (TEA, $C_6H_{15}NO_3$) in the synthesis, showing that synthesis with TEA exhibits lower polarization, better cycling ability, and significantly improved rate performance [268].

It has been shown that different carbonate precipitators affect its synthesis and consequently battery performance, the best discharge capacity value being obtained for sodium bicarbonate ($NaHCO_3$) [198].

A new synthesis process based on the sodium carbonate (Na_2CO_3) coprecipitation method was developed for this active material. The obtained morphology and elemental distribution are shown in Fig. 3.8a [80]. It has been also shown that using Na_2CO_3 as a precipitant improves the electronic conductivity of the electrode, which contributes to improved rate capability [283]. Calcium carbonate ($CaCO_3$) template was also used for the synthesis of this active material [276].

The chemical vapor deposition (CVD) method with sucrose as a source of carbon for the coating of this active material was developed, allowing better uniformity, higher quality, and better electrochemical performance in comparison to the thermal evaporation method [82].

Nanoplates of this active material with exposed [010] active facets were synthesized in a polyol medium (ethylene glycol, EG, $C_2H_6O_2$), and these facets can lead to more channels for Li^+-ion migration, resulting in an improvement in their electrochemical performance [128].

The impact of the different morphologies on the electrochemical performance was studied, in which highly dense materials exhibit the best power performances in terms of volumetric figures as shown in Fig. 3.8b [22].

$xLi_2MnO_{3.(1-x)}LiNi_{1/3}Co_{1/3}Mn_{1/3}O_2$ ($x = 1/4$, $1/3$, and $1/2$) hollow microspheres have been fabricated (Fig. 3.8c) with high specific capacity and capacity retention for $x = 1/3$ [129]. Further, $xLi_2MnO_{3.(1-x)}LiNi_{1/3}Co_{1/3}Mn_{1/3}O_2$ was also synthesized by the molten salt method, and, in this case, $x = 0.4$ presents better electrochemical performance [267].

Structural stability and consequently battery performance by reducing capacity fade have been improved by doping/modifying with different elements such as sodium (Na) [27, 67, 133] as illustrated in Fig. 3.8d, aluminum (Al) and iron (Fe) [142], silicon

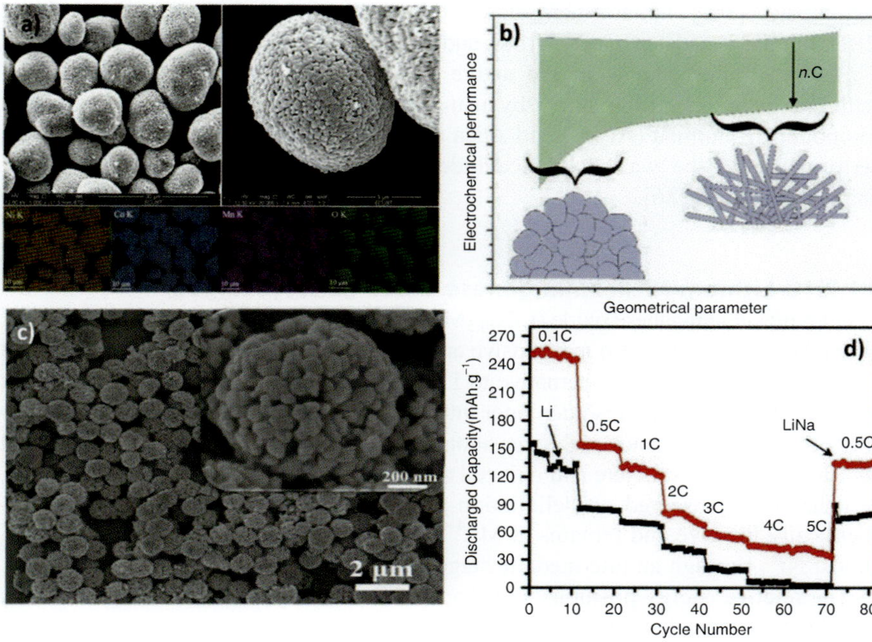

Fig. 3.8 (**a**) SEM-EDS analysis and elemental maps of this compound. (Reprinted from Ceramics International, 44, Li-Po He, Shu-Ying Sun, Jian-Guo Yu, Performance of $LiNi_{1/3}Co_{1/3}Mn_{1/3}O2$ prepared from spent lithium ion batteries by a carbonate co-precipitation method, 7, 2018, with permission from Elsevier [80]), (**b**) different morphologies versus electrochemical performance. (Reprinted from Journal of Power Sources, 346, Pierre-Etienne Cabelguen, David Peralta, Mikael Cugnet, Pascal Maillet, Impact of morphological changes of $LiNi_{1/3}Mn_{1/3}Co_{1/3}O2$ on lithium-ion cathode performances, 11, 2017, with permission from Elsevier [22]), and (**c**) SEM image of $xLi_2MnO_{3.(1-x)}LiNi_{1/3}Co_{1/3}Mn_{1/3}O_2$ microspheres. (Reproduced from Ref. [129] with permission of The Royal Society of Chemistry (RSC) on behalf of the Centre National de la Recherche Scientifique (CNRS) and the RSC. [129]) and (**d**) cycling performance of this active material doped with Na element. (Reprinted from Journal of Power Sources, 247, Chunxia Gong, Weixin Lv, Limin Qu, Oluwatosin Emmanuel Bankole, Guanghua Li, Rui Zhang, Meng Hu, Lixu Lei, Syntheses and electrochemical properties of layered $Li0.95Na0.05Ni_{1/3}Co_{1/3}Mn_{1/3}O2$ and $LiNi1/3Co_{1/3}Mn_{1/3}O2$, 5, 2014, with permission from Elsevier [67])

(Si) [179], phosphate $(PO_4)^{3-}$ [38], chromium(III) oxide (Cr_2O_3) [79], ZrO_2 [85], lanthanum oxide (La_2O_3) [94, 224], alumina (Al_2O_3) [110, 190], titania (TiO_2) [122], aluminum phosphate ($AlPO_4$) [130], manganese oxide (MnO_x) [258], lithium manganese oxide ($LiMn_2O_4$) [262], zirconia (ZrO_2) [263], fluoroborate glass ($MO–MF_2–B_2O_3$, where M = metal like Pb) [193], lithium titanium phosphate [$LiTi_2(PO_4)_3$] [279], and europium (III) oxide (Eu_2O_3) [290] coatings. These different coatings also suppress the increase of charge transfer resistance (Rct). Improved performance has been also achieved in cathode composites of this active material with lithium iron phosphate ($LiFePO_4$) [150], vanadium(V) oxide V_2O_5 to improve electrochemical performance [162], lithium manganese oxide ($LiMn_2O_4$) to improve rate capability [181], lithium vanadate ($LiVO_3$) [184], and polypyrrole (PPy, $[C_4H_2NH)]_n$) [273].

Beyond improvements in the active material, scientific advances have been achieved in the electrode slurry by using water in its production. Thus, for this compound, γ-alumina (Al_2O_3) and silica (SiO_2) significantly enhance the electrode performance of the water-based cathode [159] as well as new binders such as polyacrylic latex [286].

3.1.2.10 Lithium Nickel Manganese Oxide ($LiNi_{0.5}Mn_{0.5}O_2$)

Lithium nickel manganese oxide ($LiNi_{0.5}Mn_{0.5}O_2$) shows a high theoretical capacity of 280 $mAh.g^{-1}$, and it is also inexpensive and nontoxic and shows thermal stability and cycle life being a possible alternative to $LiCoO_2$ [161]. It shows some disadvantages such as substantial Li/Ni disorder, poor intrinsic rate capability, and structural impurity [102]. The oxidation states of the transition metals in this active material are Mn^{4+} and Ni^{2+} in which their structure and unit cell are shown in Fig. 3.9 [44].

It has been observed on delithiation of $Li_{1-x}Ni_{0.5}Mn_{0.5}O_2$ that Mn is electrochemically inactive and remains at Mn^{4+}, whereas the Ni is oxidized from Ni^{2+} to almost Ni^{4+} through an intermediate stage of Ni^{3+} [44].

The relative energy positions and shape of the density of states of O_{2p} and Ni_{3d} have been investigated for this active material, and it was verified that O_{2p} electrons are itinerant and exist in the vicinity of the Fermi energy more than Ni_{3d} electrons [207].

The electrode/electrolyte interface of this active material was investigated, and after one electrochemical cycle, it demonstrated the strong influence of the variation of potential [53].

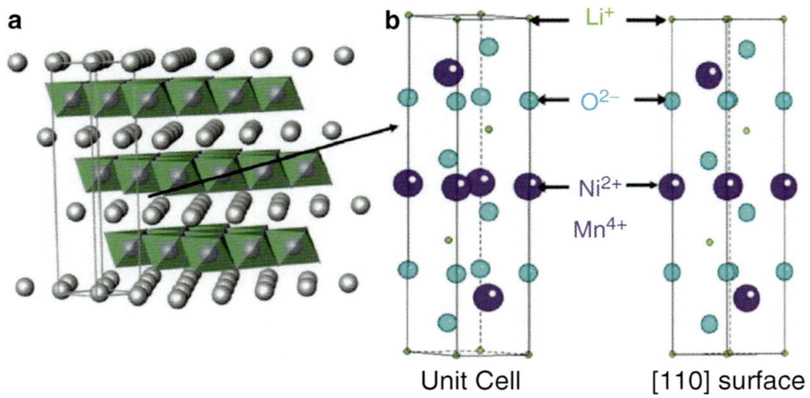

Fig. 3.9 (**a**) Structure and (**b**) unit cell of $LiNi_{0.5}Mn_{0.5}O_2$. (Reprinted from Journal of Applied Physics, Local structure of $LiNi_{0.5}Mn_{0.5}O2$ cathode material probed by in situ x-ray absorption spectroscopy, Aniruddha Deb, Uwe Bergmann, Stephen P. Cramer, et al, Mar 15, 2006, Volume 99, Issue with the permission of AIP Publishing [44])

Lithium nickel manganese oxide ($LiNi_{0.5}Mn_{0.5}O_2$) has been synthesized by different methods such as solid solution [161], freeze drying [214], in situ conversion route [148], coprecipitation [51] [208], combination of hydroxide and molten salt methods [92], wet chemical [222], solid-state reaction [246], hydrothermal method [281], and ultrasonic irradiation in coprecipitation method [272].

For the coprecipitation method, a morphology composed of hexagonal cubes was obtained, which improves the electrochemical performance [51].

The synthesis of this active material through ball milling was proposed, leading to materials with very small capacity fading [253]. This structure was doped with fluorine (F) element to reduce cation mixing and improve the poor rate performance [86]. The same was observed by ruthenium (Ru) doping of $LiNi_{0.5}Mn_{0.5}O_2$ [222].

Composites of $LiNi_{0.5}Mn_{0.5}O_2$ with lithium cobalt oxide ($LiCoO_2$) and lithium manganese oxide (Li_2MnO_3) were prepared by spray drying in which the discharge capacity increases with increasing cobalt (Co) content [219]. Similarly, a composite cathode with lithium manganese alumina ($LiMn_{1.9}Al_{0.1}O_4$) was developed, leading to the improvement in the electrochemical performance through structural stabilization [229].

3.1.2.11 Lithium Vanadium Oxide (LiV_2O_5)

Lithium vanadium oxide (LiV_2O_5) has the advantages of higher energy density (~450 $Wh.kg^{-1}$) [47], long cycle, and low environmental impact, but it has as a disadvantage a high capacity fading resulting from structural modifications [10]. Their structure was reported to be the space group $Pna2_1$ [C_{2V}] but refined in $Pn2_1a$ where the Li atom has a crude octahedral coordination with one distance [to 0 (6)] considerably lower than the others [2].

The electronic properties of α-$Li_xV_2O_5$ ($x = 0.5$ and 1) were investigated through density functional theory (DFT) in which the intercalation of Li into V_2O_5 does not change the electron transition property of V_2O_5, which is an indirect band gap semiconductor [120].

The ω-phase of LiV_2O_5 consists in the stability of its tetragonal structure ($\Delta V/V < 1\%$) as the Li insertion/extraction proceeds and, at a higher rate above of C/5, the capacity decreases through a significant dissolution process of the vanadium oxide, leading to the presence of V^{4+} and V^{3+} species in the electrolyte [118].

Through NMR spectra of the fully discharged $Li_xV_2O_5$ cathode, it is observed a changing of V ion from V^{+5} to V^{+4}. The lithium intercalated $Li_xV_2O_5$ system has been studied using ab initio calculations in which different phases were obtained during the reaction ((omega) ω- $Li_3V_2O_5$ and (zeta) ζ-$Li_2V_2O_5$) [199]. For intercalation amounts larger than one, the (delta) δ-phase is transformed into a (gamma) γ-one via an irreversible reconstruction mechanism [48].

The lithium insertion/extraction of this active material in the 4–2.15 V potential range was analyzed through Raman micro-spectrometry, being detected the emergence of various $Li_xV_2O_5$ phases in the wide Li composition range $0 < x < 2$ [10].

This electrode was prepared using the plasma-enhanced chemical vapor deposition (PECVD) technique in which the optimized substrate temperature and radio-frequency (RF) power are 250 °C and 50 W, respectively [141].

The carbothermal reduction method was used for synthesizing this active material that demonstrated a high reversible specific capacity of 130 $mAh.g^{-1}$ [14].

Further, this active material was synthesized by the sol-gel method; it shows nanostructure and is capable of reversible cycling [23].

(gamma) γ-LiV_2O_5 was synthesized via a simple solvothermal method and exhibits a better electrochemical performance [127].

Three-layered $Al_2O_3/Li_xV_2O_5/LiV_3O_8$ was prepared, and the $Li_xV_2O_5$ middle layer contributes to the improvement of interfacial electrochemical properties of the hybrid electrode [221].

3.1.2.12 LiV_3O_8

Lithium vanadium(V) oxide (LiV_3O_8) is an active material very interesting for rechargeable batteries considering its high specific energy, long cycle, and low cost [238]. In addition to its properties and application as cathodic material, it has been also investigated for the anode, but with low capacity value [113].

Its specific capacity of about 280 $mAh.g^{-1}$ has been achieved corresponding to the reversible intercalation of three equivalents of Li per mol $Li_{1+x}V_3O_8$ [157].

The atomic structures and the phase transition for $Li_{1-x}V_3O_8$ were calculated through the local density functional theory as a function of lithiation and the structure of $1 + x = 1.2$ and $1 + x = 4$ coincide with the predicted low-energy configurations [16]. Its lithium insertion kinetics was affected largely by the small diffusivity of Li^+ in the $Li_4V_3O_8$ phase formed for $x > 1.5$ upon lithiation [109].

The diffusion coefficient for this active material is relatively low ($\sim 10^{-12}$–10^{-13} cm^2 s^{-1}), and it is observed that it decreases with the intercalation of Li^+, which slows down the mobility of the Li^+ ion [20]. During the first lithium electrochemical insertion/extraction process in the electrode surface, it is observed the formation of LiF as well as compounds containing phosphorus originated by solid electrolyte interface (SEI) formation [34, 37].

The effect of oxygen defects in the structure has been analyzed and it has been observed the formation of new sites available for the occupation of Li^+ ions [108].

Their final properties of the active material depend on the conditions and methods of synthesis, which include sol-gel (citric acid, $C_6H_8O_7$, and (hydrogen) peroxide, H_2O_2) [57, 238], solution reaction [54], solid state [57], hydrothermal [57, 144], aqueous precipitation [57], mechanochemical [115], microwave [248], soft chemical via a low-temperature heating process [58], spray pyrolysis [96], and solid-state routine with quenching in freezing atmosphere [147].

The influence of the synthesis method on its electrochemical performance was evaluated, and it has been verified that the active material produced from the sol-gel citric acid method and the aqueous precipitation route presents excellent results [57].

The polymer polyvinylpyrrolidone was used in the polymer precursor method for synthesizing this active material, leading to good cycling stability [203]. Another polymer such as polyethylene glycol (PEG, $C_{2n}H_{4n}$+$2O_{n+1}$) was used in its synthesis [220].

The obtained morphologies are nanobelts (Fig. 3.10a) [70], nanorods [287], spheres [96] (Fig. 3.10b), and flake like [58].

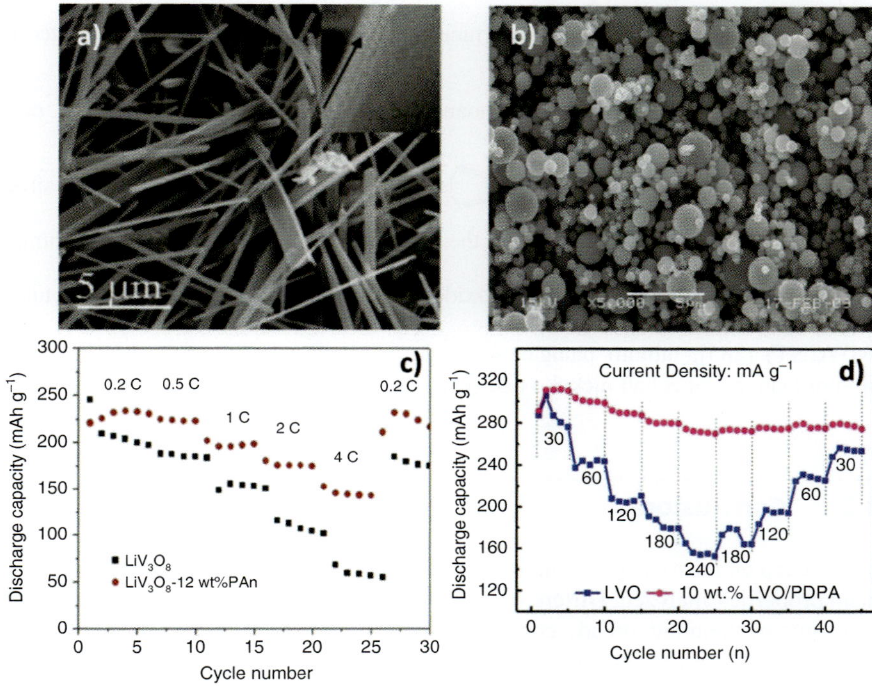

Fig. 3.10 (**a**) SEM image of $Li_{1+x}V_3O_8$ nanobelts. (Reprinted by permission from Springer Nature, Journal of Sol-Gel Science and Technology Facile preparation and electrochemical properties of large-scale $Li_{1x}V_3O_8$ nanobelts, Yuanxiang Gu, Fangfang Jian, Volume 46, Issue 2, (2008) [70]) and (**b**) spheres. (Reprinted from Materials Chemistry and Physics, 126, S.H. Ju, Y.C. Kang, Effect of preparation temperature on the morphology, crystal structure and electrochemical properties of LiV_3O_8 powders prepared by spray pyrolysis, 5, 2011 with permission from Elsevier [96]). Cycling performance of LiV_3O_8 with (**c**) polyaniline . (Reprinted from Journal of Power Sources, 220, Xuan-Wen Gao, Jia-Zhao Wang, Shu-Lei Chou, Hua-Kun Liu, Synthesis and electrochemical performance of LiV_3O_8/polyaniline as cathode material for the lithium battery, 7, 2012 with permission from Elsevier [64]) and (**d**) polydiphenylamine . (Reprinted with permission from Ref [292]. Copyright (2018) American Chemical Society)

It has been demonstrated that the nanorod morphology with high crystallinity greatly improves the stability of the crystallographic structure during cycling [146].

For improving electrical conductivity and cycling performance, this compound was doped of with manganese [107], molybdenum (Mo) [217], chromium [59], chlorine [125], and titanium [99] and prepared in composites with polypyrrole [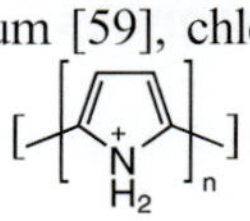] [56], polyaniline [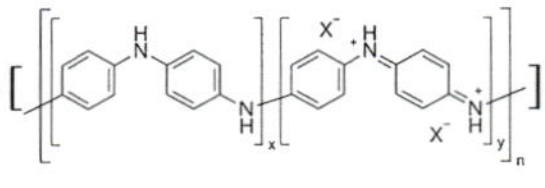] [64] (as it is shown in Fig. 3.10c), polythiophene

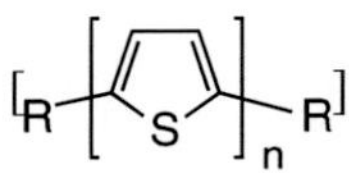 [138], polydiphenylamine 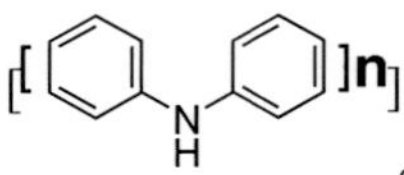 (capacity presented in Fig. 3.10d [293], poly(1,5-diaminoanthraquinone) 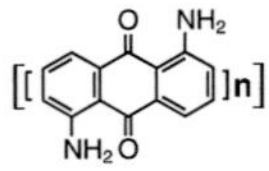 [131], poly (3, 4-ethylenedioxythiophene) 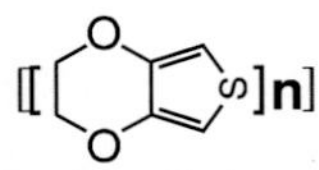 [291], carbon nanosheet [88], silver (Ag) nanoparticles anchored in their nanobelts [138], polytriphenylamine 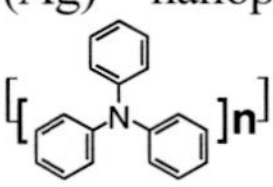 [131], reduced graphene oxide (GO) [167], multiwalled carbon nanotubes (MWCNT) [285], lithium manganese oxide ($LiMn_2O_4$) [280], and core-shell with lithium manganese cobalt nickel oxide ($Li_{1.2}Mn_{0.54}Co_{0.13}Ni_{0.13}O_2$) [223].

3.2 Conclusion

In this chapter, the main promising cathode active materials were presented and their main properties analyzed. Different kinds of materials are used as a cathode material, and some of them are already commercialized. A variety of active materials with a high range of specific capacities can be used on lithium-ion batteries for a different application. Different synthesis routes have been used to fabricate these materials including hydrothermal, sol-gel, precipitation, solid-state method, mechanochemical, microwave, soft chemical via a low-temperature heating process, and spray pyrolysis, among others. For each synthesis procedure, it is possible to control one or more specific particle properties and therefore reaching a better and improved cathode material. It is also possible to state that nanosize and carbon coating has a crucial importance in the electrochemical performance. Another possible treatment is doping of the active material, allowing to overcome some intrinsic drawbacks of the pure active materials.

Acknowledgments This work was supported by the Portuguese Foundation for Science and Technology (FCT) in the framework of the Strategic Funding UID/FIS/04650/2013 project PTDC/CTM-ENE/5387/2014, project no. PTDC/FISMAC/28157/2017, and grant SFRH/BPD/112547/2015 (C.M.C.). The authors thank the Basque Government Industry Department under the ELKARTEK and HAZITEK Programs for its financial support.

References

1. D.S. Adipranoto et al., Crystal-local structure analyses for cathode LIBs LiNi1−xCoxO2 (0 ≤ x ≤ 1) by neutron diffraction, in *Proceedings of the 2nd International Symposium on Science at J-PARC? Unlocking the Mysteries of Life, Matter and the Universe? JPS Conference Proceedings*, vol. 8. Journal of the Physical Society of Japan (2015). https://doi.org/10.7566/JPSCP.8.031017

2. D.N. Anderson, R.D. Willett, Refinement of the structure of LiV_2O_5. Acta Crystallogr. B **27**, 1476–1477 (1971). https://doi.org/10.1107/S0567740871004199
3. B. Andriyevsky, K. Doll, T. Jacob, Electronic and transport properties of $LiCoO_2$. Phys. Chem. Chem. Phys. **16**, 23412–23420 (2014). https://doi.org/10.1039/c4cp03052d
4. V. Aravindan, J. Gnanaraj, Y.-S. Lee, S. Madhavi, $LiMnPO_4$ – a next-generation cathode material for lithium-ion batteries. J. Mater. Chem. A **1**, 3518–3539 (2013). https://doi.org/10.1039/C2TA01393B
5. A.R. Armstrong, P.G. Bruce, Synthesis of layered $LiMnO_2$ as an electrode for rechargeable lithium batteries. Nature **381**, 499–500 (1996). https://doi.org/10.1038/381499a0
6. T. Asano, S. Yubuchi, A. Sakuda, A. Hayashi, M. Tatsumisago, Electronic and ionic conductivities of $LiNi_{1/3}Mn_{1/3}Co_{1/3}O_2$-$Li_3PS_4$ positive composite electrodes for all-solid-state lithium batteries. J. Electrochem. Soc. **164**, A3960–A3963 (2017). https://doi.org/10.1149/2.1501714jes
7. J. Auvergniot, A. Cassel, J.-B. Ledeuil, V. Viallet, V. Seznec, R. Dedryvère, Interface stability of argyrodite Li6PS5Cl toward $LiCoO_2$, $LiNi_{1/3}Co_{1/3}Mn_{1/3}O_2$, and $LiMn_2O_4$ in bulk all-solid-state batteries. Chem. Mater. **29**, 3883–3890 (2017). https://doi.org/10.1021/acs.chemmater.6b04990
8. T. Azib, F. Le Cras, H. Porthault, Direct fabrication of $LiCoO_2$ thin-films in water-ethanol solutions by electrochemical-hydrothermal method. Electrochim. Acta **160**, 145–151 (2015). https://doi.org/10.1016/j.electacta.2015.01.091
9. J.P. Baboo et al., Facile redox synthesis of layered $LiNi_{1/3}Co_{1/3}Mn_{1/3}O_2$ for rechargeable Li-ion batteries. Electrochim. Acta **224**, 243–250 (2017). https://doi.org/10.1016/j.electacta.2016.12.050
10. R. Baddour-Hadjean, A. Marzouk, J.P. Pereira-Ramos, Structural modifications of $Li_xV_2O_5$ in a composite cathode ($0 \leq x < 2$) investigated by Raman micro spectrometry. J. Raman Spectrosc. **43**, 153–160 (2012). https://doi.org/10.1002/jrs.2984
11. N. Bai, H. Chen, W. Zhou, K. Xiang, Y. Zhang, C. Li, H. Lu, Preparation and electrochemical performance of $LiFePO_4$/C microspheres by a facile and novel co-precipitation. Electrochim. Acta **167**, 172–178 (2015). https://doi.org/10.1016/j.electacta.2015.03.163
12. T. Balakrishnan, N. Sankarasubramanian, A. Kavitha, A. Kathalingam, Studies on structural, optical, electrical and morphological properties of $LiCoO_2$ thin films prepared by sol-gel method. Mater. Res. Innov., 1–6 (2018). https://doi.org/10.1080/14328917.2018.1431190
13. L. Bao et al., Hydrothermal synthesis of flower-like $LiMnPO_4$ nanostructures self-assembled with (010) nanosheets and their application in Li-ion batteries. CrystEngComm **17**, 6399–6405 (2015). https://doi.org/10.1039/C5CE01253H
14. J. Barker, M.Y. Saidi, J.L. Swoyer, Performance evaluation of the electroactive material, γ LiV_2O_5, made by a carbothermal reduction method. J. Electrochem. Soc. **150**, A1267–A1272 (2003). https://doi.org/10.1149/1.1600462
15. P.T. Barton, P.Y. Daniel, P.A. Chater, S.J. Ram, M.J. Rosseinsky, Chemical inhomogeneity, short-range order, and magnetism in the $LiNiO_2$-NiO solid solution. Chem. Eur. J. **19**, 14521–14531 (2013). https://doi.org/10.1002/chem.201301451
16. R. Benedek, M.M. Thackeray, L.H. Yang, First-principles calculation of atomic structure and electrochemical potential of Li_1+xV_3O_8. J. Power Sources **81–82**, 487–490 (1999). https://doi.org/10.1016/S0378-7753(99)00208-6
17. S. Beninati, L. Damen, M. Mastragostino, Fast sol-gel synthesis of $LiFePO_4$/C for high power lithium-ion batteries for hybrid electric vehicle application. J. Power Sources **194**, 1094–1098 (2009). https://doi.org/10.1016/j.jpowsour.2009.06.035
18. X. Bie et al., Spin glass transition in the rhombohedral $LiNi_{1/3}Co_{1/3}Mn_{1/3}O_2$. J. Alloys Compd. **572**, 79–83 (2013). https://doi.org/10.1016/j.jallcom.2013.03.242
19. G.E. Blomgren, The development and future of lithium-ion batteries. J. Electrochem. Soc. **164**, A5019–A5025 (2017). https://doi.org/10.1149/2.0251701jes
20. F. Bonino, S. Panero, M. Pasquali, G. Pistoia, Rechargeable lithium batteries based on Li_1+xV_3O_8 thin films. J. Power Sources **56**, 193–196 (1995). https://doi.org/10.1016/0378-7753(95)80033-D
21. A. Burukhin, O. Brylev, P. Hany, B.R. Churagulov, Hydrothermal synthesis of $LiCoO_2$ for lithium rechargeable batteries. Solid State Ionics **151**, 259–263 (2002). https://doi.org/10.1016/S0167-2738(02)00721-X

22. P.-E. Cabelguen, D. Peralta, M. Cugnet, P. Maillet, Impact of morphological changes of $LiNi_{1/3}Co_{1/3}Mn_{1/3}O_2$ on lithium-ion cathode performances. J. Power Sources **346**, 13–23 (2017). https://doi.org/10.1016/j.jpowsour.2017.02.025
23. S. Caes et al., Mesoporous lithium vanadium oxide as a thin film electrode for lithium-ion batteries: comparison between direct synthesis of LiV_2O_5 and electrochemical lithium intercalation in V_2O_5. J. Mater. Chem. A **2**, 5809–5815 (2014). https://doi.org/10.1039/C4TA00090K
24. Ş.E. Can Şener, J.L. Sharp, A. Anctil, Factors impacting diverging paths of renewable energy: a review. Renew. Sust. Energ. Rev. **81**, 2335–2342 (2018). https://doi.org/10.1016/j.rser.2017.06.042
25. R.J. Cava, A. Santoro, D.W. Murphy, S.M. Zahurak, R. Fleming, P. Marsh, R.S. Roth, The structure of the lithium-inserted metal oxide LiV_2O_5. J. Solid State Chem. **65**, 63–71 (1986). https://doi.org/10.1016/0022-4596(86)90089-7
26. G. Ceder, S.K. Mishra, Stability of orthorhombic and monoclinic-layered $LiMnO_2$. Electrochem. Solid-State Lett. **2**, 550–552 (1999). https://doi.org/10.1149/1.1390900
27. Z. Chen, T. Xie, L. Li, M. Xu, H. Zhu, W. Wang, Characterization of Na-substituted $LiNi_{1/3}Co_{1/3}Mn_{1/3}O_2$ cathode materials for lithium-ion battery. Ionics **20**, 629–634 (2014). https://doi.org/10.1007/s11581-013-1022-y
28. Z. Chen, M. Xu, B. Du, H. Zhu, T. Xie, W. Wang, Morphology control of lithium iron phosphate nanoparticles by soluble starch-assisted hydrothermal synthesis. J. Power Sources **272**, 837–844 (2014). https://doi.org/10.1016/j.jpowsour.2014.09.019
29. B. Chen, L. Ben, H. Yu, Y. Chen, X. Huang, Understanding surface structural stabilization of the high-temperature and high-voltage cycling performance of Al^{3+}-modified $LiMn_2O_4$ cathode material. ACS Appl. Mater. Interfaces **10**, 550–559 (2018). https://doi.org/10.1021/acsami.7b14535
30. C. Cherkouk, T. Nestler, Cathodes – a technological review. AIP Conf. Proc. **1597**, 134–145 (2014). https://doi.org/10.1063/1.4878484
31. J. Cho, K. Yong Jeong, T.J. Kim, B. Park, Enhanced structural stability of o-$LiMnO_2$ by a sol-gel coating of Al_2O_3. Chem. Mater. **13**, 18–20 (2001). https://doi.org/10.1021/cm000759+
32. J. Cho, T.J. Kim, B. Park, The effect of a metal-oxide coating on the cycling behavior at 55°C in orthorhombic $LiMnO_2$ cathode materials. J. Electrochem. Soc. **149**, A288–A292 (2002). https://doi.org/10.1149/1.1446870
33. E. Cho, S.-W. Seo, K. Min, Theoretical prediction of surface stability and morphology of $LiNiO_2$ cathode for Li-ion batteries. ACS Appl. Mater. Interfaces **9**, 33257–33266 (2017). https://doi.org/10.1021/acsami.7b08563
34. H.C. Choi, S.B. Kim, Y.M. Jung, Application of soft X-ray absorption spectroscopy and two-dimensional correlation spectroscopy to the electrochemical reaction in the $Li_1{+}xV_3O_8$/Li cell. J. Mol. Struct. **799**, 91–95 (2006). https://doi.org/10.1016/j.molstruc.2006.03.022
35. D. Choi, X. Li, W.A. Henderson, Q. Huang, S.K. Nune, J.P. Lemmon, V.L. Sprenkle, $LiCoPO_4$ cathode from a $CoHPO_4{\cdot}xH_2O$ nanoplate precursor for high voltage Li-ion batteries. Heliyon **2**(2), e00081 (2016). https://doi.org/10.1016/j.heliyon.2016.e00081
36. T.L. Christiansen, E.D. Bojesen, M. Sondergaard, S. Birgisson, J. Becker, B.B. Iversen, Crystal structure, microstructure and electrochemical properties of hydrothermally synthesized $LiMn_2O_4$. CrystEngComm **18**, 1996–2004 (2016). https://doi.org/10.1039/c5ce02358k
37. H. Chul Choi, M. Kyu Lee, H. Joon Shin, S. Bin Kim, Investigation of the $Li_1{+}xV_3O_8$ electrode–organic electrolyte interface by scanning photoelectron microscopy. J. Electron Spectrosc. Relat. Phenom. **130**, 85–96 (2003). https://doi.org/10.1016/S0368-2048(03)00087-2
38. L. Cong et al., $(PO_4)_3^-$ polyanions doped $LiNi_{1/3}Co_{1/3}Mn_{1/3}O_2$: an ultrafast-rate, long-life and high-voltage cathode material for Li-ion rechargeable batteries. Electrochim. Acta **201**, 8–19 (2016). https://doi.org/10.1016/j.electacta.2016.03.088
39. K.R. Crompton, M.P. Hladky, J.W. Staub, B.J. Landi, Enhanced overdischarge stability of $LiCoO_2$ by a solution deposited $AlPO_4$ coating. J. Electrochem. Soc. **164**, A3214–A3219 (2017). https://doi.org/10.1149/2.1171713jes

40. Y. Cui, C. Yang, Z. Zhuang, M. Wang, Q. Zhuang, Synthesis and electrochemical performance of spheroid $LiNi_{1/3}Co_{1/3}Mn_{1/3}O_2$ in the electrolyte modified by ethylene sulfate and methylene methanedisulfonate. J. Inorg. Organomet. Polym. Mater. **28**, 731–737 (2018). https://doi.org/10.1007/s10904-017-0722-6
41. R. Dang et al., Lithium-ion conductor and electronic conductor Co-coating modified layered cathode material $LiNi_{1/3}Co_{1/3}Mn_{1/3}O_2$. Electrochim. Acta **247**, 443–450 (2017). https://doi.org/10.1016/j.electacta.2017.07.041
42. C. Daniel, D. Mohanty, J. Li, D.L. Wood, Cathode materials review. AIP Conf. Proc. **1597**, 26–43 (2014). https://doi.org/10.1063/1.4878478
43. I.J. Davidson, R.S. McMillan, J.J. Murray, J.E. Greedan, Lithium-ion cell based on orthorhombic $LiMnO_2$. J. Power Sources **54**, 232–235 (1995). https://doi.org/10.1016/0378-7753(94)02074-D
44. A. Deb, U. Bergmann, S.P. Cramer, E.J. Cairns, Local structure of $LiNi_{1/2}Mn_{1/2}O_2$ cathode material probed by in situ x-ray absorption spectroscopy. J. Appl. Phys. **99**, 063701 (2006). https://doi.org/10.1063/1.2179198
45. C. Delacourt, C. Wurm, P. Reale, M. Morcrette, C. Masquelier, Low-temperature preparation of optimized phosphates for Li-battery applications. Solid State Ionics **173**, 113–118 (2004). https://doi.org/10.1016/j.ssi.2004.07.061
46. C. Delmas, I. Saadoune, Electrochemical and physical properties of the $Li_xNi_{1-y}Co_yO_2$ phases. Solid State Ionics **53–56**, 370–375 (1992). https://doi.org/10.1016/0167-2738(92)90402-B
47. C. Delmas, S. Brèthes, M. Ménétrier, ω-$Li_xV_2O_5$ – a new electrode material for rechargeable lithium batteries. J. Power Sources **34**, 113–118 (1991). https://doi.org/10.1016/0378-7753(91)85030-Z
48. C. Delmas, H. Cognac-Auradou, J.M. Cocciantelli, M. Ménétrier, J.P. Doumerc, The $LixV_2O_5$ system: an overview of the structural modifications induced by the lithium intercalation. Solid State Ionics **69**, 257–264 (1994). https://doi.org/10.1016/0167-2738(94)90414-6
49. X.-r. Deng, D.K. Hu G-r, Z.-D. Peng, X.-G. Gao, Y.-N. Yang, Synthesis and electrochemical properties of Co, Mn-coated $LiNiO_2$ lithium-ion battery cathode materials. Mater. Chem. Phys. **109**, 469–474 (2008). https://doi.org/10.1016/j.matchemphys.2007.12.025
50. B. Diouf, R. Pode, Potential of lithium-ion batteries in renewable energy. Renew. Energy **76**, 375–380 (2015). https://doi.org/10.1016/j.renene.2014.11.058
51. S. Dou, Q. Li, Synthesis and electrochemical properties of hexagonal sliced $LiNi_{1/2}Mn_{1/2}O_2$ as cathode materials for Li-ion batteries. Asian J. Chem. **26**(9), 2632–2636 (2014). https://doi.org/10.14233/ajchem.2014.15771
52. J. Dou, X. Kang, T. Wumaier, H. Yu, N. Hua, Y. Han, G. Xu, Effect of a lithium boron oxide glass coating on the electrochemical performance of $LiNi_{1/3}Co_{1/3}Mn_{1/3}O_2$. J. Solid State Electrochem. **16**, 1481–1486 (2012). https://doi.org/10.1007/s10008-011-1550-1
53. N. Dupré, J.-F. Martin, D. Guyomard, A. Yamada, R. Kanno, Characterization of interphases appearing on $LiNi_{1/2}Mn_{1/2}O_2$ using 7Li MAS NMR. J. Power Sources **189**, 557–560 (2009). https://doi.org/10.1016/j.jpowsour.2008.10.017
54. L. En-Hui, L. Xin-Hai, H. Zhao-Hui, H. Ze-Qiang, D. Ling-Feng, Chemical diffusion behaviors of lithium-ion in the LiV_3O_8 prepared by wet method reaction. Acta Phys.-Chim. Sin. **20**, 377–381 (2004). https://doi.org/10.3866/pku.whxb20040409
55. J. Fan, Studies of 18650 cylindrical cells made with doped $LiNiO_2$ positive electrodes for military applications. J. Power Sources **138**, 288–293 (2004). https://doi.org/10.1016/j.jpowsour.2004.06.010
56. C.Q. Feng, S.Y. Chew, Z.P. Guo, J.Z. Wang, H.K. Liu, An investigation of polypyrrole–LiV_3O_8 composite cathode materials for lithium-ion batteries. J. Power Sources **174**, 1095–1099 (2007). https://doi.org/10.1016/j.jpowsour.2007.06.190
57. J. Feng, X. Liu, X. Zhang, J. Jiang, J. Zhao, M. Wang, Effects of synthesis methods on Li_1+xV_3O_8 as cathodes in lithium-ion batteries. J. Electrochem. Soc. **156**, A768–A771 (2009a). https://doi.org/10.1149/1.3174386
58. Y. Feng, F. Hou, Y. Li, A new low-temperature synthesis and electrochemical properties of LiV_3O_8 hydrate as cathode material for lithium-ion batteries. J. Power Sources **192**, 708–713 (2009b). https://doi.org/10.1016/j.jpowsour.2009.02.079

59. Y. Feng, Y. Li, F. Hou, Preparation and electrochemical properties of Cr doped LiV_3O_8 cathode for lithium-ion batteries. Mater. Lett. **63**, 1338–1340 (2009c). https://doi.org/10.1016/j.matlet.2009.03.009
60. C.P. Fonseca, R.M. Paula, E.M.J.A. Pallone, S. Neves, A new approach to obtain lithium nickel cobalt oxide porous films. Electrochim. Acta **51**, 6419–6425 (2006). https://doi.org/10.1016/j.electacta.2006.04.026
61. L. Fransson, T. Eriksson, K. Edström, T. Gustafsson, J.O. Thomas, Influence of carbon black and binder on Li-ion batteries. J. Power Sources **101**, 1–9 (2001). https://doi.org/10.1016/S0378-7753(01)00481-5
62. Y. Fujii, H. Miura, N. Suzuki, T. Shoji, N. Nakayama, Structural and electrochemical properties of $LiNi_{1/3}Co_{1/3}Mn_{1/3}O_2$: calcination temperature dependence. J. Power Sources **171**, 894–903 (2007). https://doi.org/10.1016/j.jpowsour.2007.06.017
63. D. Fujimoto, Y. Lei, Z.-H. Huang, F. Kang, J. Kawamura, Synthesis and electrochemical performance of $LiMnPO_4$ by hydrothermal method. Int. J. Electrochem. **2014**, 9 (2014). https://doi.org/10.1155/2014/768912
64. X.-W. Gao, J.-Z. Wang, S.-L. Chou, H.-K. Liu, Synthesis and electrochemical performance of LiV_3O_8/polyaniline as cathode material for the lithium battery. J. Power Sources **220**, 47–53 (2012). https://doi.org/10.1016/j.jpowsour.2012.07.114
65. S. Gao et al., Sol-gel synthesis and electrochemical properties of c-axis oriented $LiCoO_2$ for lithium-ion batteries. RSC Adv. **5**, 51483–51488 (2015). https://doi.org/10.1039/c5ra06571b
66. H. Geaney, C. O'Dwyer, Examining the role of electrolyte and binders in determining discharge product morphology and cycling performance of carbon cathodes in Li–O_2 batteries. J. Electrochem. Soc. **163**, A43–A49 (2016). https://doi.org/10.1149/2.1011514jes
67. C. Gong et al., Syntheses and electrochemical properties of layered $Li_{0.95}Na_{0.05}Ni_{1/3}Co_{1/3}Mn_{1/3}O_2$ and $LiNi_{1/3}Co_{1/3}Mn_{1/3}O_2$. J. Power Sources **247**, 151–155 (2014). https://doi.org/10.1016/j.jpowsour.2013.08.081
68. J.B. Goodenough, K.-S. Park, The Li-ion rechargeable battery: a perspective. J. Am. Chem. Soc. **135**, 1167–1176 (2013). https://doi.org/10.1021/ja3091438
69. A. Gören, C.M. Costa, M.M. Silva, S. Lanceros-Méndez, State of the art and open questions on cathode preparation based on carbon-coated lithium iron phosphate. Compos. Part B **83**, 333–345 (2015). https://doi.org/10.1016/j.compositesb.2015.08.064
70. Y. Gu, F. Jian, Facile preparation and electrochemical properties of large-scale Li_1+xV_3O_8 nanobelts. J. Sol-Gel Sci. Technol. **46**, 161–165 (2008). https://doi.org/10.1007/s10971-008-1711-6
71. H.-B. Gu, B. Jin, D.-K. Jun, Z. Han, Improved electrochemical performance of $LiCoPO_4$ nanoparticles for lithium-ion batteries. J. Nanosci. Nanotechnol. **7**(11), 4037–4040 (2007). https://doi.org/10.1166/jnn.2007.078
72. J. Guan et al., Mechanochemical process enhanced cobalt and lithium recycling from wasted lithium-ion batteries. ACS Sustain. Chem. Eng. **5**, 1026–1032 (2017). https://doi.org/10.1021/acssuschemeng.6b02337
73. X. Guo, S. Greenbaum, F. Ronci, B. Scrosati, X-ray diffraction and 7Li nuclear magnetic resonance studies of iron- and cobalt-substituted $LiNiO_2$ prepared from inorganic transition metal nitrates. Solid State Ionics **168**, 37–49 (2004). https://doi.org/10.1016/j.ssi.2004.02.006
74. B.T. Habte, F. Jiang, Microstructure reconstruction and impedance spectroscopy study of $LiCoO_2$, $LiMn_2O_4$ and $LiFePO_4$ Li-ion battery cathodes. Microporous Mesoporous Mater. **268**, 69–76 (2018). https://doi.org/10.1016/j.micromeso.2018.04.001
75. X. Han, Q. Meng, T. Sun, J. Sun, Preparation and electrochemical characterization of single-crystalline spherical $LiNi_{1/3}Co_{1/3}Mn_{1/3}O_2$ powders cathode material for Li-ion batteries. J. Power Sources **195**, 3047–3052 (2010). https://doi.org/10.1016/j.jpowsour.2009.11.007
76. Y.-m. Han, Z.-f. Zhang, L.-b. Zhang, J.-h. Peng, M.-b. Fu, C. Srinivasakannan, J. Du, Influence of carbon coating prepared by microwave pyrolysis on properties of $LiNi_{1/3}Co_{1/3}Mn_{1/3}O_2$. Trans. Nonferrous Metals Soc. China **23**, 2971–2976 (2013). https://doi.org/10.1016/S1003-6326(13)62822-8

77. L. Han, C. Yu, Q. Wang, S. Guo, K. Xu, W. Zhang, R. Yang, Mechanism of the enhanced electrochemical properties of $LiNi_{1/3}Co_{1/3}Mn_{1/3}O_2$ cathode materials by a pre-sintering process. Russ. J. Electrochem. **54**, 49–55 (2018). https://doi.org/10.1134/s1023193518010044
78. A.M. Hashem, A.E. Abdel-Ghany, H.M. Abuzeid, H. Ehrenberg, A. Mauger, H. Groult, C.M. Julien, $LiNi_{1/3}Mn_{1/3}Co_{1/3}O_2$ synthesized by sol-gel method: structure and electrochemical properties. ECS Trans. **50**, 91–96 (2013). https://doi.org/10.1149/05024.0091ecst
79. R. He, L. Zhang, M. Yan, Y. Gao, Z. Liu, Effects of Cr_2O_3-modified $LiNi_{1/3}Co_{1/3}Mn_{1/3}O_2$ cathode materials on the electrochemical performance of lithium-ion batteries. J. Mater. Sci. **52**, 4599–4607 (2017). https://doi.org/10.1007/s10853-016-0704-z
80. L.-P. He, S.-Y. Sun, J.-G. Yu, Performance of $LiNi_{1/3}Co_{1/3}Mn_{1/3}O_2$ prepared from spent lithium-ion batteries by a carbonate co-precipitation method. Ceram. Int. **44**, 351–357 (2018). https://doi.org/10.1016/j.ceramint.2017.09.180
81. P.S. Herle, B. Ellis, N. Coombs, L.F. Nazar, Nano-network electronic conduction in iron and nickel olivine phosphates. Nat. Mater. **3**, 147–152 (2004). https://doi.org/10.1038/nmat1063
82. Q. Hou, G. Cao, P. Wang, D. Zhao, X. Cui, S. Li, C. Li, Carbon coating nanostructured-$LiNi_{1/3}Co_{1/3}Mn_{1/3}O_2$ cathode material synthesized by chemical vapor deposition method for high performance lithium-ion batteries. J. Alloys Compd. **747**, 796–802 (2018). https://doi.org/10.1016/j.jallcom.2018.03.115
83. Y. Hou et al., Highly [010]-oriented self-assembled $LiCoPO_4$/C nanoflakes as high-performance cathode for lithium-ion batteries. Nano Res. **11**, 2424–2435 (2018). https://doi.org/10.1007/s12274-017-1864-0
84. G.R. Hu, G. Liao, Z.D. Peng, J. Xiao, X.L. Zhang, X.Y. Yu, Structure and electrochemical properties of $LiCoO_2$ synthesized by microwave heating. J. Cent. S. Univ. Technol. **11**, 261–264 (2004). https://doi.org/10.1007/s11771-004-0053-y
85. S.-K. Hu, G.-H. Cheng, M.-Y. Cheng, B.-J. Hwang, R. Santhanam, Cycle life improvement of ZrO_2-coated spherical $LiNi_{1/3}Co_{1/3}Mn_{1/3}O_2$ cathode material for lithium ion batteries. J. Power Sources **188**, 564–569 (2009). https://doi.org/10.1016/j.jpowsour.2008.11.113
86. C. Hu, J. Guo, J. Wen, Y. Peng, Y. Chen, Preparation and electrochemical performance of $LiNi_{1/2}Mn_{1/2}O_2-xF_x$ $(0 \leqslant x \leqslant 0.04)$ cathode material synthesized with hydroxide co-precipitation for lithium-ion batteries. J. Alloys Compd. **581**, 121–127 (2013). https://doi.org/10.1016/j.jallcom.2013.05.223
87. B. Huang, Y.-I. Jang, Y.-M. Chiang, D.R. Sadoway, Electrochemical evaluation of $LiCoO_2$ synthesized by decomposition and intercalation of hydroxides for lithium-ion battery applications. J. Appl. Electrochem. **28**, 1365–1369 (1998). https://doi.org/10.1023/a:1003443108681
88. N.H. Idris, M.M. Rahman, J.-Z. Wang, Z.-X. Chen, H.-K. Liu, Synthesis and electrochemical performance of LiV_3O_8/carbon nanosheet composite as cathode material for lithium-ion batteries. Compos. Sci. Technol. **71**, 343–349 (2011). https://doi.org/10.1016/j.compscitech.2010.11.025
89. M. Inagaki, H. Konno, O. Tanaike, Carbon materials for electrochemical capacitors. J. Power Sources **195**, 7880–7903 (2010). https://doi.org/10.1016/j.jpowsour.2010.06.036
90. B.J. Jacas, F. Cristina, S. Laura, M.J. Ramon, Static and dynamic studies on $LiNi_{1/3}Co_{1/3}Mn_{1/3}O_2$-based suspensions for semi-solid flow batteries. ChemSusChem **9**, 1938–1944 (2016). https://doi.org/10.1002/cssc.201600285
91. Y.I. Jang, B. Huang, H. Wang, D.R. Sadoway, Y.M. Chiang, Electrochemical cycling-induced spinel formation in high-charge-capacity orthorhombic $LiMnO_2$. J. Electrochem. Soc. **146**, 3217–3223 (1999). https://doi.org/10.1149/1.1392457
92. K.H. Jeong, H.-W. Ha, N.J. Yun, M.Z. Hong, K. Kim, Zr-doped $Li[Ni_{1/2}-xMn_{1/2}-xZr_{2x}]O_2$ (x=0, 0.025) as cathode material for lithium ion batteries. Electrochim. Acta **50**, 5349–5353 (2005). https://doi.org/10.1016/j.electacta.2005.03.014
93. Q. Jiang, D. Liu, H. Zhang, S. Wang, Plasma-assisted sulfur doping of $LiMn_2O_4$ for high-performance lithium-ion batteries. J. Phys. Chem. C **119**, 28776–28782 (2015). https://doi.org/10.1021/acs.jpcc.5b10298

94. Q. Jiang, P. Lang, J. Li, J. Tang, Improving the elevated-temperature behaviors of $LiNi_{1/3}Co_{1/3}Mn_{1/3}O_2$ by surface modification with Nano-La_2O_3. J. Alloys Compd. **742**, 549–554 (2018). https://doi.org/10.1016/j.jallcom.2018.01.354
95. E.M. Jin, B. Jin, Y.S. Jeon, K.H. Park, H.B. Gu, Electrochemical properties of $LiMnO_2$ for lithium polymer battery. J. Power Sources **189**, 620–623 (2009). https://doi.org/10.1016/j.jpowsour.2008.09.102
96. S.H. Ju, Y.C. Kang, Effect of preparation temperature on the morphology, crystal structure and electrochemical properties of LiV_3O_8 powders prepared by spray pyrolysis. Mater. Chem. Phys. **126**, 133–137 (2011). https://doi.org/10.1016/j.matchemphys.2010.11.050
97. C. Julien, A. Mauger, A. Vijh, K. Zaghib, *Lithium Batteries: Science and Technology* (Springer, New York, 2015)
98. C. Julien, A. Mauger, K. Zaghib, H. Groult, Optimization of layered cathode materials for lithium-ion batteries. Materials **9**, 595 (2016)
99. S. Junli et al., Synthesis and electrochemical performance of Ti^{4+} doped LiV_3O_8. Chin. J. Chem. **27**, 863–867 (2009). https://doi.org/10.1002/cjoc.200990145
100. M. Kageyama, D. Li, K. Kobayakawa, Y. Sato, Y.-S. Lee, Structural and electrochemical properties of $LiNi_{1/3}Co_{1/3}Mn_{1/3}O_2 - xF_x$ prepared by solid-state reaction. J. Power Sources **157**, 494–500 (2006). https://doi.org/10.1016/j.jpowsour.2005.08.002
101. H.-K. Kang, B.-Y. Jung, M.-G. Kim, Y. Lee, D.-M. Kim, K. Ryu, K.-S. Han, Direct thermal reaction for preparation of lithiated cathode materials without any artificial mixing procedures. Solid State Ionics **169**, 151–159 (2004). https://doi.org/10.1016/j.ssi.2003.08.059
102. K. Kang, Y.S. Meng, J. Bréger, C.P. Grey, G. Ceder, Electrodes with high power and high capacity for rechargeable lithium batteries. Science **311**, 977–980 (2006). https://doi.org/10.1126/science.1122152
103. W. Karino, Order of the transition metal layer in $LiNi_{1/3}Co_{1/3}Mn_{1/3}O_2$ and stability of the crystal structure. Ionics **22**, 991–995 (2016). https://doi.org/10.1007/s11581-016-1706-1
104. S. Karthik Prabhu, G. Hirankumar, A. Maheswaran, R.S. Daries Bella, C. Sanjeeviraja, Structural and electrical studies on Zn^{2+}doped $LiCoPO_4$. J. Electrost. **72**, 181–186 (2014). https://doi.org/10.1016/j.elstat.2014.02.001
105. J. Kasnatscheew et al., Learning from overpotentials in lithium-ion batteries: a case study on the $LiNi_{1/3}Co_{1/3}Mn_{1/3}O_2$ (NCM) cathode. J. Electrochem. Soc. **163**, A2943–A2950 (2016). https://doi.org/10.1149/2.0461614jes
106. J. Kasnatscheew, M. Evertz, B. Streipert, R. Wagner, S. Nowak, I. Cekic Laskovic, M. Winter, Changing established belief on capacity fade mechanisms: thorough investigation of $LiNi_{1/3}Co_{1/3}Mn_{1/3}O_2$ (NCM111) under high voltage conditions. J. Phys. Chem. C **121**, 1521–1529 (2017). https://doi.org/10.1021/acs.jpcc.6b11746
107. J. Kawakita, H. Katagiri, T. Miura, T. Kishi, Lithium insertion behaviour of manganese or molybdenum substituted Li_1+xV_3O_8. J. Power Sources **68**, 680–685 (1997). https://doi.org/10.1016/S0378-7753(96)02596-7
108. J. Kawakita, M. Majima, T. Miura, T. Kishi, Preparation and lithium insertion behaviour of oxygen-deficient Li_1+x$V_3O_8 - \delta$. J. Power Sources **66**, 135–139 (1997). https://doi.org/10.1016/S0378-7753(96)02540-2
109. J. Kawakita, T. Miura, T. Kishi, Lithium insertion and extraction kinetics of Li_1+xV_3O_8. J. Power Sources **83**, 79–83 (1999). https://doi.org/10.1016/S0378-7753(99)00270-0
110. Y. Kim, H.S. Kim, S.W. Martin, Synthesis and electrochemical characteristics of Al_2O_3-coated $LiNi_{1/3}Co_{1/3}Mn_{1/3}O_2$ cathode materials for lithium-ion batteries. Electrochim. Acta **52**, 1316–1322 (2006). https://doi.org/10.1016/j.electacta.2006.07.033
111. B.-H. Kim, J.-H. Kim, I.-H. Kwon, M.-Y. Song, Electrochemical properties of $LiNiO_2$ cathode material synthesized by the emulsion method. Ceram. Int. **33**, 837–841 (2007). https://doi.org/10.1016/j.ceramint.2006.01.021
112. A. Kinoshita, K. Yanagida, A. Yanai, Y. Kida, A. Funahashi, T. Nohma, I. Yonezu, Electrochemical characteristics of $LiNi_{1-x}Co_xO_2$ as positive electrode materials for lithium secondary batteries. J. Power Sources **102**, 283–287 (2001). https://doi.org/10.1016/S0378-7753(01)00806-0

113. J. Köhler, H. Makihara, H. Uegaito, H. Inoue, M. Toki, LiV_3O_8: characterization as anode material for an aqueous rechargeable Li-ion battery system. Electrochim. Acta **46**, 59–65 (2000). https://doi.org/10.1016/S0013-4686(00)00515-6
114. F. Kong, C. Liang, R.C. Longo, Y. Zheng, K. Cho, Atomic-scale understanding of non-stoichiometry effects on the electrochemical performance of Ni-rich cathode materials. J. Power Sources **378**, 750–758 (2018). https://doi.org/10.1016/j.jpowsour.2018.01.008
115. N.V. Kosova, S.V. Vosel, V.F. Anufrienko, N.T. Vasenin, E.T. Devyatkina, Reduction processes in the course of mechanochemical synthesis of Li_1+xV_3O_8. J. Solid State Chem. **160**, 444–449 (2001). https://doi.org/10.1006/jssc.2001.9272
116. K.J. Kreder, G. Assat, A. Manthiram, Aliovalent substitution of $V^{3+}$for Co^{2+}in $LiCoPO_4$ by a low-temperature microwave-assisted solvothermal process. Chem. Mater. **28**, 1847–1853 (2016). https://doi.org/10.1021/acs.chemmater.5b05042
117. Y.S. Lee, M. Yoshio, Preparation of orthorhombic $LiMnO_2$ material by quenching. Electrochem. Solid-State Lett. **4**, A166–A169 (2001). https://doi.org/10.1149/1.1399879
118. C. Leger, S. Bach, P. Soudan, J.-P. Pereira-Ramos, Structural and electrochemical properties of ωLi×V_2O_5 (0.4⩽x⩽3) as rechargeable cathodic material for lithium batteries. J. Electrochem. Soc. **152**, A236–A241 (2005). https://doi.org/10.1149/1.1836155
119. W. Li, J.C. Currie, Morphology effects on the electrochemical performance of $LiNi_{1-x}Co_xO_2$. J. Electrochem. Soc. **144**, 2773–2779 (1997). https://doi.org/10.1149/1.1837894
120. Z.Y. Li, Q.H. Wu, Electronic structures of $Li_xV_2O_5$ (x=0.5 and 1): a theoretical study. ChemPhysChem **9**, 300–304 (2008). https://doi.org/10.1002/cphc.200700560
121. W. Li, J.C. Currie, J. Wolstenholme, Influence of morphology on the stability of $LiNiO_2$. J. Power Sources **68**, 565–569 (1997). https://doi.org/10.1016/S0378-7753(96)02566-9
122. D. Li, Y. Kato, K. Kobayakawa, H. Noguchi, Y. Sato, Preparation and electrochemical characteristics of $LiNi_{1/3}Co_{1/3}Mn_{1/3}O_2$ coated with metal oxides coating. J. Power Sources **160**, 1342–1348 (2006). https://doi.org/10.1016/j.jpowsour.2006.02.080
123. D. Li, Y. Sasaki, K. Kobayakawa, H. Noguchi, Y. Sato, Preparation, morphology and electrochemical characteristics of $LiNi_{1/3}Co_{1/3}Mn_{1/3}O_2$ with LiF addition. Electrochim. Acta **52**, 643–648 (2006). https://doi.org/10.1016/j.electacta.2006.05.044
124. D. Li, Z. Peng, H. Ren, W. Guo, Y. Zhou, Synthesis and characterization of $LiNi_{1-x}Co_xO_2$ for lithium batteries by a novel method. Mater. Chem. Phys. **107**, 171–176 (2008). https://doi.org/10.1016/j.matchemphys.2007.06.069
125. L. Li, J. Lifang, S. Junli, L. Sichen, Y. Huatang, W. Yongmei, Electrochemical performance of LiV3O8−xClx cathode materials synthesized by a low-temperature solid state method. Chin. J. Chem. **27**, 1093–1098 (2009). https://doi.org/10.1002/cjoc.200990182
126. J. Li, C. Cao, X. Xu, Y. Zhu, R. Yao, $LiNi_{1/3}Co_{1/3}Mn_{1/3}O_2$ hollow nano-micro hierarchical microspheres with enhanced performances as cathodes for lithium-ion batteries. J. Mater. Chem. A **1**, 11848–11852 (2013). https://doi.org/10.1039/C3TA12375H
127. N. Li, G. Hx, Q. Yt, Orthorhombic γLiV_2O_5 as cathode materials in lithium-ion batteries: synthesis and property. Chin. J. Chem. Phys. **26**, 597–600 (2013). https://doi.org/10.1063/1674-0068/26/05/597-600
128. J. Li, R. Yao, C. Cao, $LiNi_{1/3}Co_{1/3}Mn_{1/3}O_2$ nanoplates with {010} active planes exposing prepared in polyol medium as a high-performance cathode for Li-ion battery. ACS Appl. Mater. Interfaces **6**, 5075–5082 (2014). https://doi.org/10.1021/am500215b
129. J. Li, M. Li, L. Zhang, J. Wang, General synthesis of xLi_2MnO_3[middle dot]$(_{1-x})$ $LiNi_{1/3}Co_{1/3}Mn_{1/3}O_2$ $(x = 1/4, 1/3, \text{and } 1/2)$ hollow microspheres towards enhancing the performance of rechargeable lithium-ion batteries. J. Mater. Chem. A **4**, 12442–12450 (2016). https://doi.org/10.1039/C6TA04219H
130. L. Li, Y. Cao, H. Zheng, C. Feng, $AlPO_4$ coated $LiNi_{1/3}Co_{1/3}Mn_{1/3}O_2$ for high performance cathode material in lithium batteries. J. Mater. Sci. Mater. Electron. **28**, 1925–1930 (2017). https://doi.org/10.1007/s10854-016-5745-1

131. W. Li, L. Zhu, Z. Yu, L. Xie, X. Cao, LiV_3O_8/polytriphenylamine composites with enhanced electrochemical performances as cathode materials for rechargeable lithium batteries. Materials **10**, 344 (2017)
132. X. Li et al., Improved rate capability of a $LiNi_{1/3}Co_{1/3}Mn_{1/3}O_2$/CNT/graphene hybrid material for Li-ion batteries. RSC Adv. **7**, 24359–24367 (2017). https://doi.org/10.1039/C7RA03438E
133. Y. Li, J. Liu, Y. Lei, C. Lai, Q. Xu, Enhanced electrochemical performances of Na-doped cathode material $LiNi_{1/3}Co_{1/3}Mn_{1/3}O_2$ for lithium-ion batteries. J. Mater. Sci. **52**, 13596–13605 (2017). https://doi.org/10.1007/s10853-017-1449-z
134. J. Li, S. Luo, X. Ding, Q. Wang, P. He, Three-dimensional honeycomb-structural $LiAlO_2$-modified $LiMnPO_4$ composite with superior high rate capability as Li-ion battery cathodes. ACS Appl. Mater. Interfaces **10**, 10786–10795 (2018). https://doi.org/10.1021/acsami.7b17597
135. L. Li, Y. Bian, X. Zhang, Q. Xue, E. Fan, F. Wu, R. Chen, Economical recycling process for spent lithium-ion batteries and macro- and micro-scale mechanistic study. J. Power Sources **377**, 70–79 (2018). https://doi.org/10.1016/j.jpowsour.2017.12.006
136. X. Li, Z. Su, Y. Wang, Electrochemical properties of monoclinic and orthorhombic $LiMnO_2$ synthesized by a one-step hydrothermal method. J. Alloys Compd. **735**, 2182–2189 (2018). https://doi.org/10.1016/j.jallcom.2017.11.384
137. F. Liang, Y. Yao, Y. Dai, B. Yang, W. Ma, T. Watanabe, Preparation of porous structure LiFePO4/C composite by template method for lithium-ion batteries. Solid State Ionics **214**, 31–36 (2012). https://doi.org/10.1016/j.ssi.2012.02.043
138. S. Liang et al., LiV_3O_8/Ag composite nanobelts with enhanced performance as cathode material for rechargeable lithium batteries. J. Alloys Compd. **583**, 351–356 (2014). https://doi.org/10.1016/j.jallcom.2013.08.124
139. D. Linden, T.B. Reddy, *Handbook of Batteries* (McGraw-Hill, New York, 2002)
140. G.W. Ling et al., Structural and thermal stabilities of spinel $LiMn_2O_4$ materials under commercial power batteries cycling and abusive conditions. Int. J. Electrochem. Sci. **7**, 2455–2467 (2012)
141. P. Liu, J.-G. Zhang, J.A. Turner, C.E. Tracy, D.K. Benson, R.N. Bhattacharya, Fabrication of LiV_2O_5 thin-film electrodes for rechargeable lithium batteries. Solid State Ionics **111**, 145–151 (1998). https://doi.org/10.1016/S0167-2738(98)00184-2
142. D. Liu, Z. Wang, L. Chen, Comparison of structure and electrochemistry of Al- and Fe-doped $LiNi_{1/3}Co_{1/3}Mn_{1/3}O_2$. Electrochim. Acta **51**, 4199–4203 (2006). https://doi.org/10.1016/j.electacta.2005.11.045
143. H. Liu, Q. Cao, L.J. Fu, C. Li, Y.P. Wu, H.Q. Wu, Doping effects of zinc on $LiFePO_4$ cathode material for lithium ion batteries. Electrochem. Commun. **8**, 1553–1557 (2006). https://doi.org/10.1016/j.elecom.2006.07.014
144. H. Liu, H. Yang, T. Huang, Synthesis, structure and electrochemical properties of one-dimensional nanometer materials LiV_3O_8. Mater. Sci. Eng. B **143**, 60–63 (2007). https://doi.org/10.1016/j.mseb.2007.07.010
145. H. Liu, P. Zhang, G.C. Li, Q. Wu, Y.P. Wu, $LiFePO_4$/C composites from carbothermal reduction method. J. Solid State Electrochem. **12**, 1011–1015 (2008). https://doi.org/10.1007/s10008-007-0478-y
146. H. Liu, Y. Wang, K. Wang, Y. Wang, H. Zhou, Synthesis and electrochemical properties of single-crystalline LiV_3O_8 nanorods as cathode materials for rechargeable lithium batteries. J. Power Sources **192**, 668–673 (2009). https://doi.org/10.1016/j.jpowsour.2009.03.066
147. Y. Liu, X. Zhou, Y. Guo, Structure and electrochemical performance of LiV_3O_8 synthesized by solid-state routine with quenching in freezing atmosphere. Mater. Chem. Phys. **114**, 915–919 (2009). https://doi.org/10.1016/j.matchemphys.2008.10.064
148. Y. Liu, B. Chen, F. Cao, X. Zhao, J. Yuan, Synthesis of nano-architectured $LiNi_{1/2}Mn_{1/2}O_2$ spheres for high-performance rechargeable lithium-ion batteries via an in situ conversion route. J. Mater. Chem. **21**, 10437–10441 (2011). https://doi.org/10.1039/C1JM10408J
149. Z. Liu, H. Zhen, Y. Kim, C. Liang, Synthesis of $LiNiO_2$ cathode materials with homogeneous Al doping at the atomic level. J. Power Sources **196**, 10201–10206 (2011). https://doi.org/10.1016/j.jpowsour.2011.08.059

150. L. Liu et al., Studies of the electrochemical properties and thermal stability of $LiNi_{1/3}Co_{1/3}Mn_{1/3}O_2/LiFePO_4$ composite cathodes for lithium-ion batteries. Ionics **20**, 1087–1093 (2014). https://doi.org/10.1007/s11581-013-1055-2
151. C. Liu, Z.G. Neale, G. Cao, Understanding electrochemical potentials of cathode materials in rechargeable batteries. Mater. Today **19**, 109–123 (2016). https://doi.org/10.1016/j.mattod.2015.10.009
152. Y. Liu et al., Formation of size-dependent and conductive phase on lithium iron phosphate during carbon coating. Nat. Commun. **9** (2018). https://doi.org/10.1038/s41467-018-03324-7
153. Y. Long, Z. Zhang, Z. Wu, J. Su, X. Lv, Y. Wen, Microwave-assisted polyol synthesis of $LiMnPO_4$/C and its use as a cathode material in lithium-ion batteries. Particuology **33**, 42–49 (2017). https://doi.org/10.1016/j.partic.2016.10.006
154. Q. Lu, G.S. Hutchings, Y. Zhou, H.L. Xin, H. Zheng, F. Jiao, Nanostructured flexible Mg-modified $LiMnPO_4$ matrix as high-rate cathode materials for Li-ion batteries. J. Mater. Chem. A **2**, 6368–6373 (2014). https://doi.org/10.1039/C4TA00654B
155. J. Ludwig, T. Nilges, Recent progress and developments in lithium cobalt phosphate chemistry – syntheses, polymorphism and properties. J. Power Sources **382**, 101–115 (2018). https://doi.org/10.1016/j.jpowsour.2018.02.038
156. J. Ludwig, C. Marino, D. Haering, C. Stinner, H.A. Gasteiger, T. Nilges, Morphology-controlled microwave-assisted solvothermal synthesis of high-performance $LiCoPO_4$ as a high-voltage cathode material for Li-ion batteries. J. Power Sources **342**, 214–223 (2017). https://doi.org/10.1016/j.jpowsour.2016.12.059
157. V. Manev, A. Momchilov, A. Nassalevska, G. Pistoia, M. Pasquali, A new approach to the improvement of Li_1+xV_3O_8 performance in rechargeable lithium batteries. J. Power Sources **54**, 501–507 (1995). https://doi.org/10.1016/0378-7753(94)02135-P
158. S.K. Martha et al., $LiMnPO_4$ as an advanced cathode material for rechargeable lithium batteries. J. Electrochem. Soc. **156**, A541–A552 (2009). https://doi.org/10.1149/1.3125765
159. M. Memm, A. Hoffmann, M. Wohlfahrt-Mehrens, Water-based $LiNi_{1/3}Co_{1/3}Mn_{1/3}O_2$-cathodes with good electrochemical performance by use of additives. Electrochim. Acta **260**, 664–673 (2018). https://doi.org/10.1016/j.electacta.2017.12.014
160. Y.S. Meng, G. Ceder, C.P. Grey, W.-S. Yoon, Y. Shao-Horn, Understanding the crystal structure of layered $LiNi_{1/2}Mn_{1/2}O_2$ by electron diffraction and powder diffraction simulation. Electrochem. Solid-State Lett. **7**, A155–A158 (2004). https://doi.org/10.1149/1.1718211
161. X. Meng, S. Dou, W.-l. Wang, High power and high capacity cathode material $LiNi_{1/2}Mn_{1/2}O_2$ for advanced lithium-ion batteries. J. Power Sources **184**, 489–493 (2008). https://doi.org/10.1016/j.jpowsour.2008.04.015
162. X. Meng, H. Cao, J. Hao, P. Ning, G. Xu, Z. Sun, Sustainable preparation of $LiNi_{1/3}Co_{1/3}Mn_{1/3}O_2 - V_2O_5$ cathode materials by recycling waste materials of spent lithium-ion battery and vanadium-bearing slag. ACS Sustain. Chem. Eng. **6**, 5797–5805 (2018). https://doi.org/10.1021/acssuschemeng.7b03880
163. H.-N. Migeon, M. Zanne, C. Gleitzer, J. Aubry, The Li_2O–NiO–O_2 system at 670°C and the consequences of non-stoichiometry on the magnetic properties of the $Li_xNi_{1-x}O_{1\pm y}$ phases. J. Mater. Sci. **13**, 461–466 (1978). https://doi.org/10.1007/bf00541793
164. Y. Mizuno et al., Electrospinning synthesis of wire-structured $LiCoO_2$ for electrode materials of high-power Li-ion batteries. J. Phys. Chem. C **116**, 10774–10780 (2012). https://doi.org/10.1021/jp2123482
165. Y. Mizuno et al., Fabrication of $LiCoO_2$ crystal layers using a flux method and their application for additive-free lithium-ion rechargeable battery cathodes. Cryst. Growth Des. **14**, 1882–1887 (2014). https://doi.org/10.1021/cg5000217
166. K. Mizushima, P.C. Jones, P.J. Wiseman, J.B. Goodenough, LixCoO2 ($0<x<-1$): A new cathode material for batteries of high energy density. Mater. Res. Bull. **15**(6), 783–799 (1980). https://doi.org/10.1016/0025-5408(80)90012-4
167. R. Mo, Y. Du, D. Rooney, G. Ding, K. Sun, Ultradispersed nanoarchitecture of LiV_3O_8 nanoparticle/reduced graphene oxide with high-capacity and long-life lithium-ion battery

cathodes. Sci. Rep. **6**, 19843 (2016). https://doi.org/10.1038/srep19843. https://www.nature.com/articles/srep19843#supplementary-information
168. M. Mogi et al., Combined in-situ X-ray absorption spectroscopy and first-principles calculation studies on local structural and electronic structural alternations of $LiNi_1/_3Co_1/_3Mn_1/_3O_2$. ECS Trans. **50**, 47–58 (2013). https://doi.org/10.1149/05026.0047ecst
169. P. Mohan, G. Paruthimal Kalaignan, Electrochemical behaviour of surface modified SiO_2-coated $LiNiO_2$ cathode materials for rechargeable lithium-ion batteries. J. Nanosci. Nanotechnol. **13** (2013). https://doi.org/10.1166/jnn.2013.7379
170. P. Mohan, G. Paruthimal Kalaignan, P.V.S. Muralidharan, Improved the electrochemical performance of $LiCe_{(1-x)}Ni_{(x)}O_2$ cathode material for rechargeable lithium-ion battery. Journal of Nanoscience and Nanotechnology **12** (2012). https://doi.org/10.1166/jnn.2012.6510
171. J. Mohtasham, Review article-renewable energies. Energy Procedia **74**, 1289–1297 (2015). https://doi.org/10.1016/j.egypro.2015.07.774
172. J. Molenda, P. Wilk, J. Marzec, Structural, electrical and electrochemical properties of $LiNiO_2$. Solid State Ionics **146**, 73–79 (2002). https://doi.org/10.1016/S0167-2738(01)00992-4
173. A. Momchilov, V. Manev, A. Nassalevska, A. Kozawa, Rechargeable lithium battery with spinel-related MnO_2 II. Optimization of the $LiMn_2O_4$ synthesis conditions. J. Power Sources **41**, 305–314 (1993). https://doi.org/10.1016/0378-7753(93)80048-T
174. A. Moradabadi, P. Kaghazchi, Mechanism of Li intercalation/deintercalation into/from the surface of $LiCoO_2$. Phys. Chem. Chem. Phys. **17**, 22917–22922 (2015). https://doi.org/10.1039/c5cp02246k
175. A.W. Moses, H.G.G. Flores, J.-G. Kim, M.A. Langell, Surface properties of $LiCoO_2$, $LiNiO_2$ and $LiNi_{1-x}Co_xO_2$. Appl. Surf. Sci. **253**, 4782–4791 (2007). https://doi.org/10.1016/j.apsusc.2006.10.044
176. R.V. Moshtev, P. Zlatilova, B. Puresheva, V. Manev, Material balance of petroleum coke/$LiNiO_2$ lithium-ion cells. J. Power Sources **56**, 137–144 (1995). https://doi.org/10.1016/0378-7753(95)80025-C
177. K. Mukai et al., Microscopic magnetism in lithium insertion materials of $LiNi_{1-x}Co_xO_2$ (x=0, ¼, ½ ½, ¾, and 1). J. Power Sources **174**, 843–846 (2007). https://doi.org/10.1016/j.jpowsour.2007.06.105
178. S. Muto, K. Tatsumi, Y. Kojima, H. Oka, H. Kondo, K. Horibuchi, Y. Ukyo, Effect of Mg-doping on the degradation of $LiNiO_2$-based cathode materials by combined spectroscopic methods. J. Power Sources **205**, 449–455 (2012). https://doi.org/10.1016/j.jpowsour.2012.01.071
179. S.-H. Na, H.-S. Kim, S.-I. Moon, The effect of Si doping on the electrochemical characteristics of $LiNi_xMn_yCo(_{1-x-y})O_2$. Solid State Ionics **176**, 313–317 (2005). https://doi.org/10.1016/j.ssi.2004.08.016
180. K. Nakamura, H. Ohno, K. Okamura, Y. Michihiro, I. Nakabayashi, T. Kanashiro, On the diffusion of Li^+ defects in $LiCoO_2$ and $LiNiO_2$. Solid State Ionics **135**, 143–147 (2000). https://doi.org/10.1016/S0167-2738(00)00293-9
181. K.-W. Nam, W.-S. Yoon, H. Shin, K.Y. Chung, S. Choi, X.-Q. Yang, In situ X-ray diffraction studies of mixed $LiMn_2O_4$–$LiNi_{1/3}Co_{1/3}Mn_{1/3}O_2$ composite cathode in Li-ion cells during charge-discharge cycling. J. Power Sources **192**, 652–659 (2009). https://doi.org/10.1016/j.jpowsour.2009.02.088
182. H. Nie et al., $LiCoO_2$: recycling from spent batteries and regeneration with solid state synthesis. Green Chem. **17**, 1276–1280 (2015). https://doi.org/10.1039/c4gc01951b
183. T. Ohzuku, Y. Makimura, Formation of solid solution and its effect on lithium insertion schemes for advanced lithium-ion batteries: X-ray absorption spectroscopy and X-ray diffraction of $LiCoO_2$, $LiCo_1/_2Ni_1/_2O_2$ and $LiNiO_2$. Res. Chem. Intermed. **32**, 507–521 (2006). https://doi.org/10.1163/156856706777973745
184. T. Onodera, J. Kawaji, A. Sato, T. Okumura, Electrochemical performance of a newly-designed all-solid-state Li-ion battery with a $LiNi_{1/3}Co_{1/3}Mn_{1/3}O_2$–$LiVO_3$ mono-particle

layered cathode and a lamellar $LiVO_3$ anode. J. Ceram. Soc. Jpn. **125**, 494–499 (2017). https://doi.org/10.2109/jcersj2.16316
185. C.Y. Ouyang, S.Q. Shi, M.S. Lei, Jahn–Teller distortion and electronic structure of $LiMn_2O_4$. J. Alloys Compd. **474**, 370–374 (2009). https://doi.org/10.1016/j.jallcom.2008.06.123
186. H. Ozawa, T. Ogihara, I. Mukoyama, K. Myojin, H. Aikiyo, T. Okawa, A. Harada, Synthesis of lithium manganate powders by spray pyrolysis and its application to lithium ion batteries for trams. World Electr. Veh. J. **1**, 19 (2007)
187. A.K. Padhi, K.S. Nanjundaswamy, J.B. Goodenough, Phospho-olivines as positive-electrode materials for rechargeable lithium batteries. J. Electrochem. Soc. **144**, 1188–1194 (1997). https://doi.org/10.1149/1.1837571
188. S.-C. Park, Y.-M. Kim, S.-C. Han, S. Ahn, C.-H. Ku, J.-Y. Lee, The elevated temperature performance of $LiMn_2O_4$ coated with $LiNi_{1-X}Co_XO_2$ (X=0.2 and 1). J. Power Sources **107**, 42–47 (2002). https://doi.org/10.1016/S0378-7753(01)00978-8
189. M. Park, X. Zhang, M. Chung, G.B. Less, A.M. Sastry, A review of conduction phenomena in Li-ion batteries. J. Power Sources **195**, 7904–7929 (2010). https://doi.org/10.1016/j.jpowsour.2010.06.060
190. Z.-D. Peng, X.-r. Deng, K. Du, G.-r. Hu, X.-g. Gao, Y.-x. Liu, Coating of $LiNi_{1/3}Mn_{1/3}Co_{1/3}O_2$ cathode materials with alumina by solid-state reaction at room temperature. J. Cent. S. Univ. Technol. **15**, 34–38 (2008). https://doi.org/10.1007/s11771-008-0008-9
191. H. Peng, Z.H. Zhu, P.X. Huang, X. Li, Effects of synthesis temperature and holding time on the electrochemical properties of $LiNi_{1/3}Co_{1/3}Mn_{1/3}O_2$ cathode material using rheological phase method. Mater. Sci. Forum **814**, 351–357 (2015). https://doi.org/10.4028/www.scientific.net/MSF.814.351
192. G. Pistoia, *Lithium-Ion Batteries: Advances and Applications* (Elsevier Science, Amsterdam, 2013)
193. Z. Qibing et al., Enhanced electrochemical capability of $LiNi_{1/3}Co_{1/3}Mn_{1/3}O_2$ cathode materials coated with fluoroborate glass for lithium-ion batteries. ChemElectroChem **4**, 1199–1204 (2017). https://doi.org/10.1002/celc.201600828
194. M.D. Radin et al., Narrowing the gap between theoretical and practical capacities in Li-ion layered oxide cathode materials. Adv. Energy Mater. **7** (2017). https://doi.org/10.1002/aenm.201602888
195. K. Rajammal, D. Sivakumar, N. Duraisamy, K. Ramesh, S. Ramesh, Effect of sintering temperature on structural properties of $LiMnPO_4$ cathode materials obtained by sol-gel method. J. Sol-Gel Sci. Technol. **80**, 514–522 (2016). https://doi.org/10.1007/s10971-016-4111-3
196. M.M. Rao, C. Liebenow, M. Jayalakshmi, H. Wulff, U. Guth, F. Scholz, High-temperature combustion synthesis and electrochemical characterization of $LiNiO_2$, $LiCoO_2$ and $LiMn_2O_4$ for lithium-ion secondary batteries. J. Solid State Electrochem. **5**, 348–354 (2001). https://doi.org/10.1007/s100080000157
197. H. Ren, Y. Wang, D. Li, L. Ren, Z. Peng, Y. Zhou, Synthesis of $LiNi_{1/3}Co_{1/3}Mn_{1/3}O_2$ as a cathode material for lithium battery by the rheological phase method. J. Power Sources **178**, 439–444 (2008). https://doi.org/10.1016/j.jpowsour.2007.12.004
198. H. Ren, Y. Huang, Y. Wang, Z. Li, P. Cai, Z. Peng, Y. Zhou, Effects of different carbonate precipitators on $LiNi_{1/3}Co_{1/3}Mn_{1/3}O_2$ morphology and electrochemical performance. Mater. Chem. Phys. **117**, 41–45 (2009). https://doi.org/10.1016/j.matchemphys.2009.01.002
199. X. Rocquefelte, F. Boucher, P. Gressier, G. Ouvrard, First-principle study of the intercalation process in the $Li_xV_2O_5$ system. Chem. Mater. **15**, 1812–1819 (2003). https://doi.org/10.1021/cm021254h
200. S. Rodrigues, N. Munichandraiah, A.K. Shukla, Novel solution-combustion synthesis of $LiCoO_2$ and its characterization as cathode material for lithium-ion cells. J. Power Sources **102**, 322–325 (2001). https://doi.org/10.1016/S0378-7753(01)00770-4
201. P. Rosaiah, O.M. Hussain, J. Zhu, Y. Qiu, Spectroscopic and electrochemical properties of lithium-rich $LiFePO_4$ cathode synthesized by solid-state reaction. J. Electron. Mater. **46**, 4865–4874 (2017). https://doi.org/10.1007/s11664-017-5474-0

202. A. Rougier, I. Saadoune, P. Gravereau, P. Willmann, C. Delmasa, Effect of cobalt substitution on cationic distribution in $LiNi_1 - yCo_yO_2$ electrode materials. Solid State Ionics **90**, 83–90 (1996). https://doi.org/10.1016/S0167-2738(96)00370-0
203. A. Sakunthala, M.V. Reddy, S. Selvasekarapandian, B.V.R. Chowdari, H. Nithya, P. Christopher Selvin, Synthesis and electrochemical studies on LiV_3O_8. J. Solid State Electrochem. **14**, 1847–1854 (2010). https://doi.org/10.1007/s10008-010-1044-6
204. P. Samarasingha, D.-H. Tran-Nguyen, M. Behm, A. Wijayasinghe, $LiNi_{1/3}Mn_{1/3}Co_{1/3}O_2$ synthesized by the Pechini method for the positive electrode in Li-ion batteries: material characteristics and electrochemical behaviour. Electrochim. Acta **53**, 7995–8000 (2008). https://doi.org/10.1016/j.electacta.2008.06.003
205. T. Sasaki et al., Capacity-fading mechanisms of $LiNiO_2$-based lithium-ion batteries: I. Analysis by electrochemical and spectroscopic examination. J. Electrochem. Soc. **156**, A289–A293 (2009). https://doi.org/10.1149/1.3076136
206. M. Sathiya, A.S. Prakash, K. Ramesha, A.K. Shukla, Rapid synthetic routes to prepare $LiNi_{1/3}Mn_{1/3}Co_{1/3}O_2$ as a high voltage, high-capacity Li-ion battery cathode material. Mater. Res. Bull. **44**, 1990–1994 (2009). https://doi.org/10.1016/j.materresbull.2009.06.007
207. Y. Satou, S. Komine, S. Shimizu, Orbital-Specific observation of O_{2p} and Ni_3d electrons in $LiNi_{1/2}Mn_{1/2}O_2$, a cathode material for lithium-ion batteries. AIP Adv. **7**, 095117 (2017). https://doi.org/10.1063/1.4996705
208. S.B. Schougaard, J. Bréger, M. Jiang, C.P. Grey, J.B. Goodenough, $LiNi_{1/2}+\delta Mn_{1/2}-\delta O_2$ – a high-rate, high-capacity cathode for lithium rechargeable batteries. Adv. Mater. **18**, 905–909 (2006). https://doi.org/10.1002/adma.200500113
209. B. Scrosati, J. Garche, Lithium batteries: status, prospects and future. J. Power Sources **195**, 2419–2430 (2010). https://doi.org/10.1016/j.jpowsour.2009.11.048
210. M. Shandilya, R. Rai, J. Singh, Review: hydrothermal technology for smart materials. Adv. Appl. Ceram. **115**, 354–376 (2016). https://doi.org/10.1080/17436753.2016.1157131
211. Y. Shao-Horn, S.A. Hackney, A.R. Armstrong, P.G. Bruce, R. Gitzendanner, C.S. Johnson, M.M. Thackeray, Structural characterization of layered $LiMnO_2$ electrodes by electron diffraction and lattice imaging. J. Electrochem. Soc. **146**, 2404–2412 (1999). https://doi.org/10.1149/1.1391949
212. B.-J. Shen, J.-S. Ma, H.-C. Wu, C.-H. Lu, Microwave-mediated hydrothermal synthesis and electrochemical properties of $LiNi_{1/3}Co_{1/3}Mn_{1/3}O_2$ powders. Mater. Lett. **62**, 4075–4077 (2008). https://doi.org/10.1016/j.matlet.2008.05.034
213. S.P. Sheu, I.C. Shih, C.Y. Yao, J.M. Chen, W.M. Huang, Studies of $LiNiO_2$ in lithium-ion batteries. J. Power Sources **68**, 558–560 (1997). https://doi.org/10.1016/S0378-7753(97)02526-3
214. O.A. Shlyakhtin, Y.S. Yoon, S.H. Choi, Y.-J. Oh, Freeze-drying synthesis of $LiNi_{1/2}Mn_{1/2}O_2$ cathode materials. Electrochim. Acta **50**, 505–509 (2004). https://doi.org/10.1016/j.electacta.2004.05.051
215. L. Singh, A. Kumar, H. Lee, J. Lee, M. Ji, Y. Lee, Ultrafast auto flame synthesis for the mass production of $LiCoO_2$ as a cathode material for Li-ion batteries. J. Solid State Electrochem. **22**(8), 2561–2568 (2018). https://doi.org/10.1007/s10008-018-3972-5
216. M.Y. Song, H.U. Kim, H.R. Park, Electrochemical properties of $LiNiO_2$ cathode after TiO_2 or ZnO addition. Ceram. Int. **40**, 4219–4224 (2014). https://doi.org/10.1016/j.ceramint.2013.08.084
217. H. Song, Y. Liu, C. Zhang, C. Liu, G. Cao, Mo-doped LiV_3O_8 nanorod-assembled nanosheets as a high performance cathode material for lithium ion batteries. J. Mater. Chem. A **3**, 3547–3558 (2015). https://doi.org/10.1039/C4TA05616G
218. Y. Sun, P. Wan, J. Pan, C. Xu, X. Liu, Low temperature synthesis of layered $LiNiO_2$ cathode material in air atmosphere by ion exchange reaction. Solid State Ionics **177**, 1173–1177 (2006). https://doi.org/10.1016/j.ssi.2006.04.050
219. Y. Sun, Y. Xia, Y. Shiosaki, H. Noguchi, Preparation and electrochemical properties of $LiCoO_2$–$LiNi_{1/2}Mn_{1/2}O_2$–Li_2MnO_3 solid solutions with high Mn contents. Electrochim. Acta **51**, 5581–5586 (2006). https://doi.org/10.1016/j.electacta.2006.02.041

220. J. Sun, L. Jiao, X. Wei, W. Peng, L. Liu, H. Yuan, Effect of PEG molecular weight on the crystal structure and electrochemical performance of LiV_3O_8. J. Solid State Electrochem. **14**, 615–619 (2010). https://doi.org/10.1007/s10008-009-0829-y
221. D. Sun et al., Multi-layered $Al_2O_3/Li_xV_2O_5/LiV_3O_8$ nanoflakes with superior cycling stability as cathode material for Li-ion battery. Electrochim. Acta **157**, 211–217 (2015). https://doi.org/10.1016/j.electacta.2015.01.081
222. C.-F. Sun, N. Amruthnath, J.-S. Yu, W.-J. Li, Enhanced electrochemical performances of layered $LiNi_{0.5}Mn_{0.5}O_2$ as cathode materials by Ru doping for lithium-ion batteries. Ionics **22**, 1501–1508 (2016). https://doi.org/10.1007/s11581-016-1751-9
223. K. Sun et al., Hybrid LiV_3O_8/carbon encapsulated $Li_{1.2}Mn_{0.54}Co_{0.13}Ni_{0.13}O_2$ with improved electrochemical properties for lithium-ion batteries. RSC Adv. **6**, 28729–28736 (2016). https://doi.org/10.1039/C6RA02688E
224. G. Sun et al., Synergistic effects of ion doping and surface-modifying for lithium transition-metal oxide: synthesis and characterization of La_2O_3-modified $LiNi_{1/3}Co_{1/3}Mn_{1/3}O_2$. Electrochim. Acta **272**, 11–21 (2018). https://doi.org/10.1016/j.electacta.2018.03.175
225. C.-Y. Tang, L. Feng, R.T. Haasch, S.J. Dillon, Surface redox on $Li\left[LiNi_{1/3}Co_{1/3}Mn_{1/3}\right]O_2$ characterized by insitu X-ray photoelectron spectroscopy and insitu Auger electron spectroscopy. Electrochim. Acta **277**, 197–204 (2018). https://doi.org/10.1016/j.electacta.2018.04.199
226. M.M. Thackeray, P.J. Johnson, L.A. de Picciotto, P.G. Bruce, J.B. Goodenough, Electrochemical extraction of lithium from $LiMn_2O_4$. Mater. Res. Bull. **19**, 179–187 (1984). https://doi.org/10.1016/0025-5408(84)90088-6
227. R. Thirunakaran, R. Ravikumar, S. Gopukumar, A. Sivashanmugam, Electrochemical evaluation of dual-doped $LiMn_2O_4$ spinels synthesized via co-precipitation method as cathode material for lithium rechargeable batteries. J. Alloys Compd. **556**, 266–273 (2013). https://doi.org/10.1016/j.jallcom.2012.12.053
228. M. Tian, Effects of sintering time on the performance of $LiNi_{1/3}Co_{1/3}Mn_{1/3}O_2$ synthesized by high temperature ball milling method. Int. J. Electrochem. Sci. **12**, 7166–7173 (2017). https://doi.org/10.20964/2017.08.42
229. M. Tian et al., Phase structure and electrochemical performance of layered-spinel integrated $LiNi_{1/2}Mn_{1/2}O_2$–$LiMn_{1.9}Al_{0.1}O_4$ composite cathodes for lithium-ion batteries. Ceram. Int. **42**, 16916–16926 (2016). https://doi.org/10.1016/j.ceramint.2016.07.191
230. K. Vediappan et al., Stirring effect in hydrothermal synthesis of nano C-$LiFePO_4$. J. Power Sources **266**, 99–106 (2014). https://doi.org/10.1016/j.jpowsour.2014.05.005
231. C. Venkateswara Rao, A. Leela Mohana Reddy, Y. Ishikawa, P.M. Ajayan, $LiNi_{1/3}Co_{1/3}Mn_{1/3}O_2LiNi_{1/3}Co_{1/3}Mn_{1/3}O_2$–Graphene composite as a promising cathode for lithium-ion batteries. ACS Appl. Mater. Interfaces **3**, 2966–2972 (2011). https://doi.org/10.1021/am200421h
232. M. Wang, A. Navrotsky, Enthalpy of formation of $LiNiO_2$, $LiCoO_2$ and their solid solution, $LiNi1-xCoxO_2$. Solid State Ionics **166**, 167–173 (2004). https://doi.org/10.1016/j.ssi.2003.11.004
233. J. Wang, X. Sun, Olivine $LiFePO_4$: the remaining challenges for future energy storage. Energy Environ. Sci. **8**, 1110–1138 (2015). https://doi.org/10.1039/c4ee04016c
234. L. Wang, F. Zhou, Y.S. Meng, G. Ceder, First-principles study of surface properties of $LiFePO_4$: surface energy, structure, Wulff shape, and surface redox potential. Phys. Rev. B Condens. Matter Mater. Phys. **76**, 165435 (2007). https://doi.org/10.1103/PhysRevB.76.165435
235. C. Wang, X. Ma, L. Zhou, J. Cheng, J. Sun, Y. Zhou, Study on the rapid synthesis of $LiNi_{1-x}Co_xO_2$ cathode material for lithium secondary battery. Electrochim. Acta **52**, 3022–3027 (2007). https://doi.org/10.1016/j.electacta.2006.09.047
236. K. Wang, R. Cai, T. Yuan, X. Yu, R. Ran, Z. Shao, Process investigation, electrochemical characterization and optimization of $LiFePO_4$/C composite from mechanical activation using sucrose as carbon source. Electrochim. Acta **54**, 2861–2868 (2009). https://doi.org/10.1016/j.electacta.2008.11.012
237. F. Wang, J. Yang, Y. NuLi, J. Wang, Highly promoted electrochemical performance of 5 V $LiCoPO_4$ cathode material by addition of vanadium. J. Power Sources **195**, 6884–6887 (2010). https://doi.org/10.1016/j.jpowsour.2010.04.071

238. D. Wang, L. Cao, J. Huang, J. Wu, Synthesis and electrochemical properties of LiV_3O_8 via an improved sol-gel process. Ceram. Int. **38**, 2647–2652 (2012). https://doi.org/10.1016/j.ceramint.2011.11.030
239. R. Wang et al., Effect of different binders on the electrochemical performance of metal oxide anode for lithium-ion batteries. Nanoscale Res. Lett. **12**, 575 (2017). https://doi.org/10.1186/s11671-017-2348-6
240. Y. Wang et al., Layered oxides – $LiNi_{1/3}Co_{1/3}Mn_{1/3}O_2$ as anode electrode for symmetric rechargeable lithium-ion batteries. J. Power Sources **378**, 516–521 (2018). https://doi.org/10.1016/j.jpowsour.2017.12.043
241. T. Wenming et al., Synthesis of mesoporous orthorhombic $LiMnO_2$ cathode materials via a one-step flux method for high performance lithium-ion batteries. Mater. Res. Express **5**, 065511 (2018)
242. J. Wolfenstine, Electrical conductivity of doped $LiCoPO_4$. J. Power Sources **158**, 1431–1435 (2006). https://doi.org/10.1016/j.jpowsour.2005.10.072
243. J. Wolfenstine, J. Read, J.L. Allen, Effect of carbon on the electronic conductivity and discharge capacity $LiCoPO_4$. J. Power Sources **163**, 1070–1073 (2007). https://doi.org/10.1016/j.jpowsour.2006.10.010
244. Y. Wu, *Lithium-Ion Batteries: Fundamentals and Applications* (CRC Press, Boca Raton, 2015)
245. M. Wu, A. Chen, R. Xu, Y. Li, Nanocrystalline $LiNi_{1-x}Co_xO_2$ cathode materials prepared by a gel-combustion process. Mater. Sci. Eng. B **99**, 336–339 (2003). https://doi.org/10.1016/S0921-5107(02)00484-1
246. Q. Wu, X. Li, M. Yan, Z. Jiang, Electrochemical properties of submicron-sized layered $LiNi_{1/2}Mn_{1/2}O_2$. Electrochem. Commun. **5**, 878–882 (2003). https://doi.org/10.1016/j.elecom.2003.08.011
247. H.M. Wu, J.P. Tu, Y.F. Yuan, Y. Li, X.B. Zhao, G.S. Cao, Structural, morphological and electrochemical characteristics of spinel $LiMn_2O_4$ prepared by spray-drying method. Scr. Mater. **52**, 513–517 (2005). https://doi.org/10.1016/j.scriptamat.2004.10.037
248. F. Wu, L. Wang, C. Wu, Y. Bai, F. Wang, Study on Li_1+xV_3O_8 synthesized by microwave sol-gel route. Mater. Chem. Phys. **115**, 707–711 (2009). https://doi.org/10.1016/j.matchemphys.2009.02.007
249. F. Wu, M. Wang, Y. Su, L. Bao, S. Chen, A novel method for synthesis of layered $LiNi_{1/3}Mn_{1/3}Co_{1/3}O_2$ as cathode material for lithium-ion battery. J. Power Sources **195**, 2362–2367 (2010). https://doi.org/10.1016/j.jpowsour.2009.10.043
250. B. Wu et al., Controlled solvothermal synthesis and electrochemical performance of $LiCoPO_4$ submicron single crystals as a cathode material for lithium ion batteries. J. Power Sources **304**, 181–188 (2016). https://doi.org/10.1016/j.jpowsour.2015.11.023
251. X. Wu et al., Analysis of the effects of different carbon coating strategies on structure and electrochemical behavior of $LiCoPO_4$ material as a high-voltage cathode electrode for lithium-ion batteries. Electrochim. Acta **279**, 108–117 (2018). https://doi.org/10.1016/j.electacta.2018.05.067
252. X. Wu, S.-X. Zhao, L. Wei, H. Deng, C.-W. Nan, Improving the structure stability and electrochemical performance of Li_2MnSiO_4/C cathode materials by Ti-doping and porous microstructure. J. Alloys Compd. **735**, 1158–1166 (2018). https://doi.org/10.1016/j.jallcom.2017.11.230
253. H. Xia, S.B. Tang, L. Lu, Novel synthesis and electrochemical behavior of layered $LiNi_{1/2}Mn_{1/2}O_2$. J. Alloys Compd. **449**, 296–299 (2008). https://doi.org/10.1016/j.jallcom.2006.02.097
254. H. Xia, Z. Luo, J. Xie, Nanostructured $LiMn_2O_4$ and their composites as high-performance cathodes for lithium-ion batteries. Prog. Nat. Sci.: Mater. Int. **22**, 572–584 (2012). https://doi.org/10.1016/j.pnsc.2012.11.014
255. X. Xiao, L. Wang, D. Wang, X. He, Q. Peng, Y. Li, Hydrothermal synthesis of orthorhombic $LiMnO_2$ nano-particles and $LiMnO_2$ nanorods and comparison of their electrochemical performances. Nano Res. **2**, 923–930 (2009). https://doi.org/10.1007/s12274-009-9094-8
256. L. Xiao, X. Liu, X. Zhao, H. Liang, H. Liu, Preparation of $LiNi_{1/2}Mn_{1/2}O_2$ cathode materials by urea hydrolysis coprecipitation. Solid State Ionics **192**, 335–338 (2011). https://doi.org/10.1016/j.ssi.2010.06.039

257. H. Xiao, Y. Wang, K. Xie, S. Cheng, X. Cheng, High capacitance $LiMn_2O_4$ microspheres with different microstructures as cathode material for aqueous asymmetric supercapacitors. J. Alloys Compd. **738**, 25–31 (2018). https://doi.org/10.1016/j.jallcom.2017.12.143
258. X. Xiong, Z. Wang, G. Yan, H. Guo, X. Li, Role of V_2O_5 coating on $LiNiO_2$-based materials for lithium-ion battery. J. Power Sources **245**, 183–193 (2014). https://doi.org/10.1016/j.jpowsour.2013.06.133
259. Z. Xu, L. Gao, Y. Liu, L. Li, Review – recent developments in the doped $LiFePO_4$ cathode materials for power lithium ion batteries. J. Electrochem. Soc. **163**, A2600–A2610 (2016). https://doi.org/10.1149/2.0411613jes
260. F. Xu, H. Yan, J. Chen, Z. Zhang, C. Fan, Improving electrochemical properties of $LiNi_{1/3}Co_{1/3}Mn_{1/3}O_2$ by enhancing thermal decomposition of carbonates to synthesize ultrafine powders. J. Electroanal. Chem. **820**, 118–122 (2018). https://doi.org/10.1016/j.jelechem.2018.04.069
261. S. Yamada, M. Fujiwara, M. Kanda, Synthesis and properties of $LiNiO_2$ as cathode material for secondary batteries. J. Power Sources **54**, 209–213 (1995). https://doi.org/10.1016/0378-7753(94)02068-E
262. J. Yan, H. Liu, Y. Wang, X. Zhao, Y. Mi, B. Xia, Enhanced high-temperature cycling stability of $LiNi_{1/3}Co_{1/3}Mn_{1/3}O_2$-coated $LiMn_2O_4$ as cathode material for lithium ion batteries. Ionics **21**, 1835–1842 (2015). https://doi.org/10.1007/s11581-015-1371-9
263. A. Yano, A. Ueda, M. Shikano, H. Sakaebe, Z. Ogumi, Surface structure and high-voltage charging/discharging performance of low-content Zr-oxide-coated $LiNi_{1/3}Co_{1/3}Mn_{1/3}O_2$. J. Electrochem. Soc. **163**, A75–A82 (2016). https://doi.org/10.1149/2.0211602jes
264. J. Yao, C. Shen, P. Zhang, D.H. Gregory, L. Wang, Enhanced cyclability of spinel $LiMn_2O_4$ by controlling the phase purity and structural strain. J. Phys. Chem. Solids **73**, 1390–1395 (2012). https://doi.org/10.1016/j.jpcs.2012.07.006
265. T.F. Yi, C.L. Hao, C.B. Yue, R.S. Zhu, J. Shu, A literature review and test: structure and physicochemical properties of spinel $LiMn_2O_4$ synthesized by different temperatures for lithium-ion battery. Synth. Met. **159**, 1255–1260 (2009). https://doi.org/10.1016/j.synthmet.2009.02.026
266. F. Yu, S.H. Lim, Y. Zhen, Y. An, J. Lin, Optimized electrochemical performance of three-dimensional porous $LiFePO_4$/C microspheres via microwave irradiation assisted synthesis. J. Power Sources **271**, 231–238 (2014). https://doi.org/10.1016/j.jpowsour.2014.08.009
267. T. Yuan, H. Liu, Y. Gu, H. Cui, Y. Wang, The phase structure and electrochemical performance of $xLi_2MnO_3{\cdot}(_{1-x})$ $LiNi_{1/3}Co_{1/3}Mn_{1/3}O_2$ during the synthesis and charge-discharge process. Appl. Phys. A **122**, 812 (2016). https://doi.org/10.1007/s00339-016-0306-2
268. J. Zeng et al., Facile triethanolamine-assisted combustion synthesized layered $LiNi_{1/3}Co_{1/3}Mn_{1/3}O_2$ cathode materials with enhanced electrochemical performance for lithium-ion batteries. J. Alloys Compd. **735**, 1977–1985 (2018). https://doi.org/10.1016/j.jallcom.2017.11.321
269. W.-J. Zhang, Structure and performance of $LiFePO_4$ cathode materials: a review. J. Power Sources **196**, 2962–2970 (2011). https://doi.org/10.1016/j.jpowsour.2010.11.113
270. Y. Zhang, C.-Y. Wang, Cycle-life characterization of automotive lithium-ion batteries with $LiNiO_2$ cathode. J. Electrochem. Soc. **156**, A527–A535 (2009). https://doi.org/10.1149/1.3126385
271. Z. Zhang, S.S. Zhang, *Rechargeable Batteries: Materials, Technologies and New Trends* (Springer, Berlin, 2015)
272. B. Zhang, G. Chen, P. Xu, Z. Lv, Effect of ultrasonic irradiation on the structure and electrochemical properties of cathode material $LiNi_{1/2}Mn_{1/2}O_2$ for lithium batteries. Solid State Ionics **178**, 1230–1234 (2007). https://doi.org/10.1016/j.ssi.2007.06.010
273. P. Zhang, L. Zhang, X. Ren, Q. Yuan, J. Liu, Q. Zhang, Preparation and electrochemical properties of $LiNi_{1/3}Co_{1/3}Mn_{1/3}O_2$ – PPy composites cathode materials for lithium-ion battery. Synth. Met. **161**, 1092–1097 (2011). https://doi.org/10.1016/j.synthmet.2011.03.021
274. X. Zhang, A. Mauger, W. Jiang, H. Groult, C.M. Julien, Diffusion of Li+ ions in $LiNi_{1/3}Co_{1/3}Mn_{1/3}O_2$. ECS Trans. **35**, 89–94 (2011). https://doi.org/10.1149/1.3654205

275. Z. Zhang, W. He, G. Li, J. Xia, H. Hu, J. Huang, Ultrasound-assisted hydrothermal renovation of $LiCoO_2$ from the cathode of spent lithium-ion batteries. Int. J. Electrochem. Sci. **9**, 3691–3700 (2014)
276. J. Zhang, Z. Wu, W. Hua, H. Liu, B. Zhong, High-performance porous spherical cathode materials based on $CaCO_3$-template synthesis of $LiNi_{1/3}Co_{1/3}Mn_{1/3}O_2$ for lithium-ion batteries. Ionics **21**, 3151–3158 (2015). https://doi.org/10.1007/s11581-015-1501-4
277. Y. Zhang et al., Sacrificial template strategy toward a hollow $LiNi_{1/3}Co_{1/3}Mn_{1/3}O_2$ nanosphere cathode for advanced lithium-ion batteries. ACS Omega **2**, 7593–7599 (2017). https://doi.org/10.1021/acsomega.7b00764
278. C. Zhang et al., Significant improvement on electrochemical performance of $LiMn_2O_4$ at elevated temperature by atomic layer deposition of TiO_2 nanocoating. ACS Sustain. Chem. Eng. **6**, 7890–7901 (2018). https://doi.org/10.1021/acssuschemeng.8b01081
279. L.-L. Zhang, J.-Q. Wang, X.-L. Yang, G. Liang, T. Li, P.-L. Yu, D. Ma, Enhanced electrochemical performance of fast ionic conductor $LiTi_2(PO_4)_3$-coated $LiNi_{1/3}Co_{1/3}Mn_{1/3}O_2$ cathode material. ACS Appl. Mater. Interfaces **10**, 11663–11670 (2018). https://doi.org/10.1021/acsami.7b19692
280. M. Zhao, Q. Zheng, F. Wang, W. Dai, X. Song, Electrochemical performance of high specific capacity of lithium-ion cell LiV_3O_8//$LiMn_2O_4$ with $LiNO_3$ aqueous solution electrolyte. Electrochim. Acta **56**, 3781–3784 (2011). https://doi.org/10.1016/j.electacta.2011.02.057
281. H. Zhao, Z.B. Hu, X.L. Xiao, X.F. Liu, Hydrothermal synthesis and electrochemical performance of $LiNi_{1/2}Mn_{1/2}O_2$ as lithium-ion battery cathode. Appl. Mech. Mater. **672–674**, 638–645 (2014). https://doi.org/10.4028/www.scientific.net/AMM.672-674.638
282. J. Zhao, Z. Wang, H. Guo, X. Li, Enhanced electrochemical properties of $LiNiO_2$-based cathode materials by nanoscale manganese carbonate treatment. Appl. Surf. Sci. **403**, 426–434 (2017). https://doi.org/10.1016/j.apsusc.2017.01.089
283. X. Zheng, X. Li, Z. Wang, H. Guo, Z. Huang, G. Yan, D. Wang, Investigation and improvement on the electrochemical performance and storage characteristics of $LiNiO_2$-based materials for lithium-ion battery. Electrochim. Acta **191**, 832–840 (2016). https://doi.org/10.1016/j.electacta.2016.01.142
284. H. Zheng, X. Chen, Y. Yang, L. Li, G. Li, Z. Guo, C. Feng, Self-assembled $LiNi_{1/3}Co_{1/3}Mn_{1/3}O_2$ nanosheet cathode with high electrochemical performance. ACS Appl. Mater. Interfaces **9**, 39560–39568 (2017). https://doi.org/10.1021/acsami.7b10264
285. X. Zhong Ren, C. Shi, P. Xin Zhang, J. Hong Liu, Synthesis and electrochemical performance of LiV_3O_8/MWCNTs cathode material for lithium-ion batteries. **454** (2012). https://doi.org/10.4028/scientific5/AMR.454.105
286. H. Zhong, M. Sun, Y. Li, J. He, J. Yang, L. Zhang, The polyacrylic latex: an efficient water-soluble binder for the $LiNi_{1/3}Co_{1/3}Mn_{1/3}O_2$ cathode in Li-ion batteries. J. Solid State Electrochem. **20**, 1–8 (2016). https://doi.org/10.1007/s10008-015-2967-8
287. C. Zhongxue, X. Fei, C. Shunan, L. Zhengfeng, Y. Hanxi, A. Xinping, C. Yuliang, High rate, long lifespan LiV_3O_8 nanorods as a cathode material for lithium-ion batteries. Small **13**, 1603148 (2017). https://doi.org/10.1002/smll.201603148
288. H. Zhou, Y. Li, J. Zhang, W. Kang, D.Y.W. Yu, Low-temperature direct synthesis of layered m-$LiMnO_2$ for lithium-ion battery applications. J. Alloys Compd. **659**, 248–254 (2016). https://doi.org/10.1016/j.jallcom.2015.11.038
289. X. Zhu, H. Zhan, H. Liu, Y. Zhou, Synthesis and characterization of $LiNi_{0.95}$–$xCo_xAl_{0.05}O_2$ for lithium-ion batteries. Rare Metals **25**, 303–308 (2006). https://doi.org/10.1016/S1001-0521(06)60058-2
290. W. Zhu, Z. Zhuang, Z. Lin, Y. Yang, Y. Lin, Z. Huang, Enhanced electrochemical properties and thermal stability of $LiNi_{1/3}Co_{1/3}Mn_{1/3}O_2$ by surface modification with Eu_2O_3. Ionics **22**, 1533–1540 (2016). https://doi.org/10.1007/s11581-016-1681-6
291. L. Zhu, W. Li, Z. Yu, L. Xie, X. Cao, Synthesis and electrochemical performances of LiV_3O_8/poly (3,4-ethylene dioxythiophene) composites as cathode materials for rechargeable lithium batteries. Solid State Ionics **310**, 30–37 (2017). https://doi.org/10.1016/j.ssi.2017.08.002

292. C. Zhu, Z. Wu, J. Xie, Z. Chen, J. Tu, G. Cao, X. Zhao, Solvothermal-assisted morphology evolution of nanostructured $LiMnPO_4$ as high-performance lithium-ion batteries cathode. J. Mater. Sci. Technol. (2018). https://doi.org/10.1016/j.jmst.2018.04.017
293. L. Zhu, L. Xie, X. Cao, LiV_3O_8/polydiphenylamine composites with significantly improved electrochemical behavior as cathode materials for rechargeable lithium batteries. ACS Appl. Mater. Interfaces **10**, 10909–10917 (2018). https://doi.org/10.1021/acsami.8b00364
294. D.A. Ziolkowska, J.B. Jasinski, B. Hamankiewicz, K.P. Korona, S.H. Wu, A. Czerwinski, In situ XRD and TEM studies of sol-gel-based synthesis of $LiFePO_4$. Cryst. Growth Des. **16**, 5006–5013 (2016). https://doi.org/10.1021/acs.cgd.6b00575

Rechargeable Lithium Metal Batteries

4

Bin Liu and Huilin Pan

Contents

Abstract

Along with the state-of-the-art lithium-ion (Li-ion) batteries approaching their limitation in specific energy density, "beyond Li-ion" battery technologies have become alternative energy storage solutions due to their higher specific energy density. Among them, rechargeable lithium metal battery (LMB) has been considered as one of the most promising battery technologies that promises a great increase in energy density. Among the known anode materials, the Li metal has some unique and attractive features, such as an ultrahigh theoretical capacity (3,860 mAh g^{-1}), a lowest negative electrochemical potential (−3.04 vs SHE), and a low gravimetric density (0.534 g cm^{-1}). Therefore, Li metal has rendered as

Author Contribution: Dr. H. Pan designed the manuscript. Dr. B. Liu and Dr. H. Pan wrote the manuscript together. The authors further acknowledge that there is no financial relationship with the editors or publisher and have contributed original work in this chapter, other than what was acknowledged or appropriately cited with copyright permission.

B. Liu · H. Pan (✉)
Energy and Environment Directorate, Pacific Northwest National Laboratory, Richland, WA, USA
e-mail: Bin.liu@pnnl.gov; Huilin.pan@pnnl.gov

Q. Zhen et al. (eds.), *Nanostructured Materials for Next-Generation Energy Storage and Conversion*, https://doi.org/10.1007/978-3-662-58675-4_4

a potential ultimate anode material in rechargeable batteries. Furthermore, a combination of Li metal with oxygen (O_2) or sulfur (S) cathodes brings more viable options for the next-generation high-energy rechargeable batteries (3,505 Wh kg^{-1} for Li-O_2, 2,600 Wh kg^{-1} for Li-S batteries) with a great reduction in battery cost due to the high abundance and broad distribution of O_2 and S sources. However, the use of Li metal anode in LMBs still causes some critical issues, including the well-known Li dendritic growth, uncontrolled interfacial reactions with electrolytes, and large volumetric change during Li plating/stripping process. The "short circuit" of the battery by dendrite formation and the continuous depletion of electrolytes/active bulk lithium have been significantly challenging the practical application of rechargeable LMBs. In this chapter, we will discuss the fundamental challenges for both Li anode and cathodes and the proposed strategies for rechargeable LMBs. A perspective on the future research direction is also presented to initiate more helpful thoughts and promote to solve the critical issues in this research field.

4.1 Introduction

The growing markets for portable electronics, electric vehicles, and stationary energy storage systems have already triggered a rapid increase in the ever-changing demands for energy storage devices [1, 2]. Among various energy storage systems, rechargeable lithium (Li)-ion batteries (LIBs), as one of the most promising power sources, have successfully revolutionized communications and the transportation style of our modern life, driving mobile phones/laptops, and more recently electric vehicles [3, 4]. However, the state-of-the-art Li-ion batteries using graphite anodes with a low theoretical specific capacity of 372 mAh g^{-1} have almost reached their theoretical specific energy density (~350 Wh kg^{-1}), which still cannot satisfy the fast-growing requirement of the higher-energy-density batteries for wide applications, particularly long-lasting electric vehicles.

To address the low capacity issue of an aforementioned graphite anode, researchers around the world have been conducting extensive research on exploring alternative anodes for LIBs, such as silicon [5–8], oxides [9–14], sulfides [15–17], etc. Unfortunately, many issues existing in those reported anodes are as follows: (a) excess volume change during battery cycling, (b) high working voltage plateau, and (c) complicated material synthesis steps and high cost, which cause low practical energy density of Li-ion batteries based on above anodes. Besides, Li metal has been highly considered as one of most ideal anode candidates due to its ultrahigh theoretical specific capacity (3860 mAh g^{-1}), lowest negative electrochemical potential (−3.040 V versus standard hydrogen electrode), and lightweight (0.534 g cm^{-3}) [18, 19]. In particular, Li metal batteries (LMBs) have recently become one of the most promising alternatives for future high-energy battery technologies when further development of low-capacity LIBs used currently has been in dilemma. The combination of high-capacity oxygen (O_2) or sulfur

(S) cathode with the Li metal anode attracted intensive interest from the worldwide research community due to their high theoretical energy compared with the traditional LIBs, e.g., 3505 Wh kg^{-1} for Li-O_2 and 2600 Wh kg^{-1} for Li-S batteries. Further, Fig. 4.1 represents a clear comparison of specific energy densities among different battery systems including Li-air battery, Li-S battery, zinc (Zn)-air battery, advanced Li-ion battery, the existing Li-ion battery, and nickel metal hydride (Ni-MH) battery [20]. Although the estimation of practical energy densities of energy storage devices is complicated, the specific energy values of different battery chemistries clearly map the development perspective of high-energy-density batteries in the future.

Back to the development history of LMBs, the non-rechargeable lithium metal battery (LMBs) was first applied in digital watches, calculators, and medical devices by Exxon in the 1970s [21]. However, such a primary lithium-metal battery inevitably resulted in high cost and environmental pollution. To solve these issues, Moli Energy developed the first commercial rechargeable lithium metal battery (LMBs) through using molybdenum sulfide (MoS_2) cathode and excess Li anode in the later 1980s, but it presented a serious safety risk because of the short circuit induced by lithium dendrite growth, thermal runaway, and potential explosion hazards. Thus all of the lithium-metal packs were recalled in 1989. *Nippon Denki Kabushiki Gaisha* (NEC) and Tadiran (a merge of Tadi Batteries and Ran, today controlled by Saft

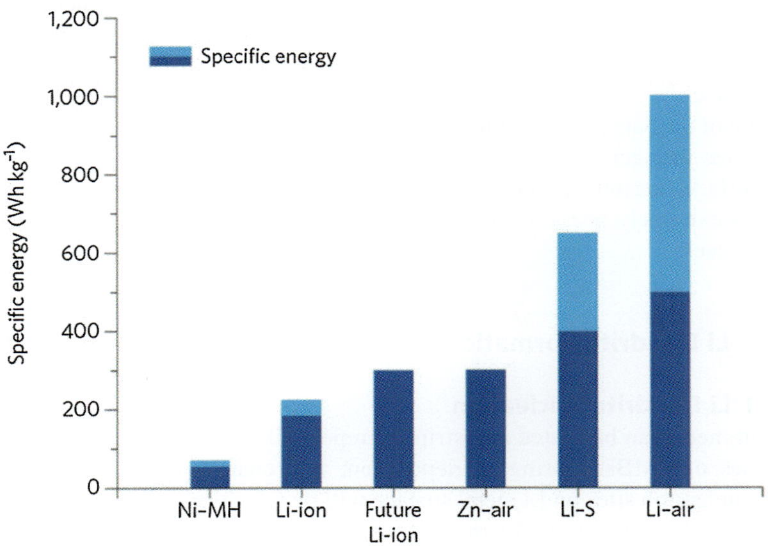

Fig. 4.1 A bar chart representing the comparison of specific energy among various battery systems. Battery materials, separators, electrolytes, and casings are taken into account. Dark blue denotes practical specific energy densities. Light blue indicates the potential changing range of those specific energy densities affected by current collectors, cell packaging, and system overheads. (Reproduced with the permission from Aurbach et al. [20]. Copyright 2016, Nature Publishing Group)

Groupe SA) tried to improve the battery design later but with limited success, whereas very few companies are able to realize rechargeable LMBs. Herein, safety concerns hindered the opportunity of the further commercialization of LMBs at that time. Over the last 40 years, LMBs have been extensively investigated to eliminate safety concern of LMBs.

In this chapter, we first present an overview of Li-metal batteries, and then summarize issues/challenges and recently proposed strategies to develop Li metal anodes and promising cathodes in LMBs. Finally, a general summary and some perspectives for Li metal battery chemistry will be provided. We aim to provide a schematic of the fundamentals of lithium metal battery (LMBs), illustrate key progress in LMBs, and give an inspiration for future developments of next-generation Li-metal battery technology.

4.2 Li Metal Anode

Lithium is the lightest one in the alkali metal group and has the smallest atomic radius of all metals. These characteristics enable Li metal with ultrahigh specific capacity and quick Li^+ ion transfer. Li metal anode with an extremely high capacity of 3860 mAh g^{-1} has the most negative potential of all the currently known electrode materials, which enables high discharge voltage and high-energy density in LMBs. However, there are two main barriers facing LMBs. One is the growth of Li dendrites during repeated plating/stripping processes, and the other is low Coulombic efficiency (CE) of these processes. These two critical issues for Li metal anode cause the safety hazard: internal short circuits led by dendrites and limited cycle life of the batteries due to irreversible and continuous reactions between Li and electrolytes that generate thick solid electrolyte interphase (SEI) layers on the Li metal surface, consume Li and electrolyte, as well as increase the internal resistance. Thus it's extremely necessary to clearly understand and consider Li dendrites and related issues.

4.2.1 Li Dendrite Formation

4.2.1.1 Li Dendrite Nucleation

Lithium metal can be plated and stripped repeatedly during the charge/discharge processes of LMBs. During Li deposition, i.e., charge process of LMBs, the initial nucleation sites on Li metal surface will be a dominant factor for governing subsequent Li deposition. Li metal thermodynamically easily grows dendritic morphology during such a Li deposition process. In contrast, some multivalent ion metal such as magnesium (Mg) tends to be dendrite-free with two (2D) – and three-dimensional (3D) morphology rather than one-dimensional (1D) dendritic morphology. This is because that Mg has a higher free energy difference between high-dimensional and low-dimensional phases than Li [3, 22, 23]. In addition, the surface diffusion of the Mg^{2+} ion on Mg anode surface is obviously faster than that on Li or Na surface, which makes Mg metal deposit toward the surrounding sites instead of on some lumped area to grow undesirable dendrites. Overall, in terms of thermodynamic view, both lower surface energy and higher migration energy of Li metal primarily contribute to dendrite formation during Li metal deposition.

During the initial stage of Li-ion plating process, the nucleation models can be mainly classified as follows: (1) heterogeneous nucleation model [24, 25], (2) space-charge model [26], and (3) SEI-induced nucleation model [27]. Li nucleation occurrence on the alien current collectors is identified as a heterogeneous nucleation behavior. As demonstrated by Cui's group, such a heterogeneous nucleation is demonstrated by the selective deposition of Li on the gold (Au) substrates rather than copper (Cu, Fig. 4.2) [25]. In addition, Chazalviel reported a space-charge model for Li dendrite nucleation [26]. Briefly, during Li-ion deposition at high rates, the concentration of anions in dilute electrolyte decreases near the electrode surface, which leads to the establishment of a very large space charge an electric field near the electrolyte/electrode interface. This facilitates a subsequent ramified growth of Li [26]. Furthermore, Li metal is thermodynamically unstable with an organic solvent; thus both can react to form solid-electrolyte interphase on Li metal anode [28]. This SEI layer will continuously break down and re-form during Li deposition/stripping processes, further promoting dendritic Li growth. However, most models in the reported literature do not focus on the effect of the SEI on the Li dendrite growth mechanisms. So, this can be classified as an SEI-induced nucleating model. The Sacci's group revealed direct visualization of SEI formation during lithium deposition using in situ transmission electron microscopy (TEM) [27]. It can be seen that the dark dashed curve indicates the gradual formation of SEI, prior to Li deposition process, as shown in Fig. 4.3. Nevertheless, the actual formation sequence of SEI formation and Li deposition process would be immediately varied by changing electrolytes in a practical condition.

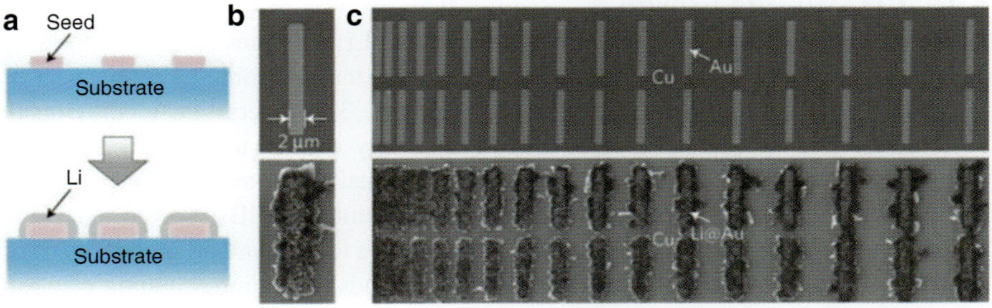

Fig. 4.2 (**a**) Schematic of patterned Li deposition. Li metal can be preferably plated on seeds without an overpotential due to the different overpotentials during Li nucleation. (**b**) SEM images of a gold strip before and after Li deposition at a current density of 0.5 mA cm^{-2} for 0.1 mAh cm^{-2}. (**c**) SEM images of a gold strip array with various separations before and after Li deposition. (Reproduced with the permission from Yan et al. [25]. Copyright 2016, Nature Publishing Group)

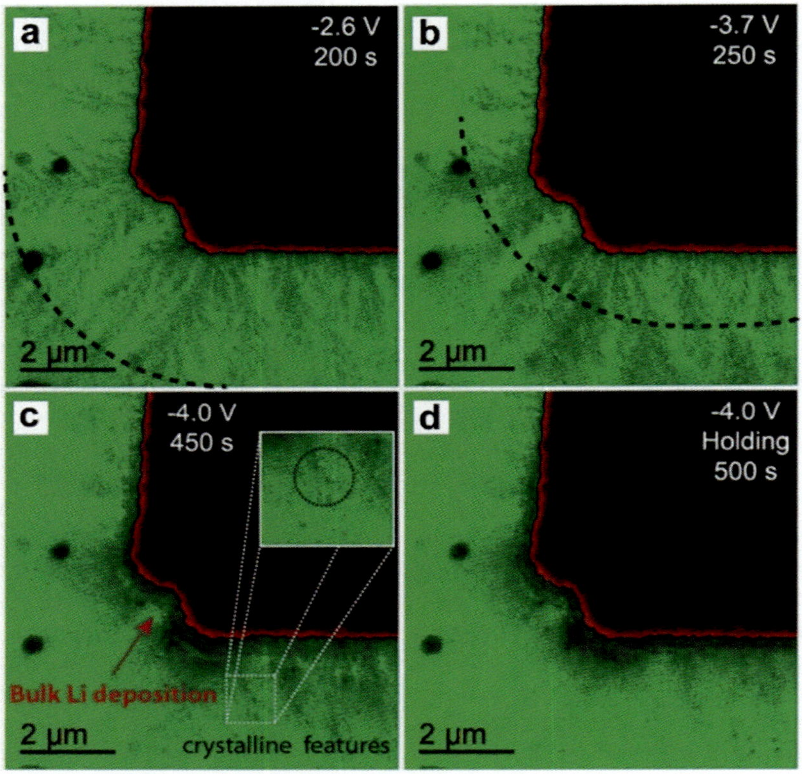

Fig. 4.3 Selected individual images revealing dendrite evolution during different times: (**a**) 200 s (potential: −2.6 V), (**b**) 250 s (potential: −3.7 V), (**c**) 450 s (potential: −4.0 V), and (**d**) 500 s (potential: −4.0 V). Inset: the formation of crystalline features. (Reproduced with the permission from Sacci et al. [27]. Copyright 2014, Royal Society of Chemistry)

4.2.1.2 Li Dendrite Growth

After nucleation of initial-stage Li dendrite growth, Li continues to grow into large deposits. Li dendrite growth is mainly determined by the electronic field, the electrolyte concentration, and properties of SEI films [29, 30]. According to previously reported results, morphologies of Li dendrites mainly contain needlelike, mosslike, treelike structures [31–33]. Needlelike one-dimensional dendrites without numerous branches hold the most threat to cause the short circuit in LMBs. So far, extensive efforts have been made to characterize and analyze the formation and growth steps of Li dendrites by using different characterization methods, including scanning electron microscopy (SEM), atomic force microscopy (AFM), transmission electron microscopy (TEM), optical microscopy, nuclear magnetic resonance (NMR), and Fourier transform infrared spectroscopy (FTIR) [34–39]. For example, Mehdi et al. directly observed real-time SEI formation and Li dendrite growth evolution at Li anode/electrolyte interface during the first three charge-discharge cycles (Fig. 4.4) [40].

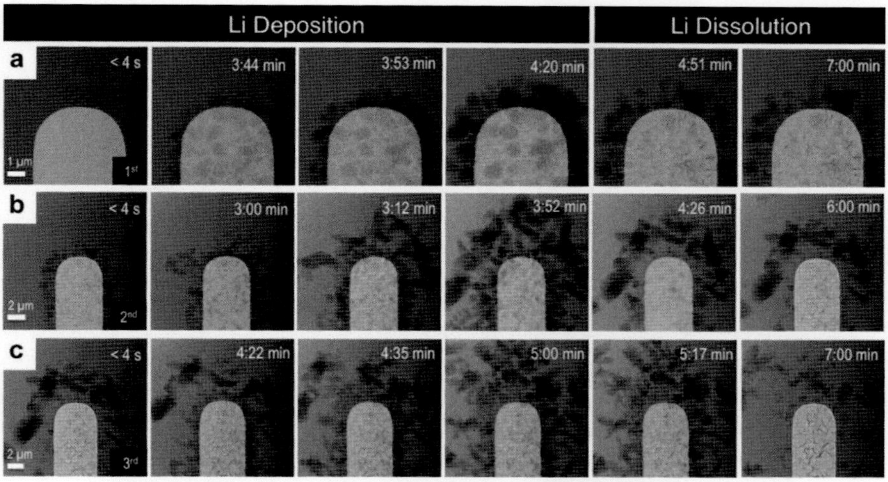

Fig. 4.4 High-angle annular dark-field (HAADF) images of Li deposition and dissolution at the interface between the Pt working electrode and the lithium hexafluorophosphate [$LiPF_6$] solution in propylene carbonate [$C_4H_6O_3$] electrolyte during the first (**a**), second (**b**), and third (**c**) charge/discharge cycles of the liquid cell. (Reproduced with the permission from Mehdi et al. [40]. Copyright 2015, American Chemical Society)

Also, some inactive "dead" Li residues detach from the platinum (Pt) electrode after Li stripping process, and it's one of the main reasons for the low Coulombic efficiency (CE) in rechargeable LMBs. In recent work [41], an *operando* observation of the Li electrode surface were time-synchronized with the corresponding potential traces of the designed cell in 1 M lithium hexafluorophosphate solution ($LiPF_6$) in ethylene carbonate and dimethyl carbonate (EC, $C_3H_4O_3$/DMC, $(CH_3O)_2CO$) electrolyte in the first three half-cycles from an even larger electrode area scale (Fig. 4.5). The formation/dissolution of new lithium was entangled with the formation of "dead" lithium during repeated cycles across the whole electrode surface. This causes a continuous displacing of "dead" lithium in the electrode and the consequent uncontrollable growth of Li upon cycling.

Some considerable external influences (i.e., current density, pressure, and temperature) on the Li dendrite growth behaviors have been investigated as well. An early work has revealed the influence of the current density and Li plating capacity on the dendrite formation on Li electrode [42]. It has been found that the total amount of discharge capacity enabling no dendrite formation increases with increasing current density up to 1 mA cm^{-2} and then decreases with further increasing current density, less than 2 mA cm^{-2} (Fig. 4.6). An optimal condition of a current rate of 1 mA cm^{-2} with a total amount of discharge of 12 C cm^{-2} for Li metal deposition was found to prevent the heterogeneous Li dendrite growth. In addition, the external pressure applied on lithium metal battery (LMBs) will have an immediate and partial suppression of Li dendrite growth occurred on Li metal anode. In a pioneering research in 2003, researchers found that dendrite growth could be

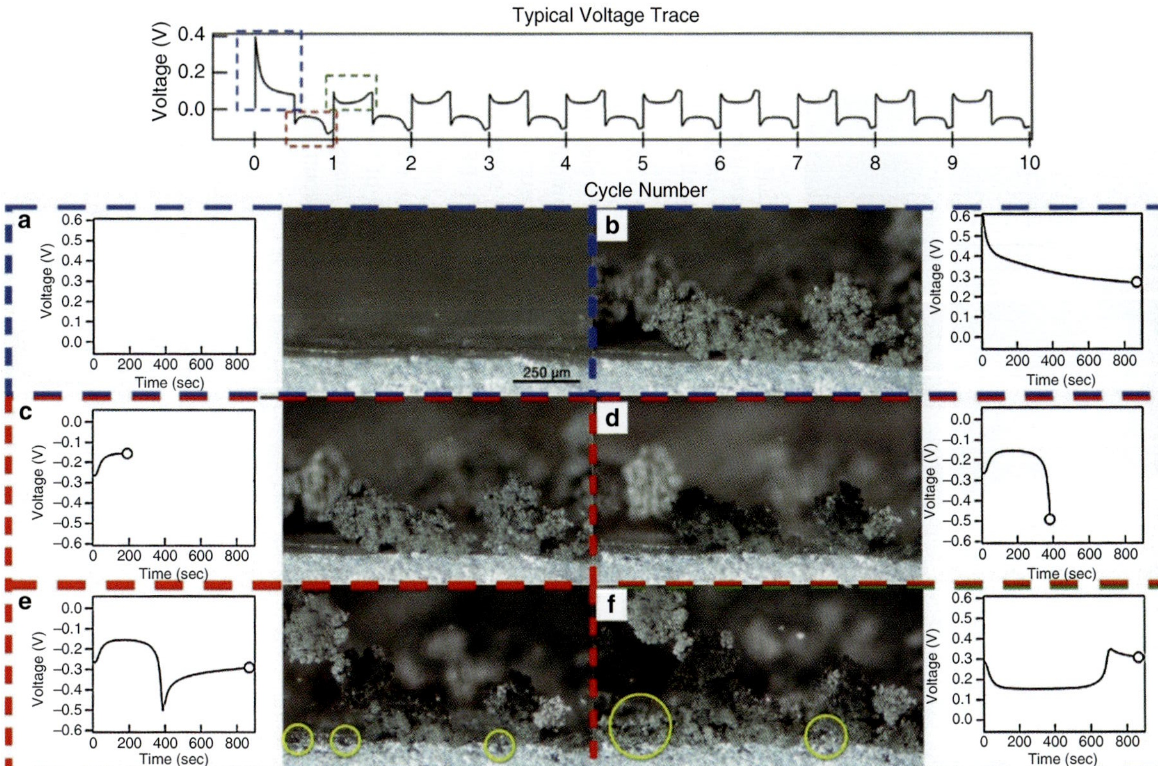

Fig. 4.5 Li formation evolution and corresponding voltage trace are recorded during different steps. (**a**) Before cycling; (**b**) after first half-cycle; (**c**) at cell polarization minimum; (**d**) at cell polarization maximum; (**e**) at the end of half-cycle; (**f**) at the end of third half-cycle. Some pitting sites are highlighted in yellow circles. (Reproduced with the permission from Wood et al. [41]. Copyright 2016, American Chemical Society)

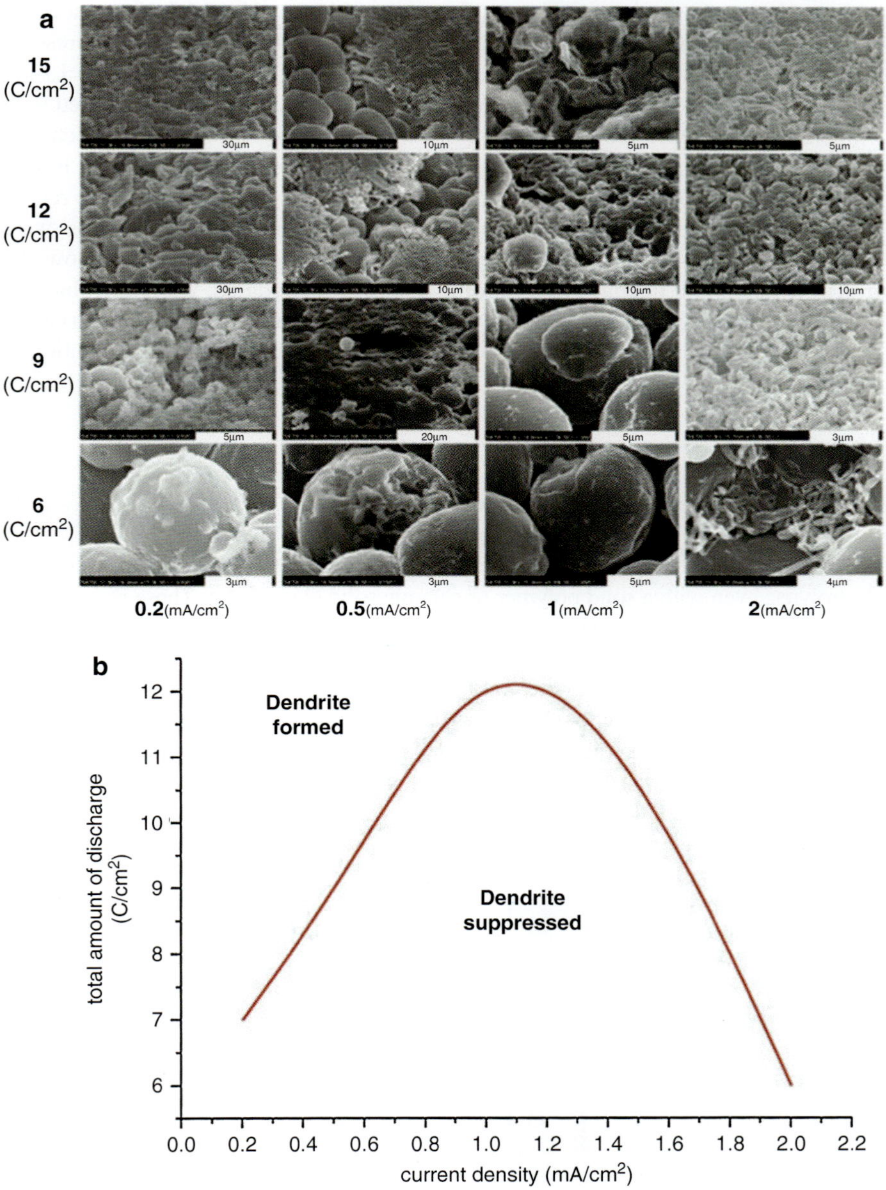

Fig. 4.6 (**a**) Scanning electron microscope (SEM) images of the lithium after one cycle of dissolution/deposition under various conditions of current densities and total amounts of discharge. (**b**) The illustrated curve of dendrite formation/dendrite suppression as a function of current densities and the total amount of discharge. (Reproduced with the permission from Seong et al. [42]. Copyright 2008, Elsevier)

inhibited by mechanical ways when utilizing electrolytes with a high shear modulus of more than two times larger than that of Li metal [43]. This critical role of pressure particularly in lithium metal battery (LMB) should arouse more attention. Li-ion diffusion, electrolyte viscosity, and reaction kinetics on Li metal are associated with temperature. Akolkar reported modeling dendrite growth during lithium electrodeposition at a sub-ambient temperature, giving a good guideline to define the lowest temperature for safe operation of Li-metal batteries [44]. A key temperature region required to suppress dendrite formation under different current densities were suggested in Fig. 4.7. The red curve indicates a critical temperature. Below the critical temperature, uncontrolled dendrite growth was initiated, while the blue one shows a much lower dendrite growth rate ratio. The region above the blue curve can be identified as the region for suppression of dendrite growth and stable operation of batteries.

4.2.2 Low Coulombic Efficiency Issue

The high reactivity of Li metal results in the parasitic reactions with almost all liquid electrolytes, even with a solid polymer or ceramic solid electrolytes. Those reactions are generally hard to control, which continuously consumes both electrolyte and Li metal, leading to low Coulombic efficiency (CE) during LMBs. The aforementioned Li dendrite problem could intensify the low CE issue of Li metal due to the increased exposed surface area of Li to the electrolytes. Effective methods are required to minimize the side reactions between the electrolyte and Li metal to address the above issue.

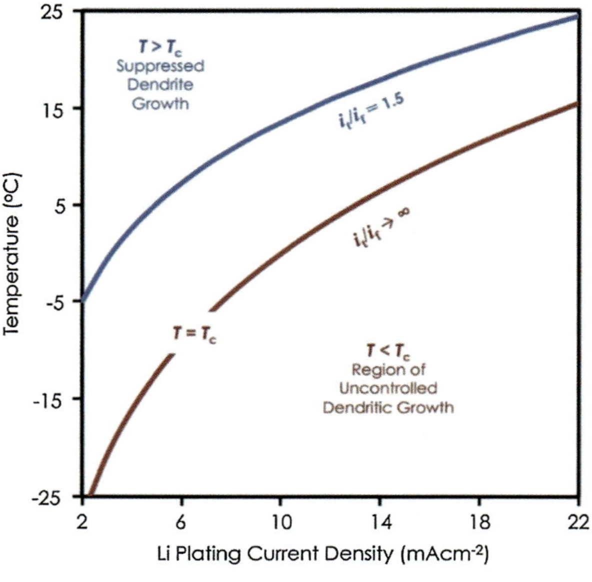

Fig. 4.7 Variation of the dendrite growth rate ratio as a function of temperature and a current density of Li plating process. (Reproduced with the permission from Akolkar [44]. Copyright 2014, Elsevier)

4.2.2.1 Evaluation of CE of Lithium Plating/Stripping

In most literature, different calculation methods of CE often result in different values of CE. In addition, CE is dependent on cycling conditions (i.e., current density, temperature, cycling time, and aging). Thus, the value of the CE for a given electrolyte could be misleading and not comparable if the calculation of CE is not associated with detailed testing conditions [28]. Here, four kinds of evaluation methods of CE for Li plating/stripping processes are listed below:

Method 1: CE of Li is calculated through:

$$CE = \frac{Q_S}{Q_D} \tag{4.1}$$

where Q_S and Q_D are charged for Li stripping and plating determined for each cycle. This method is first defined by Aurbach in 1999 [45], which is applicable when Li is deposited on the inert electrode surfaces, such as nickel (Ni), copper (Cu), and tungsten (W).

Method 2: CE of Li metal is calculated through [46]

$$CE = \frac{nQ_C}{nQ_C + Q_D} \tag{4.2}$$

where n denotes the total deposition/stripping times, Q_D is for originally known amount of deposition charge for Li metal on the working electrodes (Ni or Cu), Q_D is usually determined by the thickness and density of Li metal electrode, and Q_C is a fraction of charge cycled across the cell in order to advance lithium striping-deposition cycles. The stripping overpotential is monitored upon cycling. Above process will be completed once the anodic overpotential reaches more than 100 mV.

Method 3: Moreover, some literature also used a figure-of-merit (FOM) value rather than CE. For example [47],

$$FOM_{Li} = \frac{nQ_D}{Q_T} = \frac{1}{1 - CE} \tag{4.3}$$

where nQ_D is the total accumulated cell capacity, while Q_T is the theoretical Li capacity or Li turnover number.

Method 4: Besides the above methods, Zhang's group developed more reliable methods for measuring CE to precisely quantify the amount of Li consumed during cycling and estimate the cycle life of LMBs [48]. In this method, a given amount of charge (Q_T) is used to deposit Li on the copper (Cu) as a Li reservoir,

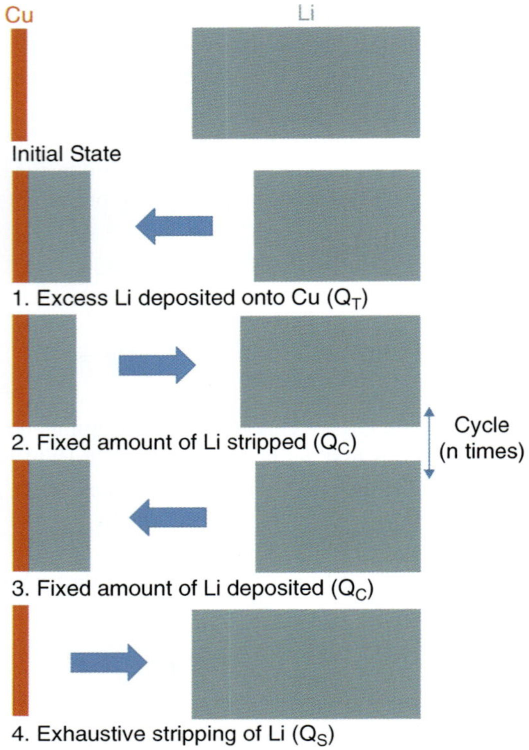

Fig. 4.8 Schematic of cycling Li||Cu cells for measurement of Li CE using method 4 with the constant current protocol in the 4 M lithium bis(fluorosulfonyl) imide (LiFSI, $[F_2LiNO_4S_2]$)-1,2-dimethoxyethane (DME, $[CH_3OCH_2CH_2OCH_3]$) electrolyte. (Reproduced with the permission from Adams et al. [48]. Copyright 2018, WILEY-VCH)

then a smaller portion of the charge (Q_C) is used to cycle Li for n cycles (Fig. 4.8). After n cycles, a final exhaustive stripping process, corresponding to the stripping charge (Q_S), is conducted to completely remove all of Li residues on Cu. So, the average CE (CE_{avg}) can be calculated below:

$$CE_{avg} = \frac{nQ_C + Q_S}{nQ_C + Q_T} \tag{4.4}$$

The advantage of this method is to minimize the effects of Cu current collector because Li plating/stripping occurs on Li metal substrate rather than the external substrate in a practical LMB [48].

4.2.2.2 Solid Electrolyte Interphase on Li Metal Surface

The formation of complicated solid electrolyte interphase through the reaction between Li and electrolytes was first found by Dey in 1970 [49], then named as solid electrolyte interphase (SEI) film by Peled in 1979 [50]. Basically, any electrolytes can be reduced at the Li metal surface due to the highly negative electrochemical potential of Li. The passivation effect from SEI facilitates the cell operation in a reductive environment at a voltage window of more than 4 V. Different from the well-known graphite anode in current LIBs, Li anode has more restrictive requirements on the SEI properties to be rechargeable. The composition, morphology, ionic conductivity, flexibility, and elasticity are strongly related to Li plating behavior and CE of Li metal. In practice, the formation of SEI on Li metal surface is originated from the intrinsic reaction between electrolyte and Li metal. A broad range of factors from various electrolyte solvents and salts and their concentrations, etc. will dramatically change the lowest unoccupied molecular orbital (LUMO)/highest occupied molecular orbital (HOMO), ion diffusion, and viscosity of electrolytes. The SEI formed on Li metal is therefore tailored by thermodynamic and kinetics of reactions between electrolyte and Li metal. Ideally, the SEI should possess the following features: (1) high ionic conductivity to minimize the diffusion resistance of Li^+ through this SEI film, (2) dense layer with appropriate thickness to prevent continuous direct contact between electrolyte and Li metal, (3) strong mechanical and elastic function to greatly accommodate the volume and morphological change of Li metal during cycles, and (4) excellent stability in terms of chemistry/electrochemistry in the long-lasting cycles.

Influence of Electrolyte Solvent

Organic carbonate-based electrolytes, widely used in today's commercial LIBs, are currently applied for the LMBs. The SEI composition from carbonate electrolytes is mainly Li alkyl carbonates [$ROCO_2Li$] produced from one-electron reduction of alkyl carbonates [39], and composition of inorganic components (lithium oxide [Li_2O], lithium carbonate [CLi_2O_3], as well as Li halides [LiI]) [51]. Unfortunately, the SEI film generated from traditional carbonate electrolyte usually has poor flexibility to stably accommodate repeated Li plating and stripping [52]. Compared to carbonate solvents, ethers generally presented higher stability against Li metal anode for both CE and suppressed dendrite growth [53], which could be attributed to the strong binding affinity to Li surfaces and good flexibility of as-formed oligomers in ether-based electrolytes. However, common ethers have low anodic decomposition voltage (<4 V vs Li^+/Li) and high flammability [53].

Influence of Electrolyte Salt

Selection of electrolyte salts also plays an important role in the SEI formation, further affecting CE of Li metal that is strongly dependent on the nature of SEI. Kanamura et al. studied the effect of different electrolyte salts (lithium hexafluoroarsenate [$LiAsF_6$], lithium hexafluorophosphate [$LiPF_6$], lithium perchlorate [$LiClO_4$], or lithium tetrafluoroborate [$LiBF_4$]) on SEI on Li electrode [54, 55]. The SEI composition and resistance can be accordingly adjusted by using different salts and thus influence the CE of Li metal. Ding et al. gave an evident difference for the CE of Li electrodes with different electrolyte salts as follows: lithium hexafluoroarsenate (95% [$LiAsF_6$]) > lithium bis(oxalate)borate (LiBOB, 93% [$LiB(C_2O_4)_2$]) > lithium difluoro(oxalato)borate (LiDFOB, 86% [$LiBF_2(C_2O_4)$] >

lithium hexafluorophosphate (77% [$LiPF_6$]) > lithium trifluoromethanesulfonate (73% [$LiCF_3SO_3$]) > lithium perchlorate (72% [$LiClO_4$]) lithium bis(trifluoromethanesulfonyl)imide (LiTFSI, 72% [$CF_3SO_2NLiSO_2CF_3$]) ~ lithium tetrafluoroborate (72% [$LiBF_4$]) > lithium iodide (69% [LiI])} [56].

4.2.3 Strategies to Stabilize Li Anode

To deal with above two critical issues of Li metal anode, i.e., dendrite growth and low CE, researchers have developed numerous methods, such as optimizing electrolytes, designing three-dimensional hosts for Li metal deposition, tailoring interfacial engineering, etc. Those considerable strategies already showed encouraging results to realize the practical application of Li metal anodes in LMBs, especially high-energy density Li-S and Li-O_2 batteries in the near future.

4.2.3.1 Electrolyte Engineering

Electrolyte stability has been regarded as a most important element for any battery systems, not only limited to LMBs. Basically, the anions of the Li salts and the solvent molecules can decompose to produce SEI films on Li metal surface either during the initial contact between electrolyte and Li or through the subsequent electrochemical processes [51, 56–58]. Herein, electrolyte engineering toward a stable interface between Li and electrolyte was proven to be one fundamental approach to stabilize Li metal anode.

Liquid Electrolyte Additives

Researchers have demonstrated that the use of the efficient additives in the electrolytes can suppress the Li dendrite growth. In a typical 1 M lithium hexafluorophosphate [$LiPF_6$]/ethylene carbonate (EC, [$C_3H_4O_3$])-ethyl methyl carbonate (EMC, [$C_4H_8O_3$]) electrolyte, the deposited Li on copper (Cu) substrate generally grows in a random way to form needlelike and mosslike morphologies (also called "dendrites") [59]. Qian et al. found that the addition of a trace amount of water (50 ppm) to 1 M lithium hexafluorophosphate ($LiPF_6$)/propylene carbonate (PC, [$C_4H_6O_3$]) can help the formation of dendrite-free compact and dense Li-deposited film composed of well-aligned nanorods (Fig. 4.9a) [58]. A trace amount of hydrogen fluoride [HF] originated from the decomposition reaction of $LiPF_6$ with water [H_2O] can be reduced during the initial Li deposition process by forming a dense and uniform lithium fluoride [LiF]-rich SEI layer on the substrate. The uniform SEI layer led to a uniform distribution of the electric field, enabling subsequent dendrite-free Li deposition. 0.05 M cesium hexafluorophosphate [$CsPF_6$] has also been reported as an effective addition in 1 M $LiPF_6$-PC electrolyte, leading to a highly compact and very smooth Li deposited layer on Cu substrate (Fig. 4.9b) [60, 61]. Cesium ion (Cs^+) has been proposed to play a critical role in generating a self-healing electrostatic shield effect on Li metal surface. With electrostatic shielding of Cs^+, the Li tends to deposit on the adjacent regions of initial Li growth tips and thus eliminates the Li dendrite growth. In addition, the $LiPF_6$ was

Fig. 4.9 (**a**) Scanning electron microscope (SEM) images of Li deposits with different morphologies: well-aligned nanorods in the 1.0 M lithium hexafluorophosphate [$LiPF_6$]/propylene carbonate (PC, [$C_4H_6O_3$]) electrolyte with 50 ppm water [H_2O] additive. (Reproduced with the permission from Qian et al. [58]. Copyright 2015, Elsevier) (**b**) Compact and very smooth film in the 1 M $LiPF_6$/PC electrolyte with 0.05 M cesium hexafluorophosphate [$CsPF_6$]. (Reproduced with the permission from Ding et al. [60]. Copyright 2013, American Chemical Society). (**c**) Large wires in the 0.05 M $LiPF_6$-added LiTFSI [$CF_3SO_2NLiSO_2CF_3$]-LiBOB [$LiB(C_2O_4)_2$]/EC [$C_3H_4O_3$]-EMC [$C_4H_8O_3$] (4:6 by wt.) electrolyte. (Reproduced with the permission from Zheng et al. [59]. Copyright 2017, Nature Publishing Group)

also reported as an effective additive in the LiTFSI [$CF_3SO_2NLiSO_2CF_3$]-LiBOB [$LiB(C_2O_4)_2$]/EC [$C_3H_4O_3$]-EMC [$C_4H_8O_3$] dual-salt electrolyte, which promotes the formation of more compact wiry Li deposits with conductive SEI layer on Li surface (Fig. 4.9c) [59]. The $LiPF_6$ additive can facilitate the polymerization of EC solvent molecules to produce polycarbonates on the Li surface and protect cathode and Al current collector. Though the aforementioned additives have been proven to be effective to suppress Li dendrite, some additives (i.e., H_2O and $CsPF_6$) are not able to improve CE of Li anodes much. Recent work presented the use of mixed additives of 2% $LiAsF_6$ and 2% vinylene carbonate (VC, [$C_3H_2O_3$]) or fluoroethylene carbonate (FEC, [$C_3H_3FO_3$]) is beneficial for both formation of well-aligned nanorod Li deposits and the increase of Li CE from 73.2% to 96.5% [62]. The increase in CE was due to the electrochemical reduction of $LiAsF_6$ additive, i.e., lithium fluoride [LiF] and lithium arsenate [Li_xAs], which passivates copper (Cu) substrate and serves for nanoseeds for aligned Li growth. Meanwhile, the polymerization products of vinylene carbonate (VC) or fluoroethylene carbonate (FEC) protects Li metal and promotes the cycling stability of LMBs.

Concentrated Liquid Electrolyte

Jeong et al. pioneered the study in high concentration propylene carbonate (PC, [$C_4H_6O_3$]) electrolytes to suppress Li dendrite deposition in 2008 [63]. After that, different kinds of high concentrated electrolytes (HCE) have been confirmed to be effective to the protection of Li metal and suppression of Li dendrite growth. For examples, 7 M bis(trifluoromethane)sulfonimide lithium salt (LiTFSI, [$CF_3SO_2NLiSO_2CF_3$]) in 1,3-dioxolane (DOL, [$C_3H_6O_2$]) and dimethyl ether (DME, [$(CH_3)_2O$]) [64], 4.9 mol kg^{-1} lithium bis(fluorosulfonyl)imide (LiFSI, [$F_2LiNO_4S_2$]) in fluorosulfonylimide (FSI)-based ionic liquids [65], 4.2 M lithium bis(trifluoromethanesulfonyl)amide (LiTFSA, $[(CH_3)_3Si]_2NLi]$) in acetonitrile (AN, $[CH_3CN]$) [66], 4 M LiFSI in DME [67], 4 M lithium nitrate $[LiNO_3]$ in dimethyl sulfoxide (DMSO, $[(CH_3)_2SO]$) [68], 3 M LiTFSI in DME [69], and LiTFSI in 3 M DMSO electrolytes [70] have been reported (Fig. 4.10). According to calculation analysis in those works, the excess free solvents in dilute electrolytes react easily and directly with Li metal anodes, thus leading to continuous corrosion of Li metal. But highly concentrated electrolytes with optimal salt-solvent coordination are able to enhance the stability of electrolyte upon cycling. In an ideal concentrated electrolyte, all solvents and anions coordinate to Li ions to create a fluid polymeric network of Li^+ cations and anions to suppress potential parasitic reactions.

Although high concentration electrolytes have already demonstrated attractive features, such as high reductive/oxidative/thermal stability, high carrier density, fast electrode reaction, and low volatility, the greatly increased cost from the high concentration of salt has become a big concern for the high concentration electrolytes. Recently, Zhang's group reported a new concept of localized high concentration electrolyte (LHCE) by adding inert diluent into an high concentrated electrolyte (HCE) [71]. Briefly, the selected diluent solvent does not solvate salt but is miscible with solvent as well as Li^+-solvent solvates in the highly concentrated electrolytes (HCE). Consequently, the LHCE lowers the overall concentration of salt but retains a similar local coordination environment of salt and solvent to those in the HCE. For example, an LHCE of 1.2 M lithium bis(fluorosulfonyl)imide (LiFSI, $[F_2NO_4S_2Li]$) in 1,2-dimethoxyethane (DME, $[CH_3OCH_2CH_2OCH_3]$) – bis(2,2,2-trifluoroethyl ether (BTFE, $[CF_3CH_2)_2O]$) (1:2 by mol/mol) was reported by mixing an HCE of 5.5 M LiFSI-DMC electrolyte and diluent BTFE. Figure 4.11 shows the scanning electron microscope (SEM) images of Li-deposited layers on Cu at a current density of 1.0 mA cm^{-2} with HCE (Fig. 4.11a, b)

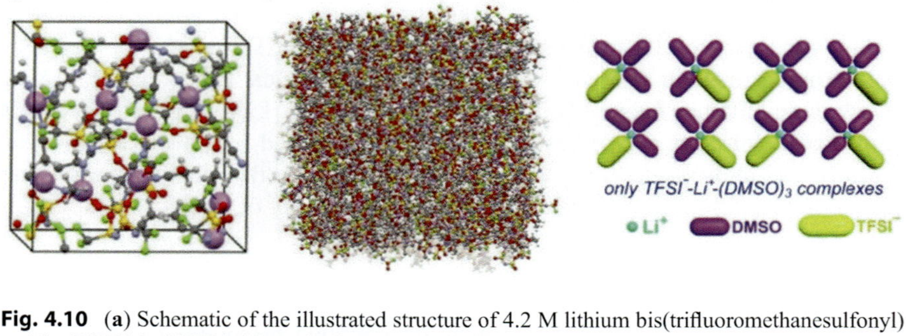

Fig. 4.10 (**a**) Schematic of the illustrated structure of 4.2 M lithium bis(trifluoromethanesulfonyl) amide (LiTFSA, $[(CH_3)_3Si]_2NLi]$) in acetonitrile (AN, $[CH_3CN]$) electrolyte. Atom color: Li, purple; C, dark gray; H, light gray; O, red; S, yellow; and F, green. (Reproduced with the permission from Yamada et al. [66]. Copyright 2014, American Chemical Society) (**b**) Schematic of the illustrated structure of 4 M bis(trifluoromethane)sulfonimide lithium salt (LiTFSI, $[CF_3SO_2NLiSO_2CF_3]$ in dimethyl ether (DME, $[CH_3)_2O]$) electrolyte. Atom color: Li, purple; O, red; N, blue; S, yellow; and F, green. (Reproduced with the permission from Qian et al. [67]. Copyright 2015, Nature Publishing Group) (**c**) Schematic of the illustrated structure of concentrated LiTFSI-3 M dimethyl sulfoxide (DMSO, $[CH_3)_2SO]$ electrolyte. (Reproduced with the permission from Liu et al. [70]. Copyright 2017, WILEY-VCH)

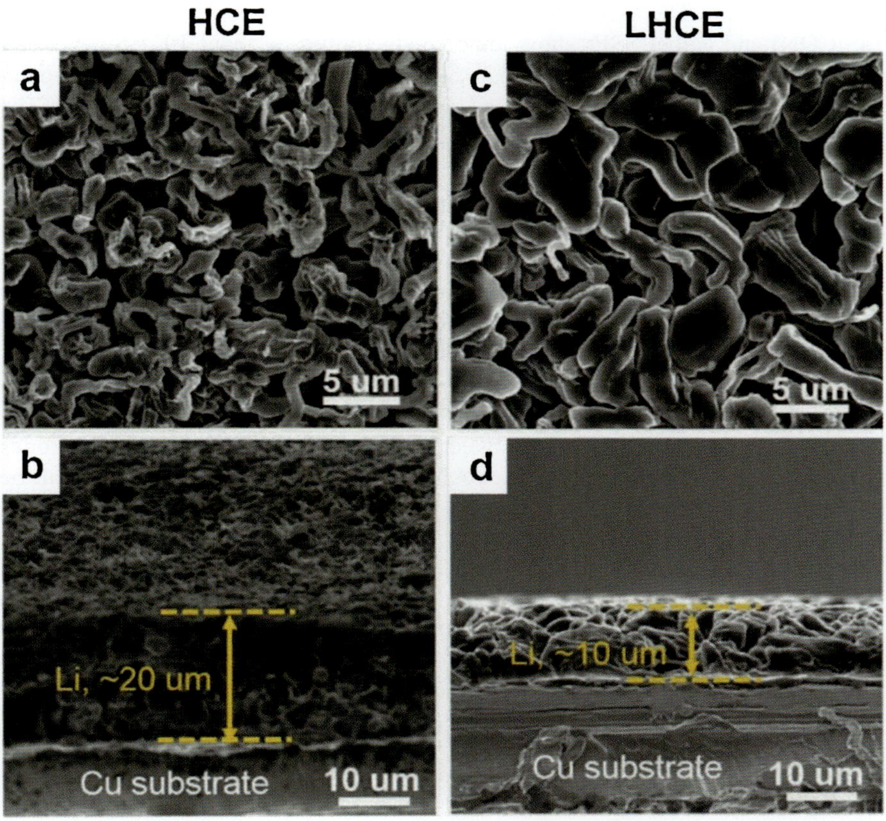

Fig. 4.11 Top and cross-sectional views of SEM images of Li deposits (plating amount: 1.5 mAh cm^{-2}) on Cu substrate at a current density of 1.0 mA cm^{-2} in the HCE of 5.5 M LiFSI/DMC (**a**, **b**), and the LHCE of 1.2 M lithium bis(fluorosulfonyl)imide (LiFSI, [$F_2NO_4S_2Li$]) in 1,2-dimethoxyethane (DME, [$CH_3OCH_2CH_2OCH_3$]) – bis(2,2,2-trifluoroethyl ether (BTFE, [$(CF_3CH_2)_2O$]) (1:2 by mol) (**c**, **d**). (Reproduced with the permission from Chen et al. [71]. Copyright 2018, WILEY-VCH)

and LHCE of 1.2 M LiFSI/DMC-BTFE (Fig. 4.11c, d). Clearly, the thickness of Li deposition from HCE of 5.5 M LiFSI/DMC is higher than that from LHCE of 1.2 M LiFSI/DMC-BTFE under a same Li deposition amount. The development of LHCE could be a more practical approach to stabilize the interfacial reactions on Li metal in LMB systems and other battery systems [72–74].

Solid-State Electrolyte

In contrast to liquid organic electrolytes, solid-state electrolytes (SSEs) have been considered as a final solution for the safety issues in the battery system, such as flammability, poor chemical stability, leakage, etc. The low ionic conductivity had been one significant limitation of SSEs before. In the recent years, it has been improved a lot. The oxygen-based SSEs (i.e., lithium super ionic conductor, LiSICON) and garnet SSEs show ionic conductivities of about 10^{-3} S cm^{-2} and 10^{-4} S cm^{-1}, respectively [18, 75–77]. Sulfide-based SSEs has reached a comparable value of that in a liquid electrolyte (10^{-2} S cm^{-1} at room temperature) [78]. The use of high-modulus SSEs in LMBs replacing liquid electrolytes has been found to suppress Li dendrite growth [19].

Besides the ionic conductivity issue of SSE, there are still grand challenges in the applications of SSEs in LMBs: (a) poor interfacial stability against Li metal anodes, especially for the sulfide SSEs, (b) large interfacial resistance between the SSE and Li metal, and (c) poor stability in the atmospheric environment. For example, interfacial side-reaction species can be detected on the interface between Li and SSE after simple physical contact [79]. Further, a clear comparison of electrochemical stability window among different SSE materials has been well summarized by Ceder's group on basis of the thermodynamics of formation of resistive interfacial phases between solid-state electrolyte and electrodes (Fig. 4.12) [80]. The effective electrochemical window of certain SSEs can be extended after reacting with electrode materials by forming barrier layers. In order to improve the interfacial contact, atomic layer deposition (ALD) coating layer on the SSEs surface represented the positive effect to improve wetting of EES surface and minimize the interfacial resistance on Li metal surface [81]. Meanwhile, the interfacial problems in the cathode are as important as that in the anode part; however, very few attentions were paid to such area so far. Overall, there is still a long way to improve the interfacial properties before applying SSEs in a practical LMB.

4.2.3.2 Stable Host for Li Metal

The volume change of electrode materials during the charge and discharge process is commonly observed, and it plays an important role in affecting the lifetime for various kinds of battery systems. For example, a relatively low volumetric expansion rate such as ~10% in graphite anode can be accommodated by porous electrode itself (Fig. 4.13a), while up to 300% volume expansion in silicon anode has become a critical concern for its cycle life due to the increasing challenges of SEI layer stability (Fig. 4.13b) [82, 83]. Despite those challenges, the modifications of electrode architecture and electrolyte additive have been proven to be on the right way to solve volume expansion problems in recent years [82, 84]. In contrast to the conventional electrode materials, the Li metal anode suffers virtually infinite volume change during plating and stripping process due to the "hostless" nature of Li metal anode (Fig. 4.13c). In addition, the huge volumetric expansion causes repeated breaking and re-forming of the SEI layer on the lithium metal surface and speeds up the degradation of Li metal anode and depletion of electrolyte during cycling [85, 86]. Very recently, the critical issue of volumetric change has been tackled by introducing a stable host material for Li metal [87]. Such approaches exhibit significant

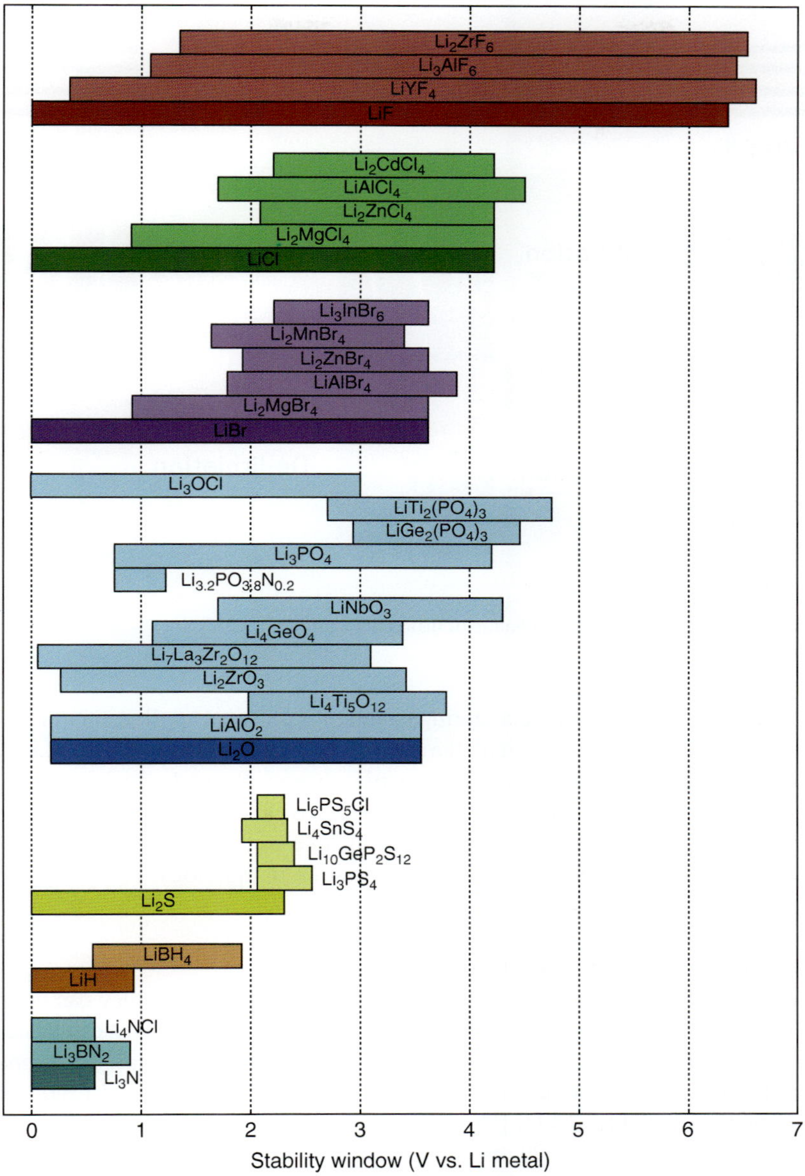

Fig. 4.12 Electrochemical stability (ECS) ranges of different electrolyte materials grouped by anions, with their corresponding binary. The high voltage stability of these materials is determined by the anions. (Reproduced with the permission from Richards et al. [80]. Copyright 2015, American Chemical Society)

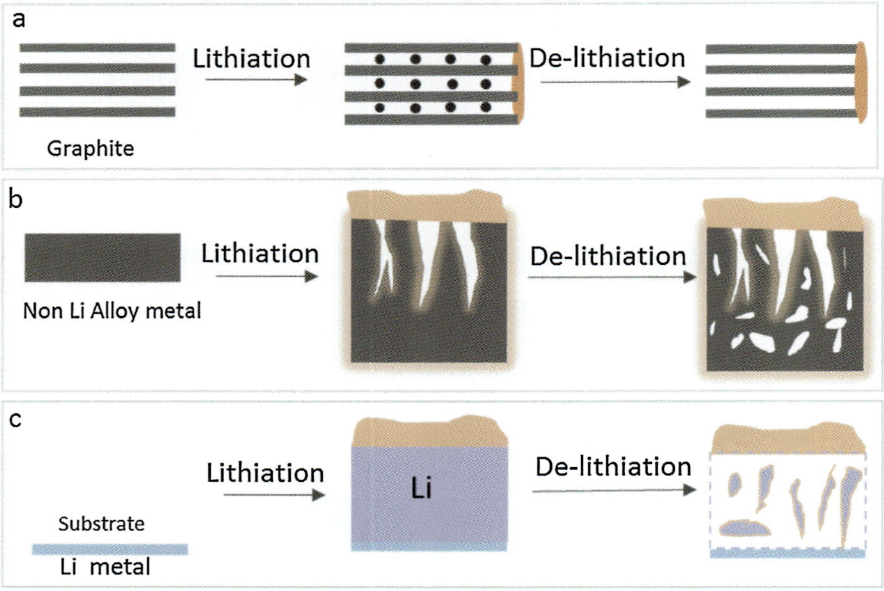

Fig. 4.13 Scheme of the volume change of different electrode materials. (**a**) Graphite, (**b**) silicon, and (**c**) Li metal

advantages in stabilizing Li metal during cycling for practical application of Li metal as the high-capacity anode. First, the host materials reserve space for Li metal deposition and reduce the overall volumetric change. Second, the host materials have higher surface area than planer Li metal, which dramatically reduces the effective local current density and increases the availability of more Li nuclei sites to guide a uniform and dendrite-proof Li deposition morphology.

Electronically Conductive Carbonaceous Host

Carbon-based materials have high conductivity and lightweight and they can be made with different shapes and structures, which make them very suitable host candidates for Li metal deposition. Yi and co-workers first reported a lithium-scaffold composite electrode fabricated by lithium melt infusion into a three-dimensional (3D) porous carbon fiber matrix (Fig. 4.14a) [88]. The composite electrode showed stable cycling with a small overpotential of 90 mV at a large current density of 3 mA cm^{-2} during Li plating/striping over 80 cycles. The wettability of Li metal (or lipophilicity) was also found to be of great importance for the spontaneous Li infusion into the carbon host since most carbon materials can hardly wet the molten Li metal. With a thin silicon (Si) coating layer by chemical vapor deposition (CVD) method on the carbon fiber scaffold surface, the carbon matrix changed from "lithiophobicity" to "lipophilicity" by increasing the bonding interaction between the carbon host surface and molten Li, such as forming a binary Li-Si alloy interfacial phase [88]. Other compounds, such as zinc

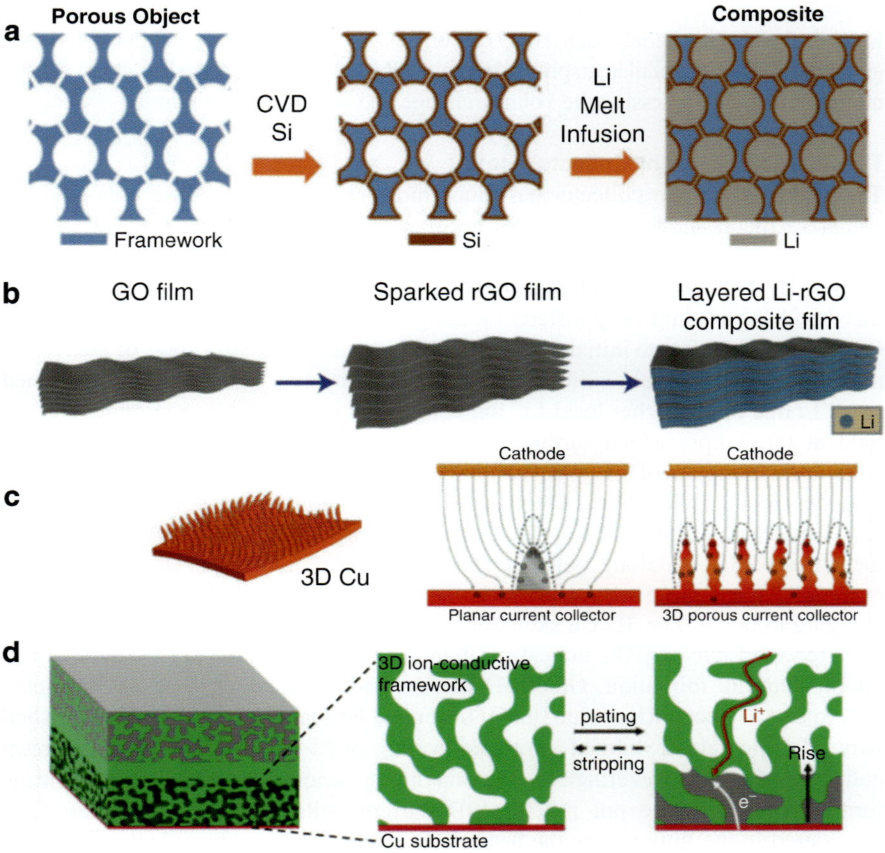

Fig. 4.14 Conductive 3D hosts for Li metal. (**a**) Schematic illustration of the design of a porous Li-scaffold composite with a lipophilicity treatment on host surface. (Reproduced with the permission from Liang et al. [88]. Copyright 2016, National Academy of Sciences) (**b**) Sparked r-GO film as host for Li metal. (Reproduced with the permission from Lin et al. [89]. Copyright 2016, Springer Nature) (**c**) Illustration of the electrochemical deposition processes of Li metal on a planar current collector and 3D copper (Cu) current collector. (Reproduced with the permission from Yang et al. [90]. Copyright 2015, Springer Nature) (**d**) The schematic of 3D ion-conductive host for Li plating and stripping. (Reproduced with the permission from Yang et al. [91]. Copyright 2017, National Academy of Sciences)

oxide (ZnO), that show strong chemical adsorption and reactions with Li metal may also be good candidates as a coating layer onto carbon host [92, 93].

Besides surface coating, surface functional groups that show strong affinity to Li metal may also help. Layered sparked reduced graphene oxide (r-GO) was found to possess excellent Li wettability due to the multifunctional hydrophilic groups on the surface of r-GOs (Fig. 4.14b). The Li-rGOs composite anode showed ~89% of the theoretical capacity of Li metal (3390 mAh $g_{\text{composite Li}}^{-1}$ anode vs 3800 mAh g_{Li}^{-1}) and significant reduction in the volume change to below 20%, thus resulting in significant improvement in cycling, reduction in polarization, and dendrite-free properties [89]. Other types of functional carbon materials such as 3D carbon nanotubes (CNT) [94], 3D fluorine-doped graphene [95], and cellular graphene scaffold [96] have been reported to show similar improvement in suppressing the volume change and dendrite growth during cycling.

The 3D Metal Current Collector Host

The 3D metal current collector has much more electroactive surface area than the planar Li metal anode, which could be directly used as a conducting host for Li deposition. Submicron-sized copper (Cu) skeleton synthesized by reduction of copper (II) hydroxide [$Cu(OH)_2$] fibers on Cu foil was proven to accommodate Li metal and show completely different plating behavior of Li metal [90]. On a planar Cu foil, the Li prefers to initially form small Li islands at the early stage of nucleation process. Then, Li metal tends to deposit on the sharp tips of the previously formed small Li due to the higher local Li^+ ion concentration induced by a stronger electric field at those tips, which further amplifies Li dendrite growth. In contrast, the submicron skeleton of 3D Cu with numerous protuberant tips on the Cu fiber provides more homogenous electric field distribution and produces uniform Li deposition within the 3D Cu current collector (Fig. 4.14c). Lithium gradually fills the pores of 3D Cu along with the increasing of Li amount on the 3D host and eventually forms an even surface. The filled Li can be completely stripped after charging to 0.5 V. The 3D Cu skeleton and dendrite-free feature remain unchanged after repeated cycles [90], indicating high reliable stability of 3D Cu skeleton to inhibit dendrite formation. Other 3D metal current collectors, such as 3D nickel (Ni) foam [97], cobalt(II) oxide [CoO] decorated Ni foam [98], 3D Cu-Ni core-shell wire [99], etc., also exhibited improved cycling of Li embedded in the 3D current collector network. Nevertheless, the maximum amount of Li without dendrite formation that can be put into the 3D current collector needs to be carefully considered in the future since the heavy and inactive metal current collector significantly reduces the achievable energy density of composite Li anode for practical application. In addition, the potential risk of piercing of 3D metal current collector through the separator during practical cell assembling also needs to be considered.

3D Ionic Conducting Host

The solid-state electrolyte presents higher interfacial stability toward Li metal compared with the liquid electrolyte due to limited reactivity between Li metal and solid-state electrolyte [100, 101]. Recently, solid-state electrolytes have been considered as a fundamental strategy to eliminate all the issues of Li metal such as dendrite growth and low coulombic efficiency for the ultimate application of Li metal batteries [102, 103]. A 3D Li^+ ion-conductive framework made of an ionic conductor can serve as a good candidate for the Li metal battery because it not only provides the stable host with stable interface against Li metal but also allows fast Li^+ ion transport through the host. Nevertheless, the solid electrolyte is usually hard to be processed as a thin film or 3D porous framework in practice; therefore, 3D Li^+ ion-conductive framework applied in Li metal is rarely reported in the literature. Very recently, Hu's group demonstrated a solid-state composite Li metal anode through integrating a 3D solid-state Li^+ ion-conductive host with Cu current collector [91, 104]. The 3D ion-conductive framework was fabricated by using a lithium lanthanum calcium zirconium niobium oxide [$Li_7La_{2.75}Ca_{0.25}Zr_{1.75}Nb_{0.25}O_{12}$] garnet-type Li^+ ion conductor with porous-dense-porous structure (Fig. 4.14d). As shown in Fig. 4.14d, the ion-conductive host consists of a dense layer as the separator and two porous layers for hosting cathode-active materials and Li metal anode. The 3D solid-state ion-conductive host shows a safe and dendrite-free behavior for Li metal during plating/stripping process. In addition, the Li starts to rise from the bottom near the Cu current collector during Li plating process and shows a reverse process during stripping, which further reduces the possibility of penetration of Li dendrite through the middle dense separator. The 3D ion-conductive host for Li metal can be cycled with good stability for 300 h at 1 mAh cm^{-2} at 0.5 mA cm^{-2} without failure of cell short or significant increase in overpotential.

Guided Li Deposition in 3D Host

A 3D host provides a favored choice to reduce the volume change of Li metal and improve its cyclability. However, it is of equal or more importance to guide the deposition of Li metal preferentially within the predesigned 3D host during repeated cycles, especially under the large cycled capacity of Li and at high current densities. It was found that Li deposition shows different overpotentials on different metallic substrates due to different nucleation barriers [25]. The commonly used Cu substrate for Li exhibits a substantial nucleation barrier because of negligible solubility of Li in Cu, whereas no barrier is shown for metals with a certain amount of Li solubility. A controlled Li nucleation and growth was demonstrated to effectively engineer the Li deposition process by a favored nuclei seed (such as gold (Au), silver (Ag), and zinc (Zn)) for Li metal [25, 105]. For example, the Au seeds located in the hollow carbon sphere facilitate the Li solely deposited inside the carbon shell and also act with a suitable host to accommodate Li metal and protect the Li metal from reacting with electrolytes (Fig. 4.15a) [25]. A much improved coulombic efficiency (>98%) was obtained over 300 cycles in an alkyl carbonate electrolyte for carbon sphere host with Au seeds, while a low coulombic efficiency of 90% with fast decay in 50 cycles was observed in a traditional Cu substrate. In addition, the Li deposition can also be guided by tuning the surface chemistry of host materials and 3D patterned substrates to direct Li deposition at specific positions with modified electric field distribution on the substrates (Fig. 4.15b) [106–108].

4.2.3.3 Interfacial Engineering for Li Metal

The interface/interphase between Li metal and electrolyte is the place where the Li plating/striping occurs in an LMB. The stability of interface/interphase, typically the SEI layer, has the immediate impact on the behavior of Li plating/striping and safety of Li metal batteries. The interfacial stability for Li metal is actually much more critical than that in conventional LIB technologies due to the high reactivity of Li metal with various kinds of electrolytes. Additional suitable protection layer on Li metal could be helpful to block the continuous reaction between electrolyte and Li

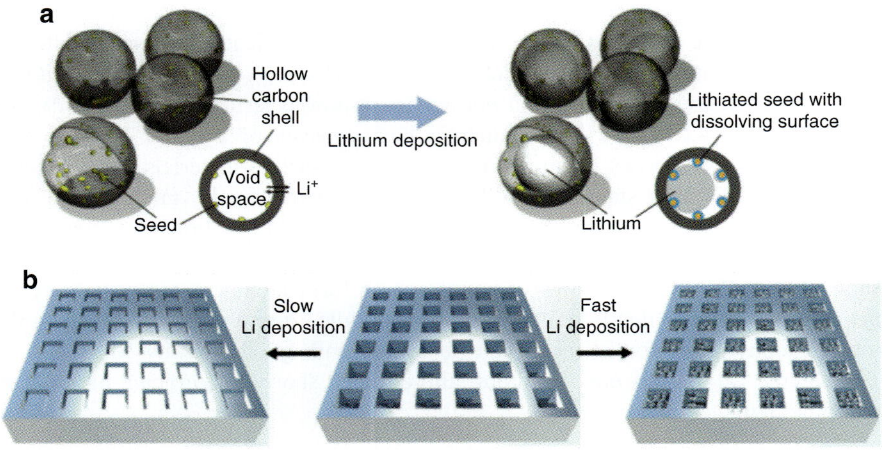

Fig. 4.15 (**a**) Schematic of induced Li metal deposition with Au seed in hollow carbon spheres. (Reproduced with the permission from Yan et al. [25]. Copyright 2016, Springer Nature) (**b**) Schematic of the illustration the surface patterned Li metal deposition. (Reproduced with the permission from Park et al. [106]. Copyright 2016, WILEY-VCH)

metal and provide a physical barrier layer to suppress Li dendrite growth and penetration. The interfacial protection layer can be either artificial or electrochemically formed layer on the Li metal surface, and this layer is highly expected to have the high mechanical strength and electrochemical stability upon cycling. In situ-formed SEI layers with tailored morphology, composition, and function largely depend on the formula of electrolytes, which have been discussed in Sect. 4.2.3.1 of this chapter. Herein, this section mainly focuses on the discussion of artificial protection layers on Li metal surface.

Solid/Gel Electrolyte Layer

An artificial solid/gel electrolyte interphase layer between Li metal and the liquid electrolyte can function as the effective protection layer for Li metal that can suppress Li corrosion and dendrite growth and provide fast Li^+ ion transport through the interphase layer. For example, a uniform artificial ionic conductive SEI layer primarily composed of lithium phosphorous sulfur [Li_3PS_4] (space group Pnma) prepared by in situ reaction of polyphosphoric acid (PPA, $H_n + 2P_nO3_{n+1}$) with Li metal layer has been found to be stable without a break/repair and short circuit during cycling of Li metal (Fig. 4.16a) [109]. Highly ionic conductive lithium nitride [Li_3N] protective layer obtained by direct reaction between Li and nitrogen (N_2) gas at room temperature was also demonstrated to suppress the nonuniform deposition of Li, improving the safety of the battery without sacrificing the migration of Li^+ ions [113]. Besides the ionic conducting compounds, in situ-formed polymer electrolyte layers deprived of the polarization or cross-linking reaction of organic electrolyte

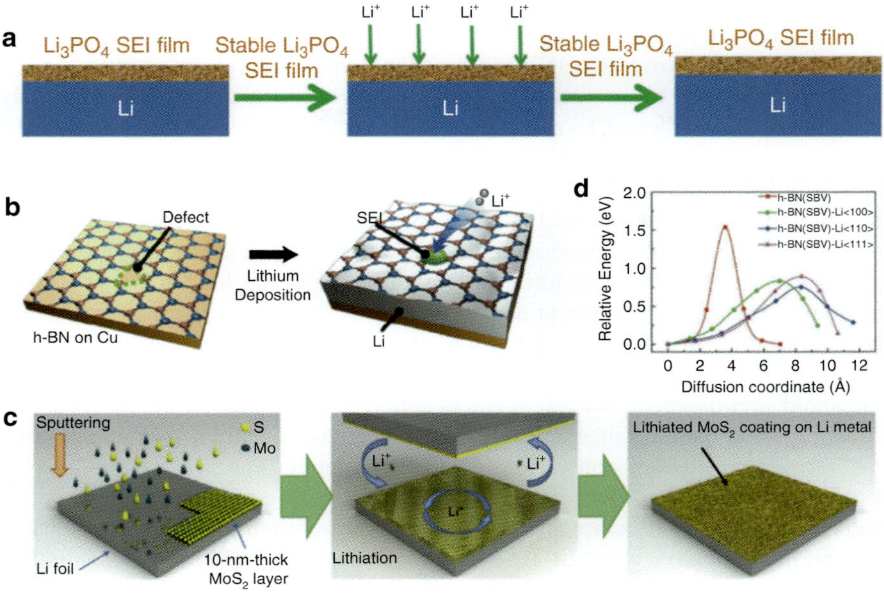

Fig. 4.16 (**a**) The lithium phosphate [Li_3PO_4]-modified Li metal anodes during SEI formation and cycling. (Reproduced with the permission from Li et al. [109]. Copyright 2015, WILEY-VCH) (**b**) The subnanometer defects in h-BN film grown on copper serve as channels for Li ions during lithium deposition. (Reproduced with the permission from Yan et al. [110]. Copyright 2014, American Chemical Society). (**c**) Schematics illustrating the fabrication method for a molybdenum sulfide (MoS_2)-coated Li anode via sputtering and subsequent lithiation. (Reproduced with the permission from Cha et al. [111]. Copyright 2018, Springer Nature). (**d**) Comparisons on Li^+ ion diffusion through lithiated and unlithiated materials of h-BN with single boron vacancy. (Reproduced with the permission from Tian et al. [112]. Copyright 2017, WILEY-VCH)

ingredients by the ultraviolet radiation-curing method were also effective in stabilizing Li metal [114, 115]. A combination of soft polymer electrolyte with inorganic ceramic particles can further improve the mechanical strength of artificial solid electrolyte layer to suppress Li dendrite piercing through the separator. For example, a composite layer of poly(vinylidene fluoride (PVdF)-hexafluoropropylene) (HFP) $\{(–CH_2CF_2–)_x[–CF_2CF(CF_3)–]_y\}$and aluminum oxide [$Al_2O_3$] particles demonstrated significantly improved elastic and shear modulus compared with the PVdF-HFP polymer and delivered a stable cycling under a current density up to 10 mAh cm^{-2} [116].

Layered Compound

The 2D layered compounds have wide applications in the electronic device, catalysis, and energy storage areas due to unique electronic structures [117, 118]. Recently, 2D layered materials were also extended to their potential application in protecting Li metal anode with a function of artificially tailoring of the top SEI layer properties and its Young's modulus, while the Li^+ ions can also freely pass through the layer in the vertical direction. For example, ultrathin 2D layered hexagonal boron nitride (h-BN) and graphene were used to cover onto the Cu current collector to form a 2D layer-Li-Cu sandwich structure, which effectively stabilized the interface of Li metal (Fig. 4.16b) and suppressed the growth of Li dendrites through the stiff B-N bond and chemical stability [110]. A 2D molybdenum sulfide (MoS_2) layer with a thickness of ~10 nm by the sputtering method was reported to function as a protection layer for Li metal anodes and greatly improve cell performance in Li-S batteries by stabilization of the electrodeposition of Li metal and suppression of dendrite nucleation (Fig. 4.16c) [111]. A low-voltage hysteresis even at a high current density of 10 mA cm^{-2} was observed, which has been attributed to the strong adhesion of the lithiated MoS_2 layer to Li and favorable Li transport through the layer (small energy barrier of 0.155 eV). A threshold improvement in cycle life was achieved compared with base Li metal, showing stable performance and low impedance upon cycling [111]. In addition, theoretical investigations on various 2D layered materials as a protection layer on Li metal have recently been carried out [112]. It has been found that the defect patterns, crystalline structure, bond length, and metal proximity effect play the key role on the protective effect of 2D layered materials in terms of ion diffusion and mechanical properties, which are based on the fundamental of the electronic interaction or redistribution inside the materials systems [112]. Although introducing defects and proximity effect may facilitate the Li^+ ion diffusion through the layer, it will weaken the mechanical strength of the protective layer (Fig. 4.16d). Therefore, a trade-off between ion diffusion and protective effect should be considered for practical applications.

Li-Rich Composite Alloy Film

Lithium metal has a certain amount of solubility in various metals such as aluminum (Al), zinc (Zn), indium (In), bismuth (Bi), etc., which makes Li alloy achievable in practice [119]. The reactivity of Li metal can be potentially and fundamentally adjusted due to the modification of the electronic structure of Li by forming Li alloys. Meanwhile, using the alloy approach, the properties of the SEI layer on Li metal, i.e., composition, structure, and function, therefore, can be tuned correspondingly [120]. Li alloy can be made by direct reactions with the other metals or chemically/electrochemically lithiation method. For example, the Li_xM alloy-protection layer can be chemically bound onto the Li metal foil surface through the following in situ reactions of metal chlorides with Li [120]:

$$\mathrm{xLi} + \mathrm{MCl}_x \rightarrow \mathrm{M} + \mathrm{xLiCl}\ (\mathrm{M} = \mathrm{ln,Zn,Bi,\ and\ As,etc}) \quad (4.5)$$

The as-formed metal layer will immediately undergo reactions with the underlying Li metal, forming a single or multiphases of Li alloy (it depends on the phase diagram of Li-M alloy):

$$yLi + zM \rightarrow Li_yM_z \quad (4.6)$$

In the cases of $Li_{13}In_{13}$ and LiZn, the in situ-formed composite Li alloy protection layer showed smooth Li deposition under the alloy protection layer in $Li_{13}In_{13}|Li$ and LiZn|Li cell, where the insulating LiCl in the alloy composite layer prevents the reduction of Li^+ ion on the top surface. The composite alloy layer retained its composition during cycling and demonstrated an extended cycling life of 1400 h in a symmetrical cell at 2 mA cm^{-2} [120]. In addition, a composite of Li_xM (M can be Si, tin (Sn), Al) alloy particles encapsulated by large graphene sheet further improved stability in different air conditions due to the completely covered Li_xM alloy clusters by graphene sheet. The Li_xM/graphene maintained a stable structure and improved cyclability over 400 cycles with 98% capacity retention in a half cell [121]. Moreover, the environmental stability of Li-containing anode toward the atmosphere, electrode materials, and electrolyte can be further improved by atomic layer deposition (ALD) of oxides or fluorides in the future.

4.2.3.4 Others

Except for the direct approaches of the chemical and/or physical engineering of Li metal anode, some external parameters such as the pressure will also fundamentally change the electrochemical behavior of Li metal thermodynamically. The Gibbs free energy of Li metal electrode will be tailored by external force p through $G(p, T) = U + pV–TS$, where G is Gibbs free energy, (kJ/mol); U is the internal energy, (kJ/mol); p is the pressure, (atm); V is the volume, (L); T is the thermodynamic temperature, (K); and S is the system entropy, (JK^{-1}). The pressure, therefore, affects the growth mode and morphology of Li metal during the Li plating/stripping process intrinsically. For example, it was recently found that the electrochemical deposition of Li metal can be affected by the external pressure [122]. Instead of highly porous Li deposition, a much more compact Li deposition was achieved under a pressure condition. The improved Li morphology under pressure leads to fivefold longer cycle life and 5% higher coulombic efficiency with the same electrolytes [122]. Though the external parameters are important to Li metal, this area is currently still almost blank. Experimental results combined with theoretical simulation of external pressure will be helpful to further understand the effect of pressure on the Li deposition behavior and improve safety for lithium metal battery (LMBs) in the future.

4.3 Cathodes for Li Metal Batteries

As mentioned at the beginning of this chapter, the LIBs have been approaching their limitation in energy density, which has become insufficient for their practical needs, especially for the use in the electrification of transport [123]. Rechargeable LMBs have been the most promising solution that can go beyond the horizon of LIB technologies due to the use of high-energy Li metal as the anode. In this section, we are going to discuss a few options of cathodes for the future development of next-generation high-energy rechargeable Li metal batteries.

4.3.1 Oxygen Cathode

Li-air or Li-O_2 battery has attracted worldwide interest over the last two decades due to their high theoretical specific energy (3505 Wh kg^{-1} for nonaqueous battery; 3582 Wh kg^{-1} for aqueous one) [124–128]. The first nonaqueous Li-O_2 battery was proposed by Abraham and Jiang in 1996 [129]. Since then, the extensive worldwide efforts have already been devoted to building an ideal Li-O_2 battery system to realize its practical application [130–138]. In general, a typical Li-O_2 battery includes an air electrode composed of the porous gas diffusion layer and a catalyst/carbon conductive agent, a separator soaked with electrolyte, and a Li metal anode. During the discharge process, gaseous O_2 first diffuses into the porous air electrode/electrolyte interface and then participates in an electrochemical reduction reaction ($2Li^+ + O_2 + 2e^- \rightarrow Li_2O_2$, $E_0 = 2.96$ V vs Li/Li^+). Upon the subsequent charge process, as-formed Li_2O_2 as discharge products can be decomposed, and meanwhile, O_2 is also released during an electrochemical oxidation ($Li_2O_2 \rightarrow 2Li + 2e^- + O_2$) [139].

Although a number of key considerable progress in Li-O_2 battery have been made to overcome the barriers in the practical large-scale application of Li-O_2 batteries, there are three main challenges in a rechargeable Li-O_2 battery: (1) The reduced oxygen species generated during the discharge process or oxygen reduction reaction (ORR), especially the superoxide radical anion, are very reactive with electrolyte, oxygen cathode, and Li metal anode [140]; (2) a high overpotential is required for oxygen evolution reaction (OER) during charge process [141]; and (3) continuous accumulation of undecomposed Li_2O_2 and other side reaction products at the oxygen-cathode surface during cycling leads to increased battery impedance [142]. Above factors cause limited reversible capacity, poor cycle life, and low CE of Li-O_2 batteries.

4.3.1.1 Catalysts for Oxygen Cathode

As the electrochemical reactions on the oxygen cathode of a Li-O_2 battery is indeed a kind of catalytic electrochemical process, thus those catalysts decorated on the oxygen cathode largely determine battery performance [138]. Rechargeable Li-O_2 batteries already benefited from many proposed catalysts, including noble metals, carbon materials, oxides, carbides, and polymers.

Noble Metals and Alloys

Some precious metals or alloys (e.g., platinum (Pt) [143, 144], gold (Au) [145], PtAu [146], silver (Ag) [147, 148], palladium (Pd) [149, 150], and iridium (Ir) [151]) have been proven to show exceptional catalytic capability during oxygen reduction reaction (ORR) and oxygen evolution reaction (OER) processes in Li-O_2 batteries. Primary purpose to introduce noble metal catalysts to oxygen cathodes is to decrease overpotential upon cycling, especially the OER process. Shao-horn's group presented bifunctional PtAu alloy catalysts by combining Pt with good catalytic performance during OER and Au with good catalytic capability during ORR, which significantly decreased the overpotential in their $Li\text{-}O^{-2}$ batteries and increased the round-trip efficiency [146]. Nevertheless, the high cost of those catalysts already hinders their future scale-up application in $Li\text{-}O_2$ batteries.

Functional Carbon Materials

Compared to the high cost of noble metals mentioned above, the carbonaceous material should be one of the most inexpensive alternatives. Importantly, different carbon materials provide unique features: low cost, wettability, large surface area, high electrical conductivity, and good stability under harsh environments [138]. Basically, porous carbon is the critical component for architecture design of gas diffusion layer in $Li\text{-}O_2$ batteries. Meanwhile, carbon can serve as both a substrate/conductive agent and a catalytic material [133]. Carbon materials with some defects or dopants can improve oxygen reduction reaction (ORR) kinetics in $Li\text{-}O_2$ batteries [152–157]. For example, it was reported that the $Li\text{-}O_2$ battery based on only functionalized graphene sheets with high-density reactive sites for $Li\text{-}O_2$ reaction and microporous channels for oxygen gas diffusion delivered an ultrahigh capacity of 15,000 mAh g^{-1} [153]. However, later subsequent research indicated that bare carbon catalyst without any protection (e.g., from catalyst's or other artificial films' protection) will suffer severe corrosion and oxidation by the attack from reduced oxygen species (especially superoxide radical anion). In addition, carbon materials have limited catalytic effect for the oxygen evolution reaction (OER) completely removing Li_2O_2 during charge process of $Li\text{-}O_2$ batteries, resulting in increased charge overpotential and poor cycle life. To address these challenges, Liu et al. recently proposed a new concept of the one-step in situ pre-charging process to simultaneously promote thin protective films on carbon nanotubes oxygen-cathode and Li metal anode in an inert atmosphere prior to the $Li\text{-}O_2$ battery cycling process (Fig. 4.17a) [158]. The pretreatment was conducted in inert atmosphere from the open-circuit voltage (OCV) to 4 .3V at 0.1 mA cm^{-2} and then holding the cells at 4.3 V for 0, 5, 10, 15, and 20 min, respectively. After the above optimal pre-treatment, $Li\text{-}O_2$ batteries demonstrated significantly extended cycle life of 110 cycles under the capacity-limited protocol of 1000 mAh g^{-1} (limited cycles of 43 times for untreated cell) in Fig. 4.17b. This is because the protective films induced by the decomposition of a small amount electrolyte can significantly mitigate the oxidation of CNT air electrodes and suppress corrosion of Li metal anode upon cycling.

Carbides

As aforementioned, carbon-based oxygen cathodes still have stability issue such as decomposition caused by attaching from oxygen species and promoting electrolyte decomposition upon cycling [126, 159]. Non-carbon material as a catalyst in oxygen cathodes becomes an alternative to improve the stability of air electrodes during cycling. The non-carbon catalysts themselves should present some required features, such as sufficient electronic conductivity, low density, low cost, and chemical and

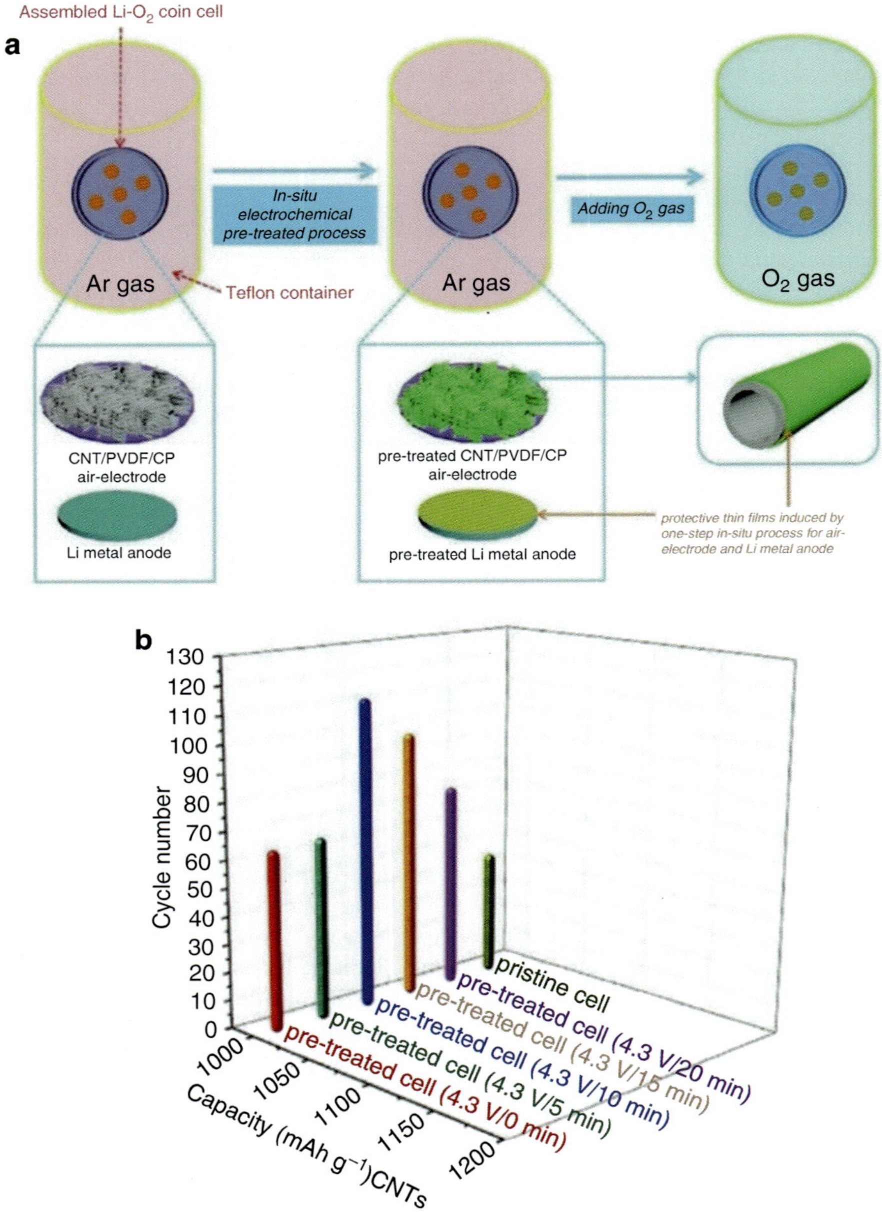

Fig. 4.17 (**a**) Schematic of the one-step electrochemical pre-charging treatment process of a Li-O_2 coin cell in Ar atmosphere to in situ produce thin protective films on both carbon electrode and Li metal anode prior to regular cycling in the O_2 atmosphere. (**b**) Comparison of stable cyclic life of Li-O_2 cells with pristine carbon nanotubes (CNTs) electrode and pretreated CNTs electrodes at 4.3 V/0 min, 4.3 V/5 min, 4.3 V/10 min, 4.3 V/15 min, and 4.3 V/20 min, at a current density of 0.1 mA cm^{-2} between 2.0 and 4.5 V. (Reproduced with the permission from Liu et al. [158]. Copyright 2018, WILEY-VCH)

electrochemical stability [160]. Titanium carbide (TiC) is one example that has been reported as a non-carbon catalyst to indicate stable cycling performance in a dimethyl sulfoxide (DMSO, $[(CH_3)_2SO]$)-based electrolyte. The improved cycle life relies on the high electronic conductivity and protective oxide surface layer of TiC catalyst [159]. TiC showed lower stability than carbon-based materials in tetraglyme (tetraethylene glycol dimethyl ether $[C_{10}H_{22}O_5]$)-based electrolytes [161]. While boron carbide (B_4C) is highly resistive against chemical attack, it serves as an electrode material for batteries and fuel cells [162, 163]. Recently, Zhang et al. reported that Li-O_2 batteries using B_4C catalyst-based oxygen electrode exhibited a good discharge/charge performance in the tetraglyme-based electrolyte for over 250 cycles under a capacity-limited protocol of 100 mAh g^{-1}. This could be attributed to the catalytic capability of B_4C, the absence of surface layer, and relatively consistent surface composition [160]. However, further improvement for deep discharge/charge of Li-O_2 batteries still needs more investigation.

Metal Oxides

Metal oxides can also function as catalyst in Li-O_2 batteries, such as cobalt(II,III) oxide $[Co_3O_4]$ [164–166], nickel cobaltite $[NiCo_2O_4]$ [124, 167], manganese (IV) oxide $[MnO_2]$ [168–170], manganese cobaltite $[MnCo_2O_4]$ [171, 172], zinc cobaltite $[ZnCo_2O_4]$ [125, 173], ruthenium(IV) oxide $[RuO_2]$ [174, 175], etc. The intrinsic advantages of oxide metal catalysts contain high abundance, low cost, and environmental friendliness. Metal oxides usually need to be mixed with conductive agents or prepared on conductive substrates due to the low electronic conductivity. Interestingly, metal oxides usually deliver an impressive catalytic performance with active sites located at the surface of the catalysts. In an early work by Bruce's group in 2008, (alpha manganese(IV) oxide) α-MnO_2 nanowire catalyst delivered high capacities for Li-O_2 batteries and attracted intensive research interest [176]. Later on, more metal oxide candidates have been explored in the last few years, and their catalytic activities and cycling stability have been progressively improved. In a recent work [174], researchers tried to put a ruthenium(IV) oxide (RuO_2) shell on the surface of core carbon nanotubes (CNTs) to form RuO_2/CNTs composite catalyst by using a common sol-gel method (Fig. 4.18a–d). The Li-O_2 batteries based on such a composite catalyst showed lowered discharge and charge overpotential of 0.21 V and 0.51 V, respectively, compared to only CNTs-based Li-O_2 batteries (Fig. 4.18e). No decay on the specific capacity of Li-O_2 batteries was observed with RuO_2/CNTs oxygen cathodes for more than 100 cycles at a cutoff capacity of 300 mAh g^{-1} (Fig. 4.18f).

4.3.1.2 Mechanism Analysis on Oxygen Cathode

Oxygen Reduction Reaction (ORR)

As mentioned above, in a typical Li-O_2 battery, Li ions in electrolytes diffuse to the surface of oxygen to react with it to first form LiO_2 intermediates ($Li^+ + O_2 + e^- \rightarrow LiO_2$), which can be gradually converted to Li_2O_2 through either

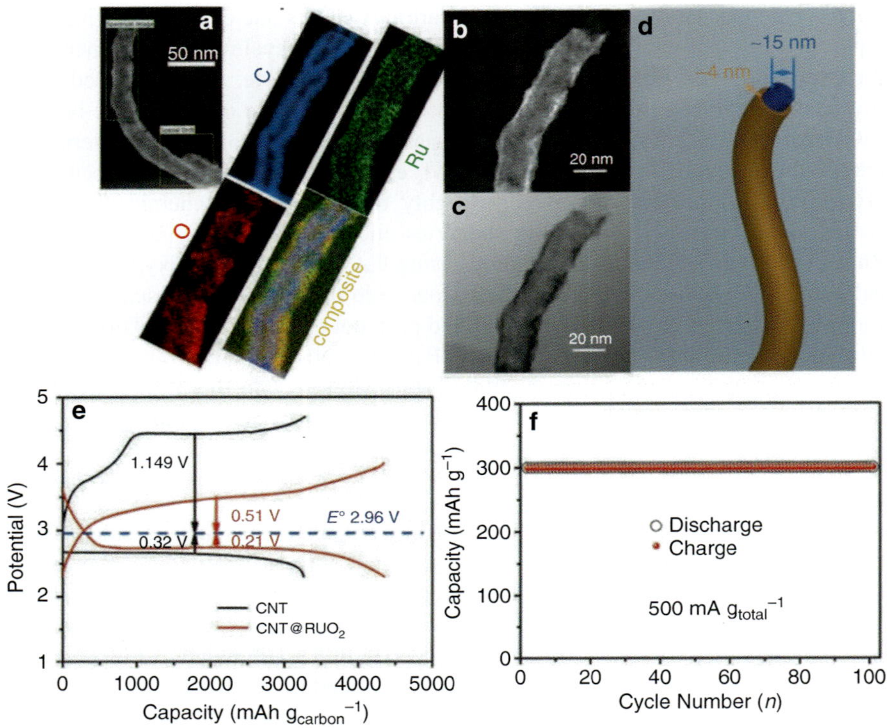

Fig. 4.18 (**a**) High-angle annular dark-field scanning transmission electron microscopy (HAADF-STEM) image and energy-dispersive X-ray (EDX) maps of the carbon (C), oxygen (O), ruthenium (Ru), and their composite. (**b, c**) HAADF-STEM and bright field scanning transmission electron microscopy (BF-STEM) images of a single CNT/RuO_2 composite. (**d**) Schematic of a single CNT/RuO_2 structure. (**e**) Discharge/charge profiles of the Li-O_2 batteries with the only CNTs (2.3–4.7 V) and CNT/RuO_2 composite (2.3–4.0 V) at a current density of 385 $mAg^{-1}{}_{carbon}$. (**f**) Cycling stability of the Li-O_2 battery at a current density of 500 $mAg^{-1}{}_{total}$. (Reproduced with the permission from Jian et al. [174]. Copyright 2014, WILEY-VCH)

continued electrochemical process ($LiO_2 + Li^+ + e^- \rightarrow Li_2O_2$) or chemical disproportionation process ($2LiO_2 \rightarrow Li_2O_2 + O_2$) [177]. Two different models of O_2 reduction have been presented. One is an O_2 reduction to form Li_2O_2 occurred on the electrode surface [178, 179], and another is a Li_2O_2 formation in electrolyte solution after O_2 reduction [180, 181]. Bruce's group pioneered to prove the importance of LiO_2 solubility for oxygen reduction in different donor-number (DN) electrolyte solutions (Fig. 4.19) [182]. In high donor-number (DN) electrolyte solvents, O_2 reduction at high potentials and low potentials belong to the solution pathway (blue) and the surface pathway (red), respectively. In low DN solvents, O_2 reduction follows the surface pathway at all potentials (red). The solubility of LiO_2 should be regarded as a dominant factor for the formation pathway of discharge products. In addition, the O_2 reduction

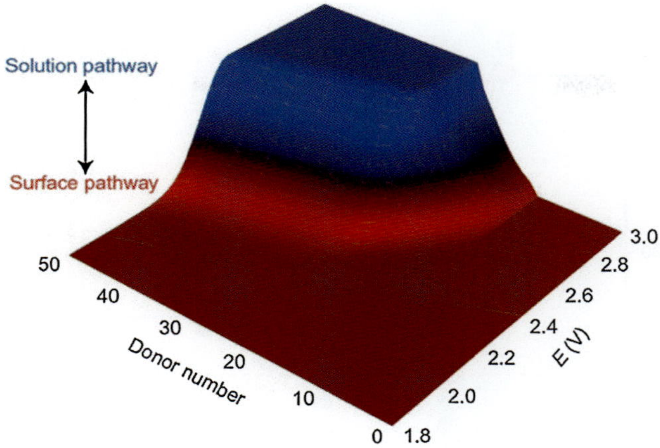

Fig. 4.19 The schematic of the O_2 reduction mechanism dominated by donor number (DN) and operating potential. (Reproduced with the permission from Johnson et al. [182]. Copyright 2014, Nature Publishing Group)

pathway can also be affected by temperature in the nonaqueous Li-O_2 batteries [177]. The lifetime of superoxide and solution pathway plays a dominant role during O_2 reduction process in the temperature range of −20 °C to 0 °C. The electrochemical kinetics of O_2 reduction and the surface pathway dominate discharge behavior in the temperature range of 0 °C to 40 °C (Fig. 4.20). In addition to traditional Li_2O_2 as primary discharge products, some other species (e.g., lithium oxide (LiO_2) [151] and lithium hydroxide (LiOH) [161]) as primary reversible discharge products have been proposed, but more investigation is still needed.

Oxygen Evaluation Reaction (OER)

The charging Li-O_2 battery is the oxidation of previously formed lithium oxide [Li_2O_2] discharge products to release O_2, so-called oxygen evaluation reaction (OER). A relatively high applied voltage (i.e., >4 V) is required to achieve the full oxidation of Li_2O_2 (Fig. 4.21) [183]. Therefore, to reduce charge overpotential, effective catalysts are highly desired in the OER process. The OER catalysts include the solid catalysts on the electrode surface and the soluble redox mediators in the electrolytes. In general, mediators should have high stability, high diffusion coefficient, fast charge transfer kinetics. These mediators can be oxidized at a potential slightly above the equilibrium potential of the Li_2O_2 formation. The oxidized mediators can freely diffuse near the oxygen cathode surface to participate in the oxidation of Li_2O_2 particles and remarkably reduce charge overpotential for the oxidation of Li_2O_2 consequently [20, 184–187]. Nevertheless, there are still three critical concerns for the use of redox mediators: (1) how to find an appropriate mediator with a good compatibility with Li-O_2

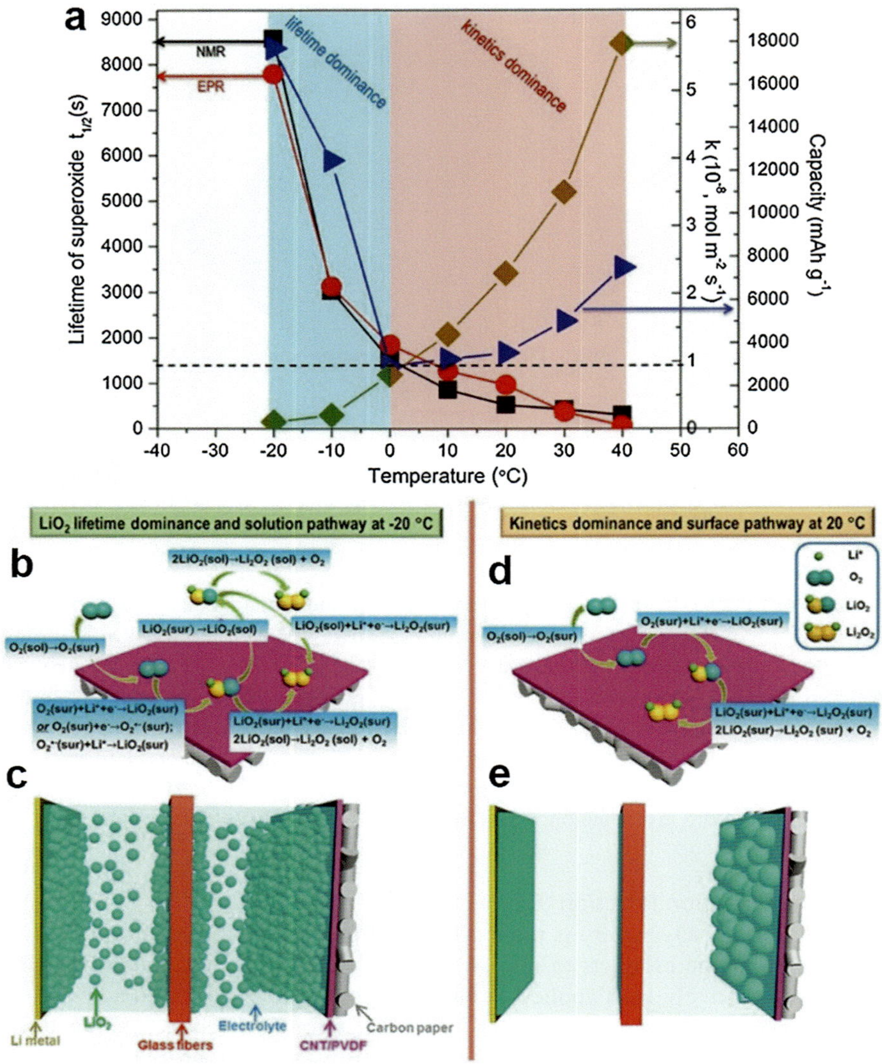

Fig. 4.20 (**a**) Temperature dependence of the experimental lifetime of superoxide from nuclear magnetic resonance (NMR) and electron paramagnetic resonance (EPR), the calculated electrochemical kinetics, and the discharge capacities at different temperature points. Schematic of LiO_2 evolution during discharge process in Li-O_2 batteries at −20 °C (**b, c**) and 20 °C (**d, e**). Note: "sol" and "sur" mean solution and surface, respectively. (Reproduced with the permission from Liu et al. [177]. Copyright 2017, American Chemical Society)

chemistry; (2) how much battery capacity is contributed by the reactions of redox mediator itself, rather than Li-O_2 battery chemistry; and (3) how to efficiently suppress the mediator's diffusion to Li anode surface and the consequent side reactions with Li metal.

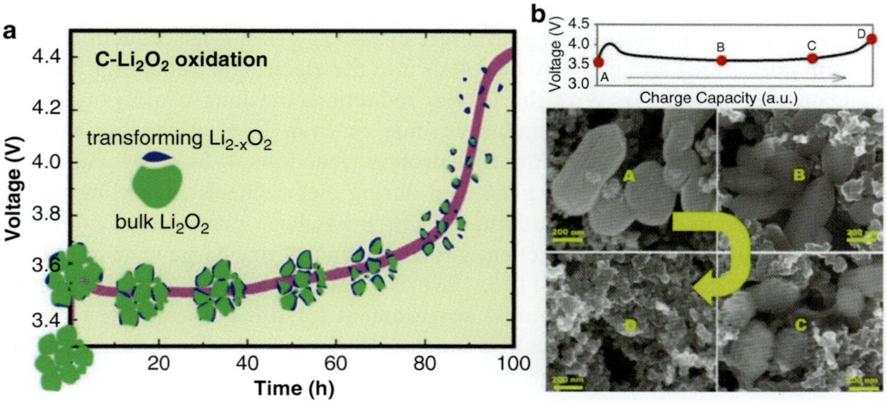

Fig. 4.21 (**a**) Oxidation curve of lithium oxide (Li_2O_2) during the charging process in Li-O_2 batteries. (**b**) Scanning electron microscopy (SEM) images recorded at different stages of oxidation of Li_2O_2 correspond to the panel. (Reproduced with the permission from Ganapathy et al. [183]. Copyright 2014, American Chemical Society)

4.3.2 Sulfur Cathode

Lithium-sulfur (Li-S) battery has a capacity of 1675 mAh g^{-1} on the basis of a reaction of $2Li + S \leftrightarrow Li_2S$, which is almost ten times more than the capacity of the conventional lithium-ion cathodes with the same mass such as lithium cobaltite [$LiCoO_2$]. The use of sulfur cathode in lithium metal battery (LMB) brings a high theoretical energy density of 2600 Wh kg^{-1} [188]. In addition, sulfur is extremely abundant in earth crust with wide distribution and a very low cost [189], which makes more sense for the sulfur as electrode materials than the currently used expensive transition metal oxides cathodes. In fact, sulfur has been widely explored as a cathode material combined with alkaline metal as the anode for the energy storage devices for more than 50 years. A high-temperature molten salt sodium sulfur [Na-S] battery was first developed by Ford in the late1960s [190]. Room temperature lithium-sulfur [Li-S] primary battery was patented by D. Herbert and J. Ulam in 1962 [191]. Earlier in the 1970s, due to the technical limits of nanoengineering, the sulfur was existed as the polysulfide solution (Li_2S_x) in various polarized electrolytes, from dimethylformamide [$HCON(CH_3)_2$], tetrahydrofuran [C_4H_8O] and dimethyl sulfoxide [$(CH_3)_2SO$], to ensure the well-contacts of active species and the conductive matrix. The most important pioneer study was conducted by E. Peled's group in the 1980s, introducing the linear and cyclone ether as the electrolyte solvent, which still states as the "standard" electrolyte nowadays [188, 192].

4.3.2.1 Li-S Battery Chemistry

The working mechanism of the Li-S battery involves a combination of the electrochemical reduction and chemical disproportion of the lithium polysulfide and represents typical two plateaus at 2.35 V and 2.1 V vs Li in the ether-based electrolyte, respectively, with multi-reaction pathways (Fig. 4.22). Generally, the reaction of sulfur proposes a ring-structure cyclic-S_8's ring-open reduction to the lithium polysulfide and partially dissolves into the electrolyte at the high voltage plateau (reactions 1–3). With the further reduction, the long-chain polysulfide would split into the short-chain polysulfide or even Li_2S at the low-voltage plateau and deposited out from the electrolyte (reactions 4–5) [194]. Activated trisulfur [S_3^{*-}] radicals can be generated during electrochemical cycling due to the disproportion reactions (reaction 3), which has been proposed to affect the chemical equilibrium of sulfur chemistry at the end of discharge state and drive the reverse process of charging [195, 196]. The reaction pathway of sulfur can be tuned by the physical and chemical environment for sulfur species in a certain condition. For example, the sulfur-encapsulated-microporous carbon cathode presents a slop discharge/charge behavior in carbonate electrolyte instead of the commonly seen two-plateau behavior in ether-based electrolyte [197].

4.3.2.2 The Fundamental Challenge of Li-S Batteries Chemistry

In a Li-S cell, there are two of the main fundamental challenges. First, the intermediate lithium polysulfide species has a high solubility in the electrolyte such as the ether-based electrolyte [194]. The dissolution of polysulfides causes active material loss and therefore fast capacity decay during cycling. The dissolved polysulfide may migrate into the Li metal side and get reduced by Li metal and then migrate to the cathode side again, causing the well-known "shuttle effect" and low coulombic efficiency (inset of Fig. 4.22). Ether-based electrolytes with lithium nitrate [$LiNO_3$] additive have been reported to effectively restrict the "shuttle" of polysulfide and present good stability with sulfur, but it is still challenging due to the gradual consumption of $LiNO_3$ during long cycling [198]. Second, sulfur and its discharge product Li_2S are highly insulating material. The redistribution of sulfur species

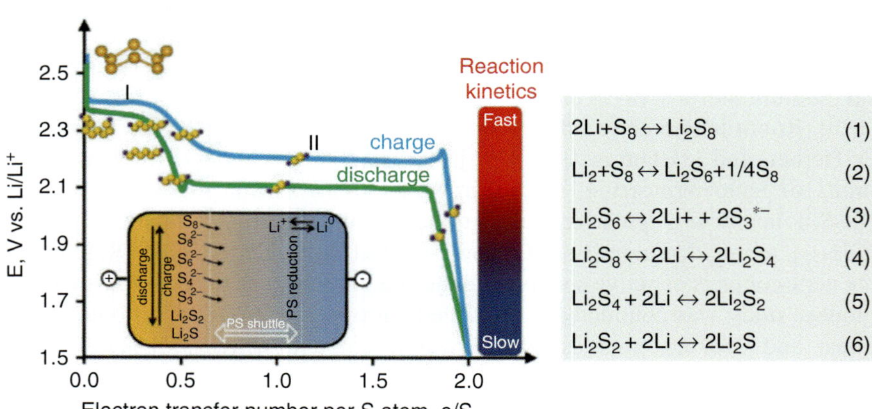

Fig. 4.22 The typical charge/discharge curve and reactions in Li-S batteries. Inset: polysulfide shuttle. (Reproduced with the permission from Wang et al. [193]. Copyright 2013, The Royal Society of Chemistry)

during cycling causes uncontrollable passivation and aggregation of insulating species on the sulfur cathode, thus leading to low sulfur utilization and shortened cycle life. Most of the critical issues in Li-S batteries are originated from the aforementioned two challenges and have been hindering their further practical application.

4.3.2.3 Strategies in Li-S Batteries

Phase I: Encapsulating Sulfur in Conducting Porous Host Materials

In the past decades, most efforts in Li-S battery research field have been devoted to the designing of functional cathode architecture to capture polysulfides and promote sulfur reactions during charge/discharge process. The most investigated strategy is to encapsulate sulfur into the porous conducting host matrix. Nazar and co-workers pioneered the work of highly ordered nanostructured carbon-sulfur composite cathode by infusing melted sulfur into nanopores at 155 °C (Fig. 4.23a) [199]. The conductive mesoporous carbon (CMK3) constrains sulfur within its mesopores and generates essential electronic contact with the insulating sulfur species. The nanostructure of carbon-sulfur composite cathode also provides access for Li^+ ions to ingress/egress the carbon-sulfur composite cathode for the reactivity of sulfur and aids the traping of polysulfide by the adsorption properties of nanomaterials. A high capacity of 1320 mAh g^{-1} was obtained with the carbon-sulfur composite cathode in ether-based electrolytes [199]. Later on, various porous host materials for sulfur have been comprehensively investigated to confine sulfur within the pores and/or high surface area surface, such as mesoporous carbon [200], hollow carbon sphere (point group I_h) [201], graphene (monolayer conforms to point group P6/mmm) [202, 203], hollow carbon nanofibers (CNF) [204, 205], transition metal oxide yolk shells (Fig. 4.23b) [206] or particles [207], carbide compounds [208], and conductive microporous polymers (Fig. 4.23c) [209], etc. In addition, the optimization of cell configuration such as with a porous interlayer on the top of the porous sulfur cathode can help to further mitigate the polysulfide diffusion into electrolyte and function as a top current collector, leading to improved cyclic stability (Fig. 4.23d) [210].

Phase II: Surface Chemistry Modification of Conducting Matrix

In addition to the architecture of porous cathode structure, the chemical interaction between polysulfide and conducting framework is another important factor that determines the capability of capturing polysulfides within the sulfur cathode during long-term cycling of Li-S batteries. Tailored surface chemistry of the porous nanostructures has attracted increasing interest recently. For example, carbon nanofibers (CNF) with tin-doped indium oxide nanoparticles decorating the surface (indium tin oxide, ITO [In_2O_3/SnO_2]-carbon hybrid) presented remarkable improvement in capacity and cycling stability compared with the bare carbon nanofiber electrodes (Fig. 4.24a) [211]. It was proposed that the ITO-carbon hybrid material shows a stronger binding ability to the soluble polysulfide and exhibits better spacial control of polysulfide deposition on the electrode surface than the bare carbon electrodes. The polar surface of ITO-carbon hybrid enhances the redox kinetics of polysulfides

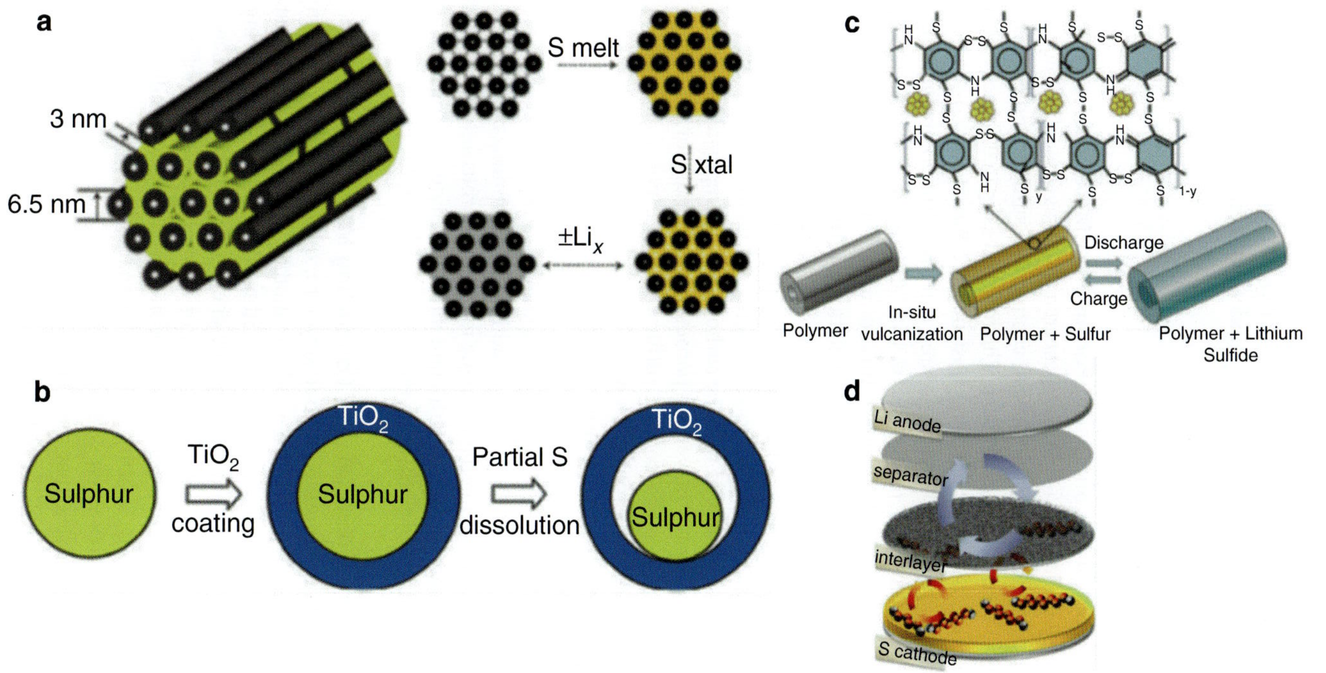

Fig. 4.23 (**a**) The schematic of the sulfur (yellow) confined in the interconnected structure of mesoporous carbon, CMK-3 by impregnation of molten sulfur, and the subsequent discharge-charge process with Li. (Reproduced with the permission from Ji et al. [199]. Copyright 2009, Springer Nature) (**b**) Schematic of the synthetic process that involves coating of sulfur nanoparticles with titania [TiO_2] to form sulfur-TiO_2 core-shell nanostructures, followed by partial dissolution of sulfur in toluene to achieve the yolk-shell morphology. (Reproduced with the permission from Seh et al. [206]. Copyright 2013, Springer Nature) (**c**) Schematic illustration of the construction and discharge/charge process of the SPANI-NT/S composite. (Reproduced with the permission from Xiao et al. [209]. Copyright 2012, WILEY-VCH) (**d**) The schematic of the nanoporous carbon interlayer for Li-S batteries. (Reproduced with the permission from Su and Manthiram. Copyright 2012, The Royal Society of Chemistry)

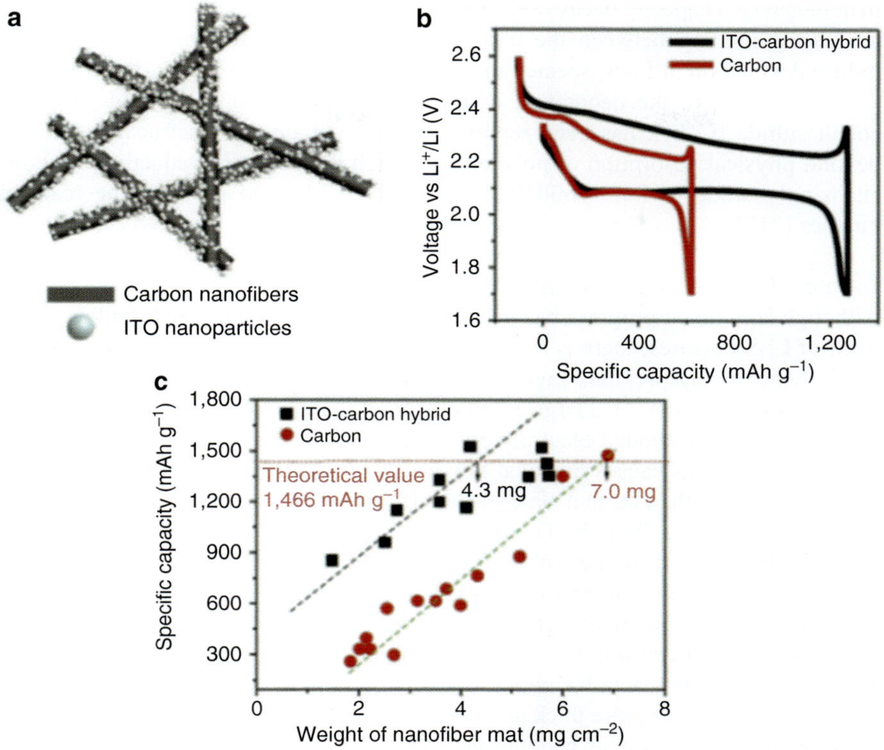

Fig. 4.24 (**a**) The schematic of indium tin oxide (ITO)-carbon hybrid. (**b**) The charge and discharge initial discharge/charge voltage profiles at a C/5 current rate of ITO-C and carbon nanofiber electrode, respectively. (**c**) Plot of initial discharge capacities of ITO-C and carbon nanofiber electrodes at a C/5 current rate as the function of the mass of the used electrodes. (Reproduced with the permission from Yao et al. [211]. Copyright 2014, Springer Nature)

and mediates the reduction of polysulfides to form solid Li_2S_x on the electrode and therefore improves sulfur utilization (Fig. 4.24b, c) [211]. In addition, other type of transition metal oxides, sulfides, and carbides, such as titanium oxide [Ti_4O_7] [209], titanium sulfide [TiS_2] [212], and titanium carbide [Ti_2C] [213], also exhibited improved performance of sulfur cathode because of the strong interaction of the polysulfide species with the surface of transition metal atoms. While some work further indicated that only transition metal-based compounds with a redox potential in a certain targeted voltage window (2.4 V < E_{redox} < 3.05 V) react with polysulfides to form active surface-bound polythionate species that can engage in the surface redox chemistry of sulfur [214]. For example, the manganese(IV) oxide [MnO_2] nanosheets (3.05 V) can oxidize the polysulfides [Li_2S_x] to form thiosulfate [$Li_2S_2O_3$]/polythionate [$S_nO_6^{2-}$] groups chemically bound to reduced metal oxide surface and function as a mediator for the sulfur redox on the surface [215]. The MnO_2 nanosheet sulfur cathode presented a stable cycle life with sulfur host materials with a capacity decay rate of 0.036% per cycle over 2000 cycles. However, too strong binding between the transition metal and the polysulfide, which can induce destruction of Li_2S_x species and potentially larger polarization, is not favored [216]. In addition, the decoration of transition metal-based compounds, such as cobalt sulfide [CoS_2], has been reported to demonstrate the electrocatalyst effect beyond physical adsorption of polysulfide, which enhances the reduction of short-chain polysulfide to form solid lithium sulfide [Li_2S] and promote the reaction kinetics [217].

Toward a Practical Application of Sulfur Cathodes

Although significant progress has been made on the cathode research and development of Li-S batteries, there is still a big gap for their real-world application so far [218]. Most of the previous investigations were performed using thin film sulfur electrodes (<2 mg cm^{-2}), a large amount of conducting porous carbon (30–40 wt % in electrodes) and flooded electrolyte usage, typically electrolyte/sulfur (E/S; μlE/gS) ratio >20. This causes practical low-energy density of Li-S batteries, which is even much lower than current state-of-the-art lithium nickel cobalt aluminum oxide [$LiNiCoAlO_2$, NCA-18650B. The battery is 18 mm wide and is 65 mm in length, and 0 is the division marker (650/10 = 65), and B is the battery chemistry with lithium as a cathode, an organic electrolyte, and carbon monofluoride as anode] cells [218]. To achieve a practical high-energy density in Li-S batteries, it is imperative to find solutions to increasing the sulfur loading and reducing the electrolyte without sacrificing the sulfur reaction kinetics for high energy in Li-S system [219]. Basically, it is much harder to prepare thick and dense sulfur cathode than conventional Li-ion cathodes due to the use of nanoporous carbon/sulfur composite materials, which causes electrode peer-off from current collector easier. On the other hand, it is remarkably challenging to reduce electrolyte usage in the Li-S system due to the highly porous carbon (C)/sulfur (S) composite cathode materials and poor reaction kinetics of insulating sulfur species. Nevertheless, we believe advanced sulfur architecture combined with functional electrolyte will be effective to address the above issues in Li-S batteries. Several useful approaches to increase sulfur loading and reduce electrolyte usage will be introduced, respectively.

Thick Sulfur Loading Electrodes

Integrating the sulfur/carbon nanocomposite into large secondary particles is an effective way to reduce void space and surface area of composite materials, which is beneficial for preparing thick sulfur electrodes while retaining the good electronic contact and fast ion diffusion. For example, a facile and effective integrated approach of synthesizing micro-size particles from commercial potassium boride [KB] nanoparticles was reported for sulfur cathodes. Uniform and crack-free coating with high loading of 2–8 mg cm^{-2} sulfur have successfully been achieved with the integrated particles (Fig. 4.25) [220]. Binder is another important factor that directly affects the quality of sulfur cathodes. An in situ cross-linked binder was reported to construct stable and high volumetric energy density in Li-S batteries, by coupling multifunctional and hierarchically

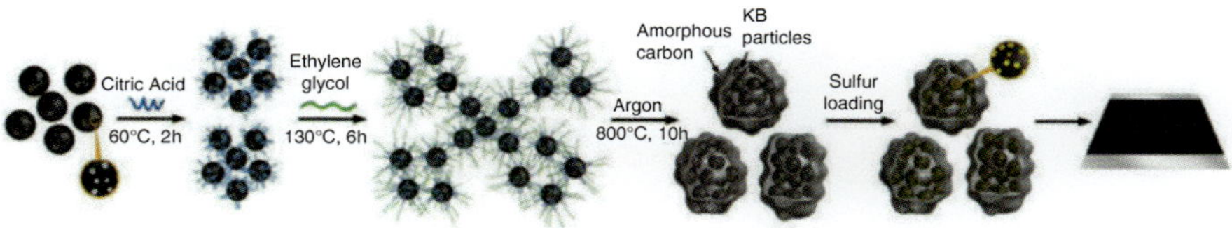

Fig. 4.25 Schematic illustration of the synthesis process of integrated potassium boride [KB] for thick electrode coating. (Reproduced with the permission from Lv et al. [220]. Copyright 2016, WILEY-VCH)

structured sulfur composite. The cross-linked elastomeric binder empowers the hierarchical sulfur composites with crack-free and high-loading electrodes using traditional slurry processing. Using this approach, electrodes with up to 14.9 mg cm^{-2} sulfur loading [221]. In addition, the 3D conducting framework constructed porous carbon such as graphene were demonstrated to be helpful to increase sulfur loading, accommodate the sulfur volume change, and provide an excellent conducting network for electrons and ions [222]. However, the highly porous 3D cathodes absorb too much electrolyte in practice, which should be carefully considered for real application in batteries.

Reducing Electrolyte Usage in Sulfur Cathodes

There are basically two main challenges to reduce their electrolyte usage to achieve high energy in Li-S batteries. First, the currently used highly porous carbon/S composite cathode heavily requires a large amount of electrolyte to wet the whole electrode. Second, lean electrolyte condition causes poor sulfur reaction kinetics and reversibility (especially for the transportation from long-chain polysulfide [Li_2S_x] to lithium sulfide [Li_2S_2/Li_2S], large polarization, low sulfur utilization, and limited cycle life) [223]. Reducing the porosity of sulfur cathodes will directly favor the reduction of the needed electrolyte volume in sulfur cathodes. A low electrolyte/sulfur (E/S) ratio of 3.5 ml_E g_S^{-1} was realized in coin cells for the high packing sulfur electrodes with integrated sulfur cathodes [221]. In contrast to the encapsulation approach for sulfur cathodes, Pan et al. revealed a new approach for sulfur cathodes that do not rely on sulfur encapsulation within high surface area carbon, called "non-encapsulation" approach (Fig. 4.26a and b) [224]. They used a low surface area, open carbon fiber architecture to control the nucleation and growth of the sulfur species by manipulating the carbon surface chemistry and the solvent properties, realizing ~100% sulfur utilization and high-energy density by in situ forming large sulfur sphere agglomerates (Fig. 4.26c and d). The large sulfur agglomerate well fills the void space in the carbon framework in the cathode and leads to high electrode density and volumetric energy of sulfur cathode and, therefore, benefits low electrolyte usage in practice. This finding offers an alternative approach for designing high-energy and low-cost Li-S batteries through controlling sulfur reaction on low-surface-area carbon.

On the other hand, considering the poor reaction kinetics of sulfur, it is extremely important to retain a good ionic conducting network under lean electrolyte condition. The limited electrolyte usually hardly distributes uniformly in conventional sulfur

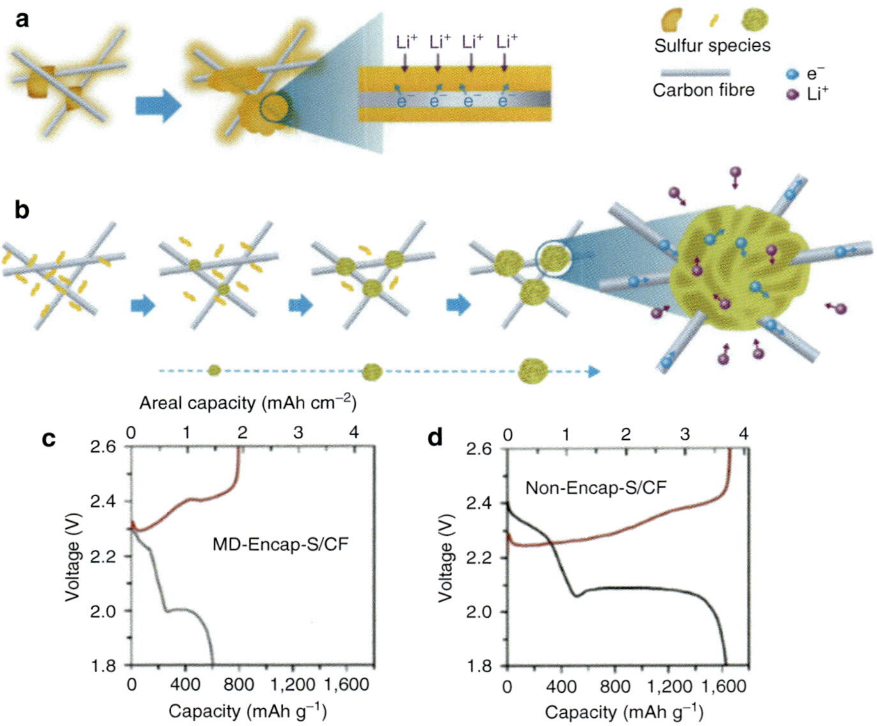

Fig. 4.26 Two different growth pathways of sulfur species during the electrochemical process in Li-S batteries. (**a**) In the traditional melt-diffusion encapsulation approach (melt-diffusion [MD]-encapsulation approach), the sulfur species are involved in a 2D surface electrochemical reaction mechanism that produces a continuous insulating sulfur/lithium sulfide [S/Li_2S] film, which limits the charge transport across the insulating film and causes passivation. (**b**) In the non-encapsulation approach the sulfur species are deposited onto the carbon fibers from an electrochemical precipitation process from the Li_2S_8 catholyte. The heterogeneous nucleation leads to the formation of 3D "flowerlike" particles that generate a mixed conducting network for Li^+ and electrons. The arrows indicate the possible moving directions of Li^+ and electrons. (**c, d**) The charge/discharge curve of MD-Encap-sulfur/carbon fibers [S/CF] and non-encap S/CF electrodes in the second cycle. (Reproduced with the permission from Pan et al. [224]. Copyright 2017, Springer Nature)

cathodes, which causes a large amount of inactive sulfur without contacting electrolyte, and therefore significantly reduces sulfur utilization and cycle life. Chen et al. recently reported a preformed ion conducting network for sulfur electrodes by using ionic conducting poly(ethylene oxide) (PEO_{10}, [$(CH_2CH_2O)_{100}$], lithium bis-trifluoromethanesulfonimide, LiTFSI) polymer as a binder [225]. The PEO_{10}LiTFSI polymer gel immobilizes the electrolyte and confines polysulfides within the ion conducting phase and thus produces a stable sulfur/electrolyte reaction interface with a very thin layer of gel electrolyte surrounding, enabling smooth charge transfer on the interface. A very lower E/S ratio of 3.3 $ml_E\ g_S^{-1}$ was achieved with a high capacity of 1200 mAh g^{-1} (one magnitude higher sulfur utilization than nonpolymer cathodes) and stable cycling. Such improvement is closely correlated with the excellent conductivity network across the cathode with very limited electrolyte volume [225].

Until now, reducing electrolyte usage in Li-S batteries is still challenging. Further understanding of the lean electrolyte system, i.e., reaction and failure mechanisms, is required to fully address the challenges under lean electrolyte condition and achieve high energy in Li-S batteries. One example is the limited cycling life of Li-S batteries under lean electrolyte condition [223]. The passivation by an accumulation of insulating Li_2S in sulfur cathode during cycling was reported to be one important reason [223, 225]. In the future, combining the advanced sulfur cathode promoting high sulfur utilization with the optimized electrolyte addressing cathode passivation could promise a remarkable improvement to significantly extend the cycle life of high-energy Li-S batteries.

4.3.2.4 Solid-State Li-S Batteries

Replacing the liquid organic electrolyte by solid-state electrolyte could fundamentally eliminate the critical issues of "shuttle effect" and the redistribution of sulfur species due to non-soluble intermediated species in the solid electrolyte, which improves the stability of the Li-metal anode by preventing continuous reaction between Li metal and electrolyte. For example, the sulfide-based electrolytes such as lithium-sulfur phosphorus sulfide [Li_2S-P_2S_5] were reported to present high ionic conductivity and excellent compatibility with sulfur cathodes, showing a stable cycling over 1000 cycles with an alternative indium-lithium [In-Li] alloy anode [226, 227]. The use of In-Li anode increases the cost significantly and sacrifices the energy density due to the higher voltage (~0.6 V vs Li) of alloy anode. On the other hand, the active sulfur content (typically $<50\%$) in the sulfur cathode and applied current densities in solid-state Li-S batteries are currently quite low due to the large interfacial resistance of solid-solid interface. In the future, the active material ratio in the solid-state electrode should be increased for a practical high energy of solid batteries. The interfacial properties, i.e., resistance, stability, and flexibility, in solid-state batteries will be critical to the realization of solid-state Li-S batteries.

4.4 Conclusion

Lithium metal batteries offer a sustainable increase in theoretical energy density compared with current Li-ion batteries. Extensive efforts have been devoted to the research on Li metal anode and cathodes (mainly oxygen electrode and sulfur electrode) for high-energy Li-metal batteries. Though encouraging progress has been made in such areas, such as designing of functional electrolytes, 3D conducting host for Li metal and advanced air and sulfur cathodes; however, there is still the huge gap in their practical applications. The practically achievable energy densities of Li-metal batteries, such as Li-air and Li-S, are still quite modest due to the current use of a large amount of inactive materials such as extra Li metal, host materials, electrolytes, etc. Reducing the gap between a research and the practical application are critical to effectively transferring the scientific discoveries to a practical use. Another relative candidate – the solid-state Li-metal battery – is just at the infant stage of the investigation and shows great potential to address some urgent challenges in lithium metal battery (LMB), especially for a safety concerns.

The rechargeability and safety issue of lithium metal anode are the ultimate challenges for the practical application of lithium metal battery (LMB) as high-energy rechargeable battery technology. The fundamental solutions to the development of LMBs still rely on solving the problems in Li metal anode, i.e., the dendritic growth of Li, the huge volumetric change and the depletion of Li and electrolyte upon cycling, the corresponding continuous interfacial resistance buildup in the cell, etc. Furthermore, the grand challenges of Li metal are entangled with each other, which makes the Li metal anode complicated. In the future, the advanced characterization tool that can quantify the Li evolution and distribution for both active and "dead" lithium-based materials is highly required. We believe that the cycling stability and safety issue of LMBs could be finally solved through the solid step-by-step development from materials to cell-level design with the necessary support of advanced characterization tools. Finally, the practical application of LMB technology will step into its golden age with a promise of high energy and improved safety.

Acknowledgments This work is partially supported by US Department of Energy (DOE), Office of Basic Energy Sciences, Division of Materials Sciences and Engineering, under Award KC020105-FWP12152 for the design and execution of experiments, and the Assistant Secretary for Energy Efficiency and Renewable Energy, Vehicle Technologies Office, Advanced Battery Materials Research (BMR) program of the US Department of Energy (DOE) under Contract No. DE-AC02-05CH11231. The authors also gratefully acknowledge the helpful comments and suggestions of the reviewers, which have improved the presentation.

References

1. J.-M. Tarascon, M. Armand, Issues and challenges facing rechargeable lithium batteries. Nature **414**, 359–367 (2001). https://doi.org/10.1038/35104644
2. B. Lee, Y. Ko, G. Kwon, S. Lee, K. Ku, J. Kim, K. Kang, Exploiting biological systems: toward eco-friendly and high-efficiency rechargeable batteries. Joule **2**, 61–75 (2018). https://doi.org/10.1016/j.joule.2017.10.013
3. X.B. Cheng, R. Zhang, C.Z. Zhao, Q. Zhang, Toward safe lithium metal anode in rechargeable batteries: a review. Chem. Rev. **117**, 10403–10473 (2017). https://doi.org/10.1021/acs.chemrev.7b00115
4. H. Pan, Y. Shao, P. Yan, Y. Cheng, K.S. Han, Z. Nie, C. Wang, J. Yang, X. Li, P. Bhattacharya, K.T. Mueller, J. Liu, Reversible aqueous zinc/manganese oxide energy storage from conversion reactions. Nat. Energy **1**, 16039 (2016). https://doi.org/10.1038/nenergy.2016.39
5. H. Wu, G. Chan, J.W. Choi, I. Ryu, Y. Yao, M.T. McDowell, S.W. Lee, A. Jackson, Y. Yang, L. Hu, Y. Cui, Stable cycling of double-walled silicon nanotube battery anodes through solid-

electrolyte interphase control. Nat. Nanotechnol. **7**, 310–315 (2012). https://doi.org/10.1038/nnano.2012.35

6. J.C. Hyesun Kim, Superior lithium electroactive mesoporous Si@carbon core-shell nanowires for lithium battery anode material. Nano Lett. **8**, 3688–3691 (2008). https://doi.org/10.1021/nl801853x
7. H. Wu, Y. Cui, Designing nanostructured Si anodes for high energy lithium-ion batteries. Nano Today **7**, 414–429 (2012). https://doi.org/10.1016/j.nantod.2012.08.004
8. B. Liu, X. Wang, H. Chen, Z. Wang, D. Chen, Y.B. Cheng, C. Zhou, G. Shen, Hierarchical silicon nanowires-carbon textiles matrix as a binder-free anode for high-performance advanced lithium-ion batteries. Sci. Rep. **3**, 1622 (2013). https://doi.org/10.1038/srep01622
9. M.-S. Park, Y.-M. Kang, G.-X. Wang, S.-X. Dou, H.-K. Liu, The effect of morphological modification on the electrochemical properties of SnO_2 nanomaterials. Adv. Funct. Mater. **18**, 455–461 (2008). https://doi.org/10.1002/adfm.200700407
10. X. Hou, B. Liu, X. Wang, Z. Wang, Q. Wang, D. Chen, G. Shen, SnO_2-microtube-assembled cloth for fully flexible self-powered photodetector nanosystems. Nanoscale **5**, 7831–7837 (2013). https://doi.org/10.1039/c3nr02300a
11. B. Liu, X. Wang, B. Liu, Q. Wang, D. Tan, W. Song, X. Hou, D. Chen, G. Shen, Advanced rechargeable lithium-ion batteries based on bendable $ZnCo_2O_4$-urchins-on-carbon-fibers electrodes. Nano Res. **6**, 525–534 (2013). https://doi.org/10.1007/s12274-013-0329-3
12. M.-S. Wu, P.-C.J. Chiang, J.-T. Lee, J.-C. Lin, Synthesis of manganese oxide electrodes with interconnected nanowire structure as an anode material for rechargeable lithium-ion batteries. J. Phys. Chem. B **109**, 23279–23284 (2005). https://doi.org/10.1021/jp054740b
13. B. Liu, J. Zhang, X. Wang, G. Chen, D. Chen, C. Zhou, G. Shen, Hierarchical three-dimensional $ZnCo(_2)O(_4)$ nanowire arrays/carbon cloth anodes for a novel class of high-performance flexible lithium-ion batteries. Nano Lett. **12**, 3005–3011 (2012). https://doi.org/10.1021/nl300794f
14. L. Wang, B. Liu, S. Ran, H. Huang, X. Wang, B. Liang, D. Chen, G. Shen, Nanorod-assembled Co_3O_4 hexapods with enhanced electrochemical performance for lithium-ion batteries. J. Mater. Chem. **22**, 23541 (2012). https://doi.org/10.1039/c2jm35617a
15. R.D. Apostolova, E.M. Shembel', I. Talyosef, J. Grinblat, B. Markovsky, D. Aurbach, Study of electrolytic cobalt sulfide Co9S8 as an electrode material in lithium accumulator prototypes. Russ. J. Electrochem. **45**, 311–319 (2009). https://doi.org/10.1134/s1023193509030112
16. X. Meng, Y. Cao, J.A. Libera, J.W. Elam, Atomic layer deposition of aluminum sulfide: growth mechanism and electrochemical evaluation in lithium-ion batteries. Chem. Mater. **29**, 9043–9052 (2017). https://doi.org/10.1021/acs.chemmater.7b02175
17. X. Wang, Q. Xiang, B. Liu, L. Wang, T. Luo, D. Chen, G. Shen, TiO_2 modified FeS nanostructures with enhanced electrochemical performance for lithium-ion batteries. Sci. Rep. **3**, 2007 (2013). https://doi.org/10.1038/srep02007
18. B. Liu, J.-G. Zhang, W. Xu, Advancing lithium metal batteries. Joule **2**, 833–845 (2018). https://doi.org/10.1016/j.joule.2018.03.008
19. W. Xu, J. Wang, F. Ding, X. Chen, E. Nasybulin, Y. Zhang, J.-G. Zhang, Lithium metal anodes for rechargeable batteries. Energy Environ. Sci. **7**, 513–537 (2014). https://doi.org/10.1039/c3ee40795k
20. D. Aurbach, B.D. McCloskey, L.F. Nazar, P.G. Bruce, Advances in understanding mechanisms underpinning lithium-air batteries. Nat. Energy **1**, 16128 (2016). https://doi.org/10.1038/nenergy.2016.128
21. M.S. Whittingham, Electrical energy storage and intercalation chemistry. Science **192**, 1126–1127 (1976). https://doi.org/10.1126/science.192.4244.1126
22. C. Ling, D. Banerjee, M. Matsui, Study of the electrochemical deposition of Mg in the atomic level: why it prefers the non-dendritic morphology. Electrochim. Acta **76**, 270–274 (2012). https://doi.org/10.1016/j.electacta.2012.05.001

23. T.D. Gregory, R.J. Hoffman, R.C. Winterton, Nonaqueous electrochemistry of magnesium: applications to energy storage. J. Electrochem. Soc. **137**, 775–780 (1990). https://doi.org/10.1149/1.2086553
24. A. Pei, G. Zheng, F. Shi, Y. Li, Y. Cui, Nanoscale nucleation and growth of electrodeposited lithium metal. Nano Lett. **17**, 1132–1139 (2017). https://doi.org/10.1021/acs.nanolett.6b04755
25. K. Yan, Z. Lu, H.-W. Lee, F. Xiong, P.-C. Hsu, Y. Li, J. Zhao, S. Chu, Y. Cui, Selective deposition and stable encapsulation of lithium through heterogeneous seeded growth. Nat. Energy **1**, 16010 (2016). https://doi.org/10.1038/nenergy.2016.10
26. J.N. Chazalviel, Electrochemical aspects of the generation of ramified metallic electrodeposits. Phys. Rev. A **42**, 7355–7367 (1990). https://doi.org/10.1103/PhysRevA.42.7355
27. R.L. Sacci, N.J. Dudney, K.L. More, L.R. Parent, I. Arslan, N.D. Browning, R.R. Unocic, Direct visualization of initial SEI morphology and growth kinetics during lithium deposition by *in situ* electrochemical transmission electron microscopy. Chem. Commun. (Camb.) **50**, 2104–2107 (2014). https://doi.org/10.1039/c3cc49029g
28. J.-G. Zhang, W. Xu, W.A. Henderson, *Lithium Metal Anodes and Rechargeable Lithium Metal Batteries* (Springer, Switzerland, 2016). https://doi.org/10.1007/978-3-319-44054-5
29. O. Crowther, A.C. West, Effect of electrolyte composition on lithium dendrite growth. J. Electrochem. Soc. **155**, A806 (2008). https://doi.org/10.1149/1.2969424
30. C. Brissot, M. Rosso, J.-N. Chazalviel, P. Baudry, S. Lascaud, *In situ* study of dendritic growth in lithium/PEO-salt/lithium cells. Electrochim. Acta **43**, 1569–1574 (1998). https://doi.org/10.1016/S0013-4686(97)10055-X
31. C.-Y. Tang, S.J. Dillon, *In situ* scanning electron microscopy characterization of the mechanism for Li dendrite growth. J. Electrochem. Soc. **163**, A1660–A1665 (2016). https://doi.org/10.1149/2.0891608jes
32. F. Sun, R. Moroni, K. Dong, H. Markötter, D. Zhou, A. Hilger, L. Zielke, R. Zengerle, S. Thiele, J. Banhart, I. Manke, Study of the mechanisms of internal short circuit in a Li/Li cell by synchrotron x-ray phase contrast tomography. ACS Energy Lett. **2**, 94–104 (2016). https://doi.org/10.1021/acsenergylett.6b00589
33. M.S. Park, S.B. Ma, D.J. Lee, D. Im, S.G. Doo, O. Yamamoto, A highly reversible lithium metal anode. Sci. Rep. **4**, 3815 (2014). https://doi.org/10.1038/srep03815
34. D. Aurbach, I. Weissman, H. Yamin, E. Elster, The correlation between charge/discharge rates and morphology, surface chemistry, and performance of Li electrodes and the connection to cycle life of practical batteries. J. Electrochem. Soc. **145**, 1421–1426 (1998). https://doi.org/10.1149/1.1838498
35. X.H. Liu, L. Zhong, L.Q. Zhang, A. Kushima, S.X. Mao, J. Li, Z.Z. Ye, J.P. Sullivan, J.Y. Huang, Lithium fiber growth on the anode in a nanowire lithium ion battery during charging. Appl. Phys. Lett. **98**, 183107 (2011). https://doi.org/10.1063/1.3585655
36. T. Osaka, T. Homma, T. Momma, H. Yarimizu, *In situ* observation of lithium deposition processes in solid polymer and gel electrolytes. J. Electroanal. Chem. **421**, 153–156 (1997). https://doi.org/10.1016/S0022-0728(96)04870-X
37. R. Bhattacharyya, B. Key, H. Chen, A.S. Best, A.F. Hollenkamp, C.P. Grey, *In situ* NMR observation of the formation of metallic lithium microstructures in lithium batteries. Nat. Mater. **9**, 504–510 (2010). https://doi.org/10.1038/nmat2764
38. Y.S. Cohen, Y. Cohen, D. Aurbach, Micromorphological studies of lithium electrodes in alkyl carbonate solutions using *in situ* atomic force microscopy. J. Phys. Chem. B **104**, 12282–12291 (2000). https://doi.org/10.1021/jp002526b
39. D. Aurbach, M.L. Daroux, P.W. Faguy, E. Yeager, Identification of surface films formed on lithium in propylene carbonate solutions. J. Electrochem. Soc. **134**, 1611–1620 (1987). https://doi.org/10.1149/1.2100722
40. B.L. Mehdi, J. Qian, E. Nasybulin, C. Park, D.A. Welch, R. Faller, H. Mehta, W.A. Henderson, W. Xu, C.M. Wang, J.E. Evans, J. Liu, J.G. Zhang, K.T. Mueller, N.D. Browning, Observation and quantification of nanoscale processes in lithium batteries by operando electrochemical (S)TEM. Nano Lett. **15**, 2168–2173 (2015). https://doi.org/10.1021/acs.nanolett.5b00175

41. K.N. Wood, E. Kazyak, A.F. Chadwick, K.H. Chen, J.G. Zhang, K. Thornton, N.P. Dasgupta, Dendrites and pits: untangling the complex behavior of lithium metal anodes through operando video microscopy. ACS Cent. Sci. **2**, 790–801 (2016). https://doi.org/10.1021/acscentsci.6b00260
42. I.W. Seong, C.H. Hong, B.K. Kim, W.Y. Yoon, The effects of current density and amount of discharge on dendrite formation in the lithium powder anode electrode. J. Power Sources **178**, 769–773 (2008). https://doi.org/10.1016/j.jpowsour.2007.12.062
43. C. Monroe, J. Newman, Dendrite growth in lithium/polymer systems. J. Electrochem. Soc. **150**, A1377 (2003). https://doi.org/10.1149/1.1606686
44. R. Akolkar, Modeling dendrite growth during lithium electrodeposition at sub-ambient temperature. J. Power Sources **246**, 84–89 (2014). https://doi.org/10.1016/j.jpowsour.2013.07.056
45. D. Aurbach, The electrochemical behavior of active metal electrodes in nonaqueous solutions, in *Nonaqueous Electrochemistry*, (Marcel Dekker, Boca Raton 1999), pp. 289–411. https://doi.org/10.1201/9780824741389
46. G.B. Appetecchi, F. Croce, F. Ronci, B. Scrosati, F. Alessandrini, M. Carewska, P.P. Prosini, Electrochemical characterization of a composite polymer electrolyte with improved lithium metal electrode interfacial properties. Ionics **5**, 59–63 (1999). https://doi.org/10.1007/BF02375904
47. K.M. Abraham, J.L. Goldman, The use of the reactive ether, tetrahydrofuran (THF), in rechargeable lithium cells. J. Power Sources **9**, 239–245 (1983). https://doi.org/10.1016/0378-7753(83)87024-4
48. B.D. Adams, J. Zheng, X. Ren, W. Xu, J.-G. Zhang, Accurate determination of coulombic efficiency for lithium metal anodes and lithium metal batteries. Adv. Energy Mater. **8**, 1702097 (2018). https://doi.org/10.1002/aenm.201702097
49. A. Dey, Film formation on lithium anode in propylene carbonate. J. Electrochem. Soc. **117**, C248 (1970). https://doi.org/10.1149/1.2407470
50. E. Peled, The electrochemical behavior of alkali and alkaline earth metals in nonaqueous battery systems-the solid electrolyte interphase model. J. Electrochem. Soc. **126**, 2047–2051 (1979). https://doi.org/10.1149/1.2128859
51. D. Aurbach, Review of selected electrode–solution interactions which determine the performance of Li and Li-ion batteries. J. Power Sources **89**, 206–218 (2000). https://doi.org/10.1016/S0378-7753(00)00431-6
52. D. Lin, Y. Liu, Y. Cui, Reviving the lithium metal anode for high-energy batteries. Nat. Nanotechnol. **12**, 194–206 (2017). https://doi.org/10.1038/nnano.2017.16
53. K. Xu, Nonaqueous liquid electrolytes for lithium-based rechargeable batteries. Chem. Rev. **104**, 4303–4418 (2004). https://doi.org/10.1021/cr030203g
54. K. Kanamura, H. Tamura, S. Shiraishi, Z.-I. Takehara, XPS analysis for the lithium surface immersed in γ-butyrolactone containing various salts. Electrochim. Acta **40**, 913–921 (1995). https://doi.org/10.1016/0013-4686(93)E0020-M
55. K. Kanamura, S. Shiraishi, H. Tamura, Z.-I. Takehara, X-ray photoelectron spectroscopic analysis and scanning electron microscopic observation of the lithium surface immersed in nonaqueous solvents. J. Electrochem. Soc. **141**, 2379–2385 (1994). https://doi.org/10.1149/1.2055129
56. F. Ding, W. Xu, X. Chen, J. Zhang, M.H. Engelhard, Y. Zhang, B.R. Johnson, J.V. Crum, T.A. Blake, X. Liu, J.-G. Zhang, Effects of carbonate solvents and lithium salts on morphology and coulombic efficiency of lithium electrode. J. Electrochem. Soc. **160**, A1894–A1901 (2013). https://doi.org/10.1149/2.100310jes
57. R. Younesi, G.M. Veith, P. Johansson, K. Edström, T. Vegge, Lithium salts for advanced lithium batteries: Li–metal, Li–O2, and Li–S. Energy Environ. Sci. **8**, 1905–1922 (2015). https://doi.org/10.1039/c5ee01215e
58. J. Qian, W. Xu, P. Bhattacharya, M. Engelhard, W.A. Henderson, Y. Zhang, J.-G. Zhang, Dendrite-free Li deposition using trace-amounts of water as an electrolyte additive. Nano Energy **15**, 135–144 (2015). https://doi.org/10.1016/j.nanoen.2015.04.009

59. J. Zheng, M.H. Engelhard, D. Mei, S. Jiao, B.J. Polzin, J.-G. Zhang, W. Xu, Electrolyte additive enabled fast charging and stable cycling lithium metal batteries. Nat. Energy **2**, 17012 (2017). https://doi.org/10.1038/nenergy.2017.12
60. F. Ding, W. Xu, G.L. Graff, J. Zhang, M.L. Sushko, X. Chen, Y. Shao, M.H. Engelhard, Z. Nie, J. Xiao, X. Liu, P.V. Sushko, J. Liu, J.G. Zhang, Dendrite-free lithium deposition via self-healing electrostatic shield mechanism. J. Am. Chem. Soc. **135**, 4450–4456 (2013). https://doi.org/10.1021/ja312241y
61. M.L. Gordin, F. Dai, S. Chen, T. Xu, J. Song, D. Tang, N. Azimi, Z. Zhang, D. Wang, Bis(2,2,2-trifluoroethyl) ether as an electrolyte co-solvent for mitigating self-discharge in lithium-sulfur batteries. ACS Appl. Mater. Interfaces **6**, 8006–8010 (2014). https://doi.org/10.1021/am501665s
62. X. Ren, Y. Zhang, M.H. Engelhard, Q. Li, J.-G. Zhang, W. Xu, Guided lithium metal deposition and improved lithium coulombic efficiency through synergistic effects of LiAsF6 and cyclic carbonate additives. ACS Energy Lett. **3**, 14–19 (2017). https://doi.org/10.1021/acsenergylett.7b00982
63. S.-K. Jeong, H.-Y. Seo, D.-H. Kim, H.-K. Han, J.-G. Kim, Y.B. Lee, Y. Iriyama, T. Abe, Z. Ogumi, Suppression of dendritic lithium formation by using concentrated electrolyte solutions. Electrochem. Commun. **10**, 635–638 (2008). https://doi.org/10.1016/j.elecom.2008.02.006
64. L. Suo, Y.S. Hu, H. Li, M. Armand, L. Chen, A new class of Solvent-in-Salt electrolyte for high-energy rechargeable metallic lithium batteries. Nat. Commun. **4**, 1481 (2013). https://doi.org/10.1038/ncomms2513
65. H. Yoona, P.C. Howlett, A.S. Best, M. Forsyth, D.R. MacFarlane, Fast charge/discharge of Li metal batteries using an ionic liquid electrolyte. J. Electrochem. Soc. **160**, A1629–A1637 (2013). https://doi.org/10.1149/2.022310jes
66. Y. Yamada, K. Furukawa, K. Sodeyama, K. Kikuchi, M. Yaegashi, Y. Tateyama, A. Yamada, Unusual stability of acetonitrile-based superconcentrated electrolytes for fast-charging lithium-ion batteries. J. Am. Chem. Soc. **136**, 5039–5046 (2014). https://doi.org/10.1021/ja412807w
67. J. Qian, W.A. Henderson, W. Xu, P. Bhattacharya, M. Engelhard, O. Borodin, J.G. Zhang, High rate and stable cycling of lithium metal anode. Nat. Commun. **6**, 6362 (2015). https://doi.org/10.1038/ncomms7362
68. N. Togasaki, T. Momma, T. Osaka, Enhanced cycling performance of a Li metal anode in a dimethylsulfoxide-based electrolyte using highly concentrated lithium salt for a lithium−oxygen battery. J. Power Sources **307**, 98–104 (2016). https://doi.org/10.1016/j.jpowsour.2015.12.123
69. B. Liu, W. Xu, P. Yan, X. Sun, M.E. Bowden, J. Read, J. Qian, D. Mei, C.-M. Wang, J.-G. Zhang, Enhanced cycling stability of rechargeable Li-O_2 batteries using high-concentration electrolytes. Adv. Funct. Mater. **26**, 605–613 (2016). https://doi.org/10.1002/adfm.201503697
70. B. Liu, W. Xu, P. Yan, S.T. Kim, M.H. Engelhard, X. Sun, D. Mei, J. Cho, C.-M. Wang, J.-G. Zhang, Stabilization of Li metal anode in DMSO-based electrolytes via optimization of salt-solvent coordination for Li-O_2 batteries. Adv. Energy Mater. **7**, 1602605 (2017). https://doi.org/10.1002/aenm.201602605
71. S. Chen, J. Zheng, D. Mei, K.S. Han, M.H. Engelhard, W. Zhao, W. Xu, J. Liu, J.G. Zhang, High-voltage lithium-metal batteries enabled by localized high-concentration electrolytes. Adv. Mater. **30**, e1706102 (2018). https://doi.org/10.1002/adma.201706102
72. J. Zheng, S. Chen, W. Zhao, J. Song, M.H. Engelhard, J.-G. Zhang, Extremely stable sodium metal batteries enabled by localized high-concentration electrolytes. ACS Energy Lett. **3**, 315–321 (2018). https://doi.org/10.1021/acsenergylett.7b01213
73. X. Ren, S. Chen, H. Lee, D. Mei, M.H. Engelhard, S.D. Burton, W. Zhao, J. Zheng, Q. Li, M.S. Ding, M. Schroeder, J. Alvarado, K. Xu, Y.S. Meng, J. Liu, J.-G. Zhang, W. Xu, Localized high-concentration sulfone electrolytes for high-efficiency lithium-metal batteries. Chem (2018). https://doi.org/10.1016/j.chempr.2018.05.002

74. S. Chen, J. Zheng, L. Yu, X. Ren, M.H. Engelhard, C. Niu, H. Lee, W. Xu, J. Xiao, J. Liu, J.-G. Zhang, High-efficiency lithium metal batteries with fire-retardant electrolytes. Joule (2018). https://doi.org/10.1016/j.joule.2018.05.002
75. S. Ohta, S. Komagata, J. Seki, T. Saeki, S. Morishita, T. Asaoka, All-solid-state lithium-ion battery using garnet-type oxide and Li_3BO_3 solid electrolytes fabricated by screen-printing. J. Power Sources **238**, 53–56 (2013). https://doi.org/10.1016/j.jpowsour.2013.02.073
76. Z. Lin, C. Liang, Lithium-sulfur batteries: from liquid to solid cells. J. Mater. Chem. A **3**, 936–958 (2015). https://doi.org/10.1039/c4ta04727c
77. V. Thangadurai, S. Narayanan, D. Pinzaru, Garnet-type solid-state fast Li ion conductors for Li batteries: critical review. Chem. Soc. Rev. **43**, 4714–4727 (2014). https://doi.org/10.1039/c4cs00020j
78. N. Kamaya, K. Homma, Y. Yamakawa, M. Hirayama, R. Kanno, M. Yonemura, T. Kamiyama, Y. Kato, S. Hama, K. Kawamoto, A. Mitsui, A lithium superionic conductor. Nat. Mater. **10**, 682–686 (2011). https://doi.org/10.1038/nmat3066
79. S. Wenzel, S. Randau, T. Leichtweiß, D.A. Weber, J. Sann, W.G. Zeier, J. Janek, Direct observation of the interfacial instability of the fast ionic conductor Li10GeP2S12 at the lithium metal anode. Chem. Mater. **28**, 2400–2407 (2016). https://doi.org/10.1021/acs.chemmater.6b00610
80. W.D. Richards, L.J. Miara, Y. Wang, J.C. Kim, G. Ceder, Interface stability in solid-state batteries. Chem. Mater. **28**, 266–273 (2015). https://doi.org/10.1021/acs.chemmater.5b04082
81. X. Han, Y. Gong, K.K. Fu, X. He, G.T. Hitz, J. Dai, A. Pearse, B. Liu, H. Wang, G. Rubloff, Y. Mo, V. Thangadurai, E.D. Wachsman, L. Hu, Negating interfacial impedance in garnet-based solid-state Li metal batteries. Nat. Mater. **16**, 572–579 (2017). https://doi.org/10.1038/nmat4821
82. H. Wu, G. Chan, J.W. Choi, I. Ryu, Y. Yao, M.T. McDowell, S.W. Lee, A. Jackson, Y. Yang, L. Hu, Y. Cui, Stable cycling of double-walled silicon nanotube battery anodes through solid–electrolyte interphase control. Nat. Nanotechnol. **7**, 310 (2012). https://doi.org/10.1038/nnano.2012.35
83. S. Schweidler, L. de Biasi, A. Schiele, P. Hartmann, T. Brezesinski, J. Janek, Volume changes of graphite anodes revisited: a combined operando x-ray diffraction and *in situ* pressure analysis study. J. Phys. Chem. C **122**, 8829–8835 (2018). https://doi.org/10.1021/acs.jpcc.8b01873
84. N. Liu, Z. Lu, J. Zhao, M.T. McDowell, H.-W. Lee, W. Zhao, Y. Cui, A pomegranate-inspired nanoscale design for large-volume-change lithium battery anodes. Nat. Nanotechnol. **9**, 187 (2014). https://doi.org/10.1038/nnano.2014.6
85. D. Lu, Y. Shao, T. Lozano, W.D. Bennett, G.L. Graff, B. Polzin, J. Zhang, M.H. Engelhard, N.T. Saenz, W.A. Henderson, Failure mechanism for fast-charged lithium metal batteries with liquid electrolytes. Adv. Energy Mater. **5** (2015). https://doi.org/10.1002/aenm.201400993
86. N. Nitta, F. Wu, J.T. Lee, G. Yushin, Li-ion battery materials: present and future. Mater. Today **18**, 252–264 (2015). https://doi.org/10.1016/j.mattod.2014.10.040
87. D. Lin, Y. Liu, Y. Cui, Reviving the lithium metal anode for high-energy batteries. Nat. Nanotechnol. **12**, 194–206 (2017). https://doi.org/10.1038/nnano.2017.16
88. Z. Liang, D. Lin, J. Zhao, Z. Lu, Y. Liu, C. Liu, Y. Lu, H. Wang, K. Yan, X. Tao, Composite lithium metal anode by melt infusion of lithium into a 3D conducting scaffold with lithophilic coating. Proc. Natl. Acad. Sci. **113**, 2862–2867 (2016). https://doi.org/10.1073/pnas.1518188113
89. D. Lin, Y. Liu, Z. Liang, H.-W. Lee, J. Sun, H. Wang, K. Yan, J. Xie, Y. Cui, Layered reduced graphene oxide with nanoscale interlayer gaps as a stable host for lithium metal anodes. Nat. Nanotechnol. **11**, 626–632 (2016)
90. C.-P. Yang, Y.-X. Yin, S.-F. Zhang, N.-W. Li, Y.-G. Guo, Accommodating lithium into 3D current collectors with a submicron skeleton towards long-life lithium metal anodes. Nat. Commun. **6**, 8058 (2015). https://doi.org/10.1038/ncomms9058

91. C. Yang, L. Zhang, B. Liu, S. Xu, T. Hamann, D. McOwen, J. Dai, W. Luo, Y. Gong, E.D. Wachsman, Continuous plating/stripping behavior of solid-state lithium metal anode in a 3D ion-conductive framework. Proc. Natl. Acad. Sci. **115**, 3770–3775 (2018). https://doi.org/10.1073/pnas.1719758115
92. Y. Liu, D. Lin, Z. Liang, J. Zhao, K. Yan, Y. Cui, Lithium-coated polymeric matrix as a minimum volume-change and dendrite-free lithium metal anode. Nat. Commun. **7**, 10992 (2016). https://doi.org/10.1038/ncomms10992
93. C. Jin, O. Sheng, J. Luo, H. Yuan, C. Fang, W. Zhang, H. Huang, Y. Gan, Y. Xia, C. Liang, 3D lithium metal embedded within lithophilic porous matrix for stable lithium metal batteries. Nano Energy **37**, 177–186 (2017). https://doi.org/10.1016/j.nanoen.2017.05.015
94. S. Matsuda, Y. Kubo, K. Uosaki, S. Nakanishi, Lithium-metal deposition/dissolution within internal space of CNT 3D matrix results in prolonged cycle of lithium-metal negative electrode. Carbon **119**, 119–123 (2017). https://doi.org/10.1016/j.carbon.2017.04.032
95. Z. Li, X. Li, L. Zhou, Z. Xiao, S. Zhou, X. Zhang, L. Li, L. Zhi, A synergistic strategy for stable lithium metal anodes using 3D fluorine-doped graphene shuttle-implanted porous carbon networks. Nano Energy **49**, 179–185 (2018). https://doi.org/10.1016/j.nanoen.2018.04.040
96. W. Deng, X. Zhou, Q. Fang, Z. Liu, Microscale lithium metal stored inside cellular graphene scaffold toward advanced metallic lithium anodes. Adv. Energy Mater. **8**, 1703152 (2018). https://doi.org/10.1002/aenm.201703152
97. S.S. Chi, Y. Liu, W.L. Song, L.Z. Fan, Q. Zhang, Prestoring lithium into stable 3D nickel foam host as dendrite-free lithium metal anode. Adv. Funct. Mater. **27**, 1700348 (2017). https://doi.org/10.1002/adfm.201700348
98. X.-Y. Yue, W.-W. Wang, Q.-C. Wang, J.-K. Meng, Z.-Q. Zhang, X.-J. Wu, X.-Q. Yang, Y.-N. Zhou, CoO nanofiber decorated nickel foams as lithium dendrite suppressing host skeletons for high energy lithium metal batteries. Energy Storage Mater. **14**, 335–344 (2018). https://doi.org/10.1016/j.ensm.2018.05.017
99. L.-L. Lu, Y. Zhang, Z. Pan, H.-B. Yao, F. Zhou, S.-H. Yu, Lithiophilic Cu–Ni core-shell nanowire network as a stable host for improving lithium anode performance. Energy Storage Mater. **9**, 31–38 (2017). https://doi.org/10.1016/j.ensm.2017.06.004
100. M. Kotobuki, K. Kanamura, Y. Sato, T. Yoshida, Fabrication of all-solid-state lithium battery with lithium metal anode using Al2O3-added Li7La3Zr2O12 solid electrolyte. J. Power Sources **196**, 7750–7754 (2011). https://doi.org/10.1016/j.jpowsour.2011.04.047
101. J.-M. Tarascon, M. Armand, *Materials for Sustainable Energy: A Collection of Peer-Reviewed Research and Review Articles from Nature Publishing Group* (World Scientific, Singapore, 2011), pp. 171–179. https://doi.org/10.1142/9789814317665_0024
102. X. Yao, D. Liu, C. Wang, P. Long, G. Peng, Y.-S. Hu, H. Li, L. Chen, X. Xu, High-energy all-solid-state lithium batteries with ultralong cycle life. Nano Lett. **16**, 7148–7154 (2016). https://doi.org/10.1021/acs.nanolett.6b03448
103. Y.-S. Hu, Batteries: getting solid. Nat. Energy **1**, 16042 (2016). https://doi.org/10.1038/nenergy.2016.42
104. B. Liu, L. Zhang, S. Xu, D.W. McOwen, Y. Gong, C. Yang, G.R. Pastel, H. Xie, K. Fu, J. Dai, C. Chen, E.D. Wachsman, L. Hu, 3D lithium metal anodes hosted in asymmetric garnet frameworks toward high energy density batteries. Energy Storage Mater. **14**, 376–382 (2018). https://doi.org/10.1016/j.ensm.2018.04.015
105. C. Yang, Y. Yao, S. He, H. Xie, E. Hitz, L. Hu, Ultrafine silver nanoparticles for seeded lithium deposition toward stable lithium metal anode. Adv. Mater. **29**, 1702714 (2017). https://doi.org/10.1002/adma.201702714
106. J. Park, J. Jeong, Y. Lee, M. Oh, M.H. Ryou, Y.M. Lee, Micro-patterned lithium metal anodes with suppressed dendrite formation for post lithium-ion batteries. Adv. Mater. Interfaces **3**, 1600140 (2016). https://doi.org/10.1002/admi.201600140
107. M.H. Ryou, Y.M. Lee, Y. Lee, M. Winter, P. Bieker, Mechanical surface modification of lithium metal: towards improved Li metal anode performance by directed Li plating. Adv. Funct. Mater. **25**, 834–841 (2015). https://doi.org/10.1002/adfm.201402953

108. Q. Li, B. Quan, W. Li, J. Lu, J. Zheng, X. Yu, J. Li, H. Li, Electro-plating and stripping behavior on lithium metal electrode with ordered three-dimensional structure. Nano Energy **45**, 463–470 (2018). https://doi.org/10.1016/j.nanoen.2018.01.019
109. N.W. Li, Y.X. Yin, C.P. Yang, Y.G. Guo, An artificial solid electrolyte interphase layer for stable lithium metal anodes. Adv. Mater. **28**, 1853–1858 (2016). https://doi.org/10.1002/adma.201504526
110. K. Yan, H.-W. Lee, T. Gao, G. Zheng, H. Yao, H. Wang, Z. Lu, Y. Zhou, Z. Liang, Z. Liu, Ultrathin two-dimensional atomic crystals as stable interfacial layer for improvement of lithium metal anode. Nano Lett. **14**, 6016–6022 (2014). https://doi.org/10.1021/nl503125u
111. E. Cha, M.D. Patel, J. Park, J. Hwang, V. Prasad, K. Cho, W. Choi, 2D MoS2 as an efficient protective layer for lithium metal anodes in high-performance Li–S batteries. Nat. Nanotechnol. **13**, 337–344 (2018). https://doi.org/10.1038/s41565-018-0061-y
112. H. Tian, Z.W. Seh, K. Yan, Z. Fu, P. Tang, Y. Lu, R. Zhang, D. Legut, Y. Cui, Q. Zhang, Theoretical investigation of 2D layered materials as protective films for lithium and sodium metal anodes. Adv. Energy Mater. **7**, 1602528 (2017). https://doi.org/10.1002/aenm.201602528
113. G. Ma, Z. Wen, M. Wu, C. Shen, Q. Wang, J. Jin, X. Wu, A lithium anode protection guided highly-stable lithium-sulfur battery. Chem. Commun. **50**, 14209–14212 (2014). https://doi.org/10.1039/C4CC05535G
114. Y.M. Lee, N.-S. Choi, J.H. Park, J.-K. Park, Electrochemical performance of lithium/sulfur batteries with protected Li anodes. J. Power Sources **119**, 964–972 (2003). https://doi.org/10.1016/S0378-7753(03)00300-8
115. N.-S. Choi, Y.M. Lee, J.H. Park, J.-K. Park, Interfacial enhancement between lithium electrode and polymer electrolytes. J. Power Sources **119**, 610–616 (2003). https://doi.org/10.1016/S0378-7753(03)00305-7
116. H. Lee, D.J. Lee, Y.-J. Kim, J.-K. Park, H.-T. Kim, A simple composite protective layer coating that enhances the cycling stability of lithium metal batteries. J. Power Sources **284**, 103–108 (2015). https://doi.org/10.1016/j.jpowsour.2015.03.004
117. K. Novoselov, D. Jiang, F. Schedin, T. Booth, V. Khotkevich, S. Morozov, A. Geim, Two-dimensional atomic crystals. Proc. Natl. Acad. Sci. **102**, 10451–10453 (2005). https://doi.org/10.1073/pnas.0502848102
118. A.B.. Kaul, Two-dimensional layered materials: structure, properties, and prospects for device applications. J. Mater. Res. **29**, 348–361 (2014). https://doi.org/10.1557/jmr.2014.6
119. A.N. Dey, Electrochemical alloying of lithium in organic electrolytes. J. Electrochem. Soc. **118**, 1547–1549 (1971). https://doi.org/10.1149/1.2407783
120. X. Liang, Q. Pang, I.R. Kochetkov, M.S. Sempere, H. Huang, X. Sun, L.F. Nazar, A facile surface chemistry route to a stabilized lithium metal anode. Nat. Energy **2**, 17119 (2017). https://doi.org/10.1038/nenergy.2017.119
121. J. Zhao, G. Zhou, K. Yan, J. Xie, Y. Li, L. Liao, Y. Jin, K. Liu, P.-C. Hsu, J. Wang, H.-M. Cheng, Y. Cui, Air-stable and freestanding lithium alloy/graphene foil as an alternative to lithium metal anodes. Nat. Nanotechnol. **12**, 993 (2017). https://doi.org/10.1038/nnano.2017.129
122. X. Yin, W. Tang, I.D. Jung, K.C. Phua, S. Adams, S.W. Lee, G.W. Zheng, Insights into morphological evolution and cycling behaviour of lithium metal anode under mechanical pressure. Nano Energy **50**, 659–664 (2018). https://doi.org/10.1016/j.nanoen.2018.06.003
123. P.G. Bruce, S.A. Freunberger, L.J. Hardwick, J.-M. Tarascon, Li-O_2 and Li-S batteries with high energy storage. Nat. Mater. **11**, 172 (2011). https://doi.org/10.1038/nmat3237
124. B. Liu, P. Yan, W. Xu, J. Zheng, Y. He, L. Luo, M.E. Bowden, C.M. Wang, J.G. Zhang, Electrochemically formed ultrafine metal oxide nanocatalysts for high-performance lithium-oxygen batteries. Nano Lett. **16**, 4932–4939 (2016). https://doi.org/10.1021/acs.nanolett.6b01556
125. B. Liu, W. Xu, P. Yan, P. Bhattacharya, R. Cao, M.E. Bowden, M.H. Engelhard, C.M. Wang, J.G. Zhang, I*n situ*-grown $ZnCo_2O_4$ on single-walled carbon nanotubes as air electrode materials for rechargeable lithium-oxygen batteries. ChemSusChem **8**, 3697–3703 (2015). https://doi.org/10.1002/cssc.201500636

126. M.M. Ottakam Thotiyl, S.A. Freunberger, Z. Peng, P.G. Bruce, The carbon electrode in nonaqueous Li-O_2 cells. J. Am. Chem. Soc. **135**, 494–500 (2013). https://doi.org/10.1021/ja310258x
127. B.D. McCloskey, A. Valery, A.C. Luntz, S.R. Gowda, G.M. Wallraff, J.M. Garcia, T. Mori, L.E. Krupp, Combining accurate O_2 and Li_2O_2 assays to separate discharge and charge stability limitations in nonaqueous Li-O_2 batteries. J. Phys. Chem. Lett. **4**, 2989–2993 (2013). https://doi.org/10.1021/jz401659f
128. W. Xu, K. Xu, V.V. Viswanathan, S.A. Towne, J.S. Hardy, J. Xiao, Z. Nie, D. Hu, D. Wang, J.-G. Zhang, Reaction mechanisms for the limited reversibility of Li–O_2 chemistry in organic carbonate electrolytes. J. Power Sources **196**, 9631–9639 (2011). https://doi.org/10.1016/j.jpowsour.2011.06.099
129. K.M. Abraham, Z. Jiang, A polymer electrolyte-based rechargeable lithium/oxygen battery. J. Electrochem. Soc. **143**, 1–5 (1996). https://doi.org/10.1149/1.1836378
130. Y. Lin, B. Moitoso, C. Martinez-Martinez, E.D. Walsh, S.D. Lacey, J.W. Kim, L. Dai, L. Hu, J.W. Connell, Ultrahigh-capacity lithium-oxygen batteries enabled by dry-pressed holey graphene air cathodes. Nano Lett. **17**, 3252–3260 (2017). https://doi.org/10.1021/acs.nanolett.7b00872
131. E.N. Nasybulin, W. Xu, B.L. Mehdi, E. Thomsen, M.H. Engelhard, R.C. Masse, P. Bhattacharya, M. Gu, W. Bennett, Z. Nie, C. Wang, N.D. Browning, J.G. Zhang, Formation of interfacial layer and long-term cyclability of Li-O(2) batteries. ACS Appl. Mater. Interfaces **6**, 14141–14151 (2014). https://doi.org/10.1021/am503390q
132. A.C. Luntz, B.D. McCloskey, Nonaqueous Li-air batteries: a status report. Chem. Rev. **114**, 11721–11750 (2014). https://doi.org/10.1021/cr500054y
133. Y. Shao, F. Ding, J. Xiao, J. Zhang, W. Xu, S. Park, J.-G. Zhang, Y. Wang, J. Liu, Making Li-air batteries rechargeable: material challenges. Adv. Funct. Mater. **23**, 987–1004 (2013). https://doi.org/10.1002/adfm.201200688
134. J. Lu, Y. Lei, K.C. Lau, X. Luo, P. Du, J. Wen, R.S. Assary, U. Das, D.J. Miller, J.W. Elam, H.M. Albishri, D.A. El-Hady, Y.K. Sun, L.A. Curtiss, K. Amine, A nanostructured cathode architecture for low charge overpotential in lithium-oxygen batteries. Nat. Commun. **4**, 2383 (2013). https://doi.org/10.1038/ncomms3383
135. B.D. Adams, C. Radtke, R. Black, M.L. Trudeau, K. Zaghib, L.F. Nazar, Current density dependence of peroxide formation in the Li–O_2 battery and its effect on charge. Energy Environ. Sci. **6**, 1772 (2013). https://doi.org/10.1039/c3ee40697k
136. G.M. Veith, J. Nanda, L.H. Delmau, N.J. Dudney, Influence of lithium salts on the discharge chemistry of Li-air cells. J. Phys. Chem. Lett. **3**, 1242–1247 (2012). https://doi.org/10.1021/jz300430s
137. H.G. Jung, J. Hassoun, J.B. Park, Y.K. Sun, B. Scrosati, An improved high-performance lithium-air battery. Nat. Chem. **4**, 579–585 (2012). https://doi.org/10.1038/nchem.1376
138. F. Cheng, J. Chen, Metal-air batteries: from oxygen reduction electrochemistry to cathode catalysts. Chem. Soc. Rev. **41**, 2172–2192 (2012). https://doi.org/10.1039/c1cs15228a
139. P.G. Bruce, S.A. Freunberger, L.J. Hardwick, J.M. Tarascon, Li-O_2 and Li-S batteries with high energy storage. Nat. Mater. **11**, 19–29 (2011). https://doi.org/10.1038/nmat3191
140. Y.-C. Lu, B.M. Gallant, D.G. Kwabi, J.R. Harding, R.R. Mitchell, M.S. Whittingham, Y. Shao-Horn, Lithium-oxygen batteries: bridging mechanistic understanding and battery performance. Energy Environ. Sci. **6**, 750 (2013). https://doi.org/10.1039/c3ee23966g
141. S. Song, W. Xu, J. Zheng, L. Luo, M.H. Engelhard, M.E. Bowden, B. Liu, C.M. Wang, J.G. Zhang, Complete decomposition of Li_2CO_3 in Li-O_2 batteries using Ir/B4C as noncarbon-based oxygen electrode. Nano Lett. **17**, 1417–1424 (2017). https://doi.org/10.1021/acs.nanolett.6b04371
142. R. Black, S.H. Oh, J.H. Lee, T. Yim, B. Adams, L.F. Nazar, Screening for superoxide reactivity in Li-O_2 batteries: effect on Li_2O_2/LiOH crystallization. J. Am. Chem. Soc. **134**, 2902–2905 (2012). https://doi.org/10.1021/ja2111543

143. Y.-C. Lu, H.A. Gasteiger, M.C. Parent, V. Chiloyan, Y. Shao-Horn, The influence of catalysts on discharge and charge voltages of rechargeable Li–oxygen batteries. Electrochem. Solid-State Lett. **13**, A69 (2010). https://doi.org/10.1149/1.3363047
144. Y.-C. Lu, H.A. Gasteiger, E. Crumlin, R. McGuire, Y. Shao-Horn, Electrocatalytic activity studies of select metal surfaces and implications in Li-air batteries. J. Electrochem. Soc. **157**, A1016 (2010). https://doi.org/10.1149/1.3462981
145. Y.-C. Lu, D.G. Kwabi, K.P.C. Yao, J.R. Harding, J. Zhou, L. Zuin, Y. Shao-Horn, The discharge rate capability of rechargeable Li–O_2 batteries. Energy Environ. Sci. **4**, 2999 (2011). https://doi.org/10.1039/c1ee01500a
146. Y.-C. Lu, Z. Xu, H.A. Gasteiger, S. Chen, K. Hamad-Schifferli, Y. Shao-Horn, Platinum−gold nanoparticles: a highly active bifunctional electrocatalyst for rechargeable lithium−air batteries. J. Am. Chem. Soc. **132**, 12170–12171 (2010). https://doi.org/10.1021/ja1036572
147. A.C. Marschilok, S. Zhu, C.C. Milleville, S.H. Lee, E.S. Takeuchi, K.J. Takeuchi, Electrodes for nonaqueous oxygen reduction based upon conductive polymer-silver composites. J. Electrochem. Soc. **158**, A223 (2011). https://doi.org/10.1149/1.3527992
148. S. Lee, S. Zhu, C.C. Milleville, C.-Y. Lee, P. Chen, K.J. Takeuchi, E.S. Takeuchi, A.C. Marschilok, Metal-air electrochemical cells: silver–polymer–carbon composite air electrodes. Electrochem. Solid-State Lett. **13**, A162 (2010). https://doi.org/10.1149/1.3479660
149. A.K. Thapa, T. Ishihara, Mesoporous α-MnO_2/Pd catalyst air electrode for rechargeable lithium-air battery. J. Power Sources **196**, 7016–7020 (2011). https://doi.org/10.1016/j.jpowsour.2010.09.112
150. A.K. Thapa, K. Saimen, T. Ishihara, Pd/MnO[$_2$] air electrode catalyst for rechargeable lithium/air battery. Electrochem. Solid-State Lett. **13**, A165 (2010). https://doi.org/10.1149/1.3481762
151. J. Lu, Y.J. Lee, X. Luo, K.C. Lau, M. Asadi, H.H. Wang, S. Brombosz, J. Wen, D. Zhai, Z. Chen, D.J. Miller, Y.S. Jeong, J.B. Park, Z.Z. Fang, B. Kumar, A. Salehi-Khojin, Y.K. Sun, L.A. Curtiss, K. Amine, A lithium-oxygen battery based on lithium superoxide. Nature **529**, 377–382 (2016). https://doi.org/10.1038/nature16484
152. H.D. Lim, K.Y. Park, H. Song, E.Y. Jang, H. Gwon, J. Kim, Y.H. Kim, M.D. Lima, R. Ovalle Robles, X. Lepro, R.H. Baughman, K. Kang, Enhanced power and rechargeability of a Li-O_2 battery based on a hierarchical-fibril CNT electrode. Adv. Mater. **25**, 1348–1352 (2013). https://doi.org/10.1002/adma.201204018
153. J. Xiao, D. Mei, X. Li, W. Xu, D. Wang, G.L. Graff, W.D. Bennett, Z. Nie, L.V. Saraf, I.A. Aksay, J. Liu, J.G. Zhang, Hierarchically porous graphene as a lithium-air battery electrode. Nano Lett. **11**, 5071–5078 (2011). https://doi.org/10.1021/nl203332e
154. Y. Wang, H. Zhou, To draw an air electrode of a Li-air battery by pencil. Energy Environ. Sci. **4**, 1704 (2011). https://doi.org/10.1039/c0ee00759e
155. X. Ren, S.S. Zhang, D.T. Tran, J. Read, Oxygen reduction reaction catalyst on lithium/air battery discharge performance. J. Mater. Chem. **21**, 10118 (2011). https://doi.org/10.1039/c0jm04170j
156. Y. Li, J. Wang, X. Li, D. Geng, R. Li, X. Sun, Superior energy capacity of graphene nanosheets for a nonaqueous lithium-oxygen battery. Chem. Commun. (Camb.) **47**, 9438–9440 (2011). https://doi.org/10.1039/c1cc13464g
157. E. Yoo, H. Zhou, Li-air rechargeable battery based on metal-free graphene nanosheet catalysts. ACS Nano **5**, 3020–3026 (2011). https://doi.org/10.1021/nn200084u
158. B. Liu, W. Xu, J. Tao, P. Yan, J. Zheng, M.H. Engelhard, D. Lu, C. Wang, J.-G. Zhang, Enhanced cyclability of lithium-oxygen batteries with electrodes protected by surface films induced via *in situ* electrochemical process. Adv. Energy Mater. **8**, 1702340 (2018). https://doi.org/10.1002/aenm.201702340
159. M.M. Ottakam Thotiyl, S.A. Freunberger, Z. Peng, Y. Chen, Z. Liu, P.G. Bruce, A stable cathode for the aprotic Li-O_2 battery. Nat. Mater. **12**, 1050–1056 (2013). https://doi.org/10.1038/nmat3737
160. S. Song, W. Xu, R. Cao, L. Luo, M.H. Engelhard, M.E. Bowden, B. Liu, L. Estevez, C.-M. Wang, J.-G. Zhang, B 4 C as a stable non-carbon-based oxygen electrode material for lithium-oxygen batteries. Nano Energy **33**, 195–204 (2017). https://doi.org/10.1016/j.nanoen.2017.01.042

161. T. Liu, M. Leskes, W. Yu, A.J. Moore, L. Zhou, P.M. Bayley, G. Kim, C.P. Grey, Cycling Li-O_2 batteries via LiOH formation and decomposition. Science **350**, 530–533 (2015). https://doi.org/10.1126/science.aac7730
162. M. Minakshi, M.G. Blackford, Electrochemical characteristics of B_4C or BN added MnO_2 cathode material for alkaline batteries. Mater. Chem. Phys. **123**, 700–705 (2010). https://doi.org/10.1016/j.matchemphys.2010.05.041
163. E. Antolini, E.R. Gonzalez, Ceramic materials as supports for low-temperature fuel cell catalysts. Solid State Ionics **180**, 746–763 (2009). https://doi.org/10.1016/j.ssi.2009.03.007
164. W.H. Ryu, T.H. Yoon, S.H. Song, S. Jeon, Y.J. Park, I.D. Kim, Bifunctional composite catalysts using Co_3O_4 nanofibers immobilized on nonoxidized graphene nanoflakes for high-capacity and long-cycle Li-O_2 batteries. Nano Lett. **13**, 4190–4197 (2013). https://doi.org/10.1021/nl401868q
165. A. Riaz, K.N. Jung, W. Chang, S.B. Lee, T.H. Lim, S.J. Park, R.H. Song, S. Yoon, K.H. Shin, J.W. Lee, Carbon-free cobalt oxide cathodes with tunable nanoarchitectures for rechargeable lithium-oxygen batteries. Chem. Commun. (Camb.) **49**, 5984–5986 (2013). https://doi.org/10.1039/c3cc42794c
166. A. Débart, J. Bao, G. Armstrong, P.G. Bruce, An O_2 cathode for rechargeable lithium batteries: the effect of a catalyst. J. Power Sources **174**, 1177–1182 (2007). https://doi.org/10.1016/j.jpowsour.2007.06.180
167. B. Sun, X. Huang, S. Chen, Y. Zhao, J. Zhang, P. Munroe, G. Wang, Hierarchical macroporous/mesoporous $NiCo_2O_4$ nanosheets as cathode catalysts for rechargeable Li–O_2 batteries. J. Mater. Chem. A **2**, 12053 (2014). https://doi.org/10.1039/c4ta01888e
168. E.M. Benbow, S.P. Kelly, L. Zhao, J.W. Reutenauer, S.L. Suib, Oxygen reduction properties of bifunctional α-manganese oxide electrocatalysts in aqueous and organic electrolytes. J. Phys. Chem. C **115**, 22009–22017 (2011). https://doi.org/10.1021/jp2055443
169. L. Jin, L. Xu, C. Morein, C.-h. Chen, M. Lai, S. Dharmarathna, A. Dobley, S.L. Suib, Titanium containing γ-MnO_2 (TM) hollow spheres: one-step synthesis and catalytic activities in Li/air batteries and oxidative chemical reactions. Adv. Funct. Mater. **20**, 3373–3382 (2010). https://doi.org/10.1002/adfm.201001080
170. J. Read, Characterization of the lithium/oxygen organic electrolyte battery. J. Electrochem. Soc. **149**, A1190 (2002). https://doi.org/10.1149/1.1498256
171. S. Ma, L. Sun, L. Cong, X. Gao, C. Yao, X. Guo, L. Tai, P. Mei, Y. Zeng, H. Xie, R. Wang, Multiporous $MnCo_2O_4$ microspheres as an efficient bifunctional catalyst for nonaqueous Li–O_2 batteries. J. Phys. Chem. C **117**, 25890–25897 (2013). https://doi.org/10.1021/jp407576q
172. H. Wang, Y. Yang, Y. Liang, G. Zheng, Y. Li, Y. Cui, H. Dai, Rechargeable Li–O_2 batteries with a covalently coupled $MnCo_2O_4$–graphene hybrid as an oxygen cathode catalyst. Energy Environ. Sci. **5**, 7931 (2012). https://doi.org/10.1039/c2ee21746e
173. T.F. Hung, S.G. Mohamed, C.C. Shen, Y.Q. Tsai, W.S. Chang, R.S. Liu, Mesoporous $ZnCo_2O_4$ nanoflakes with bifunctional electrocatalytic activities toward efficiencies of rechargeable lithium-oxygen batteries in aprotic media. Nanoscale **5**, 12115–12119 (2013). https://doi.org/10.1039/c3nr04271e
174. Z. Jian, P. Liu, F. Li, P. He, X. Guo, M. Chen, H. Zhou, Core-shell-structured CNT@RuO($_2$) composite as a high-performance cathode catalyst for rechargeable Li-O($_2$) batteries. Angew. Chem. Int. Ed. Engl. **53**, 442–446 (2014). https://doi.org/10.1002/anie.201307976
175. E. Yilmaz, C. Yogi, K. Yamanaka, T. Ohta, H.R. Byon, Promoting formation of noncrystalline Li_2O_2 in the Li-O_2 battery with RuO_2 nanoparticles. Nano Lett. **13**, 4679–4684 (2013). https://doi.org/10.1021/nl4020952
176. A. Debart, A.J. Paterson, J. Bao, P.G. Bruce, Alpha-MnO_2 nanowires: a catalyst for the O_2 electrode in rechargeable lithium batteries. Angew. Chem. Int. Ed. Engl. **47**, 4521–4524 (2008). https://doi.org/10.1002/anie.200705648
177. B. Liu, W. Xu, J. Zheng, P. Yan, E.D. Walter, N. Isern, M.E. Bowden, M.H. Engelhard, S.T. Kim, J. Read, B.D. Adams, X. Li, J. Cho, C. Wang, J.-G. Zhang, Temperature dependence

of the oxygen reduction mechanism in nonaqueous Li–O_2 batteries. ACS Energy Lett. **2**, 2525–2530 (2017). https://doi.org/10.1021/acsenergylett.7b00845
178. V. Viswanathan, J.K. Norskov, A. Speidel, R. Scheffler, S. Gowda, A.C. Luntz, Li-O_2 kinetic overpotentials: Tafel plots from experiment and first-principles theory. J. Phys. Chem. Lett. **4**, 556–560 (2013). https://doi.org/10.1021/jz400019y
179. B.D. McCloskey, R. Scheffler, A. Speidel, G. Girishkumar, A.C. Luntz, On the mechanism of nonaqueous Li–O_2 electrochemistry on C and its kinetic overpotentials: some implications for Li-air batteries. J. Phys. Chem. C **116**, 23897–23905 (2012). https://doi.org/10.1021/jp306680f
180. C.J. Allen, J. Hwang, R. Kautz, S. Mukerjee, E.J. Plichta, M.A. Hendrickson, K.M. Abraham, Oxygen reduction reactions in ionic liquids and the formulation of a general ORR mechanism for Li-air batteries. J. Phys. Chem. C **116**, 20755–20764 (2012). https://doi.org/10.1021/jp306718v
181. C.O. Laoire, S. Mukerjee, K.M. Abraham, E.J. Plichta, M.A. Hendrickson, Influence of nonaqueous solvents on the electrochemistry of oxygen in the rechargeable lithium-air battery. J. Phys. Chem. C **114**, 9178–9186 (2010). https://doi.org/10.1021/jp102019y
182. L. Johnson, C. Li, Z. Liu, Y. Chen, S.A. Freunberger, P.C. Ashok, B.B. Praveen, K. Dholakia, J.M. Tarascon, P.G. Bruce, The role of LiO_2 solubility in O_2 reduction in aprotic solvents and its consequences for Li-O2 batteries. Nat. Chem. **6**, 1091–1099 (2014). https://doi.org/10.1038/nchem.2101
183. S. Ganapathy, B.D. Adams, G. Stenou, M.S. Anastasaki, K. Goubitz, X.F. Miao, L.F. Nazar, M. Wagemaker, Nature of Li_2O_2 oxidation in a Li-O_2 battery revealed by operando X-ray diffraction. J. Am. Chem. Soc. **136**, 16335–16344 (2014). https://doi.org/10.1021/ja508794r
184. T. Zhang, K. Liao, P. He, H. Zhou, A self-defense redox mediator for efficient lithium–O_2 batteries. Energy Environ. Sci. **9**, 1024–1030 (2016). https://doi.org/10.1039/c5ee02803e
185. D. Kundu, R. Black, B. Adams, L.F. Nazar, A highly active low voltage redox mediator for enhanced rechargeability of lithium-oxygen batteries. ACS Cent. Sci. **1**, 510–515 (2015). https://doi.org/10.1021/acscentsci.5b00267
186. H.D. Lim, H. Song, J. Kim, H. Gwon, Y. Bae, K.Y. Park, J. Hong, H. Kim, T. Kim, Y.H. Kim, X. Lepro, R. Ovalle-Robles, R.H. Baughman, K. Kang, Superior rechargeability and efficiency of lithium-oxygen batteries: hierarchical air electrode architecture combined with a soluble catalyst. Angew. Chem. Int. Ed. Engl. **53**, 3926–3931 (2014). https://doi.org/10.1002/anie.201400711
187. B.J. Bergner, A. Schurmann, K. Peppler, A. Garsuch, J. Janek, TEMPO: a mobile catalyst for rechargeable Li-O(2) batteries. J. Am. Chem. Soc. **136**, 15054–15064 (2014). https://doi.org/10.1021/ja508400m
188. H. Yamin, E. Peled, Electrochemistry of a nonaqueous lithium/sulfur cell. J. Power Sources **9**, 281–287 (1983). https://doi.org/10.1016/0378-7753(83)87029-3
189. https://en.wikipedia.org/wiki/Abundance_of_elements_in_Earth%27s_crust
190. J.T. Kummer, W. Neill, Google Patents, 1968
191. H. Danuta, U. Juliusz, Google Patents, 1962
192. E. Peled, Y. Sternberg, A. Gorenshtein, Y. Lavi, Lithium-sulfur battery: evaluation of dioxolane-based electrolytes. J. Electrochem. Soc. **136**, 1621–1625 (1989). https://doi.org/10.1149/1.2096981
193. D.-W. Wang, Q. Zeng, G. Zhou, L. Yin, F. Li, H.-M. Cheng, I.R. Gentle, G.Q.M. Lu, Carbon-sulfur composites for Li-S batteries: status and prospects. J. Mater. Chem. A **1**, 9382–9394 (2013). https://doi.org/10.1039/C3TA11045A
194. S.S. Zhang, Liquid electrolyte lithium/sulfur battery: fundamental chemistry, problems, and solutions. J. Power Sources **231**, 153–162 (2013). https://doi.org/10.1016/j.jpowsour.2012.12.102
195. Q. Wang, J. Zheng, E. Walter, H. Pan, D. Lv, P. Zuo, H. Chen, Z.D. Deng, B.Y. Liaw, X. Yu, X. Yang, J.-G. Zhang, J. Liu, J. Xiao, Direct observation of sulfur radicals as reaction media in lithium-sulfur batteries. J. Electrochem. Soc. **162**, A474–A478 (2015). https://doi.org/10.1149/2.0851503jes

196. J. Xiao, J.Z. Hu, H. Chen, M. Vijayakumar, J. Zheng, H. Pan, E.D. Walter, M. Hu, X. Deng, J. Feng, B.Y. Liaw, M. Gu, Z.D. Deng, D. Lu, S. Xu, C. Wang, J. Liu, Following the transient reactions in lithium–sulfur batteries using an *in situ* nuclear magnetic resonance technique. Nano Lett. **15**, 3309–3316 (2015). https://doi.org/10.1021/acs.nanolett.5b00521
197. Y. Xu, Y. Wen, Y. Zhu, K. Gaskell, K.A. Cychosz, B. Eichhorn, K. Xu, C. Wang, Confined sulfur in microporous carbon renders superior cycling stability in Li/S batteries. Adv. Funct. Mater. **25**, 4312–4320 (2015). https://doi.org/10.1002/adfm.201500983
198. S.S. Zhang, Role of $LiNO_3$ in rechargeable lithium/sulfur battery. Electrochim. Acta **70**, 344–348 (2012). https://doi.org/10.1016/j.electacta.2012.03.081
199. X. Ji, K.T. Lee, L.F. Nazar, A highly ordered nanostructured carbon-sulphur cathode for lithium-sulphur batteries. Nat. Mater. **8**, 500–506 (2009). https://doi.org/10.1038/nmat2460
200. J. Schuster, G. He, B. Mandlmeier, T. Yim, K.T. Lee, T. Bein, L.F. Nazar, Spherical ordered mesoporous carbon nanoparticles with high porosity for lithium-sulfur batteries. Angew. Chem. Int. Ed. **51**, 3591–3595 (2012). https://doi.org/10.1002/anie.201107817
201. C. Zhang, H.B. Wu, C. Yuan, Z. Guo, X.W. Lou, Confining sulfur in double-shelled hollow carbon spheres for lithium-sulfur batteries. Angew. Chem. **124**, 9730–9733 (2012). https://doi.org/10.1002/ange.201205292
202. Y. Cao, X. Li, I.A. Aksay, J. Lemmon, Z. Nie, Z. Yang, J. Liu, Sandwich-type functionalized graphene sheet-sulfur nanocomposite for rechargeable lithium batteries. Phys. Chem. Chem. Phys. **13**, 7660–7665 (2011). https://doi.org/10.1039/C0CP02477E
203. G. Zhou, S. Pei, L. Li, D.-W. Wang, S. Wang, K. Huang, L.-C. Yin, F. Li, H.-M. Cheng, A graphene–pure-sulfur sandwich structure for ultrafast, long-life lithium-sulfur batteries. Adv. Mater. **26**, 625–631 (2014). https://doi.org/10.1002/adma.201302877
204. G. Zheng, Y. Yang, J.J. Cha, S.S. Hong, Y. Cui, Hollow carbon nanofiber-encapsulated sulfur cathodes for high specific capacity rechargeable lithium batteries. Nano Lett. **11**, 4462–4467 (2011). https://doi.org/10.1021/nl2027684
205. G. Zheng, Q. Zhang, J.J. Cha, Y. Yang, W. Li, Z.W. Seh, Y. Cui, Amphiphilic surface modification of hollow carbon nanofibers for improved cycle life of lithium sulfur batteries. Nano Lett. **13**, 1265–1270 (2013). https://doi.org/10.1021/nl304795g
206. Z.W. Seh, W. Li, J.J. Cha, G. Zheng, Y. Yang, M.T. McDowell, P.-C. Hsu, Y. Cui, Sulphur–TiO_2 yolk–shell nanoarchitecture with internal void space for long-cycle lithium–sulphur batteries. Nat. Commun. **4**, 1331 (2013). https://doi.org/10.1038/ncomms2327
207. Q. Pang, D. Kundu, M. Cuisinier, L. Nazar, Surface-enhanced redox chemistry of polysulphides on a metallic and polar host for lithium-sulphur batteries. Nat. Commun. **5**, 4759 (2014). https://doi.org/10.1038/ncomms5759
208. X. Liang, A. Garsuch, L.F. Nazar, Sulfur cathodes based on conductive MXene nanosheets for high-performance lithium-sulfur batteries. Angew. Chem. 127, 3979 (2015). https://doi.org/10.1002/ange.201410174
209. L. Xiao, Y. Cao, J. Xiao, B. Schwenzer, M.H. Engelhard, L.V. Saraf, Z. Nie, G.J. Exarhos, J. Liu, A soft approach to encapsulate sulfur: polyaniline nanotubes for lithium-sulfur batteries with long cycle life. Adv. Mater. **24**, 1176–1181 (2012). https://doi.org/10.1002/adma.201103392
210. Y.-S. Su, A. Manthiram, A new approach to improve cycle performance of rechargeable lithium-sulfur batteries by inserting a free-standing MWCNT interlayer. Chem. Commun. **48**, 8817–8819 (2012). https://doi.org/10.1039/C2CC33945E
211. H. Yao, G. Zheng, P.-C. Hsu, D. Kong, J.J. Cha, W. Li, Z.W. Seh, M.T. McDowell, K. Yan, Z. Liang, Improving lithium–sulphur batteries through spatial control of sulphur species deposition on a hybrid electrode surface. Nat. Commun. **5**, 3943 (2014). https://doi.org/10.1038/ncomms4943
212. Z.W. Seh, J.H. Yu, W. Li, P.-C. Hsu, H. Wang, Y. Sun, H. Yao, Q. Zhang, Y. Cui, Two-dimensional layered transition metal disulphides for effective encapsulation of high-capacity lithium sulphide cathodes. Nat. Commun. **5**, 5017 (2014). https://doi.org/10.1038/ncomms6017

213. X. Liang, A. Garsuch, L.F. Nazar, Sulfur cathodes based on conductive MXene nanosheets for high-performance lithium-sulfur batteries. Angew. Chem. **127**, 3979–3983 (2015). https://doi.org/10.1002/ange.201410174
214. X. Liang, C.Y. Kwok, F. Lodi-Marzano, Q. Pang, M. Cuisinier, H. Huang, C.J. Hart, D. Houtarde, K. Kaup, H. Sommer, Tuning transition metal oxide–sulfur interactions for long life lithium-sulfur batteries: the "Goldilocks" principle. Adv. Energy Mater. **6**, 1501636 (2016). https://doi.org/10.1002/aenm.201501636
215. X. Liang, C. Hart, Q. Pang, A. Garsuch, T. Weiss, L.F. Nazar, A highly efficient polysulfide mediator for lithium-sulfur batteries. Nat. Commun. **6**, 5682 (2015). https://doi.org/10.1038/ncomms6682
216. Q. Zhang, Y. Wang, Z.W. Seh, Z. Fu, R. Zhang, Y. Cui, Understanding the anchoring effect of two-dimensional layered materials for lithium-sulfur batteries. Nano Lett. **15**, 3780–3786 (2015). https://doi.org/10.1021/acs.nanolett.5b00367
217. Z. Yuan, H.-J. Peng, T.-Z. Hou, J.-Q. Huang, C.-M. Chen, D.-W. Wang, X.-B. Cheng, F. Wei, Q. Zhang, Powering lithium-sulfur battery performance by propelling polysulfide redox at sulfiphilic hosts. Nano Lett. **16**, 519–527 (2016). https://doi.org/10.1021/acs.nanolett.5b04166
218. M. Hagen, D. Hanselmann, K. Ahlbrecht, R. Maça, D. Gerber, J. Tübke, Lithium–sulfur cells: the gap between the state-of-the-art and the requirements for high energy battery cells. Adv. Energy Mater. **5**, 1401986 (2015). https://doi.org/10.1002/aenm.201401986
219. D. Eroglu, K.R. Zavadil, K.G. Gallagher, Critical link between materials chemistry and cell-level design for high energy density and low cost lithium-sulfur transportation battery. J. Electrochem. Soc. **162**, A982–A990 (2015). https://doi.org/10.1149/2.0611506jes
220. D. Lv, J. Zheng, Q. Li, X. Xie, S. Ferrara, Z. Nie, L.B. Mehdi, N.D. Browning, J.G. Zhang, G.L. Graff, High energy density lithium-sulfur batteries: challenges of thick sulfur cathodes. Adv. Energy Mater. **5**, 1402290 (2015). https://doi.org/10.1002/aenm.201402290
221. Q. Pang, X. Liang, C.Y. Kwok, J. Kulisch, L.F. Nazar, A comprehensive approach toward stable lithium-sulfur batteries with high volumetric energy density. Adv. Energy Mater. **7**, 1601630 (2017). https://doi.org/10.1002/aenm.201601630
222. G. Zhou, E. Paek, G.S. Hwang, A. Manthiram, Long-life Li/polysulphide batteries with high sulphur loading enabled by lightweight three-dimensional nitrogen/sulphur-codoped graphene sponge. Nat. Commun. **6**, 7760 (2015). https://doi.org/10.1038/ncomms8760
223. H. Pan, K.S. Han, M.H. Engelhard, R. Cao, J. Chen, J.G. Zhang, K.T. Mueller, Y. Shao, J. Liu, Addressing passivation in lithium–sulfur battery under lean electrolyte condition. Adv. Funct. Mater., 1707234 (2018). https://doi.org/10.1002/adfm.201707234
224. H. Pan, J. Chen, R. Cao, V. Murugesan, N.N. Rajput, K.S. Han, K. Persson, L. Estevez, M.H. Engelhard, J.-G. Zhang, K.T. Mueller, Y. Cui, Y. Shao, J. Liu, Non-encapsulation approach for high-performance Li–S batteries through controlled nucleation and growth. Nat. Energy **2**, 813–820 (2017). https://doi.org/10.1038/s41560-017-0005-z
225. J. Chen, W.A. Henderson, H. Pan, B.R. Perdue, R. Cao, J.Z. Hu, C. Wan, K.S. Han, K.T. Mueller, J.-G. Zhang, Y. Shao, J. Liu, Improving lithium–sulfur battery performance under lean electrolyte through nanoscale confinement in soft swellable gels. Nano Lett. **17**, 3061–3067 (2017). https://doi.org/10.1021/acs.nanolett.7b00417
226. H. Nagata, Y. Chikusa, All-solid-state lithium–sulfur battery with high energy and power densities at the cell level. Energ. Technol. **4**, 484–489 (2016). https://doi.org/10.1002/ente.201500297
227. T. Yamada, S. Ito, R. Omoda, T. Watanabe, Y. Aihara, M. Agostini, U. Ulissi, J. Hassoun, B. Scrosati, All solid-state lithium-sulfur battery using a glass-type P2S5–Li2S electrolyte: benefits on anode kinetics. J. Electrochem. Soc. **162**, A646–A651 (2015). https://doi.org/10.1149/2.0441504jes

High-Energy All-Solid-State Lithium-Metal Batteries by Nanomaterial Designs

5

Wei Liu

Contents

Abstract

Replacing flammable organic liquid electrolytes by solid-state electrolytes (SSEs) provides a fundamental solution to safety issue for lithium-ion batteries (LIBs), driven by the increasing demand of high energy density, stable cycling performance, low cost, and abuse tolerance for energy storage systems in the widespread applications of electric vehicles (EVs) and smart grids. In addition, SSEs

Author Contribution: The author further acknowledges that there is no financial relationship with the editors or publisher and have contributed original work in this chapter, other than what acknowledged or appropriately cited with copyright permission.

W. Liu (✉)
School of Physical Science and Technology, ShanghaiTech University, Shanghai, China
e-mail: liuwei1@shanghaitech.edu.cn

Q. Zhen et al. (eds.), *Nanostructured Materials for Next-Generation Energy Storage and Conversion*, https://doi.org/10.1007/978-3-662-58675-4_5

could in principle suppress the lithium dendrite growth and enable the utilization of lithium metal anode that is regarded as a "holy grail" of lithium batteries, in all-solid-state lithium metal batteries (ASSLMBs). However, the problems including sluggish ion transport in SSEs and the interfacial incompatibility of SSEs with electrodes significantly limit the overall performances of ASSLMBs. In this chapter, we first highlight the critical scientific challenges as well as technological progresses for the currently studied SSEs including organic polymers, inorganic electrolytes, and the composites. The fundamental understanding and characterizations on the lithium nucleation, dendrite growth, and formation of solid electrolyte interphase (SEI) during Li electrochemical deposition is discussed systematically. The strategies to address the large interfacial impedances between SSE and Li metal anode attributed to low ionic conductivity and poor instability of SEI, and degraded physical contact are also overviewed. The approaches based on nanostructured materials and nanotechnology advances of great promise to improve the overall performances of ASSLMBs are discussed in detail. Finally, the remaining issues and future opportunities of ASSLMBs for next-generation high-energy storages are prospected.

5.1 Introduction

Without a doubt, lithium-ion batteries (LIBs), with the advantages of high energy and power densities, long cycle life, and no memory effect, have exclusively occupied the market of consumer electronics and showed increasing promise in the applications of electric vehicles (EVs) and smart grid. However, the fundamental understandings as well as battery technologies are criticized for the slow advancement and development of LIBs over the past 30 years [39, 155]. The current commercialized LIBs ($LiCoO_2$ or $LiFePO_4$ as cathode and graphite as anode) couldn't satisfy the strong demand of widespread deployment for large-scale energy storages, which motivates the urgent exploration of new materials and structures for the next-generation batteries with high energy and power density as well as low cost and high safety. All-solid-state lithium batteries (ASSLBs) using solid-state electrolytes (SSEs) (Fig. 5.1) are of great interest to improve the overall performances, such as fundamentally solving the safety issue related to the flammable organic liquid electrolyte, absence of electrolyte leakage, wider temperature and electrochemical voltage tolerance, and ease of miniaturization [52, 135, 143]. SSEs could be classified into three types generally: organic polymers, inorganic electrolytes, and the composites. Solid polymer electrolytes (SPEs) are easy to be synthesized with flexible shaped structure, easy lamination stacking, and low cost. In contrast, rigid inorganic ceramics with high elastic moduli are more stable in aggressive environments and are conceivable to be potentially fast charging without the concentration gradients of ions due to the almost unit Li-ion transference number (LTN).

Moreover, the adoption of SSEs could prevent the Li dendrite growth, resolve the shuttle effect of dissoluble polysulfide for sulfur cathode, improve the stability of

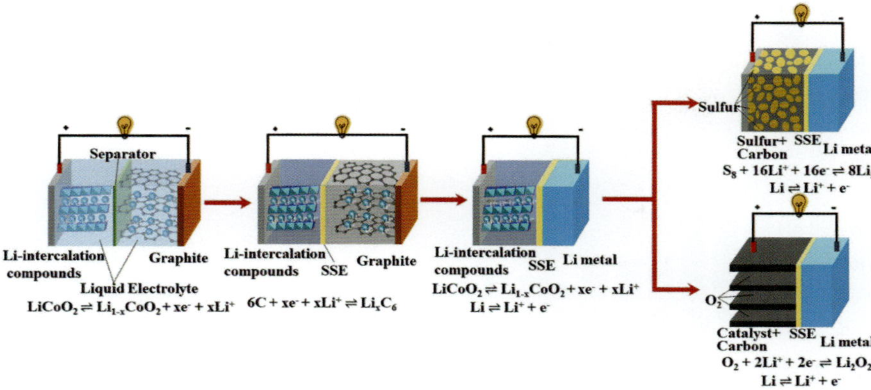

Fig. 5.1 Schematic diagrams of the development trend of lithium batteries from LIBs using conventional liquid electrolyte to ASSLIBs based on SSEs and then to Li metal-based advanced batteries including ASSLMBs, ASSLSBs, and ASSLABs [221]

battery components due to open surroundings for air electrode, and subsequently promote the batteries based on Li metal anode coupling with high-capacity cathode materials, such as Li-S and Li-O_2 batteries shown in Fig. 5.1. Lithium metal is the ideal anode material, on the fact that it has a high theoretical specific capacity (3860 mAh g^{-1}), the lowest negative electrochemical potential (−3.040 V vs. the standard hydrogen electrode), and a low density (0.59 g cm^{-3}) [120, 132, 226]. As a result, Li metal has been first used in primary Li batteries since the 1970s and was used in Li-MoS_2 batteries in the late 1980s by Moli Energy, but was abandoned and replaced by insertion carbonaceous anode later in 1991 by Sony Corporation, due to the uncontrolled Li dendrite growth that could penetrate the separator and cause short circuit and thermal runaway [190]. Meanwhile, scientists have also tried to replace liquid electrolyte by polymeric electrolytes to block dendrite growth, resulting in the Li metal polymer batteries [10]. Nevertheless, history repeats itself in recent years. Driven by the development of cathode chemistries with high specific capacities to meet the potentially high energy density of 500 Wh kg^{-1} for batteries, Li metal anode has become attractive again with extensive investigations. As indicated in Fig. 5.2a, b, the numbers of the literatures of the keyword of "Li metal anode" and "all-solid-state lithium battery" show a quick increase in recent 5 years. The most studied keywords in the publications from 2015 to 2018 can be found in Fig. 5.2c, such as lithium metal anode, Li-S batteries, conductivity, mechanism, Li-O_2 batteries, solid-state, and LLZO, which reveal the related issues and studied approaches. Growing researches and various strategies have been proposed to circumvent the Li dendritic growth and the low CE during Li deposition/dissolution cycling. All-solid-state lithium metal batteries (ASSLMBs) based on SSEs have been paid increasing attention recently.

Despite considerate efforts have been devoted to study batteries based on SSEs since the year of 1960s, ASSLBs haven't been fully commercialized, except the PEO-based

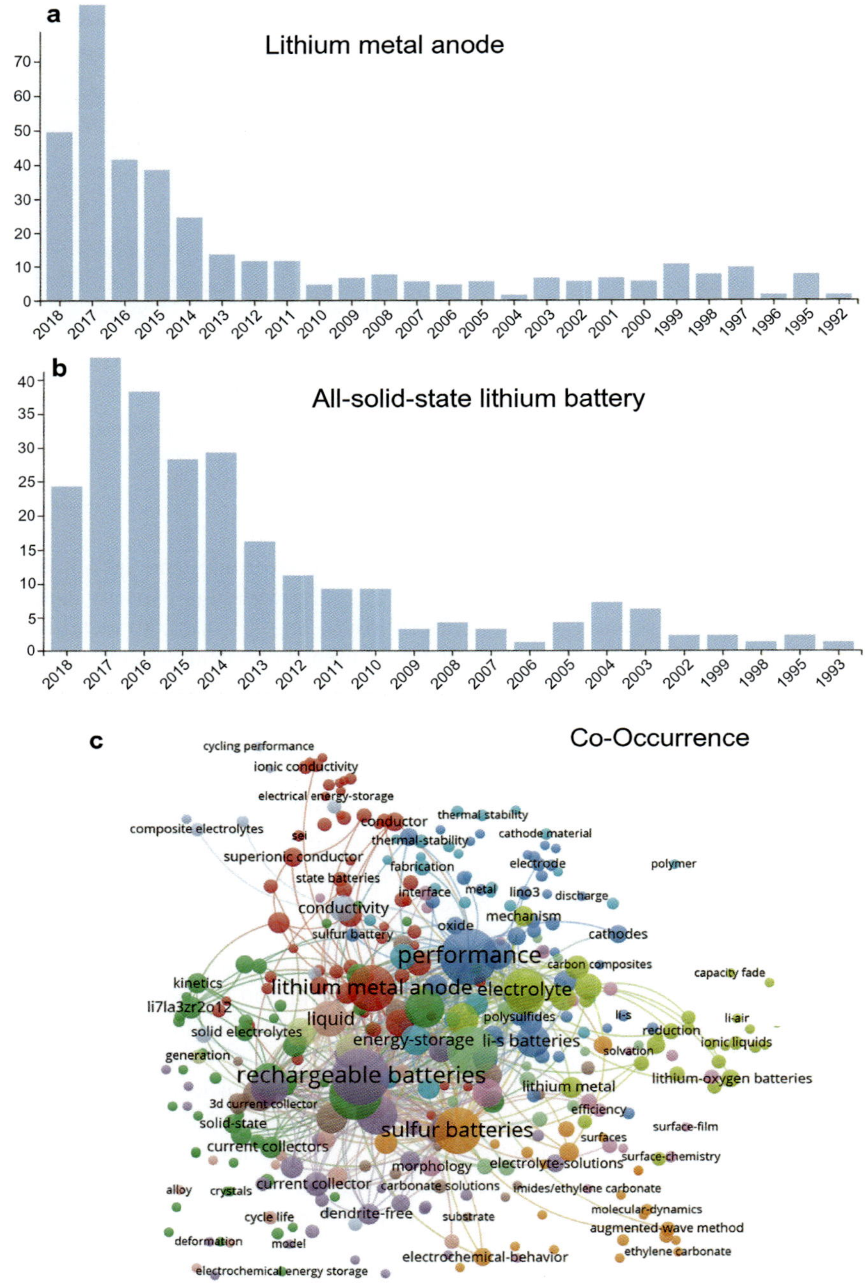

Fig. 5.2 The analysis on the related publications from 1990 to current. The number of the publications of the keyword of a "Li metal anode" (**a**) and "all-solid-state lithium battery" (**b**). (**c**) The co-occurrence network of the keywords related to ASSLMBs

batteries operated at the temperatures higher than 60 °C. The utilization of SSEs usually limits the electrochemical performances of the resultant batteries, especially the power density, which is mainly attributed to the low ionic conductivities, the high interfacial resistances, and the poor incompatibility with electrodes. To realize the practical applications of ASSLBs, SSEs need to satisfy the following requirements: (1) high ionic conductivity $>10^{-4}$ S cm^{-1} and low electronic conductivity at room temperature (RT); (2) low interfacial electrolyte/electrode impedances and good compatibility with electrodes; (3) high chemical, electrochemical, thermal, and physical stabilities; (4) good mechanical strength; (5) high LTN closest to 1; and (6) easy fabrication with low cost. Disappointedly, till now, a SSE that could satisfy all the above requirements hasn't explored. But, continuous progresses have been achieved attributed to many scientists and engineers' great efforts. Along with the quick development of nanomaterials and nanotechnologies, various approaches have been proposed to solve the issues with SSEs, including fine processing, nanoscale interfacial engineering, nanocomposites and nanostructures, etc.

In this chapter, we provide a comprehensive evaluation of various SSEs in ASSLBs from fundamental ion-transport mechanisms to fabrication technologies. We then highlight the mechanism, modeling, and observation of Li nucleation and dendrite growth during Li plating/stripping cycling. The most important interfacial stability of SSEs with Li metal has been discussed systematically. Strategies based on nanoscaled chemistries and nanotechnologies to address the Li dendrite growth and the incompatibility of SSEs with Li metal anode have also been overviewed. Last, current challenges and future prospects of the next-generation ASSLMBs with high energy and power density, long cycling performance, low cost, high safety, and power grid compatibility toward commercialization are summarized.

5.2 Solid-State Electrolytes

As the performance-limiting factor, SSE plays a key role in battery, the current progresses and challenges of which are discussed in detail in this section. For solid ionic conductors, the ion transfer generally follows a random hopping, referred to as "universal dynamic response" (UDR) containing AC conductivity $\sigma(\omega)$ and DC conductivity σ_{DC} as below [61]:

$$\sigma(\omega) = \sigma_{DC} + A\omega^c \tag{5.1}$$

where ω_p is characteristic frequency and c is a constant. DC conductivity could be determined according to a random walk theory:

$$\sigma_{DC} = f_h \left(\frac{ne^2 d^2 \gamma}{6kT} \right) \tag{5.2}$$

where $f_h = \omega_p$ is hopping rate, d is characteristic hopping distance, n is carrier concentration, e is charge, γ is a geometrical factor, and k is the Boltzmann constant.

Hence, it can be seen that fast hopping rate, longer hopping distance, and larger ion concentration can result in higher ionic conductivity.

Ion transport in ionic solids is a thermally activated process, in which an energy barrier must be overcome. The dependence on temperature of the ionic conductivity for SSEs typically follows two dominant conduction mechanisms: Arrhenius relationship and Vogel-Tamman-Fulcher (VTF) type. The crystalline polymer-salt complexes and inorganic ceramic electrolytes prefer to follow the Arrhenius behavior, while the ion transport in amorphous phase of SPEs showing strong coupling between ions with polymer segmental more satisfies the VTF equation [84]. The Arrhenius equation is described as follows:

$$\sigma = \sigma_0 \exp\left(-\frac{E_a}{kT}\right) \tag{5.3}$$

where σ_0 is the pre-exponential factor related to the number of charge carriers and E_a is the activation energy that can be calculated from the slope of log σ vs. $1/T$ plot. The VTF type can be expressed according to the following empirical equation [163]:

$$\sigma = \sigma_0 T^{-\frac{1}{2}} \exp\left(-\frac{B}{k(T - T_0)}\right) \tag{5.4}$$

where B is the pseudo-activation energy related to the segmental mobility of polymer chain and T_0 is the equilibrium glass transition temperature ($T_0 \approx T_g - 50$ K). The linear Arrhenius relationship for SSEs indicates ion conduction is via a simple hopping mechanism unlike the VTF behavior. The plot of conductivity vs. $1/T$ for VTF is generally nonlinear due to the ion-conduction mechanism involving ion hopping associated with the segmental motion of polymeric chains.

5.2.1 Solid Polymer Electrolytes

Since the system of poly(ethylene oxide) (PEO) complexing alkali metal salt showing high ionic conductivity was discovered in 1973 by Wright and co-workers [51], and later was developed by Armand et al. [9], it has been intensively studied for various applications, including batteries, fuel cells, sensors, and electrochromic devices [8]. In the following past two decades, SPE has been one of the most challenging and passionately adored topics in the field of material science and solid-state electrochemistry. Other polymer hosts satisfying the requirements of flexible polymer chains and sufficient electron-donor atoms or groups to couple with cations by a suitable distance between coordinating centers have been also developed, such as polyacrylonitrile (PAN), poly(vinylidene fluoride) (PVDF), poly(vinylidene fluoride-co-hexafluoropropylene) (PVDF-co-HFP), poly(methyl methacrylate) (PMMA), and polyvinylchloride (PVC). To date, PEO-based electrolytes receive the most attention and are the most promising applicable candidates for ASSLIBs, due to their high power of solvating various kinds of salts, low glass transition temperature, good electrochemical stability, low cost and low toxicity.

The mobility of the cations dissolved in the polymer matrix is related to hopping from the coordination site to the neighboring one associated with the ether oxygen atoms in the polymer backbone, which is assisted by the motions of the segments of the polymer chain (i.e., PEO monomer is –(CH_2–CH_2–O)–), as shown in Fig. 5.3a. Berthier and co-workers demonstrated that ion conduction preferentially took place in the amorphous region of PEO that is a semicrystalline polymer [20]. A consequence of the correlation indicates that the ionic motion is limited by the reluctant local segmental motion of the polymer host at ambient temperatures, which leads to a low ionic conductivity of 10^{-7} S cm^{-1} according to the following basic expression:

$$\sigma = \sum n_i e_i \mu_i \tag{5.5}$$

where n, e, and μ are the effective number of mobile ions, the electric charge, and the mobility of the i species, respectively. High ionic conductivity of 10^{-4} S cm^{-1} could be only achieved at the temperatures higher than 60 °C [13]. Therefore, various approaches on reducing glass transition temperature with lower crystallinity have been proposed [52], including linear chain polymers [152], comb-branched polymers [71], adding plasticizers [17, 99] or solvents, polymer blends [200, 209], cross-linking [140],

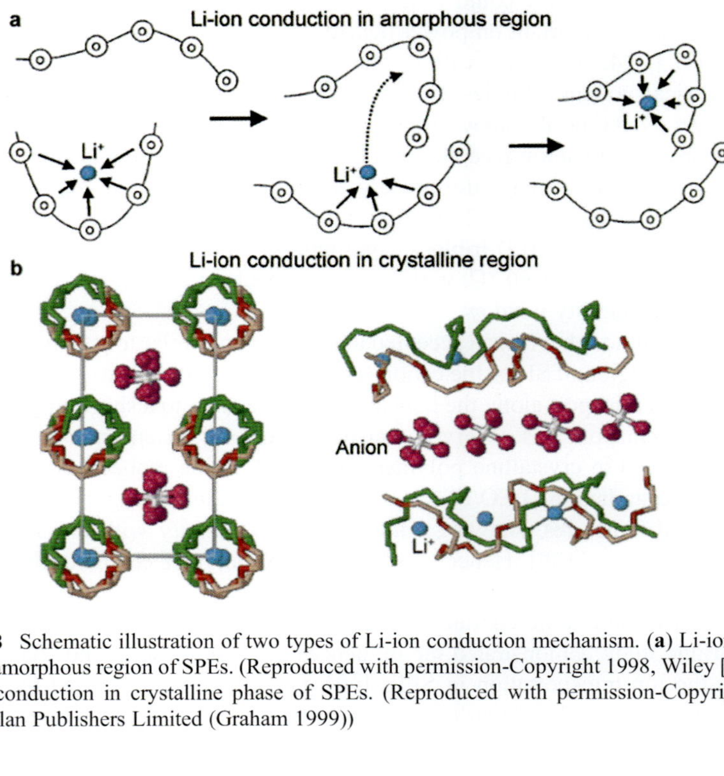

Fig. 5.3 Schematic illustration of two types of Li-ion conduction mechanism. (**a**) Li-ion conduction in amorphous region of SPEs. (Reproduced with permission-Copyright 1998, Wiley [147]); (**b**) Li-ion conduction in crystalline phase of SPEs. (Reproduced with permission-Copyright 1999, Macmillan Publishers Limited (Graham 1999))

copolymers [54, 186], and hyperbranched derivatives [196]. Archer and Coates demonstrated a cross-linked polyethylene/poly(ethylene oxide) (PE/PEO) of very high ionic conductivity of $>1.0 \times 10^{-4}$ S cm^{-1} at RT, with a same order of magnitude shear moduli as lithium, which could be used in ASSLMBs [98].

Along with the reduction the crystalline phase, the number of free ions depending on the degree of salt dissociation in the polymer host could be increased via the above approaches. Adding of plasticizer or fillers in SSEs may promote dissociation of ions, which results in more free ions for transport and subsequently improves ionic conductivity. The salts with large anions can easily be dissociated in the polymer matrix releasing free Li-ions and also prevent polymer crystallization process with increased amorphous region, which consequently promote the Li-ion conductivity. Lithium bis (trifluoromethane sulfonimide) (LiTFSI) is a promising lithium salt but relatively expensive that has a high solubility and good electrochemical stability, which results in a high ionic conductivity of 10^{-5} S cm^{-1} for SPEs [94, 107, 197]. However, the interfacial stability of the PEO-LiTFSI electrolyte with Li metal anode is relatively poor. Lithium bis(oxalate)borate (LiBOB) is also a good lithium salt with large anions and the ability to delocalize charge and reduce anion-cation reactions, via the bulky anion BOB. Scrosati et al. [7, 42, 224] reported the PEO-LiBOB electrolyte, showing high ionic conductivity of 10^{-4} S cm^{-1} at 40 °C.

Furthermore, alternative approaches to enhance ionic conductivity by enhancing LTN (the ratio of Li^+ to the total cationic and anionic carrier ions) have been proposed. Improving LTN could reduce the concentration gradients built by the anions that is also important on power density of LIBs [43]. But, LTN is generally lower than 0.3–0.4, since both cations and anions are mobile in the polymer host under electrical voltage. Single-ion conductors with the ability of reduced anion mobility can overcome the above issue. For example, affixing anions to polymer backbone and adding anion receptor are effective approaches to obtain a single-ion conducting polymer, which unfortunately usually leads to ionic conductivity of lower than 10^{-5} S cm^{-1} [153, 216]. Bouchet et al. [24] demonstrated a single-ion P(STFSILi)-PEO-P(STFSILi) triblock copolymers, showing an ionic conductivity of 1.3×10^{-5} S cm^{-1} at 60 °C, with improved mechanical strength and enlarged electrochemical stability window.

In addition, although it is well accepted that the cations prefer to transport through the amorphous region of SPEs, the cations enclosed within the helices of PEO-based electrolytes can transport along the helical axis, but leading to a short-range motion. In 1999, ion transport was observed that occur not only in amorphous region of polymers but also in crystalline polymers (Fig. 5.4). Some pioneers such as Peter G. Bruce have studied the PEO/Li salt complex crystals indicating that PEO chains formed helical structures, showing comparable conductivity with the amorphous polymer electrolytes [139]. Hence, manipulation of polymer chains to the desired orientation is of great importance to achieve a long-range ordered pathway, which results in anisotropic ionic conductivity as well as mechanical property for SPEs. Later, various study demonstrated the influences of crystallization and chain orientation on the ion transportation in SPEs [60]. For the commonly used method of casting from solution, the planar orientation of helices is preferred to be parallel to

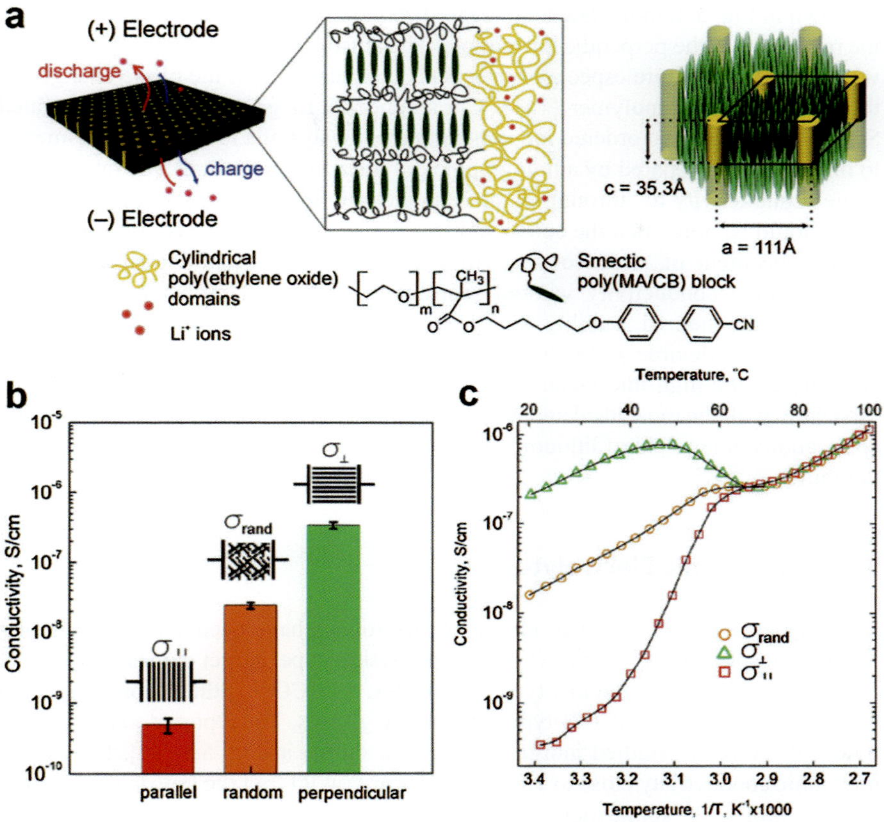

Fig. 5.4 Structure and electrical property of block copolymer with aligned PEO cylindrical domains. (**a**) Structure of block copolymer with aligned PEO cylindrical domains. (**b**) The in-plane and through-plane conductivities at RT for block copolymer with aligned domains, compared with one with random domains. (**c**) Arrhenius plots of block copolymers with and without aligned domains. (Reproduced with permission-Copyright 2010, American Chemical Society [141])

the substrate plane. However, for practical applications, the preferred order along perpendicular direction is more crucial [40, 67]. Anodic oxidation of aluminum (AAO) was used to confine PEO, showing higher ionic conductivity in smaller pores, which is due to the reduced crystallinity and the stretching and ordering of the polymer backbone [23]. Golodnitsky et al. [68] demonstrated an orientation of the helices along the perpendicular direction for PEO-based electrolytes, prepared by applying a magnetic field during polymer solution casting and drying. The ordering resulted in one order magnitude enhancement on ionic conductivity. Majewski et al. [141] reported a more detailed study on the ion transportation of PEO-based electrolytes with respect to chain orientation. The ionic conductivity comparison for the aligned PEO cylinders in two orthogonal directions and the random polymer is shown in Fig. 5.4, indicating an increased conductivity by the order of the parallel, the random, and the perpendicular sample. The anisotropy in conductivity decreased with rising temperature especially over 60 °C, indicating that the phase changed to the disordered melt polymer [141]. Christopher Y. Li group [181] demonstrated SPEs with tunable and ordered ion-conducting pathways of a few tens of nanometers to micrometers, prepared by a holographic polymerization method. The ratio of "in-plane" conductivity to "through-plane" conductivity ionic could reach up to 37.

It should be noted that the considerable and sustained efforts have been made on the enhancement of ionic conductivities of SPEs, but a qualified material with satisfied ionic conductivity without sacrificing mechanical property and electrochemical and thermal stability still needs to be exploited. Moreover, although SPEs are more flexible and volume comfortable compared with the rigid inorganic electrolytes, the high interfacial resistance and poor compatibility with electrodes, especially with the most ideal anode, lithium metal, remain the limitation on the wide applications in the field of lithium batteries with high energy density and long cycle durability.

5.2.2 Inorganic Electrolytes

For decades, a large number of inorganic electrolytes have been studied, including lithium phosphorous oxynitride (LiPON), perovskite type, garnet oxides, NASICON (sodium superionic conductor)-type phosphates, LISICON (lithium-ion superionic conductor) and thio-LISICON-type, and sulfide glasses. The reported ionic conductivities of currently studied inorganic SSEs are shown in Fig. 5.5 [14]. Li_3N has a high ionic conductivity close to 1.0×10^{-3} S cm^{-1} at RT, but the low decomposition voltage limits its applications [5, 108]. LiPON generally prepared by radio-frequency magnetron sputtering showing a low conductivity of 2×10^{-6} S cm^{-1} at RT has been studied for thin-film ASSLIBs [82].

$Li_{3x}La_{2/3-x}\square_{1/3-2x}TiO_3$ ($0.04 < x < 0.17$) is perovskite (ABO_3) structured solid ion conductors deriving from $La_{2/3}\square_{1/3}TiO_3$, which shows grain conductivity up to 10^{-3} S cm^{-1} (x~0.1) and total ionic conductivity of 10^{-5} S cm^{-1} due to highly resistive grain boundary [97, 185]. LLTO generally shows two types of crystal structure: cubic α-LLTO (space group Pm3m) of randomly arranged Li^+ and La^{3+} over the A sites and tetragonal β-LLTO with (P4/mmm) with ordering distribution of Li^+ and La^{3+} on A sites. In the tetragonal LLTO, the alternate stacking of La-rich layer and La-poor layer along the c-axis with a square planar bottleneck formed between A sites dominates the Li-ion migration by the vacancy mechanism, which shows lower ionic conductivity than cubic LLTO (Fig. 5.5b). To further improve ionic conductivity of LLTO, A, B, or A and B substitutions have been intensely investigated. Dopants that could stabilize the cubic structure and expand the "bottleneck" of LLTO are favorable for the conductivity enhancement [1]. A-site substitution of Al and Sr in LLTO could improve bulk ionic conductivity from 1.32×10^{-3} to 2.95×10^{-3} and 2.54×10^{-3} S cm^{-1}, respectively [161]. Zr^{4+} doping increased the total ionic conductivity of LLTO slightly from 3.08×10^{-5} to

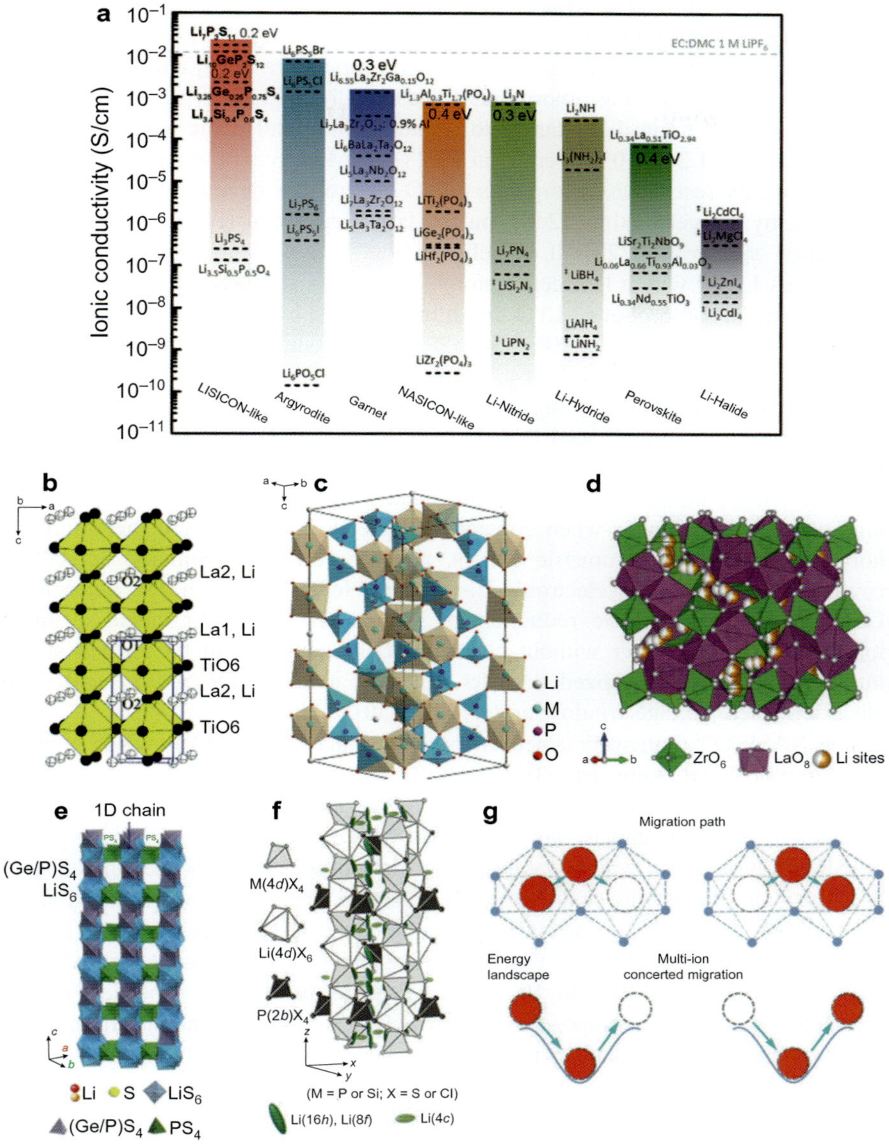

Fig. 5.5 Ionic conductivities and crystal structure of currently studied inorganic electrolytes. (**a**) Ionic conductivities at RT for the inorganic electrolytes with activation energy. (Reproduced with permission-Copyright 2010, American Chemical Society [14] Reproduced with permission-Copyright 2014, the Royal Society of Chemistry [178]); (**b–f**) Crystal structure of LLTO, NASICON, LLZO, LGPS, and LPS. (Images reproduced with permission: Copyright 2003, 2015, 2010, American Chemical Society [65, 166, 185] Copyright 2011, 2016, Macmillan Publishers Limited [90, 96]); (**g**) Schematic illustration of concerted migrations of multiple ions in the superionic conductors. (Reproduced with permission-Copyright 2017, Macmillan Publishers Limited [78])

5.84×10^{-5} S cm^{-1} [122]. Furthermore, many strategies have been proposed to further improve the total ionic conductivity by increasing the Li-ion conduction along grain boundaries, including high sinter temperature [106] and adding additives. The additives such as amorphous silica and $Li_7La_3Zr_2O_{12}$ (LLZO) have been used to modify grain boundaries to enhance the total ionic conductivity to 1.0×10^{-4} and 1.2×10^{-4} S cm^{-1} at RT [145]. Moreover, to overcome the high grain boundary resistance, amorphous LLTO electrolytes have been prepared successfully, by sol-gel method [249], pulsed laser deposition (PLD) [144], and RF magnetron sputtering [223]. It is believed that the local arrangement of Ti in amorphous LLTO with a less dense atomic packing is similar to that in crystalline counterpart. The amorphous LLTO thin films prepared by a sol-gel method showed an ionic conductivity of 4.5×10^{-6} S cm^{-1} at 30 °C, which is due to the open space and disordered structure [249]. It also found that the amorphous LLTO thin film prepared by PLD showed a higher ionic conductivity of 8.98×10^{-4} S cm^{-1} [142]. Ahn et al. demonstrated that the amorphous LLTO prepared by PLD with an ionic conductivity of 2.0×10^{-5} S cm^{-1} at RT showed a good stability with Li metal [3]. In addition, Zheng et al. found that the amorphous LLTO remained an ionic conductor behavior when contact with metallic lithium from the results of normalized chronoamperometric curves, which is different from the crystalline counterpart showing an electronic conductor. Hence, although the amorphous LLTO also undergoes the reduction of Ti^{4+} to Ti^{3+} with Li metal, a stable ion-conductive SEI layer without electronic conduction could be formed at the interface due to the localized disorder, showing a high LTN > 0.82 and wide electrochemical voltage window up to 12 V [250].

The NASICON framework generally has the formula of $A_xB_y(PO_4)_3$, where A and B are monovalent and multivalent (Ti, Ge, Hf, Zr, Sn) cations, with a rhombohedral unit cell and space group of $R\bar{3}c$ (Fig. 5.5c). The composition of $LiTi_2(PO_4)_3$- and $LiGe_2P_3O_{12}$-based materials are the most promising electrolytes among the NASICON-type SSEs, which was originally developed by Fu and co-workers [55, 56]. The ionic conductivity could be greatly improved by partial aliovalent substitution, such as Al^{3+}, Sc^{3+}, Zn^{2+}, and Ca^{2+}. Among them, $Li_{1+x}Al_xGe_{2-x}P_3O_{12}$ (LAGP, $0 \leq x \leq 1.2$) and $Li_{1+x}Al_xTi_{1-x}(PO_4)_3$ (LATP) systems are the most studied with RT ionic conductivities of 10^{-3} to 10^{-5} S cm^{-1}, which is strongly dependent on the preparation method and composition. Compared with polycrystalline materials, glass-ceramics with low grain boundary resistance are more easily manufactured into a desired shape and size [126]. The $Li_{1.4}Al_{0.4}Ti_{1.6}(PO_4)_3$ glass-ceramic prepared by spark plasma sintering (SPS) technique was reported to show a high ionic conductivity of 1.12×10^{-3} S cm^{-1} at RT [225]. But, similar as perovskites LLTO, LATP is not stable with lithium metal due to the reduction of Ti^{4+} to Ti^{3+} leading to a high electronic conductivity. LAGP is more stable in comparison to LATP. A LATP/LAGP double layer with no defects and good stability with lithium metal showed an electrical conductivity of 3.4×10^{-4} S cm^{-1} and an electronic conductivity of 9.6×10^{-9} S cm^{-1} at RT [246]. The composition of $Li_{1.4}Al_{0.4}Ge_{1.6}P_3O_{12}$ prepared by a sol-gel process demonstrated a total conductivity of 1.22×10^{-3} cm^{-1} at RT [242]. LAGP glass-ceramic with $x = 0.5$ prepared by the melt-quenching technique showed an ionic conductivity as high as 5.08×10^{-3} S cm^{-1} at 27 °C [193]. However, the dopant of Al^{3+} is effective to improve ion mobility in NASICON-type electrolytes, but it should be noted that the impurity phase of less conducting phase $AlPO_4$ was observed at the surfaces in both LATP- and LAGP-based materials [192]. In addition, it should be noted that the raw material of germanium Ge is expensive to enhance the battery cost according to a real practical application.

Recently, the most studied inorganic Li-ion conductors is garnet-based electrolytes with the formula of $Li_xLa_3M_2O_{12}$ (M = Ta, Nb, Zr). Among the garnet oxides, $Li_7La_3Zr_2O_{12}$ (LLZO) has been received the most interest, since the first report in 2007 by Murugan et al., which showed an ionic conductivity of 2.44×10^{-4} S cm^{-1} and a good stability with lithium metal and a wide electrochemical window [151]. LLZO was observed to have two crystal polymorphs: high-temperature cubic phase and low-temperature tetragonal phase [65]. The former with a space group of *I*41/*acd* exhibits two orders of magnitude higher in ionic conductivity than the latter with completely ordered distribution of lithium-ions of *Ia*3*d*, as shown in Fig. 5.5d. To stabilize the cubic phase and enhance ionic conductivity of LLZO, various cation dopants, such as Al^{3+} [65], Ga^{3+} [219], Nb^{5+} [130], and W^{6+} [109], have been proposed to have a positive effect on increasing the number of vacancies and the disordered degree of Li sublattice. Xin Guo group [219] demonstrated the highest ionic conductivity of 1.46×10^{-3} S cm^{-1} at RT with a low activation energy of 0.25 eV for the garnet-based electrolytes with the composition of $Li_{7-3x}Ga_xLa_3Zr_2O_{12}$ at $x = 0.25$ prepared by solid-state reaction. They believed that the high ionic conductivity is due to a high Li-ion mobility from the Coulombic repulsion between Ga^{3+} and Li^+. The Li-ion transport varies with crystal direction, and recently it is found that self-textured Ga-LLZO with a strong (420) preferred orientation showed a high ionic conductivity of 2.06×10^{-3} S cm^{-1} at RT that is six times higher than of the non-textured one, which is due to a shorter and less tortuous ion conduction pathways [162]. It is also interesting to note that although garnet-based ion conductors were reported to have good stability compared with other SSEs, it was found in further literatures that LLZO electrolyte could be aged by reacting with CO_2/H_2O in air forming Li_2CO_3 and LiOH. It was shown that protons entered the lattice of LLZO and replaced the less mobile Li-ions in the tetrahedral sites. The H^+/Li^+ exchange reaction could degrade the sintered bulk integrity and reduce ionic conductivity [86, 220]. Nb and Y have been used to co-doped LLZO electrolyte to improve the stability in air, which showed an ionic conductivity 6.91×10^{-4} S cm^{-1} at 30 °C without change after exposure in air for 1.5 months [62]. The addition of 2 wt.% LiF was able to improve the stability of garnet ceramic against moist air [111, 112].

To date, sulfide-type electrolytes have been reported to show the highest ionic conductivities up to 10^{-2} S cm^{-1} among SSEs, comparable to liquid electrolyte. Crystalline, amorphous, and partially crystalline structures could be observed in sulfides. Compared with oxygen, sulfur possesses larger ionic radius, which results in a more open ion channels in the structure and a higher ionic mobility. In 1981, the first sulfide-based glass as SSE was reported (Mercier et al. 1981). In 1984, Tachez and co-workers reported a crystalline sulfide electrolyte with the composition of

Li_3PS_4, showing a conductivity of 3×10^{-7} S cm^{-1} with an activation energy of 0.46 eV. Following that, various Li-P-S systems have been studied with respect to ionic conductivity [202]. As the pioneers to study the sulfide-based SSEs, Kanno and co-workers first reported the crystalline sulfide of thio-LISICON-type structure $Li_{4-x}Ge_{1-x}P_xS_4$ with ionic conductivities $>10^{-3}$ S cm^{-1} in 2001 [93]. In 2011, Kamaya et al. [90] reported a breakthrough work on a solid Li-ion conductor $Li_{10}GeP_2S_{12}$ (LGPS) with a 3D framework structure and 1D Li-ion transport pathway formed by the PS_4 and $(P/Ge)S_4$ tetrahedra tunnels along c-axis, as shown in Fig. 5.5e. LGPS has a super high ionic conductivity of 1.2×10^{-2} S cm^{-1}, comparable to liquid electrolyte, a very low electronic conductivity of 5.7 nS cm^{-1} at RT, and a low activation energy of 24 kJ mol^{-1}. Very recently, a glass-ceramic $Li_7P_3S_{11}$ (LPS) electrolyte with even higher ionic conductivity of 1.7×10^{-2} S cm^{-1} at RT and a lower activation energy of 17 kJ mol^{-1} was demonstrated, due to the reduced grain boundary resistance [178]. It believed that LPS is triclinic with the space group of *P1*, showing Li-ion transport along zigzag chains between P_2S_7 di-tetrahedra and slightly distorted PS_4 tetrahedra in the open space, as illustrated in Fig. 5.5f [212]. The fundamental understanding on the high ionic conductivity of SSEs compared to liquid electrolyte was reported by Ceder and co-workers [203]. They calculated and found that body-centered cubic (BCC)-like anion frameworks delivered a fast Li-ion transport pathway via a network of interconnected tetrahedral sites, which is in agreement with the reported experiment results. To study the common feature for the superionic conductors, Yifei Mo group [78] used ab initio modeling to predict that Li-ion diffusion behavior and found that Li-ion occupied high-energy sites and transport via concerted migrations of multiple ions, with low migration energy barriers, rather than the model of isolated ion hopping that is typical in ionic solid conductors, as illustrated in Fig. 5.5g.

5.2.3 Composite Solid Electrolytes

Over the last few decades, solid composite polymer electrolytes (CPEs) with incorporated inorganic fillers have been attracted much attention, which could improve the overall performance, including increased ionic conductivity, LTN, voltage window, electrolyte/electrode compatibility, and mechanical strength. Extensive investigations on complexion in SSEs have been reported since the first demonstration on Li-ion conductivity improvement by adding Al_2O_3 in LiI with a large rations in 1973 [116]. Due to the fast ion transport along interfaces, a percolation effect was used to explain the conductivity enhancement in the composite electrolytes. Later, Maier et al. found that a dramatically large lateral ionic conductivity of heterolayered CaF_2/BaF_2 thin films with a proportion to the number of interfaces [173]. Similar discover was found in the Y_2O_3-doped ZrO_2 (YSZ)/$SrTiO_3$ epitaxial multilayers, showing an eight orders of magnitude enhancement in total conductivity [64].

In 1982, Weston and Steele reported the PEO-$LiClO_4$ CPE with α-Al_2O_3 fillers for the first time, showing improved ionic conductivity as well as mechanical properties (Weston and Steele 1982). Since then, extensive investigations on inorganic fillers in

CPEs have been carried out, especially the nanoscale ceramic fillers with large specific surface area, including passive fillers such as Al_2O_3 [201], SiO_2 [100, 154], TiO_2 ([118]; Croce et al. 1998), ZrO_2 [194], Sm_2O_3 [38], BN [104], and clay [121] and active ones (Table 5.1) such as Li_3N [50], $LiAlO_2$ [89], LAGP [87], LLTO [123], LLZO [244], and LGPS [248]. Nan group [204] demonstrated a PVDF-based CPE with $Li_{6.75}La_3Zr_{1.75}Ta_{0.25}O_{12}$ (LLZTO) ceramic nanoparticles (NPs), showing a high ionic conductivity of 5×10^{-4} S cm^{-1} at RT with good mechanical strength and thermal stability [244]. The PEO-$LiClO_4$ composite electrolyte with 10 wt.% LATP nanoparticles exhibited an ionic conductivity of 1.70×10^{-4} S cm^{-1} at 20 °C. It was suggested that two contributions play the dominated role on the enhanced ionic conductivity, including reduced crystallinity degree of polymer host and highly conductive interface between organic polymer and inorganic filler. Scrosati et al. [177] demonstrated that the addition of these nanofillers could hinder the polymer reorganization and result in a lower crystallinity degree with long-term stability of the amorphous phase, allowing Li-ions to move more freely on the interfaces. The pioneers Wieczorek and co-workers suggested that the surface groups of the ceramic nanofillers associated with polymer chain and lithium salts by Lewis acid-base type interactions, which resulted in the promotion of salt dissociation and increase of the concentration of free Li-ions (Wieczorek et al. 1989; [215]).

Table 5.1 Conductivities of solid composite electrolytes with Li-ion conductive component

Li-ion filler	Weight percent (%)	Polymer	Lithium salt	σ_{RT} (S cm^{-1})	References
LLZO NPs	10	PVDF	$LiClO_4$	5×10^{-4}	[244]
LLZO NPs	85	PEO- poly(ethylene glycol) (PEG)	LiTFSI	6.24×10^{-5}	[32]
Random LLZO NWs	5	PAN	$LiClO_4$	1.31×10^{-4}	[231]
LLZO NW framework	80	PEO	LiTFSI	2.5×10^{-4}	[57]
LLTO NPs	15	PMMA	$LiNO_3$	1.13×10^{-4}	[169]
Random LLTO NWs	15	PAN	$LiClO_4$	2.4×10^{-4}	[123]
Random LLTO NWs	15	PEO	LiTFSI	2.4×10^{-4}	[253]
LLTO NP framework	44	PEO	LiTFSI	8.8×10^{-5}	[15]
Aligned LLTO NWs	3	PAN	$LiClO_4$	3.0×10^{-5}	[130]
LATP NPs	10	PEO	$LiClO_4$	2.0×10^{-4}	[204]
Aligned LATP NPs	60	PEO	$LiClO_4$	5.2×10^{-5}	[240]
LATP NPs	70	PEO-succinonitrile (SN)	$LiClO_4$	1.1×10^{-4}	[89]
LAGP NPs	20	PEO	LiTFSI	6.76×10^{-4}	[247]
LGPS NPs	1	PEO	LiTFSI	1.18×10^{-5}	[248]

Most recently, compared with isolated nanoparticles, randomly dispersed one-dimensional (1D) ceramic nanofillers could build a continuous three-dimensional (3D) ion-conducting network, which provides a significant enhancement of ionic conductivity. Liu et al. [123] originally reported PAN-based CPEs with Li-ion conductive nanowires with an extremely high ionic conductivity of 2.4×10^{-4} S cm^{-1} at RT, which showed three and two orders magnitude enhancement of ionic conductivity in comparison to the filler-free electrolyte and the corresponding NP-filled polymer electrolyte. In this work, 1D inorganic Li-ion conductor ($Li_{0.33}La_{0.557}TiO_3$) was prepared for the first time by electrospinning. Later, LLZO nanowires were reported to effectively improve ionic conductivity of SPEs [57, 230, 231]. To achieve a high ionic conductivity according to percolation of nanofillers, some ceramic scaffolds have been proposed. Liangbing Hu group [57] demonstrated a CPE with 3D inorganic matrix of garnet nanofibers and PEO polymer with lithium salt, which showed an ionic conductivity of 2.5×10^{-4} S cm^{-1} at RT. Guihua Yu group [15] synthesized a hydrogel-derived LLTO framework composited with PEO-based SPE, showing an ionic conductivity of 8.8×10^{-5} S cm^{-1} at RT. Goodenough group [32] proposed a concept of from "ceramic-in-polymer" to "polymer-in-ceramic" with the ability of polymer host to ceramic matrix, as shown in Fig. 5.6a. The PEO-LLZO-PEG (10:85:5) with 60 wt.% LiTFSI prepared by hot pressing showed an ionic conductivity of 6.24×10^{-5} at RT and a good stability against Li metal anode. It is worth noting that the ionic conductivity of CPEs could be further improved via the approach of aligning the NWs that attributed to continuous transport pathways without resistive junctions like NPs and random distributed NWs. After the development of random NW-filled SPE, Liu et al. [130] demonstrated a CPE with aligned LLTO NWs that were parallel to electrode direction (Fig. 5.6b), which showed one order magnitude enhancement of ionic conductivity, compared with the one with random NWs. The approaches of controlling Li-ion transport pathway with the help of aligned nanofillers were also realized in several other publications [240, 245]. Yuan yang group [240] developed a CPE with vertically aligned and connected LATP NPs via ice templating method, which showed an ionic conductivity of 5.2×10^{-5} S cm^{-1} at RT, 3.6 times higher than that of CPE with randomly dispersed LATP NPs. Yi Cui group [245] exploited vertically aligned nanoscale ceramic/polymer interfaces with densely packing, via a surface-modified anodized aluminum oxide with filled PEO-LiTFSI, showing a high ionic conductivity of 10^{-3} S cm^{-1} at 0 °C. An extremely high conductivity of 5.82×10^{-4} S cm^{-1} was achieved in the composite electrolyte at RT. Hence, it is important to control the orientation of nanofillers to improve electrical properties of the composite electrolytes.

5.2.4 3D Solid-State Electrolyte Structures

In addition to the study on the material synthesis with modification of composition, it is also critical to develop unconventional configurations for battery structures based on the novel fabrication technologies [69, 198]. Planar structured LIBs have been occupying the main market share, but the energy density and capacity are limiting. Various 3D structures with a focus on electrode/electrolyte interfaces for LIBs have been studied to

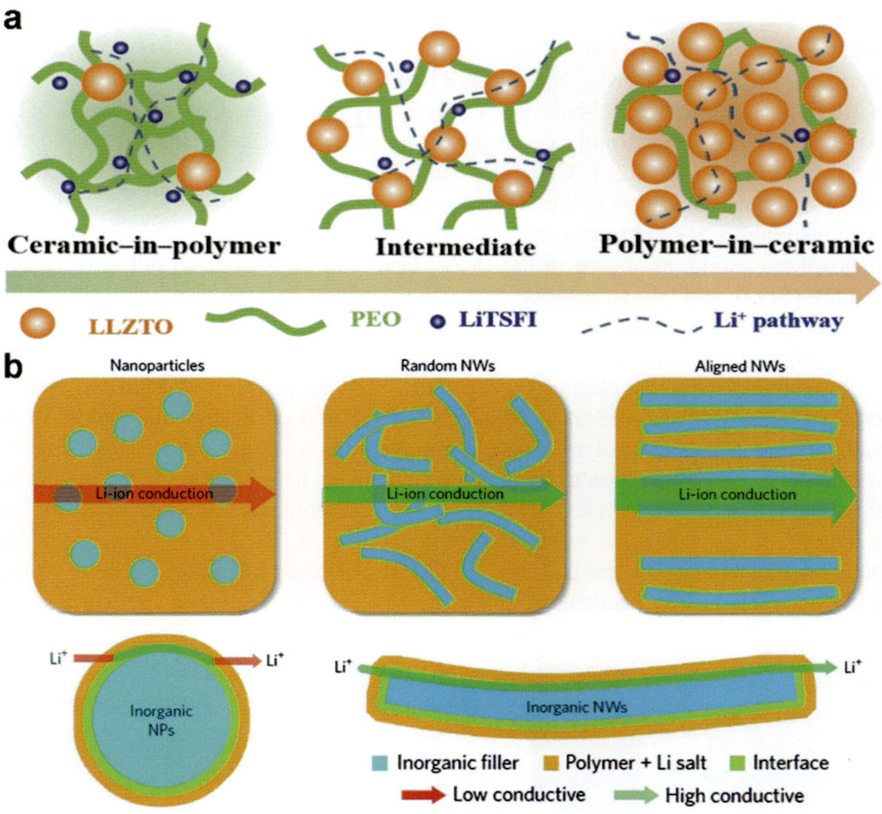

Fig. 5.6 Schematic illustration for various Li-ion conduction mechanisms in composite solid electrolytes. (**a**) Schematic illustration for composite solid electrolytes with three different mechanisms of "ceramic-in-polymer," "intermediate," and "polymer-in-ceramic." (Reproduced with permission-Copyright 2017, Elsevier Ltd. [32]); (**b**) Schematic illustration of Li-ion transport mechanism for CPEs with nanoparticles, random nanowires, and aligned nanowires. (Reproduced with permission-Copyright 2017, Macmillan Publishers Limited [130])

improve the electrochemical performances, especially the energy density, due to not only large contact area between electrode and electrolyte but also high loading of active materials and modified electron/ion pathways. Several configurations of 3D batteries have been proposed, including array of interdigital array of electrode pillars, interdigital array of electrode plates, rod array of cylindrical anodes with a thin electrolyte coating and filled cathode material in the remaining free space, and sponge-type cathode with thin electrolyte coating and filled anode in the remaining free space, as illustrated in Fig. 5.7a [134]. Moreover, as shown in Fig. 5.7b, compared with liquid electrolyte that could wet the electrodes, the solid electrolyte leads to a poor contact with the electrodes, leading to a large interfacial resistance. Hence, the large surface area in 3D battery could improve the contact between solid electrolyte with electrodes [103].

Sun and co-workers built 3D microbatteries (3D-MBs) based on a poly(trimethylene carbonate) (PTMC)-based electrolyte with a high LTN >0.6 and a high ionic conductivities of ~10^{-5} S cm^{-1} at RT. A 3D current collect was first prepared, active material and PTMC-based electrolyte was then coated on it. The 3D-MBs based on PTMC could stably cycling with an areal capacity of 0.2 mAh cm^{-2} at ambient temperature 22 ± 1 °C [187]. In addition, the membrane of Cu_2O-coated Cu nanopillars has also been as the 3D supporter in the above work. Ergang et al. [49] demonstrated a 3D interpenetrating electrochemical cell based on ordered carbon nanosphere anode, as shown in Fig. 5.7c. The solid electrolyte of poly(phenylene oxide) (PPO) and V_2O_5 cathode were later coated on the carbon anode. The reversible specific capacity was increased from 0.7 to 350 μAh g^{-1}, by forming the V_2O_5 cathode from xerogel to ambigel, due to improved Li^+ diffusion.

In addition, inorganic electrolytes have also been adopted in 3D ASSLBs. Notten et al. demonstrated an ASS 3D-integrated battery based on Si thin film anode, LCO cathode, and LiPON electrolyte, from a silicon substrate with high surface area using the micro-etching and deposition technologies. 3D ASSLBs have also been fabricated from the support of solid electrolytes. For example, LLTO membrane with micro-sized holes, a honeycomb structure, has been prepared for 3D ASSLIBs. The mixtures including active materials of LCO and $Li_4Mn_5O_{12}$ (LMO) were then filled into the holes of the LLTO membrane. The resultant 3D solid LCO/LLTO/LMO cell could deliver a small

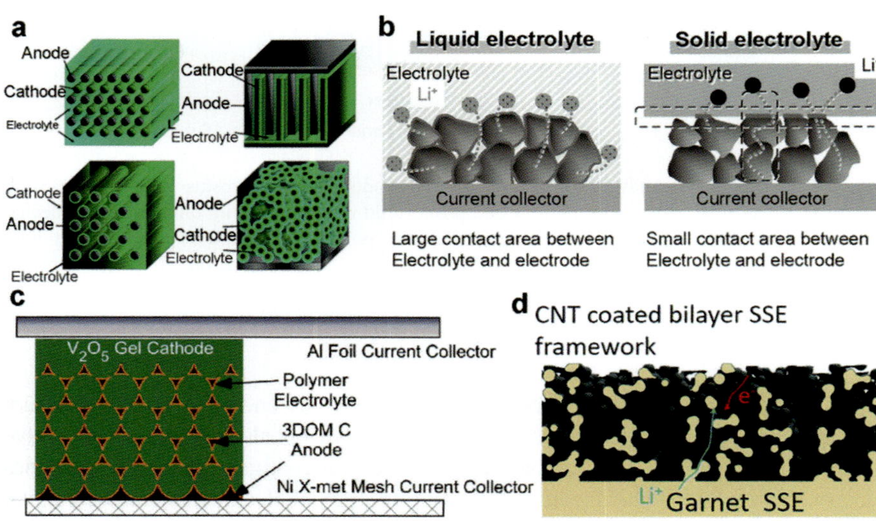

Fig. 5.7 Illustrations of 3D battery structures. (**a**) Prospective four different configurations of 3D batteries. (Reproduced with permission-Copyright 2004, American Chemical Society [134]); (**b**) Schematic illustration of electrode/electrolyte contact using liquid electrolyte and solid electrolyte. (Reproduced with permission-Copyright 2013, Elsevier Inc. [103]); (**c**) Illustration of SPE based 3D batteries. (Reproduced with permission-Copyright 2007, the Electrochemical Society [49]); (**d**) Illustration of inorganic electrolyte-based 3D batteries. (Reproduced with permission-Copyright 2017, the Royal Society of Chemistry [59])

discharge capacity of 7.3 μAh cm^{-2} [102]. To address the high interfacial resistance, ion-conducting components are needed to mix with active electrode materials, such as polymer-Li salt and inorganic electrolyte NPs, while the inorganic nanoparticles are distributed isolately in the composite electrode, showing sluggish ionic Li-ion conduction pathway. Recently, Liangbing Hu group [59] demonstrated a 3D bilayer garnet-based SSE framework (Fig. 5.7d), consisting of a thin dense LLZO layer with a thickness of 20 μm and a thick porous LLZO layer with a thickness of 50–100 μm. The sulfur cathode was filled into the pores of this bilayer electrolyte. The ASSLSB with a high sulfur loading of >7 mg cm^{-2} showed a high discharge capacity is around 645 mAh g^{-1} at the current density of 0.2 mA cm^{-2} at the first cycle, and high CE >99% for the subsequent cycles, indicating that the "shuttle effect" was blocked by the dense layer. A high energy density of 248.2 Wh kg^{-1} based on the total mass of the electrolyte with sulfur cathode and Li metal anode was achieved. Moreover, Li metal could also be filled into the pores to achieve a composite Li metal anode without volume change during cycling. The $Li(Ni_{0.5}Mn_{0.3}Co_{0.2})O_2$ (NMC)/garnet/Li cells, with a high NMC mass loading of 32 mg cm^{-2}, could deliver a high energy densities of 329 Wh kg^{-1} and 972 Wh L^{-1} [133].

5.3 Lithium Metal Anode

With the ever-increasing demand on high-energy-density energy storages, lithium metal anode as the "holy grail" in the field of rechargeable lithium batteries has been revived since the most recent years. Lithium metal is the ideal anode material attributed to the high theoretical specific capacity (3860 mAh g^{-1}), the low redox potential (−3.040 V vs. standard hydrogen electrode), and the low relative density [4, 120, 125, 131, 195, 243]. In addition, as the development of high-capacity chemistries taking place of conventional intercalation cathodes, such as sulfur cathode and oxygen electrode for Li-S and Li-O_2 batteries, lithium metal as anode is highly needed. The advantages of lithium metal anode are very clear, but the problems with failure mechanisms it confront with are also critical to limit its practical applications in the field of batteries. The most serious and detrimental issue is the dendrite growth at the lithium surface during repeated charging/discharging, which could penetrate the separator, cause internal short circuit, and lead to thermal runaway. It is possible that the dendrites break off leading to "dead lithium" floated on the lithium surface and the pulverizing of lithium metal. The second issue is the formation of solid electrolyte interphase (SEI) on Li metal surface with electrolyte, due to the highly reactive feature of lithium metal with high Fermi energy level, leading to the consumption of lithium metal as well as electrolyte, increase of internal resistance, and the subsequently low Coulombic efficiency (CE) and poor cycling performance [35, 115]. The SEI layer is easy to break upon the infinite volume change during Li plating/stripping, resulting in the exposure of fresh lithium metal and continuous generation of thick SEI. The large and repeatable volume change also creates the accommodation of large strain and reduces the electrochemical performance of lithium metal anode.

5.3.1 Understanding on Lithium Nucleation and Dendrite Growth

The fundamental understanding on Li nucleation and growth during Li platting/stripping cycling is critical to guide experiments for stabilizing Li metal anode [87, 217]. According to the fundamentals of materials science, the formation of a new phase in a parent phase results in a Gibbs free energy change ΔG with respect to several contributions [27]:

$$\Delta G = \Delta G_{\text{chem}} + \Delta G_{\text{int}} + \Delta G_{\text{strain}} = \frac{4}{3}\pi r^3 \left(\Delta G^v_{\text{chem}} + \Delta G^v_{\text{strain}}\right) + 4\pi r^2 \gamma \tag{5.6}$$

where ΔG_{chem} (<0) is the chemical energy change driving the phase transformation, ΔG_{int} (>0) is the interface/surface energy change due to the nuclei formation of new phase, and ΔG_{strain} (>0) is the strain energy derived from the volume difference of precipitates from the volume they occupied. ΔG^v_{chem} and $\Delta G^v_{\text{strain}}$ are the chemical energy and the strain energy change per unit volume, respectively, and γ is the interfacial energy per area. The nuclei could continuously grow with decreased free energy when the critical radius could be reached by the energy fluctuation to overcome the energy barrier.

For the last two decades, various theory models of electrochemical systems based on the well-established phase field method numerically and analytically have been proposed to study the behavior of Li nucleation and dendrite growth, by charge mass transfer through the binary electrolyte (salt dissolved in solvent) to the negative and positive electrodes and the cation reduction on the anode surface under polarization, as shown in Fig. 5.8a [31]. The open circuit voltage (OCV) is dependent on the chemical potential difference between the anode (μ_A) and the cathode (μ_C) and working voltage also limited by the electrolyte electrochemical window that is the energy gap between the highest occupied molecular orbital (HOMO) and the lowest unoccupied molecular orbital (LUMO) (Fig. 5.8b) [70]. The classical Butler-Volmer kinetics could describe the Li metal deposition and the deposited rate, with the dependences on the cation concentration at the anode surface and the overpotential, as follows [117, 149]:

$$i_n = i_{0,\text{ref}} \exp\left[\frac{(1 - D_a)\Delta\mu_{e^-}}{RT}\right]\left[\exp\left(\frac{D_a F \eta_s}{RT}\right) - \exp\left(\frac{D_c F \eta_s}{RT}\right)\right] \tag{5.7}$$

where i_n is current density at the deformed interface, $i_{0,\text{ref}}$ is the exchange current density at an undeformed interface, D_a and D_c are the anodic and cathodic diffusion coefficients, $\Delta\mu_{e^-}$ is the change of electrochemical potential in the electrons, η_s is the surface overpotential, R is the gas constant, F is the Faraday's constant, and T is the temperature. Generally, a diffusion limited growth system resulting in a local concentration gradient is proposed that could lead to unstable dendrite growth at the negative electrode surface. Under high current densities exceeding diffusion limitation, the cation concentration could drop to zero resulting in a local space

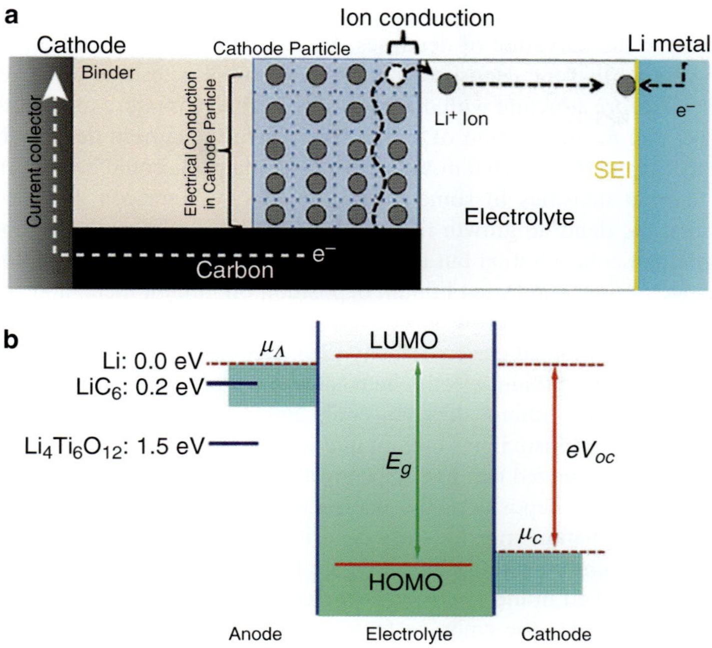

Fig. 5.8 Structure and electrochemical potentials in a battery. (**a**) Battery structure and ion/electron conduction pathway. (Reproduced with permission-Copyright 2010, Elsevier B.V. [156]); (**b**) Electrode electrochemical potentials versus electrolyte window. (Reproduced with permission-Copyright 2009, American Chemical Society [70])

charge at the electrode surface, leading to the transition from charge transfer control to mass transfer process and the subsequently ramified deposits. The corresponding time is Sand's time according to the following equations [170, 172]:

$$\tau_S = \pi D\left(\frac{z_c e C_{c0}}{2 i_n}\right)^2 \left(\frac{\mu_a + \mu_c}{\mu_a}\right)^2 \tag{5.8}$$

$$D = \frac{\mu_a D_c + \mu_c D_a}{\mu_a + \mu_c} \tag{5.9}$$

where D is the ambipolar diffusion constant, μ_a and μ_c the anion and the cation mobilities, z_c the cation charge number, and C_{c0} the initial cation concentration. Therefore, it can be seen that low current densities prefer to a stable Li plating/stripping cycling and high current densities promote the heterogeneous growth. The pioneer Chazalviel demonstrated that at high current densities, the velocity of dendrite growth appeared is proportional to the anion mobility in the electrolyte and the applied electric field [31]. Meanwhile, Monroe and Newman [148] indicated that the dendrite growth velocity is proportional to the overpotential and inversely proportional to the curvature of dendrites. Kanamura et al. [94] conducted experiments and found that the deposited lithium metal on Ni substrate changed from a smooth hemisphere structure with diameter of 0.3 μm to a mixture of smooth Li and Li dendrite, and to most region of Li dendrite, when the current density increased from 0.2 to 2.0 and then to 10.0 mA cm^2, while Li filaments could be appeared even at small current densities in some investigations. JM Tarascon and co-workers believed that the dendrite growth model could not simply be studied according to lithium gradient concentration but the Li surface state should be taken into consideration as well. They conducted lithium deposition on lithium metal and found that lithium metal preferred to grow in deep defects with higher electric field, where induced variations in local current density [28]. Dendrite-free mossy lithium metal could be obtained by lithium electrochemical plating and PLD on polished lithium foil, but the dendrite lithium deposits could not be unavoidable upon extended cycling due to the intrinsic inner lithium microstructure [66].

Recently, it summarized that five steps toward the final lithium dendrite structure during homogeneous deposition: the nucleation suppression, the long incubation time, the short incubation time, the early dendrite growth, and the late growth regime [47]. The lithium embryos in the first step are thermodynamically unstable and are easy to be redissolved in the electrolyte. The critical overpotential could determine the incubation time for the embryos to grow kinetically. The time for nucleation incubation was determined to be proportional to Sand's time [25]. In the early dendrite growth regime, the thermodynamically and kinetically favored nuclei could grow to reach the same growth velocity, and in the late dendrite growth regime, uniformed deposited structure with localized electric fields could control the next morphology of the deposit [47].

The fundamental understanding on the Li nuclei and dendrite growth is critical to design new material and structure to stable lithium metal anode. It is clear that current densities, salt concentration, surface defects, and substrate feature have strong influences on lithium plating/stripping. Moreover, the interfacial stress generated at the solid/solid interface upon the Li nucleation could also affect the lithium dendrite growth [30]. Compressive stress during Li plating could provide the driving force to push out Li filaments at localized diffusional creep. Recently, it was found that Li dendrite was be suppressed by the use of a soft substrate of polydimethylsiloxane (PDMS), as a result of the surface-wrinkling-induced stress relaxation [206]. More obviously, it is found that the morphology of deposited lithium metal at higher applied pressure is more uniform and dense [66]. assembled the Li/V_2O_5-P_2O_5 cells and then coated the gasket with ethylene-propylene-diene monomer (EPDM)/toluene emulsion. A hydraulic hand press fitted with an iron cylinder with the same diameter as the Li metal anode was applied on the test cells. The SEM images indicated that dense and uniform Li metal could be deposited under pressure, which is due to the good electrical path between deposited Li particles and anode substrate.

Moreover, surface diffusion prefers to smooth out the cavities or protrusions on Li surface, with physically meaningful rates at temperature range of $0.5T_M$ to

$0.75T_M$ (T_M is the melting temperature) of the metal [165]. Therefore, temperature is also a factor that can influence the morphology of deposited lithium metal. It is reported that mushroom-shaped lithium deposit was obtained at temperatures lower than 5 °C with a shorter incubation time and needlelike lithium was found at temperatures higher than 5 °C [136]. In addition, it was observed that in the Li/PEO/Li symmetric cell, several Li dendrites growth with a proximately same velocity and only one first contact with the cathode that acted as a fuse and burnt under high current densities [170]. While and recently, Li et al. [113] demonstrated a dendrite healing mechanism with respect to temperature under even higher current densities. They found that Li dendrite protrusion could be disappeared and return to smooth morphology under current densities >9 mA cm^{-2}, which is due to the surface migration dependent on temperature by self-heating under high current densities.

5.3.2 Characterization and Observation on Lithium Dendrite

In the past two decades, morphology observations and technologies on the Li nuclei and growth using advanced facilities, including optical microscopy (OM) [182], scanning electron microscopy (SEM) [189], atomic force microscopy (AFM), transmission electron microscopy (TEM) [144, 171, 205, 236, 238], X-ray microtomography [75], nuclear magnetic resonance (NMR) spectroscopy [21], and X-ray photoelectron spectroscopy (XPS) [199], have been utilized to analyze the behavior of Li electrochemical deposition/dissolution [226]. Lithium dendrite was first observed by SEM in 1980 [48], and later in situ morphological observations on lithium deposition under galvanostatic cycling were studied since the late 1980s [18]. Lithium dendrite growth based on tip mode that lithium metal deposits on the surface of pre-deposited lithium nuclei under the SEI layer has been widely observed and accepted. But, other growth behaviors have also been observed. It is believed that if the adhesion between deposited lithium and substrate is strong, a surface growth mode is preferred; if the contact is weak, a root growth mode that lithium deposits on the electrode substrate is more likely to occur [241], while, as shown in Fig. 5.9a, b, Bai et al. observed the deposited lithium structure in liquid electrolyte of 1 M $LiPF_6$ in EC/DMC using a glass capillary cell by an OM and found two different mechanisms with respect to the applied current and capacity. Mossy lithium from the roots was observed below Sand's capacity with a reaction limitation, and then it transferred to dendritic lithium growth at the tips above Sand's time with a transport limitation growth [16]. In addition to the tip mode and root mode, lithium growth in between kinks have also been observed [182, 183].

Recently, Bazant and Li groups [105] observed that by in situ TEM in liquid cell (Fig. 5.9c), the surface growth with a morphology of densely packed "buds" and the root growth with a morphology of loosed "whiskers" showed strong dependence on the SEI growth rate and overpotential. In addition, an operando TEM was carried out to confirm the growth of lithium dendrite formation of "dead lithium" with less

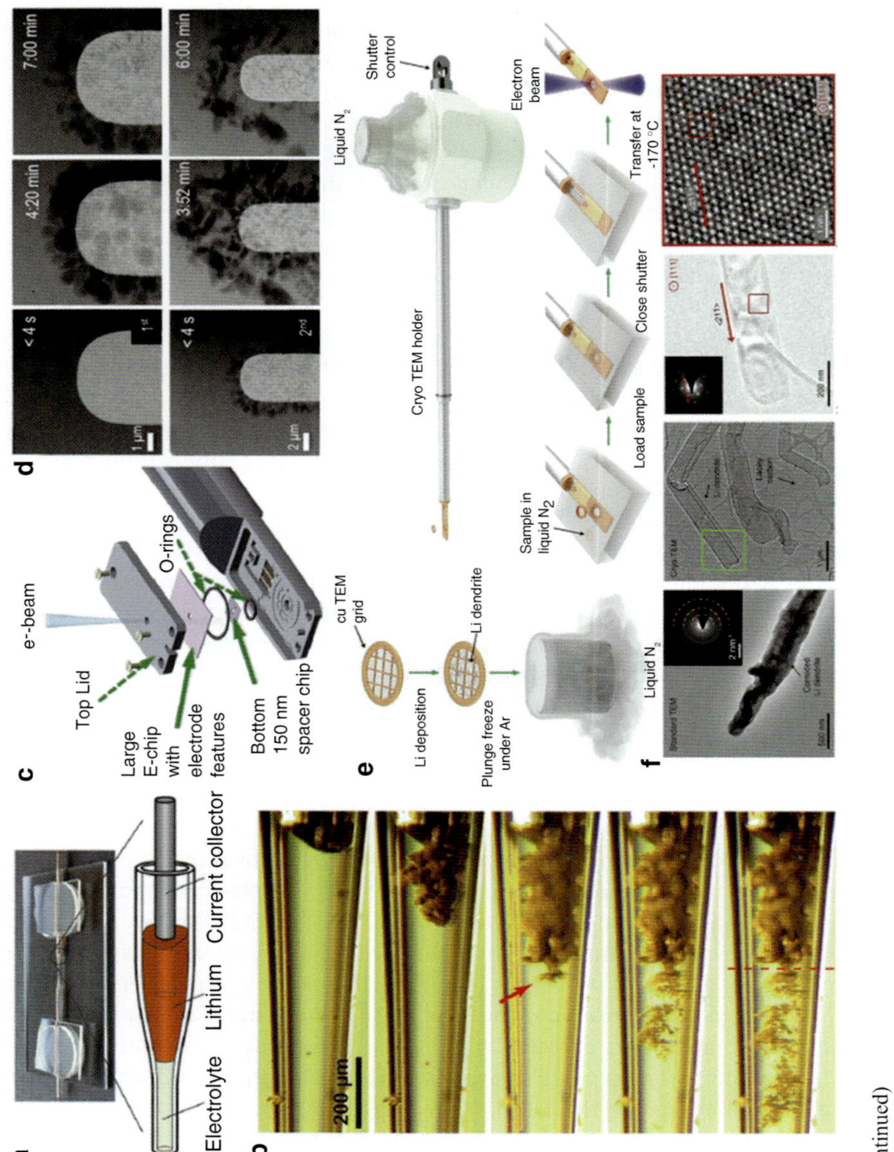

Fig. 5.9 (continued)

density during cycling (Fig. 5.9d) [144]. It is worth noting that the atomic-resolution imaging of lithium metal that is beam-sensitive was obtained for the first time using cryo-TEM from Yi Cui group, indicating a big breakthrough, as illustrated in Fig. 5.9e [112]. They demonstrated that Li metal dendrites prepared by electrochemically deposited onto a Cu TEM grid were single-crystalline nanowires with preferred crystal faces (Fig. 5.9f). And, the SEI layer was found to be Li carbonate and Li oxide without the trace of LiF. While, Ying Shirley Meng and co-workers [205] found that the electrochemically deposited Li metal is amorphous, with organic species and crystalline LiF on its surface. Further investigations by the utilization of TEM are needed to study the atomic-resolution imaging of electrochemically deposited lithium metal. In addition, lithium dendrite growth was also observed by real-time AFM under a galvanostatic condition [101]. Transmission grazing incidence small angle X-ray scattering transmission (GISAXS) analyses by a homemade Li electrodeposition cell was used to detect the real-time evolution from primary Li nuclei to dendrites [88]. In situ NMR was also utilized to determine the structure of lithium dendrite. Clare P. Grey group [29] demonstrated the ability of in situ $^{6}Li/^{7}Li$ NMR to study the morphologies of deposited lithium metal quantitatively.

5.3.3 Solid Electrolyte Interphases on Lithium Metal

Lithium metal is highly reactive and thermodynamically unstable, which results in the formation of a passivation layer of SEI with thickness of several nanometers on its surface immediately when contact with liquid electrolyte in time constants of milliseconds. SEI consists of insoluble and partially soluble Li-ion conductive components with high electronic resistivity, which determines the overall performances of the batteries [22]. Upon repeatable Li electrodissolution/electrodeposition with large volume change, SEI layer is easy to crack with fresh Li metal exposure, leading to the inhomogeneities on the electrode surface where are more likely to concentrate Li-ion flux ("hot spots") and the subsequent Li dendrite growth. Moreover, during long cycling, the uncontrolled formation of thick SEI results in the consumption of both the liquid electrolyte and lithium metal, which leads to low CE [158, 208]. It should be noted that the lithium metal is easy to be covered by a native film which consisted of Li_2CO_3, Li_2O, and LiOH [93].

Fig. 5.9 Morphology observations on the Li nuclei and growth during Li electrochemical deposition/dissolution. (**a**) Illustration of the glass capillary cell for in situ OM setup. (**b**) The morphology of the deposited lithium in liquid electrolyte. Dendritic lithium growth at the tips was observed above Sand's time. (Reproduced with permission-Copyright 2016, the Royal Society of Chemistry [16]); (**c**) Setup for operando TEM. (**d**) TEM images for the lithium dendrite growth and the "dead lithium." (Reproduced with permission-Copyright 2015, American Chemical Society [144]); (**e**) Experiment procedure for cryo-TEM. (**f**) TEM image and high-resolution TEM image of Li dendrite. (Reproduced with permission-Copyright 2017, American Association for the Advancement of Science [111])

The composition and structure of SEI is quite complex with a heterogeneous morphology that is dependent on the cations and anions, solvent molecules and impurities in the liquid electrolyte, as well as current density, temperature, and so on. The models on the SEI formation have been studied since 1979 by Paled [158, 159]. He predicted a mosaic structure SEI with Li_2O, LiF, Li_2CO_3 and semi-carbonates and polyolefins, as shown in Fig. 5.10a. Another SEI model predicted by Doron Aurbach is multilayer structure: inner region is thin and compact inorganic materials blocking the diffusion of solvent molecules or anions through it, and the outer region is more porous (Fig. 5.10b). The reduction products of both insoluble inorganic components reacted with salts (e.g., lithium hexa-fluorophosphate ($LiPF_6$), lithium bis(trifluoromethanesulfonyl)imide (LiTFSI), lithium perchlorate ($LiClO_4$)) and partially soluble semi-carbonates and polymers with solvents. ROLi, RCOOLi, and $ROCOLi_2$ (R = alkyl groups) species were found for the commonly used solvents of ethers (e.g., 1, 2-dimethoxyethane (DME), tetrahydrofuran (THF), tetraethylene glycol dimethyl ether (TEGDME)), esters (e.g., ethylene carbonate (EC), dimethyl carbonate (DMC), propylene carbonate (PC)), and alkyl carbonate, respectively [11]. It has been proposed in 1997 that the artificial SEI layer consisted of LiF and Li_2O on lithium metal surface was effective to stabilize lithium deposition/dissolution [188]. The use of 1,3-dioxolane (DOL) solvent could provide a uniform Li metal deposition with high CE, which is due to the partial polymerization via an anionic mechanism and the formation of a more flexible SEI layer with the ability of accommodation the volume changes, as illustrated in Fig. 5.10b [12]. Very recently, Yi Cui group [111]discovered two different mechanisms by cryo-TEM (Fig. 5.10c), besides the observation of atomic image of beam-sensitive lithium metal. They found SEI in carbonate-based electrolyte consisted of small crystalline domains identified to be Li_2O and Li_2CO_3 with diameter of 3 nm dispersed randomly in an amorphous organic polymer matrix formed by the decomposition of carbonate electrolyte. When adding 10 volume % fluoroethylene carbonate in the carbonate-based electrolyte, a multilayer structure with a more order arrangement was observed, different from Aurbach's demonstration. The inner layer is an amorphous polymer matrix, and the outer is Li_2O with large grain size of ~15 nm, without the trace of LiF that are generally believed the main composition of SEI.

Variety of experiments has been carried out to investigate the structure and compositions of the SEI layer on Li metal upon Li deposition/dissolution cycling. XPS that is widely used to probe surface chemistry has been carried out to confirm that SEI formed on the graphite surface is multilayers composed of LiF and Li_2CO_3 close to electrode/SEI and a porous polymeric layer at SEI/electrolyte (Andersson et al. 2001). Kanamura et al. found based on XPS that the SEI layer on the surface of deposited lithium metal on Ni substrate consisted of LiF and Li_2O under low current density of 0.2 mA cm^{-2} and a thick inner Li_2O layer with an outer layer of LiOH, Li_2CO_3, or $LiOCO_2R$ under 10.0 mA cm^{-2} [92]. Fourier-transform infrared (FTIR) spectroscopy was also utilized to determine the SEI composition. Morigaki [150] found that the reduction products of lithium alkyl carbonates on lithium metal surface were observed by in situ FTIR. It is found by electrochemical impedance spectroscopy (EIS) that the interfacial resistances corresponding to SEI and the charge diffusion at Li metal anode are much larger than that of electrolyte, which plays a key role on the Li deposition/dissolution cycling [22, 191]. After several cycles, the interfacial resistances decreased, which is because the reduction of the overpotentials with increased surface area of Li metal anode due to dendritic growth. However, the EIS data analysis using equivalent circuit sometimes is empirical and is not accurate.

Hence, the morphology, compositions, and electrochemical and thermal stability [46] of SEI are highly dependent on salt and solvent types. In addition, the behavior of SEI also has a strong dependence on the factors that influence lithium metal deposition/dissolution. Beside current density, temperature is also important. It is found that low temperature (−20 °C) could result in an improved cycling performance of lithium metal anode, due to the formation of compact and stable SEI of LiF with low resistance at low temperature that maintained at RT after subsequent cycling [83], while Andersson et al. found that $LiOCO_2R$ formed on graphite was not stable and decomposed at elevated temperature by XPS, which leaded to capacity decay during cycling. LiF on the surface was observed to break into crystallites upon increasing storage temperature, due to decomposition of salt in the electrolyte. And, the amount of polymeric carbon in SEI increased with time, which is because the continuous decomposition of the solvent [6].

It is undoubted that SEI plays a critical role on determining the electrochemical performances and safety of LMBs. Over the past two decades, since the first report on SEI formed on alkali and alkaline earth metals with electrolyte [157], various approaches have been proposed to introduce an ideal SEI layer with reasonable thickness of a few nanometers to block the diffusion of solvated electrons, that is, high cation conductivity with low electronic conductivity, good electrochemical stability, high mechanical strength and tolerance to the volume change during cycling, and a wide range of temperature.

5.4 Interface Between Solid-State Electrolyte and Li Metal Anode

5.4.1 Understanding on Solid/Solid Interface Stability

With considerable effort on exploring novel materials, various SSEs with high ionic conductivities larger than 10^{-3} S cm^{-1} have been discovered, while less attention has been focused on the SSE/electrode interfaces. It should be noted that the fundamental behavior for lithium plating and dendrite growth through SSEs is different from liquid electrolytes. Two decades ago, Rishi Raj proposed a kinetic approach to understand the Li nucleation at cavities in grain boundaries with lower ionic conductivity than grains of SSEs [164, 165]. During the Li metal nucleation, two states are involved: a supersaturation of Li with an electro-chemo-mechanical potential μ_{Li}^{*} and an embryo of Li metal formed at the grain boundary with the

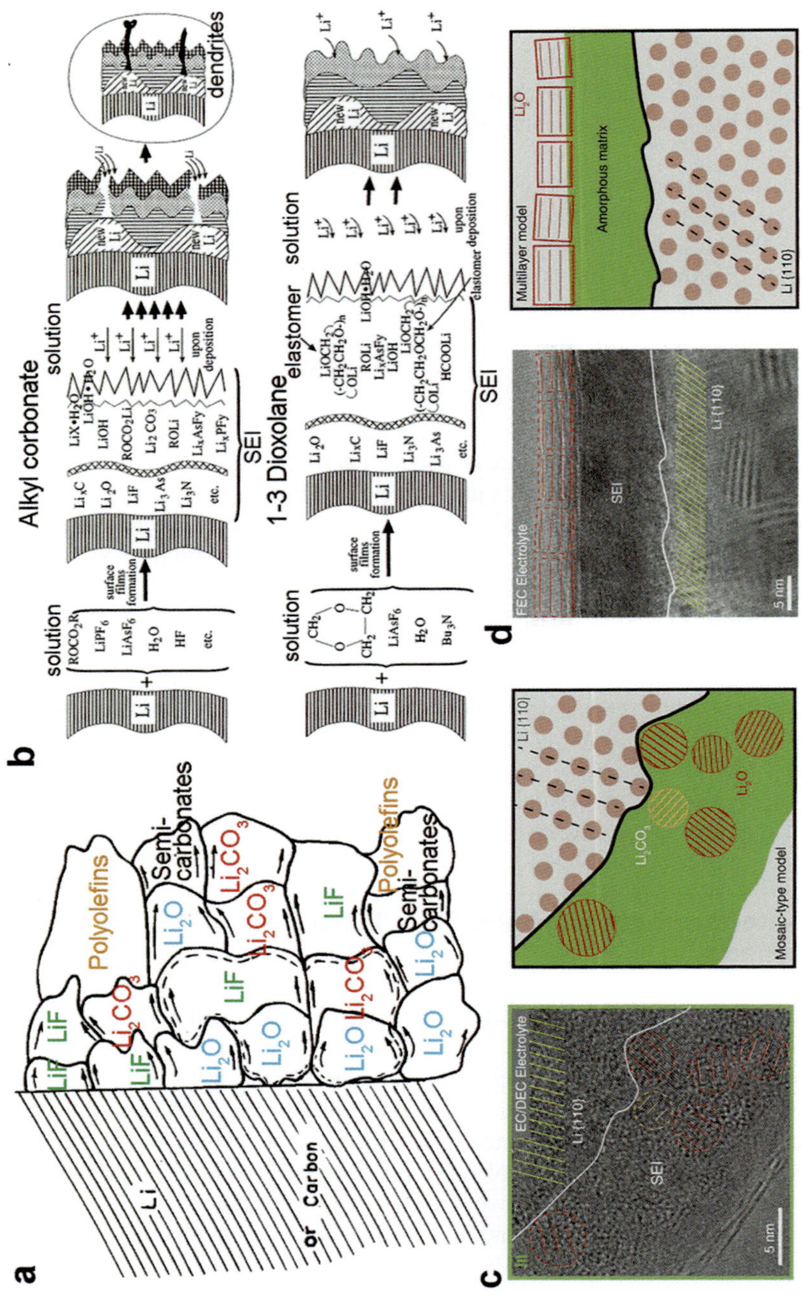

Fig. 5.10 (continued)

reference potential of pure lithium μ_0^*. The difference ($\Delta\mu_{\mathrm{Li}}^* = \mu_{\mathrm{Li}}^* - \mu_0^*$) is the excess chemical potential that provides the driving force for the Li metal nucleation, which is against with the excess surface energy due to the embryo formation. Differentiating of ΔG with respect to radius and equating to zero could determine the critical radius r_c with the free energy barrier to nucleation. r_c could be determined as follows:

$$r_c = \frac{2\gamma}{n_V \Delta\mu_{\mathrm{Li}}^*} \tag{5.10}$$

where n_V is the number of Li atoms per unit volume. Hence, it can be seen that the diameter of Li nuclei is dependent on the surface energy, the density of Li atoms and the applied overpotential for Li electrochemical plating. When a current is applied on the solid-state battery, the voltage drops across the grains and the grain boundaries in SSE and the SEI layer at the interface. The voltage drop attributed to the grain boundary close to the interface with Li metal can provide an overpotential (ΔV), that is, the driving force for nucleation of Li dendrites (State 1), as illustrated in Fig. 5.11. A flux of Li-ions equal to the applied current density (j) then forms at the grain boundary due to the formation of a space charge layer. Li precipitates of embryos appear at the grain boundary of SSE by the expense of the supersaturation (State 2). The Li embryo created into the grain boundary produces a mechanical compression that must be overcome for continuous Li nucleation. The total driving force for the Li nucleation at the boundary could be expressed as follows:

$$\Delta\mu_{\mathrm{Li}}^* = j\rho_B\delta_B\left(1 - \frac{d'}{d}\right)|e| - \frac{\sigma_F}{n_V} \tag{5.11}$$

where d is grain size, ρ_B is resistivity of grain boundary, δ_B is effective boundary width, $|e|$ is the positive charge of Li-ion, and σ_F is the fracture stress of SSE. Hence, if the driving force is lower than the energy barrier (combination Eqs. 5.10 and 5.11), the Li dendrite could be unlikely to form according to the following expression:

$$j\rho_B\delta_B\left(1 - \frac{d'}{d}\right)|e| \le \frac{2\gamma}{r_c n_V} + \frac{\sigma_F}{n_V} \tag{5.12}$$

Therefore, from the above expression, it can be seen that the Li dendrite formation could be avoided in terms of (1) current density lower than critical current density, (2) grain boundary of SSE with high ionic conductivity, and (3) SSE with high fracture stress.

Fig. 5.10 Schematic presentation of four different surface chemistry distributions for the SEI layer on Li metal anode. (**a**) Mosaic-like structure. (Reproduced with permission-Copyright 1997, the Electrochemical Society, Inc. [159]); (**b**) Multilayer structure with inorganic material Li_2O in the inner region to Li metal. (Reproduced with permission-Copyright 2000, Elsevier Science S.A. [11]); (**c**) Small crystalline domains dispersed randomly in an amorphous polymer matrix. (**d**) Multilayer structure with inorganic material Li_2O on the surface of SEI. (Reproduced with permission-Copyright 2017, American Association for the Advancement of Science [111, 112])

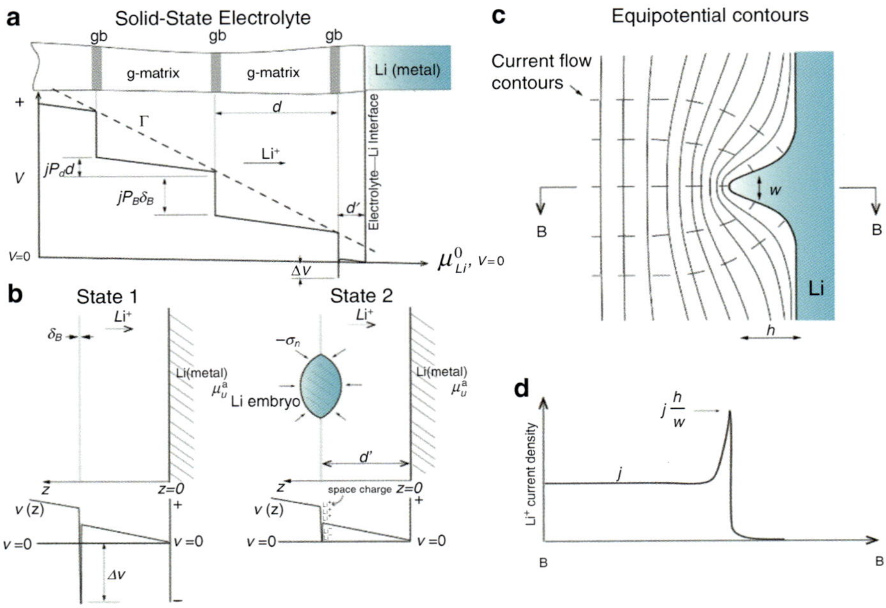

Fig. 5.11 Schematic illustration of Li nucleation at the interface of solid electrolyte and Li metal. (**a**) Schematic illustration of SSE with Li metal anode and the voltage drop across SSE and interface. (**b**) The two states for Li nucleation. (**c**) Current flow and equipotential contours near the tip of Li protrusion. (**d**) Li-ion current density near the interface depending on the aspect ratio of the protrusion. (Reproduced with permission-Copyright 2017, Elsevier B.V. [165])

In addition, similar with the Li dendrite formation in liquid electrolyte, the condition of Li metal is also a key factor. A protrusion on the surface of the Li metal surface can concentrate the local Li-ion flux and promote the growth of lithium dendrite. The current density near the tip of the protrusion is shown as follows according to the Saint-Venant's principle:

$$j_{\text{int}} = j\frac{h}{w} \tag{5.13}$$

where w and h are the width and the height of the bump tip on Li metal surface. Therefore, smooth Li metal surface takes the advantage of stable Li electrochemical deposition.

In general, the stiffness for SSEs with high elastic modulus is required to suppress Li dendrite growth. On the other hand, the volume change in anode and cathode associated with Li-ion transfer in and out during charge/discharge could damage the mechanical integrity of the electrodes as well as the electrode/electrolyte interface, which leads to the electrode cracks and delamination of SSE from electrodes. The conformal ability for SSEs to volume change during Li plating/stripping could avoid the detachment from electrodes and maintain a good contact with low interfacial resistance. It seems like that the high modulus for SSEs is more important than the flexibility, as predicted by Monroe and Newman [53, 149] that SSEs with at least twice shear modulus of Li metal (Li metal: ~3.4 GPa) are able to block Li dendrite growth. But, one requirement is needed for implementing this prediction is that the contact between the electrolyte and electrode must be tight [184]. Most SPEs and inorganic electrolytes have been proved to be penetrated by lithium dendrite growth [25, 44, 170]. Brissot et al. [25] found that for the symmetrical Li cells based on PEO electrolyte, like the cells using liquid electrolyte, Li dendrites started to grow at Sand's time at high current density, when the ionic concentrations dropped to zero at the anode. However, at low current density, Li dendrites were also been observed attributed to the local inhomogeneities at the electrode surface, which is different from Chazalviel's model [31]. In addition, Ce-Wen Nan group [167] reported that although LLZO has a high enough modulus of approximately 60 GPa, Li dendrite could penetrate it and cause short circuit under a medium current density of 0.5 mA cm^{-1}, while, a soft passivating interface layer coated on SSEs or Li metal could prevent dendrite grow, as reported by some related literatures [63, 114]. The onset of Li infiltration depends on SSE surface morphology, in particular the defect size and density [160].

Although the solid/solid interface is not as dynamic as the liquid/solid interface, it contributes to the major voltage drop compared to the resistance of SSE in the battery, which plays a critical role in degrading electrochemical performance of ASSLBs [26]. The ion transport over the interface is more rate-determining than that in bulk for a battery, which is the major bottleneck for practical application of ASSLBs [235]. Most of SSEs are not thermodynamically stable species with limited voltage window against Li metal anode and high-voltage cathode. A SEI layer with a mixed ionic-electronic conductivity can allow for continuous flow of Li-ions and electrons through the mixed conducting interphase (MCI), leading to the complete lithiation of the electrolyte, as shown in Fig. 5.12a [210]. Hence, a kinetically stabilized ion-conducting SEI layer without electronic conduction is required for stable Li deposition/dissolution cycling. It could be seen from Fig. 5.12b, c that the formation of interface layer at the Li|SSE and SSE|cathode interfaces increases the impedances largely and this subsequently leads to potential drops, which limit the electrochemical performance of the batteries, especially the power density. The interfacial impedances attributed to space charge layer, electrochemical side reaction, mechanical incompatibility, and so on [168, 218]. The space charge layer with a thickness of around 10 nm at SSE|cathode interface is associated with the formation of Li-ion deficient region due to the different bonding ability that reduces ion conductivity. A Li-reduced decomposition layer could be formed at the Li|SSE interface. It revealed that SPEs could be react with Li metal, forming a passivation layer, according to the time evolution of interfacial resistance in symmetric Li [41]. For inorganic electrolytes, Gerbrand Ceder group [168] and Yifei Mo group [72, 252] have predicted by computational methodology the electrochemical stability

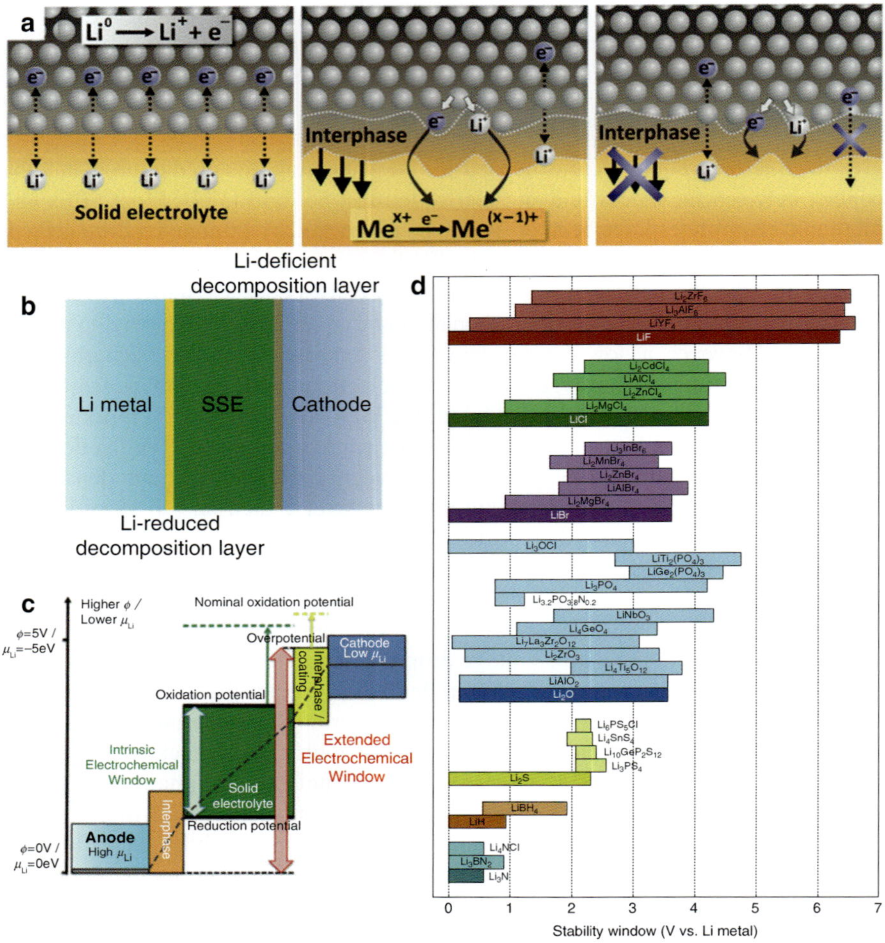

Fig. 5.12 Mechanism illustration for the interfaces of electrolyte with electrode. (**a**) Schematic illustration of the three different interfaces between SSE and metallic lithium: unreactive and stable interface, reactive and unstable interface with a MCI layer, and reactive and metastable interface with a SEI layer. (Reproduced with permission-Copyright 2015, the Royal Society of Chemistry [210]); (**b**) Schematic of a full cell. (**c**) Schematic diagram about the electrochemical window (color bars) and the Li chemical potential profile (black line) in ASSLIB. (Reproduced with permission-Copyright 2015, American Chemical Society [252]); (**d**) Electrochemical stability window of various SSE materials. (Reproduced with permission-Copyright 2015, American Chemical Society [168])

window of the most studied inorganic SSEs (Fig. 5.12d), which is also supported by experiment results. The ion-conducting SEI layer formed on the electrode surface due to the electrochemical instability increases interfacial resistance. The interfacial resistance that impedes ion conduction is also as a result of the chemical mixing due to the interdiffusion of Li-ions through the interfaces.

Moreover, the mechanical instability, due to the large volume change of electrode during lithiation/delithiation processes, is also one of the most challenging issues to be solved, which leads to the loss of contact between the electrolyte and electrode and degrades the cycling performance. In this review, the stability of Li|SSE interface and the approaches to suppress dendrite growth will be discussed in the following session.

Various experiments have been carried to confirm the instability of SSEs with metallic Li, including XPS [76, 175, 211], NMR [235], SEM [2, 222], TEM [138], EIS, etc. Even LiPON that is recognized as one of the most stable SSEs was found to be reacted with lithium metal [233]. It is found the decomposition chemicals consisted of Li_3PO_4, Li_3P, and Li_3N by a highly surface-sensitive XPS [175]. It is generally recognized that LGPS could react with lithium metal due to the chemical instability of Ge^{4+}, leading to an interphase layer formed at the interface. It is found that by in situ XPS (Fig. 5.13a), the new chemical species, such as Li_2S, Li_sP, and $Li_{15}Ge_4$ alloy (or Ge), were formed due to the decomposition of LGPS at the interface according to the following equation:

$$Li_{10}GeP_2S_{12} + 23.75Li \rightarrow 12Li_2S + 2Li_3P + 1/4Li_{15}Ge_4 \tag{5.14}$$

The interfacial resistance associated with the resultant chemicals was also measured by time-resolved EIS. As shown in Fig. 5.13b, c, the interfacial impedance increased with time was found, due to the poor conductive decompositions of LGPS when in contact with Li metal anode. The thickness of the SEI layer was deduced to be ~20 nm after 30 h, by assuming the main component was Li_2S with a low ionic conductivity of 8×10^{-9} S cm^{-1}. It also could calculate that the SEI thickness was about 370 nm for LGPS and 23 nm for $Li_7P_3S_{11}$ after 1 year, which is corresponding to an interfacial resistance of 4.6 and 0.28 kΩ cm^2, respectively [211]. It was believed that LPS with the highest ionic conductivity for SSEs to date was stable against Li metal [77], but later it is found by in situ XPS technique that LPS could react with Li metal with the decomposed interphase layer consisted of Li_2S and Li_3P [212].

Direct morphology observation visually on the formation of interphase species when SSE contacts with Li metal is clearer on the mechanism understanding. Miaofang Chi group [138] an in situ aberration-corrected scanning transmission electron microscopy (STEM) to detect the morphology change at the interface when cubic LLZO contacting with Li metal, as exhibited in Fig. 5.13d, e. They revealed that the surface of cubic LLZO was reduced to a tetragonal-like LLZO interphase layer of around five unit cells with a thickness of 6 nm by electron energy loss spectroscopy (EELS) and this SEI layer was stable over time to prevent the further reaction of LLZO with Li metal. Moreover, the NASICON-type inorganic electrolytes, such as LAGP or LATP, have also been confirmed to react with Li metal leading to the generation of a MCI layer [76]. It also has been proven that the perovskite-type LLTO could be reduced by lithium insertion to produce reduced Ti-ions and Ti metal due to the reduction of Li metal, based on the ex situ XPS results [83, 210, 230].

To detect the morphology or the phase using a microscope or spectroscopy technology, the applied beam irradiation may cause damage on the samples,

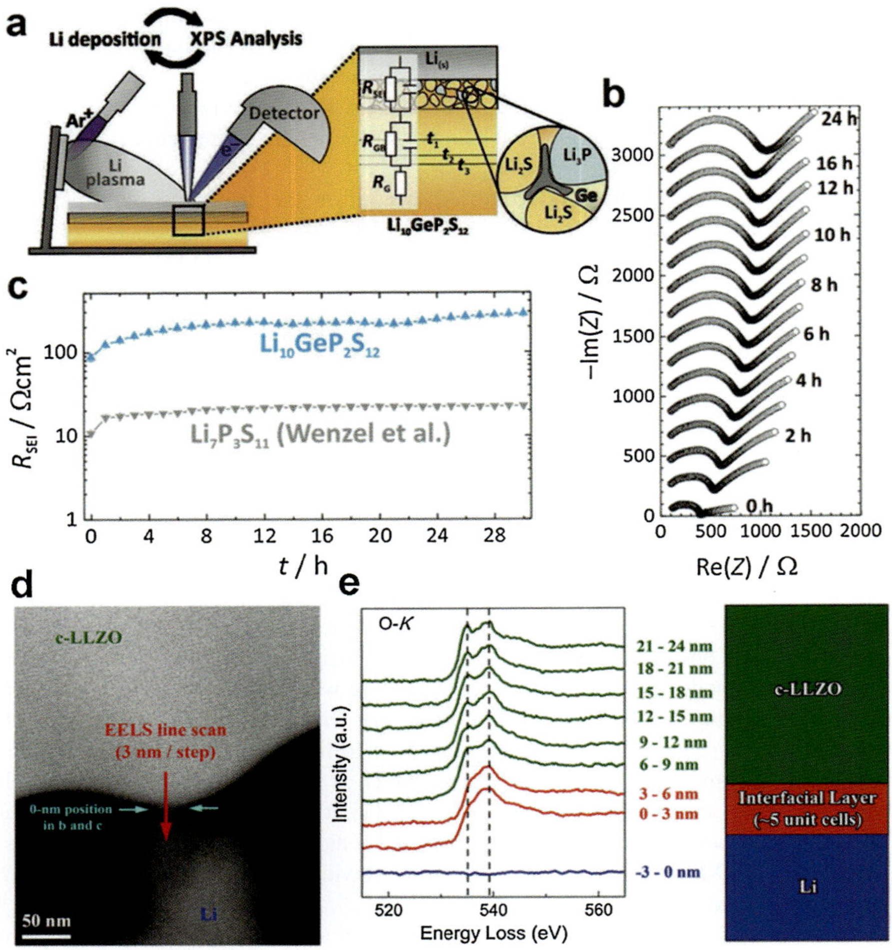

Fig. 5.13 Observations on the electrolyte/Li metal interface. (**a**) Schematic of setup for XPS measurement and the structure of the battery. (**b**) Impedance spectra for LGPS contacting with Li metal with time evolution. (**c**) Interfacial resistances for LGPS and $Li_7P_3S_{11}$ with time. (Reproduced with permission-Copyright 2016, American Chemical Society [211]); (**d**) HAADF-STEM image of cubic LLZO contacted with metallic Li. (**e**) EELS line scan of O-K edges according to panel **d**, with the schematic illustration of the interfacial layer. (Reproduced with permission-Copyright 2016, American Chemical Society [138])

especially beam-sensitive species, which could lead to an error or misunderstanding on the results [121]. Xiangxin Guo group [222] found that metallic lithium could be expulsed on the surface of garnet-based ceramic electrolyte under the beam irradiation by SEM, due to the injection of electrons into the garnet. Hence, this behavior needs to be considered when using SEM or TEM to probe garnet oxides.

5.4.2 Li Metal Anode-Solid Polymer Electrolyte Interface

One of the most exciting advantages of SSEs is the ability to prevent lithium dendrite growth, which is highly dependent on the composition of SSEs, cycling parameters, and temperature. However, in the late 1980s, it was reported by Scrosati that PEO itself couldn't block dendrite growth [176, 228]. Jean-Marie Tarascon group [44] found that without any external force, the SPE membrane could be easily pushed up by lithium dendrite growth, which leads to the delamination of SPE from Li metal anode. An early work on ASSLIB was reported by the pioneers Zaghib, Armand, and Gauthier, which demonstrates that the ASSLIB using solvent-free PEO-based electrolyte, graphite, or $Li_4Ti_5O_{12}$ anode coupled with LCO cathode delivered charge-discharge curves only for several cycling [237]. Hence, various strategies have been proposed to improve the interface stability of SPEs against Li metal anode toward practical applications, including electrolyte additives, cross-linked polymer networks, polymer blends, block copolymer electrolytes, and hyperbranched polymer.

5.4.2.1 Electrolyte Additives

Various components have been studied as additives in SPEs to improve stability of Li plating/stripping cycling. The most promising is adding inorganic nanomaterials into SPEs to form CPEs, which improves ionic conductivity as well as mechanical properties and electrochemical stability. Zhang and Armand group [50] reported PEO-based CPEs with lithium azide (LiN_3) (2 wt.%) for all-solid-state Li-S batteries (ASSLSBs), showing a polarization voltage as low as 5 mV for the Li symmetric cells at the current density of 0.1 mA cm^{-2} at 70 °C. The improved stability with Li metal is due to the formation of a thin and compact SEI layer with highly ionic conductivity passivation layer on Li metal anode, which could prevent dendrite formation as well as polysulfide shuttling. ASSLSBs using PEO-based CPEs with LiN_3 showed a high discharge capacity of 800 mAh g^{-1} after 30 cycles at 70 °C. The promising inorganic electrolyte LLZO has been widely studied as fillers in SPEs. Ce-Wen Nan and Yang Shen group [244] reported PVDF-based composite electrolytes with $Li_{6.75}La_3Zr_{1.75}Ta_{0.25}O_{12}$ (LLZTO) ceramics, showing a high ionic conductivity of about 5×10^{-4} S cm^{-1} at RT due to the acid-base interaction of LLZTO nanoparticles with the N atom and C=O group of polymer and solvent molecules. The LCO|CPE|Li cell delivered a specific capacity of 147 mAh g^{-1} at 0.4 C after 120 cycles with a high-capacity retention of 98%. The addition of LLZTO in the Li symmetric cells showed an improved cycling stability at 25 °C under various current densities. Candace K. Chan group [231] confirmed the ability of incorporation of inorganic ceramic to improve ionic conductivity and Li dendrite resistance, by the study of cubic-phase LLZO NW-filled PAN-$LiClO_4$ CPE, which showed prolonged time of short circuit for the Li symmetric cells. Liangbin Hu group [57] demonstrated a PEO-based CPE with 3D Li-ion conducting network made of LLZO nanowires, showing a stable Li plating/stripping under the current density of 0.2 mA cm^{-2} for 500 h and 0.5 mA cm cm^{-2} over 300 h.

5.4.2.2 Interphase Layer

The adhesion between SSE and electrodes is critical for cycling performance of ASSLBs. Although SPEs are flexible relatively, the contact with Li metal anode gets worse with cycling due to the poor wettability and the large volume change upon Li platting/stripping. Hence, introducing an interphase layer with strong adhesion with Li metal at the interface is promising. Yi Cui group [128] developed a 3D composite Li metal anode with a flowable interfacial layer (Fig. 5.14), guaranteeing a good adhesion with the solid electrolyte during cycling by reducing the interfacial fluctuation to local scale. For the composite 3D Li metal anode without volume change, Li metal was fused into the pores of layered reduced graphene oxide (rGO) host, and PEG with a molecular weight of 10,000 showing a viscous semiliquid state was then infiltrated into the 3D composite Li metal anode, so that the resultant flowable interphase could improve the adhesion of the electrolyte with Li metal anode. The Li/LFP cells using CPE with silica NPs showed a high specific capacity of 110 mAh g^{-1} at 5 C after 300 cycles at 80 °C.

5.4.2.3 Cross-Linked Polymer Networks

The ionic conductivities as well as mechanical properties of SPEs can be enhanced by cross-linking. Khurana et al. [98] reported a cross-linked polyethylene (PE)/PEO electrolyte showing the longest short-circuit time among to date, which is due to the

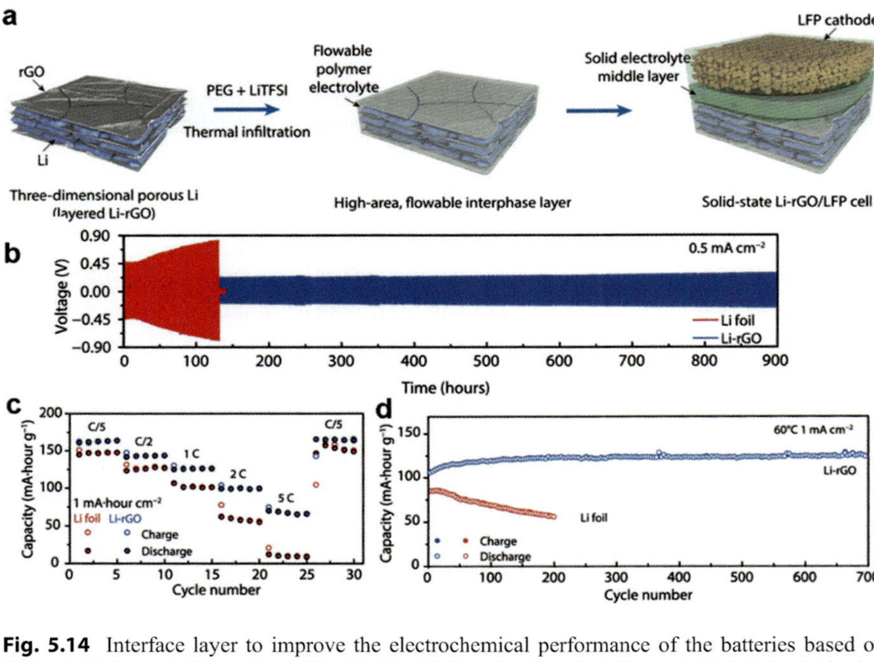

Fig. 5.14 Interface layer to improve the electrochemical performance of the batteries based on SPEs. (**a**) Schematic illustration of the 3D Li with flowable interfacial layer. (**b**) Li plating/stripping performance of the Li symmetric cells, (**c**) the rate capability, and (**d**) cycling performance of the LFP/Li cells. (Reproduced with permission-Copyright 2017, American Association for the Advancement of Science [128])

high ionic conductivity of $>10^{-4}$ S cm^{-1} at RT and the large shear modulus as the same order of magnitude as lithium metal. With reduced crystallinity, PEO were cross-linked with poly(ethyleneglycol) diacrylate (PEGDA) and divinylbenzene (DVB) prepared by UV-induced photopolymerization. The cross-linked SPEs showed a high voltage window of 4.3 V and an ionic conductivity of 1.4×10^{-4} S cm^{-1} at 70 °C. The ASS Li/LFP cells showed a specific capacity of 138 mAh g^{-1} at 70 °C after 20 cycles. In addition, for ASSLSB, the specific discharge capacity decreased from 375 to 175 mAh g^{-1} after 50 cycles [19]. Lynden A. Archer group [37] developed CPEs with hairy NPs as cross-linkers. But to improve ionic conductivity, the cross-linked polymer electrolytes are not totally dry, which is gel polymer electrolytes with mechanical modulus of 1 MPa and ionic conductivity of 10^{-5} S cm^{-1}. The NP surfaces with a tethered short hydrophilic oligomer to cross-link with rigid polymer providing ion conduction pathway and improves mechanical strength to suppress lithium dendrite. And the hairy nanoparticles are able to prevent particle agglomeration.

5.4.2.4 Polymer Blends

Polymer blending is an effect approach method to improve the overall performances of SPEs. Recently, Yu-Guo Guo and Li-Jun Wan group [239] reported a flexible SPE based on an interpenetrating network of poly(ether-acrylate) (ipn-PEA) by photopolymerization of PEO and branched acrylate, showing a high mechanical strength of ca. 12 GPa and high ionic conductivity of 2.2×10^{-4} S cm^{-1} at RT (Fig. 5.4a). The rigid network of PEA could be softened by PEO to improve the contact between electrolyte and electrodes, and meanwhile the PEA network could prevent the crystallization of PEO to enhance ionic conductivity. The Li/LFP cells based on the ipn-PEA electrolyte showed a high specific capacity of 141 mA h g^{-1} at 0.5 C and 66 mA h g^{-1} at 5 C. A capacity retention of 85% was achieved after 200 cycles at 1 C.

5.4.2.5 Block Copolymer Electrolytes

Alternative approach to improve the SPE modulus is by the adoption of block copolymers, such as combining PEO-based electrolytes with a rigid polymer like polystyrene (PS). The ratio of hard non-conductive to soft conductive component is critical to the mechanical strength and ionic conductivity. However, the shear moduli of SPEs are complex, with in-phase and out-of-phase components strongly depending on temperature and frequency, which is less understood fundamentally [73], ZnO [133], Al [58], and amorphous Si. Schauser et al. [172] studied the dendrite growth based on PS-block-PEO solid electrolytes by X-ray microtomography at the temperature range of 90–120 °C. They found the lithium dendrite location changed from within the lithium metal anode to the electrolyte, which is because the shear modulus decreased significantly with temperature in the low-frequency regime.

It is recognized that in SPEs the anion concentration gradients can limit the power density of LIBs [43]. Later, the Li|P(STFSILi)-PEO-P(STFSILi)|LFP cells based on a single-ion polymer electrolyte showed a specific capacity of 138 mAh g^{-}1 at 2C at

80 °C. The cyclability at 60 C with more than 50 stable cycles is particularly noticeable. Capacity retention of >85% at 0.5 C at the temperature range of 60–80 °C was achieved in solvent-free polymer-based LIBs [24]. However, SPEs based on nanostructured lamellar block copolymer show solid-like behavior in the bulk, as a result of a randomly oriented granular structure. The requirement of high shear modulus of at least 6 GPa for SSEs as Monroe and Newman suggested is inconsistent with the need of electrolyte like a chewing gum resulting in the good contact between electrolyte and electrode. Stone et al. [184] show a self-assembled lamellar block copolymer with a perpendicular orientation showing a liquid crystalline symmetry, which satisfied the conflicting requirements of high modulus and good interfacial adhesion at the interface. It also found the improved prevention to Li dendrite formation as well as the ability of strong interface contact.

5.4.3 Li Metal Anode-Inorganic Electrolyte Interface

Compared with SPEs, inorganic electrolytes usually possess almost unity LTN, resulting in a low concentration polarization with a possibility of high power density [252]. However, as discussed above, the electrochemical and chemical instability of inorganic electrolytes with Li metal, narrow voltage window, and the poor interface contact lead to high interfacial resistances and limit the practical application of inorganic SSEs in ASSLBs [168]. Even the garnet LLZO-based oxides that is recognized as one of the most stable SSEs were found to be reacted with Li metal [167]. The instability of garnet electrolytes with Li metal was revealed by in situ STEM. It was found that a tetragonal-like LLZO interface layer with about five unit cell thickness was formed after the cubic LLZO contacted with Li metal. The SEI layer was able to prevent further reaction between the cubic LLZO and Li metal [138]. Additionally, although SSEs have enough high modulus to suppress Li dendrite as Monroe and Newman suggested, it was found by experiments that even inorganic ceramic couldn't block the growth of lithium dendrite at the interface [160]. Both experimental and calculated results indicated that LLZO are highly stiff with shear moduli of about 60 GPa [234]. However, Ishiguro and co-workers [84] found that the Li symmetrical cells based on garget electrolyte showed a short-circuit time of 280 s at a slightly high current density of 0.5 mA cm^{-2}. Li deposition is stable without short circuit under the current densities lower than 0.1 mA cm^{-2}. Yet-Ming Chiang group [160] conducted rationally designed experiments to monitor Li metal penetration into inorganic electrolytes, including Li_2S-P_2S_5, β-Li_3PS_4, polycrystalline, and single-crystalline garnet-based oxide. They found a different mechanism of Li dendrite growth during Li electrochemical deposition for SSEs from the amplification of kinetic variation at the Li metal surface for liquid electrolyte, which is strongly dependent on the surface morphology of SSEs. It was believed that Li metal plating was started from Li metal surface (Fig. 5.15a) and grew to fill the space between Li metal and SSE, especially the pre-existing flaws (i.e., cracks, voids, pinholes) or wetting incompatibility on SSE where the localized current

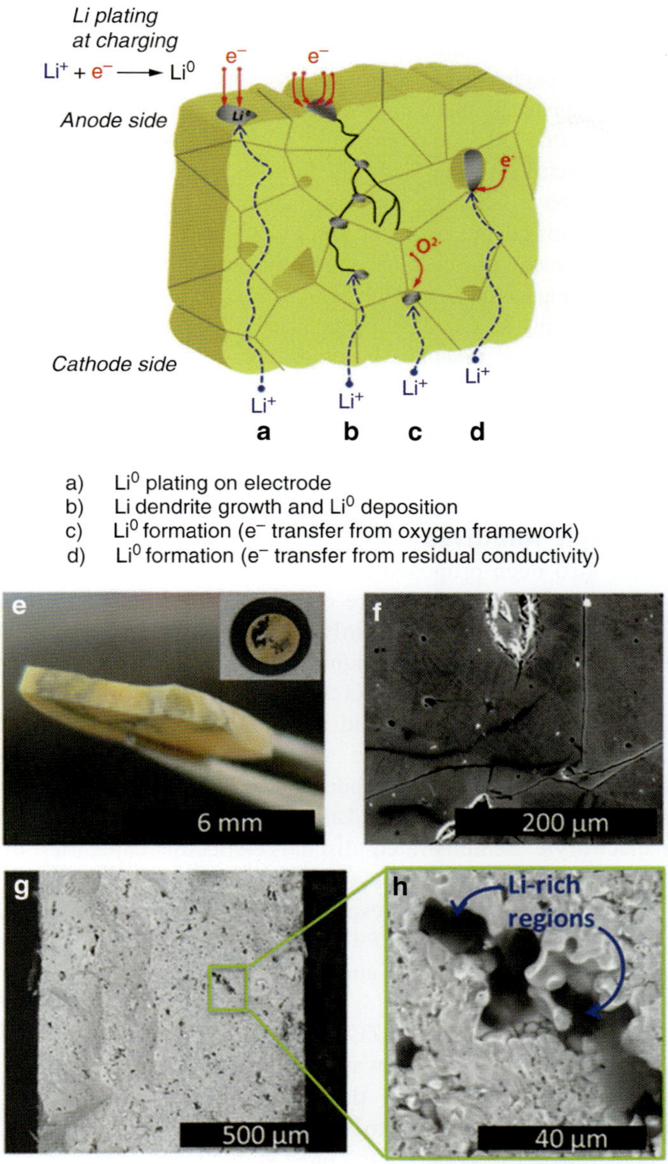

Fig. 5.15 Mechanism illustration of lithium dendrite growth with inorganic SSEs. (**a–d**) Schematic illustration of a cross section in LLZO-based ceramic electrolyte indicating the mechanism of the formation of Li metal at the interface and the interior of the electrolyte during Li plating. (**e**) Photo of the ceramic electrolyte penetrated by Li dendrite. (**f**) SEM image of the surface morphology and (**g**, **h**) the cross-sectional view of the ceramic. (Reproduced with permission-Copyright 2017, American Chemical Society [2])

concentrated. The deposited Li metal then grows to drive crack propagation and penetrates through grain boundaries and interconnected pores of the ceramic electrolyte, leading to the electromechanical failure of the electrolyte (Fig. 5.15b). In addition, it is demonstrated that isolated Li metal agglomerates could be found in the inorganic solid electrolyte, which is due to Li-ion reduction by electrons from the oxygen backbone of the LLZO-based electrolyte (Fig. 5.15c) or from the residual electronic conductivity (Fig. 5.15d). The above behavior was also supported by Aguesse and co-workers [2]. Figure 5.15e showed the photo of the LLZO-based ceramic penetrated by Li dendrite. The SEM image of Fig. 5.15f shows the surface morphology of the LLZO-based electrolyte facing Li metal, and the SEM images in Fig. 5.14g, h indicate that the Li metal corresponding to the dark precipitates filled in the pores of electrolyte. Furthermore, the interfacial instability becomes more critical by repeatable Li plating/stripping cycles, especially for a larger amount of Li metal. Hence, to prevent Li dendrite growth in ASSLMBs, scientists have raised various approaches, including electrolyte additives, artificial SEI layer, nanoscale interfacial engineering, microstructure modifications and high pressure, etc.

5.4.3.1 Electrolyte Additives

Additives have also been proposed to reduce the interfacial resistance of inorganic electrolytes with Li metal. More recently, Feng Pan group [207] demonstrated incorporated with LLZO electrolytes, a porous metal-organic framework (MOF) host impregnated with ionic liquid (Fig. 5.14), which enabled the direct contact between ionic liquid and LLZO through the open pores in the MOF host, resulting in a "nanowetted" solid-solid interfaces and improved interfacial ion transport. The composite electrolyte showed a high ionic conductivity of 1.0×10^{-4} S cm^{-1}, a wide electrochemical window up to 5.2 V, and a good stability with Li metal anode. The Li/LCO and Li/LFP cells based on the composite electrolyte demonstrated discharge-specific capacities of 130 and 140 mAh g^{-1} at 0.1 C at first cycle and decreased to 33 and 37 mAh g^{-1} when rising the current rate to 0.8 C, respectively. The Li/LCO and Li/LFP cells achieved a high-capacity retention of 97% after 150 cycles at 0.1 C and delivered energy densities of 196.9 Wh kg^{-1} and 377.0 Wh L^{-1} and 151.3 Wh kg^{-1} and 304.1 Wh L^{-1}.

By addition of Li_3PO_4 into LLZO-based oxide electrolyte, a self-limiting ion-conductive SEI layer of Li_3P was formed due to the reaction of Li_3PO_4 with Li metal, which results in the interfacial resistance from 2080 to 1008 Ω cm^{-2} and then to 454 Ω cm^{-2} after stable cycling for 60 h for the Li symmetric cells [227]. LiF is one of the most stable Li compounds with Li metal with large voltage window, which is introduced into stable garnet oxide to improve its stability in moisture air. The LLZO-based electrolyte with 2 wt.% LiF indicated less Li_2CO_3 on the surface, showing smaller interfacial impedances compared with the one without LiF. The Li/LFP cells based on the modified electrolyte with an extra polymer coating facing Li metal showed discharge capacities of 142 mAh g^{-1} at 80 mA cm^{-2} at 65 °C for the first cycle and maintained 120 mAh g^{-1} after 100 cycles with high CE of 99.8–100%. ASSLSB was also demonstrated, showing discharge capacities of 1137, 1074, and 1042 mAh g^{-1} at 100, 200, and 300 mA cm^{-2} for the first cycle and near 100% CE for 100 cycles [111, 112].

5.4.3.2 Design of Artificial SEI Layer

Designing an artificial SEI layer that is ion-conducting and electron-insolating as well as can conformally contact with Li metal anode at the electrolyte/electrode interface is an effect approach to reduce the interfacial impedance and suppress Li dendrite growth [36]. Liangbin Hu group proposed an approach of coating "lithiophilic" layer by in situ reaction with Li metal (Fig. 5.16a), including Al_2O_3 [73], ZnO [133], Al [58, 59], and amorphous Si [137] to form a stable artificial SEI layer at the interface of garnet electrolytes with Li metal anode, which could decrease interfacial resistance. The introduction of Al_2O_3 layer with several nanometer thickness lithiated by Li metal dramatically decreased the interfacial impedance from 1710 to 1 Ω cm^{-2} at RT (Fig. 5.16b), due to the resultant conformal garnet/Li interface. The lithiated layer could also prevent the further reaction of LLZO with Li metal. The Li symmetrical cells using Al_2O_3-coated garnet electrolyte delivered a stable cycling with a polarization voltage of 22 mV under the current density of 0.2 $mAcm^{-2}$ for 30 min (Fig. 5.16c) [73]. Another approach to improve the contact of LLZO with Li metal to remove the surface contaminants (Li_2CO_3 and LiOH) that are formed in the moisture air [34, 179, 180]. Sharafi et al. [179] demonstrated that the interfacial resistance was reduced to only 2 Ω cm^{-2}, due to the good wettability of clean LLZO with Li metal. The Li symmetrical cells could run at 0.2 mA cm^{-2} for 100 cycles stably.

In addition to introducing a stable artificial SEI layer via an extra interphase layer or mixing with other components, an interfacial layer in situ formed between the solid electrolytes with Li metal is cost-effective. Goodenough group [110] developed an inorganic electrolyte $LiZr_2(PO_4)_3$ with a high ionic conductivity of 2×10^{-4} S cm^{-1} at RT and good electrochemical voltage window up to 5.5 V vs. Li^+/Li. A Li-ion conducting SEI layer as a strong adhesion was formed at the interface between this electrolyte and Li metal, which is consisted of Li_3P and Li_8ZrO_6. The ASS Li/LFP cells showed a specific capacity of 140 and 120 mAh g^{-1} at the current density of 50 and 100 μA cm^{-2}, respectively. For sulfide electrolytes that are not stable with Li metal, an interphase layer was introduced to stable the Li deposition/dissolution cycling. Au was proposed effectively to prevent the reaction between Li_2S-P_2S_5 solid electrolyte and Li metal due to the alloy formation with Li metal and the good compatibility with sulfide electrolyte [95].

5.4.3.3 Nanoscale Interfacial Engineering

Co-firing electrolyte and electrode is an effect approach to improve the interfacial contact for the ASSLBs using an inorganic electrolyte. But the common used electrical conducting component such as carbon black is oxidized during heating at high temperatures, so that an inorganic conducting alternative of In_2O_5Sn (ITO) shows advantage. Ce-Wen Nan group [124] demonstrated an ASSLB based on LLZO electrolyte. The

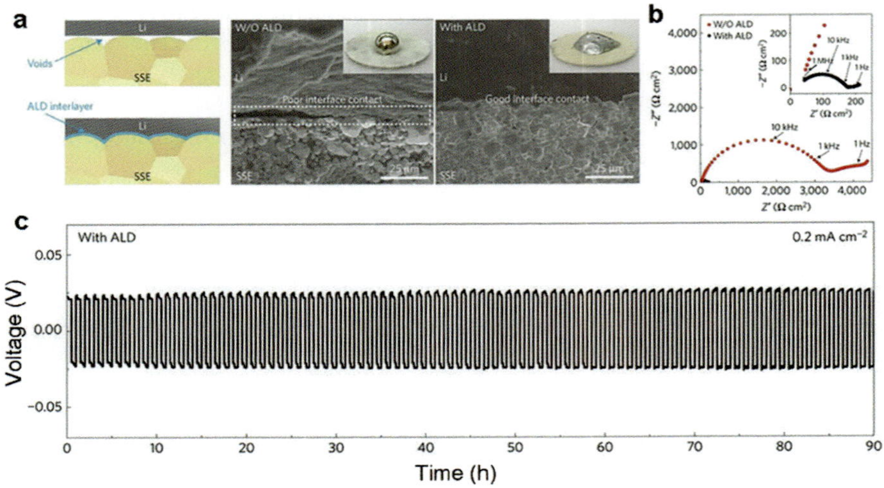

Fig. 5.16 Artificial SEI layer to improve the electrochemical performance of the batteries based on inorganic solid electrolytes. (**a**) Schematic illustration of the artificial SEI layer coated ceramic electrolyte and the cross-sectional SEM images for garnet-based electrolyte with and without interphase layer when contacting with Li metal. (**b**) Impedance spectra for the Li symmetric cells based on garnet electrolyte with and without interphase layer. (**c**) Cycling performance for the Li symmetric cell based on garnet electrolyte with interphase layer. (Reproduced with permission-Copyright 2017, Macmillan Publishers Limited [73])

composite LCO cathode consisted of LCO, LTO, and Li_3BO_3 (LBO) that is a solid Li-ion conductor with low melting temperature of 700 °C. The electrolyte and composite electrode were co-fired at 700 °C. The resultant battery showed a specific discharge capacity of 101.3 mAh g^{-1} at the first cycle at 0.025 C at RT. Recently, Xiangxin Guo and Chunsheng Wang et al. [74] demonstrated an all-ceramic lithium battery with a very low interfacial resistance by interphase engineering, which is attributed to thermal soldering the LLZO electrolyte and the LCO cathode with $Li_{2.3-x}C_{0.7+x}B_{0.3-x}O_3$ SEI via the reaction of $Li_{2.3}C_{0.7}B_{0.3}O_3$ solder with their conformally Li_2CO_3 coating (Fig. 5.17). Li_2CO_3 is also able to prevent the element diffusion at high temperature. The Li/LLZO/LCO ASSLIBs delivered a high specific capacity of 83 mAh g^{-1} after 100 cycles at RT, and good rate capability, which is the best performance for all-ceramic Li batteries to date. In addition to the inorganic interphase layer, polymer was also proposed to improve the contact between the garnet-based electrolyte and electrode and suppress Li dendrite growth. An asymmetric SSE consisted of LLZO covered with polymer electrolyte of 7.5 nm thickness was able to prevent Li dendrite growth [45].

Sulfide electrolytes such as LGPS, with extremely high ionic conductivities, process poor compatibility with metallic lithium, showing a limited application in ASSLMBs. A lithium compatible interphase layer of 75%Li_2S-24%P_2S_5–1%P_2O_5 was adopted to prevent the contact of LGPS with Li metal anode in ASSLSBs [232]. To further improve the electrochemical performance through reducing the interfacial resistance and the strain of sulfur cathode, a composite sulfur cathode was synthesized by coating sulfur

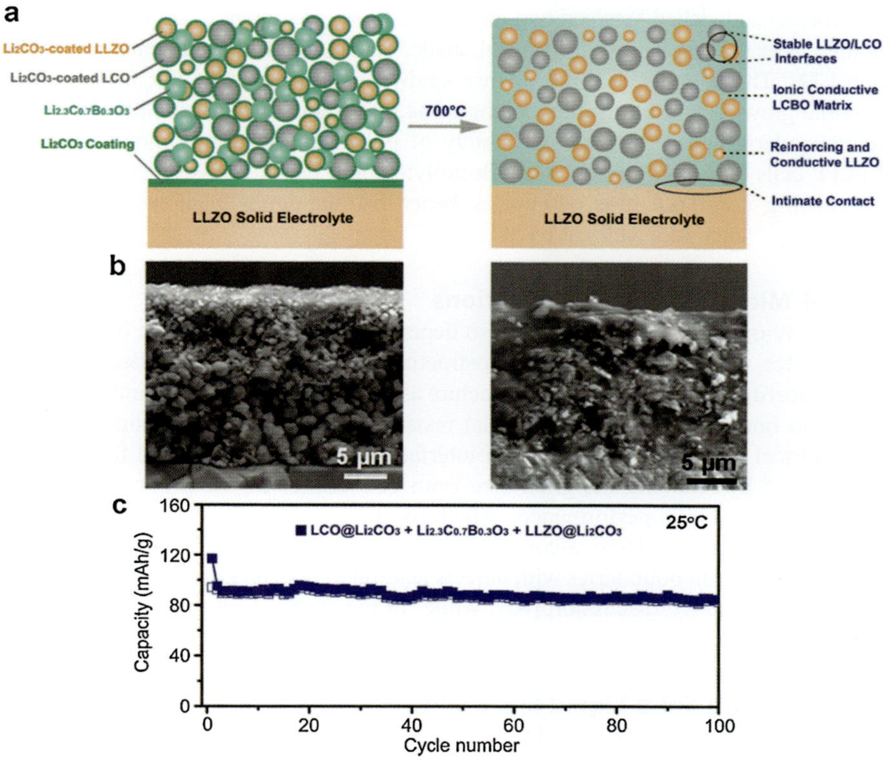

Fig. 5.17 Nanoscale interfacial engineering to improve the electrochemical performance of the batteries based on inorganic solid electrolytes. (**a**) Schematic illustration of the interphase-engineered all-ceramic electrode/cathode. (**b**) Cross-sectional SEM images of the electrolyte with composite cathode before and after sintering. (**c**) Cycling performance of Li/LLZO/LCO cell at 0.05 C at RT. (Reproduced with permission-Copyright 2018, Elsevier Inc. [74])

with thickness of ~ 2 nm onto rGO. This composite cathode was then coated on the LGPS side of the bilayer sulfide electrolyte. The resultant ASSLSBs showed a high discharge-specific capacity of 1629 mA h g^{-1} at 60 °C at 0.05 C for the first cycle, and a specific capacity of 830 mA g^{-1} at 1.0 C after 750 cycles, indicating a good long cycling performance.

In addition to the stable inorganic interface layer, polymer-based layers have also been proposed to improve the contact between the inorganic electrolyte and metallic Li anode. A gel electrolyte was used to protect the garnet-based electrolyte and improve the wettability. The interfacial resistance reduced from 6.5×10^4 to 248 Ω cm^{-2} and from 1.4×10^4 to 214 Ω cm^{-2} for the interface with cathode and Li metal anode, respectively. The Li/LFP cell based on the gel electrolyte coated with solid electrolyte showed a high specific capacity of 140 mAh g^{-1} after 70 cycles at RT [127]. John B. Goodenough group [251] demonstrated a polymer/ceramic/polymer sandwich structured composite solid electrolytes for ASSLIBs. The multilayer electrolyte consisted of a LATP ceramic electrolyte sandwiched between two cross-linked SPE membranes, showing a reduced double-layer electric field near Li metal anode, due to the block of ceramic layer with unit LTN. The polymer/ceramic/polymer sandwich electrolyte was able to prevent Li dendrite growth, attributed to the homogenous Li-ion flux distribution at the polymer/lithium interface and the good wettability of polymer. The results showed that the Li/LFP cells based on the sandwich electrolyte could deliver a discharge capacity of 102 mAh g^{-1} at 0.6 C after 640 cycles, better than the one without ceramic layer of 70 mAh g^{-1} after 325 cycles at 65 °C.

5.4.3.4 Microstructure Modifications

The behavior of Li plating/striping also depends on the microstructure of inorganic electrolytes, namely, the surface microstructure. Cheng et al. [34] studied the effect of engineering of the surface microstructure associated with grain size and especially the grain boundaries, on the interfacial resistance and cycling performance for Li symmetrical cells. It is found that the interfacial impedance decreased from 130 to 37 Ω cm^{-2} by reducing the grain size from 100–200 to 20–40 μm, resulting in a more stable cycling performance, which is due to the larger amount of grain boundaries at surface. Here, a confusion probably arises that Li nucleation location prefers to the grain boundaries with defects leading to uneven Li plating and dendrite growth in the above discussion [82], while, in the study, it revealed that plenty of homogenous grain boundaries are benefit for stable Li electrochemical deposition.

5.4.3.5 Hybrid Solid-Liquid Electrolyte

Inspired by the Tai Chi philosophy, to utilize the both advantages of SSE with good mechanical property and liquid electrolyte with high ionic conductivity and wettability, the approach of hybrid solid-liquid electrolyte is generally accepted before the real high-performance SSEs come out. Chen et al. [33] demonstrated a hybrid electrolyte consisted of PVDF-HFP framework, LiTFSI-Py13TFSI ionic liquid, and TiO_2 NPs, showing a high ionic conductivity of 7.4×10^{-3} S cm^{-1} at RT and a wide electrochemical window up to 5.5 V. The Li symmetrical cells could run for 800 h under the current density of 0.1 mA cm^{-2} without short circuit. The Li/LFP cells based on the hybrid electrolyte delivered a high specific capacity of 145, 122, 50, and 14 mAh g^{-1} at 0.2, 0.5, 1.0, and 2.0 C, respectively. It also indicated that the Li/LFP cells could stably run at 0.1 C without obvious voltage hysteresis incensement for 500 days.

5.5 Conclusion

Nowadays, overwhelming majority of the commercial LIBs are based on liquid electrolytes or gel polymer electrolytes, confronting with safety concern as well as limited energy density. However, many scientists and engineers have devoted their efforts on the development of SSEs toward the practical applications of ASS batteries since the middle of the twentieth century, even though only high-temperature Na-S batteries based on SSEs are viable. But it can be prospected that although liquid electrolyte is promising for near-term applications for LIBs, SSE as an alternative is of great interest in long-term technologies for post-LIBs. Meanwhile, to adopt the ideal anode material of lithium metal in the batteries coupled with high-capacity chemistries such sulfur and oxygen, SSEs with high elastic moduli are of great promise to suppress Li dendrite growth and subsequently improve safety.

SSEs with high ionic conductivities ($>10^{-3}$ S cm^{-1} at RT), comparable to that of liquid electrolyte, have been discovered continuously. Sulfide electrolytes (e.g., $Li_{10}GeP_2S_{12}$ and $Li_7P_3S_{11}$) and garnet oxides (e.g., $Li_{6.25}Ga_{0.25}La_3Zr_2O_{12}$) could achieve ionic conductivities of higher than 10^{-3} S cm^{-1} at RT, but both are sensitive to moisture in air and unstable with Li metal anode. The large interfacial impedances, derived from the chemical and electrochemical instability, poor incompatibility, and physical contact of SSEs with electrodes, remain the critical issues to limit the commercialization of SSEs as well as lithium metal anode in ASSLMBs. The Li-reduced decomposition layer formed at the Li|SSE interface is highly resistive, which especially restricts the rate capability of the batteries, while, in principle, inorganic electrolytes and single-ion SPEs with near unit LTN show great potential to achieve high power density. Advanced image- and spectrum-based techniques have been adopted to understand the behavior of lithium nucleation and dendrite growth at the interface of SSE and Li metal anode during Li deposition/dissolution cycling. Moreover, various approaches have proposed to improve the stability of SSEs including organic polymers and inorganic electrolytes with Li metal, which are highlighted in detail in this chapter. Introducing a stable artificial SEI layer with ionically conducting and electronically insulating at the Li|SSE interface is a promising approach. In addition, a good conformal contacting for SSEs with Li metal critically benefits for stable cycling, due to the infinite volume change for Li metal anode over repeated charge/discharge cycles. Hence, the structure of a composite lithium metal anode which consisted of a lithiophilic and ionic conductive porous host filled with Li metal, connected tightly with a thin SSE layer of high ionic conductivity, is of great promise for high-performance ASSLMBs. Although the current electronic devices and EVs are powered by LIBs exclusively based on liquid electrolyte and graphite, and there is a long way to go for scientists and engineers before the widely practical applications, ASSLMBs will increase the chance of a full commercialization of the next-generation large-scaled energy storages with high energy and power density, long cycling life, high safety, and durability.

Acknowledgments The author thanks Dr. Qing Lu for the analysis on the related publications and Jingbo Louise Liu and Dr. Sajid (Bashir) Liu for the invitation.

References

1. M. Abreu-Sepúlveda, D.E. Williams, A. Huq, et al., Synthesis and characterization of substituted garnet and perovskite-based lithium-ion conducting solid electrolytes. Ionics **22**, 317–325 (2015). https://doi.org/10.1007/s11581-015-1556-2
2. F. Aguesse, W. Manalastas, L. Buannic, et al., Investigating the dendritic growth during full cell cycling of garnet electrolyte in direct contact with Li metal. ACS Appl. Mater. Interfaces **9**, 3808–3816 (2017). https://doi.org/10.1021/acsami.6b13925

3. J.-K. Ahn, S.-G. Yoon, Characteristics of amorphous lithium lanthanum titanate electrolyte thin films grown by PLD for use in rechargeable lithium microbatteries. Electrochem. Solid-State Lett. **8**, A75–A78 (2005). https://doi.org/10.1149/1.1843571
4. P. Albertus, S. Babinec, S. Litzelman, et al., Status and challenges in enabling the lithium metal electrode for high-energy and low-cost rechargeable batteries. Nat. Energy **3**, 16–21 (2017). https://doi.org/10.1038/s41560-017-0047-2
5. U.V. Alpen, A. Rabenau, G.H. Talat, Ionic conductivity in Li_3N single crystals. Appl. Phys. Lett. **30**, 621–623 (1977). https://doi.org/10.1063/1.89283
6. A.M. Andersson, K. Edström, Chemical composition and morphology of the elevated temperature SEI on graphite. J. Electrochem. Soc. **148**, A1100–A1109 (2001). https://doi.org/10.1149/1.1397771
7. G.B. Appetecchia, D. Zaneb, B. Scrosati, PEO-based electrolyte membranes based on $LiBC_4O_8$ salt. J. Electrochem. Soc. **151**, A1369–A1374 (2004). https://doi.org/10.1149/1.1774488
8. M.B. Armand, The history of polymer electrolytes. Solid State Ionics **69**, 309–319 (1994). https://doi.org/10.1016/0167-2738(94)90419-7
9. M.B. Armand, J.M. Chabagno, M.J. Duclot, in *Second International Meeting on Solid. Electrolytes*, St. Andrews, Scotland, 1978
10. M.B. Armand, J.M. Chabagno, M.J. Duclot, *Fast Ion Transport in Solids Electrodes and Electrolytes* (North Holland, Amsterdam, 1979)
11. D. Aurbach, Review of selected electrode–solution interactions which determine the performance of Li and Li ion batteries. J. Power Sources **89**, 206–218 (2000). https://doi.org/10.1016/S0378-7753(00)00431-6
12. D. Aurbach, E. Zinigrad, H. Teller, et al., Factors which limit the cycle life of rechargeable lithium (metal) batteries. J. Electrochem. Soc. **147**, 1274–1279 (2000). https://doi.org/10.1149/1.1393349
13. S. Austin Suthanthiraraj, M. Johnsi, Nanocomposite polymer electrolytes. Ionics **23**, 2531–2542 (2016). https://doi.org/10.1007/s11581-016-1924-6
14. J.C. Bachman, S. Muy, A. Grimaud, et al., Inorganic solid-state electrolytes for lithium batteries: mechanisms and properties governing ion conduction. Chem. Rev. **116**, 140–162 (2016). https://doi.org/10.1021/acs.chemrev.5b00563
15. J. Bae, Y. Li, J. Zhang, et al., A 3D nanostructured hydrogel-framework-derived high-performance composite polymer lithium-ion electrolyte. Angew. Chem. Int. Ed. Engl. **57**, 2096–2100 (2018). https://doi.org/10.1002/anie.201710841
16. P. Bai, J. Li, F.R. Brushett, et al., Transition of lithium growth mechanisms in liquid electrolytes. Energy Environ. Sci. **9**, 3221–3229 (2016). https://doi.org/10.1039/c6ee01674j
17. L.R.A.K. Bandara, M.A.K.L. Dissanayake, B.-E. Mellander, Ionic conductivity of plasticized (PEO)-LiCF3SO3 electrolytes. Electrochim. Acta **43**, 1447–1451 (1998). https://doi.org/10.1016/S0013-4686(97)10082-2
18. P. Baudry, M. Armand, M. Gauthier, et al., In situ observation by SEM of positive composite electrodes during discharge of polymer lithium batteries. Solid State Ionics **28**, 1567–1571 (1988). https://doi.org/10.1016/0167-2738(88)90421-3
19. H. Ben youcef, O. Garcia-Calvo, N. Lago, et al., Cross-linked solid polymer electrolyte for All-solid-state rechargeable lithium batteries. Electrochim. Acta **220**, 587–594 (2016). https://doi.org/10.1016/j.electacta.2016.10.122
20. C. Berthier, W. Gorecki, M. Minier, et al., Microscopic investigation of ionic conductivity in alkali metal salts-poly(ethylene oxide) adducts. Solid State Ionics **11**, 91–95 (1983). https://doi.org/10.1016/0167-2738(83)90068-1
21. R. Bhattacharyya, B. Key, H. Chen, et al., In situ NMR observation of the formation of metallic lithium microstructures in lithium batteries. Nat. Mater. **9**, 504–510 (2010). https://doi.org/10.1038/nmat2764
22. G. Bieker, M. Winter, P. Bieker, Electrochemical in situ investigations of SEI and dendrite formation on the lithium metal anode. Phys. Chem. Chem. Phys. **17**, 8670–8679 (2015). https://doi.org/10.1039/c4cp05865h

23. C. Bishop, D. Teeters, Crystallinity and order of poly(ethylene oxide)/lithium triflate complex confined in nanoporous membranes. Electrochim. Acta **54**, 4084–4088 (2009). https://doi.org/10.1016/j.electacta.2009.02.043
24. R. Bouchet, S. Maria, R. Meziane, et al., Single-ion BAB triblock copolymers as highly efficient electrolytes for lithium-metal batteries. Nat. Mater. **12**, 452–457 (2013). https://doi.org/10.1038/nmat3602
25. C. Brissot et al., Dendritic growth mechanisms in lithium/polymer cells. J. Power Sources **81**, 925–929 (1999). https://doi.org/10.1016/S0378-7753(98)00242-0
26. M.R. Busche, T. Drossel, T. Leichtweiss, et al., Dynamic formation of a solid-liquid electrolyte interphase and its consequences for hybrid-battery concepts. Nat. Chem. **8**, 426–434 (2016). https://doi.org/10.1038/nchem.2470
27. W.D. Callister, *Fundamentals of Materials Science and Engineering* (Wiley, London, 2000)
28. J. Cannarella, C.B. Arnold, The effects of defects on localized plating in lithium-ion batteries. J. Electrochem. Soc. **162**, A1365–A1373 (2015). https://doi.org/10.1149/2.1051507jes
29. H.J. Chang, N.M. Trease, A.J. Ilott, et al., Investigating Li microstructure formation on Li anodes for lithium batteries by in situ $^{6}Li/^{7}Li$ NMR and SEM. Phys. Chem. Chem. Phys. **119**, 16443–16451 (2015). https://doi.org/10.1021/acs.jpcc.5b03396
30. E. Chason, A. Engwall, F. Pei, et al., Understanding residual stress in electrodeposited Cu thin films. J. Electrochem. Soc. **160**, D3285–D3289 (2013). https://doi.org/10.1149/2.048312jes
31. J.N. Chazalviel, Electrochemical aspects of the generation of ramified metallic electrodeposits. Phys. Rev. A **42**, 7355–7367 (1990). https://doi.org/10.1103/PhysRevA.42.7355
32. L. Chen, Y. Li, S.-P. Li, et al., PEO/garnet composite electrolytes for solid-state lithium batteries: from "ceramic-in-polymer" to "polymer-in-ceramic". Nano Energy **46**, 176–184 (2018a). https://doi.org/10.1016/j.nanoen.2017.12.037
33. N. Chen, Y. Xing, L. Wang, et al., "Tai Chi" philosophy driven rigid-flexible hybrid ionogel electrolyte for high-performance lithium battery. Nano Energy **47**, 35–42 (2018b). https://doi.org/10.1016/j.nanoen.2018.02.036
34. L. Cheng, W. Chen, M. Kunz, et al., Effect of surface microstructure on electrochemical performance of garnet solid electrolytes. ACS Appl. Mater Interfaces **7**, 2073–2081 (2015). https://doi.org/10.1021/am508111r
35. X.B. Cheng, R. Zhang, C.Z. Zhao, et al., A review of solid electrolyte interphases on lithium metal anode. Adv. Sci. **3**, 1500213 (2016). https://doi.org/10.1002/advs.201500213
36. X.B. Cheng, R. Zhang, C.Z. Zhao, et al., Toward safe lithium metal anode in rechargeable batteries: a review. Chem. Rev. **117**, 10403–10473 (2017). https://doi.org/10.1021/acs.chemrev.7b00115
37. S. Choudhury, R. Mangal, A. Agrawal, et al., A highly reversible room-temperature lithium metal battery based on crosslinked hairy nanoparticles. Nat. Commun. **6**, 10101 (2015). https://doi.org/10.1038/ncomms10101
38. P. Chu, Sm_2O_3 composite PEO solid polymer electrolyte. J. Power Sources **115**, 288–294 (2003). https://doi.org/10.1016/s0378-7753(02)00717-6
39. S. Chu, Y. Cui, N. Liu, The path towards sustainable energy. Nat. Mater. **16**, 16–22 (2016). https://doi.org/10.1038/nmat4834
40. S.H. Chung, Y. Wang, S.G. Greenbaum, Uniaxial stress effects in poly (ethylene oxide) -LiI polymer electrolyte film: a 7Li nuclear magnetic resonance study. Electrochem. Solid-State Lett. **2**, 553–555 (1999). https://doi.org/10.1149/1.1390901
41. F. Croce, L. Persi, F. Ronci, et al., Nanocomposite polymer electrolytes and their impact on the lithium battery technology. Solid State Ionics **135**, 47–52 (2000). https://doi.org/10.1016/S0167-2738(00)00329-5
42. F. Croce, L. Settimi, B. Scrosati, et al., Nanocomposite, PEO-LiBOB polymer electrolytes for low temperature, lithium rechargeable batteries. J. New Mater. Electrochem. Syst. **9**, 3–9 (2006)

43. K.M. Diederichsen, E.J. McShane, B.D. McCloskey, Promising routes to a high Li+ transference number electrolyte for lithium ion batteries. ACS Energy Lett. **2**, 2563–2575 (2017). https://doi.org/10.1021/acsenergylett.7b00792
44. M. Dollé, L. Sannier, B. Beaudoin, et al., Live scanning electron microscope observations of dendritic growth in lithium/polymer cells. Electrochem. Solid-State Lett. **5**, A286–A289 (2002). https://doi.org/10.1149/1.1519970
45. H. Duan, Y.X. Yin, Y. Shi, et al., Dendrite-Free li-metal battery enabled by a thin asymmetric solid electrolyte with engineered layers. J. Am. Chem. Soc. **140**, 82–85 (2018). https://doi.org/10.1021/jacs.7b10864
46. K. Edströma, A.M. Anderssona, A. Bishop, Carbon electrode morphology and thermal stability of the passivation layer. J. Power Sources **97–98**, 87–91 (2001). https://doi.org/10.1016/S0378-7753(01)00593-6
47. D.R. Ely, R.E. Garcia, Heterogeneous nucleation and growth of lithium electrodeposits on negative electrodes. J. Electrochem. Soc. **160**, A662–A668 (2013). https://doi.org/10.1149/1.057304jes
48. I. Epelboin et al., Behavior of secondary lithium and aluminum-lithium electrodes in propylene carbonate. J. Electrochem. Soc. **127**, 2100–2104 (1980). https://doi.org/10.1149/1.2129354
49. N.S. Ergang, M.A. Fierke, Z. Wang, et al., Fabrication of a fully infiltrated three-dimensional solid-state interpenetrating electrochemical cell. J. Electrochem. Soc. **160**(154), A1135–A1139 (2007). https://doi.org/10.1149/1.2794288
50. G.G. Eshetu, X. Judez, C. Li, et al., Lithium azide as an electrolyte additive for all-solid-state lithium-sulfur batteries. Angew. Chem. Int. Ed. Engl. **56**, 15368–15372 (2017). https://doi.org/10.1002/anie.201709305
51. D.E. Fenton, J.M. Parker, P.V. Wright, Complexes of alkali metal ions with poly(ethylene oxide). Polymer **14**, 589–590 (1973). https://doi.org/10.1016/0032-3861(73)90146-8
52. J.W. Fergus, Ceramic and polymeric solid electrolytes for lithium-ion batteries. J. Power Sources **195**, 4554–4569 (2010). https://doi.org/10.1016/j.jpowsour.2010.01.076
53. A. Ferrese, P. Albertus, J. Christensen, et al., Lithium redistribution in lithium-metal batteries. J. Electrochem. Soc. **159**, A1615–A1623 (2012). https://doi.org/10.1149/2.027210jes
54. C.P. Fonseca, S. Neves, Characterization of polymer electrolytes based on poly (dimethyl siloxane-co-ethylene oxide). J. Power Sources **104**, 85–89 (2002). https://doi.org/10.1016/S0378-7753(01)00902-8
55. J. Fu, Superionic conductivity of glass-ceramics in the system Li_2O-Al_2O_3-TiO_2-P_2O_5. Solid State Ionics **96**, 195–200 (1997a). https://doi.org/10.1016/s0167-2738(97)00018-0
56. J. Fu, Fast Li^+ ion conducting glass-ceramics in the system Li_2O-Al_2O_3-GeO_2-P_2O_5. Solid State Ionics **104**, 191–194 (1997b). https://doi.org/10.1016/S0167-2738(97)00434-7
57. K.K. Fu, Y. Gong, J. Dai, et al., Flexible, solid-state, ion-conducting membrane with 3D garnet nanofiber networks for lithium batteries. Proc. Natl. Acad. Sci. U. S. A. **113**, 7094–7099 (2016). https://doi.org/10.1073/pnas.1600422113
58. K. Fu, Y. Gong, B. Liu, et al., Toward garnet electrolyte–based Li metal batteries: an ultrathin, highly effective, artificial solid-state electrolyte/metallic Li interface. Sci. Adv. **3**, e1601659 (2017a). https://doi.org/10.1126/sciadv.1601659
59. K. Fu, Y. Gong, G.T. Hitz, et al., Three-dimensional bilayer garnet solid electrolyte based high energy density lithium metal–sulfur batteries. Energy Environ. Sci. **10**, 1568–1575 (2017b). https://doi.org/10.1039/c7ee01004d
60. X.B. Fu, G. Yang, J.Z. Wu, et al., Fast lithium-ion transportation in crystalline polymer electrolytes. ChemPhysChem **19**, 45–50 (2018). https://doi.org/10.1002/cphc.201701092
61. K. Funke, Jump relaxation in solid electrolytes. Prog. Solid State Chem. **22**, 111–195 (1993). https://doi.org/10.1016/0079-6786(93)90002-9
62. J. Gai, E. Zhao, F. Ma, et al., Improving the Li-ion conductivity and air stability of cubic $Li_7La_3Zr_2O_{12}$ by the co-doping of Nb, Y on the Zr site. J Eur. Ceram. Soc. **38**, 1673–1678 (2018). https://doi.org/10.1016/j.jeurceramsoc.2017.12.002

63. Y. Gao, Y. Zhao, Y.C. Li, et al., Interfacial chemistry regulation via a skin-grafting strategy rnables high-performance lithium-metal batteries. J. Am. Chem. Soc. **139**, 15288–15291 (2017). https://doi.org/10.1021/jacs.7b06437
64. J. Garcia-Barriocanal, A. Rivera-Calzada, M. Varela, et al., Colossal ionic conductivity at interfaces of epitaxial ZrO_2: $Y_2O_3/SrTiO_3$ heterostructures. Science **321**, 676–680 (2008). https://doi.org/10.1126/science.1156393
65. C.A. Geiger, E. Alekseev, B. Lazic, et al., Crystal chemistry and stability of "$Li_7La_3Zr_2O_{12}$" garnet: a fast lithium-ion conductor. Inorg. Chem. **50**, 1089–1097 (2011). https://doi.org/10.1021/ic101914e
66. L. Gireaud, S. Grugeon, S. Laruelle, et al., Lithium metal stripping/plating mechanisms studies: a metallurgical approach. Electrochem. Commun. **8**, 1639–1649 (2006). https://doi.org/10.1016/j.elecom.2006.07.037
67. D. Golodnitsky, E. Livshits, A. Ulus, Fast ion transport phenomena in oriented semicrystalline LiI-P(EO)n-based polymer electrolytes. J. Phys. Chem. A **105**, 10098–10106 (2001). https://doi.org/10.1021/jp012547l
68. D. Golodnitsky, E. Livshits, R. Kovarsky, et al., New generation of ordered polymer electrolytes for lithium batteries. Electrochem. Solid-State Lett. **7**, A412–A415 (2004). https://doi.org/10.1149/1.1803434
69. D. Golodnitsky, M. Nathan, V. Yufit, et al., Progress in three-dimensional (3D) Li-ion microbatteries. Solid State Ionics **177**, 2811–2819 (2006). https://doi.org/10.1016/j.ssi.2006.02.048
70. J.B. Goodenough, Y. Kim, Challenges for rechargeable Li batteries. Chem. Mater. **22**, 587–603 (2010). https://doi.org/10.1021/cm901452z
71. P.G. Hall, G.R. Davis, J.E. McIntyre, Ion conductivity in polysiloxane comb polymers with ethylene glycol teeth. Polym. Comm. **27**, 98–100 (1986)
72. F. Han, Y. Zhu, X. He, et al., Electrochemical stability of $Li_{10}GeP_2S_{12}$ and $Li_7La_3Zr_2O_{12}$ solid electrolytes. Adv. Energy Mater. **6** (2016). https://doi.org/10.1002/aenm.201501590
73. X. Han, Y. Gong, K.K. Fu, et al., Negating interfacial impedance in garnet-based solid-state Li metal batteries. Nat. Mater. **16**, 572–579 (2017). https://doi.org/10.1038/nmat4821
74. F. Han, J. Yue, C. Chen, et al., Interphase engineering enabled all-ceramic lithium battery. Joule **2**, 497–508 (2018). https://doi.org/10.1016/j.joule.2018.02.007
75. K.J. Harry, D.T. Hallinan, D.Y. Parkinson, et al., Detection of subsurface structures underneath dendrites formed on cycled lithium metal electrodes. Nat. Mater. **13**, 69–73 (2014). https://doi.org/10.1038/nmat3793
76. P. Hartmann, T. Leichtweiss, M.R. Busche, et al., Degradation of NASICON-type materials in contact with lithium metal: formation of mixed conducting interphases (MCI) on solid electrolytes. J. Phys. Chem. C **117**, 21064–21074 (2013). https://doi.org/10.1021/jp4051275
77. A. Hayashi, Characterization of Li_2S-P_2S_5 glass-ceramics as a solid electrolyte for lithium secondary batteries. Solid State Ionics **175**, 683–686 (2004). https://doi.org/10.1016/j.ssi.2004.08.036
78. X. He, Y. Zhu, Y. Mo, Origin of fast ion diffusion in super-ionic conductors. Nat. Commun. **8**, 15893 (2017). https://doi.org/10.1038/ncomms15893
79. W.A. Henderson, S. Passerini, Ionic conductivity in crystalline–amorphous polymer electrolytes-$P(EO)_6$:LiX phases. Electrochem. Commun. **5**, 575–578 (2003). https://doi.org/10.1016/s1388-2481(03)00131-0
80. A.M. Hussain, F.A. Ghaffar, S.I. Park, et al., Metal/polymer based stretchable antenna for constant frequency far-field communication in wearable electronics. Adv. Funct. Mater. **25**, 6565–6575 (2015). https://doi.org/10.1002/adfm.201503277
81. Y. Inaguma, High ionic conductivity in lithium lanthanum titanate. Solid State Commun. **86**, 689–693 (1993). https://doi.org/10.1016/0038-1098(93)90841-A
82. K. Ishiguro, Y. Nakata, M. Matsui, et al., Stability of Nb-doped cubic $Li_7La_3Zr_2O_{12}$ with lithium metal. J. Electrochem. Soc. **160**, A1690–A1693 (2013). https://doi.org/10.1149/2.036310jes
83. M. Ishikawa, Y. Tasaka, N. Yoshimoto, et al., Optimization of physicochemical characteristics of a lithium anode interface for high-efficiency cycling: an effect of electrolyte temperature. J. Power Sources **97–98**, 262–264 (2001). https://doi.org/10.1016/S0378-7753(01)00621-8

84. P. Jeevanandam, S. Vasudevan, Conductivity of a confined polymer electrolyte: lithium-polypropylene glycol intercalated in layered $CdPS_3$. J. Phys. Chem. B **102**, 4753–4758 (1998). https://doi.org/10.1021/jp980357d
85. S. Jiao, J. Zheng, Q. Li, et al., Behavior of lithium metal anodes under various capacity utilization and high current density in lithium metal batteries. Joule **2**, 110–124 (2018). https://doi.org/10.1016/j.joule.2017.10.007
86. Y. Jin, P.J. McGinn, $Li_7La_3Zr_2O_{12}$ electrolyte stability in air and fabrication of a Li/Li7La3Zr2O12/Cu0.1V2O5 solid-state battery. J. Power Sources **239**, 326–331 (2013). https://doi.org/10.1016/j.jpowsour.2013.03.155
87. Y.-C. Jung, M.-S. Park, C.-H. Doh, et al., Organic-inorganic hybrid solid electrolytes for solid-state lithium cells operating at room temperature. Electrochim. Acta **218**, 271–277 (2016). https://doi.org/10.1016/j.electacta.2016.09.141
88. H. Jung, B. Lee, M. Lengyel, et al., Nanoscale in situ detection of nucleation and growth of Li electrodeposition at various current densities. J. Mater. Chem. A **6**, 4629–4635 (2018). https://doi.org/10.1039/c8ta00343b
89. N.T. Kalyana Sundaram, A. Subramania, Nano-size LiAlO2 ceramic filler incorporated porous PVDF-co-HFP electrolyte for lithium-ion battery applications. Electrochim. Acta **52**, 4987–4993 (2007). https://doi.org/10.1016/j.electacta.2007.01.066
90. N. Kamaya, K. Homma, Y. Yamakawa, et al., A lithium superionic conductor. Nat. Mater. **10**, 682–686 (2011). https://doi.org/10.1038/nmat3066
91. K. Kanamura, H. Tamura, S. Shiraishi, et al., XPS analysis for the lithium surface immersed in γ-butyrolactone containing various salts. Electrochim. Acta **40**, 913–921 (1995). https://doi.org/10.1016/0013-4686(93)E0020-M
92. K. Kanamura, S. Shiraishi, Z.I. Takehara, Electrochemical deposition of very smooth lithium using nonaqueous electrolytes containing HF. J. Electrochem. Soc. **143**, 2187–2197 (1996). https://doi.org/10.1149/1.1836979
93. R. Kanno, M. Murayama, Lithium ionic conductor thio-LISICON: the Li_2S-GeS_2-P_2S_5 system. J. Electrochem. Soc. **148**, A742–A746 (2001). https://doi.org/10.1149/1.137902
94. N. Karan, D. Pradhan, R. Thomas, et al., Solid polymer electrolytes based on polyethylene oxide and lithium trifluoro-methane sulfonate (PEO–$LiCF_3SO_3$): ionic conductivity and dielectric relaxation. Solid State Ionics **179**, 689–696 (2008). https://doi.org/10.1016/j.ssi.2008.04.034
95. A. Kato, A. Hayashi, M. Tatsumisago, Enhancing utilization of lithium metal electrodes in all-solid-state batteries by interface modification with gold thin films. J. Power Sources **309**, 27–32 (2016a). https://doi.org/10.1016/j.jpowsour.2016.01.068
96. Y. Kato, S. Hori, T. Saito, et al., High-power all-solid-state batteries using sulfide superionic conductors. Nat. Energy **1** (2016b). https://doi.org/10.1038/nenergy.2016.30
97. H. Kawai, J. Kuwano, Lithium ion conductivity of a-site deficient perovskite solid solution $La_{0.67-x}Li_{3x}TiO_3$. J. Electrochem. Soc. **141**, L78–L79 (1994). https://doi.org/10.1149/1.2055043
98. R. Khurana, J.L. Schaefer, L.A. Archer, et al., Suppression of lithium dendrite growth using cross-linked polyethylene/poly(ethylene oxide) electrolytes: a new approach for practical lithium-metal polymer batteries. J. Am. Chem. Soc. **136**, 7395–7402 (2014). https://doi.org/10.1021/ja502133j
99. Y.-T. Kim, The effect of plasticizers on transport and electrochemical properties of PEO-based electrolytes for lithium rechargeable batteries. Solid State Ionics **149**, 9 (2002). https://doi.org/10.1016/S0167-2738(02)00130-3
100. J.-W. Kim, K.-S. Ji, J.-P. Lee, et al., Electrochemical characteristics of two types of PEO-based composite electrolyte with functional SiO_2. J. Power Sources **119–121**, 415–421 (2003). https://doi.org/10.1016/s0378-7753(03)00263-5
101. M. Kitta, H. Sano, Real-time observation of Li deposition on a Li electrode with operand atomic force microscopy and surface mechanical imaging. Langmuir **33**, 1861–1866 (2017). https://doi.org/10.1021/acs.langmuir.6b04651

102. M. Kotobuki et al., Fabrication of three-dimensional battery using ceramic electrolyte with honeycomb structure by sol–gel process. J. Electrochem. Soc. **157**, A493–A498 (2010). https://doi.org/10.1149/1.3308459
103. M. Kotobuki, H. Munakata, K. Kanamura, All-solid-state Li battery for future energy technology. Handbook Adv. Ceram. (2013). https://doi.org/10.1016/b978-0-12-385469-8.00019-8
104. J. Kumar, S.J. Rodrigues, B. Kumar, Interface-mediated electrochemical effects in lithium/polymer-ceramic cells. J. Power Sources **195**, 327–334 (2010). https://doi.org/10.1016/j.jpowsour.2009.06.098
105. A. Kushima, K.P. So, C. Su, et al., Liquid cell transmission electron microscopy observation of lithium metal growth and dissolution: root growth, dead lithium and lithium flotsams. Nano Energy **32**, 271–279 (2017). https://doi.org/10.1016/j.nanoen.2016.12.001
106. W.J. Kwon, H. Kim, K.-N. Jung, et al., Enhanced Li+ conduction in perovskite $Li_{3x}La_{2/3-x}\square_{1/3-2x}TiO_3$ solid-electrolytes via microstructural engineering. J. Mater. Chem. A **5**, 6257–6262 (2017). https://doi.org/10.1039/c7ta00196g
107. J. Li, X. Huang, L. Chen, X-ray diffraction and vibrational spectroscopic studies on PAN-LiTFSI polymer electrolytes. J. Electrochem. Soc. **147**, 2653–2657 (2000). https://doi.org/10.1149/1.1393585
108. W. Li, G. Wu, C.M. Araújo, et al., Li^+ ion conductivity and diffusion mechanism in α-Li3N and β-Li3N. Energy Environ. Sci. **3**, 1524–1530 (2010). https://doi.org/10.1039/c0ee00052c
109. Y. Li, Z. Wang, Y. Cao, et al., W-doped $Li_7La_3Zr_2O_{12}$ ceramic electrolytes for solid state Li-ion batteries. Electrochim. Acta **180**, 37–42 (2015). https://doi.org/10.1016/j.electacta.2015.08.046
110. Y. Li, W. Zhou, X. Chen, et al., Mastering the interface for advanced all-solid-state lithium rechargeable batteries. Proc. Natl. Acad. Sci. U. S. A. **113**, 13313–13317 (2016). https://doi.org/10.1073/pnas.1615912113
111. Y. Li, Y. Li, A. Pei, et al., Atomic structure of sensitive battery materials and interfaces revealed by cryo-electron microscopy. Science **358**, 506–510 (2017a). https://doi.org/10.1126/science.aam6014
112. Y. Li, B. Xu, H. Xu, et al., Hybrid polymer/garnet electrolyte with a small interfacial resistance for lithium-ion batteries. Angew. Chem. Int. Ed. Engl. **56**, 753–756 (2017b). https://doi.org/10.1002/anie.201608924
113. L. Li, S. Basu, Y. Wang, et al., Self-heating–induced healing of lithium dendrites. Science **359**, 1513–1516 (2018). https://doi.org/10.1126/science.aap8787
114. N.W. Li, Y. Shi, Y.X. Yin, et al., A fexible solid electrolyte interphase layer for long-life lithium metal anodes. Angew. Chem. Int. Ed. Engl. **57**, 1505–1509 (2018a). https://doi.org/10.1002/anie.201710806
115. S. Li, M. Jiang, Y. Xie, et al., Developing high-performance lithium metal anode in liquid electrolytes: challenges and progress. Adv. Mater. **30**, e1706375 (2018b). https://doi.org/10.1002/adma.201706375
116. C.C. Liang, Conduction characteristics of the lithium iodide-aluminum oxide solid electrolytes. J. Electrochem. Soc. **120**, 1289–1292 (1973). https://doi.org/10.1149/1.2403248
117. L. Liang, L.-Q. Chen, Nonlinear phase field model for electrodeposition in electrochemical systems. Appl. Phys. Lett. **105**, 263903 (2014). https://doi.org/10.1063/1.4905341
118. C.W. Lin, C.L. Hung, M. Venkateswarlu, et al., Influence of TiO_2 nano-particles on the transport properties of composite polymer electrolyte for lithium-ion batteries. J. Power Sources **146**, 397–401 (2005). https://doi.org/10.1016/j.jpowsour.2005.03.028
119. F. Lin, I.M. Markus, M.M. Doeff, et al., Chemical and structural stability of lithium-ion battery electrode materials under electron beam. Sci. Rep. **4**, 5694 (2014). https://doi.org/10.1038/srep05694
120. D. Lin, Y. Liu, Y. Cui, Reviving the lithium metal anode for high-energy batteries. Nat. Nanotechnol. **12**, 194–206 (2017a). https://doi.org/10.1038/nnano.2017.16
121. Y. Lin, X. Wang, J. Liu, et al., Natural halloysite nano-clay electrolyte for advanced all-solid-state lithium-sulfur batteries. Nano Energy **31**, 478–485 (2017b). https://doi.org/10.1016/j.nanoen.2016.11.045

122. M. Ling, X. Zhu, Y. Jiang, et al., Comparative study of solid-state reaction and sol-gel process for synthesis of Zr-doped $Li_{0.5}La_{0.5}TiO_3$ solid electrolytes. Ionics **22**, 2151–2156 (2016). https://doi.org/10.1007/s11581-016-1744-8
123. W. Liu, N. Liu, J. Sun, et al., Ionic conductivity enhancement of polymer electrolytes with ceramic nanowire fillers. Nano Lett. **15**, 2740–2745 (2015). https://doi.org/10.1021/acs.nanolett.5b00600
124. T. Liu, Y. Ren, Y. Shen, et al., Achieving high capacity in bulk-type solid-state lithium ion battery based on $Li_{6.75}La_3Zr_{1.75}Ta_{0.25}O_{12}$ electrolyte: interfacial resistance. J. Power Sources **324**, 349–357 (2016a). https://doi.org/10.1016/j.jpowsour.2016.05.111
125. W. Liu, D. Lin, A. Pei, et al., Stabilizing lithium metal anodes by uniform Li-ion flux distribution in nanochannel confinement. J. Am. Chem. Soc. **138**, 15443–15450 (2016b). https://doi.org/10.1021/jacs.6b08730
126. Z. Liu, S. Venkatachalam, H. Kirchhain, et al., Study of the glass-to-crystal transformation of the NASICON-type solid electrolyte $Li_{1+x}Al_xGe_{2-x}(PO_4)_3$. Solid State Ionics **295**, 32–40 (2016c). https://doi.org/10.1016/j.ssi.2016.07.006
127. X. Liu, Y. Li, T. Yang, et al., High lithium ionic conductivity in the garnet-type oxide $Li_{7-2x}La_3Zr_{2-x}Mo_xO_{12}$ (x = 0–0.3) ceramics by sol-gel method. J. Am. Ceram. Soc. **100**, 1527–1533 (2017a). https://doi.org/10.1111/jace.14736
128. Y. Liu, D. Lin, Y. Jin, et al., Transforming from planar to three-dimensional lithium with flowable interphase for solid lithium metal batteries. Sci. Adv. **3**, eaao0713 (2017b). https://doi.org/10.1126/sciadv.aao0713
129. B. Liu, Y. Gong, K. Fu, et al., Garnet solid electrolyte protected Li-metal batteries. ACS Appl. Mater. Interfaces **9**, 18809–18815 (2017c). https://doi.org/10.1021/acsami.7b03887
130. W. Liu, S.W. Lee, D. Lin, et al., Enhancing ionic conductivity in composite polymer electrolytes with well-aligned ceramic nanowires. Nat. Energy **2**, 17035 (2017d). https://doi.org/10.1038/nenergy.2017.35
131. W. Liu, W. Li, D. Zhuo, et al., Core-shell nanoparticle coating as an interfacial layer for dendrite-free lithium metal anodes. ACS Cent. Sci. **3**, 135–140 (2017e). https://doi.org/10.1021/acscentsci.6b00389
132. B. Liu, J.-G. Zhang, W. Xu, Advancing lithium metal batteries. Joule **2**, 833–845 (2018a). https://doi.org/10.1016/j.joule.2018.03.008
133. B. Liu, L. Zhang, S. Xu, et al., 3D lithium metal anodes hosted in asymmetric garnet frameworks toward high energy density batteries. Energy Storage Mater. **14**, 376–382 (2018b). https://doi.org/10.1016/j.ensm.2018.04.015
134. J.W. Long, B. Dunn, D.R. Rolison, et al., Three-dimensional battery architectures. Chem. Rev. **104**, 4463–4492 (2004). https://doi.org/10.1021/cr020740l
135. B.V. Lotsch, J. Maier, Relevance of solid electrolytes for lithium-based batteries: a realistic view. J. Electroceram. **38**, 128–141 (2017). https://doi.org/10.1007/s10832-017-0091-0
136. C.T. Love, O.A. Baturina, K.E. Swider-Lyons, Observation of lithium dendrites at ambient temperature and below. ECS Electrochem. Lett. **4**, A24–A27 (2014). https://doi.org/10.1149/2.0041502eel
137. W. Luo, Y. Gong, Y. Zhu, et al., Transition from superlithiophobicity to superlithiophilicity of garnet solid-state electrolyte. J. Am. Chem. Soc. **138**, 12258–12262 (2016). https://doi.org/10.1021/jacs.6b06777
138. C. Ma, Y. Cheng, K. Yin, et al., Interfacial stability of Li metal-solid electrolyte elucidated via in situ electron microscopy. Nano Lett. **16**, 7030–7036 (2016). https://doi.org/10.1021/acs.nanolett.6b03223
139. G.S. MacGlashan, Y.G. Andreev, G.B. Peter, Structure of the polymer electrolyte poly(ethylene oxide)$_6$:$LiAsF_6$. Nature **398**, 792–794 (1999). https://doi.org/10.1038/19730
140. A. Magistris, K. Singh, PEO-based polymer electrolytes. Poly Interna. **28**, 277–280 (1992). https://doi.org/10.1016/S0167-2738(98)00114-3
141. P.W. Majewski, M. Gopinadhan, W.S. Jang, et al., Anisotropic ionic conductivity in block copolymer membranes by magnetic field alignment. J. Am. Chem. Soc. **132**, 17516–17522 (2010). https://doi.org/10.1021/ja107309p

142. O. Maqueda, F. Sauvage, L. Laffont, et al., Structural, microstructural and transport properties study of lanthanum lithium titanium perovskite thin films grown by pulsed laser deposition. Thin Solid Films **516**, 1651–1655 (2008). https://doi.org/10.1016/j.tsf.2007.05.004
143. A. Mauger, M. Armand, C.M. Julien, et al., Challenges and issues facing lithium metal for solid-state rechargeable batteries. J. Power Sources **353**, 333–342 (2017). https://doi.org/10.1016/j.jpowsour.2017.04.018
144. B.L. Mehdi, J. Qian, E. Nasybulin, et al., Observation and quantification of nanoscale processes in lithium batteries by operando electrochemical (S)TEM. Nano Lett. **15**, 2168–2173 (2015). https://doi.org/10.1021/acs.nanolett.5b00175
145. A. Mei, X.-L. Wang, J.-L. Lan, et al., Role of amorphous boundary layer in enhancing ionic conductivity of lithium–lanthanum–titanate electrolyte. Electrochim. Acta **55**, 2958–2963 (2010). https://doi.org/10.1016/j.electacta.2010.01.036
146. R. Mercier, J P. Malugani, B. Fahys, et al., Superionic conduction in Li2S-P2S5-LiI-glasses[J]. Solid State Ionics, **5**, 663–666 (1981).
147. W.H. Meyer, Polymer electrolytes for lithium-ion batteries. Adv. Mater. **10**, 439–448 (1998). https://doi.org/10.1002/(SICI)1521-4095(199804)10:6<439::AID-ADMA439>3.0.CO;2-I
148. C. Monroe, J. Newman, Dendrite growth in lithium/polymer systems a propagation model for liquid electrolytes under galvanostatic conditions. J. Electrochem. Soc. **150**, A1377–A1384 (2003). https://doi.org/10.1149/1.1606686
149. C. Monroe, J. Newman, The impact of elastic deformation on deposition kinetics at lithium/polymer interfaces. J. Electrochem. Soc. **152**, A396–A404 (2005). https://doi.org/10.1149/1.1850854
150. K. Morigaki, Analysis of the interface between lithium and organic electrolyte solution. J. Power Sources **104**, 13–23 (2002). https://doi.org/10.1016/S0378-7753(01)00871-0
151. R. Murugan, V. Thangadurai, W. Weppner, Fast lithium ion conduction in garnet-type $Li_7La_3Zr_2O_{12}$. Angew. Chem. Int. Ed. Engl. **46**, 7778–7781 (2007). https://doi.org/10.1002/anie.200701144
152. K. Nagaoka, H. Naruse, I. Shinohara, High ionic conductivity in poly(dimethyl siloxane-co-ethylene oxide) dissolving lithium perchlorate. J. Polym. Sci. Polym. Lett. Ed. **22**, 659–663 (1984). https://doi.org/10.1002/pol.1984.130221205
153. Y. Nakai, K. Ito, H. Ohno, Ion conduction in molten salts prepared by terminal-charged PEO derivatives. Solid State Ionics **113**, 199–204 (1998). https://doi.org/10.1016/S0167-2738(98)00286-0
154. C.W. Nan, L. Fan, Y. Lin, et al., Enhanced ionic conductivity of polymer electrolytes containing nanocomposite SiO_2 particles. Phys. Rev. Lett. **91**, 266104 (2003). https://doi.org/10.1103/PhysRevLett.91.266104
155. N. Nitta, F. Wu, J.T. Lee, et al., Li-ion battery materials: present and future. Mater. Today **18**, 252–264 (2015). https://doi.org/10.1016/j.mattod.2014.10.040
156. M. Park, X. Zhang, M. Chung, et al., A review of conduction phenomena in Li-ion batteries. J. Power Sources **195**, 7904–7929 (2010). https://doi.org/10.1016/j.jpowsour.2010.06.060
157. E. Peled, The electrochemical behavior of alkali and alkaline earth metals in nonaqueous battery systems-the solid electrolyte interphase model. J. Electrochem. Soc. **126**, 2047–2051 (1979). https://doi.org/10.1149/1.2128859
158. E. Peled, S. Menkin, Review-SEI: past, present and future. J. Electrochem. Soc. **164**, A1703–A1719 (2017). https://doi.org/10.1149/2.1441707jes
159. E. Peled, D. Golodnitsky, G. Ardel, Advanced model for solid electrolyte interphase electrodes in liquid and polymer electrolytes. J. Electrochem. Soc. **144**, L208–L210 (1997). https://doi.org/10.1149/1.1837858
160. L. Porz, T. Swamy, B.W. Sheldon, et al., Mechanism of lithium metal penetration through inorganic solid electrolytes. Adv. Energy Mater. **7**, 1701003 (2017). https://doi.org/10.1002/aenm.201701003
161. D. Qian, B. Xu, H.-M. Cho, et al., Lithium lanthanum titanium oxides: a fast ionic conductive coating for lithium-ion battery cathodes. Chem. Mater. **24**, 2744–2751 (2012). https://doi.org/10.1021/cm300929r

162. S. Qin, X. Zhu, Y. Jiang, et al., Growth of self-textured Ga^{3+}-substituted Li7La3Zr2O12 ceramics by solid state reaction and their significant enhancement in ionic conductivity. Appl. Phys. Lett. **112**, 113901 (2018). https://doi.org/10.1063/1.5019179
163. E. Quartarone, P. Mustarelli, Electrolytes for solid-state lithium rechargeable batteries: recent advances and perspectives. Chem. Soc. Rev. **40**, 2525–2540 (2011). https://doi.org/10.1039/c0cs00081g
164. R. Raj, Nucleation of cavities at second phase particles in grain boundaries. Acta Metall. **26**, 995–1006 (1978). https://doi.org/10.1016/0001-6160(78)90050-0
165. R. Raj, J. Wolfenstine, Current limit diagrams for dendrite formation in solid-state electrolytes for Li-ion batteries. J. Power Sources **343**, 119–126 (2017). https://doi.org/10.1016/j.jpowsour.2017.01.037
166. Y. Ren, K. Chen, R. Chen, et al., Oxide electrolytes for lithium batteries. J. Am. Ceram. Soc. **98**, 3603–3623 (2015a). https://doi.org/10.1111/jace.13844
167. Y. Ren, Y. Shen, Y. Lin, et al., Direct observation of lithium dendrites inside garnet-type lithium-ion solid electrolyte. Electrochem. Commun. **57**, 27–30 (2015b). https://doi.org/10.1016/j.elecom.2015.05.001
168. W.D. Richards, L.J. Miara, Y. Wang, et al., Interface stability in solid-state batteries. Chem. Mater. **28**, 266–273 (2015). https://doi.org/10.1021/acs.chemmater.5b04082
169. M. Romero, R. Faccio, S. Vázquez, et al., Enhancement of lithium conductivity and evidence of lithium dissociation for LLTO-PMMA nanocomposite electrolyte. Mater. Lett. **172**, 1–5 (2016). https://doi.org/10.1016/j.matlet.2016.02.128
170. M. Rosso, C. Brissot, A. Teyssot, et al., Dendrite short-circuit and fuse effect on Li/polymer/Li cells. Electrochim. Acta **51**, 5334–5340 (2006). https://doi.org/10.1016/j.electacta.2006.02.004
171. R.L. Sacci, J.M. Black, N. Balke, et al., Nanoscale imaging of fundamental li battery chemistry: solid-electrolyte interphase formation and preferential growth of lithium metal nanoclusters. Nano Lett. **15**, 2011–2018 (2015). https://doi.org/10.1021/nl5048626
172. H.J.S. Sand, III. On the concentration at the electrodes in a solution, with special reference to the liberation of hydrogen by electrolysis of a mixture of copper sulphate and sulphuric acid. London, Edinburgh, Dublin Philosop. Philos. Maga. J. Sci. **1**, 45–79 (1901). https://doi.org/10.1080/14786440109462590
173. N. Sata, K. Eberman, K. Eberl, Mesoscopic fast ion conduction in nanometre-scale planar heterostructures. Nature **408**, 946–949 (2000). https://doi.org/10.1038/35050047
174. N.S. Schauser, K.J. Harry, D.Y. Parkinson, et al., Lithium dendrite growth in glassy and rubbery nanostructured block copolymer electrolytes. J. Electrochem. Soc. **162**, A398–A405 (2014). https://doi.org/10.1149/2.0511503jes
175. A. Schwöbel, R. Hausbrand, W. Jaegermann, Interface reactions between LiPON and lithium studied by in-situ X-ray photoemission. Solid State Ionics **273**, 51–54 (2015). https://doi.org/10.1016/j.ssi.2014.10.017
176. B. Scrosati, A. Selvaggi, F. Croce, et al., The Li/LiV_3O_8 polymer electrolyte lithium battery III. Investigation of the electrode interfaces. J. Power Sources **24**, 287–294 (1988). https://doi.org/10.1016/0378-7753(88)80104-6
177. B. Scrosati, F. Croce, L. Persi, Impedance spectroscopy study of PEO-based nanocomposite polymer electrolytes. J. Electrochem. Soc. **147**, 1718–1721 (2000). https://doi.org/10.1149/1.1393423
178. Y. Seino, T. Ota, K. Takada, et al., A sulphide lithium super ion conductor is superior to liquid ion conductors for use in rechargeable batteries. Energy Environ. Sci. **7**, 627–631 (2014). https://doi.org/10.1039/c3ee41655k
179. A. Sharafi, E. Kazyak, A.L. Davis, et al., Surface chemistry mechanism of ultra-low interfacial resistance in the solid-state electrolyte $Li_7La_3Zr_2O_{12}$. Chem. Mater. **29**, 7961–7968 (2017a). https://doi.org/10.1021/acs.chemmater.7b03002
180. A. Sharafi, S. Yu, M. Naguib, et al., Impact of air exposure and surface chemistry on $Li–Li_7La_3Zr_2O_{12}$ interfacial resistance. J. Mater. Chem. A **5**, 13475–13487 (2017b). https://doi.org/10.1039/c7ta03162a

181. D.M. Smith, B. Dong, R.W. Marron, et al., Tuning ion conducting pathways using holographic polymerization. Nano Lett. **12**, 310–314 (2012). https://doi.org/10.1021/nl203599y
182. J. Steiger, D. Kramer, R. Mönig, Mechanisms of dendritic growth investigated by in situ light microscopy during electrodeposition and dissolution of lithium. J. Power Sources **261**, 112–119 (2014a). https://doi.org/10.1016/j.jpowsour.2014.03.029
183. J. Steiger, D. Kramer, R. Mönig, Microscopic observations of the formation, growth and shrinkage of lithium moss during electrodeposition and dissolution. Electrochim. Acta **136**, 529–536 (2014b). https://doi.org/10.1016/j.electacta.2014.05.120
184. G.M. Stone, S.A. Mullin, A.A. Teran, et al., Resolution of the modulus versus adhesion dilemma in solid polymer electrolytes for rechargeable lithium metal batteries. J. Electrochem. Soc. **159**, A222–A227 (2012). https://doi.org/10.1149/2.030203jes
185. S. Stramare, V. Thangadurai, W. Weppner, Lithium lanthanum titanates: a review. Chem. Mater. **15**, 3974–3990 (2003). https://doi.org/10.1021/cm0300516
186. J. Sun, X. Liao, A.M. Minor, et al., Morphology-conductivity relationship in crystalline and amorphous sequence-defined peptoid block copolymer electrolytes. J. Am. Chem. Soc. **136**, 14990–14997 (2014). https://doi.org/10.1021/ja5080689
187. B. Sun, H.D. Asfaw, D. Rehnlund, et al., Toward solid-state 3D-microbatteries using functionalized polycarbonate-based polymer electrolytes. ACS Appl. Mater. Inter. **10**, 2407–2413 (2018). https://doi.org/10.1021/acsami.7b13788
188. Z.I. Takehara, Future prospects of the lithium metal anode. J. Power Sources **68**, 82–86 (1997). https://doi.org/10.1016/S0378-7753(96)02546-3
189. C.-Y. Tang, S.J. Dillon, In situ scanning electron microscopy characterization of the mechanism for Li dendrite growth. J. Electrochem. Soc. **163**, A1660–A1665 (2016). https://doi.org/10.1149/2.0891608jes
190. J.M. Tarascon, The Li-ion battery: 25 years of exciting and enriching experiences. Electrochem. Soc. Inter. **25**, 79–83 (2016). https://doi.org/10.1149/2.F08163if
191. J.G. Thevenin, R.H. Muller, Impedance of lithium electrodes in a propylene carbonate electrolyte. J. Electrochem. Soc. **134**, 273–280 (1987). https://doi.org/10.1149/1.2100445
192. J.S. Thokchom, B. Kumar, The effects of crystallization parameters on the ionic conductivity of a lithium aluminum germanium phosphate glass–ceramic. J. Power Sources **195**, 2870–2876 (2010). https://doi.org/10.1016/j.jpowsour.2009.11.037
193. J.S. Thokchom, N. Gupta, B. Kumar, Superionic conductivity in a lithium aluminum germanium phosphate glass–ceramic. J. Electrochem. Soc. **155**, A915–A920 (2008). https://doi.org/10.1149/1.2988731
194. L. TianKhoon, N.H. Hassan, M.Y.A. Rahman, et al., One-pot synthesis nano-hybrid ZrO_2–TiO_2 fillers in 49% poly(methyl methacrylate) grafted natural rubber (MG49) based nano-composite polymer electrolyte for lithium ion battery application. Solid State Ionics **276**, 72–79 (2015). https://doi.org/10.1016/j.ssi.2015.03.034
195. M.D. Tikekar, S. Choudhury, Z. Tu, et al., Design principles for electrolytes and interfaces for stable lithium-metal batteries. Nat. Energy **1**, 16114 (2016). https://doi.org/10.1038/nenergy.2016.114
196. T. Uno, H. Sano, M. Matsumoto, et al., Properties of composite solid polymer electrolyte using hyperbranched polymer with ether-linkage. J. Solid State Electrochem. **14**, 2161–2167 (2010). https://doi.org/10.1007/s10008-009-0999-7
197. A. Vallée, S. Besner, J. Prud'Homme, Comparative study of poly(ethylene oxide) electrolytes made with $LiN(CF_3SO_2)_2$, $LiCF_3SO_3$ and $LiClO_4$: thermal properties and conductivity behavior. Electrochim. Acta **37**, 1579–1583 (1992). https://doi.org/10.1016/0013-4686(92)80115-3
198. A. Vlad, N. Singh, C. Galande, et al., Design considerations for unconventional electrochemical energy storage architectures. Adv. Energy Mater. **5**, 1402115 (2015). https://doi.org/10.1002/aenm.201402115
199. G. Wan, F. Guo, H. Li, et al., Suppression of dendritic lithium growth by in situ formation of a chemically stable and mechanically strong solid electrolyte interphase. ACS Appl. Mater. Inter. **10**, 593–601 (2018). https://doi.org/10.1021/acsami.7b14662

200. S. Wang, K. Min, Solid polymer electrolytes of blends of polyurethane and polyether modified polysiloxane and their ionic conductivity. Polymer **51**, 2621–2628 (2010). https://doi.org/10.1016/j.polymer.2010.04.038
201. Z. Wang, X. Huang, L. Chen, Understanding of effects of nano-Al_2O_3 particles on ionic conductivity of composite polymer electrolytes. Electrochem. Solid-State Lett. **6**, E40–E44 (2003). https://doi.org/10.1149/1.1615352
202. X. Wang, L. Dong, H. Zhang, et al., Recent progress in electronic skin. Adv. Sci. **2**, 1500169 (2015a). https://doi.org/10.1002/advs.201500169
203. Y. Wang, W.D. Richards, S.P. Ong, et al., Design principles for solid-state lithium superionic conductors. Nat. Mater. **14**, 1026–1031 (2015b). https://doi.org/10.1038/nmat4369
204. W. Wang, E. Yi, A.J. Fici, et al., Lithium ion conducting poly(ethylene oxide)-based solid electrolytes containing active or passive ceramic nanoparticles. J. Phys. Chem. C **121**, 2563–2573 (2017a). https://doi.org/10.1021/acs.jpcc.6b11136
205. X. Wang, M. Zhang, J. Alvarado, et al., New insights on the structure of electrochemically deposited lithium metal and its solid electrolyte interphases via cryogenic TEM. Nano Lett. **17**, 7606–7612 (2017b). https://doi.org/10.1021/acs.nanolett.7b03606
206. X. Wang, W. Zeng, L. Hong, et al., Stress-driven lithium dendrite growth mechanism and dendrite mitigation by electroplating on soft substrates. Nat. Energy **3**, 227–235 (2018a). https://doi.org/10.1038/s41560-018-0104-5
207. Z. Wang, Z. Wang, L. Yang, et al., Boosting interfacial Li^+ transport with a MOF-based ionic conductor for solid-state batteries. Nano Energy **49**, 580–587 (2018b). https://doi.org/10.1016/j.nanoen.2018.04.076
208. S. Wei, S. Choudhury, Z. Tu, et al., Electrochemical interphases for high-energy storage using reactive metal anodes. Acc. Chem. Res. **51**, 80–88 (2018). https://doi.org/10.1021/acs.accounts.7b00484
209. Z. Wen, T. Itoh, Y. Ichikawa, et al., Blend-based polymer electrolytes of poly(ethylene oxide) and hyperbranched poly[bis(triethylene glycol)benzoate] with terminal acetyl groups. Solid State Ionics **134**, 281–289 (2000). https://doi.org/10.1016/S0167-2738(00)00707-4
210. S. Wenzel, T. Leichtweiss, D. Krüger, et al., Interphase formation on lithium solid electrolytes—An in situ approach to study interfacial reactions by photoelectron spectroscopy. Solid State Ionics **278**, 98–105 (2015). https://doi.org/10.1016/j.ssi.2015.06.001
211. S. Wenzel, S. Randau, T. Leichtweiß, et al., Direct observation of the interfacial instability of the fast ionic conductor $Li_{10}GeP_2S_{12}$ at the lithium metal anode. Chem. Mater. **28**, 2400–2407 (2016a). https://doi.org/10.1021/acs.chemmater.6b00610
212. S. Wenzel, D.A. Weber, T. Leichtweiss, et al., Interphase formation and degradation of charge transfer kinetics between a lithium metal anode and highly crystalline $Li_7P_3S_{11}$ solid electrolyte. Solid State Ionics **286**, 24–33 (2016b). https://doi.org/10.1016/j.ssi.2015.11.034
213. J.E. Weston, B.C.H. Steele, Effects of inert fillers on the mechanical and electrochemical properties of lithium salt-poly (ethylene oxide) polymer electrolytes[J]. Solid State Ionics **7**(1), 75–79 (1982).
214. W. Wieczorek, K. Such, H. Wyciślik, J. Płocharski, Modifications of crystalline structure of PEO polymer electrolytes with ceramic additives. Solid State Ionics **36**, 255–257 (1989).
215. W. Wieczorek, Z. Florjancyk, J.R. Stevens, Composite polyether based solid electrolytes. Electrochim. Acta **40**, 2251–2258 (1995). https://doi.org/10.1016/0013-4686(95)00172-B
216. D.H. Wong, J.L. Thelen, Y. Fu, et al., Nonflammable perfluoropolyether-based electrolytes for lithium batteries. Proc. Natl. Acad. Sci. U. S. A. **111**, 3327–3331 (2014). https://doi.org/10.1073/pnas.1314615111
217. K.N. Wood, M. Noked, N.P. Dasgupta, Lithium metal anodes: toward an improved understanding of coupled morphological, electrochemical, and mechanical behavior. ACS Energy Lett. **2**, 664–672 (2017). https://doi.org/10.1021/acsenergylett.6b00650
218. B. Wu, S. Wang, W.J. Evans Iv, et al., Interfacial behaviours between lithium ion conductors and electrode materials in various battery systems. J. Mater. Chem. A **4**, 15266–15280 (2016). https://doi.org/10.1039/c6ta05439k

219. J.F. Wu, E.Y. Chen, Y. Yu, et al., Gallium-doped $Li_7La_3Zr_2O_{12}$ garnet-type electrolytes with high lithium-ion conductivity. ACS Appl. Mater. Inter. **9**, 1542–1552 (2017). https://doi.org/10.1021/acsami.6b13902
220. W. Xia, B. Xu, H. Duan, et al., Ionic conductivity and air stability of Al-Doped $Li_7La_3Zr_2O_{12}$ sintered in alumina and Pt crucibles. ACS Appl. Mater. Inter. **8**, 5335–5342 (2016). https://doi.org/10.1021/acsami.5b12186
221. S. Xia, X. Wu, W. Liu, et al., Practical challenges and future perspectives on all-solid-state lithium metal batteries. Chem. **5**, 1–33 (2019). https://doi.org/10.1016/j.chempr.2018.11.013
222. X. Xie, J. Xing, D. Hu, et al., Lithium expulsion from the solid-state electrolyte $Li_{6.4}La_3Zr_{1.4}Ta_{0.6}O_{12}$ by controlled electron injection in a SEM. ACS Appl. Mater. Inter. **10**, 5978–5983 (2018). https://doi.org/10.1021/acsami.7b17276
223. Y. Xiong, H. Tao, J. Zhao, et al., Effects of annealing temperature on structure and opt-electric properties of ion-conducting LLTO thin films prepared by RF magnetron sputtering. J. Alloys Compd. **509**, 1910–1914 (2011). https://doi.org/10.1016/j.jallcom.2010.10.086
224. K. Xu, S. Zhang, T.R. Jow, et al., LiBOB as salt for lithium-ion batteries a possible solution for high temperature operation. Electrochem. Solid-State Lett. **5**, A26–A29 (2002). https://doi.org/10.1149/1.1426042
225. X. Xu, Z. Wen, X. Yang, et al., Dense nanostructured solid electrolyte with high Li-ion conductivity by spark plasma sintering technique. Mater. Res. Bull. **43**, 2334–2341 (2008). https://doi.org/10.1016/j.materresbull.2007.08.007
226. W. Xu, J. Wang, F. Ding, et al., Lithium metal anodes for rechargeable batteries. Energy Environ. Sci. **7**, 513–537 (2014). https://doi.org/10.1039/c3ee40795k
227. B. Xu, W. Li, H. Duan, et al., Li_3PO_4-added garnet-type $Li_{6.5}La_3Zr_{1.5}Ta_{0.5}O_{12}$ for Li-dendrite suppression. J. Power Sources **354**, 68–73 (2017). https://doi.org/10.1016/j.jpowsour.2017.04.026
228. R. Xue, H. Huang, X. Huang, et al., Storage study of lithium battery with PEO/PC/$LiClO_4$ electrolyte. Solid State Ionics **74**, 133–136 (1994). https://doi.org/10.1016/0167-2738(94)90202-X
229. K.-Y. Yang, I.-C. Leu, K.-Z. Fung, et al., Mechanism of the interfacial reaction between cation-deficient La0.56Li0.33TiO3 and metallic lithium at room temperature. J. Mater. Res. **23**, 1813–1825 (2011). https://doi.org/10.1557/jmr.2008.0255
230. T. Yang, Z.D. Gordon, Y. Li, et al., Nanostructured garnet-type solid electrolytes for lithium batteries: electrospinning synthesis of $Li_7La_3Zr_2O_{12}$ nanowires and particle size-dependent phase transformation. J. Phys. Chem. C **119**, 14947–14953 (2015). https://doi.org/10.1021/acs.jpcc.5b03589
231. T. Yang, J. Zheng, Q. Cheng, et al., Composite polymer electrolytes with $Li_7La_3Zr_2O_{12}$ garnet-type nanowires as ceramic fillers: mechanism of conductivity enhancement and role of doping and morphology. ACS Appl. Mater. Inter. **9**, 21773–21780 (2017). https://doi.org/10.1021/acsami.7b03806
232. X. Yao, N. Huang, F. Han, et al., High-performance all-solid-state lithium-sulfur batteries enabled by amorphous sulfur-coated reduced graphene oxide cathodes. Adv. Energy Mater. **7**, 1602923 (2017). https://doi.org/10.1002/aenm.201602923
233. X. Yu, J.B. Bates, G.E. Jellison, et al., A Stable thin-film lithium electrolyte: lithium phosphorus oxynitride. J. Electrochem. Soc. **144**, 524–532 (1997). https://doi.org/10.1149/1.1837443
234. S. Yu, R.D. Schmidt, R. Garcia-Mendez, et al., Elastic properties of the solid electrolyte $Li_7La_3Zr_2O_{12}$ (LLZO). Chem. Mater. **28**, 197–206 (2015). https://doi.org/10.1021/acs.chemmater.5b03854
235. C. Yu, S. Ganapathy, E. Eck, et al., Accessing the bottleneck in all-solid state batteries, lithium-ion transport over the solid-electrolyte-electrode interface. Nat. Commun. **8**, 1086 (2017). https://doi.org/10.1038/s41467-017-01187-y
236. Y. Yuan, K. Amine, J. Lu, et al., Understanding materials challenges for rechargeable ion batteries with in situ transmission electron microscopy. Nat. Commun. **8**, 15806 (2017). https://doi.org/10.1038/ncomms15806

237. K. Zaghib, M. Armand, M. Gauthier, Electrochemistry of anodes in solid-state Li-ion polymer batteries. J. Electrochem. Soc. **145**, 3135–3140 (1998). https://doi.org/10.1149/1.1838776
238. Z. Zeng, W.I. Liang, H.G. Liao, et al., Visualization of electrode-electrolyte interfaces in LiPF6/EC/DEC electrolyte for lithium ion batteries via in situ TEM. Nano Lett. **14**, 1745–1750 (2014). https://doi.org/10.1021/nl403922u
239. X.X. Zeng, Y.X. Yin, N.W. Li, et al., Reshaping lithium plating/stripping behavior via bifunctional polymer electrolyte for room-temperature solid Li metal batteries. J. Am. Chem. Soc. **138**, 15825–15828 (2016). https://doi.org/10.1021/jacs.6b10088
240. H. Zhai, P. Xu, M. Ning, et al., A flexible solid composite electrolyte with vertically aligned and connected ion-conducting nanoparticles for lithium batteries. Nano Lett. **17**, 3182–3187 (2017). https://doi.org/10.1021/acs.nanolett.7b00715
241. S.S. Zhang, Problem, status, and possible solutions for lithium metal anode of rechargeable batteries. ACS Appl. Energy. Mater. **1**, 910–920 (2018). https://doi.org/10.1021/acsaem.8b00055
242. M. Zhang, K. Takahashi, N. Imanishi, et al., Preparation and electrochemical properties of $Li_{1+x}Al_xGe_{2-x}(PO_4)_3$ synthesized by a sol-gel method. J. Electrochem. Soc. **159**, A1114–A1119 (2012). https://doi.org/10.1149/2.080207jes
243. J.G. Zhang, W. Xu, W.A. Henderson, *Lithium Metal Anodes and Rechargeable Lithium Metal Batteries* (Springer, Cham, 2017a)
244. X. Zhang, T. Liu, S. Zhang, et al., Synergistic coupling between $Li_{6.75}La_3Zr_{1.75}Ta_{0.25}O_{12}$ and poly(vinylidene fluoride) induces high ionic conductivity, mechanical strength, and thermal stability of solid composite electrolytes. J. Am. Chem. Soc. **139**, 13779–13785 (2017b). https://doi.org/10.1021/jacs.7b06364
245. X. Zhang, J. Xie, F. Shi, et al., Vertically aligned and continuous nanoscale ceramic-polymer interfaces in composite solid polymer electrolytes for enhanced ionic conductivity. Nano Lett. **18**, 3829–3838 (2018). https://doi.org/10.1021/acs.nanolett.8b01111
246. E. Zhao, F. Ma, Y. Guo, et al., Stable LATP/LAGP double-layer solid electrolyte prepared via a simple dry-pressing method for solid state lithium ion batteries. RSC Adv. **6**, 92579–92585 (2016a). https://doi.org/10.1039/c6ra19415j
247. Y. Zhao, Z. Huang, S. Chen, et al., A promising PEO/LAGP hybrid electrolyte prepared by a simple method for all-solid-state lithium batteries. Solid State Ionics **295**, 65–71 (2016b). https://doi.org/10.1016/j.ssi.2016.07.013
248. Y. Zhao, C. Wu, G. Peng, et al., A new solid polymer electrolyte incorporating $Li_{10}GeP_2S_{12}$ into a polyethylene oxide matrix for all-solid-state lithium batteries. J. Power Sources **301**, 47–53 (2016c). https://doi.org/10.1016/j.jpowsour.2015.09.111
249. Z. Zheng, H. Fang, F. Yang, et al., Amorphous $LiLaTiO_3$ as solid electrolyte material. J. Electrochem. Soc. **161**, A473–A479 (2014a). https://doi.org/10.1149/2.006404jes
250. Z. Zheng, F. Hz, L. Zk, et al., A fundamental stability study for amorphous $LiLaTiO_3$ solid electrolyte. J. Electrochem. Soc. **162**, A244–A248 (2014b). https://doi.org/10.1149/2.0011503jes
251. W. Zhou, S. Wang, Y. Li, et al., Plating a dendrite-free lithium anode with a polymer/ceramic/polymer sandwich electrolyte. J. Am. Chem. Soc. **138**, 9385–9388 (2016). https://doi.org/10.1021/jacs.6b05341
252. Y. Zhu, X. He, Y. Mo, Origin of outstanding stability in the lithium solid electrolyte materials: insights from thermodynamic analyses based on first-principles calculations. ACS Appl. Mater. Inter. **7**, 23685–23693 (2015). https://doi.org/10.1021/acsami.5b07517
253. P. Zhu, C. Yan, M. Dirican, et al., $Li_{0.33}La_{0.557}TiO_3$ ceramic nanofiber-enhanced polyethylene oxide-based composite polymer electrolytes for all-solid-state lithium batteries. J. Mater. Chem. A **6**, 4279–4285 (2018). https://doi.org/10.1039/c7ta10517g

Metal-Based Chalcogenide Anode Materials for Lithium-Ion Batteries

6

Qiming Tang, Qin Jiang, Junwei Wu, and Xingjun Liu

Contents

Abstract

Rechargeable lithium-ion batteries (LIBs) have become state-of-art and are widely used in portable electronic devices and electric vehicles. With the development of materials and batteries technology, the performance of batteries has achieved enormous progress since its commercialization in the early 1990s. Recently, we saw this development slow down, as it is nearing the theoretical limit of the present system. A

Author Contribution: Q. Tang and Q. Jiang wrote this chapter. J. Wu. and X. Liu provided the relevant instruction.

Q. Tang · Q. Jiang · J. Wu (✉) · X. Liu
Shenzhen Key Laboratory of Advanced Materials, Department of Materials Science and Engineering, Harbin Institute of Technology (Shenzhen), Shenzhen, China
e-mail: qmtang_533@outlook.com; darra74@163.com; junwei.wu@hit.edu.cn; lxj@xmu.edu.cn

Q. Zhen et al. (eds.), *Nanostructured Materials for Next-Generation Energy Storage and Conversion*, https://doi.org/10.1007/978-3-662-58675-4_6

new battery system is necessary to meet the rigorous demands of high energy density. Under this circumstance, the innovation of anode materials is expected to be a promising way to remarkably increase the capacity of lithium-ion batteries. Metal-based chalcogenide (MXs, M = Sn, Mo, Fe, or Ge; X = O, S or Se) is a promising material for the anode of lithium-ion batteries (LIBs), because of their higher theoretical capacity when compared to conventional carbonaceous anodes. However, the wide application of chalcogenide anode in rechargeable battery is still hindered by the low efficiency and poor cycle performance caused by a large volume change during fully discharge and charge. Recently, there has been much interest in using nanostructured and carbon composite method in lithium-ion batteries to alleviate their inherent drawback. This chapter will provide an overview of the chalcogenide anode materials which are being studied experimentally in order to develop next-generation anodes for high-performance lithium-ion batteries.

6.1 Introduction

In the last two decades, electrical energy storage (EES) has witnessed unprecedented boom among which lithium-ion batteries are the most popular energy storage devices, powering portable electronics and electronic vehicles due to their high energy density. A great variety of cathode materials have been synthesized, and these materials can meet the needs of high capacity demand in many fields. However, research on high-performance anode materials still remains a big challenge. Historically speaking, graphite was the first type of carbon with which reversible lithiation was explored. It was found to be a good basis for the anodic reaction in rechargeable lithium (Li) batteries and then also plays a detrimental role in the commercial lithium-ion batteries (LIBs) [1]. Even though graphite has almost a perfectly layered structure, which is able to intercalate a large amount of Li, forming lithium carbide (Li_xC_6), where $0 < x < 1$, it exhibits a capacity of 372 mAh/g [2]. In order to further improve the energy density for lithium-ion batteries, massive efforts have been made to find alternative anode materials.

Anode material for lithium-ion batteries should [3]:

1. Have electrochemical potential values of lithium intercalation/de-intercalation (insertion/extraction) that are closely above lithium metal (0 V vs Li^+/Li).
2. Be able to store a large amount of lithium reversibly.
3. Have small variation of the electrode structure with lithium storage.
4. Have high electronic and ionic conductivities.
5. Be stable with the electrolyte in a wide range of electrochemical window.
6. Be abundant in the earth and nontoxic.
7. Be economical for the manufacturer.

These early studies suggested that Li intermetallics with various metals such as silicon (Si), tin (Sn), aluminum (Al), antimony (Sb), germanium (Ge), lead (Pb), bismuth (Bi), and magnesium (Mg) showed high gravimetric capacity [4–11]. But in the meantime, they also exhibit an obvious drawback that there is a large inherent increase in volume during lithiation in these alloy-based anodes. The volume expansion causes pulverization issue and aggregation of alloy-based materials which leads to the poor cycle life. Table 6.1 shows the comparison of performance between different element anode materials which can form an alloy with lithium.

Idota et al. pointed out an amorphous metal-oxide material that can store Li ions with a Coulombic capacity reaching that of hydrogen-storage alloys, ensuring protection against dendritic Li formation. The amorphous material is a metal composite oxide glass that contains tin (II) oxide (SnO) as an active center for Li adsorption [12]. The achievement is later recognized to be attributed to the application of metal composites in lithium-ion batteries. Compared with the pure metal materials, the metal chalcogenides (MXs, M = Sn, molybdenum (Mo), iron (Fe), Ge, nickel (Ni), Sb, or cobalt (Co); X = oxygen (O), sulfur (S), or selenium (Se)) have become a hot spot for anode materials in lithium-ion batteries due to their merits of mechanical, thermal, and structural stability. Generally speaking, the metal chalcogenide anode materials can be classified into two types in terms of their different structure [13, 14]:

1. Layered MXs, mainly including M sulfides or selenides such as SnS_2 $SnSe_2$, SnSeS, $SnSe_{0.5}S_{0.5}$, $MoSe_2$, GeS, MoS_2, tungsten sulfide (WS_2), etc., show an excellent ability to lithiation. According to the report by Ataca et al., single-layered MXs even display an interesting conductivity.
2. Nonlayered MXs, primarily containing FeS, FeS_2 $FeSe_2$, CoS_2, $CoSe_2$, NiS_2, Sb_2S_3, Bi_2S_3, etc. Many can be obtained from natural minerals, such as chalcopyrite, pyrite, marcasite, and sphalerite. Thus, the advantages of low price and high theoretical capacity make them very competitive compared to other anode materials.

In spite of MXs is expected to be a promising anode materials for lithium-ion batteries, it is gradually found that there is a poor cycling performance caused by relative volume change during lithiation and delithiation. After deep characterization, the mechanism of electrochemical reaction between lithium-ion and MX is

Table 6.1 Comparison of performance between different element anode materials which can form an alloy with lithium

Element	Density (g/cm^3)	Alloy	Volume change (%)	Specific capacity (mAh/g)	Specific capacity (mAh/cm^3)
Si	2.33	$Li_{4.4}Si$	300	4191	2061
Sn	7.3	$Li_{4.4}Sn$	260	992	9765
Al	2.7	LiAl	96	993	7241
Sb	6.68	Li_3Sb	200	659	4402
Ge	5.32	$Li_{4.4}Ge$	370	1621	8626
Pb	11.4	$Li_{4.4}Pb$	330	568	6478
Bi	9.8	Li_3Bi	215	384	3763
Mg	1.74	Li_3Mg	100	3350	4355

confirmed, as shown in Eqs. 6.1 and 6.2. Take SnO_2, for example, where lithiation takes place in two steps [12]:

$$SnO_2 + 4Li^+ + 4e^- \rightarrow 2Li_2O + Sn\,(\text{irreversible reaction}) \tag{6.1}$$

$$xe^- + xLi^+ + Sn \rightarrow Li_xSn\,(\text{during continuous cycling}) \tag{6.2}$$

Due to the first step of the electrochemical reaction, MXs shows the relatively low Coulombic efficiency owing to the irreversible formation of solid electrolyte interphase (SEI) films, lithium oxides, or sulfides (Li_2O, Li_2S, or Li_2S). In the second step reaction, the alloying process, the active material suffers the volume change, which will lead to a lower performance. Two approaches have been taken to solve these problems: firstly, minimizing the particle size of active materials into nanoscale could degrade the mechanical strain, so the structural integrity can be maintained during the processing of alloying and de-alloying reaction with Li-ions, and also nanoscale materials guarantee a short transport path for lithium-ions to enhance electrochemical performance [15–17]; secondly, composite materials, combining MXs with other conductive frame (such as carbon materials and conductive polymer), in which an electrochemically active phase is homogeneously dispersed within in the conductive frame. This kind of composite material can not only buffer the volume expansion during the cycling but also prevent nanoparticle aggregation and enhance the electronic conductivity [18–21].

In this chapter, research on metal chalcogenide anode materials is summarized following the line of tin-based chalcogenides, molybdenum-based chalcogenides, iron-based chalcogenides, and germanium-based chalcogenides.

6.2 Tin-Based Chalcogenides

Tin (Sn) is a promising lithium-ion battery anode material, possessing high theory charge storage capacity both by weight and by volume (with the gravimetric specific capacity of 992 mAh/g and the volume specific capacity of 9765 mAh/cm^3), which is more than two times that of commercial graphite materials employed for most commercial lithium-ion battery anodes. It is also inexpensive (roughly 20 $US/kg), relatively safe (not extremely pyrophoric), nontoxic, and highly abundant. It should be noted that pure tin is hardly used alone as an anode in lithium-ion batteries, but its composite form, Li_2X ($X = O$, S or Se), constrains the volume expansion and contraction of Sn during cycling [22].

6.2.1 Tin Oxides

Tin ($ns^2\ np^2$) oxides, mainly as SnO and SnO_2, are promising materials as the anode of lithium secondary batteries, because of its relevant higher theoretical capacity than the conventional carbonaceous anodes. In 1995, Idota et al. reported a kind of amorphous tin-based composite oxides (TCO) as anode materials for lithium-ion batteries [12]. The new Li storage mechanism attracts great attention. However, their wide application is still hampered by the severe capacity decay resulting from the irreversible formation of SEI films and the particle aggregation and pulverization caused by the huge volume changes during discharge/charge processes, as shown in Fig. 6.1 [23].

Figure 6.1 shows the key issue of SnO_2 (as well as SnS_2 or $SnSe_2$) as anode materials for lithium-ion batteries. Combining with Eqs. 6.1 and 6.2, in the first lithiation procedure, SnO_2 was reduced to Sn, and the matrix of Sn and Li_2O was coated with the SEI film that as the result of the electrolyte decomposition at around 0.8 V. The SEI film is an insulator for electrons, but a good conductor for lithium-ions, which can freely insert and de-insert through this passivation film. In the main, SEI film determines the reversibility of the cell chemistry and hence cycle life of the device. In a lithium-ion battery that employs intercalation hosts as cathode and anode, Li^+ is a limited source, and any loss of Li^+ is at the expense of capacity and energy density of the device. The formation chemistry of SEI leads to consumption of Li^+, which are converted into lithium salts insoluble in electrolyte solutions and hence "electrochemically inactive" [24]. Tin expands volumetrically by up to 260% on the full lithiation, creating two critical issues: the unstable and thick SEI films and the pulverization of active materials. As shown in Fig. 6.1a, the huge volume change of Sn particles during lithiation/delithiation cycles can cause the unstable SEI films on their surface, resulting in the continued formation of SEI films and depleting electrolyte. The large volume expansion may lead to huge stress and result in the pulverization and exfoliation of active materials, as shown in Fig. 6.1b. Thus, a wave of research has been done in order to solve or relieve these issues.

It was observed in [25] that the diameter of SnO_2 is the key factor to determine electrochemical performance. In different reaction condition, respectively, at 110 °C, 150 °C, and 200 °C, SnO_2 nanoparticles with the size of 3.0 nm, 4.0 nm, and 8.0 nm can be obtained, respectively. As shown in Fig. 6.2, X-ray diffraction (XRD) was

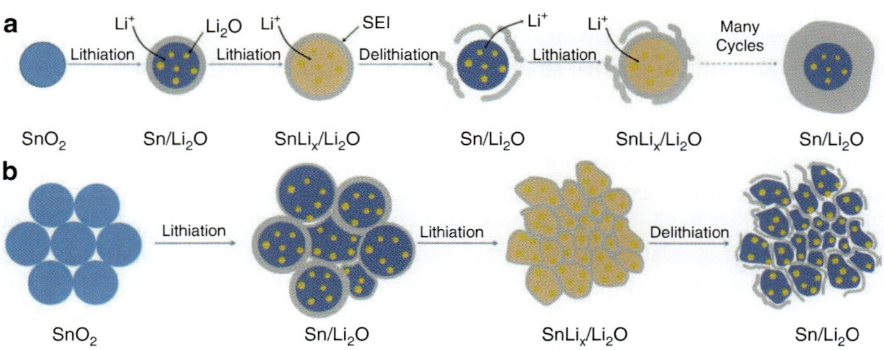

Fig. 6.1 Issues of SnO_2 as anodes for lithium-ion batteries: (**a**) the process of formation of unstable and thick solid electrolyte interphase (SEI) films; (**b**) the pulverization process of active materials. Reprinted with permission from [23]. Copyright 2015 Taylor & Francis

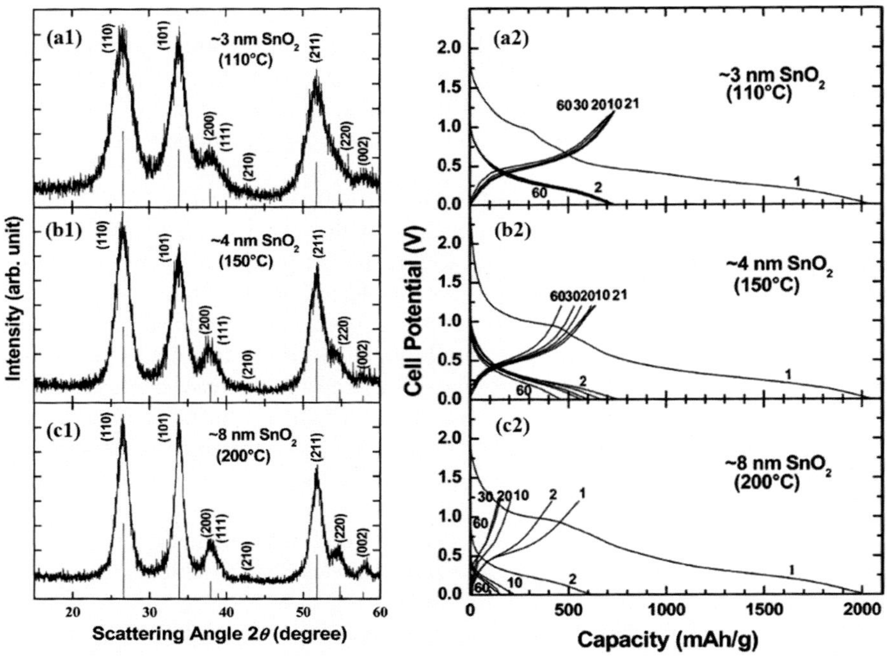

Fig. 6.2 X-ray diffraction (XRD) patterns before cycling and its corresponding voltage profiles at the rate of 60 mA/g of SnO_2 nanocrystals: (**a1** and **a2**) 3 nm, (**b1** and **b2**) 4 nm, and (**c1** and **c2**) 8 nm. Reprinted with permission from [25]. Copyright 2005 American Chemical Society

used to evaluate the structural and particle size differences between the samples obtained in different reaction condition. The electrochemical results showed that SnO_2 nanoparticles with a diameter of about 3 nm have a superior capacity and cycling stability as compared to those measuring 4 or 8 nm because the higher surface area and smaller dimensions allow the host to accommodate the lithium-ion quickly with minimal internal strain. Other nanostructured tin-oxide anodes such as SnO_2 nanowires also show higher initial Coulombic efficiency and an improved cyclic retention compared with those of SnO_2 powder, as shown in Fig. 6.3 [26].

Graphene (space group P6/*mmm*) is a two-dimensional aromatic monolayer of carbon atoms. It has excellent conductivity, the large specific surface area of over 2600 m^2/g, high flexibility, and pronounced chemical stability, and so on which can be coated to encapsulate of active nanomaterials owing to SnO_2-graphene composites had been prepared by different methods to improve electrochemical properties. For example, to optimize the cycling stability of tin-based anode materials, Mai et al. [27] have successfully synthesized the SnO_2 quantum dots @ graphene oxide (denoted as SnO_2-QDs@GO) with good dispersion and high mass loading, as shown in Fig. 6.4a1. The SnO_2-QDs@GO exhibits promising performance as an anode for lithium-ion batteries in terms of high capacity (a capacity of 1121 mAh/g at a current density of

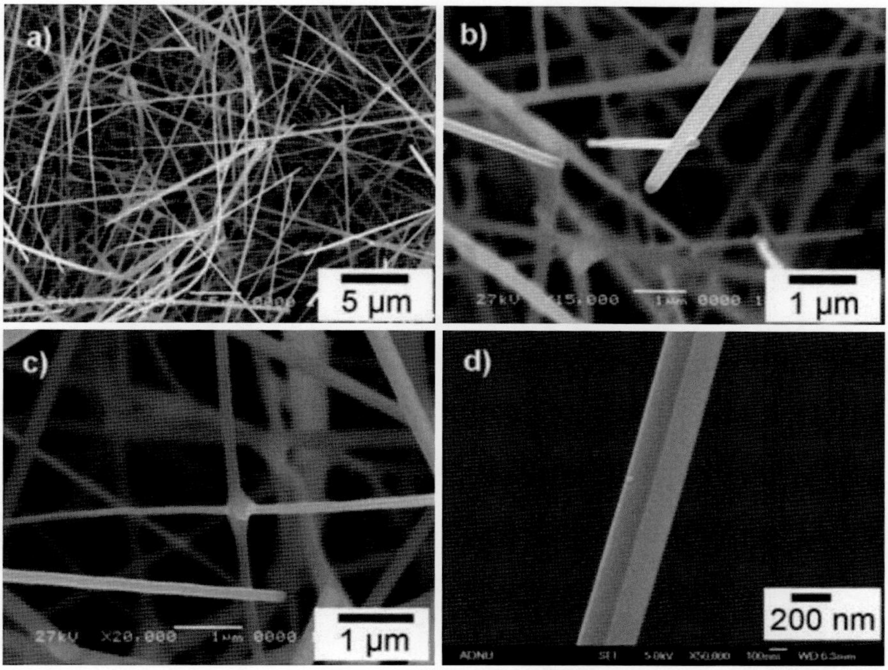

Fig. 6.3 The microstructure of SnO_2 nanowires. (**a**) Scanning electron microscope (SEM) image of SnO_2 nanowires; (**b**) SEM image of tips including Sn droplets; (**c**) SEM image of the junction; and (**d**) field emission SEM (FESEM) image of an individual nanowire stem. Reprinted with permission from [26]. Copyright 2007 John Wiley and Sons

0.1 A/g) and outstanding high-rate cycling stability, as shown in Fig. 6.4a2–a4. These results demonstrate this kind of unique design is a successful combination of high conductive graphene sheets and ultrafine quantum dots. This structure can provide a lot of active sites for lithium-ion, which in return provides high lithium storage capacity. Meanwhile, the voids between each quantum dot are able to provide space for the volume variation for lithium intercalation/de-intercalation and inhibit the aggregation of each particle. Moreover, the conductive graphene sheet endows each quantum dot with sufficient electrical conductivity, and the quantum dots offer shortened lithium diffusion length, which results in the enhanced kinetics, and, thus, high-rate cycling stability is achieved. Geng et al. [28] reported an in situ synthesized three-dimensional porous composite of fluorine tin oxides (F-SnO_2) and reduced graphene oxide (RGO) [F-SnO_2@RGO] through a one-step hydrothermal method using a F-containing Sn source, as shown in Fig. 6.4b1, b2. In the composite material, F doping results in stronger interactions between F-SnO_2 and RGO sheets, leading to more uniform and higher loading of F-SnO_2 nanoparticles on RGO sheets. As a result, the LIB cells prepared by using the F-SnO_2@RGO composite as anode material exhibited a specific

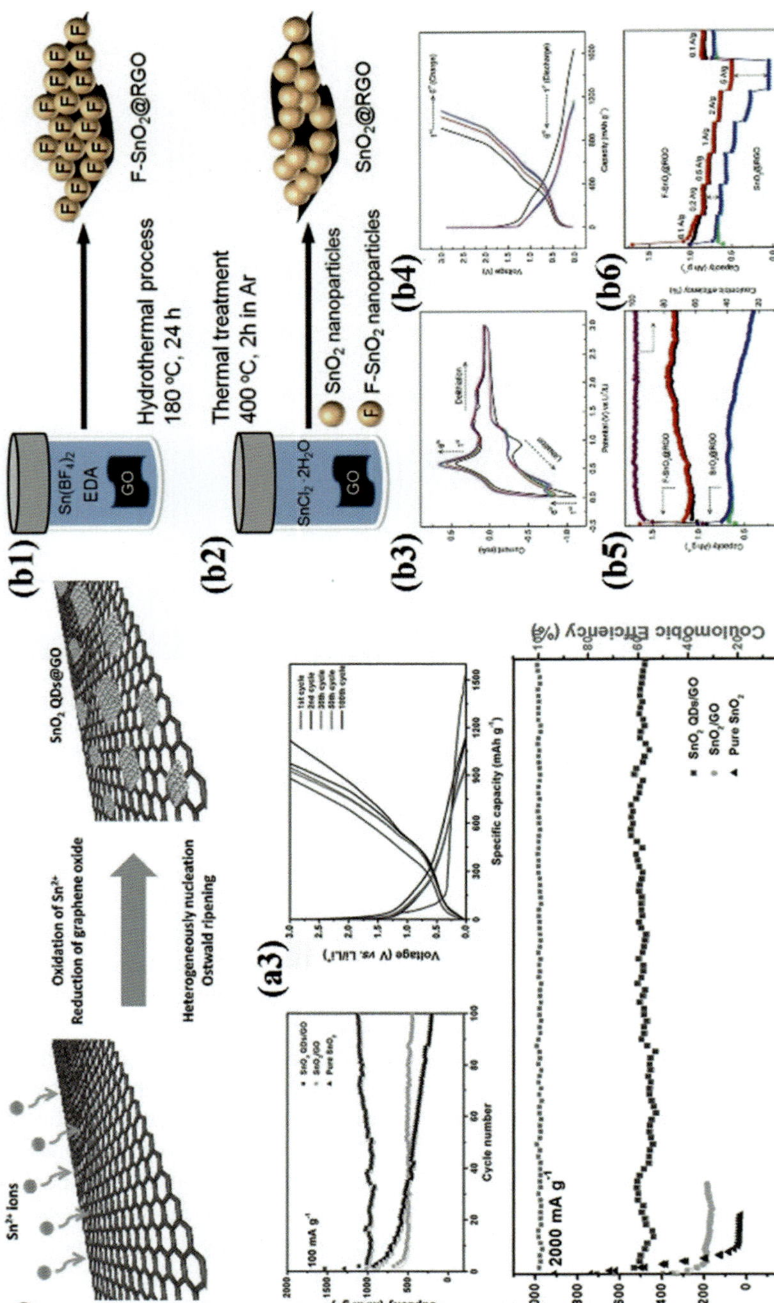

Fig. 6.4 (continued)

capacity of 1277 mAh/ g after 100 cycles. In addition, the F-SnO_2@RGO electrode also displayed remarkable performance such as long-term cycling stability and excellent rate capacity at high charge/discharge current densities, as shown in Fig. 6.4b3–b6.

6.2.2 Tin Sulfides

There is a growing interest in the applications of SnS_2 and SnS, which are inexpensive and environmentally benign and offer various properties of technological interest, especially in batteries field. Two-dimensional (2D) tin-based sulfides have been widely explored as promising electrodes for lithium-ion batteries since their two-dimensional layered structure allows lithium ions to intercalate between layers, as shown in Fig. 6.5 [29]. The structure of SnS_2 is of the cadmium iodide (CdI_2) type with the space group P-3 m1 (n8164) that consists of two layers of close-packed

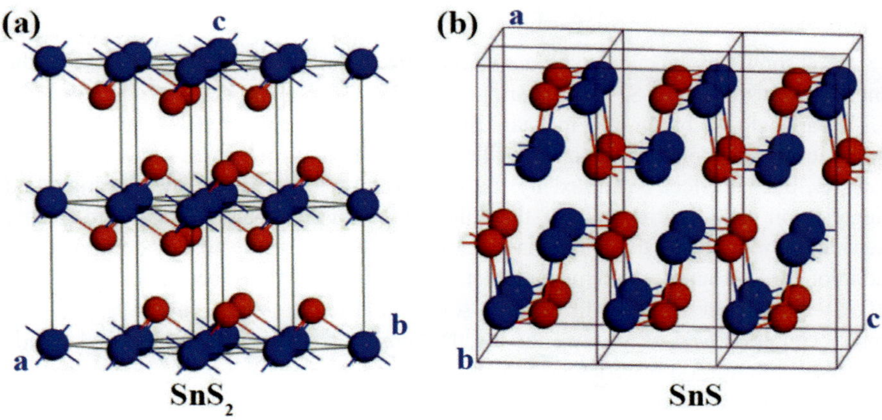

Fig. 6.5 Structures of SnS_2 (**a**) and SnS (**b**) represented by 1 × 2 × 2 and 1 × 2 × 3 supercells, respectively. The big (blue) and smaller (red) spheres represent Sn and S atoms, respectively. The unit cells are delimited by black lines. Reprinted with permission from [29]. Copyright 2016 John Wiley and Sons

Fig. 6.4 SnO_2-graphene composites anode for lithium-ion batteries. (**a1**) Schematic illustration of the formation process of SnO_2 QDs@GO; (**a2**) cycling performance of SnO_2 QDs@GO, SnO_2/GO composite, and SnO_2 particles at 100 mA/g; (**a3**) galvanostatic charge/discharge profiles of the initial two cycles, 30th cycles, 50th cycles, and 100th cycles of SnO_2 QDs@GO at 100 mA/g; (**a4**) high-rate cycling performance and the corresponding Coulombic efficiency of different electrodes; (**b1** and **b2**) schematic illustration of synthesizing the F-SnO_2@RGO and SnO_2@RGO composites; (**b3**) the first six cycles of the cyclic voltammetry (CV) curves of the F-SnO_2@RGO composite; (**b4**) the first six cycles of the galvanostatic charge/discharge profiles; (**b5**) cycling performance of the F-SnO_2@RGO and SnO_2@RGO electrodes; (**b6**) rate capability of the F-SnO_2@RGO and SnO_2@RGO electrodes. Reprinted with permission from [27, 28]. Copyright 2015 John Wiley and Sons and American Chemical Society sulfur anions with tin cations sandwiched between them in an octahedral coordination. The adjacent sulfur layers are bound by weak van der Waals interactions (Fig. 6.5a). The structure of SnS is orthorhombic with the space group Pnma (n862). Tin sulfide (SnS) can be described as a deformed sodium chloride (NaCl) lattice, where Sn and S atoms form layers stacked along the a-axis (Fig. 6.5b). Within the layers, each Sn atom is bonded to three nearest S atoms at 2.63 Å and 2 × 2.67 Å and two more distant S atoms at 3.28 Å. Thus, the appropriate layered structure with swelling tolerant hosting spaces and enhanced guest accessibility would provide enhanced diffusion for Li-ions and lead to the formation of Li-Sn alloy during the cycling and improve the cycling stability of electrode.

This is generally ascribed to the electrochemical reaction mechanism between SnS_2 and lithium-ions, which can be expressed as the following two steps: [30].

$$SnS_2 + 4Li^+ + 4e^- \rightarrow Sn + 2Li_2S \tag{6.3}$$

$$Sn + xLi^+ + xe^- \rightarrow Li_xSn\ (0 \leq x \leq 4.4) \tag{6.4}$$

As is shown by reaction (6.3), active metallic Sn is formed and embedded into the inert lithium sulfide (Li_2S) matrix during the initial discharge process. It is believed that Li_2S can help to buffer the mechanical stress from the electrode materials during subsequent lithiation/delithiation (tin alloying/de-alloying) processes between Sn and $Li_{4.4}Sn$, but its effect is limited by the huge volume variation of the materials and aggregation of Sn particles. Thus, the pulverization of electrode materials and electric separation of active particles are unavoidable, resulting in rapid capacity decay [31]. In addition, SnS_2 is an n-type semiconductor, whose conductivity is located in the range of 10^{-12}–10^{-2} S/cm [32, 33]. In order to solve/alleviate the issues, various approaches have been employed to improve the electrochemical performance of tin-based sulfide anode for lithium-ion batteries.

Su et al. [34] investigated deeply the dynamical process of the multistep reactions using in situ electron microscopy and discover the formation of an intermediate rock salt phase with the disordering of Li and Sn cations after initial two-dimensional intercalation, as shown in Fig. 6.6. The in situ observations and calculations suggest a two-phase reaction nature for intercalation, disordering, and following conversion reactions. In addition, crystalline Sn nanoparticles are well arranged within an amorphous Li_2S "matrix" in this self-assembled framework. This nanoscale framework confines the locations of individual Sn nanoparticles and prevents particle agglomeration during the subsequent cycling processes, therefore providing desired structural tolerance and warranting a sufficient electron pathway [35].

Two-dimensional layered SnS_2 nanoplates were prepared by a facile thermal decomposition method [36]. Figure 6.7 shows their structural features. It can be seen clearly that many SnS_2 nanoplates with hexagonal structure lay flat on the substrate along with the zone axis of [001]. The enhanced electrochemical properties associated with SnS_2 nanoplates can be attributed to their two-dimensional layered characteristics, including finite lateral size and enhanced open edges, which can facilitate Li-ion diffusion through

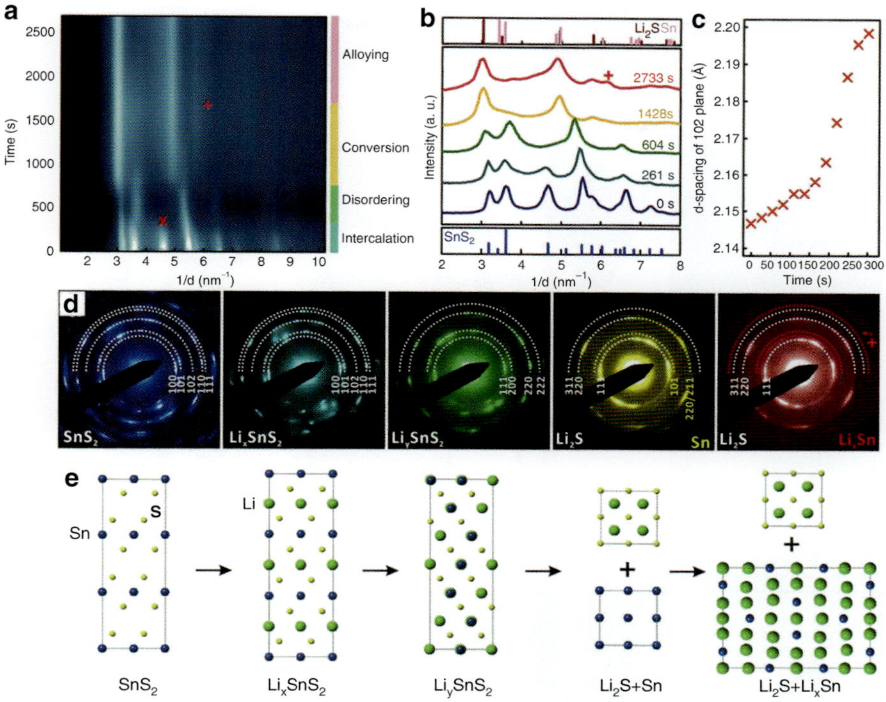

Fig. 6.6 In situ selected area diffraction (SAED) patterns of SnS_2 during lithiation. (**a**) Electron diffraction intensity profile as a function of reaction time during lithiation of SnS_2 (X indicates a 102 diffraction peak, and + denotes the Li-Sn alloy phase); (**b**) radial intensity profiles of diffraction patterns at certain times; (**c**) changes in d-spacing of 102 plane during intercalation; (**d**) diffraction patterns correspond to intensity profiles at (**b**); (**e**) atomic models representing phase evolution during lithiation. Reprinted with permission from [34]. Copyright 2018 American Chemical Society

the active materials and decrease the overvoltage for associated with the Li-Sn alloying reaction, thus driving a faster electrode reaction and providing a higher charge/discharge capacity and excellent cycling stability.

Structural phase transitions can be used to alter the properties of a material without adding any additional elements and are therefore of significant technological value. Guo et al. [37] found that the hexagonal SnS_2 phase can be transformed into the orthorhombic SnS phase after annealing in an argon atmosphere, and the transformed SnS shows enhanced energy storage performance over that of the SnS_2 due to its structural advantages, as shown in Fig. 6.8. Meanwhile, the well-developed two-dimensional nanosheets of SnS and graphene and the precise hierarchical control of various sublayers of the materials are believed to function synergistically to counterbalance the active material's large volume change; therefore, the mechanical, electrical, and electrochemical properties can be stabilized significantly.

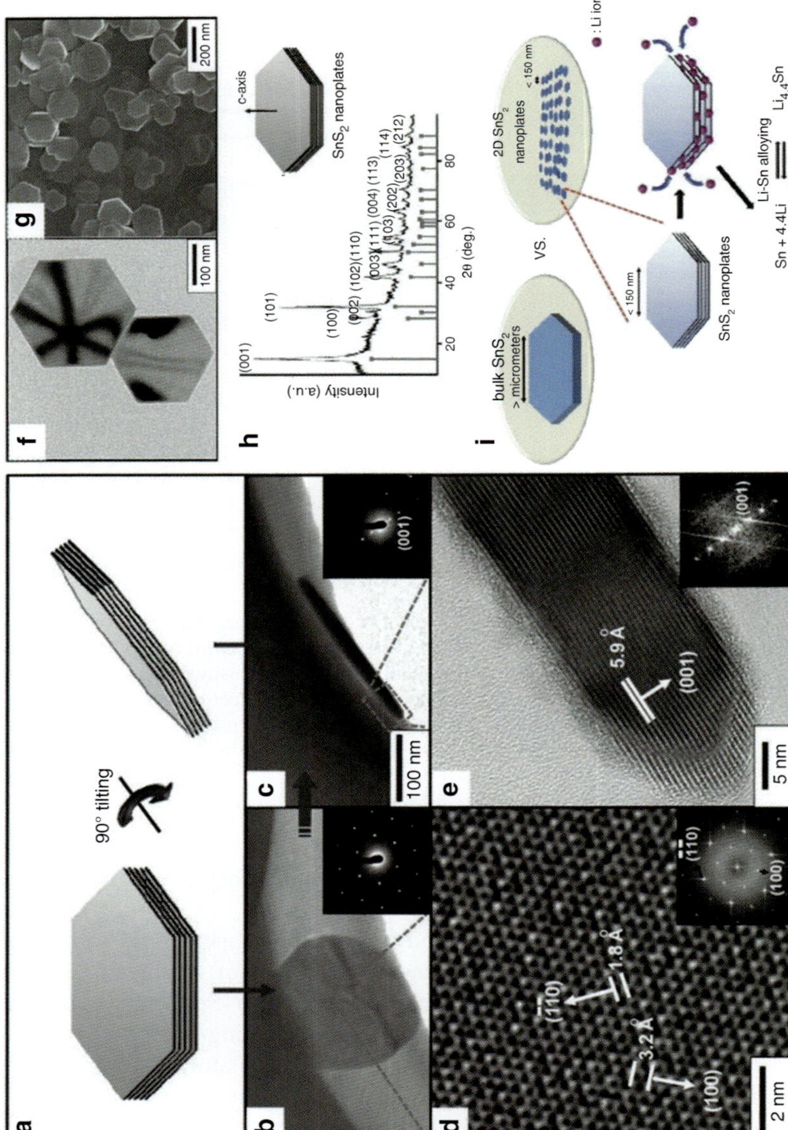

Fig. 6.7 (continued)

To enhance the cycling performance and rate capability, Zhou et al. [38] reported a simple fabrication method involving SnS and SnS_2 particles with a nitrogen-doped reduced graphene oxide (NRGO) support, which exhibits a promising cycling stability compared to other tin sulfide-based electrodes, especially with a specific capacity of 562 mAh/g at the 200th cycle at 0.2A/g. Also, they suggested the non-cycling Li_2S and nitrogen-doped reduced graphene oxide substrates are of essence for the capacity retention of Li alloyed/de-alloyed Sn nanoparticle electrodes. As recognized earlier, they reduce the volume fraction of the lithiated/delithiated tin, and they limit the mechanical stress associated with the expansion upon lithiation and shrinkage upon delithiation.

6.2.3 Tin Selenides

Se, a chalcogen group (n-1d^{10} np^4) element sharing similar chemical properties with sulfur, has a high electronic conductivity (1×10^{-3} S/m) comparing with that of S (5×10^{-28} S/m) and outstanding volumetric capacity as the electrode materials for Li-Se [39, 40]. Like tin sulfides (SnSe, $SeSe_2$), tin selenides also have a hexagonal phase and layered structural motif which are an important class of lithium storage materials because lithium-ions can be inserted and stored between the weakly interacting layers. However, as mention above, the main difficulties for using tin selenides are their dramatic volume expansion and poor cyclic performance.

Choi et al. [41] prepared $SnSe_2$ nanoplate-graphene composites through a facile method. When it was applied as anode materials in lithium-ion batteries, promising storage performances were obtained, as shown in Fig. 6.9. Figure 6.9a, b show typical TEM images of the obtained materials, basically a two-dimensional morphology with a thickness of 20–25 nm. High-resolution transmission electron microscopy (HRTEM) displays the crystal planes of the plates. According to the analysis of the side view of the nanoplates, the interplane distance was 0.61 nm which corresponds to the (001) crystal plane of hexagonal $SnSe_2$ (Fig. 6.9c). The (110) and (100) crystal planes of $SnSe_2$ with 0.29 and 0.19 nm interplanar distances were dominantly observed in the top view of the plates (Fig. 6.9d). Figure 6.9f suggests that $SnSe_2$/RGO shows a promising storage performance superior to $SnSe_2$ nanoplates or graphene alone.

Fig. 6.7 Structural features of SnS_2 nanoplates. (**a**–**c**) Transmission electron microscope (TEM) images, obtained by rotating the TEM holder stage 90°; (**d**) high-resolution TEM (HRTEM) image of the flat view; (**e**) the side view. The inset is the fast Fourier transform (FFT) images corresponding to placement parallel and with the perpendicular to the TEM substrate; (**f**) TEM image and (**g**) field emission scanning electron microscope (FESEM) image of SnS_2 nanoplates. (**h**) X-ray diffraction (XRD) patterns. These patterns are matched with the reference (vertical line, Joint committee on powder diffraction standards (JCPDS card #23–677); (**i**) A schematic of lithiation processes for bulk versus nanoplates. Facile Li-Sn alloying processes are observed for laterally confined nanoplates. Reprinted with permission from [36]. Copyright 2008 John Wiley and Sons

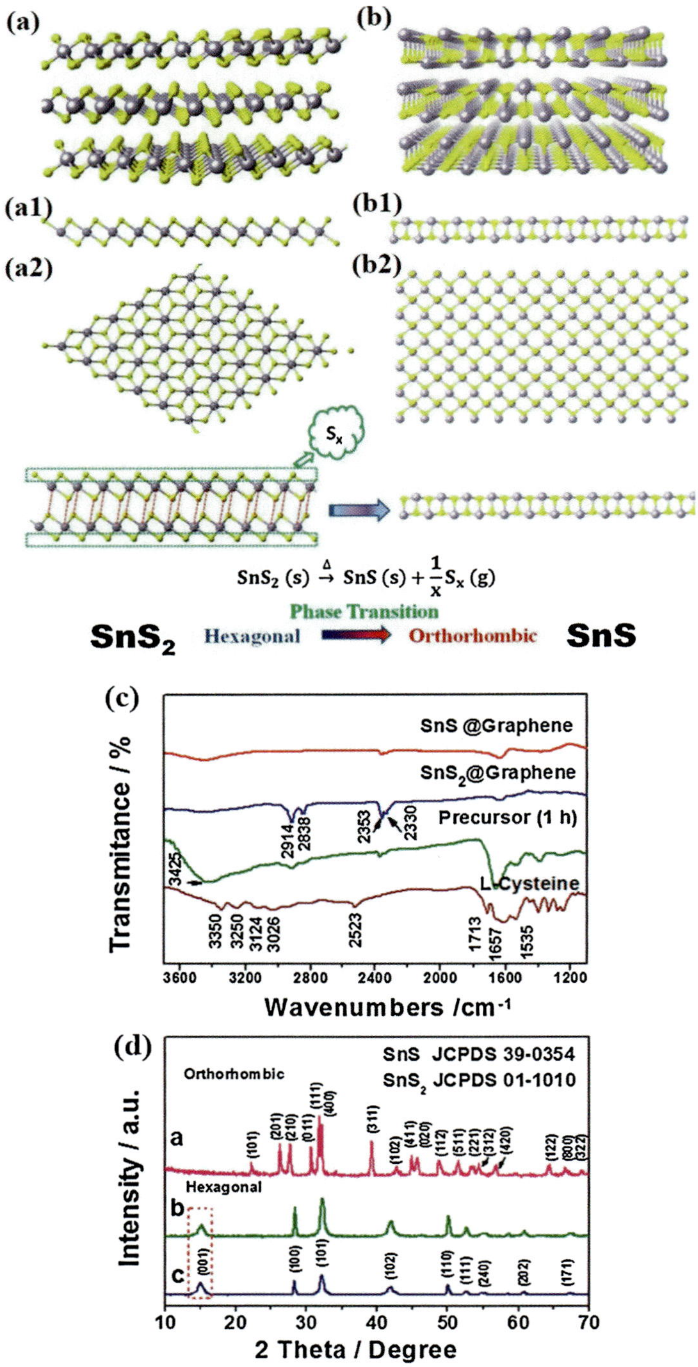

Fig. 6.8 (continued)

It was also found that combining SnSe with SnO_2 can develop superior electrochemical performance [42]. SnSe/SnO_2@Graphene heterostructure composite was prepared to form a three-dimensional (3D) hierarchical architecture, in which the SnO_2 nanospheres are homogeneously dispersed and wrapped into the graphene matrix, followed by the selenium treatment to obtain SnSe/SnO_2 heterostructure, as shown in Fig. 6.10. According to electrochemical test, SnSe/SnO_2@Gr nanocomposite yields an initial discharge of 2324 mAh/g and charge of 1185 mAh/g, which is far superior to the theoretical specific capacity of SnO_2 (1494 mAh/g) and SnSe (780 mAh/g).

6.2.4 Ternary Tin-Based Chalcogenides

Ternary tin-based chalcogenides, an emerging class of new materials, are mainly used as semiconductor due to its great carrier mobility and capability of bandgap engineering. Moreover, being earth-abundant, low cost, and environmentally friendly, tin chalcogenides are desirable for sustainable optoelectronic devices. It was found that two-dimensional ternary tin-based chalcogenides, including SnSeS and $SnSe_{0.5}S_{0.5}$, can also be a promising candidate as an electrode material in energy storage devices due to their characteristic layered structure [43–46]. In addition, pioneering research reported that partial S doping in selenides could optimize the electrical conductivity via increasing charge carrier (hole) concentration. At the same time, the surface sulfur donor states could be enhanced with the sulfidation on the selenides [47, 48]. As shown in Fig. 6.11, the schematic of various ternary tin-based chalcogenides was exhibited.

Tang et al. [44] successfully synthesized $SnSe_{0.5}S_{0.5}$/C nanocomposites by a rapid polyol method and followed by a hydrothermal treatment. Benefiting from the stable nanostructure and good conductivity, the nanocomposites can effectively restrict the large volume change and help the active materials to fully complete the electrochemical reaction, thus demonstrating its potential to be an excellent anode for lithium-ion batteries. This composite electrode exhibited high reversible discharge capacity, e.g., 989 mAh/g at 0.1 A/g and 625 mAh/g after 1000 cycles at 0.5 A/g and excellent rate performance 389 mAh/g at 5 A/g as a lithium-ion batteries anode (Fig. 6.12a–e). Then they also deeply analyze its electrochemical reaction mechanism through in situ XRD, as shown in the Fig. 6.12f–h

Fig. 6.8 Schematic diagram (bottom) illustrating the phase transition process between SnS_2 and SnS. The crystal structures of hexagonal SnS_2 (**a**, **a1**, **a2**) and orthorhombic SnS (**b**, **b1**, **b2**) are shown in different sectional views (atom color code: purple, tin; yellow, sulfur); (**c**) Fourier transform infrared (FTIR) spectra of SnS@graphene, SnS_2@graphene, SnS_2@graphene precursor after 1 h hydrothermal reaction, and L-cysteine; (**d**) X-ray diffraction (XRD) patterns of (**a**) SnS@graphene, (**b**) SnS_2@graphene architectures, and (**c**) pristine SnS_2. The XRD pattern of the SnS_2@graphene composite is similar to that of pristine SnS_2 [hexagonal SnS_2 (JCPDS No. 01-1010)], and the intensity of the (001) plane is much lower than in the case of pristine SnS_2, indicating that growth of the SnS_2 was inhibited in the hybrid due to the intervention of graphene and that a layered/sheet-like structure of SnS_2 was formed in the hybrid architecture. Reprinted with permission from [37]. Copyright 2014 American Chemical Society

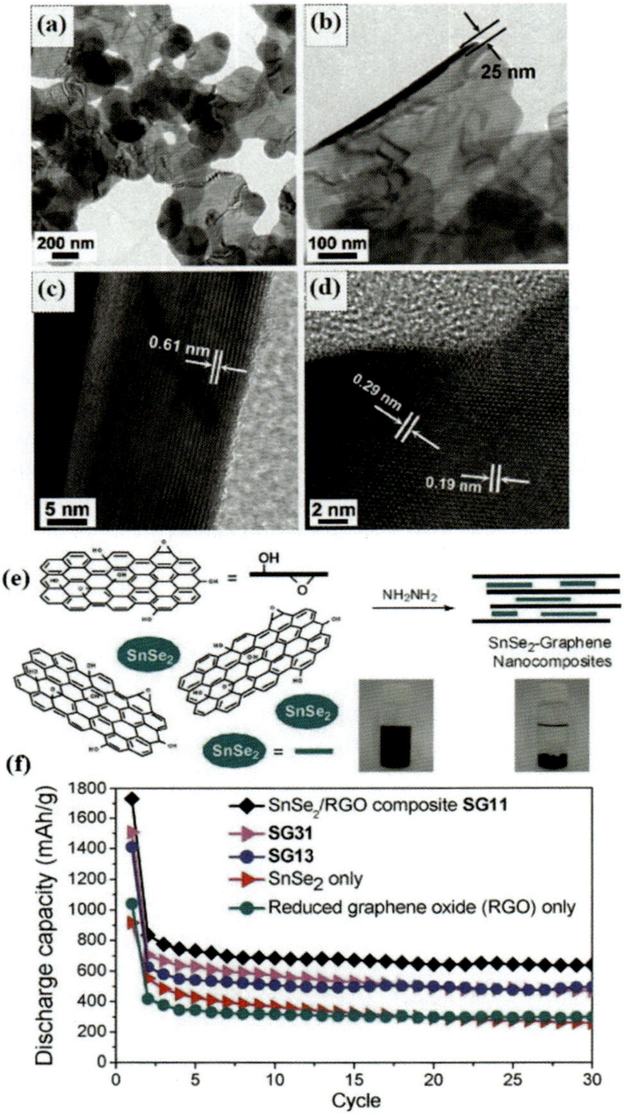

Fig. 6.9 Typical transmission electron microscope (TEM **a**, **b**) images of $SnSe_2$ nanoplates and high-resolution TEM images of side (**c**) and top views (**d**); (**e**) reaction scheme for $SnSe_2$-graphene nanocomposites made using a 1:1 weight ratio of $SnSe_2$ nanoplates and graphene oxide. Photographs of a mixture of the dispersed $SnSe_2$ nanoplates and graphene oxide and the formed nanocomposites through treatment with hydrazine; (**f**) discharge cycling performance of electrochemical cells made of reduced graphene oxide only, $SnSe_2$ nanoplates only, and $SnSe_2$ nanoplate-graphene composites (SG13, SG11, SG31) at a current density of 40 mA/g. Reprinted with permission from [41]. Copyright 2011 Royal Society of Chemistry

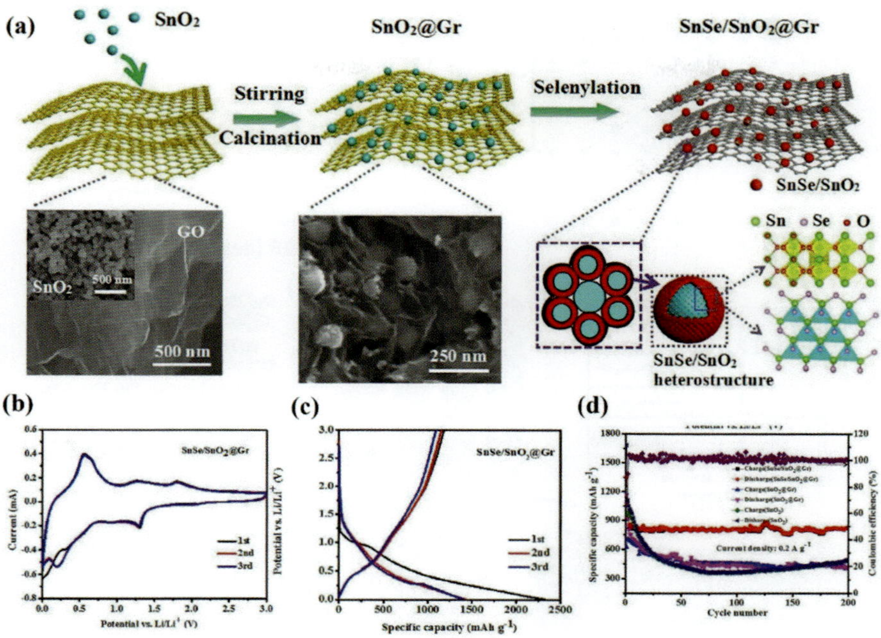

Fig. 6.10 (**a**) Schematic illustration of the formation of SnSe/SnO_2@Gr architecture. The electrochemical performance of SnSe/SnO_2@Gr electrode materials: (**b**) cyclic voltammetry (CV) curves of the first three cycles; (**c**) discharge-charge curves at 0.1 A/g in the first three cycles; (**d**) cycling performance of the SnSe/SnO_2@Gr nanocomposite, SnO_2@Gr and SnO_2 at a current density of 0.2 A/g. Reprinted with permission from [42]. Copyright 2018 Elsevier

[50]. Combining electrochemical analysis and in situ XRD results, the Li^+ storage mechanism in the $SnSe_{0.5}S_{0.5}$ anode could be summarized in three processes – intercalation, conversion, and alloying reactions:

Intercalation (2.0–0.95 V):

$$SnSe_{0.5}S_{0.5} + x\,Li^+ + xe^- \rightarrow Li_xSnSe_{0.5}S_{0.5} \tag{6.5}$$

Conversion (0.95–0.38 V):

$$Li_xSnSe_{0.5}S_{0.5} + x\,Li^+ + xe^- \rightarrow Li_xSnSe + Li_xSnS \tag{6.6}$$

Alloying (0.38–0.01 V):

$$Li_xSnSe + Li_xSnS + x\,Li^+ + xe^- \rightarrow Li_xSn + Li_xSe + Li_xS \tag{6.7}$$

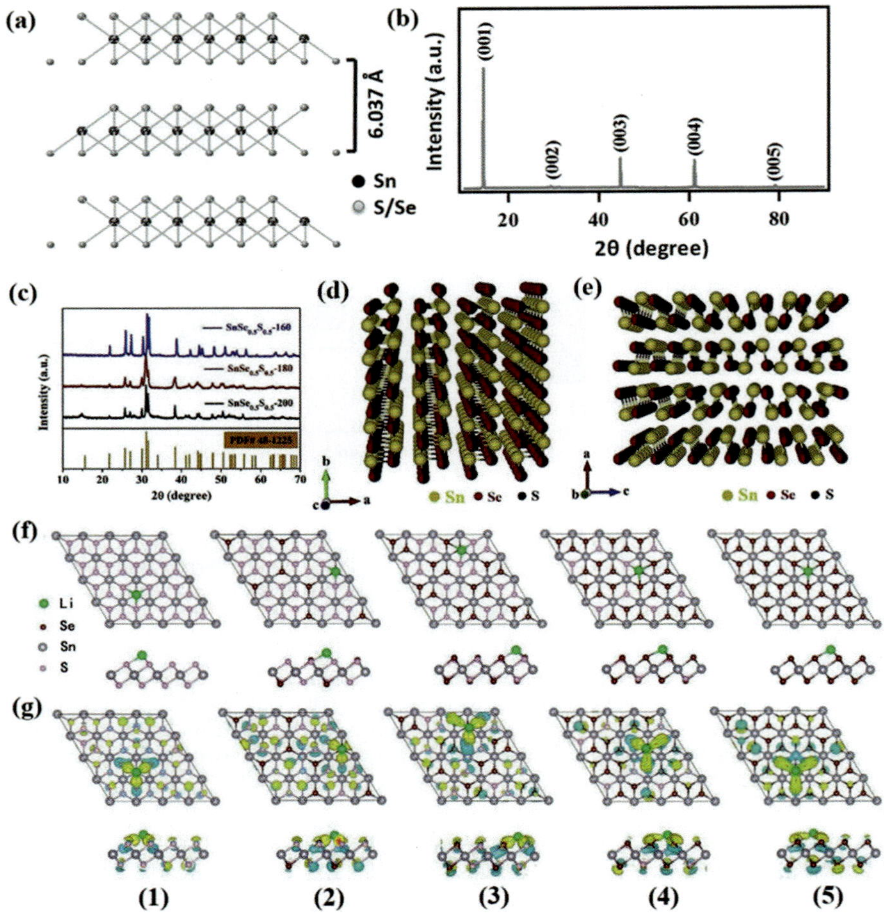

Fig. 6.11 (**a**) Hexagonal crystal structure of SnSSe; (**b**) single crystal X-ray diffraction (XRD) pattern of SnSSe; (**c**) XRD patterns of $SnSe_{0.5}S_{0.5}$ at different reaction temperature; (**d**, **e**) the schematic illustrations of the $SnSe_{0.5}S_{0.5}$ laminar structure; (**f**) the most stable lithium adsorption sites and (**g**) corresponding charge density difference plots for the optimized monolayer $SnS_{2(1-x)}Se_{2x}$ alloys with x = 0, 0.25, 0.5, 0.75, 1 marked with (1), (2), (3), (4), and (5), respectively. The atoms with the color of pink, maroon, gray, and green represent the S, Se, Sn, and Li atoms, respectively. The yellow and blue areas represent electron losses and gains. Reprinted with permission from [46, 49, 50]. Copyright 2018 and 2016 Elsevier and John Wiley and Sons

6.3 Molybdenum-Based Chalcogenides

Molybdenum and tungsten dichalcogenides (ns^1 $n\text{-}1d^5/ns^2$ $n\text{-}1d^4$, MX_2 (M = Mo, X = S and Se)) belong to the two-dimensional transition metal dichalcogenide (TMD, ns^1 nd^5) group. The unique layered structure makes them suitable for wide applications as photocatalysts and solid lubricants and in electrical energy storage devices.

6.3.1 Molybdenum Sulfides

According to the different layer-stacking sequences, molybdenum (ns^1 $n\text{-}1d^5$) sulfides or selenides, MoS_2 (as well as $MoSe_2$), can be divided into the two layer-stacked hexagonal polymorph (2H) molybdenum sulfide [2H-MoS_2], the one layer-stacked trigonal (1T) molybdenum sulfide [1T-MoS_2], and the three layer-stacked rhombohedral (3R) molybdenum sulfide [3R-MoS_2], as shown in Fig. 6.13 [14]. In such a layered structure, Mo and S atoms are covalently bonded to form two-dimensional S-Mo-S trilayers, and the adjacent planes are stacked by van der Waals interactions, which facilitate intercalation of the lithium-ions. However, the poor electronic and ionic conductivity between adjacent S-Mo-S sheets and the large volume expansion during the electrochemical process are further obstacles to the practical application of MoS_2 as an electrode material in lithium-ion batteries. In addition, the shuttle effect of "polysulfides" and loss of active materials during the charge/discharge also influence its electrochemical performance in lithium-ion batteries. Thus, various approaches have been carried out to improve the conductivity of MoS_2 and alleviate the volume change during the discharge/charge process.

Understanding the structure and phase changes associated with conversion-type materials is key to optimizing the electrochemical performance in Li-ion batteries. For Li-ion batteries, MoS_2 offers almost threefold improvement in capacity compared to currently used graphite anodes. However, in MoS_2 electrodes, from ~1.1 V versus Li^+/Li onward, Li^+ begin to react with sulfur atoms, and MoS_2 gradually changes from a trigonal prismatic (2H-MoS_2) to an octahedral (1T-Li_xMoS_2) phase [51]. After this phase transition, conversion reactions are intensified at ~0.5 V versus Li^+/Li, causing MoS_2 to fragment and disintegrate into LiS_2 and Mo nanoparticles. When these reactions coincide with electrolyte decomposition, most of the electrode turns into a gel-like matrix containing Mo nanoparticles. The overall reaction typically leads to low capacity and poor columbic efficiency. Thus, George et al. [52] found new sulfur-enriched intermediates that progressively insulate MoS_2 electrodes and cause instability from the first discharge cycle by combining ab initio density functional theory (DFT) simulation with electrochemical analysis and further confirm the electrochemical reaction mechanism of Li-MoS_2 batteries, as shown in Fig. 6.14. In the discharge curves, the sloping plateau at ~1.8 V corresponds to the onset of Li-ion insertion in MoS_2 (according to $MoS_2 + xLi^+ + xe^- \rightarrow Li_xMoS_2$). At ~1.2 V versus Li^+/Li, Li^+ begin to react with S atoms to form a LiS_2 matrix in which Mo metallic clusters are embedded (according to $2Li^+ + S + 2e^- \rightarrow Li_2S$) [53]. Below 0.5 V versus Li^+/Li is a deep conversion zone along with the formation of solid electrolyte interface (SEI). The MoS_2-double-walled carbon nanotube (DWCNT) electrodes showed a specific capacity of ~1400 mAh/g, while the MoS_2-Super P delivered ~900 mAh/g at 100 mA/g.

Zhang et al. [55] developed a simple one-pot hydrothermal route for the fabrication of MoS_2/nitrogen-doped graphene composite (MoS_2/N-G). In this composite, MoS_2 nanosheets are well grown onto the N-doped graphene, generating a unique veillike hybrid nanostructure (Fig. 6.15). The in situ nitrogen doping derived from N-containing polydopamine (PDA, $[\text{-}(C_{16}H_{12}N_2O_4)\text{-}]_n$) effectively increases the conductivity of the whole electrode. Meanwhile, the strong coupling between MoS_2 and nitrogen-doped

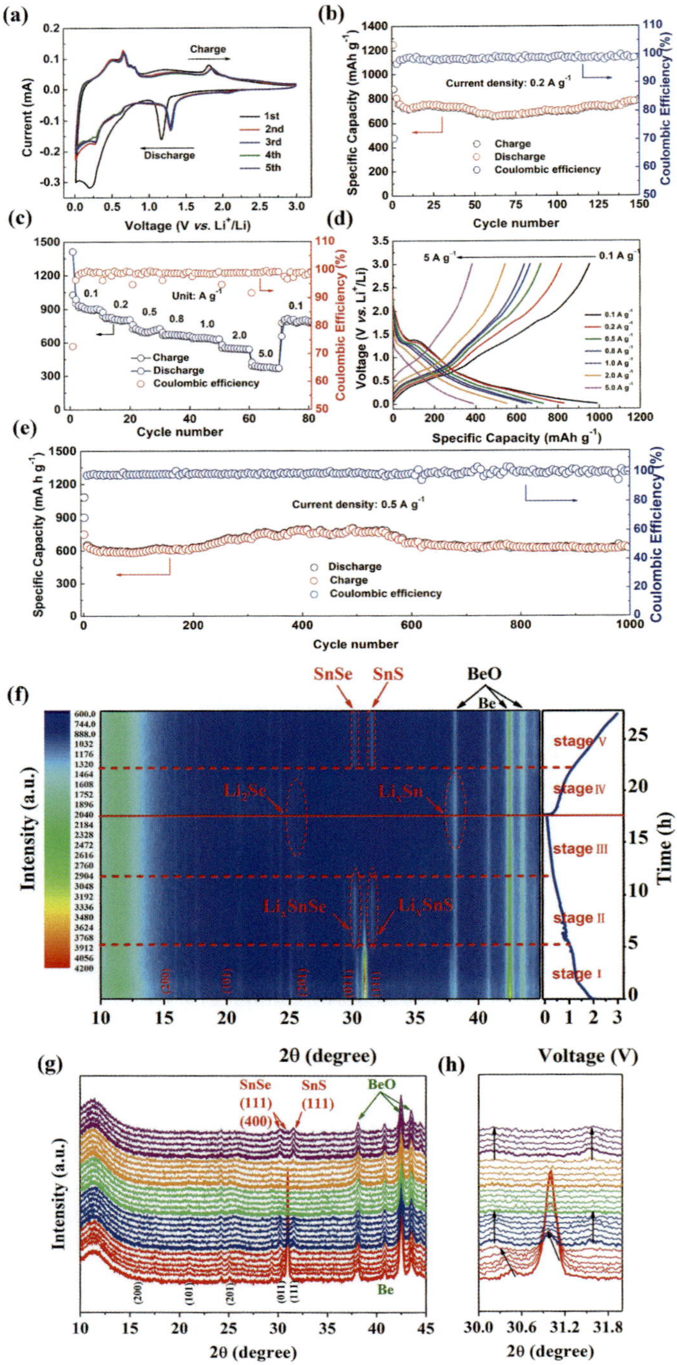

Fig. 6.12 (continued)

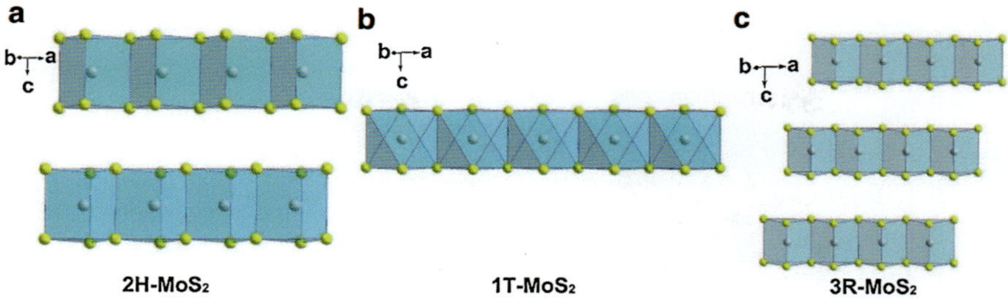

Fig. 6.13 The crystal structures of (**a**) 2H-MoS_2, (**b**) 1T-MoS_2, and (c) 3R-MoS_2. Reprinted with permission from [14]. Copyright 2017 John Wiley and Sons

graphene improves the cycling stability of the electrode material. In addition, two-dimensional MoS_2 ultrafine nanosheets aligned with graphene reduce the diffusion path of Li-ions. These contributions boost the electrochemical performance of MoS_2/N-G as the anode material for LIBs. The as-prepared MoS_2/N-G composite manifests a highly reversible capacity, superior rate performance, and superior cycling stability.

Wen et al. [56] demonstrated an effective and simple method to synthesize a three-dimensional network architecture with reduced graphene oxide (RGO) cross-linked hollow carbon spheres (HCS) as a scaffold of few-layer MoS_2 nanosheets, in which MoS_2 nanosheets are in situ grown on the surface of both RGO and HCS, forming a three-dimensional network architecture, as shown in Fig. 6.16. In this composite, MoS_2 nanosheets are tightly overlying on interconnected RGO/HCS conductive networks, which not only effectively hinders the aggregation of MoS_2 nanosheets and restacking of graphene nanosheets but also buffers the volumetric expansion during the lithium-ion insertion/extraction processes, being beneficial for improving the structural stability and cycling ability. Furthermore, the three-dimensional porous scaffolds of RGO/HCS are intimately face-to-face in contact; in this way, the few-layer MoS_2 sheets could greatly enhance the electrical conductivity and facilitate the electrolyte/ion transport, resulting in fast electrochemical reaction kinetics and much improved reversible capacity and rate capability. In addition, the expanded interlayer spacing could be able to compromise the stress caused by lithium-ion insertion and extraction and thus benefit fast ion intercalation and provide more lithium-ions' storage sites. Finally, the three-dimensional honeycomb-like network structures may enhance the affinity of the electrolyte and increase the contact sites between MoS_2 active materials and the electrolyte, leading to an enhanced Li^+ accessibility and reduce the Li-ion diffusion length.

Fig. 6.12 Electrochemical performance of $SnSe_{0.5}S_{0.5}$/C nanocomposites for lithium-ion batteries: (**a**) CV curves at the scan rate of 0.1 mV/s; (**b**) cycling performance at 0.2 A/g; (**c**) rate performance at the different current density from 0.1 to 5.0 A/g; (**d**) charge/discharge profiles at the various rate; (**e**) cycling performance and Coulombic efficiency for 1000 cycles at 0.5 A/g. In situ XRD analysis of $SnSe_{0.5}S_{0.5}$ during the first charging and discharging process: (**f**) two-dimensional Contour plots of in situ XRD patterns and corresponding discharge-charge profiles; (**g**) line plot of in situ XRD patterns; (**h**) selected 2θ regions plot corresponding with Fig. 6.12g. Reprinted with permission from [44, 50]. Copyright 2017 and 2018 Elsevier

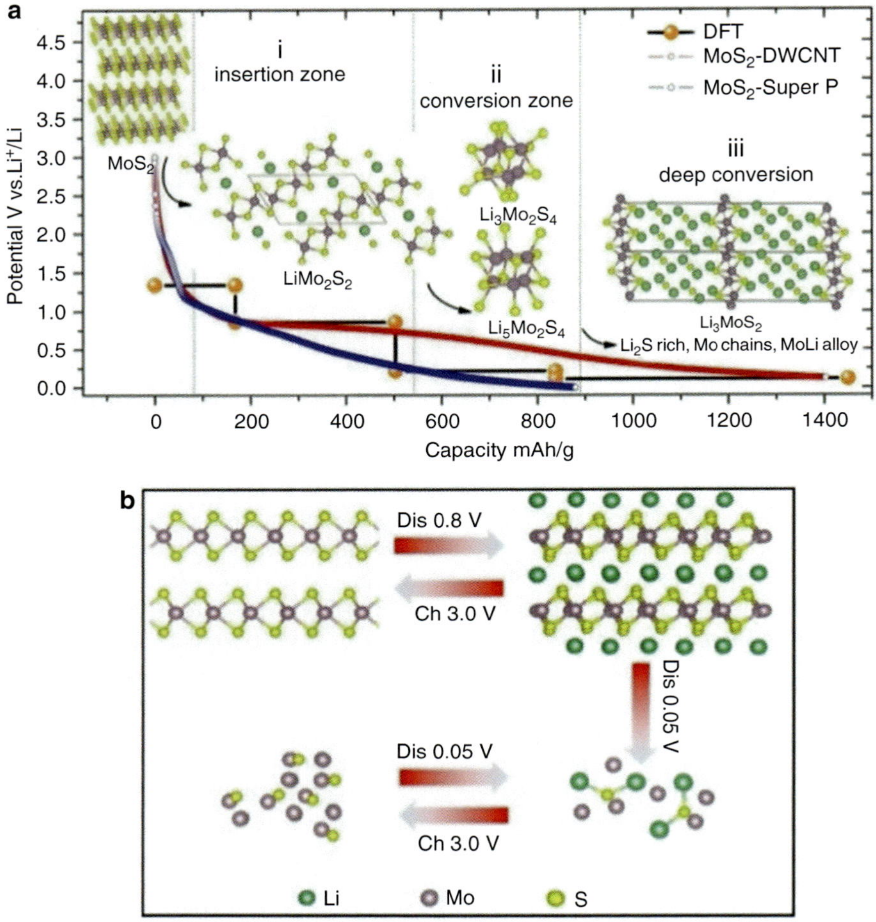

Fig. 6.14 (**a**) Representative discharge profiles from the first cycle of MoS_2-double-walled carbon nanotubes (DWCNT) and MoS_2-Super P electrodes together with the average voltages calculated from DFT energies, relative to Li^+/Li and assuming each reaction is two phase: (i) for the metastable *Cmcm* symmetry $LiMo_2S_2$ phase, the sheets are buckled. (ii) For MoS_2, two clusters were found in a sea of Li, the clusters $Li_3Mo_2S_4$ and $Li_5Mo_2S_4$, with the sulfur being pulled away from Mo and toward the Li. (iii) The $P2_1/c$ symmetry Li_3MoS_2 leading to segregated Li_2S regions; (**b**) the proposed electrochemical reaction mechanism of the MoS_2 electrode for lithium-ion batteries. Reprinted with permission from [52, 54]. Copyright 2016 and 2018 American Chemical Society

6.3.2 Molybdenum Selenides

As mentioned above, $MoSe_2$ has a similar structure to MoS_2 which are identified as three different types (unit cells), namely, one-layer tetragonal (1 T), two-layer hexagonal (2H), and three-layer rhombohedral (3R). The 2H phase MoS_2 exhibits

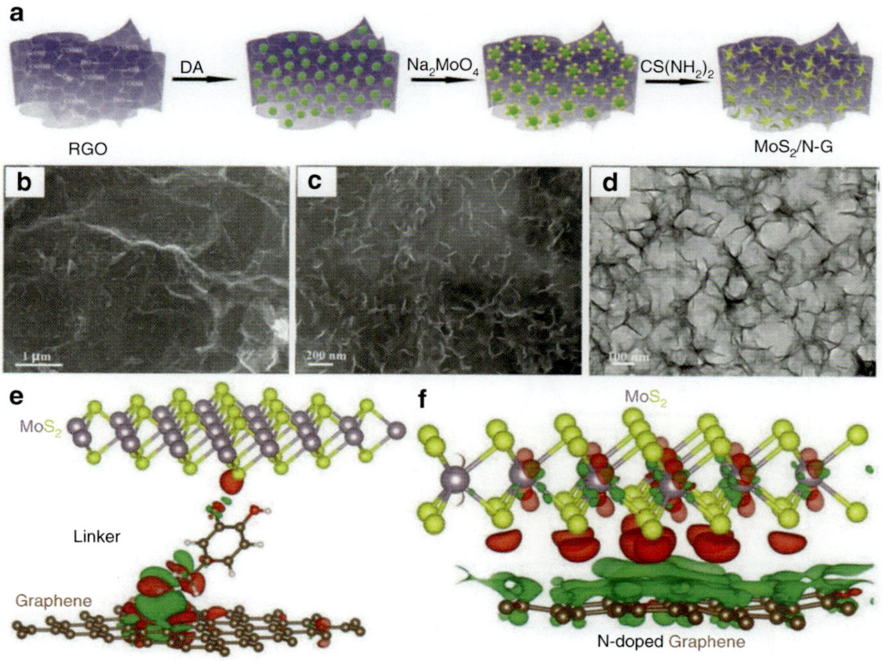

Fig. 6.15 (**a**) Schematic illustration of the preparation process; (**b** and **c**) scanning electron microscope (SEM) images; (**d**) transmission electron microscope (TEM) images; (**e**) interface charge density distribution by density functional theory (DFT) calculation. 3D charge density difference plots for the MoS_2/PDA-graphene system (**e**), and MoS_2/N-doped graphene system (**f**): red and green iso-surfaces represent charge accumulation and depletion in the 3D space with isovalues of 1.0 and 0.4 e/nm^3 for (**e**) and (**f**), respectively. Reprinted with permission from [55]. Copyright 2018 John Wiley and Sons

a monolayer band gap of ~1.9 eV, which reduces its conductivity and thus is not immediately attractive to be used as an electrode material for energy storage. In the recent, $MoSe_2$ has attracted researchers' attention owing to its adjustable the band gap and alterable the intrinsic conductivity.

Yang et al. [57] improved the conductivity by engineering the disorder of nanophase $MoSe_2$ with a rapid oxidation process. As shown in Fig. 6.17, by HRTEM and SAED analysis, the nanosheets' structure and their designed heterostructure are verified. The prepared $MoSe_2$ enables a significantly improved Li storage property and stable cycling performance.

Han et al. [58] successfully synthesized tunable $MoSe_2$ nanosheets coated homogeneously over carbon nanotubes (CNTs) via a simple solvothermal calcination process. As shown in Fig. 6.18, after a series of characterization, the three-dimensional core/shell coaxial nanotubular sponge design has the advantage

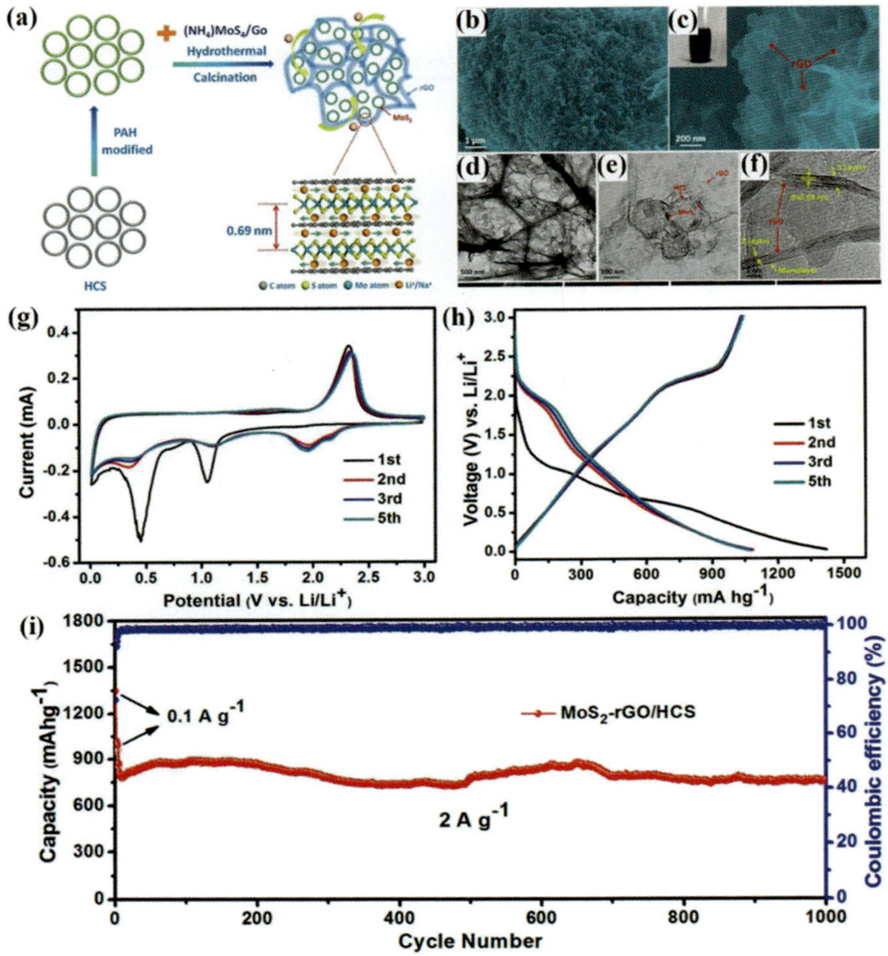

Fig. 6.16 (**a**) Schematic illustration of the formation process of MoS_2-RGO/HCS; (**b** and **c**) field emission scanning electron microscope (FESEM) images; (**d** and **e**) transmission electron microscope (TEM) images; (**f**) high-resolution (HR) TEM image; (**g**) representative cyclic voltammetry (CV) curves of the MoS_2-RGO/HCS electrode at a voltage range of 0.01 to 3.0 V and a scan rate of 0.1 mV/s; (**h**) charge-discharge voltage profiles of the MoS_2-RGO/HCS electrode at a current density of 0.1 A/g; (**i**) cycling performance and Coulombic efficiency of MoS_2-RGO/HCS for 1000 cycles at the current density of 2.0 A/g. Reprinted with permission from [56]. Copyright 2018 American Chemical Society

of being flexible, easily tunable together with an intense compaction. Thus the active material is able to yield a significantly improved areal capacity due to a large number of active sites for reversible charge storage with pathways via CNTs for a fast charge transport.

6.4 Iron-Based Chalcogenides

In addition to metal-based chalcogenide with a layered structure, non-layered MXs are also popular, especially iron-based chalcogenide (ns^2 $n\text{-}1d^6$). Due to iron's abundance, low cost, and nontoxicity, iron-based material has been extensively studied. When used as the anode for lithium-ion batteries, it shows much higher theoretical capacity than the commercial graphite electrode (theoretical capacity: 372 mAh/g). However, the rapid capacity fade seriously hinders its practical applications. A huge number of efforts had been devoted to solving the issues by synthesis methods, and various morphology or composition control techniques, designing nanostructures, carbon encapsulation, optimization of electrolytes, and selection of voltage ranges [59].

6.4.1 Iron Sulfides

Iron sulfides have inexpensive components and high capacity and are abundant in nature which results in being widely used in energy storage batteries. There are various binary iron sulfide (FeS, FeS_2, Fe_3S_4, and Fe_7S_8)-type electrodes for batteries.

The pyrite FeS_2 is a natural mineral, which is extensively used to obtain sulfuric acid. Its low price and the theoretical capacity of 894 mAh/g make it very competitive as a commercial electrode material for battery devices. As early as the 1990s, Energizer Corporation first produced commercial primary FeS_2/Li 18,650-type batteries. Unfortunately, the poor cycling performance of FeS_2 seriously hinders its wide application in rechargeable batteries. It was reported that the intercalation (Eq. 6.8) and conversion-type reaction (Eq. 6.9) coexists in the first cycle [60]. Generally, the latter reaction brings severe volume change to the electrode materials and sluggish kinetics for the reconstruction of FeS_2 at charge process, resulting in the inferior cyclic performance [61].

$$FeS_2 + xLi^+ + xe^- \rightarrow Li_xFeS_2 (0 < x < 2, 1.5\tilde{}1.7\ V) \quad (6.8)$$

$$Li_xFeS_2 + (4 - x)Li^+ + (4 - x)e^- \rightarrow Fe + 2Li_2S\ (\tilde{}1.5\ V) \quad (6.9)$$

In order to confirm the local structure evolution of this relatively reversible material, Seshadri et al. [62] employed operando studies to understand the conversion material FeS_2 (Fig. 6.19). X-ray absorption spectroscopy, pair distribution function analysis, and first-principles calculations of intermediate structures shed light on the mechanism of charge storage in the Li-FeS_2 system, with some general principles emerging for charge storage in metal-based chalcogenide materials. They also focused on the second and later charge/discharge cycles and find small, disordered domains that locally resemble iron (Fe) and Li_2S at the end of the first discharge. Upon charge, this is converted to a Li-Fe-S composition whose local structure reveals tetrahedral coordination for Fe. With the continued charge, this ternary composition displays insertion/extraction behavior at higher potentials and lower Li content. The finding of hybrid modes of charge storage, rather than simple

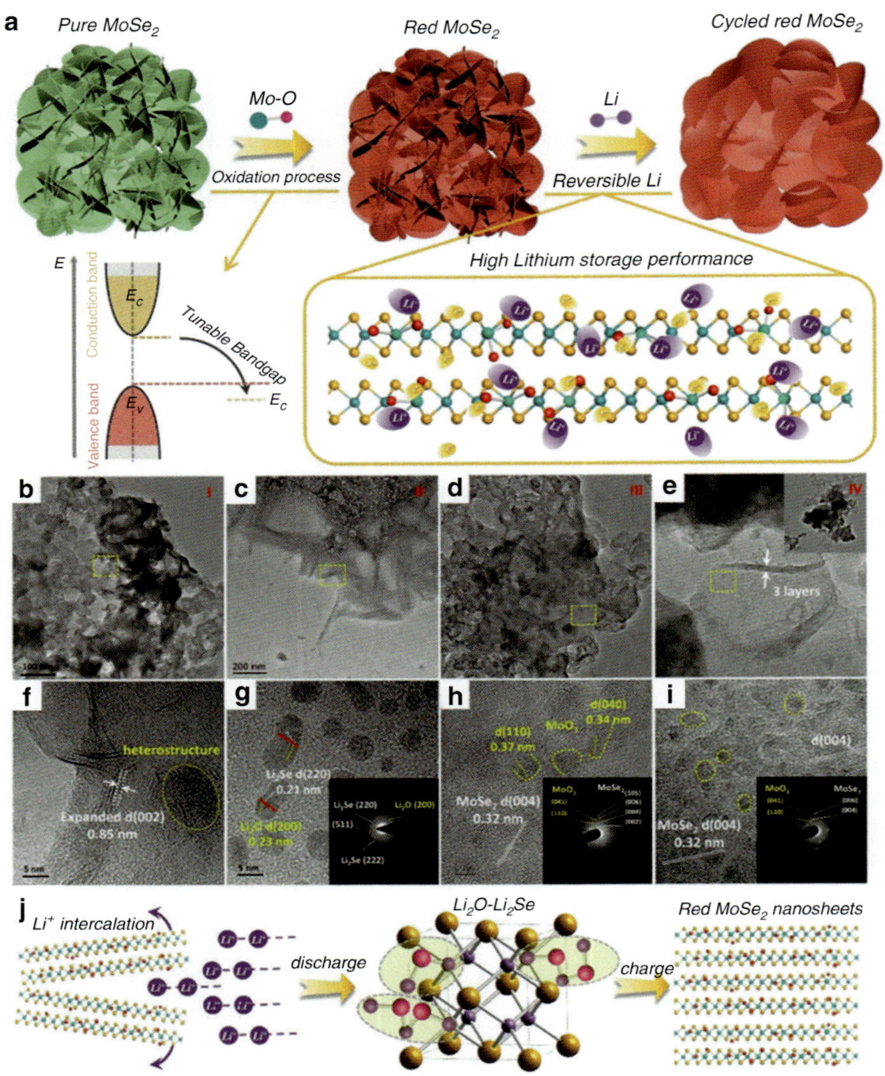

Fig. 6.17 (**a**) Schematics depicting the synthesis of red $MoSe_2$ with a tunable bandgap and its application for Li-ion batteries; (**b**–**i**) transmission electron microscope (TEM) and high-resolution (HR) TEM images of cycled red $MoSe_2$ electrodes; (**j**) schematics depicting the discharge and charge process of the red $MoSe_2$ anode. Reprinted with permission from [57]. Copyright 2018 American Chemical Society

conversion, points to the important role of intermediates that appear to store charge by mechanisms that more closely resemble intercalation.

Shen et al. [63] synthesized the composite consisting of RGO, and well-dispersed FeS_2 microparticles (FeS_2/RGO) were prepared by a one-pot

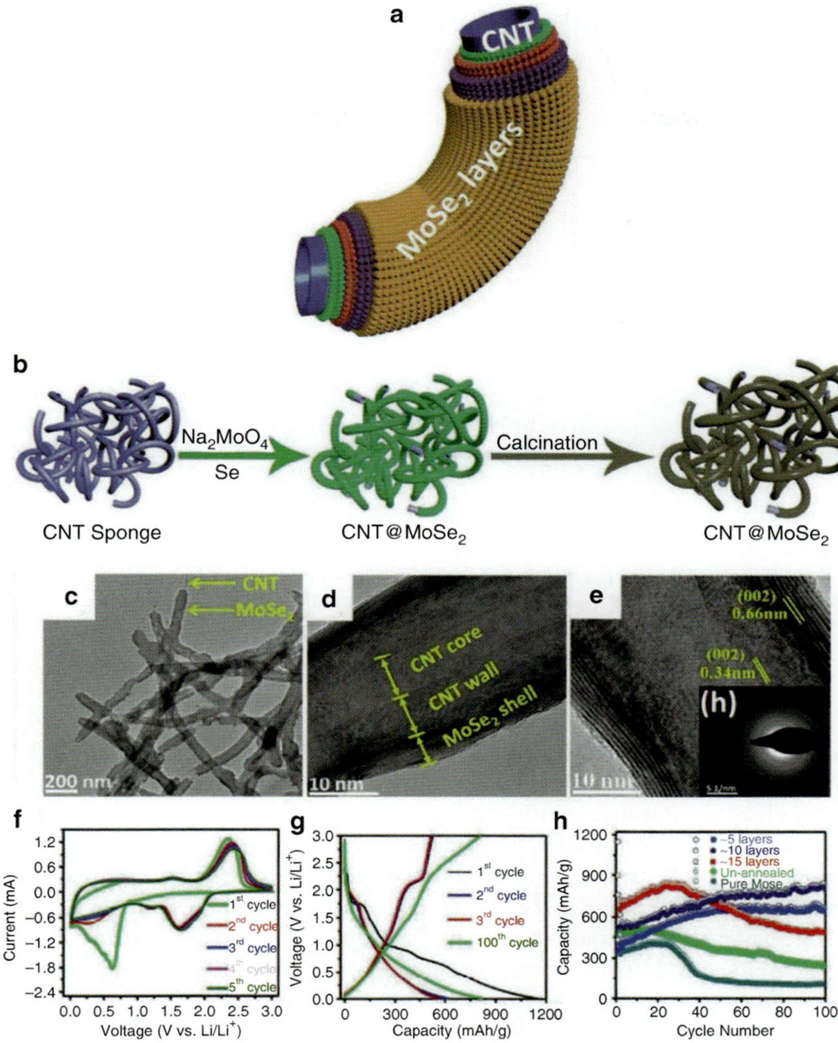

Fig. 6.18 (**a**) Schematic representation of different layers of $MoSe_2$ over carbon nanotubes (CNT); (**b**) schematic sketch of a synthesized CNT@$MoSe_2$ sponge; (**c**–**e**) transmission electron microscope (TEM) and high-resolution (HR) TEM images of CNT@$MoSe_2$ sponge; (**f**) the cyclic voltammetry (CV) curves at a scan rate of 0.3 mV/s; (**g**) galvanostatic charge/discharge voltage profiles of 1st, 2nd, 3rd, and 100th cycles at 0.1 A/g; (**h**) comparison of cycling performance of pure $MoSe_2$, CNT@$MoSe_2$-U and CNT@$MoSe_2$ sponges with various $MoSe_2$ layers at 0.1 A/g. Reprinted with permission from [58]. Copyright 2018 American Chemical Society

hydrothermal method. The characterization of the structure and morphology reveals that the FeS_2/RGO composite has interconnected networks of RGO sheets and the FeS_2 microparticle (Fig. 6.20). The unique structure of the FeS_2/RGO composite provides a high contact area between the electrolyte and the

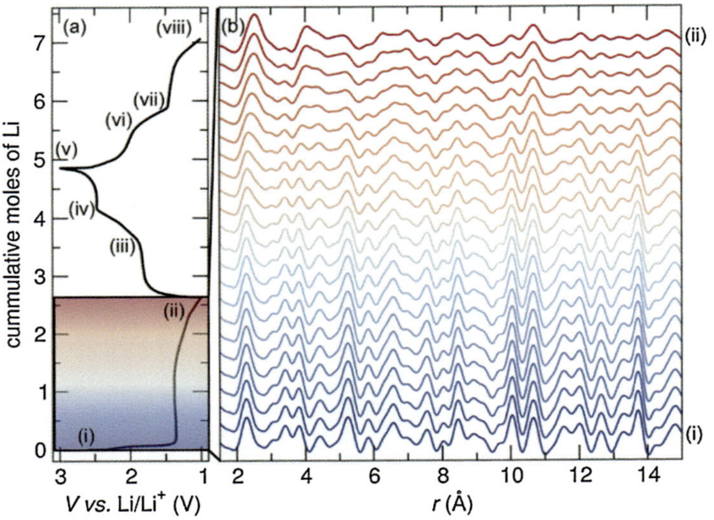

Fig. 6.19 (**a**) The first 1.5 cycles of operando electrochemistry at ≈C/17; (**b**) the corresponding powder diffraction files (PDFs) for the first discharge show conversion of (i) FeS_2 to (ii) Fe- and Li_2S-like products. The PDF also shows a decrease in crystallinity with cycling by the decreased intensity of peaks beyond 6 Å. Reprinted with permission from [62]. Copyright 2017 American Chemical Society

electrode, efficient electron conducting pathway, perfect protection against the volume change of anode materials, and excellent electrical conductivity of the overall electrode during electrochemical process.

It is known that iron sulfides, like other transition metal sulfides or oxides, suffer from low reversible capacity due to their poor electronic conductivity. Meanwhile, insulating polysulfides Li_2S_x ($2 < x < 8$) are produced during the Li storage process by conversion reactions, and these can easily dissolve in organic electrolytes and migrate to the cathode side, preventing further electrochemical Li storage reactions and resulting in poor cyclability during charge/discharge [64]. To overcome these problems, size control of iron sulfides and the restrained dissolution of polysulfides in the electrolyte are of great importance. For example, Cheng et al. [64] successfully prepared iron sulfide nanoparticle (NP)-filled CNTs (denoted Fe-S@CNTs), as shown in Fig. 6.21. Due to the unique nanoconfinement effect of the CNTs, the growth and aggregation of Fe-S NPs were efficiently suppressed during the synthesis reaction, and their huge volume expansion during lithiation was accommodated by the interior space and strong walls of the CNTs. The electrical disconnection of Fe-S and the dissolution of polysulfides in the electrolyte during the electrochemical conversion reaction were avoided by their confinement in the CNTs. As a result of this structural stability and the Li-ion transfer kinetics, the Fe-S@CNT electrode exhibits an exceptionally stable capacity retention and improved rate capability for highly reversible Li storage.

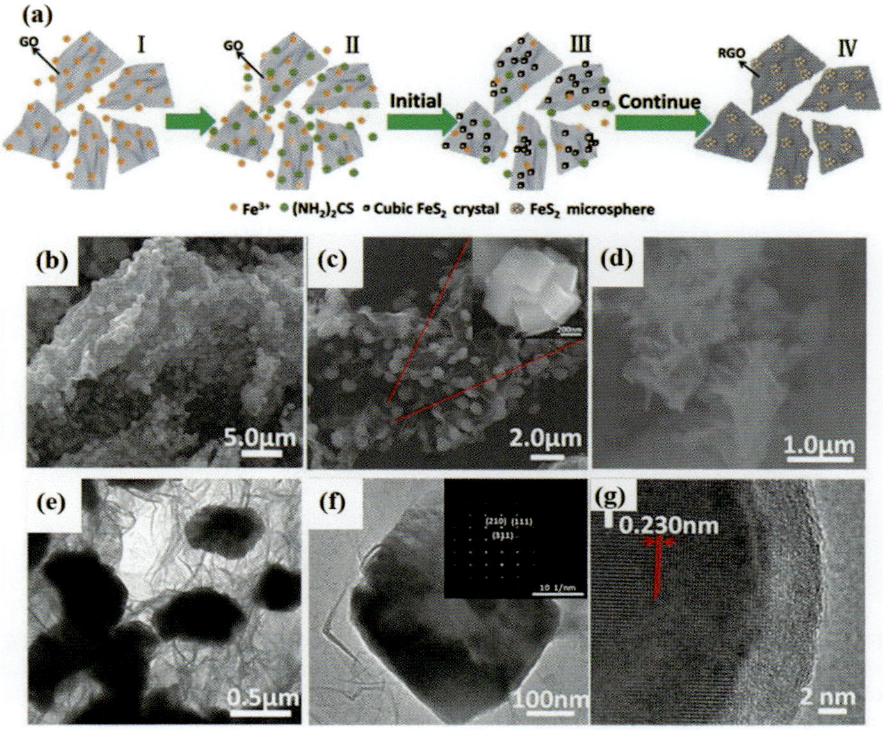

Fig. 6.20 (**a**) The illustration of the formation of the FeS_2/RGO composite (**b** and **c**) SEM images of the FeS_2/RGO composite and (**d**) pristine FeS_2; (e and f) transmission electron microscope (TEM) images, the inset of (**e**) is the electron diffraction spectrum of the FeS_2/RGO composite and (**g**) high-resolution (HR) TEM image of the FeS_2/RGO composite. Reprinted with permission from [63]. Copyright 2015 Royal Society of Chemistry

Iron sulfide (FeS), as one kind of metal sulfide, completes the lithium-ion storage via $2FeS + 2Li^+ + 2e^- \rightarrow Li_2S + Fe$ reaction, giving a theoretical specific capacity of 609 mAh/g [65]. Xue et al. [66] reported a facile one-pot solid-state method to prepare iron sulfide (FeS) confined in the three-dimensional matrix of porous carbon (PC) [FeS/PC composite]. The FeS/PC had an initial discharge capacity of 1428.8 mAh/g at 0.1 C, the highest value as far as we know, and 624.9 mAh/g capacity can still remain after 150 cycles. The systematic characterizations and electrochemical study of FeS/PC composite anode material for LIBs are presented. Wang et al. [67] combined anatase titania (TiO_2) with FeS and synthesized the FeS@TiO_2 nanowires assembled by numerous nanosheets by using a typical hydrothermal method, as shown in Fig. 6.22. The carbon-free nanocoated composite electrodes exhibit improved reversible capacity of 510 mAh/g after 100 discharge/charge cycles at 0.2 A/g, much higher than that of the pristine FeS nanostructures. Long-term cycling shows little performance degradation even after 500 discharge/charge cycles at a current density of 0.4 A/g.

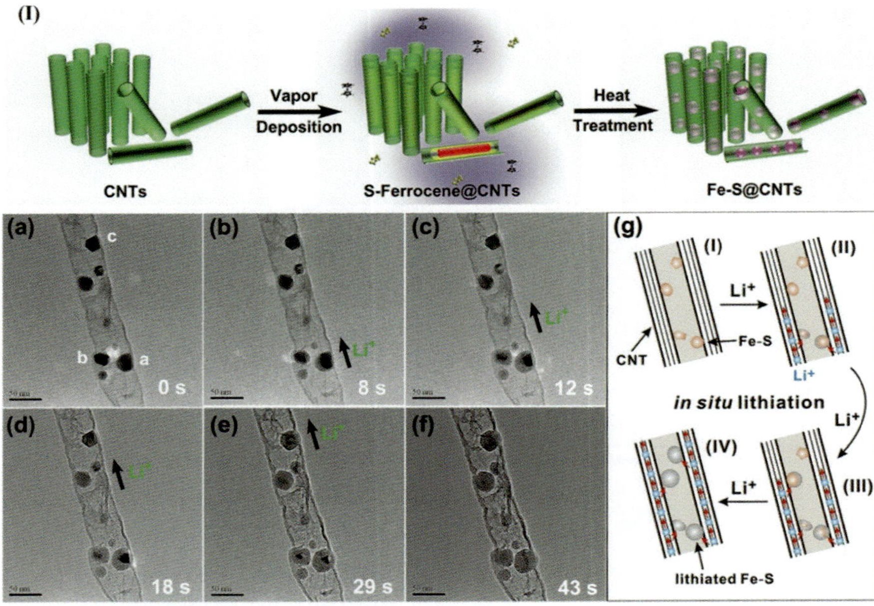

Fig. 6.21 (I) Schematic showing the preparation of the Fe-S@CNT (carbon nanotubes) material; snapshot series of the in situ lithiation of a Fe-S@CNT with densely filled Fe-S NPs; (**a**) aligned carbon nanotubes (A-CNT) with a dense filling of Fe-S NPs; (**b**) the lithiation of particle "a" starts; (**c**–**e**) sequential lithiation of different Fe-S NPs along the CNT; (**b**–**e**) The arrows in images indicate the Li transport direction and the starting lithiation sites; (**f**) transmission electron microscope (TEM) image of the Fe-S@CNT after full lithiation; (**g**) schematic showing the lithiation process of the Fe-S@CNT. Reprinted with permission from [64]. Copyright 2016 John Wiley and Sons

Greigite (Fe_3S_4), a kind of significant half-metallic magnetic material, has been investigated its application for energy storage batteries. Compared with FeS and FeS_2, Fe_3S_4 has a better electrical conductivity due to its inverse spinel structure. In addition, it exhibits a theoretical capacity of 785 mAh/g, two times higher than the conventional anode material graphite (372 mAh/g). Considering that its nontoxic and abundance, it is an optional material for high-performance lithium-ion batteries. Wang et al. [68] reported a kind of hollow nanosphere Fe_3S_4, as shown in Fig. 6.23. The growth process of the Fe_3S_4 hollow spheres has been investigated based on the observations of time-dependent experiments. For applications, these Fe_3S_4 hollow structures in LIBs exhibit stable capacity retention of 750 mAh/g over 100 cycles at a current density of 0.2 A/g.

In the recent, Fe_7S_8 also is considered as electrode materials for lithium-ion batteries due to its improved electrochemical performance with a theoretical capacity of 663 mAh/g [69]. According to the report from Zhu et al. [70], the discharge process of Fe_7S_8 anode materials is shown in the following:

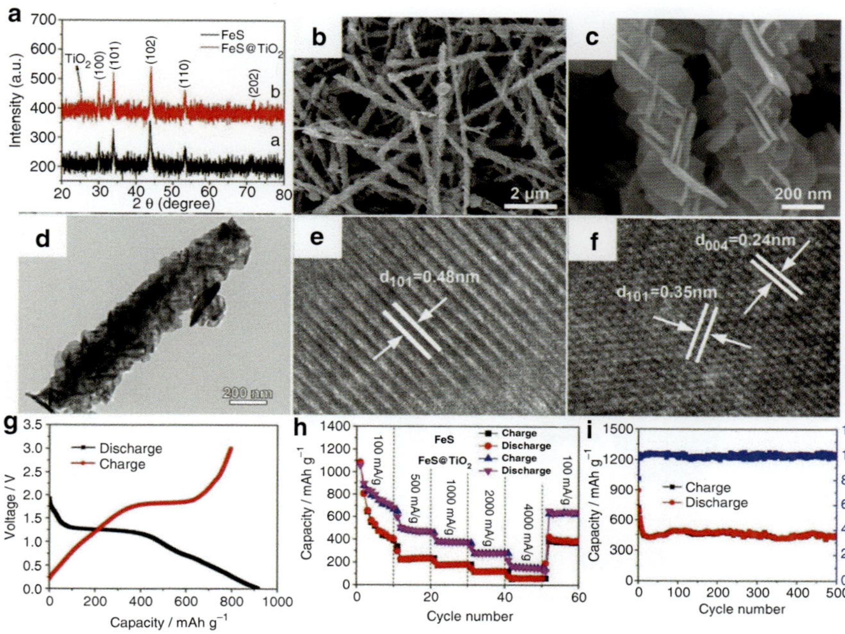

Fig. 6.22 (**a**) XRD patterns of the as-synthesized FeS and FeS@TiO_2 nanostructures; (**b**) scanning electron microscope (SEM) images; (**c**) transmission electron microscope (TEM) image and (**d**–**f**) HRTEM images of the as-prepared FeS@TiO_2 nanostructures; (**g**) discharge and charge voltage profiles of FeS@TiO_2 nanostructures electrodes; (**h**) Rate performance of FeS@TiO_2 nanostructures electrodes at various current densities of 0.1, 0.5, 1.0, 2.0, and 4.0 A/g; (**i**) long cycling performance of the electrode at a current density of 0.4 A/g. Reprinted with permission from [67]. Copyright 2013 Springer Nature

Discharge process:

$$Fe_7S_8 + 8Li^+ + 8e^- \rightarrow 4Li_2FeS_2 + 3Fe^0 \quad (6.10)$$

$$Li_2FeS_2 + 2Li^+ + 2e^- \rightarrow 2Li_2S + Fe^0 \quad (6.11)$$

Charge process:

$$2Li_2S + Fe^0 \rightarrow: Li_2FeS_2 + 2Li^+ + 2e^- \quad (6.12)$$

$$Li_2FeS_2 \rightarrow Li_{2-x}FeS_2 + xLi^+ + xe^-(0.5 < x < 0.8) \quad (6.13)$$

In order to address its big volume change and poor electrical conductivity, Zhou et al. [71] successfully prepared Fe_7S_8 nanoparticles attached on the carbon networks

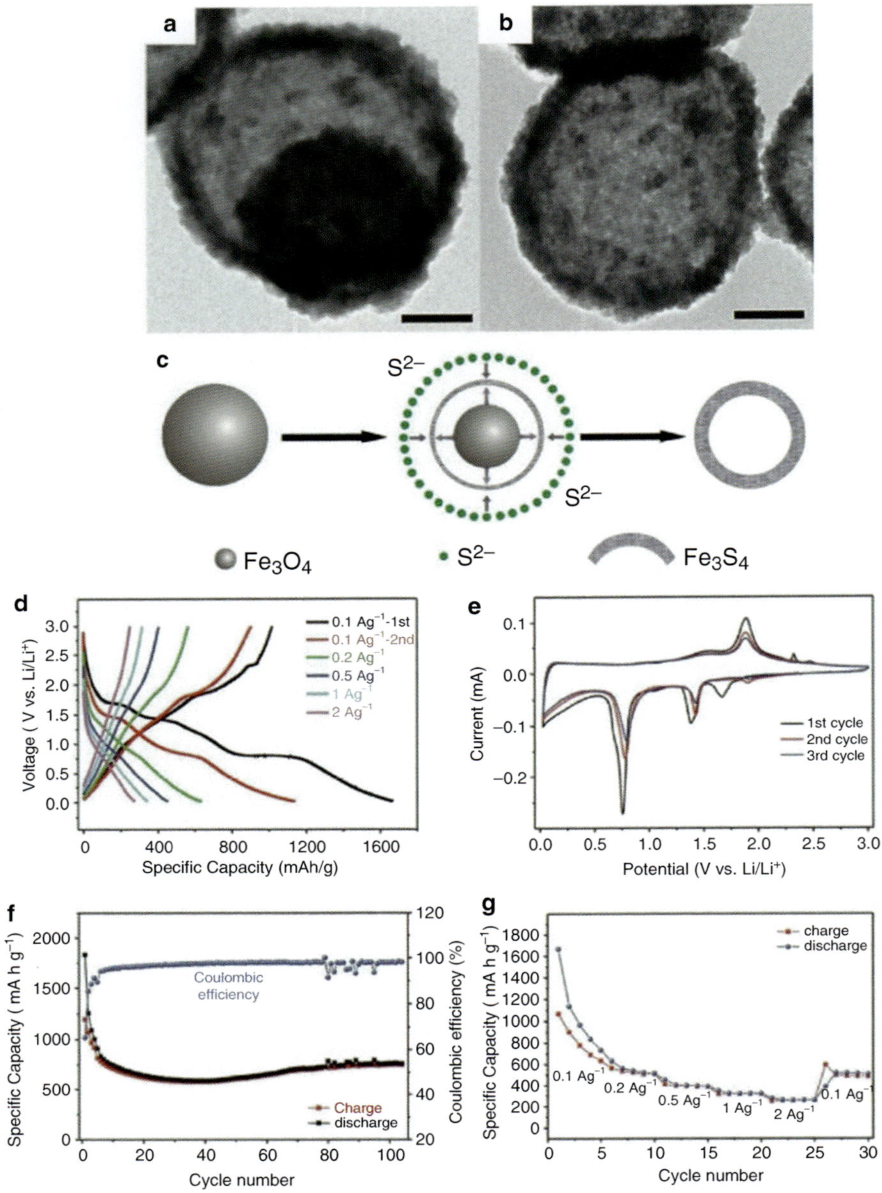

Fig. 6.23 (**a** and **b**) Transmission electron microscope (TEM) images of the products obtained at different reaction times: (**a**) 6 h, (**b**) 10 h, (all scale bars are 100 nm); (**c**) schematic illustration of the formation of Fe_3S_4 hollow spheres; (**d**) the charge-discharge curves at different current densities; (**e**) the first three consecutive cyclic voltammograms at a scan rate of 0.05 mV/s; (**f**) cycling performance and Coulombic efficiency at a current density of 0.2 A/g at room temperature; (**g**) rate performance according to the cycling rate sequence: 0.1, 0.2, 0.5, 1, 2, and 0.1 A/g. Reprinted with permission from [68]. Copyright 2014 Royal Society of Chemistry

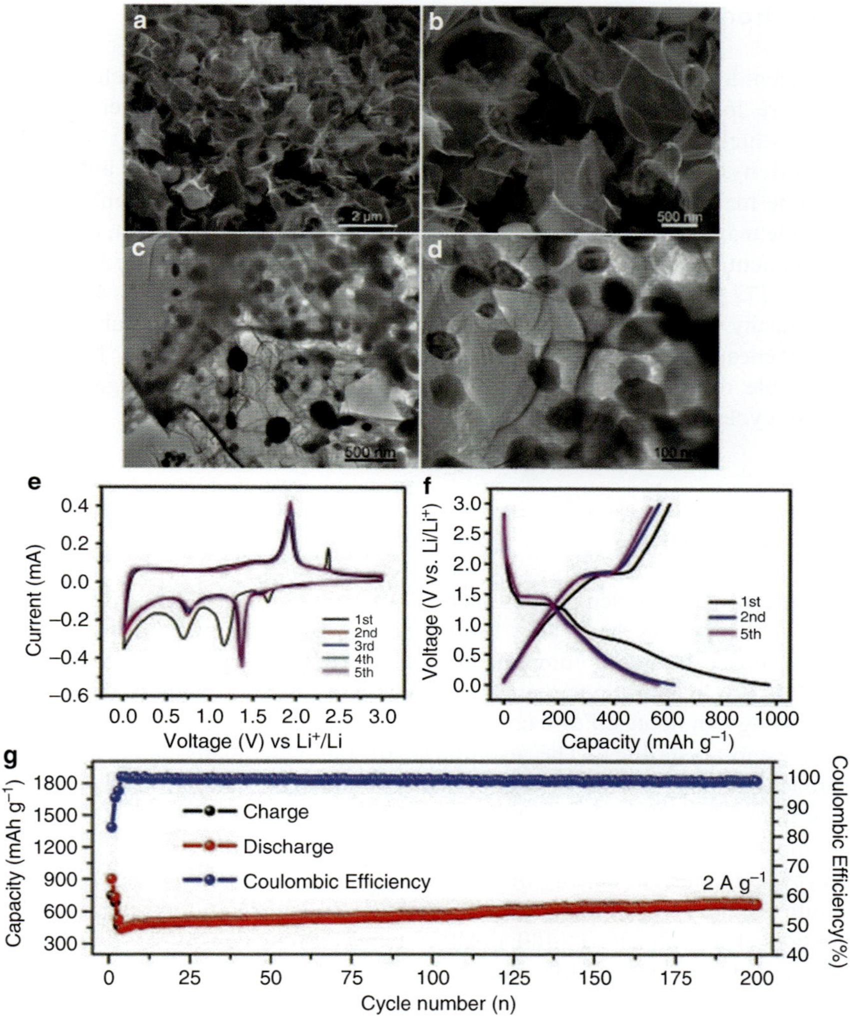

Fig. 6.24 (**a** and **b**) Scanning electron microscope (SEM) images; (**c** and **d**) transmission electron microscope (TEM) images of Fe_7S_8@C composite; (**e**) cyclic voltammetry (CV) curves; (**f**) discharge/charge profiles for the first, second, and fifth cycles; (**g**) long-term cycling performance and Coulombic efficiency of the Fe_7S_8@C composite at 2.0 A/g for lithium-ion batteries. Reprinted with permission from [71]. Copyright 2018 Elsevier

by a simple freeze-drying process followed by the subsequent calcination, forming the Fe_7S_8@C composite. As anode materials for lithium-ion batteries, its reversible capacity can maintain at 667 mAh/g at a high current density of 2 A/g after 200 cycles, as shown in Fig. 6.24.

6.4.2 Iron Selenides

Iron selenides have similar properties as iron sulfides because selenium and sulfur are located in the same group in the periodic table of elements, and $FeSe_2$, which has a comparable band gap energy ($E_g = 1.0$ eV) with good conductivity, is a promising candidate for photovoltaic devices and battery electrode materials. Moreover, Fe-based compounds are more advantageous as electrode materials for lithium-ion batteries because iron is abundant, cheap, and environmentally friendly. Mai et al. synthesized $FeSe_2$ nanowires for lithium-ion batteries [72], the reversible capacity of $FeSe_2$ in the first cycle reaches 431 mAh/g, and capacity retention rate is 45% after 25 cycles. Qian et al. [73] also reported α-FeSe encapsulated by a carbon layer as an electrode material. It shows a sustainable reversible capacity of 340 mAh/g at an average voltage of 1.5 V after 40 cycles.

6.5 Others

Other metal-based chalcogenides such as WS_2, Co_2S, $CoSe_2$, Ni_3S_4, and ternary metal chalcogenides have also been investigated, showing improved performances for Li storage in comparison with the conventional carbon-based materials [74–79]. In addition to them, GeS_2 and Sb_2S_3 were also expected to serve as outstanding candidates. For example, Yu et al. [80] reported an RGO-supported Sb_2S_3 nano-crystallites. The usefulness of the new material and synthesis approach is demonstrated by highly efficient and stable lithium battery anodes. Since both sulfur lithiation and antimony-lithium alloying reactions are reversible, they both contribute to the charge capacity, which exceeds 720 mAh/g after 50 cycles at a current density of 0.25 A/g. The very small crystallite size of the stibnite provides a minimum diffusion pathway and allows for excellent capacity retention at a high rate (>480 mAh/g at the current density of 2.0 A/g was observed). Ge is one of the most fascinating elements as an anode material because it can allow a large quantity of Li, thereby acquiring a high theoretical capacity ($Li_{3.75}Ge$: 1385 mAh/g) at room temperature. In addition, the Li-ion conductivity and electrical conductivity in Ge are ~400 and ~100 times greater, respectively, than those of Si [81]. Park et al. [82] confirmed the electrochemical Li insertion/extraction behavior of the GeS_2 electrode from extended X-ray absorption measurements as well as by cyclic voltammetry and differential capacity plots. Then an amorphous GeS_2-based composite was synthesized to develop its electrochemical performance, which shows high reversible and outstanding electrochemical performances, a highly reversible capacity (first charge capacity: 1298 mAh/g) with a high first Coulombic efficiency (83.3%), rapid rate capability (ca. 800 mAh/g at a high current rate of 0.7 A/g), and long capacity retention over 180 cycles with high capacity (1100 mAh/g), as shown in Fig. 6.25.

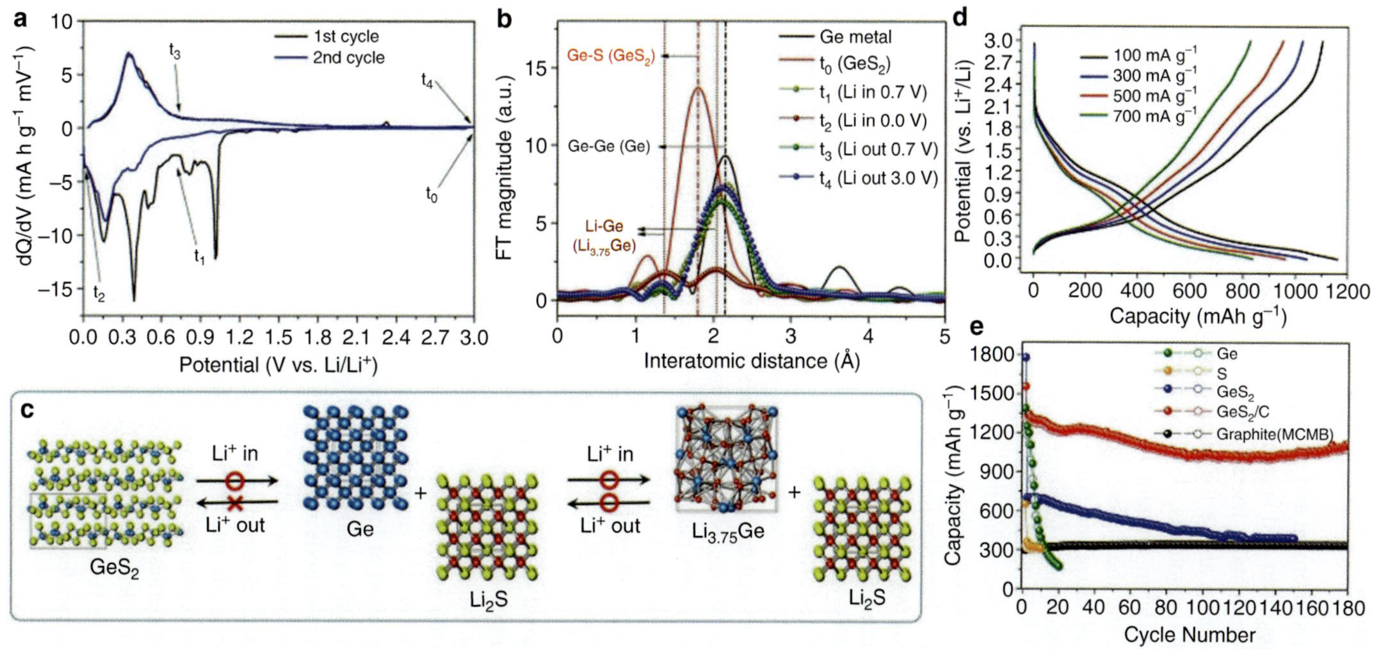

Fig. 6.25 Electrochemical Li insertion/extraction mechanism of GeS_2 electrode. (**a**) DCP for the first and second cycles; (**b**) Ge K-edge EXAFS spectra during the first cycle; (**c**) schematic diagram of phase transition mechanism during the electrochemical Li insertion/extraction in the GeS_2 electrode; (**d**) C-rate performance of the a-GeS_2/C composite electrode at the various current densities (0.1, 0.3, 0.5, and 0.7 A/g); (**e**) cycling performance of the Ge, S, GeS_2, graphite (MCMB), and a-GeS_2/C composite electrodes (cycling rate: 0.1 A/g). Reprinted with permission from [82]. Copyright 2016 American Chemical Society

6.6 Conclusions

The urgent energy requirement and environmental pollution produce much strain for scientists to develop clean and sustainable technologies that can provide abundant energy in an environmentally friendly way. Metal-based chalcogenides (MXs) are considered as promising anode for lithium-ion batteries due to its high theoretical capacities, abundant resources, low cost, etc. Nevertheless, the high capacity corresponding to a large amount of lithium-ion intercalation also brings large volume expansion, and the general conversion reaction, especially for metal-based sulfides anodes, will introduce soluble intermediates such as polysulfides, which could cause sea very capacity loss and safety problems, such as a shuttle effect that is similar to that in Li-S batteries. In addition, the poor conductivity of chalcogenides also influences their electrochemical performance. A great deal of effort has been dedicated in recent years to solving these potential issues and improving the performance of lithium-ion batteries to keep up with the demands for its applications. In this chapter, various metal-based chalcogenides and corresponding modification method, mainly including composite with carbon and nano-crystallization. Carbon modification of MXs has been considered as the most basic strategy. Carbon materials with different dimensions were selected to provide stable support for MXs and increase electrical conductivity, such as zero-dimensional carbon nanoparticles, one-dimensional carbon nanotubes, two-dimensional graphene, and three-dimensional carbon spheres. Nanocomposites show higher surface area, more active sites, and shorter transport path lengths for electrons and lithium-ion. Also, nanostructures with suitable void space (such as hollow, core/shell, or yolk/shell structure) can accommodate large volume change of electrodes during electrochemical reaction process effectively, which could improve the cycling stability of anode materials.

In the future, rational synthesis of metal-based chalcogenides anodes should be paid more attention to further develop some new structures with novel morphology, various hybrids, and doping, which could enhance the structural stability and conductivity. In addition, the following issues should be paid more attention to in the future research: Firstly, the large irreversible capacity was always observed in the first cycle, which means that a large portion of lithium-ions could not be utilized again in the following cycles when paired with a cathode in full cell. Secondly, further exploration of the interface between the electrode and the electrolyte should be carried out with a modified design of electrolyte systems that could result in enhanced performances. Thirdly, although conventional electrochemical characterization techniques, such as cyclic voltammetry, galvanostatic intermittent titration, electrochemical impedance spectroscopy, etc., are able to reveal the performance of bulk hybrid electrodes, they can hardly probe the intrinsic properties of the active material itself during ion insertion. Thus, more researches need to be dedicated to understanding the underlying mechanisms of the changes induced in the behavior of MXs by ion intercalation, especially the electronic structure and charge transfer features.

Acknowledgments This work is partially supported by Shenzhen Fundamental Research Program of Subject Distribution (JCYJ20170413102735544) and Natural Science Foundation of Guangdong (2018A030313721).

References

1. D. Aurbach, B. Markovsky, I. Weissman, et al., On the correlation between surface chemistry and performance of graphite negative electrodes for Li-ion batteries. Electrochim. Acta **45**, 67–86 (1999). https://doi.org/10.1016/S0013-4686(99)00194-2
2. R. Fong, U. Von Sacken, J.R. Dahn, Studies of lithium intercalation into carbons using nonaqueous electrochemical cells. J. Electrochem. Soc. **137**, 2009–2013 (1999). https://doi.org/10.1149/1.2086855
3. R. Yazami, *Nanomaterials for lithium-ion batteries: fundamentals and applications* (Pan Stanford, Danvers, 2014)
4. C.K. Chan, H. Peng, G. Liu, et al., High-performance lithium battery anodes using silicon nanowires. Nat. Nanotechnol. **3**, 31–35 (2008). https://doi.org/10.1038/nnano.2007.411
5. Z. Li, J. Ding, D. Mitlin, Tin and tin compounds for sodium ion battery anodes: phase transformations and performance. Acc. Chem. Res. **48**, 1657–1665 (2015). https://doi.org/10.1021/acs.accounts.5b00114
6. Y. Hamon, T. Brousse, F. Jousse, et al., Aluminum negative electrode in lithium-ion batteries. J. Power Sources **97**, 185–187 (2001). https://doi.org/10.1016/s0378-7753(01)00616-4
7. D. Larcher, S. Beattie, M. Morcrette, et al., Recent findings and prospects in the field of pure metals as negative electrodes for Li-ion batteries. J. Mater. Chem. **17**, 3759–3772 (2007). https://doi.org/10.1039/B705421C
8. C.K. Chan, X.F. Zhang, Y. Cui, High capacity Li-ion battery anodes using Ge nanowires. Nano Lett. **8**, 307–309 (2008). https://doi.org/10.1021/nl0727157
9. L. Hu, X.M. Lin, J.T. Mo, et al., Lead-based metal-organic framework with stable Lithium anodic performance. Inorg. Chem. **56**, 4289–4295 (2017). https://doi.org/10.1021/acs.inorgchem.6b02663
10. O. Crosnier, X. Devaux, T. Brousse, et al., Influence of particle size and matrix in "metal" anodes for Li-ion cells. J. Power Sources **97**, 188–190 (2001). https://doi.org/10.1016/s0378-7753(01)00617-6
11. T. Gao, F. Han, Y. Zhu, et al., Hybrid Mg^{2+}/Li^{+} battery with long cycle life and high rate capability. Adv. Energy Mater. **5**, 1401507 (2015). https://doi.org/10.1002/aenm.201401507
12. Y. Idota, T. Kubota, A. Matsufuji, et al., Tin-based amorphous oxide: a high-capacity lithium-ion-storage material. Science **276**, 1395–1397 (1997). https://doi.org/10.1126/science.276.5317.1395
13. C. Ataca, H. Sahin, S. Ciraci, Stable, single-layer MX_2 transition-metal oxides and dichalcogenides in a honeycomb-like structure. J. Phys. Chem. C **116**, 8983–8999 (2012). https://doi.org/10.1021/jp212558p
14. Z. Hu, Q. Liu, S.L. Chou, et al., Advances and challenges in metal sulfides/selenides for next-generation rechargeable sodium-ion batteries. Adv. Mater., 1700606 (2017). https://doi.org/10.1002/adma.201700606
15. K. Kravchyk, L. Protesescu, M. Bodnarchuk, et al., Monodisperse and inorganically capped Sn and Sn/SnO_2 nanocrystals for high-performance Li-ion battery anodes. J. Am. Chem. Soc. **135**, 4199–4202 (2013). https://doi.org/10.1021/ja312604r
16. J. Qin, C. He, N. Zhao, et al., Graphene networks anchored with Sn@graphene as a lithium-ion battery anode. ACS Nano **8**, 1728–1738 (2014). https://doi.org/10.1021/nn406105n
17. Z. Wang, D. Luan, F. Boey, X. Lou, Fast formation of SnO_2 nanoboxes with enhanced lithium storage capability. J. Am. Chem. Soc. **133**, 4738–4741 (2011). https://doi.org/10.1021/ja2004329

18. H. Ying, S. Zhang, Z. Meng, Z. Sun, W. Han, Ultrasmall Sn nanodots embedded inside N-doped carbon micro cages as high-performance lithium and sodium ion battery anodes. J. Mater. Chem. A **5**, 8334–8342 (2017). https://doi.org/10.1039/C7TA01480E
19. Y. Yu, L. Gu, C. Zhu, P. Van Aken, J. Maier, Tin nanoparticles encapsulated in porous multichannel carbon microtubes: preparation by single-nozzle electrospinning and application as anode material for high-performance Li-based batteries. J. Am. Chem. Soc. **131**, 15984–15985 (2009). https://doi.org/10.1021/ja906261c
20. Y. Xu, Q. Liu, Y. Zhu, Y. Liu, A. Langrock, M. Zachariah, C. Wang, Uniform nano-Sn/C composite anodes for lithium-ion batteries. Nano Lett. **13**, 470–474 (2013). https://doi.org/10.1021/nl303823k
21. L. Cui, J. Shen, F. Cheng, et al., SnO_2 nanoparticles@polypyrrole nanowires composite as anode materials for rechargeable lithium-ion batteries. J. Power Sources **196**, 2195–2201 (2011). https://doi.org/10.1016/j.jpowsour.2010.09.075
22. F. Huang, X. Lu, Y. Zhang, X. Luo, C. Yu, R. Wu, A new composite, Co–Sn metal oxide anode for lithium-ion batteries. Mater. Sci. **27**, 0137–1339 (2009)
23. M. Zhang, T. Wang, G. Cao, Promises and challenges of tin-based compounds as anode materials for lithium-ion batteries. Int. Mater. Rev. **60**, 330–352 (2015). https://doi.org/10.1179/1743280415Y.0000000004
24. K. Xu, A. Cresce, Interfacing electrolytes with electrodes in Li-ion batteries. J. Mater. Chem. **21**, 9849–9864 (2011). https://doi.org/10.1039/c0jm04309e
25. C. Kim, M. Noh, M. Choi, J. Cho, B. Park, Critical size of a nano SnO_2 electrode for a Li-secondary battery. Chem. Mater. **17**, 3297–3301 (2005). https://doi.org/10.1021/cm048003o
26. M. Park, G. Wang, Y. Kang, D. Wexler, S. Dou, H. Liu, Preparation and electrochemical properties of SnO_2 nanowires for application in lithium-ion batteries. Angew. Chem. **119**, 764–767 (2007). https://doi.org/10.1002/ange.200603309
27. K. Zhao, L. Zhang, R. Xia, Y. Dong, W. Xu, C. Niu, L. Mai, et al., SnO_2 quantum dots@graphene oxide as a high-rate and long-life anode material for lithium-ion batteries. Small **12**, 588–594 (2016). https://doi.org/10.1002/smll.201502183
28. J. Sun, L. Xiao, S. Jiang, G. Li, Y. Huang, J. Geng, Fluorine-doped SnO_2@graphene porous composite for high capacity lithium-ion batteries. Chem. Mater. **27**, 4594–4603 (2015). https://doi.org/10.1021/acs.chemmater.5b00885
29. P. Lippens, M. El Khalifi, M. Womes, Electronic structures of SnS and SnS_2. Phys. Status Solidi B **254**, 1–14 (2017). https://doi.org/10.1002/pssb.201600194
30. T. Kim, C. Kim, D. Son, M. Choi, B. Park, Novel SnS_2-nanosheet anodes for lithium-ion batteries. J. Power Sources **167**, 529–535 (2007). https://doi.org/10.1016/j.jpowsour.2007.02.040
31. M. He, L. Yuan, Y. Huang, Acetylene black incorporated three-dimensional porous SnS_2 nanoflowers with high performance for lithium storage. RSC Adv. **3**, 3374–3383 (2013). https://doi.org/10.1039/c2ra22764a
32. P. Lee, G. Said, R. Davis, T. Lim, On the optical properties of some layer compounds. J. Phys. Chem. Solids **30**, 2719–2729 (1969). https://doi.org/10.1016/0022-3697(69)90045-6
33. G. Domingo, R. Itoga, C. Kannewurf, Fundamental optical absorption in Sn S_2 and $SnSe_2$. Phys. Rev. **143**, 536–541 (1966). https://doi.org/10.1103/PhysRev.143.536
34. S. Hwang, Z. Yao, L. Zhang, M. Fu, K. He, L. Mai, et al., Multistep lithiation of tin sulfide: an investigation using in situ electron microscopy. ACS Nano **12**, 3638–3645 (2018). https://doi.org/10.1021/acsnano.8b00758
35. K. Yin, M. Zhang, Z. Hood, J. Pan, Y. Meng, M. Chi, Self-assembled framework formed during lithiation of SnS_2 nanoplates revealed by in situ electron microscopy. Acc. Chem. Res. **50**, 1513–1520 (2017). https://doi.org/10.1021/acs.accounts.7b00086
36. J. Seo, J. Jang, S. Park, C. Kim, B. Park, J. Cheon, Two-dimensional SnS_2 nanoplates with extraordinary high discharge capacity for lithium-ion batteries. Adv. Mater. **20**, 4269–4273 (2008). https://doi.org/10.1002/adma.200703122

37. T. Zhou, W. Pang, C. Zhang, J. Yang, Z. Chen, H. Liu, Z. Guo, Enhanced sodium-ion battery performance by a structural phase transition from two-dimensional hexagonal-SnS_2 to orthorhombic-SnS. ACS Nano **8**, 8323–8333 (2014). https://doi.org/10.1021/nn503582c
38. D. Youn, S. Stauffer, P. Xiao, H. Park, Y. Nam, A. Dolocan, C. Mullins, et al., Simple synthesis of nanocrystalline tin sulfide/N-doped reduced graphene oxide composites as lithium-ion battery anodes. ACS Nano **10**, 10778–10788 (2016). https://doi.org/10.1021/acsnano.6b04214
39. Q. Li, H. Liu, Z. Yao, J. Cheng, T. Li, Y. Li, V. Dravid, Electrochemistry of selenium with sodium and lithium: kinetics and reaction mechanism. ACS Nano **10**, 8788–8795 (2016). https://doi.org/10.1021/acsnano.6b04519
40. Q. Cai, Y. Li, Q. Li, J. Xu, B. Gao, X. Zhang, P. Chu, Freestanding hollow double-shell Se@CNx nanobelts as large-capacity and high-rate cathodes for Li-Se batteries. Nano Energy **32**, 1–9 (2017). https://doi.org/10.1016/j.nanoen.2016.12.010
41. J. Choi, J. Jin, I. Jung, J. Kim, H. Kim, S. Son, $SnSe_2$ nanoplate-graphene composites as anode materials for lithium-ion batteries. Chem. Commun. **47**, 5241–5243 (2011). https://doi.org/10.1039/c1cc10317b
42. K. Chen, X. Wang, G. Wang, B. Wang, X. Liu, J. Bai, H. Wang, A new generation of high-performance anode materials with semiconductor heterojunction structure of SnSe/SnO_2@Gr in lithium-ion batteries. Chem. Eng. J. **347**, 552–562 (2018). https://doi.org/10.1016/j.cej.2018.04.125
43. X. Wang, D. Chen, Z. Yang, X. Zhang, C. Wang, J. Chen, M. Xue, et al., Novel metal chalcogenide SnSSe as a high-capacity anode for sodium-ion batteries. Adv. Mater. **28**, 8645–8650 (2016). https://doi.org/10.1002/adma.201603219
44. Q. Tang, Y. Cui, J. Wu, D. Qu, A. Baker, Y. Ma, Y. Liu, et al., Ternary tin selenium sulfide ($SnSe_{0.5}S_{0.5}$) nanoalloy as the high-performance anode for lithium-ion and sodium-ion batteries. Nano Energy **41**, 377–386 (2017). https://doi.org/10.1016/j.nanoen.2017.09.052
45. Y. Zhang, J. Yang, Y. Zhang, C. Li, W. Huang, Q. Yan, X. Dong, Fe_2O_3/SnSSe hexagonal nanoplates as lithium-ion batteries anode. ACS Appl. Mater. Interfaces **10**, 12722–12730 (2018). https://doi.org/10.1021/acsami.8b01537
46. T. Wang, L. Yin, R. Zhao, C. Xia, X. Zhao, Y. An, X. Dai, et al., First-principles study of monolayer $SnS_{2(1-x)}$ Se_{2x} alloys as anode materials for Lithium-ion batteries. Appl. Surf. Sci. (2018). https://doi.org/10.1016/j.apsusc.2018.06.030
47. W. Xie, S. Populoh, X. Xiao, L. Sagarna, Y. Liu, A. Weidenkaff, Thermoelectric study of crossroads material MnTe via sulfur doping. J. Appl. Phys. **115**, 103707 (2014). https://doi.org/10.1063/1.4868584
48. R. Mehta, C. Karthik, W. Jiang, B. Singh, Y. Shi, R. Siegel, G. Ramanath, High electrical conductivity antimony selenide nanocrystals and assemblies. Nano Lett. **10**, 4417–4422 (2010). https://doi.org/10.1021/nl1020848
49. P. Perumal, R. Ulaganathan, R. Sankar, Y. Liao, T. Sun, M. Chu, Y. Chen, et al., Ultra-thin layered ternary single crystals [Sn $(S_xSe_{1-x})_2$] with bandgap engineering for high-performance phototransistors on versatile substrates. Adv. Funct. Mater. **26**, 3630–3638 (2016). https://doi.org/10.1002/adfm.201600081
50. Q. Tang, H. Su, Y. Cui, A. Baker, Y. Liu, J. Lu, D. Qu, et al., Ternary tin-based chalcogenide nanoplates as a promising anode material for lithium-ion batteries. J. Power Sources **379**, 182–190 (2018). https://doi.org/10.1016/j.jpowsour.2018.01.051
51. L. Wang, Z. Xu, W. Wang, X. Bai, Atomic mechanism of dynamic electrochemical lithiation processes of MoS_2 nanosheets. J. Am. Chem. Soc. **136**, 6693–6697. https://doi.org/10.1021/ja501686w
52. C. George, A. Morris, M. Modarres, M. De Volder, Structural evolution of electrochemically lithiated MoS_2 nanosheets and the role of carbon additive in Li-ion batteries. Chem. Mater. **28**, 7304–7310 (2016). https://doi.org/10.1021/acs.chemmater.6b02607
53. N. Sharma, G. Du, A. Studer, Z. Guo, V. Peterson, In-situ neutron diffraction study of the MoS_2 anode using a custom-built Li-ion battery. Solid State Ionics **199**, 37–43 (2011). https://doi.org/10.1016/j.ssi.2011.07.015

54. L. Zhang, D. Sun, J. Kang, J. Feng, H. Bechtel, L. Wang, J. Guo, et al., Electrochemical reaction mechanism of the MoS_2 electrode in a lithium-ion cell revealed by in situ and operando X-ray absorption spectroscopy. Nano Lett. **18**, 1466–1475 (2018). https://doi.org/10.1021/acs.nanolett.7b05246
55. Z. Lei, L. Xu, Y. Jiao, A. Du, Y. Zhang, H. Zhang, Strong coupling of MoS_2 nanosheets and nitrogen-doped graphene for high-performance pseudocapacitance lithium storage. Small, 1704410 (2018). https://doi.org/10.1002/smll.201704410
56. X. Hu, Y. Li, G. Zeng, J. Jia, H. Zhan, Z. Wen, et al., Three-dimensional network architecture with hybrid nanocarbon composites supporting few-layer MoS_2 for lithium and sodium storage. ACS Nano **12**, 1592–1602 (2018). https://doi.org/10.1021/acsnano.7b08161
57. S. Zhang, G. Wang, J. Jin, L. Zhang, Z. Wen, J. Yang, Robust and conductive red $MoSe_2$ for stable and fast lithium storage. ACS Nano **12**, 4010–4018 (2018). https://doi.org/10.1021/acsnano.8b01703
58. M. Yousaf, Y. Wang, Y. Chen, Z. Wang, W. Aftab, A. Mahmood, R. Han, et al., Tunable free-standing core-shell CNT@$MoSe_2$ anode for lithium storage. ACS Appl. Mater. Interfaces **10**, 14622–14631 (2018). https://doi.org/10.1021/acsami.7b19739
59. K. Chen, W. Zhang, L. Xue, W. Chen, X. Xiang, M. Wan, Y. Huang, Mechanism of capacity fade in sodium storage and the strategies of improvement for FeS_2 anode. ACS Appl. Mater. Interfaces **9**, 1536–1541 (2017). https://doi.org/10.1021/acsami.6b13421
60. C. Jones, P. Kovacs, R. Sharma, R. McMillan, An iron-57 Moessbauer study of the intermediates formed in the reduction of iron disulfide in the lithium/iron disulfide battery system. J. Phys. Chem. **95**, 774–779 (1991). https://doi.org/10.1021/j100155a053
61. Z. Hu, Z. Zhu, F. Cheng, K. Zhang, J. Wang, C. Chen, J. Chen, et al., Pyrite FeS_2 for high-rate and long-life rechargeable sodium batteries. Energy Environ. Sci. **8**, 1309–1316 (2015). https://doi.org/10.1039/C4EE03759F
62. M. Butala, M. Mayo, V. Doan-Nguyen, M. Lumley, C. Gobel, K. Wiaderek, G. Laurita, et al., Local structure evolution and modes of charge storage in secondary Li-FeS_2 cells. Chem. Mater. **29**, 3070–3082 (2017). https://doi.org/10.1021/acs.chemmater.7b00070
63. X. Wen, X. Wei, L. Yang, P. Shen, Self-assembled FeS_2 cubes anchored on reduced graphene oxide as an anode material for lithium-ion batteries. J. Mater. Chem. A **3**, 2090–2096 (2015). https://doi.org/10.1039/c4ta05575f
64. W. Yu, C. Liu, L. Zhang, P. Hou, F. Li, B. Zhang, H. Cheng, Synthesis and electrochemical lithium storage behavior of carbon nanotubes filled with iron sulfide nanoparticles. Adv. Sci. **3**, 1600113 (2016). https://doi.org/10.1002/advs.201600113
65. Y. Xu, W. Li, F. Zhang, X. Zhang, W. Zhang, C. Lee, Y. Tang, In situ incorporation of FeS nanoparticles/carbon nanosheets composite with an interconnected porous structure as a high-performance anode for lithium-ion batteries. J. Mater. Chem. A **4**, 3697–3703 (2016). https://doi.org/10.1039/c5ta09138a
66. S. Guo, J. Li, Z. Ma, Y. Chi, H. Xue, A facile method to prepare FeS/porous carbon composite as advanced anode material for lithium-ion batteries. J. Mater. Sci. **52**, 2345–2355 (2017). https://doi.org/10.1007/s10853-016-0527-y
67. X. Wang, Q. Xiang, B. Liu, L. Wang, T. Luo, D. Chen, G. Shen, TiO_2 modified FeS nanostructures with enhanced electrochemical performance for lithium-ion batteries. Sci. Rep. **3**, 2007 (2013). https://doi.org/10.1038/srep02007
68. J. Zheng, Y. Cao, C. Cheng, C. Chen, R. Yan, H. Huai, C. Wang, et al., Facile synthesis of Fe_3S_4 hollow spheres with high-performance for lithium-ion batteries and water treatment. J. Mater. Chem. A **2**, 19882–19888 (2014). https://doi.org/10.1039/C4TA05148C
69. K. Zhang, T. Zhang, J. Liang, Y. Zhu, N. Lin, Y. Qian, A potential pyrrhotite (Fe_7S_8) anode material for lithium storage. RSC Adv. **5**, 14828–14831 (2015). https://doi.org/10.1039/c4ra14819c
70. L. Shi, D. Li, J. Yu, H. Liu, Y. Zhao, H. Xin, C. Zhu, et al., Uniform core-shell nano biscuits of Fe_7S_8@C for lithium-ion and sodium-ion batteries with excellent performance. J. Mater. Chem. A **6**, 7967–7976 (2018). https://doi.org/10.1039/C8TA00985F

71. F. Jiang, Q. Wang, R. Du, X. Yan, Y. Zhou, Fe_7S_8 nanoparticles attached carbon networks as anode materials for both lithium and sodium ion batteries. Chem. Phys. Lett. **706**, 273–279 (2018). https://doi.org/10.1016/j.cplett.2018.06.023
72. L. Mai, Y. Gao, J. Guan, B. Hu, L. Xu, W. Jin, Formation and lithiation of ferroselite nanoflowers as high-energy Li-ion battery electrodes. Int. J. Electrochem. Sci. **4**, 755–761 (2009)
73. D. Wei, J. Liang, Y. Zhu, L. Hu, K. Zhang, J. Zhang, Y. Qian, et al., Layer structured α-FeSe: a potential anode material for lithium storage. Electrochem. Commun. **38**, 124–127 (2014). https://doi.org/10.1016/j.elecom.2013.11.021
74. K. Shiva, H. Matte, H. Rajendra, A. Bhattacharyya, C. Rao, Employing synergistic interactions between few-layer WS_2 and reduced graphene oxide to improve lithium storage, cyclability and rate capability of Li-ion batteries. Nano Energy **2**, 787–793 (2013). https://doi.org/10.1016/j.nanoen.2013.02.001
75. H. Liu, D. Su, G. Wang, S. Qiao, An ordered mesoporous WS_2 anode material with a superior electrochemical performance for lithium ion batteries. J. Mater. Chem. **22**, 17437–17440 (2012). https://doi.org/10.1039/C2JM33992G
76. Q. Wang, L. Jiao, Y. Han, H. Du, W. Peng, Q. Huan, H. Yuan, et al., CoS_2 hollow spheres: fabrication and their application in lithium-ion batteries. J. Phys. Chem. C **115**, 8300–8304 (2011). https://doi.org/10.1021/jp111626a
77. K. Zhang, M. Park, L. Zhou, G. Lee, W. Li, Y. Kang, J. Chen, Urchin-like $CoSe_2$ as a high-performance anode material for sodium-ion batteries. Adv. Funct. Mater. **26**, 6728–6735 (2016). https://doi.org/10.1002/adfm.201602608
78. N. Mahmood, C. Zhang, Y. Hou, Nickel sulfide/nitrogen-doped graphene composites: phase-controlled synthesis and high-performance anode materials for lithium-ion batteries. Small **9**, 1321–1328 (2013). https://doi.org/10.1002/smll.201203032
79. Y. Long, J. Yang, X. Gao, X. Xu, W. Fan, J. Yang, Y. Qian, et al., Solid-solution anion-enhanced electrochemical performances of metal sulfides/selenides for sodium-ion capacitors: the case of FeS_{2-x} Se_x. ACS Appl. Mater. Interfaces **10**, 10945–10954 (2018). https://doi.org/10.1021/acsami.8b00931
80. P. Prikhodchenko, J. Gun, S. Sladkevich, A. Mikhaylov, O. Lev, Y. Tay, D. Yu, Conversion of hydroperoxoantimonate coated graphenes to Sb_2S_3@graphene for a superior lithium battery anode. Chem. Mater. **24**, 4750–4757 (2012). https://doi.org/10.1021/cm3031818
81. N. Lin, T. Li, Y. Han, Q. Zhang, T. Xu, Y. Qian, Mesoporous hollow Ge microspheres prepared via molten-salt metallothermic reaction for the high-performance Li-storage anode. ACS Appl. Mater. Interfaces **10**, 8399–8404 (2018). https://doi.org/10.1021/acsami.8b00567
82. G. Sung, K. Jeon, C. Park, Highly reversible and superior Li-storage characteristics of layered GeS_2 and its amorphous composites. ACS Appl. Mater. Interfaces **8**, 29543–29550 (2016). https://doi.org/10.1021/acsami.6b10994

Cathode Material in Lithium-Ion Battery

7

Irslan Ullah Ashraf and Abdul Majid

Contents

Abstract

In the current decade, with improvements in technology, the commercial usage of rechargeable lithium-ion batteries (LIBs) has been significantly enhanced. This has not only boosted industrial interest in LIBs but also attracted major research efforts all over the world. Though all components of LIBs are crucial, their quality factors are chiefly linked with their cathode materials. That is why the majority of research is focused on upgrading the quality of cathode material. Depending on the family of materials and its coordination geometry, various families of cathode materials include chalcogenides, layered oxides, silicates, phosphates, and tavorites. In addition, there are conversion electrodes, but they lie outside the scope of this review. These classes of materials are of interest in LIB-focused research. This chapter is devoted to providing an overview of the current status and time evolution of LIBs, with a focus on cathode materials. The primary objective of this overview is to shed light on the subject matter to identify the main problems encountered by LIBs so that solutions may be sought. To present a comprehensive picture of the story, the computational and experimental literature is discussed.

I. U. Ashraf · A. Majid (✉)
Department of Physics, University of Gujrat, Gujrat, Pakistan
e-mail: Irslan.ullah@uog.edu.pk; Abdulmajid40@uog.edu.pk

Q. Zhen et al. (eds.), *Nanostructured Materials for Next-Generation Energy Storage and Conversion*, https://doi.org/10.1007/978-3-662-58675-4_7

7.1 Introduction

In the present technological age, the development of efficient and convenient sources of electrical power is one of the major scientific challenges to operate current and future appliances. To improve the quality of human life on this planet, in parallel with tackling issues related to various areas, including health, industrialization, telecommunications, agriculture, food, and climate change, for example, securing sustainable energy sources has become an issue of prime significance. The development agenda of the United Nations (UN) General Assembly has assigned several targets to member countries with a focus on a number of goals to accomplish by 2030. One of the key targets is to identify sustainable, affordable, safe, and modern energy sources. While giving opening remarks at a UN General Assembly session on climate change in 2015,[1] Secretary-General Ban Ki-Moon[2] expressed his thoughts on the importance of renewable energy sources and mentioned that global concerns on energy are due to accelerating demands based on ecological and economic factors. To meet energy needs, currently exploited energy sources are being quickly exhausted, which is threatening basic human needs.

Ecological concerns are important as energy sources like fossil fuels and even nuclear energy are not environmentally friendly, and we have to bear the deprivation in the earth environment in the form of an immense increase in Earth entropy. Keeping these considerations in view, it is high time to address energy issues and search for renewable and environmentally friendly energy sources. Developed countries and emerging economic powers are investing heavily in renewable energy sources like hydro, thermal, and biofuels. The 2050 roadmap issued by the International Renewable Energy Agency forecasted a loss of 7.4 million jobs in the nonrenewable energy sector and the creation of 19 million jobs related to renewable energy. According to the Renewables Global Futures Report 2017 (REN21),[3] employment in the renewable energy sector has increased at a rate of 1.1%, with 9.8 million people employed in 2016 in comparison to 2015. REN21 2017 estimate shows that renewable energy accounts for 62% of the total global power-generating capacity. Solar PV gave almost 47% of newly installed renewable power capacity in 2016, whereas wind and hydrothermal sources account for 34% and 15.5%, respectively. Solar energy was the leading source of power generation around the world in 2016 when market investment increased by 50%.

Since 2015, energy demands in the transport sector have increased at a 2% annual rate, resulting in an overall energy consumption rate of 28%. The main categories for the energy usage in the transportation sector include biofuels mixed with

[1]General Assembly. Transforming our world: the 2030 agenda for sustainable development. https://sustainabledevelopment.un.org/post2015/transforming our world. Accessed 21 Oct 2015.

[2]L. Steve, Renewable energy world, U.N. Secretary-General: renewables can end energy poverty. http://www.renewableenergyworld.com/articles/2011/08/u-n-secretary-generalrenewables-can-end-energy-poverty.html. Accessed 25 Aug 2011.

[3]Renewables 2018 Global Status Report. Read more at: http://www.ren21.net/status-of-renewables/global-status-report/

conventional fuels, natural gas vehicles, and electrical transport. The contribution of renewable sources is gradually increasing in this sector. Electrical systems are known to offer viable power storage in the form of batteries. These sources are increasing, although electrification of transport faces several challenges, including cost, limited driving range, short battery life, and charging infrastructure [1].

In discussions related to the electrification of vehicles and the use of portable electric appliances, the focus is on batteries and capacitors as storage systems. Conventional Electric vehicles (EVs) and portable appliances were powered by lead-acid and nickel-based batteries. These nickel cadmium and nickel-iron batteries were the first of currently used nickel metal hydride (third-generation) batteries, which are the best choice for vehicles thanks to their low operating voltage and low reduction potential for water (1.2 V) in heavy electrodes (20–70 Wh/kg). But hybrid electric vehicles (HEVs) and plugin HEVs (PHEVs) require different materials depending on their intended use. Usually, HEVs are charged by the regeneration of power during breaking and cover a few kilometers in electric mode, while PHEVs are charged by direct powering and give full support on electric power and cover comparatively larger distances in electric mode. Thus, they require different battery materials with different specifications, materials with excellent parameter values like energy density, power density, low weight, high cyclability, and safe/reliable operation. These materials were replaced by sodium nickel chloride (Na-$NiCl_2$). The materials of Li-ion batteries have evolved over time and are comprehensively discussed by M. Thackeray et al. [2]

Though other batteries may be mentioned, the current write-up core focus is on lithium-ion batteries (LIBs). Lithium-ion (Li-ion) chemistry for energy storage is the most appropriate technology of choice in current portable electronics, mechanical tools, and especially in electrified transport like HEVs and PHEVs. These have evolved enough to dominate consumer electronics and transportation. This is because of their outstanding qualities like specific energy (J/kg or Wh/kg), high energy density (Wh/I or J/I), specific power (W/kg), power density (W/I), safe operation, low cost, and long battery life [3]. Their high efficiency also allows their use in grid stations to improve energy-harvesting projects, including wind energy, solar energy, and other renewable energy sources. Yet there are gaps to be filled, like that they require long charge times and safety issues for high-temperature regimes. Hence, work is still being done in this area, and it is a hot topic of research interest around the world in both public- and private-sector funding agencies. Current issues of battery explosions, for example, the Samsung Note7 phone and Boeing's Dreamliner airplane, have raised serious safety concerns [4]. LIBs are unique compared to other water-based technologies like lead-acid and nickel-cadmium in that they have persistent issues, such as their ability to work only in a +50 °C to −50 °C temperature range, which limits their use in terrestrial environmenst; they are also heavy, and safety issues make them inferior to LIBs.

While discussing the aforementioned quality parameters of Li-ion batteries, we argue that these parameters are controlled by the characteristics that control all electrochemical reactions, including surface area, the potential difference between electrodes, ion diffusion properties, and related kinetics. All advances in batteries are linked to electrochemical processes that are inferred to evaluate the performance and stability of cells [2]. The evolution of LIBs is one of innovation, miniaturization, and development of new chemistries, although LIB development is still research in progress. Ongoing global research activities point to hopes that LIBs will be reliable and efficient energy sources into the future.

7.2 Construction and Functioning of LIBs

LIBs basically consist of electrodes, i.e., cathode and anode, separated by a solid electrolyte. These electrodes are connected externally by a circuit that allows electrons to flow, and internally these are connected by the electrolyte that allows the diffusion of Li ions but doesn't allow electrons to pass through [6]. The basic configuration remains the same throughout the historical evolution of LIBs [7], and these are very similar to conventional batteries in construction and oxidation-reduction reactions taking place at electrodes. The difference lies in the fact that in conventional batteries, an oxidation-reduction reaction occurs simultaneously at interfaces, but in LIBs these reactions occur with the mass diffusion of Li ions along the electrolyte. During the charging process, energy is provided externally and Li^+ ions are removed from the cathode and diffuse through the solid electrolyte toward the anode. During the extraction of positive Li ions, to maintain charge neutrality, an electron is lost from the electrode material and flows to the external circuit. The electron flows in the wires of the external circuit because it cannot pass through the electrolyte. In this way, the Li ion moves through the solid electrolyte, whereas electrons move from the external circuit and then meet at the anode, thereby causing Li ions to complete one cycle. The same process is reversed in the discharging process when we connect any load in the circuit. The complete process is elaborated in Fig. 7.1.

7.3 Cathode Materials in LIBs

Although all the main components including electrodes and electrolytes play an important role in electrochemical reactions and affect the quality parameters of the cell cathode, the material is of prime importance. The cathode is the transit station for ions and electrons. The material of the cathode must deal with undesirable phase transitions and chemical reactions, including parasitic reactions at the electrode/electrolyte interface and oxidation/reduction reactions. These may cause temperature-dependent destabilization, including volumetric changes during the insertion and removal of lithium ions, which has made cathode material a central subject of research [3]. The insertion of ions in a cathode occurs during the spontaneous process of battery discharge, extraction of ions take place during the charging process for which we need to provide the external electric potential. Hence, a functional material capable of efficiently and reversibly shuttling Li ions through it self and ensure the physical and chemical stability of the material can be used as cathode material for LIBs.

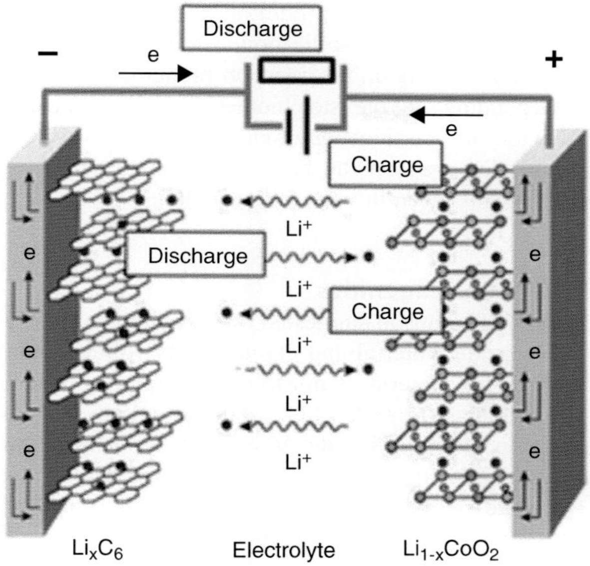

Fig. 7.1 Process of charging and discharging in LIBs. (Reprinted from Ref. [5] Copyright 2013 with permission from Elsevier)

For storing and shuttling of Li ions in some material, various electrochemical processes are involved; they can be categorized as follows: (1) intercalation, e.g., lithium manganese dioxide ($LiMnO_2$), lithium iron phosphate ($LiFePO_4$), and various subfamilies [8, 9], and (2) conversion, e.g., selenium oxide (SeO_2), tin oxide (SnO_2), and iodides (MI_x) [7]. Some also consider a third process of alloying to complete this list, but due to the generic nature of the present discussion, we will consider the most viable, applicable, and fundamental families of electrodes. Keeping these in mind, we will discuss intercalation electrodes in Sect. 1 and conversion electrodes in Sect. 2.

7.3.1 Intercalation Cathode Family

Intercalation cathodes, as the name indicates, are a family of materials that shuttle ions back and forth in a cyclical manner through their structure without any chemical change in their structure. Since electrodes have a direct interface with the electrolyte to perform activities, electrodes must possess such characteristics as electric conductivity, ionic conductivity, and a varying chemical potential for the shuttling species so that it may act as host in the intercalation mechanism. With these characteristics, if M is the metal, then the intercalation reaction must be

$$M^+ + e^- + Host \leftrightarrow < M^+Host >$$

This describes clearly that how a metal cation loses an electron (oxidized) to shuttle out from the host structure reversibly to ensure the charge–discharge process.

Obviously, the host (electrode) material plays a key dual role in the ring opening process and charge stabilization, which affects the rate of the charge–discharge process depending on the transport properties of the material.

Depending on the stoichiometry, thermodynamic kinetics, and intercalation behavior, the classification of electrodes has been clearly described by the Armand group. However, on the basis of the intercalation families, the classification of electrodes is given in what follows [10].

7.3.1.1 Chalcogenides

In 1978, Thompson suggested that titanium sulfide (TiS_2) could be considered as an electrode material and explained its voltage charge curve in Li/$LiTiS_2$ electrodes. This material is not found naturally, so it is always synthesized and was first produced in 1975 [11]. Titanium sulfide (TiS_2) is a semimetal with a very common cadmium iodide structure. It has a space group of $P3ml - d_{3d}^3$ that has point group symmetry that is mainly trigonal [11] and a layered structure where Ti is sandwiched between two S layers, with each sandwich composed of planes of titanium atoms that reside between sulfur atoms, and these layers are attached to each other by weak Van der Waals forces. Crystalline indexing can be done by a hexagonal a structure with a = 3.407 Å and c = 5.695 Å. These have a layered structure, enabling the shuttling of Li ions in the structure. We generally show this by an MX or MX_2-type structure, where M represents any metal from group 4 to group 10 and X represents chalcogen atoms, including S, Se, and Ti. The crystalline structure is shown in Fig. 7.2. A large interlayer gap facilitates Li-ion diffusion.

The Li^+ storage capacity of transition metal (TM) dichalcogenides up to the present level is 1000 mAh/g, which is much higher than currently used graphite electrodes that have a Li storage capacity of 372 mAh/g [13, 14]. A number of examples have shown excellent performance of LIBs [15–20].

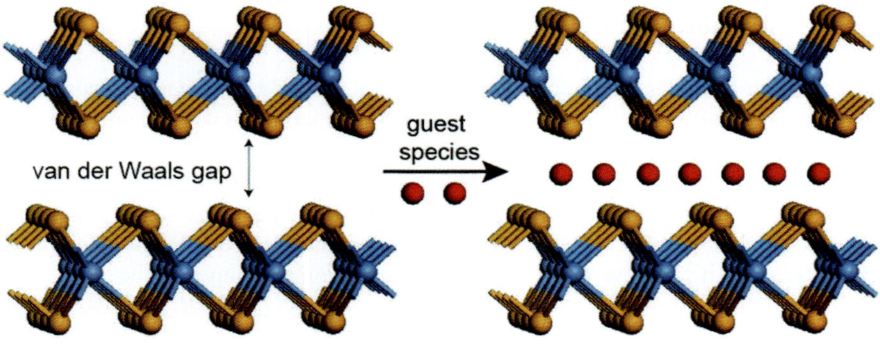

Fig. 7.2 Layered structure of transition metal (TM) chalcogenides showing intercalation of Li ion through it. (Reprinted from Ref. [12]. Copyright 2016 with permission from Royal Society of Chemistry)

The crystalline structure of Li intercalated TM chalcogenide lithium titanium sulfide ($LiTiS_2$) has been studied using diffraction analysis, and it was found that the structure could still be explained in terms of the hexagonal unit cell. However, it was found that the c-axis lattice parameter increased due to the insertion of Li ions, which is the main effect of Li insertion. No structural phase transition was found during this process, which shows that these materials could be used as an intercalation electrode [21–24]. Titanium sulfide–based electrodes still have a higher energy density and intercalation kinetics compared to layered oxide materials like lithium cobalt oxide ($LiCoO_2$) and lithium manganese oxide ($LiMn_2O_4$) [23, 25–28], as we have already described these have layered structures. When Li ions intercalate across layers, the transfer of electrons occurs from Li atom to metal cation in the structure and weak van der Waals forces are replaced by strong Coulombic interactions. The strong forces bind the structure and make it resistant to expansion. Hence, a reliable intercalation framework that supports Li kinetics in its structure is made available by elucidating the structural elastic and electronic effects during intercalation [24, 29–34].

7.3.1.2 Layered Oxides

The earliest commercialized material in LIBs were layered oxides, specifically lithium cobalt oxide ($LiCoO_2$). This material has a redox potential of 4 eV and its structural and diffusion mechanism features are described in what follows.

Depending on the stacking of oxygen, these materials have three main types of structures:

(1) ABCABC octahedral type (O_3),
(2) ABBA prismatic type (P_2), and
(3) ABBCCA prismatic type (P_3).

All these materials have layered geometries like chalcogenide materials, with a metal cation residing between two oxygen atoms, and this composes the layers of TM cations that are separated from the Li layers by oxygen; thus, overall there are tetrahedral and octahedral geometries as Li resides in an octahedral site surrounded by oxygen and it is neighbored by tetrahedral geometry from the metal cation, as shown in Fig. 7.3 for the $LiCoO_2$ structure.

The metal could be manganese (Mn), cobalt (Co), or iron (Fe), for example. In the case of a layered structure where Li-centered octahedral sites are coordinated with P-centered tetrahedral sites, Li-ion hopping or diffuses from an octahedral lattice site to another similar site via a transitional tetrahedral site [35]. It has been found that an intermediate tetrahedral site is the highest-energy-activated site throughout the transition and acts as the energy barrier during the Li hop, and this determines the hopping rate of the Li ion, which must overcome this energy barrier to complete the hop. It has been observed that a decrease in the energy barrier of 57 eV can increase by ten times the Li-ion migration. By calculating this energy barrier, the values of hopping rate and hence the power density can be found for cathode materials [35]. The diffusivity mechanism in lithium iron phosphate (Li_5FeO_4) has been studied by atomistic simulation to explore the possible paths of Li-ion migration [36]. Two channels for the migration have been identified, channel X and channel Y (Fig. 7.4), which have respective values of activation energy of 0.45 and 0.56 eV.

While increasing the power density of the electrode material, efforts are being devoted to optimizing the energy barrier during Li-ion hopping and to control other battery parameters that are linked with Li-ion hoop. For example, it has been demonstrated that doping of lithium manganese oxide ($LiMnO_2$) with zinc (Zn), chromium (Cr), cobalt (Co), Fe, and aluminum (Al) can enhance the stability of the battery material and result in better capacity retention with higher charge–discharge cycles. A Similar investigation was conducted using cationic and anionic doping in

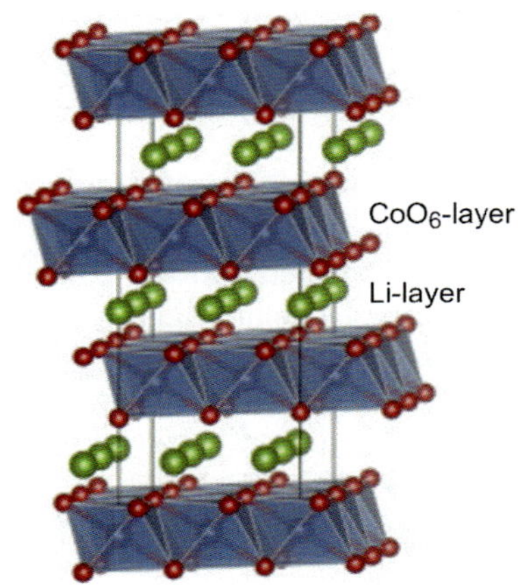

Fig. 7.3 Crystalline structure of layered oxide cathode material. (Reprinted from Ref. [5]. Copyright 2013 with permission from Elsevier)

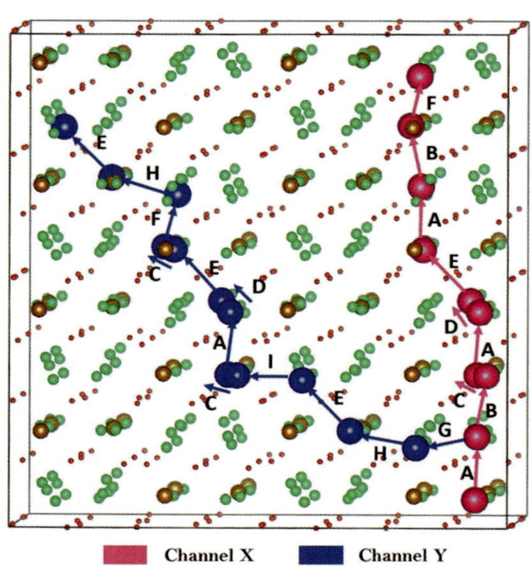

Fig. 7.4 Li migration path in Li_5FeO_4 material. Two possible paths of Li ions (blue and pink spheres) are shown, channel X and channel Y. (Reprinted from Ref. [36]. © Open access permission from Nature)

the Lithium Managnese oxide ($LiMnO_2$). in cationic doping host structure was doped with Magnesium (Mg) Titanium (Ti) Vanadium (V) Niobium (Nb), Iron (Fe) Ruthenium (Ru) Cobalt (Co), Nickel (Ni), Copper (Cu) and Aluminium (Al). While in anionic doping Nitrogen (N) and fluorine (F) doped $LiMnO_2$ was studied separately [37]. The material remained stable at all doping concentrations, and it was observed that some of the cations, like V, Nb, and Ru, significantly changed the redox potential of the host. An increase in capacity due to transition metal doping is also linked with oxygen loss in the cathode. In the charging process, electrons are compensated by the oxidation of metal cations; upon complete oxidation, for example, Mn^{4+} ($3d^3$) cannot oxidize any further, and bound oxygen get free on oxidation and leave the structure. Electrons are given during this oxidation process, and this results in an increased capacity of the cathode material. This oxygen loss is directly linked to metal-ion substitution in the material and thus plays a role in determining the specific capacity of the material. For example, the substitution of Co metal cation in the material enhances oxygen loss, while Ti inclusion reduces it. The explanation lies in the fact that mutual overlapping of Co 3d-states with oxygen 2p-states, increases the covalent character, and increased delocalization affect the stability of oxygen. This high value of covalent behavior is not observed in Ti substitution, and this does not promote oxygen loss in the material [38, 39]. The evolution of singlet oxygen followed by carbon monoxide (CO) and carbon dioxide (CO_2) at a high charge conduction state could be the result of an oxidation reaction, which is a big reason for the capacity fade problem in LIBs [40].

A recent study was conducted to explore an interesting aspect of this oxygen release at a high state of charge condition [41]. Lithium-nickel-cobalt-manganese oxide is a promising layered oxide cathode material in LIBs. It has a theoretical specific capacity of 278 mAh/g in a Li intercalation potential range of 3.0–5.0 V [41]. While the entire spectrum of specific capacity is still open to use, as the battery has a cut-off operating voltage of 4.3 V, at this specific value, only 60% of the total Li ions have been extracted, that results in a specific capacity of 160 mAh/g [42]. Li intercalation cannot be increased above a voltage of 4.3 V, at which point the material capacity decreases, reducing battery life. Other reasons for the decrease in operational lifetime include electrolyte oxidation [43], dissolution of metal cations [44–46], structural evolution during Li charge–discharge cycles [47], an increase in cathode impedance [42, 48, 49], and decay of Li ions in the material [46].

Further, different metals have varying oxidation and reduction capabilities, for example, Ni has the ability to reduce instantly, while Mn has a favorable reduction capability, so understanding these concepts helps to optimize synthesis conditions [50]. Efforts are in progress to use lithium-nickel-manganese-cobalt oxide ($LiNiCoMnO_2$) as anode material in LIBs, which may solve problems of safety and low capacity in previously used graphite and lithium titanate ($LiTiO_{12}$) anodes.

7.3.1.3 Spinal Oxides

A spinal-like structure was identified for lithium-cobalt arsenate ($LiCoAsO_4$), a novel high-voltage material. It has a reversible capacity of around 100 Ah/Kg at a cut-off voltage of 4.8 V.

Structure of Spinal Oxide

This material has an olivine structure but is unable to retain its structure under high-pressure conditions, where it undergoes a phase transition from olivine phosphate to a spinal structure [51], with Li^{+} and Co^{+} occupying an octahedral site and As^{+} occupying a tetrahedral site. Similarly, lithium-manganese oxide $LiMn_2O_4$ has spinel structure. With lithium insertion, the Li ions reside at tetrahedral sites of the structure, Mn atoms occupy octahedral 16d sites, and oxygen atoms occupy 32e sites, and the whole structure is arranged in a cubelike structure [52]. This forms a 3d repeating manganese-oxide (Mn_2O_4) framework with an edge sharing octahedral form having intermediate tunnels for Li-ion percolation, as shown in Fig. 7.5 [52].

As electrode materials, composites from the spinal family have good electronic and chemical properties with respect to energy storage, as exemplified by lithium-manganese oxide ($LiMn_2O_4$), an excellent electrode material thanks to its low cost, its environmentally friendly properties, and safety concerns [54, 55]. However, this material has a major capacity-fading problem under high flux and elevated temperature conditions. A number of reasons have been proposed for this fading [55–57], such as electrode–electrolyte chemical reactions on the surface of the electrode, manganese (Mn) dissolution in electrolyte, Jahn–Teller distortion, phase transformation during battery operation, particle size, and morphology. For example, octahedron-shaped cathode particles with 111 surface terminations have a retention capacity that is superior to that of pallet-shaped lithium-manganese oxide (LMO) with 001 surface termination. The reason for this is given in terms of dissolution of Mn that was more prominent in the case of 001 surface terminations [58].

Diffusion Mechanism in Spinal Oxides

The diffusion mechanism pathways have also been studied using experimental and theoretical techniques. The spinal structure exhibits good stability on the process of

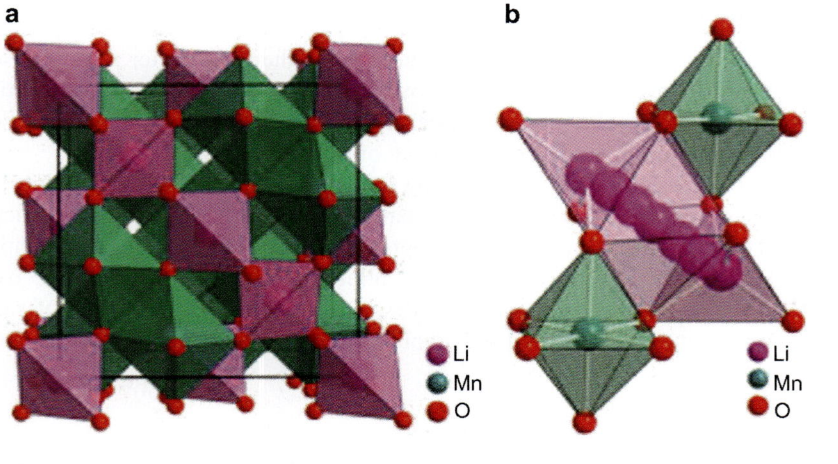

Fig. 7.5 Crystalline structure in (**a**) and migration pathway for spinal structure in (**b**) for Mn_2O_4. (Reprinted from Ref. [53]. Copyright 2013 with permission from Elsevier)

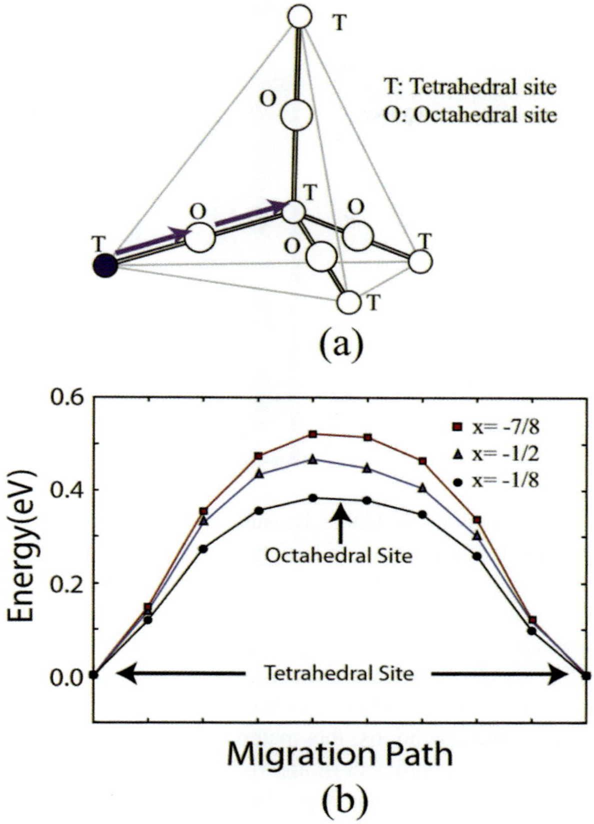

Fig. 7.6 (**a**) Li diffusion mechanism as O-T-O migration in the material, where T is a tetrahedral and O an octahedral site. (**b**) Energy changes upon hopping of ion at three different Li compositions in structure $Li_{1+x}Ti_2O_4$. (Reprinted from Ref. [60]. Copyright 2010 with permission from American Physical Society)

intercalation of Li ions, and the intercalation path is given from tetrahedral sites to octahedral sites, and this is thermodynamically at a specific Li-ion concentration of 0.250 ppm [59]. This diffusion mechanism has also been verified in lithium titanate ($Li_2Ti_2O_4$), where Li diffuses between two consecutive octahedral sites through a transitional tetrahedral site, and this energy barrier varies with Li-ion concentration inversely; the barrier variation is shown in Fig. 7.6 [60, 61].

7.3.1.4 Olivine Phosphates

The olivine phosphate material family is considered the best for use as positive electrodes in LIBs; such materials have a formula of lithium metal phosphate ($LiMPO_4$). LiMPO4s are composed of regularly arranged phosphate anion (PO_4^{3-}) tetrahedral units that actually separate the two-corner sharing octahedra that are based on metal oxide (MO_6) and lithium oxide (LiO_6) (where the metal could be iron, Fe); overall this crystallizes into an orthorhombic structure. Both of these octahedra are distinct and are of different sizes; in this way they are different from the spinal family of materials [61, 62] (Fig. 7.7).

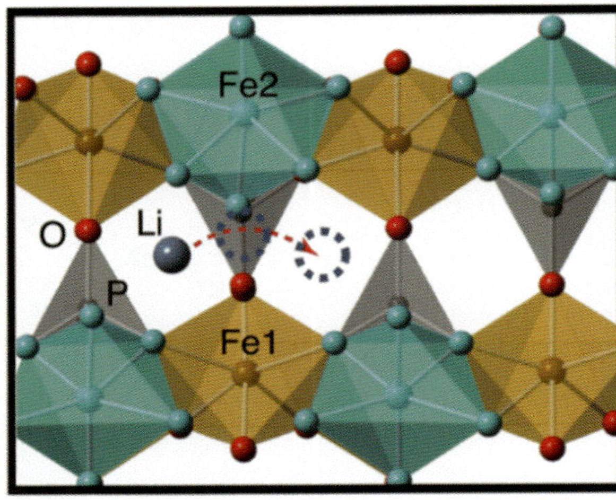

Fig. 7.7 Structure of olivine phosphate showing Li hooping path. (Reprinted from Ref. [63]. Copyright 2011 with permission from American Chemical Society)

This material is viable because of its overall behavior in each required quality parameter of low cost, good safety conditions, working potential of 3.4 V, and a capacity of 170 mAh/g [64]. These materials approach their theoretical capacity of 170 mAh/g. However, $LiFePO_4$ is considered more advantageous than other members of this family because of its good quality parameters: safety, low cost, and high capacity. However, $LiFePO_4$ also has the drawback of low electronic and ionic mobility that Limitize its application in the power electronics and automobile industry [65, 66]. At ambient temperature conditions, this material has an electronic conductivity of 10^{-9} Ohm/cm and diffusion coefficient of approximately 10^{-14}–10^{-16} cm^2/s. These parameters were calculated using various experimental and theoretical techniques. It has been seen that structural evolution plays an important role in determining Li-ion mobility in a structure, and various techniques have been developed to increase ionic mobility in structures by doping and codoping with TM cations and using different operating conditions [67–70].

The Li-ion migration pathway is once again found happening between two neighboring octahedra through an intermediate tetrahedron. A Li ion will be stable at an octahedral site having a vacant tetrahedral neighboring site. Theoretical investigation of the diffusion process of Li ions using first principle–based calculations has been performed by Henkelman et al. It has been formalized that a Li-ion diffusion barrier is strongly influenced by a crystalline environment, types of defects, and Li-ion concentration in the material. Since theoretical values do not correspond to experimental values, first, they tried to analyze the reason behind this by calculating the diffusivity at different levels, including surface diffusion, bulk diffusion, and cross-channel diffusion; they also showed that lattice strain also plays a key role in determining the Li-ion diffusion barrier [71].

7.3.1.5 Silicates

Silicate-based cathode materials with formula unit Li_2MSiO_4 (here M = Fe and Mn) have attracted attention due to their very high values of theoretical specific capacity while maintaining safety and cost values in the range of commercialized cathode families. Progress with this family of materials is being made, and they are expected to meet high-energy-density requirements. They have the ability to make Li ions shuttle back from a structure, doubling the value of the material's capacity, and Mn's rapid oxidation state of +4 plays a key role in the process, and we can achieve a theoretical capacity in a range of 330 mAh/g with a working voltage of 4.0 V [65, 72–79].

Just like $LiFePO_4$, this is also a polyanionic material that contains a distorted hexagon with an edge-sharing tetrahedron. The tetrahedral and octahedral sites are occupied by Li, Mn, and silicon (Si) variously, and it has a number of phases identified that vary upon Li extraction and insertion. The recognized phases for Li_2FeSiO_4 includes orthorhombic Pmn21, Pmnb, and monoclinic Pn, P21/n, of which the Pmn21 orthorhombic phase is considered thermodynamically most stable. It shows the lowest migration barrier and gives favorable Li-ion mobility.

$Li_2Mn_2SiO_4$ has technical challenges, such as phase variation during operating by diverting from the preferred phase of Pmn21 during synthesis and operation. This phase shifting may be due to the presence of Mn antisite defects and a Jahn–Teller distortion of Mn^{3+} ions, which lead to structural destabilization [77–79] and thereby producing Li-ion diffusion that results in a small range of theoretical capacity. Other barriers include low electron mobility that is four times lower than olivine-type phosphate materials [65, 78, 79]. That is a major barrier to Li-ion extraction and thus limits its application as an electrode material. Various techniques have been applied to eliminate these drawbacks, including carbon nanocoating [80, 81], substitution of metal cations in the structures [82, 83], and nanostructural modification [84, 85].

7.3.1.6 Tavorites

The tavorite structural compound also belongs to the polyanion cathode family which is well known because of the strong covalent bonds. In this family, lithium-iron phosphate ($LiFePO_4$) is a well-known cathode material. The tavorite structure compound has a formula unit $LiM(TO_4)X$, where M is any metal cation and X is any p-block element. The tavorite structure has the ability to shuttle back and forth two Li ions through the cathode material at the same time, thereby doubling the diffusion of Li ions as well as the power density of the material [86].

Lithium-vanadium phosphate ($LiVPO_4$) has a different crystalline structure depending upon the nature of the metal cation used in the material. However, the general structure is composed of corner-sharing vanadium oxide (VO_6) octahedron-forming chains. These chains are in the [010] direction and connect to each other via a $P(1)O_4$–$P(2)O_4$ tetrahedral configuration. The P–O distance varies from 1.50 to 1.58 Å [87]. The structure is elaborated in Fig. 7.8.

Some of the major compounds of this material family include lithium iron sulfate fluoride ($LiFeSO_4F$) [89–92], lithium vanadium sulfate fluoride ($LiVPO_4F$) [93–95], and lithium vanadium phosphate oxide ($LiVPO_4O$) [96–98]. These have been recognized as promising materials because of their important features like high operating voltages, fantastic structural stability on Li-ion intercalation, and simultaneous intercalation of two Li ions. Vanadium-based tavorites are more advantageous because V has several values of oxidation state from +2 to +5 and can donate more

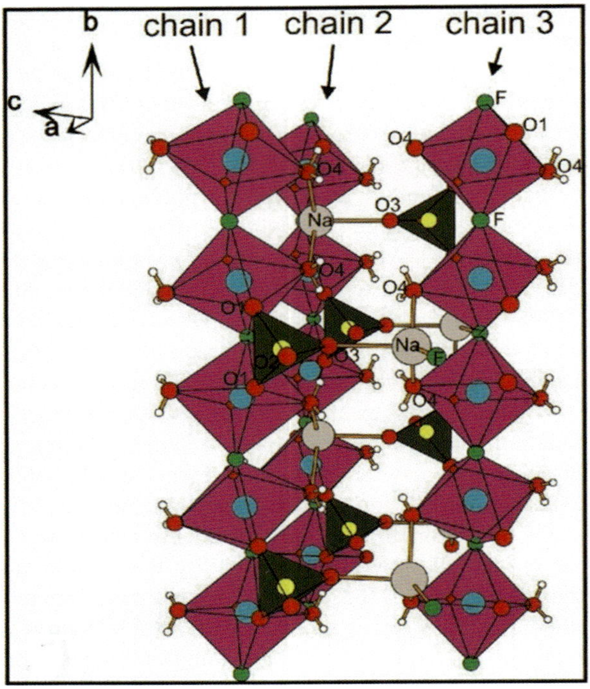

Fig. 7.8 Elaboration of tavorite structure in the form of chains. (Reprinted from Ref. [88]. Copyright 2010 with permission from American Chemical Society)

than one electron at the same time. The very high specific capacity of 312 mAh/g for $LiVPO_4F$ can be achieved by utilizing more than one redox couple at the same time, e.g., $LiVPO_4F$ has a redox potential of 4.2 V for V^{3+}/V^{4+} and redox potential of 1.8 V for the V^{2+}/V^{3+} redox couple. While a theoretical capacity of 318 mAh/g in $LiVPO_4O$ can be achieved by utilizing more than two redox couples, it has a redox potential of 3.95 V for V^{4+}/V^{5+} and redox potential of 2.3 V for V^{3+}/V^{4+}, and utilizing more than two redox couples is not practical. As we can see, the very low redox potential of 1.8 V in V^{2+}/V^{3+} for $LiVPO_4F$ makes it impractical to utilize $LiVPO_4O$, which also has to face the challenge of poor cycle stability [95, 96].

Using first principles and molecular dynamics–based techniques, the diffusion path and activation energy barrier of Li-ion transport in materials have been studied. It has been calculated that Li-ion diffusion adopts either a 1D pathway through intermediate voids along the chain [96] or a 3D pathway in the structure [99].

7.3.2 Conversion Cathode Family

Although intercalation cathodes are promising cathode materials and they are widely adopted, there is a family of widely studied cathode materials due to its benefits in comparison to intercalation electrodes. They are cheap, abundant, and environmentally friendly. They are of an AB_2 type structure with metal cations A = S, Fe, Cu,

Co, Ni, and many others as well, and anions B = O, S, or halogens. Halogens are considered poisonous in pure form, but as salts, they are stable, safe, and environmentally friendly. Conversion cathode materials can store two Li ions per chalcogen, this is not the same in case of inctercalation cathode, where ratio of Li ion per cation is even small. They form eco-friendly oxygen-based cathode chemical compounds including lithium hydroxide (LiOH) and lithium oxide (Li_2O_2) [100, 101]. Among all the families of conversion salts the oxide class exhibits the largest values of specific and volumetric capacities, e.g., Li_2O_2 has a volumetric density of 2699 mAh/cm^3 with a specific capacity of 1670 mAh/g. Among the halogens and metal halide family, metal fluorides are proving themselves promising candidates, showing high specific capacity values with a high potential at an intercalation voltage of 3.5 V, a high gravimetric density of 2196 mAh/cm^3, and a high specific capacity of 713 mAh/g^1 [102].

In the conversion cathode family, solid-state redox reactions occur at the cathode in the charging–discharging process of LIBs. During these reactions, the crystalline structure of the material changes due to the breaking and formation of new bonds during such reactions, given as follows:

$$MX_z + YLi \leftrightarrow M + zLi_{y/z}X$$

Here M is a TM ion such as Fe, Ni, Cu, or Mn, and X is a halogen atom such as F, chlorine (Cl), bromide (Br), or iodine (I), or it may be a chalcogen atom like sulfur (S) or selenium (Se) as metal salts. It could also be an oxide ion, but that results in a very low redox potential of the electrode, written

$$YLi + X \leftrightarrow Li_yX$$

Thus the whole conversion reaction [102, 103] can be divided into the following categories:

1. Chalcogens,
2. Chalcogenides,
3. Halogens, and
4. Halides.

The details of each individual family is summarized in what follows.

7.3.2.1 Chalcogens

It is commonly known that the oxygen family is a group of 16 elements in the periodic table. One of the unique elements named livermorium (Lv) also included in the chalcogens, other elements called sulfur, tellurium (Te), and selenium. The valence shell contains six electrons in the outermost shell, with general oxidation states of +2, −2, +6, and + 4 during bond formation. The structure of every element has vital importance in every respect. Many properties depend on the structure of a specific element. The chalcogens have various types of structure, for example, the monoclinic structure of oxygen, the orthorhombic structure of sulfur, and the hexagonal structure of selenium and tellurium. Depending on their composition, chalcogens may be toxic or nontoxic [104].

Chalcogens are further classified into chalcogenides, halides, and halogens.

7.3.2.2 Chalcogenides

Chalcogenides are an important class of chalcogens [105]. They differ from chalcogens in that they have one most electropositve element, unlike chalcogens. Common examples of chalcogenides are sulfides, tellurides, and polonides (Po). Mono-chalcogenides have a zinc blend structure. Chalcogenides can be of various types like dichalcogenide and Tri chalcogenides.

7.3.2.3 Chalcohalides

Just as chalcogens form chalcogenides, they form halides by combining with other elements like halogens, and these are known as chalcohalides. The halogen group includes elements such as Cl, Br, I, and fluorine. The heavier chalcohalides work efficiently in molecular interactions, but they have the drawback of being toxic. Common examples of chalcohalides include diselenium dichloride (Se_2Cl_2), tri-tellurium dichloride (Te_3Cl_2), and the less stable ditellurium dichloride (Te_2Cl_2).

Conversion cathodes are used because of their many applications in batteries thanks to their long life [3]. They include the specific family of materials like fluorides and chlorides in metallic combinations. There are many reasons to use these materials as conversion electrodes; one reason is to have an intermediate operation voltage, but nonetheless they have drawbacks, including, for example, their poor conductivity and hysteresis losses [106]. Copper chloride has an octahedral structure, but it rarely exhibits this form due to Jahn–Teller distortions; further, different materials define respective elemental properties of that material. One of two materials AgCl and $CuCl_2$ has been studied. Because of their higher oxidation state they utilize double electron transfer and used as efficient electrochemical conversion cathode. These materials are used due to the reason discussed as AgCl enhances the phase transformation reaction with the production of carbon which will be conductive [107]. It has been shown in the literature that metal chlorides can be used as efficient conversion cathodes because they have cations of relatively high oxidation states, which can be helpful to proceed phase transformation reactions which are necessary for it. The lower attraction of chlorides is due to the low the theoretical reticle capacity of chlorides compared to fluorides. Common chlorides are $CuCl_2$, $NiCl_2$, $CoCl_2$, NiC_{12}, and $FeCl_3$ which have the highest energy densities; however, one challenge remains: the corrosion of the metallic surface during their application. A typical chemical reaction is as follows [102]:

$$mLi^{+} + mCl_{m} + me^{-} \leftrightarrow mLiCl + m^{0}$$

These two types of families are commonly studied in LIB chemistries as intercalation electrodes and conversion cathodes. Research has shown that intercalation cathode material is uneconomical and highly toxic reagents, etc., with limited electrode stability, particularly under high flux or temperature, where the formulation of different chemistries is geared toward high charge storage and operational safety of the LIBs. Enhancing the capacity of intercalation cathodes above a cut-off voltage level leads to operational safety issues. To overcome these drawbacks, conversion cathode chemistries were developed, utilizing the same families of cathode materials such as chalcogenides and halides, for example. The theoretical capacity of metal halides is due to the reduction of metallic ions by the involvement of one lithium atom during conversion reactions. A comparison of the capacities of chalcogenides and chalcohalides is shown in Fig. 7.9.

Other halide composites include salts of fluorine for use in LIBs that serve as a conversion cathode material known as iron trifluoride (FeF_3), which has a high energy density and high theoretical capacity of 712 mAh/g [108]. This material has the capacity to work at its maximum operating capacity but exhibits low kinetics and poor electrochemical performance with nonoptimal characteristics. Efforts have been made to overcome these drawbacks by forming nanowires of iron trifluoride [109]. Other important fluorides are copper fluorides. Peculiarities that make them effective are their high operating potential and high ionic characteristics of the metallic composite with a high electronegativity of fluorine. The bandgap becomes

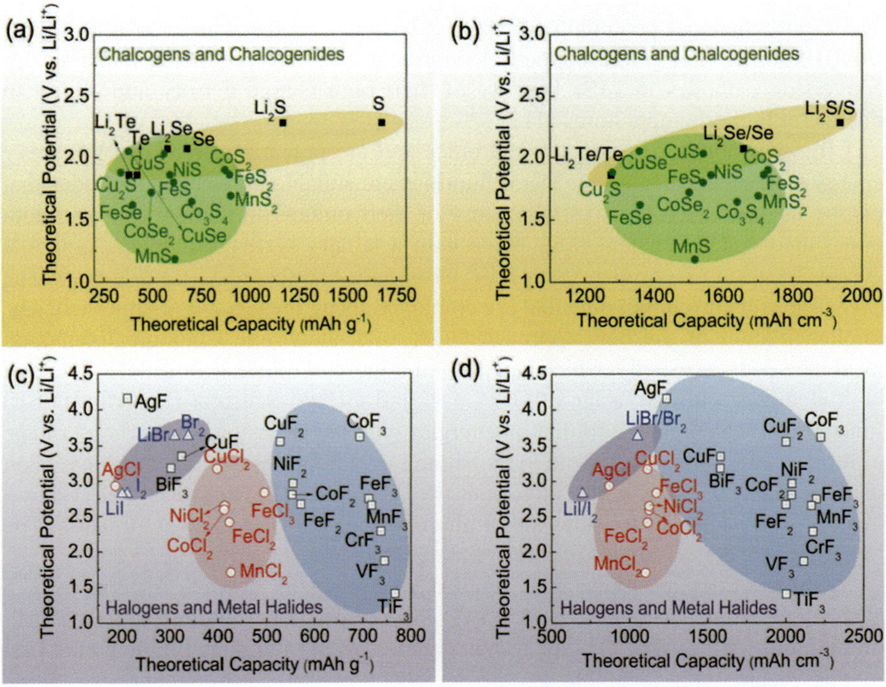

Fig. 7.9 Comparison of values of theoretical capacities and theoretical potential of chalcogenides and chalcohalides. Reprinted from Ref. [102]. Copyright 2016 with permission from Royal Society of Chemistry

wider in this case but conductivities increase with nanosizing of the formulation. With increasing energy demands, there is a need to develop materials that possess high-capacity storage with minimum loss of charge and safeguard against safety issues, such as overheating. Striking a balance between materials that are environmentally friendly and cost-effective yet exhibit high charge densities is difficult as most combinations that make sense have already been evaluated. To predict the chemical properties of binary, ternary, and higher order metal oxides, density functional theory (DFT) has been used to mine for practical formulation from an array of hypothetical possibilities with the time constraints of actual synthesis and evaluation saving time and resources and enabling the researcher to focus on practical formulations which are predicted to exhibited the required characteristics of high charge, low cost and operational safety. Iron trifluoride is a material that has been studied, at the moderate potential of 2.7 V. Iron trifluoride reacts actively with three lithium ions to deliver the maximum theoretical power, which can be shown in the following equation:

$$FeF_3 + 3Li^+ + 3e^- \leftrightarrow 3LiF + \alpha - Fe(4.5 - 1.5\text{ V } \mathit{vs\ lithium\ ion/lithium})$$

One more important aspect of fluoride is that the structure of Li-Fe-F is very important in the ternary phase when iron the fluoride readily reacts with lithium [110, 111]. The electrochemical potential of lithium determines all the voltage profiles [103]. Iron(II)fluoride (FeF_2) and copper(II) fluoride (CuF_2) are promising materials for conversion cathodes in LIBs because of their high charge density and ability to operate at higher temperatures [112].

LIBs are robust for portable electronics and are potential candidates for heavy energy storage for future use. Metal fluorides are suitable conversion cathodes that can be used effectively in LIBs because of their unique properties. Density functional theory (DFT) calculations made with a ternary system employs a potential energy function determined via a DFT basis with a Hubbard U Hamiltonian. The exchange and correlation potential are employed using the GGA + U approach. The implementation of the new approach GGA + U shows the Coulomb interaction inside the d-states exactly, which can play an important role in the orbital systems of both elements with the usage of GGA + U and a Heyd-Scuseria-Ernzerhof (HSE) functional to calculate the formation energies with the inclusion of insertion voltage of lithium ions with a composite of transition metal fluorides to study the conversion processes. The study of iron(II) fluorides (FeF_2) and iron(III) fluorides (FeF_3) shows that FeF_2 immediately degrades into lithium fluoride (LiF) and Fe via lithiation. The lithiation process completes in three phases, which are shown in Fig. 7.10.

Metal fluorides are extensively studied as conversion cathode material with a trigonal pyramidal geometry [113, 114]. An intercalation cathode transports only one lithium ion per formula unit to host lattices. Metal fluorides like FeF_2 are best in this regard with favorable geometry, reliability, and cost-effectiveness. Some metal fluorides like bismuth(III) fluoride (BiF_3), FeF_3, cobalt(II) fluoride (CoF_2), and nickel(II) fluoride (NiF_2) have the ability to show the highest voltage of hysteresis and lowest kinetics with lower reversibility of true conversion.

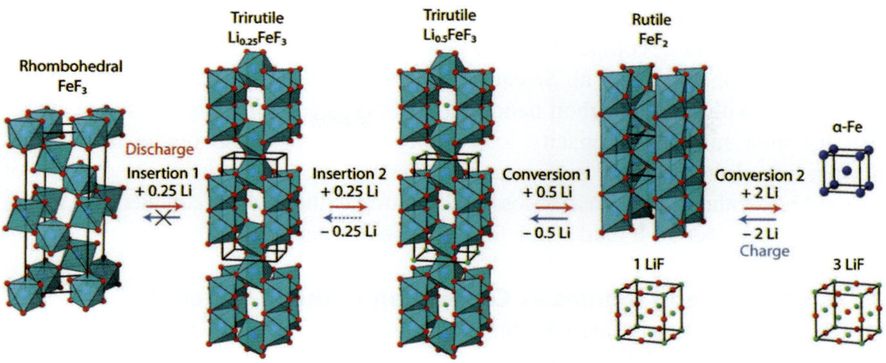

Fig. 7.10 Description of lithiation path of FeF_2 and FeF_3 using density functional theory (DFT) Heyd-Scuseria-Ernzerhof (HSE) correlation. (Reprinted from Ref. [112]. Open access. Copyright © 2016 American Chemical Society)

7.3.2.4 Metal Sulfides

Among the materials used as conversion cathode are metallic sulfides because they have a large capacity that is theoretically measured by its cost-effectiveness. The abundance of sulfur is another one of their advantages; however, one disadvantage is their low potential for lithium/lithium ions [115]. Sulfide dissolution produces polysulfides, which stimulate the loss of a percentage of sulfur under vacuum. This kind of effect can be overcome by insulating the sulfur in a hollow structure to work effectively [115]. Future storage devices demand highly efficient energy storage devices, this is the reason an extensive study is going through LIB. One advantage is doping with sulfur to enhance electrochemical processes, which in turn increases energy density [116, 117]. Diffuse toward iron sulfide, which has a cubic structure with a theoretical capacity of 894 mAh/g. The efficiency factor in conversion reactions involving LIBs is the reversible dissolution of polysulfides [118]. Advanced energy storage devices such as LIBs use metal chalcogenides because they have good characteristics like mechanical and thermal stability and conductivity. The most suitable metal dichalcogenides in LIBs as conversion cathodes are disulfides (S_2), tungsten(IV) disulfide (WS_2), and niobium(IV) sulfide (NbS_2) because they have remarkable electronic and charge density properties.

7.3.2.5 Metal Selenides and Tellurides

Sulfide-based materials exhibit certain deficiencies in terms of charge retention and operational stability and have led investigators to incorporate or substitute selenides and tellurides in conversion cathodes. The reasons for their inclusion include attractive properties, such as high theoretically calculated capacities and high electronic conductivities. The highly conductive material causes the most active utilization of energy. Selenide-based cathodes like polyselenides cause Coulombic inefficiency [119]. The great attraction is due to the conductivities of these materials but because of their low melting points, they should be insulated to enhance efficiency [120]. The electrochemical cell of selenium based materials can be enhanced by forming composites with porous carbon (PC). The composites in the form of nanosphere of Li with Se can also work efficiently [121]. Other composites of selenium with porous carbon nanofibers (PCN) as conversion cathode are attractive for their minimum capacity loss per cycle compared to others. This gives outstanding electrochemical performance. After 900 cycles it transfers the capacity of 516 mAh/g without any capacity loss. Tellurium has the highest conductivity with a larger capacity and has a binder-free form for use in electrodes.

7.3.2.6 Oxides and Iodides as Conversion Cathode Materials

Of the oxide family of materials, tin(IV) oxide (SnO_2) is an important material as conversion electrode. It has been studied using different computational tools in order to find the most efficient cathode material [122]. The important phenomenon related to it is the irreversible oxidation form of a tin oxide (SnO_2) to lithium oxide (Li_2O) during dealloying reactions; irreversible conversion reactions happen by the coating of carbon, as visualized by high-resolution electron microscopy (TEM). In contrast, unbound or free tin oxide (SnO_2) has a lower theoretical storage capacity than the composite form (SnO_2/C) [110].

Another aspect of the discussion on oxides in conversion cathodes like lithium cobalt oxide ($LiCoO_2$), lithium nickel oxide ($LiNiO_2$), and lithium iron phosphate ($LiFePO_4$) is the mechanism through which lithium ions pass, known as the rocking chair mechanism. The general reversible conversion reaction in this case when the insertion of lithium ions takes place through active electrode is given by $M^{z+}M_y{+}zLi^+ \leftrightarrow M^0{+}yLi_{z/y}X$, which mentions the charging and discharging mechanism. The conversion reaction is proportional to the bonding iconicity, but fluorides are the preferred cathodic materials in LIBs due to their high potential. When fluorides form composites with Ni, Co, Fe, Cu, and Bi, they show high electrochemical activity, which is the basic reagent of conversion reactions. A well-studied composite is carbon-doped iron fluoride, which has a theoretically calculated capacity of 600 mAh/g, whereas FeF_3 has a capacity of 712 mAh/g within a range of potential of 4.5–1.5 V. Other studies show that lithium-ion insertion occurs at a voltage of around 3.3 V.

There is another oxide of known of as of the lithium titanium dioxide (LTO) that has attracted much attention because it has the highest volumetric capacity with high life cycle and long-lasting thermal stability; these are the major properties of LTO that make it an effective material. The estimated theoretical capacity of LTO is 175 and 600 mAh/g.

7.4 Summary

A brief description of electrode materials used in LIBs in particular is given to shed light on the ongoing research efforts in the field. Since the commercialization of LIBs in 1993, a lot of research has been done in this area, which has resulted in the production of a number of cathode materials that have benefits and shortcomings. The expansion of work related to cathode materials demands a logical classification of the long list of such materials in order to move forward. These materials share similarities but have major differences that classify them according to their area of application. Two major categories of cathode materials include the intercalation cathode family and conversion cathode family, with associated subfamilies. However, knowledge related to ternary and higher-order oxides, halides, and salts of Li still needs to be further explored particularly using high-order DFTs to better model 3D crystalline structures and related electronic and chemical properties for cathode materials used in LIBs; they can be expected to gain further prominence as electric vehicles become the dominant mode of personal transportation and cell phones and tablets the dominant form of communication. This will require both basic and applied research activities in this subject area for at least the next several decades.

Acknowledgment The work has been done by the grant of HEC Pakistan under project 6509/Punjab/NRPU. All the studies and related experiments has been done in the DFT lab department of Physics.

References

1. J. Wu et al., Counter electrodes in dye-sensitized solar cells. Chem. Soc. Rev. **46**(19), 5975–6023 (2017)
2. M.M. Thackeray, C. Wolverton, E.D. Isaacs, Electrical energy storage for transportation – approaching the limits of, and going beyond, lithium-ion batteries. Energy Environ. Sci. **5**(7), 7854–7863 (2012)
3. N. Nitta et al., Li-ion battery materials: present and future. Mater. Today **18**(5), 252–264 (2015)
4. N. Williard et al., Lessons learned from the 787 Dreamliner issue on lithium-ion battery reliability. Energies **6**(9), 4682–4695 (2013)
5. R. Hausbrand et al., Fundamental degradation mechanisms of layered oxide Li-ion battery cathode materials: methodology, insights and novel approaches. Mater. Sci. Eng. B **192**, 3–25 (2015)
6. S. Kumar, N. Kumar, S. Vivekadhish, Millennium development goals (MDGS) to sustainable development goals (SDGS): Addressing unfinished agenda and strengthening sustainable development and partnership. Indian journal of community medicine: official publication of Indian Association of Preventive & Social Medicine, **41**(1), 1 (2016).
7. S. Goriparti et al., Review on recent progress of nanostructured anode materials for Li-ion batteries. J. Power Sources **257**, 421–443 (2014)
8. Y. Wang, G. Cao, Synthesis and enhanced intercalation properties of nanostructured vanadium oxides. Chem. Mater. **18**(12), 2787–2804 (2006)
9. M. Winter et al., Insertion electrode materials for rechargeable lithium batteries. Adv. Mater. **10**(10), 725–763 (1998)
10. D. Murphy, *Materials for Advanced Batteries*, vol 2 (Springer Science & Business Media, New York, 2013)
11. D. Chen et al., Double transition-metal chalcogenide as a high-performance lithium-ion battery anode material. Ind. Eng. Chem. Res. **53**(46), 17901–17908 (2014)
12. Y. Jung, Y. Zhou, J.J. Cha, Intercalation in two-dimensional transition metal chalcogenides. Inorg. Chem. Front. **3**(4), 452–463 (2016)

13. G. Guo et al., Molybdenum disulfide synthesized by the hydrothermal method as an anode for lithium rechargeable batteries. J. Mater. Sci. **40**(9–10), 2557–2559 (2005)
14. J. Xiao et al., Exfoliated MoS_2 nanocomposite as an anode material for lithium-ion batteries. Chem. Mater. **22**(16), 4522–4524 (2010)
15. K. Chang, W. Chen, L-cysteine-assisted synthesis of layered MoS_2/graphene composites with excellent electrochemical performances for lithium-ion batteries. ACS Nano **5**(6), 4720–4728 (2011)
16. K. Chang, W. Chen, In situ synthesis of MoS_2/graphene nanosheet composites with an extraordinarily high electrochemical performance for lithium ion batteries. Chem. Commun. **47**(14), 4252–4254 (2011)
17. H. Hwang, H. Kim, J. Cho, MoS_2 nanoplates consisting of disordered graphene-like layers for high rate lithium battery anode materials. Nano Lett. **11**(11), 4826–4830 (2011)
18. R. Bhandavat, L. David, G. Singh, Synthesis of surface-functionalized WS_2 nanosheets and performance as Li-ion battery anodes. J. Phys. Chem. Lett. **3**(11), 1523–1530 (2012)
19. B. Luo et al., Two-dimensional graphene–SnS_2 hybrids with superior rate capability for lithium-ion storage. Energy Environ. Sci. **5**(1), 5226–5230 (2012)
20. Y. Gong et al., Graphene-network-backboned architectures for high-performance lithium storage. Adv. Mater. **25**(29), 3979–3984 (2013)
21. M.S. Whittingham, Electrical energy storage and intercalation chemistry. Science **192**(4244), 1126–1127 (1976)
22. S. Meyer et al., Properties of intercalated 2H-$NbSe_2$, 4Hb-TaS_2, and 1T-TaS_2. J. Chem. Phys. **62**(11), 4411–4419 (1975)
23. R. Elazari et al., Li-ion cells comprising lithiated columnar silicon film anodes, TiS_2 cathodes and fluoroethylene carbonate (FEC) as a critically important component. J. Electrochem. Soc. **159**(9), A1440–A1445 (2012)
24. T. Bredow, P. Heitjans, M. Wilkening, Electric field gradient calculations for Li_xTiS_2 and comparison with ^{7}Li NMR results. Phys. Rev. B **70**(11), 115111 (2004)
25. T.A. Yersak et al., Ambient temperature and pressure mechanochemical preparation of nano-$LiTiS_2$. ECS Electrochem. Lett. **1**(1), A21–A23 (2012)
26. J.E. Trevey, C.R. Stoldt, S.-H. Lee, High power nanocomposite TiS_2 cathodes for all-solid-state lithium batteries. J. Electrochem. Soc. **158**(12), A1282–A1289 (2011)
27. B.R. Shin et al., Interfacial architecture for extra Li^+ storage in all-solid-state lithium batteries. Sci. Rep. **4**, 5572 (2014)
28. B.R. Shin et al., Comparative study of TiS_2/Li-In all-solid-state lithium batteries using glass-ceramic Li_3PS_4 and $Li_{10}GeP_2S_{12}$ solid electrolytes. Electrochim. Acta **146**, 395–402 (2014)
29. J. Fischer, H. Kim, Elastic effects in intercalation compounds: comparison of lithium in graphite and TiS_2. Phys. Rev. B **35**(7), 3295 (1987)
30. D. Clerc, R. Poshusta, A. Hess, Periodic Hartree–Fock study of Li_xTiS_2, $0 \leq x \leq 1$: the structural, elastic, and electronic effects of lithium intercalation in TiS_2. J. Phys. Chem. A **101** (47), 8926–8931 (1997)
31. F. Mendizabal, R. Contreras, A. Aizman, Molecular modeling of lithium intercalation in 1T-TiS_2. J. Phys. Condens. Matter **9**(14), 3011 (1997)
32. J. Dahn, D.C. Dahn, R. Haering, Elastic energy and staging in intercalation compounds. Solid State Commun. **42**(3), 179–183 (1982)
33. J. Wang, Model for lithium intercalation into TiS_2. Solid State Ionics **40**, 548–552 (1990)
34. C. Umrigar et al., Band structure, intercalation, and interlayer interactions of transition-metal dichalcogenides: TiS_2 and $LiTiS_2$. Phys. Rev. B **26**(9), 4935 (1982)
35. K. Kang et al., Electrodes with high power and high capacity for rechargeable lithium batteries. Science **311**(5763), 977–980 (2006)
36. N. Kuganathan, P. Iyngaran, A. Chroneos, Lithium diffusion in Li_5FeO_4. Sci. Rep. **8**(1), 5832 (2018)
37. F. Kong et al., Ab initio study of doping effects on $LiMnO_2$ and Li_2MnO_3 cathode materials for Li-ion batteries. J. Mater. Chem. A **3**(16), 8489–8500 (2015)

38. P. Xiao et al., Calculations of oxygen stability in lithium-rich layered cathodes. J. Phys. Chem. C **116**(44), 23201–23204 (2012)
39. Z. Deng, A. Manthiram, Influence of cationic substitutions on the oxygen loss and reversible capacity of lithium-rich layered oxide cathodes. J. Phys. Chem. C **115**(14), 7097–7103 (2011)
40. J. Wandt et al., Singlet oxygen evolution from layered transition metal oxide cathode materials and its implications for lithium-ion batteries. Mater. Today **21**(8), 825–833 (2018)
41. N. Yabuuchi, Y. Makimura, T. Ohzuku, Solid-state chemistry, and electrochemistry of $LiCo_{1/3}Ni_{1/3}Mn_{1/3}O_2$ for advanced lithium-ion batteries III. Rechargeable capacity and cyclability. J. Electrochem. Soc. **154**(4), A314–A321 (2007)
42. I. Buchberger et al., Aging analysis of graphite/$LiNi_{1/3}Mn_{1/3}Co_{1/3}O_2$ cells using XRD, PGAA, and AC impedance. J. Electrochem. Soc. **162**(14), A2737–A2746 (2015)
43. R. Jung et al., Chemical versus electrochemical electrolyte oxidation on NMC111, NMC622, NMC811, LNMO, and conductive carbon. J. Phys. Chem. Lett. **8**(19), 4820–4825 (2017)
44. J. Wandt et al., Transition metal dissolution and deposition in Li-ion batteries investigated by operando X-ray absorption spectroscopy. J. Mater. Chem. A **4**(47), 18300–18305 (2016)
45. D.R. Gallus et al., The influence of different conducting salts on the metal dissolution and capacity fading of NCM cathode material. Electrochim. Acta **134**, 393–398 (2014)
46. J.A. Gilbert, I.A. Shkrob, D.P. Abraham, Transition metal dissolution, ion migration, electrocatalytic reduction and capacity loss in lithium-ion full cells. J. Electrochem. Soc. **164**(2), A389–A399 (2017)
47. S.K. Jung et al., Understanding the degradation mechanisms of $LiNi_{0.5}Co_{0.2}Mn_{0.3}O_2$ cathode material in lithium-ion batteries. Adv. Energy Mater. **4**(1) (2014). https://doi.org/10.1002/aenm.201300787
48. R. Petibon et al., Effect of $LiPF_6$ concentration in $Li[Ni_{0.4}Mn_{0.4}Co_{0.2}]O_2$/graphite pouch cells operated at 4.5V. J. Power Sources **300**, 419–429 (2015)
49. K. Nelson et al., Effects of upper cutoff potential on $LaPO_4$-coated and uncoated Li $[Ni_{0.42}Mn_{0.42}Co_{0.16}]O_2$/graphite pouch cells. J. Electrochem. Soc. **163**(2), A272–A280 (2016)
50. Y. Koyama et al., Defect chemistry in layered $LiMO_2$ (M = Co, Ni, Mn, and $Li_{1/3}Mn_{2/3}$) by first-principles calculations. Chem. Mater. **24**(20), 3886–3894 (2012)
51. O. García-Moreno, O. García-Moreno, M. Alvarez-Vega, F. García-Alvarado, J. García-Jaca, J.M. Gallardo-Amores, M.L. Sanjuán, U. Amador, Influence of the structure on the electrochemical performance of lithium transition metal phosphates as cathodic materials in rechargeable lithium batteries: a new high-pressure form of $LiMPO_4$ (M = Fe and Ni). Chem. Mater. **13**, 1570 (2001)
52. M.E. Arroyo-de Dompablo, U. Amador, A computational investigation of the electrochemical properties of spinel-like $LiCoAsO_4$ as a positive electrode for lithium-ion batteries. Solid State Sci. **8**(8), 916–921 (2006)
53. T. Zhang et al., Understanding electrode materials of rechargeable lithium batteries via DFT calculations. Prog. Nat. Sci. Mater. Int. **23**(3), 256–272 (2013)
54. A. Karim, S. Fosse, K.A. Persson, Surface structure and equilibrium particle shape of the $LiMn_2O_4$ spinel from first-principles calculations. Phys. Rev. B **87**(7), 075322 (2013)
55. K. Matsuda, I. Taniguchi, Relationship between the electrochemical and particle properties of $LiMn_2O_4$ prepared by ultrasonic spray pyrolysis. J. Power Sources **132**(1–2), 156–160 (2004)
56. I. Taniguchi et al., Particle morphology and electrochemical performances of spinel $LiMn_2O_4$ powders synthesized using ultrasonic spray pyrolysis method. Solid State Ionics **146**(3–4), 239–247 (2002)
57. J.-S. Kim et al., A truncated manganese spinel cathode for excellent power and lifetime in lithium-ion batteries. Nano Lett. **12**(12), 6358–6365 (2012)
58. T. Takada et al., Structure and electrochemical characterization of $Li_{1+x}Mn_{2-x}O_4$ spinels for rechargeable lithium batteries. J. Power Sources **81**, 505–509 (1999)
59. M. Anicete-Santos et al., Intercalation processes and diffusion paths of lithium ions in spinel-type structured $Li_{1+x}Ti_2O_4$: density functional theory study. Phys. Rev. B **77**(8), 085112 (2008)

60. J. Bhattacharya, A. Van der Ven, Phase stability and nondilute Li diffusion in spinel $Li_{1+x}Ti_2O_4$. Phys. Rev. B **81**(10), 104304 (2010)
61. M.S. Islam et al., Atomic-scale investigation of defects, dopants, and lithium transport in the $LiFePO_4$ olivine-type battery material. Chem. Mater. **17**(20), 5085–5092 (2005)
62. A.K. Padhi, K.S. Nanjundaswamy, J.B. Goodenough, Phospho-olivines as positive-electrode materials for rechargeable lithium batteries. J. Electrochem. Soc. **144**(4), 1188–1194 (1997)
63. G.K.P. Dathar et al., Calculations of Li-ion diffusion in olivine phosphates. Chem. Mater. **23** (17), 4032–4037 (2011)
64. S. Novikova, A. Yaroslavtsev, Cathode materials based on olivine lithium iron phosphates for lithium-ion batteries. Rev. Adv. Mater. Sci. **49**(2), 129–139 (2017)
65. S.-Y. Chung, J.T. Blocking, Y.-M. Chiang, Electronically conductive phospho-olivines as lithium storage electrodes. Nat. Mater. **1**(2), 123 (2002)
66. N. Terada et al., Development of lithium batteries for energy storage and EV applications. J. Power Sources **100**(1–2), 80–92 (2001)
67. P.P. Prosini et al., Determination of the chemical diffusion coefficient of lithium in $LiFePO_4$. Solid State Ionics **148**(1–2), 45–51 (2002)
68. H. Liu et al., Kinetic study on $LiFePO_4$/C nanocomposites synthesized by solid state technique. J. Power Sources **159**(1), 717–720 (2006)
69. K. Tang et al., Kinetic analysis on $LiFePO_4$ thin films by CV, GITT, and EIS. Electrochim. Acta **56**(13), 4869–4875 (2011)
70. S. Zhang et al., Li-ion diffusivity and electrochemical properties of $FePO_4$ nanoparticles acted directly as cathode materials in lithium ion rechargeable batteries. Electrochim. Acta **88**, 287–293 (2013)
71. J. Yang et al., Li_2FeSiO_4 nanorods bonded with graphene for high-performance batteries. J. Mater. Chem. A **3**(18), 9601–9608 (2015)
72. J. Yang et al., Tuning structural stability and lithium-storage properties by d-orbital hybridization substitution in full tetrahedron Li_2FeSiO_4 nanocrystal. Nano Energy **20**, 117–125 (2016)
73. J. Yang et al., Graphene activated 3D-hierarchical flower-like Li_2FeSiO_4 for high-performance lithium-ion batteries. J. Mater. Chem. A **3**(32), 16567–16573 (2015)
74. R. Gummow, Y. He, Recent progress in the development of Li_2MnSiO_4 cathode materials. J. Power Sources **253**, 315–331 (2014)
75. M.S. Islam et al., Silicate cathodes for lithium batteries: alternatives to phosphates? J. Mater. Chem. **21**(27), 9811–9818 (2011)
76. V. Aravindan et al., Adipic acid assisted sol-gel synthesis of Li_2MnSiO_4 nanoparticles with improved lithium storage properties. J. Mater. Chem. **20**(35), 7340–7343 (2010)
77. Q. Cheng et al., Modification of Li_2MnSiO_4 cathode materials for lithium-ion batteries: a review. J. Mater. Chem. A **5**(22), 10772–10797 (2017)
78. R. Dominko, Li_2MSiO_4 (M = Fe and/or Mn) cathode materials. J. Power Sources **184**(2), 462–468 (2008)
79. D. Rangappa et al., Ultrathin nanosheets of Li_2MSiO_4 (M = Fe, Mn) as a high-capacity Li-ion battery electrode. Nano Lett. **12**(3), 1146–1151 (2012)
80. D. Sun et al., In-situ synthesis of carbon-coated Li_2MnSiO_4 nanoparticles with high rate performance. J. Power Sources **242**, 865–871 (2013)
81. Q. Zhang et al., Synthesis and characterization of pristine Li_2MnSiO_4 and Li_2MnSiO_4/C cathode materials for lithium-ion batteries. Ionics **18**(5), 487–494 (2012)
82. C. Hwang et al., Fast ultrasound-assisted synthesis of Li_2MnSiO_4 nanoparticles for a lithium-ion battery. J. Power Sources **294**, 522–529 (2015)
83. M. Świętosławski et al., Nanocomposite C/Li_2MnSiO_4 cathode material for lithium-ion batteries. J. Power Sources **244**, 510–514 (2013)
84. Y. Dong et al., Synthesis of La-doped Li_2MnSiO_4 nano-particle with high-capacity via the polyol-assisted hydrothermal method. Electrochim. Acta **166**, 40–46 (2015)

85. S. Zhang et al., $Li_{2+x}Mn_{1-x}P_xSi_{1-x}O_4/C$ as novel cathode materials for lithium-ion batteries. Electrochim. Acta **107**, 406–412 (2013)
86. M. Kim, S. Lee, B. Kang, High energy density polyanion electrode material: $LiVPO_4O_{1-x}F_x$ ($x \approx 0.25$) with tavorite structure. Chem. Mater. **29**(11), 4690–4699 (2017)
87. J.-M. Ateba Mba et al., Synthesis and crystallographic study of homotypic $LiVPO_4F$ and $LiVPO_4O$. Chem. Mater. **24**(6), 1223–1234 (2012)
88. M. Ati et al., Synthesis, structural, and transport properties of novel be hydrated fluorosulphates $NaMSO_4F{\cdot}2H_2O$ (M = Fe, Co, and Ni). Chem. Mater. **22**(13), 4062–4068 (2010)
89. P. Barpanda et al., A 3.90V iron-based fluorosulphate material for lithium-ion batteries crystallizing in the triplite structure. Nat. Mater. **10**(10), 772 (2011)
90. J. Dong et al., Triplite $LiFeSO_4F$ as cathode material for Li-ion batteries. J. Power Sources **244**, 716–720 (2013)
91. M. Kim, Y. Jung, B. Kang, The high electrochemical performance of 3.9V $LiFeSO_4F$ directly synthesized by a scalable solid-state reaction within 1 h. J. Mater. Chem. A **3**(14), 7583–7590 (2015)
92. R. Tripathi et al., Ultra-rapid microwave synthesis of triplite $LiFeSO_4F$. J. Mater. Chem. A **1** (9), 2990–2994 (2013)
93. R. Gover et al., $LiVPO_4F$: a new active material for safe lithium-ion batteries. Solid State Ionics **177**(26–32), 2635–2638 (2006)
94. J. Barker et al., Structural and electrochemical properties of lithium vanadium fluorophosphate, $LiVPO_4F$. J. Power Sources **146**(1–2), 516–520 (2005)
95. M. Kim, S. Lee, B. Kang, Fast-rate capable electrode material with higher energy density than $LiFePO_4$: 4.2V $LiVPO_4F$ synthesized by scalable single-step solid-state reaction. Adv. Sci. **3**(3), 1500366 (2016)
96. K.L. Harrison, A. Manthiram, Microwave-assisted solvothermal synthesis and characterization of various polymorphs of $LiVOPO_4$. Chem. Mater. **25**(9), 1751–1760 (2013)
97. L.W. Wangoh et al., Uniform second Li-ion intercalation in solid state ε-$LiVOPO_4$. Appl. Phys. Lett. **109**(5), 053904 (2016)
98. B. Wen et al., Molybdenum substituted vanadyl phosphate ε-$VOPO_4$ with enhanced two-electron transfer reversibility and kinetics for lithium-ion batteries. Chem. Mater. **28**(9), 3159–3170 (2016)
99. K.L. Harrison et al., Chemical and electrochemical lithiation of $LiVOPO_4$ cathodes for lithium-ion batteries. Chem. Mater. **26**(12), 3849–3861 (2014)
100. L. Grande et al., The lithium/air battery: still an emerging system or a practical reality? Adv. Mater. **27**(5), 784–800 (2015)
101. J. Yuan, J.-S. Yu, B. Sundén, Review of mechanisms and continuum models of multi-phase transport phenomena in porous structures of non-aqueous Li-Air batteries. J. Power Sources **278**, 352–369 (2015)
102. F. Wu, G. Yushin, Conversion cathodes for rechargeable lithium and lithium-ion batteries. Energy Environ. Sci. **10**(2), 435–459 (2016)
103. F. Wang et al., Conversion reaction mechanisms in lithium-ion batteries: a study of the binary metal fluoride electrodes. J. Am. Chem. Soc. **133**(46), 18828–18836 (2011)
104. H. Braunschweig et al., Highly strained heterocycles constructed from boron–boron multiple bonds and heavy chalcogens. Angew. Chem. Int. Ed. **55**(18), 5606–5609 (2016)
105. H.F. Franzen, Structure, and bonding in metal-rich compounds: pnictides, chalcides, and halides. Prog. Solid State Chem. **12**(1), 1–39 (1978)
106. T. Li et al., Transition-metal chlorides as conversion cathode materials for Li-ion batteries. Electrochim. Acta **68**, 202–205 (2012)
107. P.G. Bruce, B. Scrosati, J.M. Tarascon, Nanomaterialien für wiederaufladbare lithium batteries. Angew. Chem. **120**(16), 2972–2989 (2008)
108. L. Li, F. Meng, S. Jin, High-capacity lithium-ion battery conversion cathodes based on iron fluoride nanowires and insights into the conversion mechanism. Nano Lett. **12**(11), 6030–6037 (2012)

109. X. Hua et al., Comprehensive study of the CuF_2 conversion reaction mechanism in a lithium-ion battery. J. Phys. Chem. C **118**(28), 15169–15184 (2014)
110. R.E. Doe et al., First-principles investigation of the Li−Fe−F phase diagram and equilibrium and nonequilibrium conversion reactions of iron fluorides with lithium. Chem. Mater. **20**(16), 5274–5283 (2008)
111. Q. Qu et al., Strong surface-bound sulfur in conductive MoO_2 matrix for enhancing Li–S battery performance. Adv. Mater. Interfaces **2**(7), 1500048 (2015)
112. L. Li et al., Origins of large voltage hysteresis in high-energy-density metal fluoride lithium-ion battery conversion electrodes. J. Am. Chem. Soc. **138**(8), 2838–2848 (2016)
113. T. Li et al., Reversible three-electron redox behaviors of FeF_3 nanocrystals as high-capacity cathode-active materials for Li-ion batteries. J. Phys. Chem. C **114**(7), 3190–3195 (2010)
114. M. Jaishankar et al., Toxicity, mechanism and health effects of some heavy metals. Interdiscip. Toxicol. **7**(2), 60–72 (2014)
115. M. Ebner et al., Visualization and quantification of electrochemical and mechanical degradation in Li-ion batteries. Science **342**(6159), 716–720 (2013)
116. T. Evans et al., Ionic liquid enabled FeS_2 for high-energy-density lithium-ion batteries. Adv. Mater. **26**(43), 7386–7392 (2014)
117. X. Wang et al., Flexible energy-storage devices: design consideration and recent progress. Adv. Mater. **26**(28), 4763–4782 (2014)
118. M.-R. Gao et al., Nanostructured metal chalcogenides: synthesis, modification, and applications in energy conversion and storage devices. Chem. Soc. Rev. **42**(7), 2986–3017 (2013)
119. S.K. Martha et al., On the thermal stability of olivine cathode materials for lithium-ion batteries. J. Electrochem. Soc. **158**(10), A1115–A1122 (2011)
120. Z. Li et al., Confined selenium within porous carbon nanospheres as a cathode for advanced Li–Se batteries. Nano Energy **9**, 229–236 (2014)
121. L. Zeng et al., A flexible porous carbon nanofibers-selenium cathode with superior electrochemical performance for both Li–Se and Na–Se batteries. Adv. Energy Mater. **5**(4), 1401377 (2015)
122. X. Guo et al., Lithium storage in carbon-coated SnO_2 by conversion reaction. J. Power Sources **226**, 75–81 (2013)

Development of Lithium Nickel Cobalt Manganese Oxide as Cathode Material for Commercial Lithium-Ion Batteries

8

Yanbin Chen and Yafei Liu

Contents

Author Contribution: Dr. Yanbin Chen, professor, candidate of the National Talent Project in China, and candidate of the Beijing Capital Technical Leading Talent Project, is currently the vice president of Beijing Easpring Material Technology Co., Ltd. (hereinafter referred to as "Easpring"). He was awarded a Ph.D. degree from University of Science and Technology Beijing in 2001. He has been engaged in research, development, and industrialization of the key materials for lithium battery for 20 years. Up to now, he has developed more than 50 products of three series cathode materials, including lithium cobalt oxide, lithium manganese oxide, and lithium nickel cobalt manganese oxide, published more than 50 papers, obtained 16 licensed patents, and drafted 9 state and industrial standards.

Dr. Yafei Liu, professor, China State-Council Special Allowance Expert, is currently the director of Institute of Lithium-ion Battery Materials of Beijing Easpring Material Technology Co., Ltd. He was awarded a Ph.D. degree from University of Science and Technology of China in 2001. He has been engaged in research, development, and industrialization of lithium-ion cathode materials and their precursors, fuel cell materials, and gas-sensitive and functional materials for a long time. So far, he has published more than 60 papers and applied 52 patents.

Easpring is a global top supplier of cathode materials for lithium-ion batteries, providing high-end products to leading battery makers, including LG Chem, Samsung SDI, SK Innovation, Sanyo, Sony, CATL, Farasis, BYD, BAK, etc. It has been recognized as an Excellent Export Enterprise, National Technical Innovation Demonstration Enterprise, and National Accredited Enterprise Technical Center.

Y. Chen (✉) · Y. Liu
Beijing Easpring Material Technology Co., Ltd, Beijing, China

BGRIMM Technology Group, Beijing, China
e-mail: chenyanbin@easpring.com; liuyafei@easpring.com

Q. Zhen et al. (eds.), *Nanostructured Materials for Next-Generation Energy Storage and Conversion*, https://doi.org/10.1007/978-3-662-58675-4_8

Abstract

In the last three decades, the successful application of lithium-ion batteries (LIBs) for consumer electronics has laid a solid foundation for the rapid development of large-format batteries for electric vehicles (EV) and energy storage systems (ESS). By combining the merits of the high capacity of lithium nickel oxide ($LiNiO_2$), with the good rate capability of lithium cobalt oxide ($LiCoO_2$), and the thermal stability and low cost of lithium manganese oxide ($LiMnO_2$), lithium nickel cobalt manganese oxide (NCM, $LiNi_{1-x-y}Co_xMn_yO_2$) enjoys outstandingly comprehensive advantages and turns to be the major cathode material for lithium-ion batteries. This chapter is dedicated to briefly introduce the development and application of NCM materials. The problems and challenges of nickel-rich NCM are elucidated, and some corresponding solutions are put forward. In addition, examples are given to illustrate different NCM product design options for various commercial applications.

8.1 Introduction

Lithium-ion batteries have been commercialized for nearly three decades and applied predominately in consumer electronics, like a cellular phone, laptop computer, camcorder, etc. In recent years, the electric vehicle industry is booming up, and large-format lithium-ion batteries have been developed and commercialized, whose shipment will be much larger than those used in consumer electronics. Furthermore, industrialization of the energy storage system is commenced. Lithium-ion batteries are playing increasingly important roles in energy storage and conversion. Different types of electrode materials, electrolyte systems, and cell packages have been developed to meet the increasing demands of these applications. Optimizing battery design has advanced to improve the energy density, but there is no more room left. The cathode material is one of the most critical factors determining the energy density and cost of lithium-ion batteries. Therefore, developing new cathode materials with high capacity and good safety is indispensable for high-performance lithium-ion batteries [1–3].

8.2 The Road Map of Cathode Materials for Battery Electric Vehicles

With the development and commercial application of electric vehicles, the extended driving range requires the high energy of the lithium-ion battery. Up to now, in most of the commercial lithium-ion batteries (LIBs), carbon material, e.g., graphite (C), is used as anode material, while the cathode material changes from spinel lithium manganese oxide (LMO, $LiMn_2O_4$) and olivine lithium iron phosphate (LFP, $LiFePO_4$) to layer-structured material lithium nickel cobalt manganese oxide (NCM, $LiNi_{1-x-y}Co_xMn_yO_2$) and lithium nickel cobalt aluminum oxide (NCA, $LiNi_{1-x-y}Co_xAl_yO_2$), which deliver increased gravimetric and volumetric energy density, due to the high specific electrochemical capacity and high density of NCM or NCA materials.

The high volumetric energy density of lithium-ion battery requires high-density cathode materials. The bulk density of cathode is closely related to the morphology, particle size, and the distribution. Generally, irregular-shaped particles tend to involve agglomeration, bridging, and voids and thus lose some packing density. However, regular spherical particles have good fluidity, their proper size distributions allow small particles to be filled into the voids of big particles with close packing. Therefore, the spheroidization is an effective way to increase the volumetric capacity of the cathode material. Commercial NCM materials are practically prepared by chemical coprecipitation method, which not only ensures the uniform dispersion of transition metal at the atomic level but also obtains high-density spherical precursors by process tuning.

The higher density is a requirement for electric vehicles to meet customer preference for increased stamina (distance before recharging of 300–500 km) or even long distance. Such a system requires high charge density based on high-energy-density cathode materials. Practically, Ni-rich NCM materials with high packing density are preferred. Based on $LiNi_{1/3}Co_{1/3}Mn_{1/3}O_2$ (NCM111), $LiNi_{0.5}Co_{0.2}Mn_{0.3}O_2$ (NCM523), $LiNi_{0.6}Co_{0.2}Mn_{0.2}O_2$ (NCM622), and $LiNi_{0.8}Co_{0.1}Mn_{0.1}O_2$ (NCM811), the energy density of lithium-ion cells cycled in the cutoff voltages of 4.2 V–3.0 V can be achieved at 180, 210, 230, and 280 $Wh.kg^{-1}$ (Fig. 8.1).

Lithium nickel cobalt manganese oxide ($LiNi_{1-x-y}Co_xMn_yO_2$) is essentially a solid solution of lithium nickel oxide-lithium cobalt oxide-lithium manganese oxide ($LiNiO_2$-$LiCoO_2$-$LiMnO_2$) (Fig. 8.2). With the change of the relative ratio of x and y, the property changes generally corresponded to the end members. The higher the nickel (Ni) content, the higher is the capacity and more instability, the higher cobalt (Co) content means higher rate capability and higher cost, while the higher manganese (Mn) content means better thermal stability and lower cost. So the most important way to increase the energy density is to increase the Ni content, thus lowering the cobalt and manganese content. With increasing Ni content in $LiNi_{1-x-y}Co_xMn_yO_2$, the charging capacity of NCM111, NCM523, NCM622, and NCM811 increases from 175 to 186, 198, and 213 $mAh.g^{-1}$, respectively (Fig. 8.3) [4], while the cost decreases with the lowered Co content.

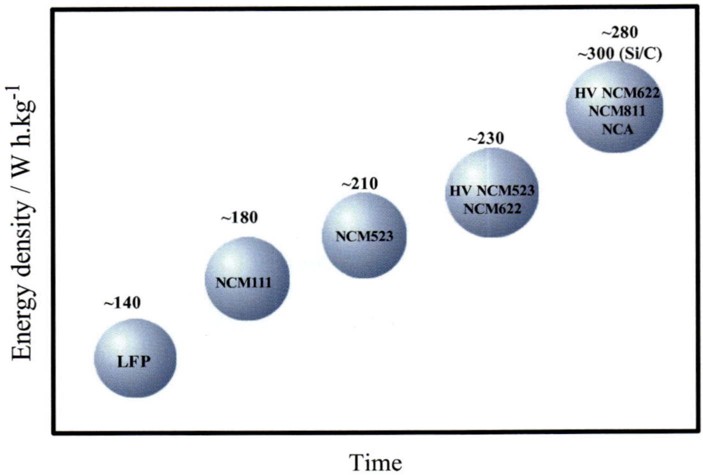

Fig. 8.1 Road map of LIB with increased energy density for electric vehicles

Fig. 8.2 Compositional phase diagram of lithium nickel oxide-nickel cobalt oxide-lithium manganese oxide ($LiCoO_2$-$LiNiO_2$-$LiMnO_2$)

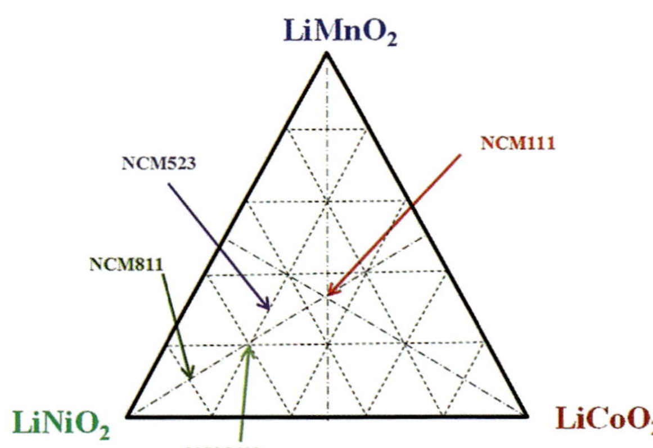

Besides increasing the Ni content, another way to get high energy density is to increase the charging cutoff voltage. As an example of NCM523, the discharging capacity increases from 162.8 to 185.3, 196.9, and 208.1 mAh.g^{-1} when the upper charging cutoff voltage is increased from 4.25 to 4.40, 4.50, and 4.60 V (Fig. 8.4); meanwhile, the cost of unit energy decreases correspondingly. During the charging process, NCM materials undergo several phase transitions: H1 → M → H2 → H3. H1, H2, and H3 represent three hexagonal phases and M a monoclinic one. Compared with H1 and H2, the H3 phase is in the deep delithiated state, which is extremely unstable in structure. As shown in Fig. 8.5, the redox onset of the H2 → H3 phase transition of NMC811 starts at >4.0 V, while those of NMC111 and NMC622 commence at >4.4 V. So raising upper cutoff voltage is generally applied to low and middle nickel NCM materials [5].

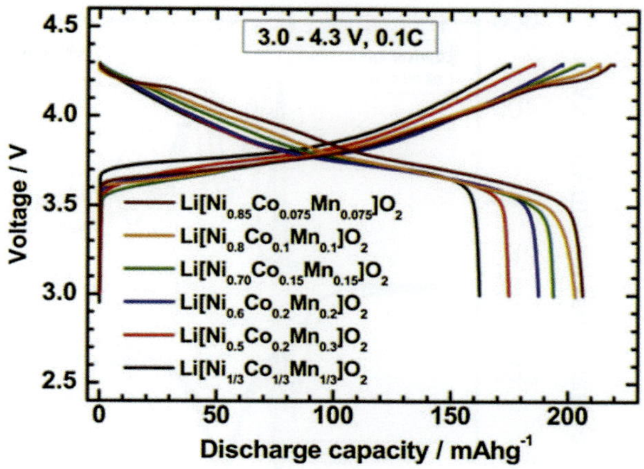

Fig. 8.3 Charge-discharge profiles of various compositions of lithium nickel cobalt manganese oxide ($LiNi_{1-x-y}Co_xMn_yO_2$) [4]

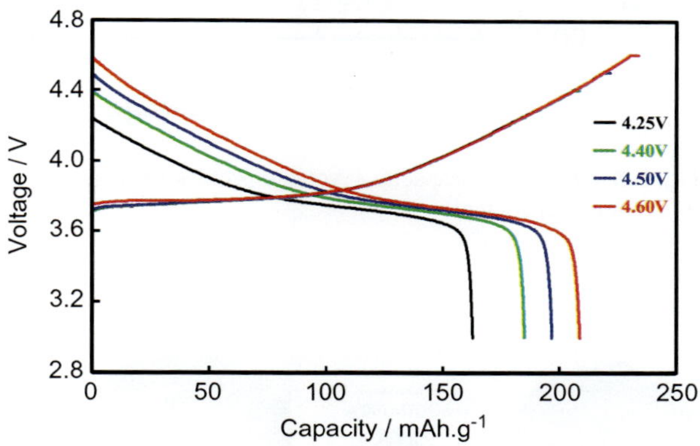

Fig. 8.4 Charge-discharge profiles of NCM523 in varied cutoff voltages

8.3 The Challenges and the Solutions of the Nickel (Ni)-Rich Lithium Nickel Cobalt Manganese Oxide Materials

8.3.1 Particle Cracking of Lithium Nickel Cobalt Manganese Oxide upon Repeated Charge-Discharge

The charging and discharging capacities of lithium nickel cobalt manganese oxide increase with the Ni content in the same cutoff voltages; therefore, the Ni-rich NCM materials are involved in a higher state of charge (SOC) with respect to the theoretical capacity, more phase transition, and larger volume change during (de)intercalation, which will further cause cracks inside the secondary particle along the grain boundary, at the triple junction where three primary particles touch. In such a way, when the particle cracking happens upon repeated charge-discharge, the structural stability deteriorates. As is mentioned in Sect. 8.2, the capacity and thus SOC increase with the increased charging cutoff voltage, and the crack becomes more serious, as illustrated in Fig. 8.6.

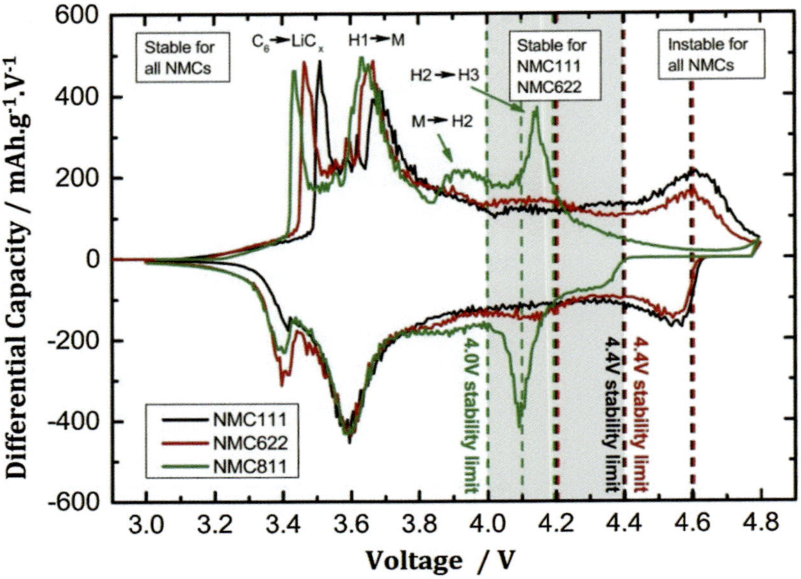

Fig. 8.5 Differential capacity – voltage profiles of lithium nickel manganese cobalt oxide with different nickel content

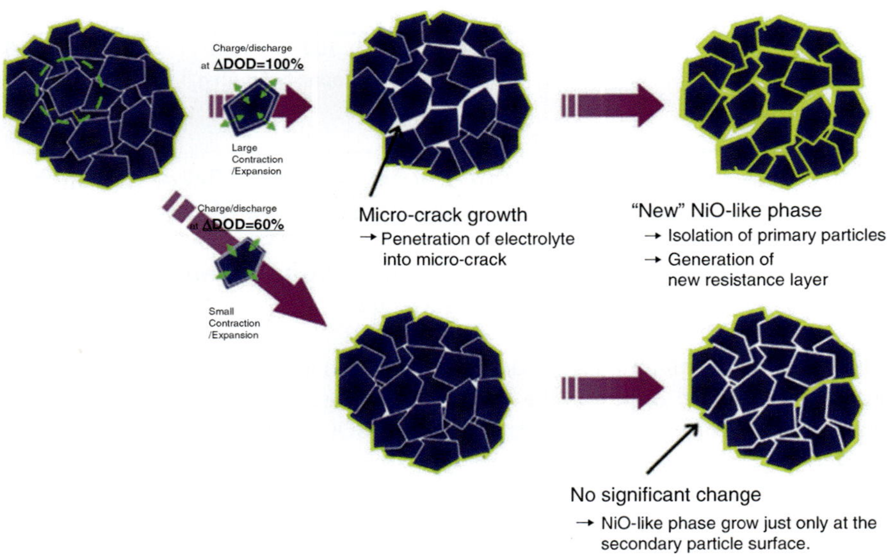

Fig. 8.6 Schematic illustration for the micro-crack growth and deterioration of lithium nickel cobalt aluminum oxide ($LiNi_{0.76}Co_{0.14}Al_{0.10}O_2$) particle during cycle tests [6]

The particle crack develops upon repeated cycling, especially at a high rate, or large depth of discharge (DOD) (Fig. 8.6) [6]. Particle cracking is supposed to be an additional but dominant failure mode of the agglomerated lithium nickel manganese cobalt oxide materials, compared to the conventional single crystal material, e.g., lithium cobalt oxide, which is extensively used as cathode material in the lithium-ion battery (LIB) of consumer electronics.

Particle cracks cause loss of inner particle connectivity and give rise to increased resistance and thus the degradation of performance. Besides, the electrolyte can penetrate along the crack into the particle, causing the increase of the interface area of NCM/electrolyte, and a relative shortage of the liquid electrolyte, the dry area inside the lithium cell may appear, and capacity may thus fade more quickly even drop sharply. On the other hand, the increase of the interface area of NCM/electrolyte may bring about the thermal instability and the risk of thermal runaway.

One solution to address the particle cracking is to make the NCM particle with high strength, resisting against the pressure during electrode pressing, and the internal stress from the repeated volume change of primary particle upon cycling. Of course, the first step is to prepare a precursor with enhanced strength, by tuning the coprecipitation conditions and even the inner configuration of the reactor. As an example of the precursor of good strength in Fig. 8.7, the particle of radial texture (Fig. 8.7a) enjoys high strength, while the other one with random texture (Fig. 8.7b)

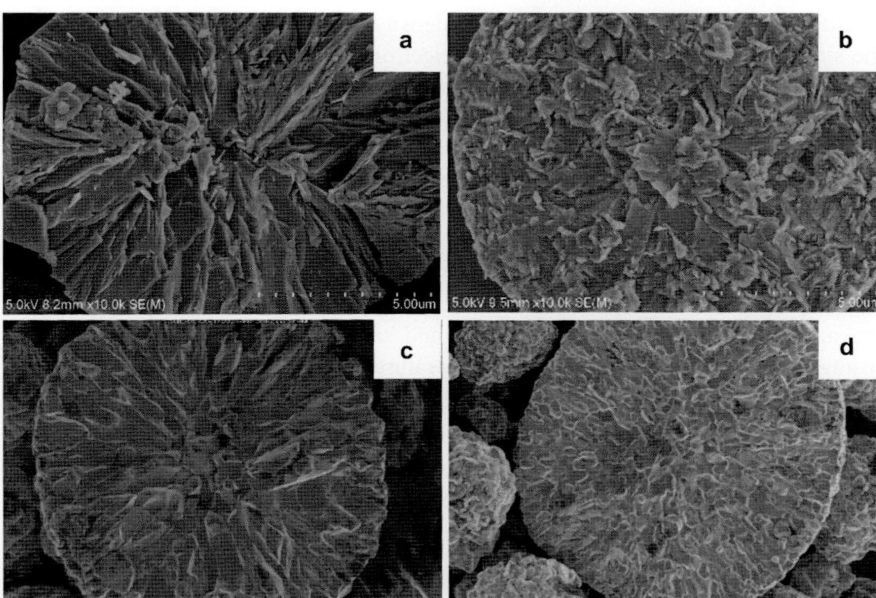

Fig. 8.7 Precursors with radial texture (**a**), random texture (**b**) and their corresponding lithium nickel manganese cobalt oxide (NCM) cathodes (**c**) and (**d**)

Table 8.1 Comparison of electrochemical performances of two types of NCM523 materials

		Single crystal	Agglomerate
Capacity in coin cell (3.0–4.3 V), $mAh.g^{-1}$	0.2C	162.0	162.0
	0.5C	157.0	157.3
	1.0C	152.0	152.6
High-temperature storage of pouch cell at 60 °C for 7 days, %	Swelling rate	3.3	9.3
	Capacity retention	93.4	87.8
SEM after 100 cycles (3.0–4.5 V)		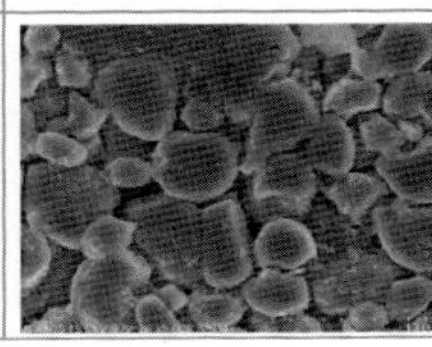	

gives low strength. NCM cathode material made from the high strength precursor generally has better cycling stability. To a much extent, the particle crack can be alleviated, although not totally avoided in such a way.

Another effective way is to prepare single crystal type lithium nickel manganese cobalt oxide materials, whose particle is strong enough to resist against physical press or internal stress. Generally, single crystal NCM material enjoys some unique attributes: (1) enhanced strength, alleviating particle crack; (2) increased pellet density, increasing the volumetric energy density; (3) lowered surface area, reducing the side reaction of NCM cathode with electrolyte; and (4) enhanced connectivity intra- and interparticles, in favor of electronic, ionic, and thermal conductivity of the electrode. Consequently, the single crystal NCM material enjoys better performance than conventional agglomerated counterpart, in terms of cycling stability, thermal storage, safety, as well as volumetric energy density. Table 8.1 gives a comparison of the performances of single crystal NCM-based LIB with those of the conventional NCM agglomerates. The single crystal particle still retains its integrity after long-term cycles.

The preparation of single crystal NCM materials can be achieved by adjusting the precursor process [7], increasing Li/Me and sintering temperature [8], or adding fusing agents for lowering sintering temperatures [9].

8.3.2 Structural and Chemical Instabilities (Related to Performance Deterioration upon Cycling and Thermal Storage)

The charging capacity increases with nickel content in lithium nickel cobalt manganese oxide ($LiNi_{1-x-y}Co_xMn_yO_2$), while the chemical stability deteriorates. The Ni-rich NCM material with higher charging capacity involves the transition metal of higher oxidation state, which tends to release active oxygen, causing side reaction of the cathode with electrolyte. There will be more serious side redox reaction between cathode and electrolyte during thermal storage and cycling at elevated temperature. In general, higher Ni content NCM material suffers from worse chemical stability and thermal stability (related to the safety issue of lithium-ion cells). Subsequently, Ni-rich NCM material-based lithium cell is confronted with big challenges in cycling, thermal storage, and safety.

Moreover, Ni-rich material behaves less stable when exposed in air, releasing lithium species impurities on the surface of the particles (this topic will be discussed in the following section).

The performance of NCM material depends fundamentally on the precursor $Ni_{1-x-y}Co_xMn_y(OH)_2$. The precursor is the predominant factor to be concerned with addressing the structural and chemical instability. Practically, $Ni_{1-x-y}Co_xMn_y(OH)_2$ precursor is prepared by mixing solutions of nickel sulfate, cobalt sulfate, and manganese sulfate with sodium hydroxide with a protective atmosphere of nitrogen (N_2). The crystallinity of $Ni_{1-x-y}Co_xMn_y(OH)_2$ depends on the precipitation conditions, such as time duration, pH, stirring intensity, temperature, and atmosphere. Well-crystallized precursor with enhanced texture, fewer surface defects, and lower impurities is naturally advantageous for preparing NCM materials. In the industrial practice, the continuous stirred-tank reactor (CSTR) or batch reactor is adopted to mass produce precursor material. Generally, the particle size distribution of precursor prepared by CSTR is wide, and that by batch mode is narrow (Fig. 8.8).

Sintering is also quite critical for the performance of NCM materials, over-sintering or less-sintering will deliver poor electrochemical performance. In the case of NCM materials prepared from CSTR precursor with wide particle size distribution, small particles are generally over-sintered with bigger primary grain size, while large particle less-sintered with smaller grain size. In other words, large particles and small ones should be sintered at different temperature for optimized performance. In a sense, precursor produced in batch mode with fairly good consistency is preferred for NCM materials with respect to electrochemical performance.

Doping is a common way for modification of functional materials. $LiNi_{1-x-y}Co_xMn_yO_2$ itself is a mutually doped solid solution consisting of component $LiNiO_2$, $LiCoO_2$, and $LiMnO_2$, enjoying better electrochemical and thermal stability than the pristine $LiNiO_2$. Besides, several other elements of part per million (ppm) level are chosen for doping to further stabilize the structure of the NCM materials, including magnesium (Mg), aluminum (Al), iron (Fe), titanium (Ti), zirconium (Zr), chromium (Cr), yttrium (Y), and gallium (Ga) [10–11]. The mechanism for improved electrochemical performance of NCM material by doping might be (1) introducing electrochemically inactive elements into the host structure, (2) preventing the undesired phase transition from the layered structure to the rock-salt-like one, and (3) promoting the lithium-ion transport due to increased lithium slab distance by the dopants. For example, Al doping improves cycling stability and thermal storage of the NCM cathode material but at the loss of capacity and rate capability. Ti doping is helpful for improved cyclability and rate capability of NCM material. Proper substitution of Cr for Mn and Y/Al for Ni can improve the cycle performance and rate capability of the materials, due to the increased lattice parameter c of the doped materials, beneficial to the diffusion of lithium ions. Most of the research on anion doping is to replace partial oxygen (O) with

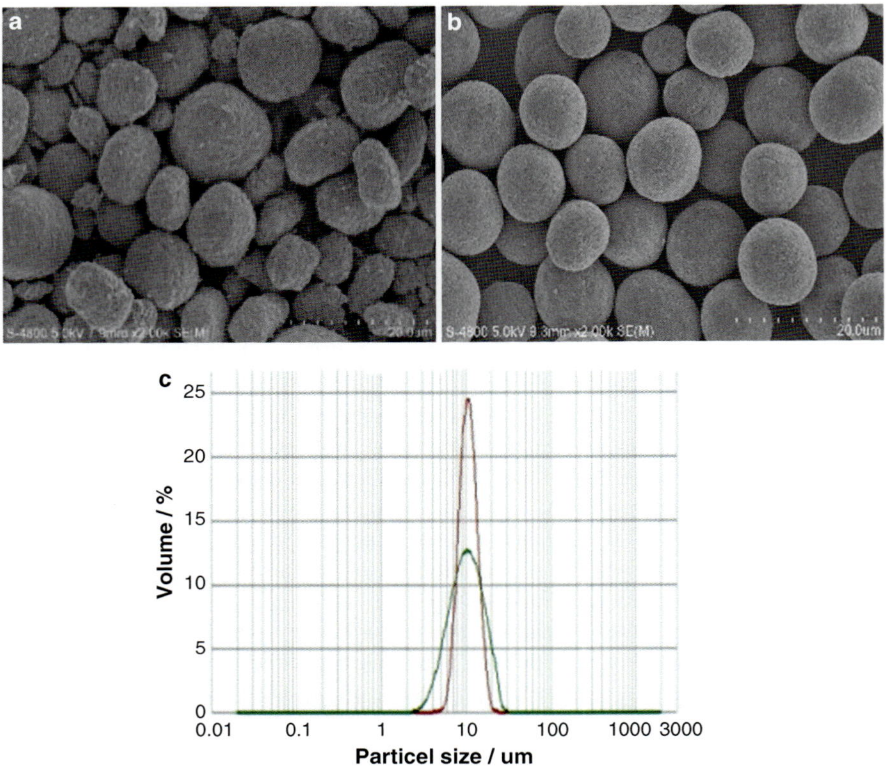

Fig. 8.8 Scanning electron microscope (SEM) of precursors precipitated by continuous stirred-tank reactor (**a**), batch process (**b**), and their particle size distribution (**c**)

fluorine (F) [12], chlorine (Cl) [13], and sulfur (S). The strong electron absorption of F enhances the structural stability of the material, and the strong bond between lithium and fluorine also has a promoting effect on the (de)intercalation of lithium ions. The co-doping of Mg and F reduces the cation mixing and improves the cycling performance and capacity of the material.

The surface coating is another effective way to improve the stability of Ni-rich cathode materials. Conventionally, one or several metal oxides or phosphates of nano-sized particles are dispersed onto the surface, followed by heat treatment, such as alumina (Al_2O_3), titania (TiO_2), ceria (CeO_2), yttria (Y_2O_3), zirconia (ZrO_2), and metal phosphate (MPO_4, where M = Al, Fe, and Y) [14–20]. The mechanism of the coating for the improvement of the materials still remains a controversy. The effect of surface coating includes (1) partial isolation of the oxidizing cathode from electrolyte solution, (2) scavenging the erosive species of hydrogen fluoride (HF) in electrolyte from attacking cathode materials, (3) consuming the surface lithium impurities, and (4) surface doping enhancement due to the diffusion of the coating element into the thin surface layer with a high concentration.

8.3.3 Surface Alkali Impurities

Layer-structured cathode materials generally have surface impurities, i.e., lithium carbonate (Li_2CO_3) and lithium hydroxide (LiOH). The higher is the Ni content, the more is the lithium impurities, nickel-rich lithium nickel cobalt manganese oxide/lithium nickel cobalt aluminum oxide (Ni-rich NCM/NCA) > nickel-poor lithium nickel cobalt manganese oxide (Ni-poor NCM) > lithium cobalt oxide (LCO). These impurities may cause gelation of the slurry during electrode coating process, resulting in gassing due to the decomposition of lithium carbonate (Li_2CO_3) at high voltage, and also are severely detrimental to cell performance in terms of both capacity retention and rate capability, especially at elevated temperature.

The surface lithium impurities consist of two kinds of sources, residual lithium and extract lithium, respectively. The former comes from the residue of lithium in the sintering process, and the latter results from the extraction of lithium from the matrix of particles, due to the chemical instability of the Ni-rich lithium nickel cobalt manganese oxide/lithium nickel cobalt aluminum oxide.

When NCM or NCA materials are prepared, extra lithium source is added to compensate for the loss of lithium during sintering; thus after sintering, there still exists some lithium residue on the surface, in the form of lithium oxide (Li_2O), at high temperature. The lithium oxide will transform to lithium hydroxide and lithium carbonate when reacting with water (H_2O) and carbon dioxide (CO_2) in the air. This part of surface impurities can be defined as "residual lithium impurities." The residual lithium impurities are a kind of enrichment of the unreacted lithium on the surface, which is related to the initial Li/Me ratio, temperature profile, atmosphere, the precursor, and lithium source used.

The other source of surface lithium impurities is the extraction of lithium species from the matrix of the particle and further conversion to lithium hydroxide and lithium carbonate, in series to the diffusion of lithium from the center to the surface layer. In a sense, lithium metal oxide ($LiMeO_2$) can be taken as Li_2O. Me_2O_3. In the existence of moisture and carbon dioxide, Li_2O is continuously extracted out of the particle and converted to LiOH and finally to Li_2CO_3; this part of lithium impurities can be defined as "extractable lithium impurities." When exposed to air, the surface residual LiOH reacts with CO_2 preferentially to form Li_2CO_3, and LiOH content decreases quickly in several hours to a lower level, while Li_2O is being extracted out and further converted to Li_2CO_3, even dominating the increase of the lithium impurities (Fig. 8.9). The more Li_2O is extracted, the less chemical stability is the NCM material. Generally, the higher Ni content NCM is covered with higher content of lithium impurities. Single crystal NCM has lower initial lithium impurities and is comparatively stable. Precursor from different process also gives NCM of different chemical stability even with the same nickel/cobalt/manganese composition. One way to restrict the lithium species extraction is to control the content of CO_2 and moisture in the atmosphere where NCM is processed. Another way is to wash NCM powders in pure water to remove the surface impurities, but the Li_2O in host structure may also be rinsed away, which degrades the cycling stability especially at high temperature.

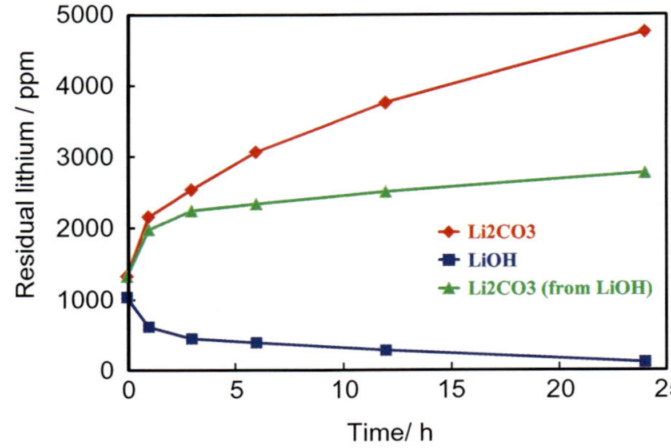

Fig. 8.9 Variation of lithium hydroxide (LiOH) and lithium carbonate (Li_2CO_3) contents in lithium nickel cobalt manganese oxide (NCM) exposed in air

8.3.4 Safety Issue

From the thermodynamic point of view, the lithium-ion battery itself is an unstable high energy system, consisting of a highly oxidizing cathode, a reductive anode, and flammable organic electrolyte. Efforts should be made to improve the safety systematically, in terms of materials, cells, modules, battery pack, car design, and even application.

There are many factors affecting the thermal stability of cathode materials. The first factor is the composition and the structure of the materials. As is well known, the thermal stability of lithium iron phosphate and lithium manganese oxide ($LiMn_2O_4$) is better than NCM and NCA. The Ni-low NCM materials are fairly stable compared to Ni-rich NCM counterparts. As is shown in Fig. 8.10, the heat flows of the charged NCM111, NCM523, NCM622, and NCM811 are 512.5, 605.7, 721.4, and 904.8 $J.g^{-1}$, and their DSC peak temperatures are 306, 290, 264, and 232 °C, respectively [4]. Accordingly, it is a possible way to improve the thermal stability of NCM with certain Ni content by increasing Mn content and lowering Co content, keeping the similar electrochemical capacity but improved thermal stability.

It will be helpful for improving thermal stability to properly increase the particle sizes and reduce the specific surface area of the cathode material. In general, single crystal materials naturally have an advantage stability due to the decreased surface area and surface defects.

Suitable doping and coating can also be adopted to minimize the exothermic reaction of the delithiated cathode material, which triggers the "thermal runaway" of LIBs.

Some case of an explosion of lithium-ion batteries has been found to be caused by metal impurities, which are unintentionally introduced during the manufacturing process of batteries and their materials. Firstly, ultra-large particles of metal scratches (Fe, Cr, Cu, and Zn) should be avoided, which might penetrate through the separator directly, resulting in internal short of the charged cathode and

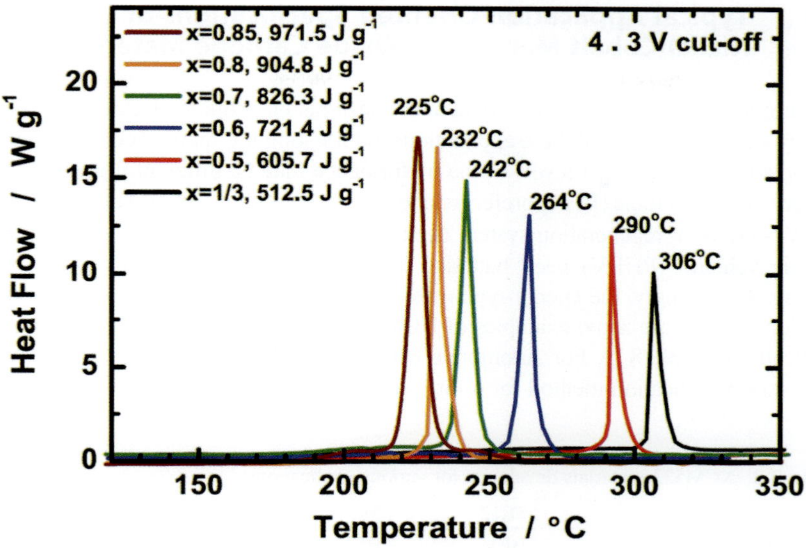

Fig. 8.10 DSC of charged lithium nickel cobalt manganese oxide (NCM) materials with different nickel (Ni) content

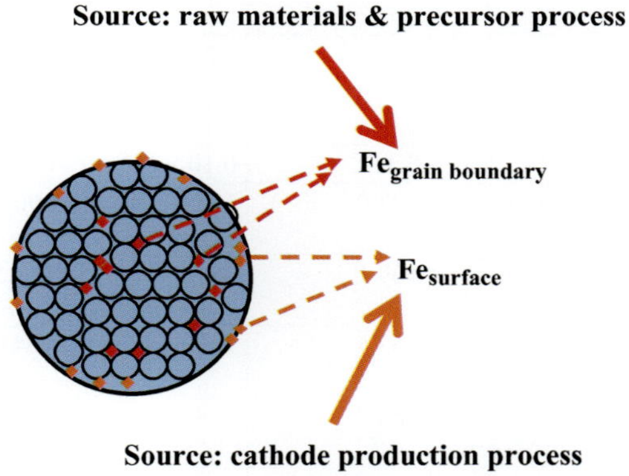

Fig. 8.11 Metal impurities existed in lithium nickel cobalt manganese oxide material

anode, causing a safety issue. Secondly, the metal impurities contaminated between particles or occluded inside the particles can be electrochemically dissolved into the electrolyte and deposited on the anode during charging, and dendrites may be formed and penetrate the separator, further cause problems of low voltage, poor safety, and short lifetime. Strict control of these metal impurities contamination should be carried out through the whole process of "raw material-cathode/anode-battery" (Fig. 8.11).

8.4 Typical Application-Oriented Specifications of Lithium Nickel Cobalt Manganese Oxide Cathode Materials

Lithium-ion batteries are extensively used in consumer electronics, electric vehicles, and energy storage system. All these applications require long life span, good reliability, and safety, although the target levels of the performance may be different from each other. Battery electric vehicle (BEV) prefers energy-type batteries, and hybrid electric vehicles (HEV) and 48 V regeneration system need power-type batteries, while plug-in hybrid electric vehicles (PHEV) need batteries with both mild energy density and power density. Accordingly, the energy-type materials and high rate-type materials should be developed. Here are some examples of NCM523 based cathode materials for different applications (Table 8.2). For automotive battery applications, doping or coating is the necessary modification method for improved reliability of cathode materials.

Table 8.2 NCM523 materials developed for various applications

Product name	5YN	5E12	5E5	5 EB-10	5SC
Application	Laptop	BEV	HEV	BEV/PHEV	BEV
Product type	Unimodal	Unimodal	Unimodal	Bimodal	Unimodal
Doping	Non	Yes	Yes	Yes	Yes
Coating	Non	Yes	Yes	Yes	Yes
D_{50}, μm	12.0	12.0	4.4	9.5	5.2
Tap density, g. cm^{-3}	2.52	2.50	1.75	2.55	2.27
Pellet density, g. cm^{-3}	3.36	3.28	2.86	3.38	3.29
Li_2CO_3, %	0.09	0.11	0.10	0.10	0.08
LiOH, %	0.11	0.10	0.10	0.10	0.06
0.2C Ch. cap., $mAh.g^{-1}$	193.0	193.0	190.0	193.0	193.7
0.2C Disch. cap., $mAh.g^{-1}$	169.1	168.8	169.5	169.0	169.3
0.5C Disch. cap., $mAh.g^{-1}$	163.5	163.6	164.0	163.7	163.0
1.0C Disch. cap., $mAh.g^{-1}$	158.7	158.7	159.6	159.0	156.8
2.0C Disch. cap., $mAh.g^{-1}$	152.0	150.0	153.2	150.7	149.6
Cap. retention. (coin cell, 25 °C, 1C/1C, 100th @4.5–3.0 V), %	91.6	95.5	95.0	95.3	95.2
SEM			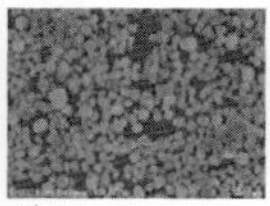		

8.5 Conclusion

Nickel-based cathode materials have been developed and commercialized in the last decade, and the application has extended from consumer electronics to electromobility, and in energy storage system, the cell configuration is diversified from 18,650 cylindrical cells of <3.0 Ah to large-sized prismatic cells and pouch cells of 50 Ah level. For small-sized cylindrical cells, like 18,650 (18 mm diameter × 65 mm length, "0" indicates cylindrical-type) or 21,700 (21 mm diameter × 70 mm length), NCM or NCA of Ni content close to 90% has been largely used, although there still remains the risk of safety issue to some extent, while for large-sized prismatic cells or pouch cells, NCM of Ni content less than 65% is being used because of more challenging issues, e.g., safety, swelling, and even cycle life. Lithium nickel cobalt manganese oxide materials applied in the high-voltage cell are initiated presently. All these applications are attributed to the understanding of the relationship between composition-process-structure-performance and the improvement in the comprehensive properties by optimizing the categories of technology including precursor, sintering, doping, coating, etc.

Obviously, efforts must be made to handle the remaining challenging issues for better reliability, especially with respect to the reuse of battery retired from electric vehicles; therefore, cutting-edge technologies are inevitably required.

Acknowledgments This work is partially supported by National Key Research and Development Plan of China (No. 2017YFB0102001).

References

1. G.E. Blomgren, The development and future of lithium-ion batteries. J. Electrochem. Soc. **164** (1), A5019–A5025 (2017)
2. M.S. Whittingham, Lithium batteries and cathode materials. Chem. Rev. **104**, 4271–4302 (2004)
3. J.B. Goodenough, Y. Kim, Challenges for rechargeable Li batteries. Chem. Mater. **22**, 587–603 (2010)
4. H.-J. Noh, S. Youn, C.S. Yoon, et al., Comparison of the structural and electrochemical properties of layered $Li[Ni_xCo_yMn_z]O_2$ (x=1/3, 0.5, 0.6, 0.7, 0.8 and 0.85) cathode material for lithium-ion batteries. J. Power Sources **233**, 121–130 (2013)
5. R. Jung, M. Metzger, F. Maglia, et al., Oxygen release and its effect on the cycling stability of $LiNi_xMn_yCo_zO_2$ (NMC) cathode materials for Li-ion batteries. J. Electrochem. Soc. **164**(7), A1361–A1377 (2017)
6. S. Watanabe, M. Kinoshita, T. Hosokawa, et al., Capacity fade of the $LiAl_yNi_{1-x-y}Co_xO_2$ cathode for lithium-ion batteries during accelerated calendar and cycle life test. J. Power Sources **258**, 210–217 (2014)
7. J. Duan, C. Wu, Y. Cao, et al., Enhanced compacting density and cycling performance of Ni-riched electrode via building mono dispersed micron scaled morphology. J. Alloys Compd. **695**, 91–99 (2017)
8. J. Li, A.R. Cameron, H. Li, et al., Comparison of single crystal and polycrystalline $LiNi_{0.5}Mn_{0.3}Co_{0.2}O_2$ positive electrode materials for high voltage Li-ion cells. J. Electrochem. Soc. **164**(7), A1534–A1544 (2017)

9. Y. Kim, Lithium nickel cobalt manganese oxide synthesized using alkali chloride flux: morphology and performance as a cathode material for lithium ion batteries. ACS Appl. Mater. Interfaces **4**, 2329–2333 (2012)
10. J. Wilcox, S. Patoux, M. Doeff, Structure and electrochemistry of $LiNi_{1/3}Co_{1/3-y}M_yMn_{1/3}O_2$ (M=Ti, Al, Fe) positive electrode materials. J. Electrochem. Soc. **156**(3), A192–A198 (2009)
11. F. Wu, M. Wang, Y.F. Su, et al., A novel layered material of $LiNi_{0.32}Mn_{0.33}Co_{0.33}Al_{0.01}O_2$ for advanced lithium-ion batteries. J. Power Sources **195**(9), 2900–2904 (2010)
12. G.H. Kim, S.T. Myung, H.J. Bang, Synthesis and electrochemical properties of $Li[Ni_{1/3}Co_{1/3}Mn_{(1/3-x)}Mg_x]O_{2-y}F_y$ via co-precipitation. Electrochem. Solid-State Lett. **7**(12), A477–A480 (2004)
13. S.X. Liu, H.L. Zhang, Effect of chlorine doping on the structure and electrochemical properties of the $LiNi_{1/3}Co_{1/3}Mn_{1/3}O_2$ cathode material. Rare Metal Mater. Eng. **42**(2), 296–300 (2013)
14. Y. Kim, H.S. Kim, S.W. Martin, Synthesis and electrochemical characteristics of Al_2O_3-coated $LiNi_{1/3}Co_{1/3}Mn_{1/3}O_2$ cathode materials for lithium-ion batteries. Electrochim. Acta **52**(3), 1316–1322 (2006)
15. F. Wu, M. Wang, Y.F. Su, et al., Effect of TiO_2-coating on the electrochemical performances of $LiCo_{1/3}Ni_{1/3}Mn_{1/3}O_2$. J. Power Sources **191**(2), 628–632 (2009)
16. F. Wu, M. Wang, Y.F. Su, et al., Modification of $LiCo_{1/3}Ni_{1/3}Mn_{1/3}O_2$ cathode material by CeO_2-coating. Sci. China Ser. E Technol. Sci. **52**(9), 2737–2741 (2009)
17. F. Wu, M. Wang, Y.F. Su, et al., Surface modification of $LiCo_{1/3}Ni_{1/3}Mn_{1/3}O_2$ with Y_2O_3 for a lithium-ion battery. J. Power Sources **189**(1), 743–747 (2009)
18. S.K. Hu, G.H. Cheng, M.Y. Cheng, et al., Cycle life improvement of ZrO_2-coated spherical $LiCo_{1/3}Ni_{1/3}Mn_{1/3}O_2$ cathode material for lithium-ion batteries. J. Power Sources **188**(2), 564–569 (2009)
19. J.H. Wang, Y. Wang, Y.Z. Guo, et al., Effect of heat-treatment on the surface structure and electrochemical behavior of $AlPO_4$-coated $LiCo_{1/3}Ni_{1/3}Mn_{1/3}O_2$ cathode materials. J. Mater. Chem. A **1**(15), 4879–4884 (2013)
20. Y. Bai, Q.J. Chang, Q. Yu, et al., A novel approach to improve the electrochemical performances of the layered $LiCo_{1/3}Ni_{1/3}Mn_{1/3}O_2$ cathode by YPO_4 surface coating. Electrochim. Acta **112**(1), 414–421 (2013)

Carbon-Based Materials for Advanced Potassium-Ion Batteries Anode

9

Xiujun Qi, Zheng Xing, and Zhicheng Ju

Contents

Abstract

Advanced potassium-ion batteries have recently captured tremendous attention in the search for cost-effective and large-scale energy storage systems. On the basis of this trend, various electrode materials are proposed and assessed for potassium-ion battery (PIB). Carbon-based materials are regarded as one of the most promising anode materials of potassium-ion batteries due to their

Author Contribution: All authors are contributing equally on the conceiving and outlook of the development of Carbon-Based Materials for Advanced Potassium-Ion Batteries Anode. The chapter context was compiled by Dr. X. Qi and Z. Xing under the supervision of Prof. Z. Ju. The results and data provided in this chapter are original and not reproduced from any other publications. The authors further acknowledge that there is no financial relationship with the editors or publisher and have contributed original work in this chapter, other than what acknowledged or appropriately cited with copyright permission.

X. Qi · Z. Xing · Z. Ju (✉)
School of Materials Science and Engineering, China University of Mining and Technology, Xuzhou, People's Republic of China
e-mail: 304849314@qq.com; xzh086@cumt.edu.cn; juzc@cumt.edu.cn

Q. Zhen et al. (eds.), *Nanostructured Materials for Next-Generation Energy Storage and Conversion*, https://doi.org/10.1007/978-3-662-58675-4_9

excellent chemical stability, high electrical conductivity, and environmental benignity. A comprehensive overview of the progress achieved within the application of carbon-based materials in potassium-ion batteries will be given in this chapter. The existing challenges and future research directions are also briefly discussed.

9.1 Introduction and Background

The rapid depletion of fossil fuel resources and global climate changes are pushing forward the development of sustainable clean energy technologies such as advanced batteries, supercapacitors, fuel cells, and solar cells. Among this energy storage and conversion devices, advanced batteries are playing a significant role in human's life. In particular, rechargeable lithium-ion batteries (LIBs) have currently been reigning over the market for portable energy storage devices including laptops, digital cameras, mobile phones, and even more electric vehicles or grid-level stationary energy storage systems (ESSs) [1–3]. However, LIBs face the challenge of lithium shortage (Table 9.1) as a result of its abundance of only 0.0017 wt % in the Earth's crust [4], uneven distribution, and the high extraction cost, which hamper the applications of LIBs for storage purposes. For this reason, it is highly urgent to search for alternative energy storage technologies with earth-abundant materials, low cost, high energy, and a long cycle life for large-scale energy storage applications.

The concerns over lithium scarcity have prompted researchers to focus on the alternative sodium-ion batteries (SIBs) system. Sodium, next to lithium in the alkali metal series, not only is more abundant (2.36 wt %) [4] and easily accessible but also possesses similar chemical properties to lithium. Based on the above advantages, SIBs have been extensively investigated [5–8]. However, the larger radius of sodium (Na^+, 1.02 Å), compared with that of Li^+ (0.76 Å), brings about great challenges in the development of SIBs. Research into the anode material of SIBs shows that sodium ions struggle to reversibly intercalate into graphite, thereby resulting in poor capacities [7, 9, 10]. In addition, the relatively high standard redox potential of Na^+/Na (−2.71 V vs. standard hydrogen electrode, SHE) leads to a lower working voltage in contrast to that of Li^+/Li (−3.04 V vs. SHE) [11] and lower energy density than Li-ion.

Currently, potassium-ion batteries (PIBs) have emerged as another prospective energy storage system with the ever-increasing demand for renewable resources [12–15]. Analogous to sodium, potassium is a relatively plentiful element in the Earth's crust (2.09 wt%) [4], which can be refined with far lower cost than lithium. More importantly, potassium (K) has a relatively lower standard redox potential (−2.93 V vs. SHE) than Na (−2.71 V vs. SHE), suggesting a broader electrochemical operation voltage window of PIBs and thereby providing more energy. Although potassium ion has an even larger radius (1.38 Å) than that of Na^+ and lithium (Li^+), K^+ as charge carrier has been proven to be viable. It's more encouraging that K^+ can intercalate/deintercalate into graphite layer reversibly, with an impressive capacity of more than 200 mAh g^{-1} [12, 16]. Furthermore, Okoshi et al. [17] revealed that K

Table 9.1 Comparison of the physical properties and economic characteristics (Li, Na, and K)

M=	Li	Na	K
Atomic number	3	11	19
Ionic radius [Å]	0.76	1.02	1.38
Atomic mass	6.941	22.99	39.10
Crust abundance [wt %]	0.0017	2.36	2.09
Redox potential vs. S.H.E [V]	−3.04	−2.71	−2.93
Melting point [°C]	180.54	97.72	63.38
Desolvation energy in EC [kJ mol^{-1}]	208.9	152.8	114.6
Cost of carbonate [US $ ton^{-1}]	23,000	200	1000
Cost of industrial-grade metal [US $ ton^{-1}]	100,000	3000	13,000
Stable graphite intercalation compounds	LiC_6	NaC_{64}	KC_8
Graphite electrode capacity [mA h g^{-1}]	372	35	279

electrolytes display higher conductivity than Li and Na electrolytes. They proposed the higher conductivity be ascribed to the smaller Stokes radius of solvated K^+ owing to its weak Lewis acidity and the small interfacial reaction resistance originating from the lower desolvation activation energy [17]. Early studies have demonstrated higher K^+ conductivity in both aqueous and nonaqueous electrolytes [18, 19]. Potassium carriers have been widely adopted in aqueous batteries [20, 21]. In organic electrolytes, such as ethylene carbonate (EC), diethyl carbonate (DEC) solvents, K^+ also show a higher ionic conductivity than Li^+ and Na^+. Considering these advantages, PIBs have rapidly attracted tremendous interest and gradually become a strong competitor to SIBs and even LIBs, and then varieties of materials have been developed and utilized as potential PIB electrodes.

In this new battery technology's very incipient stage, the research attention should be focused on potentially scalable, industrially and economically attractive, inexpensive electrode materials and the understanding of their cycle-life-property correlations. Therefore, it is the materials scientist's responsibility to develop novel, high-performance, and sustainable materials to cope with the challenge of providing renewable energy for large-scale energy conversion and storage devices while limiting environmental effects. Carbon-based materials are considered to be one of the most versatile materials for negative electrodes of the energy storage devices owing to their low cost, abundant sources, good chemical stability, high electrical conductivity, and environmental benignity [22, 23]. In particular, chemical stability at a wide temperature range whether in acidic or basic solution prompts carbon materials exceedingly attractive as electrodes in electrochemical energy devices. What's more, the significance and potential of carbon-based materials have been acknowledged by some of the highest scientific field awards such as the 1996 Nobel Prize in Chemistry (fullerenes), the 2008 Kavli Prize in Nanoscience (carbon nanotubes), and the 2010 Nobel Prize in Physics (graphene) [23]. As a result, interest in carbon-based materials and their applications is experiencing its most unprecedented development in modern materials science.

In this chapter, recent progress in carbon-based materials as anode materials of emerging nonaqueous PIBs is summarized. The class of materials is diverse and consists of graphite, graphene, hard carbon, carbon-based composites, and other carbon nanomaterials. The challenges and future research opportunities for non-aqueous PIBs are also briefly discussed.

9.2 Carbon-Based Materials for Potassium-Ion Batteries Anode

Anode materials are a critical component of alkali metal-ion batteries. The failure of the commercialized Li battery launched in the 1980s was induced by Li metal anode, mainly resulting from the dendrite formation related to security issues in the process of charge and discharge [24]. Carbon-based materials have gradually attracted researchers' considerable interest due to its unique structure and outstanding physiochemical properties. In this section, the research on carbon-based-negative electrode materials for PIBs will be divided into four parts to be discussed in detail.

9.2.1 Graphite

Among all host structures, graphite has been regarded as quite remarkable. The success of the current rechargeable LIBs is mainly ascribed to the substitute of graphite for metallic Li anode, where lithium graphite (space group P63*mc*) intercalation compounds (GICs) can form rather than Li plating/stripping, thereby effectively avoiding dendritic lithium and offering a safe operating environment. Nowadays, graphite has been employed almost exclusively in commercialized LIBs owing to its excellent overall properties. Nevertheless, a major obstacle for the development of SIBs is that sodium can't intercalate reversibly into graphite [25]. These phenomena are mainly related with sodium plating on the graphite surface before forming a stage I GIC [26], which has proven that sodium carbide (NaC_6), the only graphite intercalation compound, was not energetically favorable from a thermodynamic perspective [27]. However, K-graphite intercalated compounds (K-GICs) have been reported in the literature in the middle of the twentieth century [28]. Potassium-graphite intercalated compounds were prepared through a non-electrochemical two-zone vapor synthesis technique or alkali metal solvation method. The vapor phase synthesis technique dates from the 1960s, which contains the heating of graphite and intercalant (Li, Na, K, etc.) in two different vessels [28]. The authors also discussed the phase transformation sequence of potassium carbide KC_{36}–(stage III)–KC_{24}(stage II)–KC_8 (stage I) in the vapor pressure experiments. Furthermore, the energetics of alkali metal (Li, Na, and K) intercalation in graphite was compared by density functional theory (DFT) using van der Waals (vdW) density functionals [29]. Figure 9.1 demonstrates that the intercalation of Na is indeed energetically unfavorable, different from that of Li. More impressively, K intercalation is even more advantageous than that of Li. Their formation enthalpies

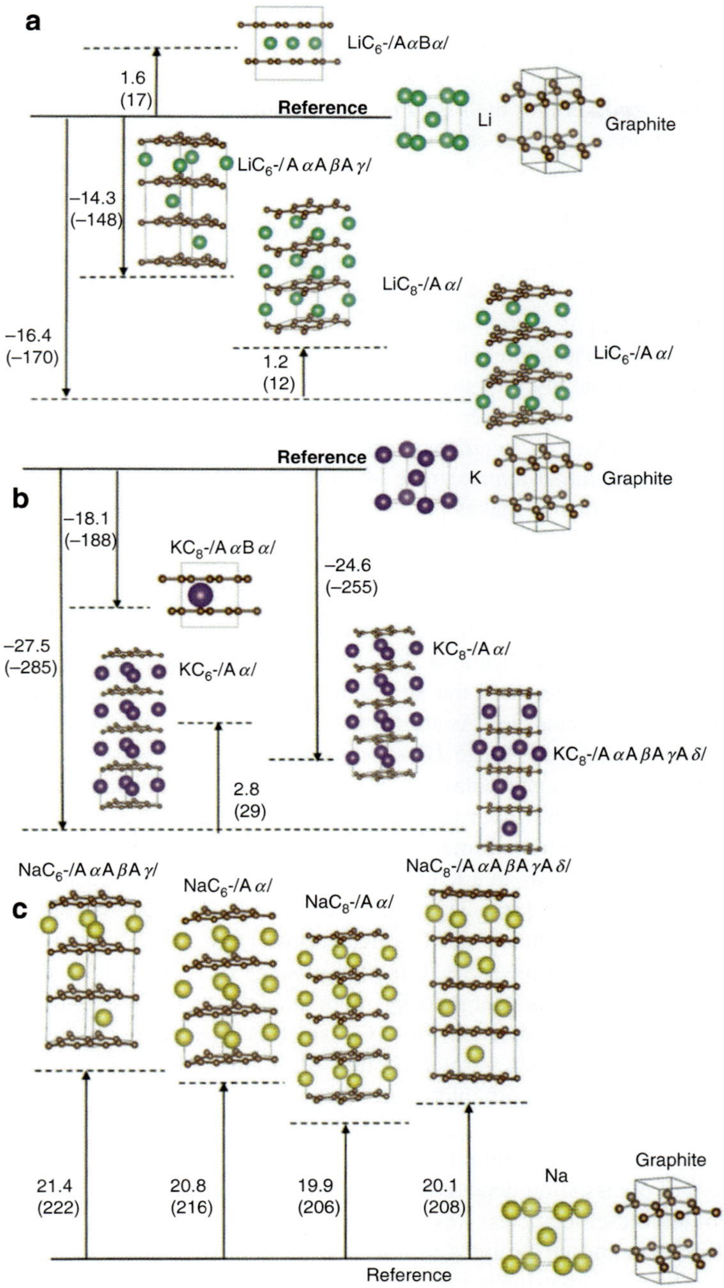

Fig. 9.1 Enthalpies of the formation of various alkali metals intercalated graphite compounds. (**a**) Li, (**b**) K, and (**c**) Na. (Adopted from Ref. [29])

for KC_8 and LiC_6 are −27.5 and − 16.4 kJ mol^{-1}, respectively. These have laid a solid theoretical foundation for future research of K ion insertion into graphite even other carbon materials.

Wang and coworkers for the first time investigated reversible (de)intercalation of K^+ ions into/from amorphous carbon nanofibers in 2014 [30]. The carbon nanofiber (CNF) electrode produces a first discharge capacity of ≈680 mA h g^{-1} and depotassiation capacity of about 300 mAh g^{-1} at a rate of 50 mAg^{-1}. However, the specific capacity decreases to 200 mAh g^{-1} on the second cycle and gradually stabilizes at about 80 mAh g^{-1} after the 20th cycle [30], implying relatively poor cycling performance.

Jian et al. [12] reported ambient-temperature electrochemical potassium (de)intercalation into graphite with a nonaqueous electrolyte of 0.8 M KPF_6 solvated in ethylene carbonate (EC)/diethyl carbonate (DEC). The graphite electrode presented a surprisingly high reversible depotassiation capacity of 273 mA h g^{-1} at C/40 (1C = 279 mAh g^{-1}, calculated using the molecular weight of KC_8 as the most potassiated phase), rather approximate to its theoretical capacity of 279 mAh g^{-1}, the excellent performance makes graphite a potential anode material for PIBs. The K/graphite cell also produced capacities of 263, 234, 172, and 80 mA h g^{-1} at current densities of C/10, C/5, C/2, and 1C [12]. As a result, although graphite can display a large capacity at low current rates, its capacity fades dramatically upon higher current rates. This implies an alternative negative electrode material might be conducive to high-rate applications. Ex situ X-ray diffraction (XRD) studies were also carried out to understand the structural evolution of K intercalation and the potassium storage mechanism. As shown in Fig. 9.2, the formation of several stages of K-GICs was observed. Stage III KC_{36}, stage II KC_{24}, and stage I KC_8 were detected in a sequence upon potassiation similar to the crystallization of the phases prepared by the two-zone vapor technique [28], while depotassiation recovers from graphite through phase transformations in an opposite sequence.

However, a recent study in a K/graphite half-cell proposes an unusual potassium intercalation mechanism [16]. This study coupled electrochemical measurements with density functional theory (DFT) calculations. They claim that while the intercalation occurs in three stages, the formation of stage III KC_{24}, followed by stage II KC_{16}, and stage I KC_8 is energetically more stable than the previously proposed KC_{36} (stage III) to KC_{24} (stage II) to KC_8 (stage I). Therefore, the mechanism of potassium insertion closely resembles that of lithium [16]. In order to clearly understand how the potassium intercalates into graphite, further research is urgently required.

The major challenge graphite faced with is the large volumetric change (60%) in the process of potassiation/depotassiation, which leads to fast capacity loss and low initial Coulombic efficiencies. Researchers have started to make efforts to choose different binders and electrolytes to alleviate the volume change to some degree. Komaba et al. [18] investigated the effect of different binders on electrochemical performance, including polyvinylidene fluoride (PVDF, $[(C_2H_2F_2)_n]$), sodium carboxymethylcellulose (CMCNa, $[(C_2H_{15}NaO_2)_n]$), and sodium polyacrylate (PANa, $[(C_3H_3NaO_2)_n]$) for their graphite/K half-cells. As observed in Fig. 9.3, the use of

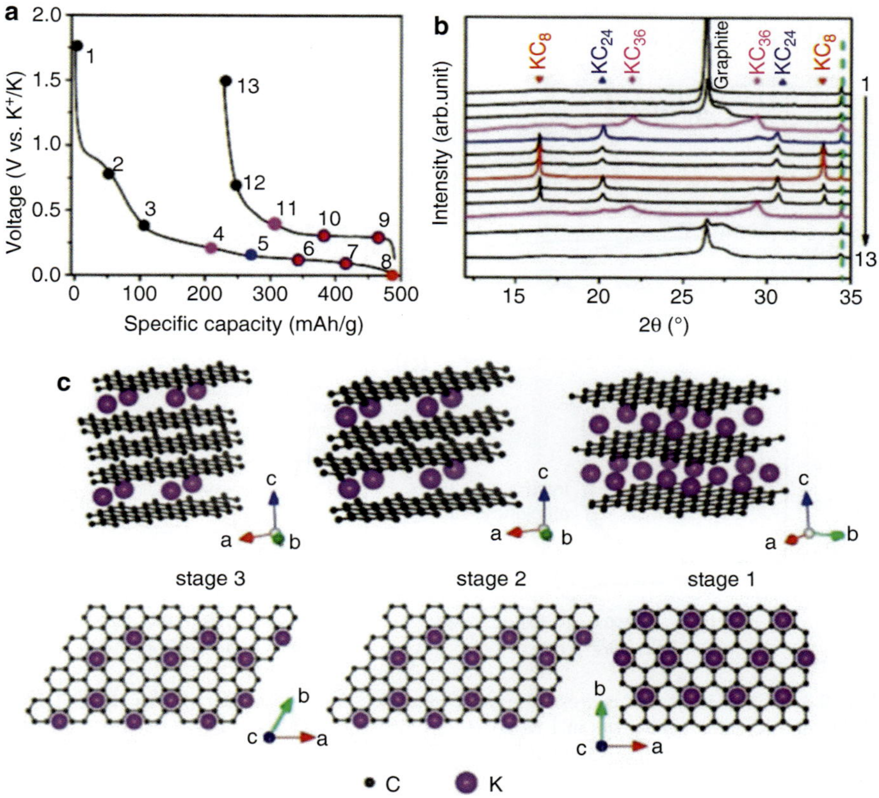

Fig. 9.2 (**a**) First-cycle graphite voltage capacity profiles at C/10. (**b**) XRD patterns of electrodes corresponding to the marked extraction points in A. (**c**) Structural diagrams of different K-GICs, side view (top row) and top view (bottom row). (Adopted from Ref. [12])

different binders causes dramatic differences in the electrochemical performance. It was revealed that higher initial Coulombic efficiencies were observed in the latter two binders for the graphite PIBs anodes compared to the PVDF binder [18]. This result was ascribed to the preformed solid electrolyte interphase (SEI) effect of the binders PANa and CMCNa; thus, less potassium is consumed for the formation of SEI layer in the first discharge cycle of the PIBs.

In addition, Zhao et al. [31] have shown the critical effects of electrolytes on the electrochemical performance of graphite in PIBs. The authors investigated three kinds of electrolytes of 1 M potassium hexafluorophosphate (KPF_6) in ethylene carbonate/propylene carbonate (EC:PC), EC/diethyl carbonate (EC:DEC), and EC/dimethyl carbonate (EC:DMC), respectively. As observed by the charge/discharge voltage profiles, graphite displayed similar reversible capacity of about 240 $mAhg^{-1}$ in the first three cycles. Nevertheless, the Coulombic efficiency and cycling stability for the three electrolytes present significant difference. It was found

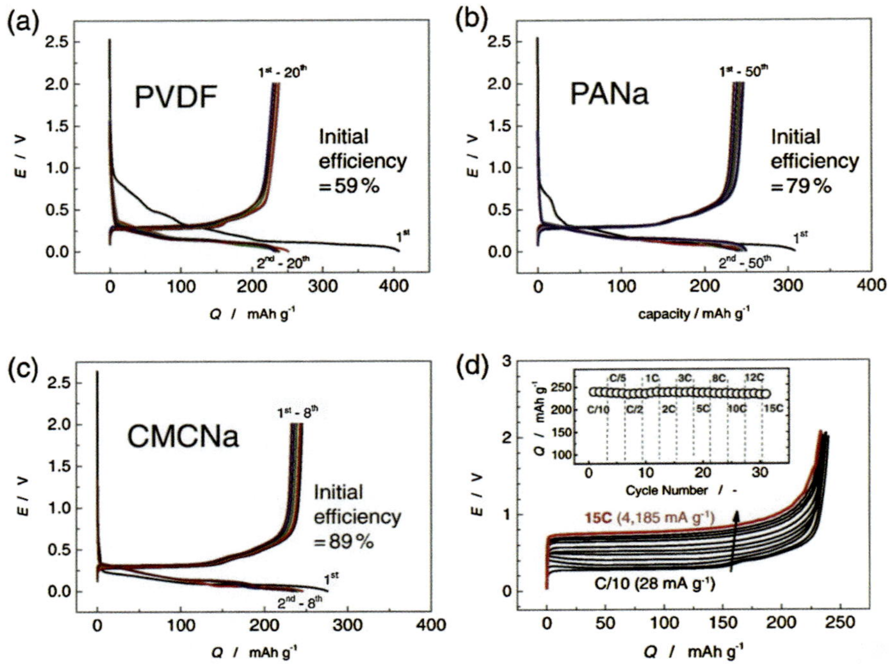

Fig. 9.3 Charge-discharge curves of graphite electrode formed on Al foil with (**a**) PVDF, (**b**) PANa, and (**c**) CMCNa binders in 1 M potassium bis(fluorosulfonyl)imide (KFSI) EC:DEC of K cells and (**d**) oxidation-rate capability for the PANa electrode. Reproduced with permission. (Adopted from Ref. [18])

that EC:DEC (used in most PIBs literature) and EC:PC worked well with the graphite anode in comparison to EC:DMC. What's worth mentioning is that EC: PC obtained the highest Coulombic efficiency in the first cycle among graphite in the other two electrolytes. It may arise from more severe decomposition of DEC and DMC that are not stable at a low potential.

To conclude, graphite is a viable and promising candidate as an anode material in PIBs. The higher average potential for K insertion into graphite as compared with Li can reduce the risk of metal plating. Even so, a more in-depth research of the material is still fairly desired so as to achieve the efficiency required from commercial batteries.

9.2.2 Graphene

While research shows some promising results with graphite for negative electrodes in PIBs, other carbon materials are still needed to be explored as alternatives due to the large volume change of graphite during the cycling process. Thus, some studies have started focusing on other carbonaceous negative electrode materials.

Graphene (space group P6/*mmm*) is composed of a monolayer of carbon atoms sp^2 bonded into a hexagonal or honeycomb lattice sheet, which was experimentally demonstrated in 2004 [32–34]. Due to its inherent strictly two-dimensional crystal structure, graphene possesses its unique electronic, thermal, mechanical, and chemical properties. For instance, graphene shows a high surface area of 2600 $m^2\ g^{-1}$, much larger than that of graphite (~10 $m^2\ g^{-1}$) and carbon nanotubes (~1300 $m^2\ g^{-1}$) [35]. Its unique electronic properties include ambipolar electric field effect, the extremely high mobility of charge carriers (~$10^5\ cm^2\ V^{-1}\ s^{-1}$), massless electron, and quantum Hall effects even at room temperature [36, 37]. These remarkable properties make graphene promising in applications of many various electrochemical energy systems. To date, there have been several methods to prepare graphene in the literature, such as mechanical exfoliation [33], ultrasonic cleavage [38], arc-discharge approach [39], chemical vapor deposition [40], epitaxial growth on silicon carbide (SiC) [41], thermal reduction [42, 43], and chemical reduction [44–47]. Based on a review of the current literature, it seems that thermal reduction and chemical reduction methods are the most commonly used in the preparation process of graphene-based materials in electrochemical devices. Especially, the chemical reduction of graphene oxide is one of the best-established routes to make graphene toward the large-scale production for commercial applications [48, 49].

Nanostructure design, composition fine-tuning, and doping have been proven to be effective approaches in improving the electrochemical performance of electrode materials. Considerable attention has been paid to heteroatom doping for tuning and enhancing carbons' unique physical and chemical properties. In particular, doped carbons with well controllable porosity and the large surface area can boost its electrochemical performance. Heteroatoms such as nitrogen (N), phosphorus (P), sulfur (S), and fluorine (F) have been exploited to modify the surface bonding state and electron distribution of carbon architecture in order to improve their wettability and electronic conductivity [50–52]. For example, N dopant enhances positive charge density, leading to the increase of electronegativity and the acceleration of charge transfer [53]. Phosphorus doping substantially changes the local structure of hexagonal carbon due to the larger diameter of P than that of C [54]. Sulfur doping generates more defective sites and expands the interlayer distance of carbon with S generally located at the edge of the carbon layer [55]. Heteroatom doping provides a useful tool for modifying carbons' unique physicochemical properties and further to boost ion storage performance.

Graphene materials have been investigated as a potential negative electrode material for PIBs. Nitrogen-doped few-layer graphene (N-FLG) was grown on nickel (Ni) foam using acetonitrile in a chemical vapor deposition (CVD) process [56]. Electrochemical tests were performed in a half-cell vs. K^+/K using CR2032 coin cells (C for ethylene carbonate, R for lithium (as anode), manganese dioxide (as cathode) and cylindrical form, 20 mm diameter, and 3.2 mm in height). N-FLG obtained an initial anodic charge (depotassiation) capacity of ~350 mAh g^{-1} at 50 mAg^{-1} and remains constant over the 5 cycles at this rate. Notably, this is well beyond the theoretical capacity (279 mAh g^{-1}) of a stage I KC_8 GIC [12], close to the theoretical Li capacity in graphite of 372 mAh g^{-1}. Galvanostatic testing indicates capacities for the N-FLG and FLG are ~200 mAhg^{-1} and ~30 mAh g^{-1} at 100 mAg^{-1}, over six-fold improved from undoped FLG. In order to elucidate the effect of N-doping sites on the enhanced performance, the mechanism for enhanced storage in N-FLG was further examined with in situ Raman spectroscopy. In situ Raman spectroscopy is a nondestructive testing technique for exploring the electronic and structural changes in graphene and graphite and has become a powerful tool to establish the understanding of Li^+ staging in carbons [57, 58]. The in situ Raman results shed light on the role of N-doping which provides active sites in the carbon lattice that can locally store K ions. Meanwhile, N-dopant sites can sustain KC_8 formation and thus provide extra charge capacity [56]. This methodology enlightens the post-researchers that atomically precise fabrication of materials can be a practical avenue to optimize the performance of electrode materials for next-generation batteries.

Ju group have also attempted some investigations on graphene-based materials in PIBs in order to make a library of carbon-based negative electrodes. A recent study reported few-layer F-doped graphene foam (FFGF) as anode materials for both LIBs and PIBs [59]. The FFGF was prepared via high-temperature solid-state method using PVDF as a single source reactant (Fig. 9.4). Fluorine doping was introduced due to F as the most electronegative element, and the introduction of F atoms into graphene makes the carbon hybridization change from sp^2 to sp^3, providing a new

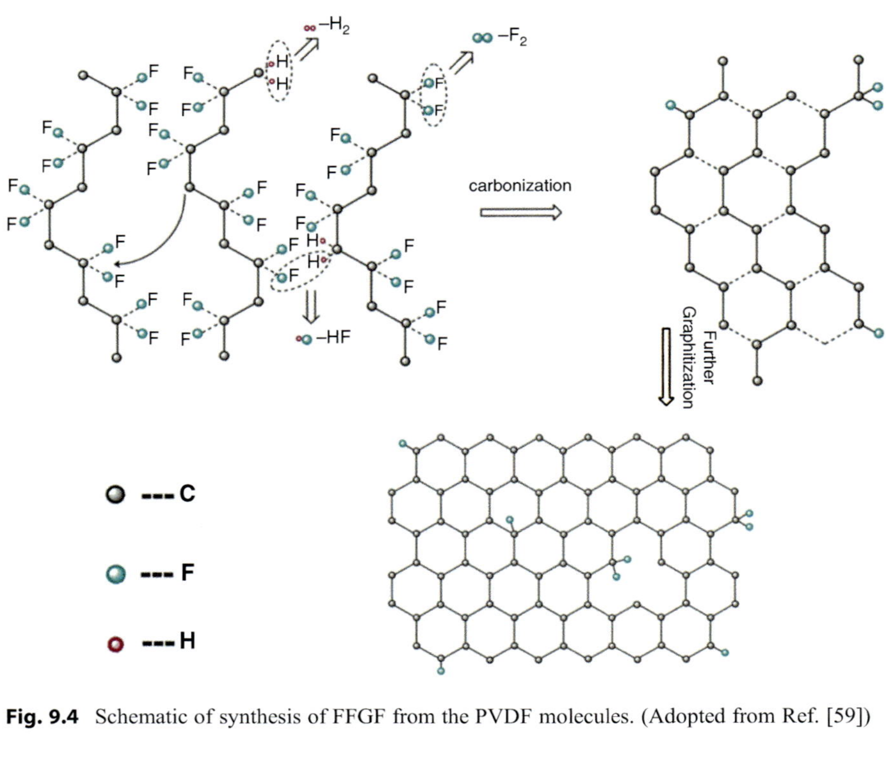

Fig. 9.4 Schematic of synthesis of FFGF from the PVDF molecules. (Adopted from Ref. [59])

possibility to modify graphene properties. Electrochemical performance of the FFGF was investigated. The cyclic voltammetry (CV) profile displays three cathodic peaks around 0.5, 0.17, and 0.05 V as well as one rather broad anodic peak around 0.68 V during the first cycle. The peak around 0.5 V in the first cycle can be attributed to the decomposition of the electrolyte and formation of the solid electrolyte interface (SEI) film on the F-doped graphene. Then, the sharp cathodic peaks situated at 0.17 V and 0.05 V are correlated with the K ions intercalation into FFGF stepwise to form KC_8. In the following anodic scan, the broad peak between 0.45 and 0.8 V corresponds to the deintercalation of K ions. From the second cycle onward, the 0.5 V peak disappears, which means that the SEI film mainly forms in the first cathodic process. Besides, the other cathodic/anodic peaks are almost overlapped, suggesting the high reversibility and stability. The galvanostatic cycling profile of FFGF vs. K^+/K at 50 mA g^{-1} shows a high initial discharge and charge capacity of 863.8 and 355.6 mA h g^{-1}, respectively. The initial Coulombic efficiency is 41.2% due to the irreversible capacity from SEI formation, but it still remained above 90% after 20 cycles. The rate cycling performance of FFGF demonstrates that this electrode delivers capacities of 326.1, 271.9, 250.7, and 212.6 mA h g^{-1} at current rates of 50, 100, 200, and 500 mA g^{-1}, respectively [59]. This material appears to be rather attractive in terms of performance. So far, further work is required to understand the mechanism of K^+ insertion/extraction and to adjust the functional groups on the graphene surface to attain optimum performance parameters.

Afterward, phosphorus and oxygen dual-doped graphene (PODG) was synthesized via a thermal annealing method using triphenylphosphine ($C_{18}H_{15}P$) and graphite oxide as precursors [60]. It exhibits fairly outstanding electrochemical performance even reaching a reversible capacity of 566.6 mAh g^{-1} at 50 mA g^{-1}, which is much higher than those of the other carbon anodes in PIBs [60]. The PODG electrode presents an ultralong cycling stability of 600 cycles at 2000 mA g^{-1} with a negligible capacity fading. Recently, few-layer nitrogen-doped graphene (FLNG) was successfully fabricated by a facile bottom-up synthesis technique with dicyandiamide and coal-tar pitch as raw materials [14]. The FLNG delivers competitive reversible capacity (320 mA h g^{-1} at 50 mA g^{-1} after 60 cycles), superior rate capability, as well as long-term cycling life at large current density (150 mA h g^{-1} at 500 mA g^{-1} after 500 cycles) [14]. This superior electrochemical performance can be ascribed to the heteroatom doping, intrinsic high surface area, and enlarged interlayer distance. These structural features and their synergistic effects would promote ion and electron transportation with nanometer-scale diffusion length as well as facilitate the penetration of electrolyte, thereby leading to the superior potassium storage performance.

9.2.3 Hard Carbon

Hard carbon, also known as nongraphitizable carbon, has been chosen as the competitive anode material in SIBs or PIBs as a result of its larger interlayer spacing. Stevens and Dahn have demonstrated that hard carbon can achieve a reversible sodium capacity of ca. 300 mAh g^{-1} [61]. However, issues of hard carbon anode in SIBs are twofold: poor rate performance and the fact that the majority of its capacity is attained at sodiation potentials close to metal plating. Jian et al. [15] for the first time reported hard carbon microspheres (HCS) as an anode material for K^+ storage and also compared their performance in SIBs and PIBs. Hard carbon microspheres are obtained via a hydrothermal synthesis of sucrose, followed by pyrolysis. More encouragingly, HCS shows advantageous performance for PIBs in two important respects compared with SIBs: (1) The ion-intercalation plateau region in the potential profiles of HCS in K cells is averaged at 0.2 V vs. K^+/K in contrast to 0.05 V vs. Na^+/Na of HCS/Na cell; this can reduce the risk of dendrite formation to a large degree in PIBs. (2) The rate capability of HCS in PIBs is much better than in SIBs. The HCS/K cells show capacities of 229, 205, 190, and 136 mAh g^{-1} at C/2, 1C, 2C, and 5C, respectively, while HCS/Na cells display capacities of 190, 127, 97, and 73 mAh g^{-1} at the same rates. In order to clarify why hard carbon exhibits a greater rate capability in PIBs than in SIBs, GITT was utilized to measure the diffusion coefficient of K-ions and Na-ions in HCS electrodes. The ionic diffusion coefficient in HCS electrodes can be determined based on Fick's second law. The results surprisingly show that the diffusion coefficient of K-ions is slightly higher than that of Na-ions in both insertion and extraction processes (Fig. 9.5). This observation may be attributed to different binding energies between K-C and Na-C and a lower charge density of larger K-ions, which may facilitate a faster K-ion diffusion with a lower activation energy. Therefore, further studies are required to obtain the clear picture of a kinetic comparison between K^+ and Na^+ in carbon structures, particularly calculation and simulation.

In addition, Jian et al. [62] demonstrated that the limited rate capability of hard carbon spheres (HCS) and the capacity loss of soft carbon (SC) can be surmounted by synthesizing hard-soft carbon composites, integrating advantages of both materials. Furthermore, the authors replaced the commonly used polyvinylidene fluoride (PVDF) binder with sodium carboxymethylcellulose (CMCNa) in the process of

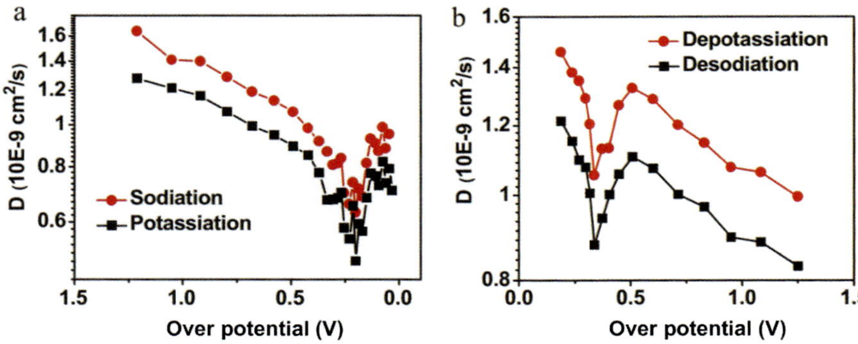

Fig. 9.5 Diffusion coefficients calculated from the GITT potential curves for the HCS/K and HCS/Na cells as a function of the state of charge during the second cycle (**a**) insertion and (**b**) extraction. (Adopted from Ref. [15])

electrode fabrication, which highly boosts the first cycle Coulombic efficiency, as it does for graphite anodes in PIBs [18]. Both HCS and HCS-SC composites exhibit much more stable cycling performance than SC, while SC and HCS-SC composites present greater rate capability than HCS. Even at 10 C, soft carbon electrode can present a high capacity of 121 mAh g^{-1} in contrast to 45 mAh g^{-1} of hard carbon. To shed light on the differences in rate capability among samples, the electronic conductivities of HCS and SC pellets were measured, which are 1.7 and 5.7 S m^{-1}, respectively. The greater electronic conductivity of SC certainly should be responsible for its outstanding rate capability. As a consequence, it can be found that different non-graphitic carbon structures can be differentiated by their performance in K-ion storage. And it also reveals that unique non-graphitic structure with a great level of disorder along the c axis is crucial to avoid capacity loss of carbonaceous anodes in PIBs.

Although these pioneering studies mentioned above achieve encouraging electrochemical results, more research should be concentrated on investigating the storage mechanisms of K^+ in hard carbon, which then provides appropriate guidance to enhance the performance of hard carbon as a PIBs anode.

Recently, Yang et al. [13] have successfully synthesized nitrogen/oxygen dual-doped hierarchical porous hard carbon (NOHPHC) applied as the anode material in PIBs by carbonizing and acidizing the NH_2-MIL-101(Al) [The Materials of Institut Lavoisier is a mesoporous framework modified by amino groups which consists of super tetrahedral building units including three-dimensional octahedral clusters [Al_3O] connected by 2-amino-1,4-benzenedicarboxylate linker] precursor (see Fig. 9.6). Mixed K-ion storage mechanisms, including capacitance and diffusion, were proposed, in which the capacitance plays a more critical part in its outstanding electrochemical performance. In particular, the existence of aluminum (Al) could serve as the catalyst to induce the graphitization of carbon [63], and the nitrogen- and oxygen-containing organic ligands could provide the N and O resources for the in situ heteroatom doping. Figure 9.7a, b shows the field-emission scanning electron microscopy (FESEM) images of the carbonized product of the NH_2-MIL-101 (Al) precursor. It can be clearly seen that the product is comprised of uniform flower-like microspheres with sizes of ca. 2 μm, which are similar to the morphology of the NH_2-MIL-101(Al) precursor. And each flower consists of plenty of primary nano-sized "petals" with thicknesses of tens of nanometers. After further removing the metal or metal oxides in the carbonized product by acid treatment, the obtained

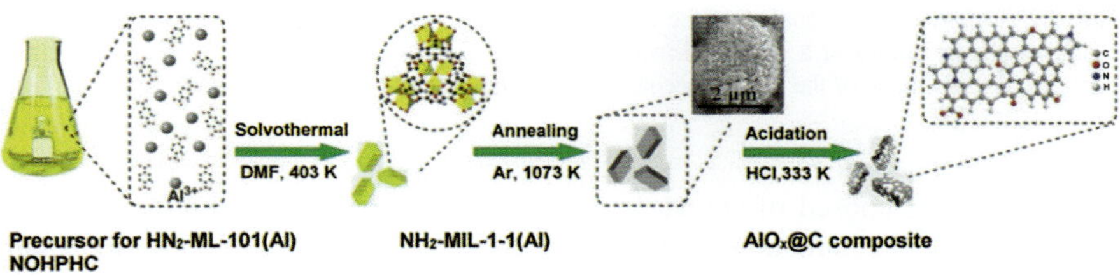

Fig. 9.6 Schematic illustration of the synthesis of NOHPHC. (Adopted from Ref. [13])

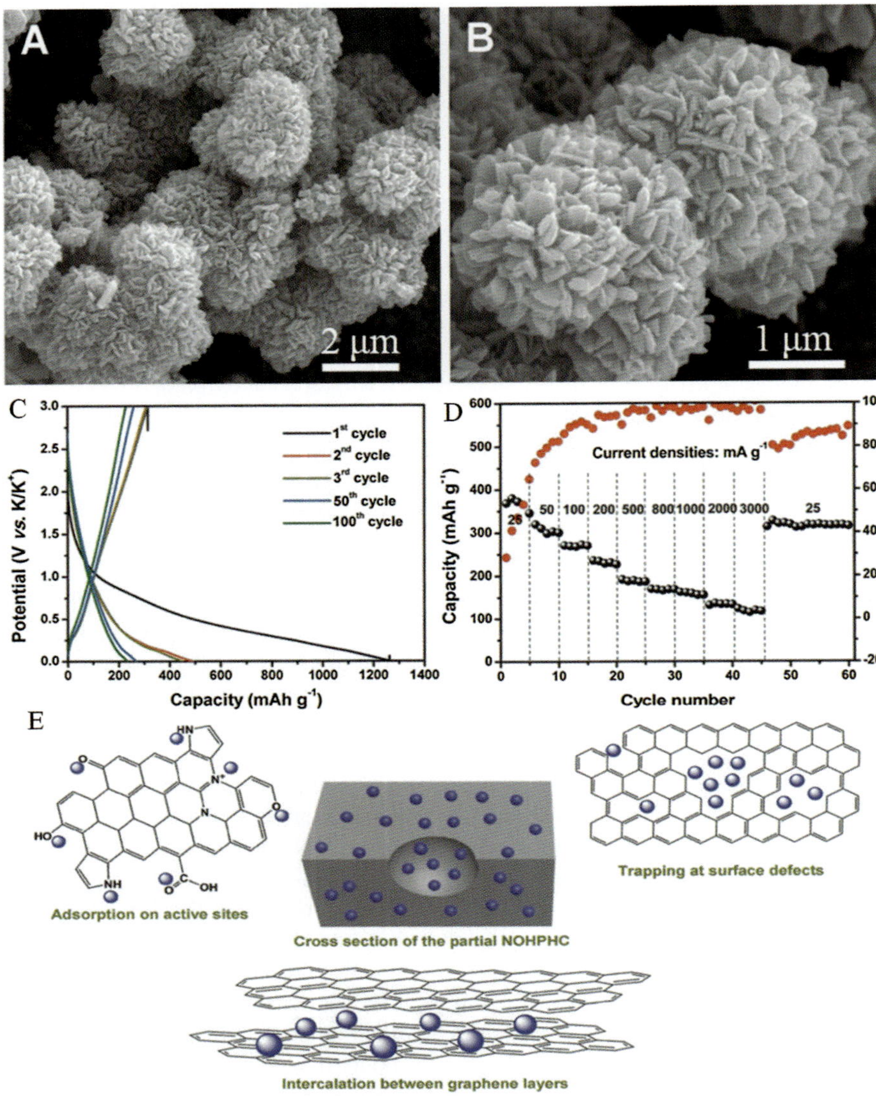

Fig. 9.7 (**a** and **b**) SEM images of the carbonized product of the NH_2-MIL-101(Al) precursor; (**c**) charge-discharge curves at a current density of 50 mA g^{-1}; (**d**) rate capability at different current densities; (**e**) schematic of the mixed mechanisms of potassium-ion storage. (Adopted from Ref. [13])

NOHPHC is composed of abundant mesoporous channels and mesopores, which could produce large amounts of defect sites to facilitate ion and electron transport. The NOHPHC electrode exhibits the high initial discharge and charge capacities of 1261 and 315 mA h g^{-1} at a current density of 50 mAg^{-1}, respectively. The low

Coulombic efficiency of 25% can be ascribed to the formation of the SEI layer on the surface of the NOHPHC. For one thing, the electrolyte used in the PIBs could be responsible for this low efficiency; for another thing, the hierarchical porous structure and the high specific surface area may also lead to the low initial efficiency and slow capacity stabilization, which need more time for the complete formation of the SEI layer. Rate performance measurements were also carried out (Fig. 9.7d). It shows the NOHPHC obtained capacities of ca. 365 and 315 mAh g^{-1} at 25 and 50 mA g^{-1}, respectively. And even at a high current density of 3000 mA g^{-1}, a capacity of 118 mAh g^{-1} can be maintained with a capacity retention of 33%. When the current density is restored to 25 mA g^{-1}, the reversible capacity nearly returns to its initial capacity which indicates the superior rate capability of the electrode. The electrode also achieved outstanding long-term cycling stability at 1050 mA g^{-1} for 1100 cycles. The capacity of the electrode fades from 174 to 130 mAh g^{-1} in the first 300 cycles, while only 4% capacity loss and nearly 100% efficiency occur in the subsequent cycles. The capacity fading in the first 300 cycles may be attributed to the fact that the fingerprint-like twisted structure was destroyed by the strain that is relevant to the intercalation of large K ions. The adsorption mechanism for capacity could account for the excellent cycling stability in the subsequent cycles. It is because of the superior rate capacity and excellent cycling performance of the NOHPHC electrode that the authors are determined to explore the contribution of two mechanisms, the surface-induced capacitive process and diffusion-controlled intercalation process, in the insertion/extraction of K^+ qualitatively and quantitatively, respectively. Figure 9.7e illustrates the mixed mechanisms of potassium-ion storage. Quantitative analysis demonstrated that both capacitance and diffusion are responsible for the excellent K-ion storage, in which capacitance accounts for the majority of the superb electrochemical performance. It will provide a useful guide on the relationship between K-ion storage mechanisms and electrochemical properties for the carbonaceous anode.

9.2.4 Carbon-Based Composites

Metals or nonmetals that undergo alloying reactions and store alkali metals are among the most appealing anode candidates for high-energy rechargeable batteries because they offer much higher capacities than the commercially employed graphite anodes. Lithium is known to form alloys (Li_xM) with various metals and nonmetals such as magnesium (Mg), Al, silicon (Si), germanium (Ge), tin (Sn), P, and antimony (Sb). For instance, the theoretical Li storage capacity is up to 4200 mA h g^{-1} for Si, 2959 mA h g^{-1} for P, 1600 mA h g^{-1} for Ge, and 990 mA h g^{-1} for Sn, respectively [64, 65]. In spite of the high theoretical capacity, the practical application of these alloy-based anode materials is still a vast challenge. A critical barrier that restricts the commercialization of these alloy-based anodes is that they suffer from intense capacity loss upon cycling. One commonly accepted reason is the aggregation and pulverization of active materials caused by severe volume changes. To date, numerous efforts have been devoted to tackling the intrinsic problem, and significant progress has been achieved. The strategies for improving the performance of alloy-type anodes mainly concentrate on engineering rational nanostructures and introducing conductive carbon substrate, both of which can effectively lead to the faster reaction kinetics and alleviate capacity decay.

So far, alloy-based anode materials in PIBs have also gradually drawn researchers' interest in terms of their high theoretical capacities and effective strategies toward large volume variation. McCulloch et al. demonstrated that nano-sized antimony carbide (Sb-C) composites exhibited an initial discharge capacity of ca. 880 mAh g^{-1} at a current density of 35 mAg^{-1} [66]. Especially, carbon can serve both as an efficient conductive pathway and a buffer layer to mitigate the large volume change of Sb during the cycling process. Due to the occurrence of some side reactions and SEI formation on the surface of the electrodes, the capacity of 880 is irreversible. Sb reversibly alloys with K upon the subsequent cycles, forming cubic K_3Sb with a capacity of $\approx$650 mAh g^{-1}.

Phosphorus (P) possesses a high theoretical capacity of 2595 mAh g^{-1} because it can form potassium phosphide (K_3P), a binary phase with K [67]. However, the low electrical conductivity and large volume changes limit its further application. In order to restrain the inherent volume expansion of phosphorus during the charge and discharge process, the phosphorus/carbon (P-C) composites are synthesized by a simple mechanical ball-milling as an anode for PIBs. The P-C electrode presents the first discharge and charge capacity of 958.9 and 482.9 mA h g^{-1}, respectively, and a low initial Coulombic efficiency of 50.3% [68]. It still maintained a rather high charge capacity of 323.5 mAh g^{-1} after 50 cycles at a rate of 50 mA g^{-1}. The outstanding electrochemical performance is mainly ascribed to a structural transformation from red P to black P and the formation of the stable P-C bond, mitigating the dramatic volume expansion upon cycling effectively. Pre-potassiation treatment was carried out to boost the low initial coulomb efficiency. The authors also conducted the X-ray diffraction characterization to further elucidate the K-ion storage mechanism. And it was found that a KP phase was detected during the alloying-de-alloying process.

Sn, as a well-known alloy anode material, has been massively studied for LIBs and SIBs. Tin (Sn) is capable of forming $Li_{4.4}Sn$ and $Na_{15}Sn_4$ alloys reversibly in these cells and displaying theoretical capacities of 991 and 845 mA h g^{-1}, respectively [64, 69]. When utilized as PIBs anode materials, the results show that an Sn atom can accept only one K atom to form KSn alloy with a capacity of 226 mA h g^{-1}. Due to the large volume changes (~197%) during the cycling process, the electrochemical performance of pure Sn is unsatisfactory as reported in previous work [70]. A common strategy to conquer this problem is to prepare Sn-carbon composites in which Sn nanoparticles are mixed with or dispersed in a carbon component. Sultana et al. [71] reported that ball-milled Sn-graphite nanocomposite anodes for PIBs delivered a capacity of 150–170 mAh g^{-1} for 10–15 cycles at 25 mA g^{-1} within the potential window of 0.01–2 V with 0.75 M KPF_6 in a mixture of EC and DEC electrolyte. Ex situ XRD characterization indicates K_2Sn_5 and K_4Sn_{23} are the likely potassiated phases. Zhang et al. [72] reported the electrochemical performance of ball-milled phosphorous/carbon (P/C) and tin phosphide/carbon (Sn_4P_3/C) composites as anode materials for K-ion storage. The P-C electrode produced over 2000 and 1000 mAh g^{-1} for the first two charge cycles, followed by a sudden capacity loss, only with a retention of 9% of the capacity after 20 cycles. In contrast, the electrochemical performance of the Sn_4P_3/C composite is more encouraging. The Sn_4P_3/C electrode displayed superior rate capability: 385 mAh g^{-1} and 222 mAh g^{-1} at 50 mA g^{-1} and 1 A g^{-1}, respectively. More importantly, this alloy anode retained a capacity of over 300 mAh g^{-1} after 50 cycles. A two-step reaction helping to buffer the stress was proposed for K^+ storage (Fig. 9.8), which explains the improved electrochemical performance clearly.

Very recently, Ju group also make some efforts in C@M (M = Sn, Sb) composites as anode materials for K-ion storage [73, 74]. Three-dimensional (3D) hierarchically porous carbon/Sn (HPCS) composite was synthesized by in situ generated sodium chloride (NaCl) crystals as templates, followed by calcination at three different temperatures. The thickness of the 3D network of carbon ultrathin films is only several nanometers. The Sn nanoparticles are uniformly distributed in the inner surface of hierarchically porous carbon materials and preserve the structure and interface well, leading to a highly reversible electrochemical reaction. Obviously, the 3D hierarchically porous framework could provide efficient space for K^+ storage and fast pathways for K^+ transportation. Electrochemical results showed that HPCS-650 exhibited both high capacity and excellent cyclability. To be specific, the electrode at 650 °C obtained a first charge and discharge capacity of 847.9 mAh g^{-1} and 384.5 mAh g^{-1} at 50 mAg^{-1}. Moreover, the electrode still maintained a reversible capacity of about 276.4 mAh g^{-1} for 100 cycles steadily with a Columbic efficiency of more than 96%. However, the sample at lower carbonization temperature (450 °C) didn't display relatively more superior performance than that of other samples. The reason may be that the lower degree of graphitization gives rise to poor conductivity and K^+ meet greater hindrance in the ion transfer process. In order to elucidate the Sn and K alloying process, the authors focus on the analysis of the third cycle of the cyclic voltammetry (CV) profiles and propose the reversible alloying and conversion reaction mechanisms. These can be summarized as the following equations:

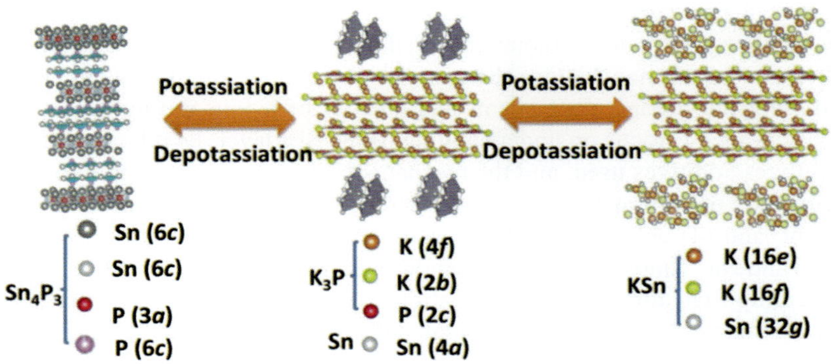

Fig. 9.8 Schematic potassiation/depotassiation reaction of Sn_4P_3/C anodes. (Adopted from Ref. [72])

$$23Sn + 4K \leftrightarrow K_4Sn_{23} \tag{9.1}$$

$$K_4Sn_{23} + K \leftrightarrow K_2Sn_5 \tag{9.2}$$

$$8C + K^+ + e^- \leftrightarrow KC_8 \tag{9.3}$$

Antimony nanoparticles uniformly embedded in the 3D porous carbon synthesized by similar methods have also been shown to offer high discharge capacities, superior rate capability, and relatively stable cycle life for K-ion storage [73]. Apparently, some metal alloying systems have displayed promising performance for K^+ storage; meanwhile, the large volume changes of K alloying reactions will pose challenges and should be further evaluated.

Compared to unsustainable carbon resources, biomass can serve as a renewable resource to prepare carbon electrodes for the applications of electrochemical energy storage [75]. It's reported that biomass carbon materials have been extensively applied as anode materials in LIBs and SIBs [76–78]. Similar to the findings reported for Li and Na, this class of materials also have been proven to intercalate K successfully. For example, some waste or scrap materials derived from seafood waste (chitin) [79], waste tires [80], or pencils [81] can be explored as carbon anodes for K^+ storage. The idea that biowaste materials are recycled to synthesize anodes for PIBs is interesting but instructive and conducive to the sustainable development of the whole society.

9.3 Conclusion and Outlook

This chapter presented an overview of the most up-to-date progress in carbon-based materials as anode materials for potassium-ion batteries. Since the intercalation of potassium ions into graphite anodes has been demonstrated to be feasible, researchers have successively conducted more investigations on other carbon anode materials; meanwhile, some advancements have been achieved. However, the research on nonaqueous PIBs is still in its infancy stage; there are some challenges to be surmounted for the future development and practical applications of this new battery chemistry, including but not limited to (1) understanding of the storage mechanisms and kinetics of K ions during the intercalation reactions; (2) focus on understanding of the SEI layer, its relationship to the electrolytes/additives used, and the mechanism of formation and growth; (3) the optimization of current electrolyte and binder systems to make it compatible and efficient; and (4) underlying safety hazards due to more chemically reactive K than Li and Na. There is still a long winding road to fulfilling viable performance of emerging PIBs for practical applications. Overall, carbon-based materials are the most prospective electrode candidate materials for the future practical applications of PIBs. And there are good reasons to believe that potassium-ion batteries have the remarkable potential to meet crucial goals for future energy storage devices.

Acknowledgments This work is supported by the Jiangsu Province Engineering Laboratory of High Efficient Energy Storage Technology and Equipment under grant no. [2017]975-18 and the Xuzhou City Key Laboratory of High Efficient Energy Storage Technology and Equipment under grant no. KH17078.

References

1. D. Larcher, J. Tarascon, Towards greener and more sustainable batteries for electrical energy storage. Nat. Chem. **7**(1), 19–29 (2015)
2. Y. Tang, Y. Zhang, W. Li, B. Ma, X. Chen, Rational material design for ultrafast rechargeable lithium-ion batteries. Chem. Soc. Rev. **44**(17), 5926–5940 (2015)
3. J.C. Pramudita, D. Sehrawat, D. Goonetilleke, N. Sharma, An initial review of the status of electrode materials for potassium–ion batteries. Adv. Energy Mater. **7**(24), 1602911 (2017)
4. S.R. Carmichael, *Practical Handbook of Physical Properties of Rocks and Minerals* (CRC Press, Boca Raton Florida, 1989)
5. H. Kim, H. Kim, Z. Ding, M.H. Lee, K. Lim, G. Yoon, K. Kang, Recent Progress in Electrode Materials for Sodium-Ion Batteries. Adv. Energy Mater. **6**(19), 1600943 (2016)
6. V. Palomares, P. Serras, I. Villaluenga, K.B. Hueso, J. Carretero-Gonzalez, T. Rojo, Na-ion batteries, recent advances and present challenges to become low cost energy storage systems. Energy Environ. Sci. **5**(3), 5884–5901 (2012)
7. H. Pan, Y.S. Hu, L. Chen, Room-temperature stationary sodium-ion batteries for large-scale electric energy storage. Energy Environ. Sci. **6**(8), 2338–2360 (2013)
8. D.A. Stevens, J.R. Dahn, High capacity anode materials for rechargeable sodium-ion batteries. J. Electrochem. Soc. **147**(4), 1271–1273 (2000)
9. S.-W. Kim, D.-H. Seo, X. Ma, G. Ceder, K. Kang, Electrode materials for rechargeable sodium-ion batteries: Potential alternatives to current lithium-ion batteries. Adv. Energy Mater. **2**(7), 710–721 (2012)
10. N. Yabuuchi, K. Kubota, M. Dahbi, S. Komaba, Research development on sodium-ion batteries. Chem. Rev. **114**(23), 11636–11682 (2014)
11. Y. Marcus, Thermodynamic functions of transfer of single ions from water to nonaqueous and mixed solvents: Part I – Gibbs free energies of transfer to nonaqueous solvents. Pure. Appl. Chem. **55**(5), 899–940 (2009)
12. Z. Jian, W. Luo, X. Ji, Carbon Electrodes for K-Ion Batteries. J. Am. Chem. Soc. **137**(36), 11566–11569 (2015)
13. J. Yang, Z. Ju, Y. Jiang, Z. Xing, B. Xi, J. Feng, S. Xiong, Enhanced capacity and rate capability of Nitrogen/Oxygen Dual-Doped Hard Carbon in capacitive potassium-ion storage. Adv. Mater. **30**(4), 1700104 (2018)
14. Z. Ju, P. Li, G. Ma, Z. Xing, Q. Zhuang, Y. Qian, Few-layer nitrogen-doped graphene with highly reversible potassium storage. Energy Storage Mater. **11**, 38–46 (2018)
15. Z. Jian, Z. Xing, C. Bommier, Z. Li, X. Ji, Hard carbon microspheres: Potassium-ion anode versus sodium-ion anode. Adv. Energy Mater. **6**(3), 1501874 (2016)
16. W. Luo, J. Wan, B. Ozdemir, W. Bao, Y. Chen, J. Dai, H. Lin, Y. Xu, F. Gu, V. Barone, L. Hu, Potassium ion batteries with graphitic materials. Nano Lett. **15**(11), 7671–7677 (2015)
17. M. Okoshi, Y. Yamada, S. Komaba, A. Yamada, H. Nakai, Theoretical analysis of interactions between potassium ions and organic electrolyte solvents: A comparison with Lithium, Sodium, and Magnesium Ions. J. Electrochem. Soc. **164**(2), A54–A60 (2017)
18. S. Komaba, T. Hasegawa, M. Dahbi, K. Kubota, Potassium intercalation into graphite to realize high-voltage/high-power potassium-ion batteries and potassium-ion capacitors. Electrochem. Commun. **60**, 172–175 (2015)
19. C.D. Wessells, S.V. Peddada, M.T. McDowell, R.A. Huggins, Y. Cui, The effect of insertion species on nanostructured open framework hexacyanoferrate battery electrodes. J. Electrochem. Soc. **159**(2), A98–A103 (2011)

20. S. Dawei, M. Andrew, Q. Shi-Zhang, W. Guoxiu, High–capacity aqueous potassium–ion batteries for large–scale energy storage. Adv. Mater. **29**(1), 1604007 (2017)
21. C.D. Wessells, S.V. Peddada, R.A. Huggins, Y. Cui, Nickel hexacyanoferrate nanoparticle electrodes for aqueous sodium and potassium ion batteries. Nano Lett. **11**(12), 5421–5425 (2011)
22. Y. Zhai, Y. Dou, D. Zhao, P.F. Fulvio, R.T. Mayes, S. Dai, Carbon materials for chemical capacitive energy storage. Adv. Mater. **23**(42), 4828–4850 (2011)
23. M.-M. Titirici, R.J. White, N. Brun, V.L. Budarin, D.S. Su, F. del Monte, J.H. Clark, M.J. MacLachlan, Sustainable carbon materials. Chem. Soc. Rev. **44**(1), 250–290 (2015)
24. X. Zou, P. Xiong, J. Zhao, J. Hu, Z. Liu, Y. Xu, Recent research progress in non-aqueous potassium-ion batteries. Phys. Chem. Chem. Phys. **19**(39), 26495–26506 (2017)
25. M.K. Datta, R. Epur, P. Saha, K. Kadakia, S.K. Park, P.N. Kumta, Tin and graphite based nanocomposites: Potential anode for sodium ion batteries. J. Power Sources **225**(3), 316–322 (2013)
26. P. Ge, M. Fouletier, Electrochemical intercalation of sodium in graphite. Solid State Ionics **28**, 1172–1175 (1988)
27. Z. Wang, A. Ratvik, T. Grande, S. Selbach, Diffusion of alkali metals in the first stage graphite intercalation compounds by vdW-DFT calculations. RSC Adv. **5**(21), 15985–15992 (2015)
28. D.E. Nixon, G.S. Parry, Formation and structure of the potassium graphites. J. Phys. D. Appl. Phys. **1**(3), 291 (1968)
29. Z. Wang, S.M. Selbach, T. Grande, Van der Waals density functional study of the energetics of alkali metal intercalation in graphite. RSC Adv. **4**(8), 4069–4079 (2014)
30. Y. Liu, F. Fan, J. Wang, Y. Liu, H. Chen, K.L. Jungjohann, Y. Xu, Y. Zhu, D. Bigio, T. Zhu, C. Wang, In situ transmission electron microscopy study of electrochemical sodiation and potassiation of carbon nanofibers. Nano Lett. **14**(6), 3445–3452 (2014)
31. J. Zhao, X. Zou, Y. Zhu, Y. Xu, C. Wang, Electrochemical intercalation of potassium into graphite. Adv. Funct. Mater. **26**(44), 8103–8110 (2016)
32. J. Hou, Y. Shao, M.W. Ellis, R.B. Moore, B. Yi, Graphene-based electrochemical energy conversion and storage: Fuel cells, supercapacitors, and lithium-ion batteries. Phys. Chem. Chem. Phys. **13**(34), 15384–15402 (2011)
33. K.S. Novoselov, A.K. Geim, S.V. Morozov, D. Jiang, Y. Zhang, S.V. Dubonos, I.V. Grigorieva, A.A. Firsov, Electric field effect in atomically thin carbon films. Science **306**(5696), 666–669 (2004)
34. Z. Xing, Z. Ju, Y. Zhao, J. Wan, Y. Zhu, Y. Qiang, Y. Qian, One-pot hydrothermal synthesis of Nitrogen-doped graphene as high-performance anode materials for lithium-ion batteries. Sci. Rep. **6**, 26146 (2016)
35. M.A. Rafiee, Graphene-based Composite Materials. Diss. Theses Gradworks **442**(2), 282–286 (2011)
36. A.K. Geim, NS the rise of graphene. Nat. Mater. **6**, 183–190 (2007)
37. M. Pumera, Graphene-based nanomaterials for energy storage. Energy Environ. Sci. **4**(3), 668–674 (2011)
38. Y. Hernandez, V. Nicolosi, M. Lotya, F. Blighe, Z. Sun, S. De, I.T. McGovern, B. Holland, M. Byrne, Y. Gunko, J. Boland, P. Niraj, G. Duesberg, S. Krishnamurti, R. Goodhue, J. Hutchison, V. Scardaci, A.C. Ferrari, J.N. Coleman, High yield production of graphene by liquid-phase exfoliation of graphite. Nat. Nanotechnol. **3**(9), 563–568 (2008)
39. N. Li, Z. Wang, K. Zhao, Z. Shi, Z. Gu, S. Xu, Large-scale synthesis of N-doped multi-layered graphene sheets by the simple arc-discharge method. Carbon **48**(1), 255–259 (2010)
40. K.S. Kim, Y. Zhao, H. Jang, S.Y. Lee, J.M. Kim, K.S. Kim, J.-H. Ahn, P. Kim, J.-Y. Choi, B.H. Hong, Large-scale pattern growth of graphene films for stretchable transparent electrodes. Nature, APS March Meeting **457**, 706 (2009)
41. C. Berger, Z. Song, X. Li, X. Wu, N. Brown, C. Naud, D. Mayou, T. Li, J. Hass, A.N. Marchenkov, Electronic confinement and coherence in patterned epitaxial graphene. Science **312**(5777), 1191–1196 (2006)

42. H.C. Schniepp, J.L. Li, M.J. Mcallister, H. Sai, M. Herrera-Alonso, D.H. Adamson, R.K. Prud'Homme, R. Car, D.A. Saville, I.A. Aksay, Functionalized single graphene sheets derived from splitting graphite oxide. J. Phys. Chem. B **110**(17), 8535–8539 (2006)
43. M.J. McAllister, J.-L. Li, D.H. Adamson, H.C. Schniepp, A.A. Abdala, J. Liu, D.L. Milius, R. Car, R.K. Prud'homme, Single sheet functionalized graphene by oxidation and thermal expansion of graphite. Chem. Mater. **19**(18), 4396–4404 (2007)
44. L. Zhang, X. Li, Y. Huang, Y. Ma, X. Wan, Y. Chen, Controlled synthesis of few-layered graphene sheets on a large scale using chemical exfoliation. Carbon **48**(8), 2367–2371 (2010)
45. S. Park, R.S. Ruoff, Chemical methods for the production of graphenes. Nat. Nanotechnol. **4**(4), 217–224 (2009)
46. S. Gilje, S. Han, M. Wang, K.L. Wang, R.B. Kaner, A chemical route to graphene for device applications. Nano Lett. **7**(11), 3394–3398 (2007)
47. W.S. Hummers, R.E. Offeman, Preparation of Graphitic Oxide. J. Am. Chem. Soc. **80**, 1339 (1958)
48. C.K. Chua, M. Pumera, Chemical reduction of graphene oxide: A synthetic chemistry viewpoint. Chem. Soc. Rev. **43**(1), 291–312 (2014)
49. C. Xu, B. Xu, Y. Gu, Z. Xiong, J. Sun, X.S. Zhao, Graphene-based electrodes for electrochemical energy storage. Energy Environ. Sci. **6**(5), 1388–1414 (2013)
50. Z.-S. Wu, W. Ren, L. Xu, F. Li, H.-M. Cheng, Doped graphene sheets as anode materials with superhigh rate and large capacity for lithium ion batteries. ACS Nano **5**(7), 5463–5471 (2011)
51. J.P. Paraknowitsch, A. Thomas, Doping carbons beyond nitrogen: An overview of advanced heteroatom doped carbons with boron, sulfur, and phosphorus for energy applications. Energy Environ. Sci. **6**(10), 2839–2855 (2013)
52. X. Qi, K. Huang, X. Wu, W. Zhao, H. Wang, Q. Zhuang, Z. Ju, Novel fabrication of N-doped hierarchically porous carbon with exceptional potassium storage properties. Carbon **131**, 79–85 (2018)
53. T. Lin, I.-W. Chen, F. Liu, C. Yang, H. Bi, F. Xu, F. Huang, Nitrogen-doped mesoporous carbon of extraordinary capacitance for electrochemical energy storage. Science **350**(6267), 1508–1513 (2015)
54. M.A. Patel, F. Luo, M.R. Khoshi, E. Rabie, Q. Zhang, C.R. Flach, R. Mendelsohn, E. Garfunkel, M. Szostak, H. He, P-Doped porous carbon as metal-free catalysts for selective aerobic oxidation with an unexpected mechanism. ACS Nano **10**(2), 2305–2315 (2016)
55. W. Kiciński, M. Szala, M. Bystrzejewski, Sulfur-doped porous carbons: Synthesis and applications. Carbon **68**, 1–32 (2014)
56. K. Share, A.P. Cohn, R. Carter, B. Rogers, C.L. Pint, Role of nitrogen-doped graphene for improved high capacity potassium ion battery anodes. ACS Nano **10**(10), 9738–9744 (2016)
57. C. Sole, N.E. Drewett, L.J. Hardwick, In situ Raman study of lithium-ion intercalation into microcrystalline graphite. Faraday Discuss. **172**(0), 223–237 (2014)
58. M. Inaba, H. Yoshida, Z. Ogumi, T. Abe, Y. Mizutani, M. Asano, In situ Raman study on electrochemical li intercalation into graphite. J. Electrochem. Soc. **142**(1), 20–26 (1995)
59. Z. Ju, S. Zhang, Z. Xing, Q. Zhuang, Y. Qiang, Y. Qian, Direct synthesis of few-layer F-Doped graphene foam and it's Lithium/Potassium storage properties. ACS Appl. Mater. Interfaces **8** (32), 20682–20690 (2016)
60. G. Ma, K. Huang, J.-S. Ma, Z. Ju, Z. Xing, Q.-c. Zhuang, Phosphorus and oxygen dual-doped graphene as superior anode material for room-temperature potassium-ion batteries. J. Mater. Chem. A **5**(17), 7854–7861 (2017)
61. D.A. Stevens, J.R. Dahn, High capacity anode materials for rechargeable sodium–ion batteries. J. Electrochem. Soc. **147**(4), 1271–1273 (2000)
62. Z. Jian, S. Hwang, Z. Li, A.S. Hernandez, X. Wang, Z. Xing, D. Su, X. Ji, Hard–soft composite carbon as a long–cycling and high–rate anode for potassium–ion batteries. Adv. Funct. Mater. **27**(26), 1700324 (2017)
63. L. Radhakrishnan, J. Reboul, S. Furukawa, P. Srinivasu, S. Kitagawa, Y. Yamauchi, Preparation of microporous carbon fibers through Carbonization of Al-Based porous coordination polymer (Al-PCP) with furfuryl alcohol. Chem. Mater. **23**(5), 1225–1231 (2011)

64. M.N. Obrovac, V.L. Chevrier, Alloy negative electrodes for Li-ion batteries. Chem. Rev. **114** (23), 11444–11502 (2014)
65. W.-J. Zhang, A review of the electrochemical performance of alloy anodes for lithium-ion batteries. J. Power Sources **196**(1), 13–24 (2011)
66. W.D. McCulloch, X. Ren, M. Yu, Z. Huang, Y. Wu, Potassium-ion oxygen battery based on a high capacity antimony anode. ACS Appl. Mater. Interfaces **7**(47), 26158–26166 (2015)
67. X. Su, S. Yang, J. Wang, N.-Y. Tang, F. Yin, Z. Li, M. Zhao, A new equation for temperature dependent solute impurity diffusivity in liquid metals. J. Phase Equilib. Diffus. **31**(4), 333–340 (2010)
68. X. Wu, W. Zhao, H. Wang, X. Qi, Z. Xing, Q. Zhuang, Z. Ju, Enhanced capacity of chemically bonded phosphorus/carbon composite as an anode material for potassium-ion batteries. J. Power Sources **378**, 460–467 (2018)
69. B. Clement, J. Xiulei, Recent development on anodes for Na–Ion batteries. Isr. J. Chem. **55**(5), 486–507 (2015)
70. Q. Wang, X. Zhao, C. Ni, H. Tian, J. Li, Z. Zhang, S.X. Mao, J. Wang, Y. Xu, Reaction and capacity-fading mechanisms of tin nanoparticles in potassium-ion batteries. J. Phys. Chem. C **121**(23), 12652–12657 (2017)
71. I. Sultana, T. Ramireddy, M.M. Rahman, Y. Chen, A.M. Glushenkov, Tin-based composite anodes for potassium-ion batteries. Chem. Commun. **52**(59), 9279–9282 (2016)
72. W. Zhang, J. Mao, S. Li, Z. Chen, Z. Guo, Phosphorus-based alloy materials for advanced potassium-ion battery anode. J. Am. Chem. Soc. **139**(9), 3316–3319 (2017)
73. H. Wang, X. Wu, X. Qi, W. Zhao, Z. Ju, Sb nanoparticles encapsulated in 3D porous carbon as anode material for lithium-ion and potassium-ion batteries. Mater. Res. Bull. **103**, 32–37 (2018)
74. K. Huang, Z. Xing, L. Wang, X. Wu, W. Zhao, X. Qi, H. Wang, Z. Ju, Direct synthesis of 3D hierarchically porous carbon/Sn composites via in situ generated NaCl crystals as templates for potassium-ion batteries anode. J. Mater. Chem. A **6**(2), 434–442 (2018)
75. Y.-P. Gao, Z.-B. Zhai, K.-J. Huang, Y.-Y. Zhang, Energy storage applications of biomass-derived carbon materials: Batteries and supercapacitors. New J. Chem. **41**(20), 11456–11470 (2017)
76. L. Wu, D. Buchholz, C. Vaalma, G.A. Giffin, S. Passerini, Apple–biowaste–derived hard carbon as a powerful anode material for Na–Ion batteries. ChemElectroChem **3**(2), 292–298 (2016)
77. Q. Jiang, Z. Zhang, S. Yin, Z. Guo, S. Wang, C. Feng, Biomass carbon micro/nano-structures derived from ramie fibers and corncobs as anode materials for lithium-ion and sodium-ion batteries. Appl. Surf. Sci. **379**, 73–82 (2016)
78. R.R. Gaddam, D. Yang, R. Narayan, K. Raju, N.A. Kumar, X.S. Zhao, Biomass-derived carbon nanoparticle as anodes for high-performance sodium and lithium-ion batteries. Nano Energy **26**, 346–352 (2016)
79. C. Chen, Z. Wang, B. Zhang, L. Miao, J. Cai, L. Peng, Y. Huang, J. Jiang, Y. Huang, L. Zhang, J. Xie, Nitrogen-rich hard carbon as a highly durable anode for high-power potassium-ion batteries. Energy Storage Mater. **8**, 161–168 (2017)
80. Y. Li, R.A. Adams, A. Arora, V.G. Pol, A.M. Levine, R.J. Lee, K. Akato, A.K. Naskar, M.P. Paranthaman, Sustainable potassium-ion battery anodes derived from waste-tire rubber. J. Electrochem. Soc. **164**(6), A1234–A1238 (2017)
81. Z. Tai, Y. Liu, Q. Zhang, T. Zhou, Z. Guo, H.K. Liu, S.X. Dou, Ultra-light and flexible pencil-trace anode for high performance potassium-ion and lithium-ion batteries. Green Energy Environ. **2**(3), 278–284 (2017)

New, Game-Changing Applications of Polymer-Based Coatings in Battery

10

Claudiu B. Bucur

Contents

10.1 Introduction

Commercial lithium-ion batteries (265 Wh/kg) provide 6.5× more energy than lead acid and more than 2× that of nickel-metal hydride (NiMH). These large gains in energy density have ushered in the wide proliferation of portable electronics and the commercialization of long-range personal electric vehicles. Car manufacturing companies such as Tesla and Volkswagen propose that large-scale production of electric vehicles will lower the battery pack cost to \$100/kWh (currently around \$150/kWh). Such large-scale projects have already started and will be more prevalent by 2020. However, the success of the electric vehicle hinges on the safety, durability, and lifetime of the battery pack. These characteristics are critically important because the lithium-ion battery operates outside its electrochemical stability window. During

C. B. Bucur (✉)
Toyota Research Institute of North America, Ann Arbor, MI, USA
e-mail: claudiu.b.bucur@outlook.com

Q. Zhen et al. (eds.), *Nanostructured Materials for Next-Generation Energy Storage and Conversion*, https://doi.org/10.1007/978-3-662-58675-4_10

charging, the electrolyte is exposed to potentials beyond its stability, so it reduces on the graphite anode and may oxidize on the cathode. The lowest unoccupied molecular orbital (LUMO) of the electrolyte is below the Fermi level of the anode, and the highest occupied molecular orbital (HOMO) of the electrolyte lies above the Fermi level of the cathode. In addition, cathode discharge products may have increased solubility in the electrolyte and may be lost during cycling. Thus, battery lifetime is governed by the degree with which these undesired reactions are mitigated. For example, the proper choice of electrolyte solvents and additives form a stable solid electrolyte interface (SEI) on the graphite anode. This thin layer of solid electrolyte is not electrically conductive, so it stops further decomposition of the electrolyte, but it allows the permeation of lithium cations (Li^+) such that battery can continue to operate. The commercial assembly process of lithium-ion batteries contains lithium in the cathode. The graphite anode is delithiated at the time of cell fabrication, so the battery is assembled in its discharged state. The electrochemically catalyzed self-assembly of the SEI during initial formation cycles irreversibly consumes lithium from the cathode which requires careful balancing of the electrodes by loading with excess cathode material. As a result, this practice reduces the energy density of the battery. In addition, the SEI decomposes at temperatures above 60 °C which further reduces the energy density due to the need for additional temperature management infrastructure. Similar chemistry occurs at the cathode.

Electrodes assembled with preformed stable interfaces may improve the cell energy density since no excess would be required at the cathode. Popular strategies for improving the interface stability of electrodes include various vapor/vacuum deposition of thin oxide layers and thin polymeric coatings. It is desirable that such coatings (1) increase the stability of the electrolyte at the electrode interface, (2) reduce dissolution of soluble species, (3) are thin enough as to not significantly increase cell impedance, and (4) are applied in a simple, fast, and cheap manner. When applied, such coatings can extend the cycle life of the battery and may allow for stability at higher charge voltages with corresponding increases in extracted capacity during discharge. In addition, advanced lithium-ion cathodes such as lithium-rich and nickel-rich cathodes also benefit from such stable coatings with reported increases in cycle life. Battery safety can also be improved by coatings applied to common polypropylene separators. In the catastrophic event of a thermal runaway, cell temperature can quickly rise above 160 °C which irreversibly shrinks commercial separators and exposes the electrodes to electric shorts. Ceramic and polymer-coated separators have been demonstrated to have improved thermal stabilities and conductivities of lithium cations and superior wettability by electrolyte solvents. The economic implementation of such coatings may foster the commercialization of advanced lithium-ion batteries with up to 20% improvement in energy density at the cell level (up to 315 Wh/kg). Over time, such advanced technology may mature to offer >10,000 cycles which may foster the introduction of electric drivetrains for vehicles used in commercial platforms (such as carrier trucks, USPS, etc.) with grueling nonstop schedules of operations. There are only a handful of batteries which go beyond lithium-ion. They commonly replace graphite anodes with alloying and ultimately pure metallic anodes (lithium, sodium, or magnesium) and pair them with cheap, high energy density cathodes such as sulfur or oxygen. While the challenges of these batteries are more numerous and more difficult to solve than those of lithium-ion, they also benefit from thin polymeric coatings to impede dissolution of discharge intermediates (sulfur is the typical example), to construct highly conductive electrode matrices, or to reign in the uncontrollable dendritic deposition of metallic lithium or sodium. As such, a lithium-sulfur battery may offer up to 3× gains in energy density (760 Wh/kg). These future, post-lithium-ion batteries may usher in the age of the long-range electric plane. For reference, an electric Tesla Model S 85 vehicle has an overall energy density of 40 Wh/kg and an average cruising power usage of 10 W/kg (C/4 operation), whereas a Cessna 172 airplane has an overall energy density of 470 Wh/kg with an average cruising power usage of 80 W/kg (if electric, C/6 operation). A rate of C corresponds to complete discharge or charge in 1 h; C/4 corresponds to full discharge in 4 h. Comprehensive reviews for the use of polymeric coatings in batteries have been published in the past 5 years [18, 25, 34, 63, 94]. The following chapter will focus on cutting edge advances in the past 3 years. Part 2 of the chapter will focus on simple polymeric coatings based on template polymerization of spin coating addition, and part 3 will demonstrate the advantages offered by the layer-by-layer multilayer technique.

10.2 Simple Polymeric Coatings

Stabilizing films which form in situ during battery operation consist in part of polymerized solvent or additive components. Common electrode-coating techniques include vapor deposition such as atomic layer deposition of stable oxides such as aluminum oxide (Al_2O_3). While this technique can control the adsorbed thickness to anatomically specified thickness, it is very slow (1 Å per cycle) and very expensive (half a million dollar instrument cost). Similarly, thin coatings can be quickly and cheaply applied from polymer solutions. The challenge is to obtain conformal and thin polymeric coatings as the most desirable conductive polymers, such as polyaniline (PANI), a mixture of poly(3,4-ethylenedioxythiophene)-polystyrene sulfonate (PEDOT:PSS) ionomers, polyacetylenes (PAC), polyphenylenes, or polypyrroles (PPY), contain conjugated double bonds and are not truly soluble in any solvent. They may, however, form good dispersions in some solvents. Table 10.1 lists many of the polymers of interest reported here.

10.2.1 Controlling Metallic Anodes

It is interesting to note that Bluecar is a three-door, all-electric car supplied by Bolloré which sports a 30 kWh (100 Wh/kg at the battery pack level) lithium polymer battery with a lithium metal anode, a polymer electrolyte, and a vanadium oxide cathode. It has an electric range of 93–60 miles. This car operates with a lithium metal anode battery and has been on the market since 2011 with no

Table 10.1 Polymers encountered in this review classified by adsorption functionality

Positively charged ions	Molecular structure
Polyaniline (PANI)	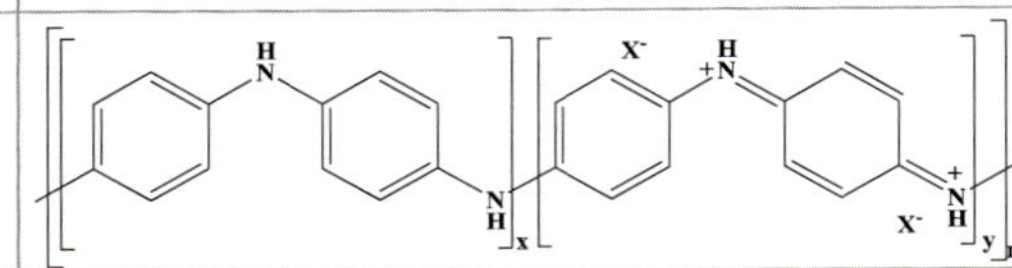
Polypyrrole (PPY)	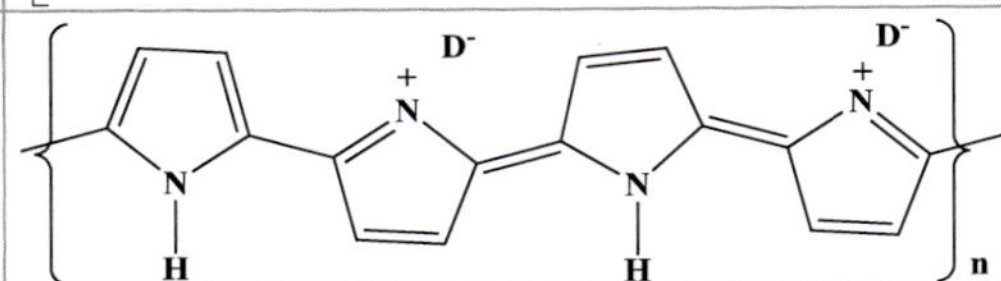
Polydiallyldimethylammonium chloride (PDADMAC)	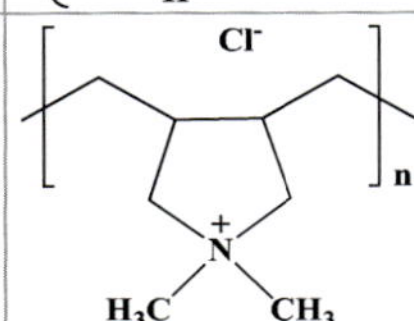
Polyethylenimine (PEI)	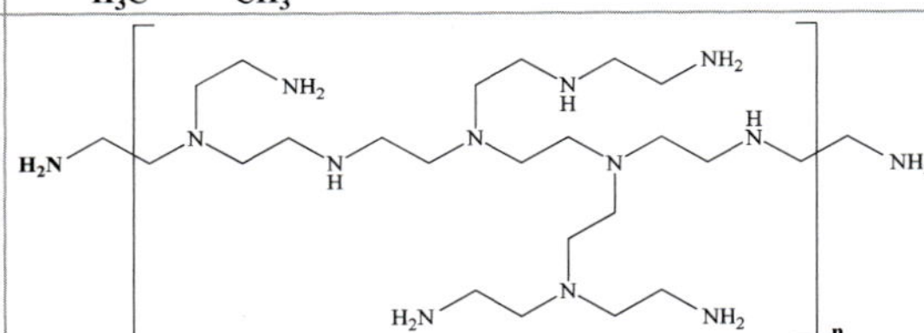
Clay flakes (Montmorillonite, Cloisite) $(Na,Ca)_{0.33}(Al,Mg)_2(Si_4O_{10})(OH)_2 \cdot nH_2O$	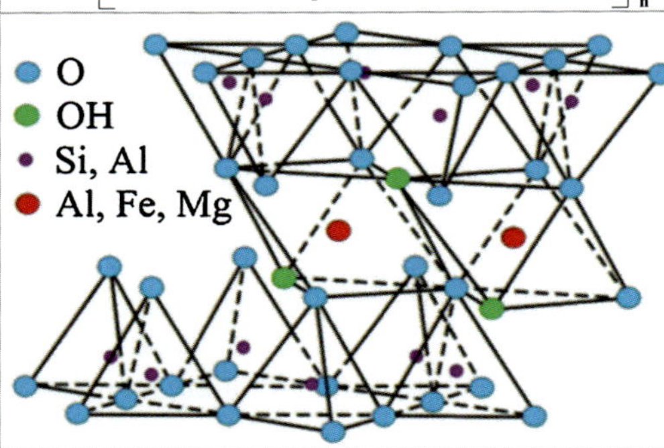
Polyamides (aramids): Kevlar®	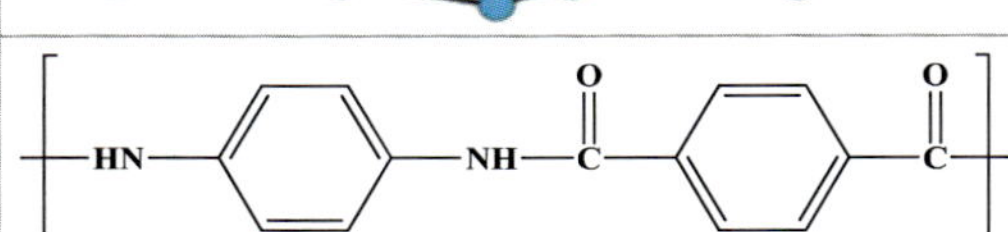
Poly(vinylsulfonic acid) (PVS)	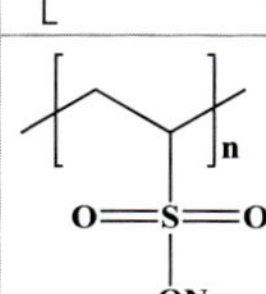
Polyvinyl alcohol (PVA)	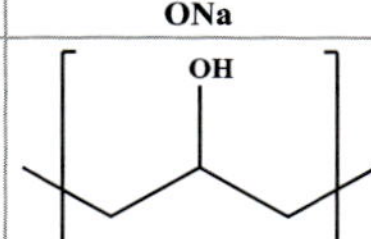

(*continued*)

Table 10.1 (continued)

Poly(allyl glycidyl ether) (PAGELS)	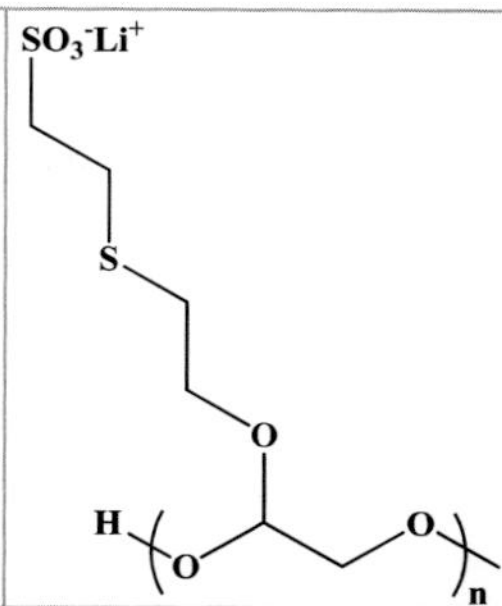
Negatively charged ions	**Molecular structure**
Nafion®	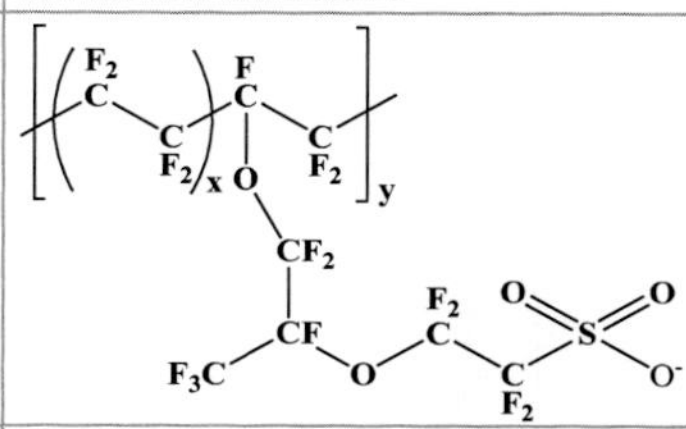
Poly(styrene sulfonate) (PSS)	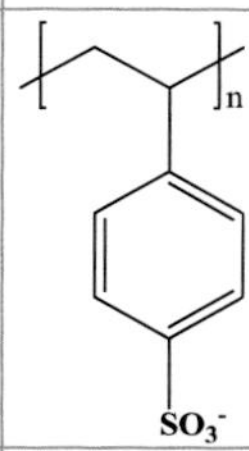
Poly (3,4-ethylenedioxythiophene)-poly(styrenesulfonate) (PEDOT: PSS)	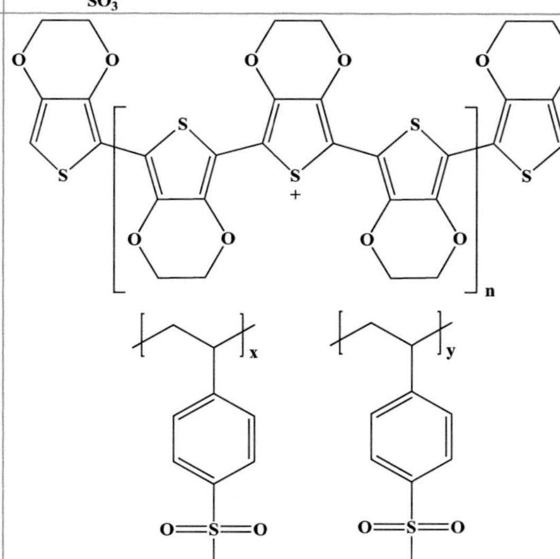

(*continued*)

Table 10.1 (continued)

Poly(acrylic acid) (PAA)	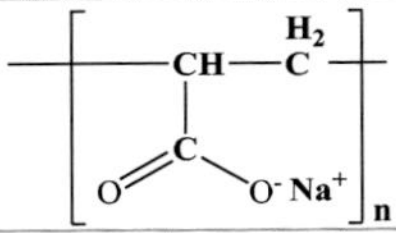
Graphene oxide (GO)	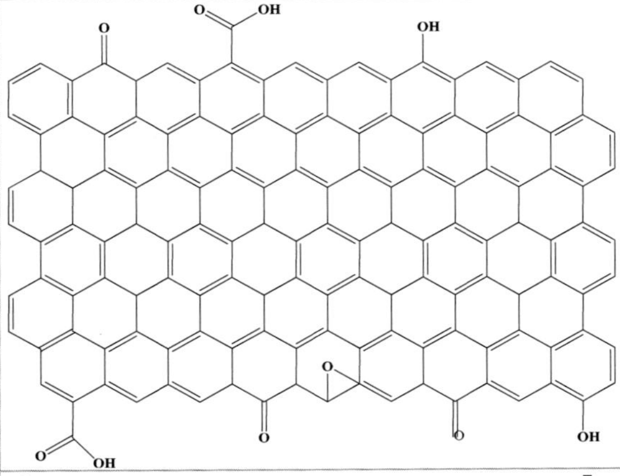
Sulfonated poly(ether ether ketone) (SPEEK)	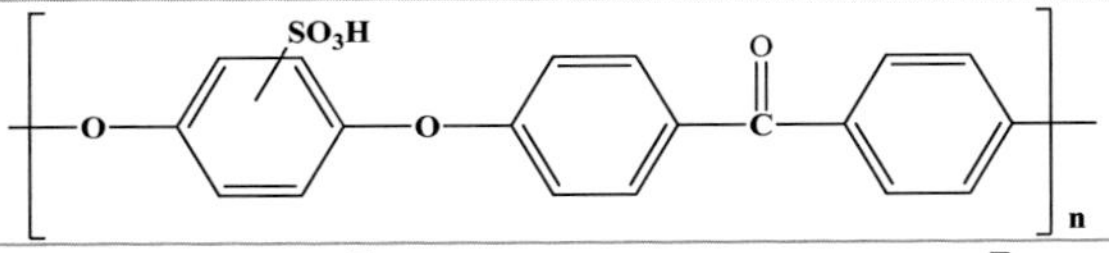
Xantham gum	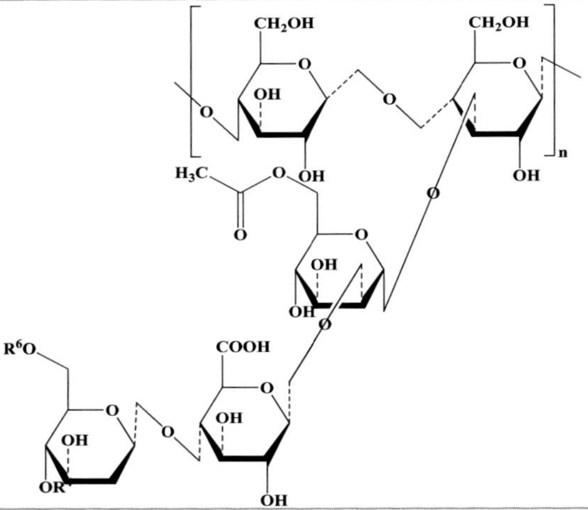
Polyvinyl alcohol (PVA)	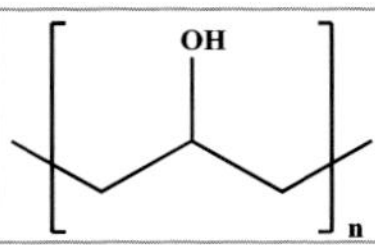

(*continued*)

Table 10.1 (continued)

Sulfonated poly(phenylene oxide) (SPPO)	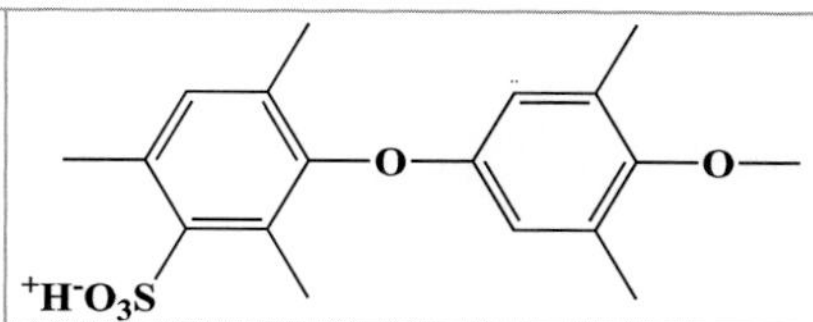
H–bonds	**Molecular structure**
Polyvinylpyrrolidone (PVP)	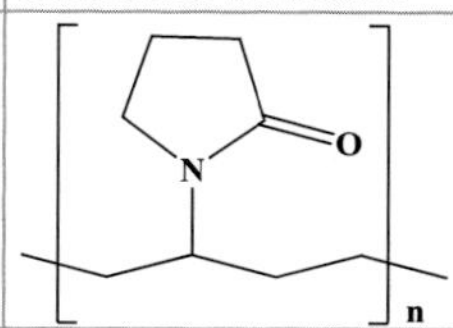
Poly(acrylic acid) (PAA)	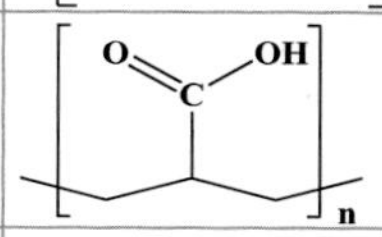
Polyethylenimine (PEI)	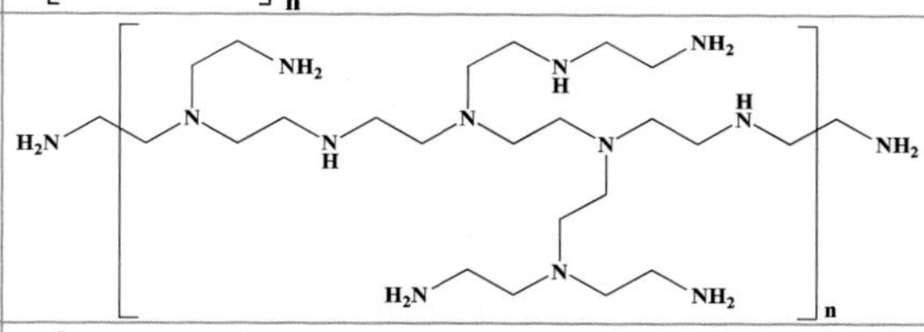
Poly(ethylene oxide) (PEO)	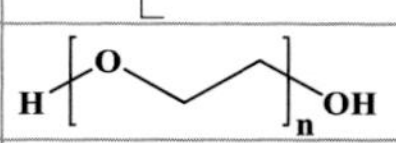
Polyethylene glycol diacrylate (PEGDA)	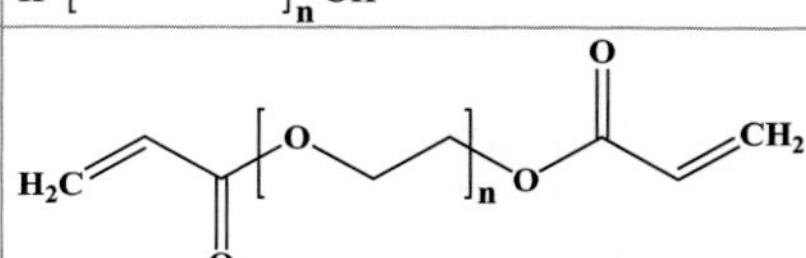
Polyamides (aramids): Kevlar®	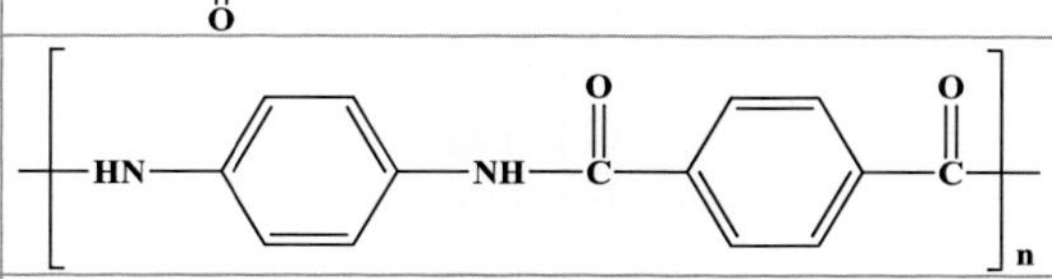
Guar gum	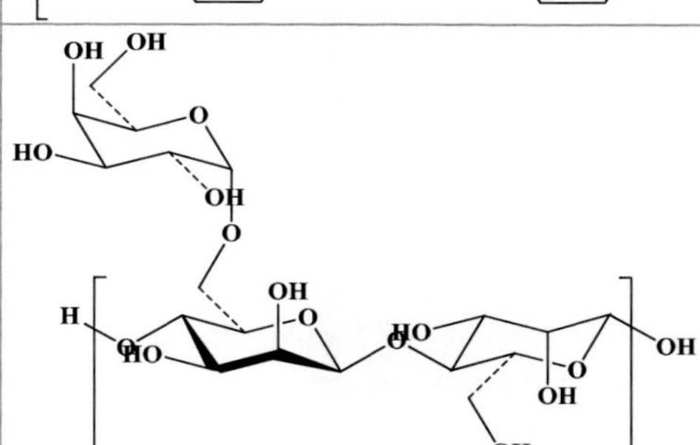

(*continued*)

Table 10.1 (continued)

Chitosan	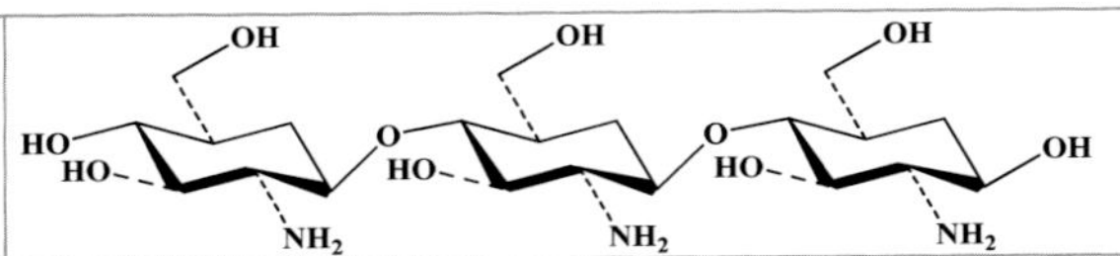
Polydopamine (PDA)	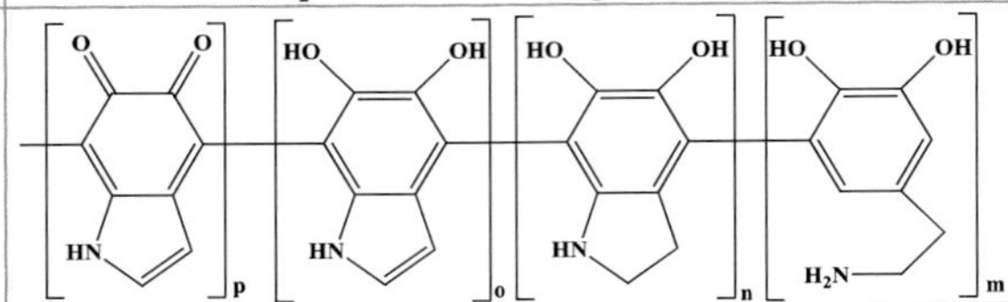
Poly (3,4-ethylenedioxythiophene), bis-poly(ethylene glycol) (PEDOT-PEG)	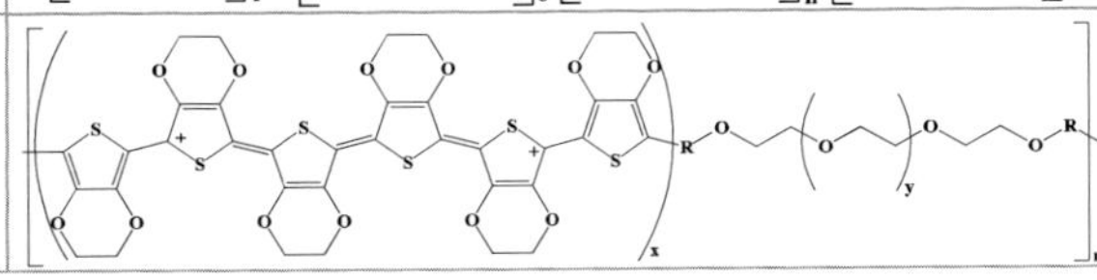

deleterious battery fires. However, other batteries with a lithium metal anode have not gained wide marketplace acceptance due to safety considerations. Several recent reports suggest that lithium metal plates smoothly from gel polymer or polymer electrolytes. Byon et al. constructed a lithium-iodine battery with a high energy density of 330 Wh/kg by pairing a high-concentration aqueous iodine solution with a water-stable solid electrolyte lithium aluminum titanium phosphate ($Li_{1+x+3z}Al_x(Ti, Ge)_{2-x}Si_{3z}P_{3-z}O_{12}$) (LATP) with a lithium-stable gel polymer electrolyte bis(trifluoromethanesulfonyl)imide (TFSI)/*N*-methyl-*N*-propylpiperidinium (PP_{13}) based on polyethylene oxide (PEO) (PEO_{20}/LiTFSI/PP_{13}TFSI/SiO_2) [103]. The gel polymer electrolyte is stable up to 200 °C, forms a stable interface with lithium metal, and promotes smooth deposition/dissolution at the anode. The total conductivity was 0.7 mS/cm at 55 °C, and a stable capacity of 200 mAh/g is reported with a flat plateau at 3.5 V (Fig. 10.1). The rate of operation was 1 mA/cm^2. For reference, typical rates of operation for commercial lithium-ion cells (Panasonic NCR18650B) at C/3 rates are 1.7 mA/cm^2 at room temperature. A similar approach of controlling the lithium metal anode was reported by Goodenough et al. who used a lithium fluoride (LiF)-enhanced garnet solid electrolyte lithium lanthanum titanate ($Li_{6.5}La_3Zr_{1.5}Ta_{0.5}O_{12}$)-(LLZT-LiF) and a cross-linked PEO/LiTFSI polymer electrolyte with a melting point of 240 °C to control the lithium metal anode [54]. With a conductivity of 0.5 mS at room temperature, lithium/LLZT/lithium iron phosphate (Li/gel/LLZT/$LiFePO_4$) and Li/gel/LLZT/sulfur batteries were constructed and cycled with a real capacity of 1 mAh/cm^2 (Fig. 10.1) and at rates of 0.2 mA/cm^2. The lithium-sulfur battery observed no capacity decay over 100 cycles with 1000 mAh/g capacities.

Nafion and polyvinylidene were cast-coated in sequence on a lithium metal foil in a lithium-sulfur battery with a liquid electrolyte. This paper by Luo et al. reports that these thin coatings result in smooth plating (Fig. 10.2) of lithium at the anode [59]. The resulting polymeric membrane had a density of 0.2 mg/cm^2 and is 3 μm

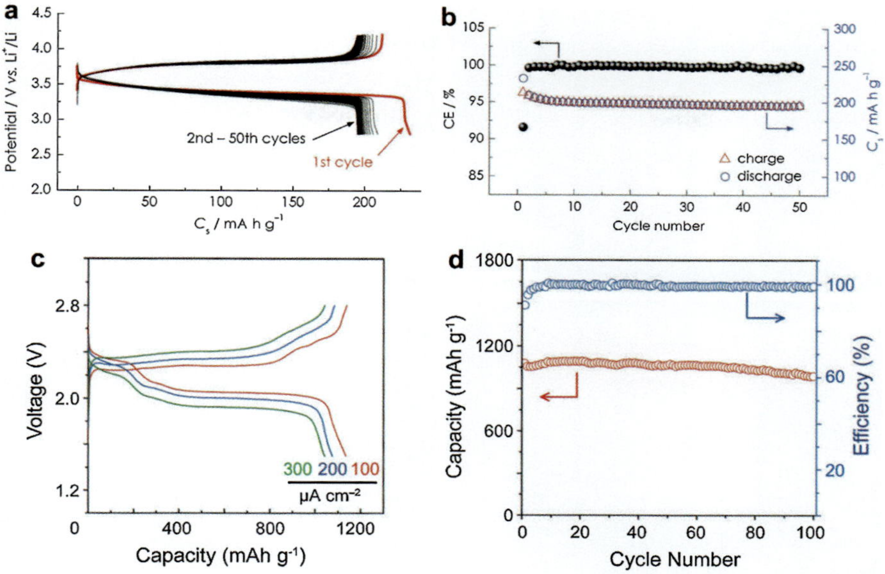

Fig. 10.1 Electrochemical performance of aqueous Li-I_2 cells at 55 °C. (**a**) Charge/discharge profiles over 50 cycles at a current rate of 1 mA/cm^2. (**b**) Corresponding Coulombic efficiency (CE) and specific capacity (C_s). (Reproduced from Zhao et al. [103]). (**c**) Charge and discharge voltage profiles of a Li-S battery with LLZT-2LiF at different current densities. (**d**) Capacity retention and cycling efficiency of the Li-S battery at 65 °C. (Reproduced from Li et al. [54])

thick. 100 cycles are reported with 700 mAh/g capacities and corresponding areal capacities of 1 mAh/cm^2. For reference, commercial lithium-ion cells can transfer 5.5 mAh of lithium per cm^2. Due to the lower voltage of a lithium-sulfur battery, it is required that even higher capacities in excess of 10 mAh have to be transferred per cm^2. In addition, direct coatings of lithium metal from solutions of polymers are problematic because there are no ideal solvents which are stable on lithium metal and can dissolve the desired polymers. For example, *N,N*-dimethylacetamide has been used in this work; however, acetamides are known to react strongly to lithium metal. Kim et al. in collaboration with Hyundai Motor Corp. have reported that poly (3,4-ethylenedioxythiophene)-co-poly(ethylene glycol) (PEDOT-PEG) can be spin cast directly on lithium metal from a nitromethane solution. In addition, dispersing aluminum fluoride (AlF_3) particulates in the PEDOT-PEG solution improves the smooth plating and stability of the lithium metal anode [43, 75]. Figure 10.2 shows the morphology of the lithium metal anode after cycling on an unprotected and a protected anode.

A highly viscoelastic self-healing polymer with a low glass transition temperature (T_g) of −26 °C was spin coated on lithium metal foils to form 1:1 solutions in chloroform and ethanol by Zheng et al. in collaboration with BASF [105]. Figure 10.3 shows the polymer structure and elasticity. This polymer is highly stretchable because of the presence of weaker hydrogen bonding cross-links in

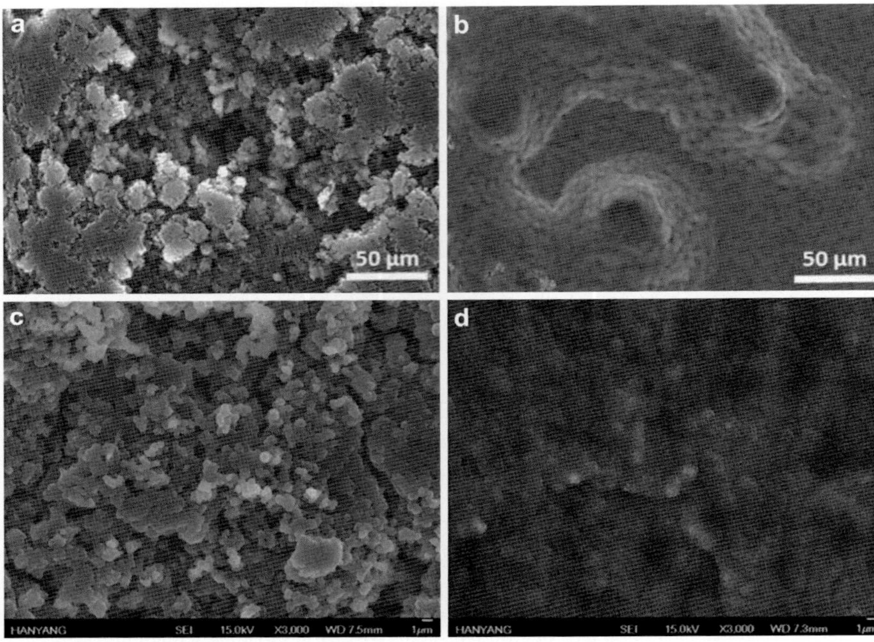

Fig. 10.2 Surface morphologies of pristine Li (**a**) and protected lithium (**b**) anodes after 20 galvanostatic cycles at 1.1 mA/cm^2 with 1 h stripping-1 h plating alternating steps. (Reproduced from Luo et al. [59]). SEM images of the lithium electrodes disassembled from the lithium-oxygen cells after 30 cycles. (**c**) Pristine lithium electrode and (**d**) surface-modified lithium electrode. (Reproduced from Kim et al. [43])

the structure. While spin coating minimizes the exposure of lithium metal to ethanol, it is not a scalable procedure. For example, solution spraying would be more desirable for large lithium foils, but prolonged exposure of lithium to ethanol would yield in rapid lithium corrosion. New solubility properties are required for direct coating of lithium by polymers. 1 mAh/cm^2 of lithium has been deposited at rates as fast as 5 mA/cm^2 with very smooth, planar deposits (Fig. 10.3).

Sodium metal is also a desirable anode because it is orders of magnitude more commonplace in the earth crust and has been historically cheaper than lithium. Unfortunately, it also deposits uncontrollably due to its high reactivity with traditional liquid electrolytes. Sodium metal has been controlled in a report by Gao et al. by the use of a low-cost gel polymer/glass fiber composite which acts as stable sodium interface as well as electrolyte/separator for the sodium salt [24]. The gel is composed of glass fiber mixed with polydopamine (PDA)-coated polyvinylidene fluoride-co-hexafluoropropylene (PVDF-HFP) which is stable to 200 °C mixed with sodium perchlorate ($NaClO_4$) in propylene carbonate. A conductivity of 4.25 mS/cm at room temperature is reported. The composite is sandwiched between electrodes and pressed in coin cells. Stable cycling is reported

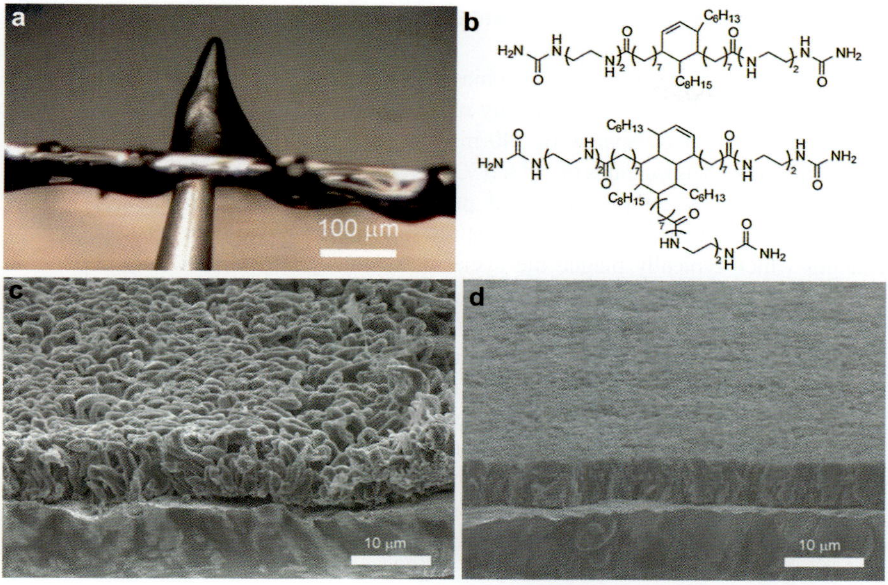

Fig. 10.3 (**a**) Physical properties of the polymer. The polymer is first laminated onto a stainless steel mesh. A needle controlled by a micromanipulator is then pushed onto the polymer. The figure shows that the viscoelastic polymer can withstand the needle without breaking. (**b**) Chemical structure of the polymer. The picture shows the molecular structure of the diacid and triacid backbones. SEM images of 1 mAh/cm^2 of lithium deposited at 1 mA/cm^2 on (**c**) a bare copper electrode and (**d**) the polymer-modified electrode at 5 mA/cm^2. (Reproduced from Zheng et al. [105])

for sodium metal with a sodium manganese ferrocyanide [$Na_2MnFe(CN_6)$] cathode at a slow rate of 0.2 mA/cm^2. Only 0.2 mAh/cm^2 is deposited at the sodium anode on each charge cycle. Gel-type polymers used as separators have also been shown to improve overcharge protection in lithium metal/$LiCoO_2$ cells by Ni et al. [65]. As such, 100% overcharge on each cycle does not reduce the stable discharge of 150 mAh/g of a $LiCoO_2$ cathode for over 50 cycles when a PVDF membrane is mixed with a heat-resistant silicone-capped electroactive polyfluorene. Poly [9,9-di-(20-ethylhexyl)fluorenyl-2,7-diyl] is endcapped with polysilsesquioxane cages. This polymer gel also forms a stable interface with a lithium metal anode. Heat-resistant, UV-curable polysilsesquioxane modifications of polypropylene (PP) separators were reported by Na et al. to promote the cycling of lithium iron phosphate ($LiFePO_4$) with metallic lithium [64] presumably due to an increase of the separator's elastic modulus. Stable plating of 2 mAh/cm^2 lithium for 250 cycles was reported to correspond to the graphene modified polyethylene (PE) separator [74]. The graphene nanosheets were co-doped with nitrogen and sulfur to induce an electrostatic interaction between the lithium metal foil and the graphene-coated separator. This interaction presumably lowered the surface tension of lithium metal and promotes smooth plating.

10.2.2 Improving Other Advanced Anodes

Other advanced anodes consist of graphite/silicon mixtures, alloying metals such as tin (Sn) or bismuth (Bi), and high-capacity metal oxides. For reference, lithiated graphite anodes have a limited capacity of 340 mAh/g, and a long cycle life can typically be obtained at low charging rates of >2 h (C/2). In addition, they require electrolytes which tend to gel at low temperatures and preclude their use in cold climates without temperature management. Polymeric coatings have been used to tame the large volume changes which typically plague the cycle life of these anodes. Sun et al. reported a stretchable, self-healing polymer to coat carbon/silicon anodes and obtain 100 stable cycles with a capacity of 800 mAh/g or 0.8 mAh/cm^2 [78]. Tin (Sn) anodes are highly desirable due to intermediate volume expansion between charge and discharge and high cyclability. PEO-coated Sn particles were reported to result in a more stable SEI during battery operation by Cao group [6]. Jiang et al. reported that polydopamine-coated SnO_2 particles can cycle stably for 300 cycles with capacities >600 mAh/g at C/3 rates [38]. Dou group reported that when Sn/Ni bimetallic nanotube arrays are coated by PEO, the interface with the electrolyte is stabilized and 200 stable cycles are obtained with capacities >800 mAh/g or 0.3 mAh/cm^2 [19]. Transition metal oxides are also promising anode materials plagued by the rapid capacity fade. Jin et al. demonstrated that polypyrrole-coated manganese/iron oxides [MnO_x/Fe_2O_3] display excellent electrochemical properties [39]. 100 stable cycles offer 1000 mAh/g.

10.2.3 Polymer Coatings for Separators

Battery separators have the important role of denying direct electrical contact between anode and cathode and forcing the flow of electrons through an outside circuit which is used to do work on a load. The flow of electrons through the outside circuit is balanced inside the battery cell by the migration of cations (e.g., Li^+) between electrodes such that the electrode charge is always neutral. For example, a loss of electrons (oxidation) is neutralized by an equimolar loss of cations. It is crucial for battery operation that the separator is thin and light and does not significantly reduce the ionic conductivity of the cell. It should also be stable over wide temperatures (it should not shrink) and potential ranges (it should not react with the electrodes). Commercial separators typically consist of polypropylene (PP) or polyethylene (PP) sheets, 9–30 μm thick with 40–90% porosities and good wettability by a liquid electrolyte.

Alternative separators. A variety of alternative separators have been recently reported to be constructed as interlayer freestanding composites of inorganic (or carbonaceous) compounds suspended in polymeric matrices. Zhai group reported a closely packed ×-polyethylene glycol diacrylate (×-PEGDA) dip-coated onto electrospun polyetherimide/polyvinylidene (PEI/PVDF) membranes [99]. This separator has improved wettability and thermal stability as compared to commercial (Celgard) separators. An ultrathin single-walled carbon nanotube (SWCNT)/polyacrylic acid (PAA) interlayer has been reported by Kim group [42] to improve the trapping of polysulfides and the cycle life of a lithium-sulfur battery with 2 mAh/cm^2 capacity (700 mAh/g sulfur) for 200 cycles. The thickness of the interlayer was 7 μm and it weighed 0.8 mg/cm^2. Zhou et al. reported the assembly of a flexible sulfur-graphene-polypropylene (PP) separator for lithium-sulfur batteries [106]. In this approach, graphene/PVDF slurry is coated onto one side of the PP separator using a slot coating machine. Silica glass films were prepared onto the surface of Celgard PP separators by impregnating a perhydropolysilazane solution in xylene for 10 min [33]. Upon heating at 100 °C, the polymer dries into silica glass films conformally coating the Celgard PP separator weaves. The modified separator is heat-resistant and does not shrink at 160 °C while maintaining the same rate of battery operation as the untreated commercial polypropylene (PP) separator. Another popular alternative Li^+-conductive separator/polymeric electrolyte is based on nonwoven polyvinylidene fluoride-hexafluoropropylene (PVDF-HFP). Shi group reported a polydopamine (PDA)-coated PVDF-HFP membrane which does not shrink even at 200 °C and has better wettability than commercial polypropylene separators [73]. The coating process is simple in aqueous solutions and consists of electrostatic interactions between the charged PDA and PVDF-HFP fibers. Seider group proposed the spraying of PVDF-HFP fibers complexed with ether-modified polysiloxanes [72]. Improvements are observed in terms of electrolyte uptake and increased ionic transport. The separator solution can be sprayed directly onto the electrode, resulting in a thin separator. An interesting approach was that of Ding group who coated PVDF-HFP fibers with a polymeric tartaric acid borate polyelectrolyte (PLTB) which results in improved thermal resistance at 150 °C and improved conductivity and strength [17]. A polyionic liquid separator based on a thiazolium polycation and a (bis)(trifluoromethanesulfonyl)imide ($TFSI^-$) anion has been reported by Grygiel group to have an improved transfer coefficient for Li^+ and better conductivity [30]. It has been demonstrated to work with lithium-ion batteries.

Modifications of commercial separators. Commercial polypropylene (PP) and polyethylene (PE) separators can be modified by conformal coatings which improve the thermal stability, wettability, and conductivity. Xu group reported an aqueous coating of PP separators with a suspension of polyvinylidene (PVDF) particles and the polyelectrolyte polyvinyl alcohol (PVA) [89]. This is achieved by simple dipping. This coating improves heat resistance, and no shrinking is observed at temperatures of 130 °C (Fig. 10.4). The polyelectrolyte PVA also improves the electrolyte uptake and wettability of the separator (Fig. 10.4). Pan et al. coats PP separators with tannic acid, which is a naturally occurring polyphenol [68]. The coating process takes place in water at pH 7 and results in dramatic improvements in electrolyte uptake. Another approach to improve the poor wettability of PP separators has been that reported by Wang et al. who used pyrogallic acid, which is cheaper than similar agents such as dopamine [81]. Lia et al. proposed the use of a cellulose aerogel and a polydopamine substrate to coat PP separators with aims at improving the thermal stability and shrink resistance at 150 °C [55]. Hwang group coated a PP separator with a polymerized ionic liquid based on 1-[(4-etheneylphenyl)methyl]-3-butylimidazolium chloride

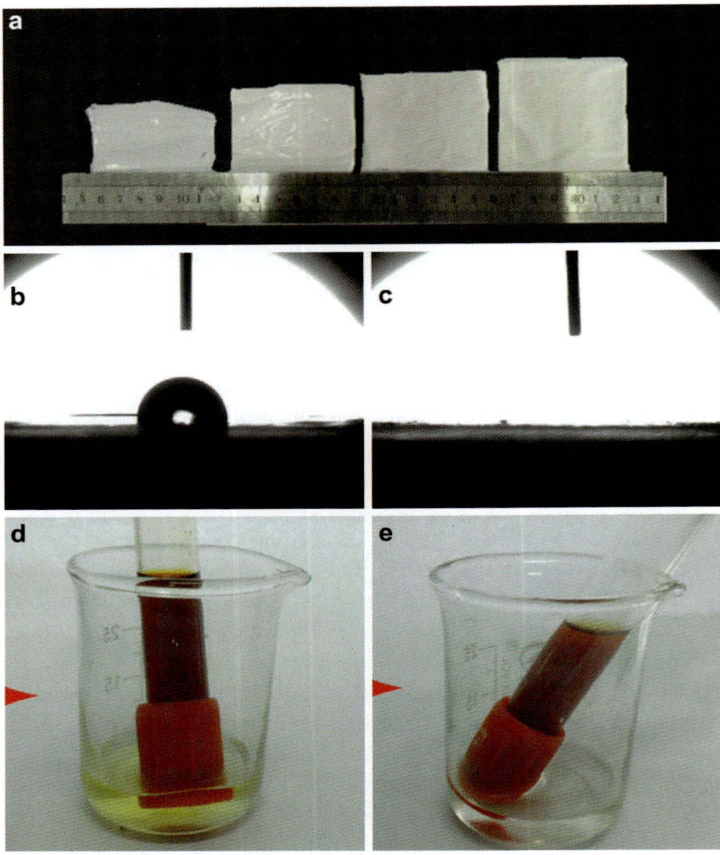

Fig. 10.4 (**a**) Shrinking resistance of PVDF-coated PP separators after being held at 130 °C for 1 h. Contact angle photographs of PVDF-coated PP separators (**c**) and PVDF-uncoated separators (**b**). (Reproduced from Xu et al. [89]). (**d**) Loss of polysulfides in PP/PE/PP separator. (**e**) No loss observed in polydopamine-coated PP/PE/PP separator. (Reproduced from Zhang et al. [101])

(EBIC) and 1-[(4-ethenylphenyl)methyl]-3-butylimidazolium hydroxide (EBIH) [35]. Improved wettability is reported. Commercial polyethylene separator also exists, especially sandwiched between PP layers to afford a shutdown mechanism at high temperatures. The melting point of PE is lower than that of PP, and once it melts, its pores close and effectively shut down the flow of ions between electrodes. Zhang group has used polydopamine to coat a PP/PE/PP commercial separator to improve polysulfide retention in a lithium-sulfur battery [101]. The coating can be assembled in an aqueous solution. Polysulfide trapping can be visualized in Fig. 10.4. Han group has grafted a PE separator induced by atmospheric plasma with charged polymers such as polyvinyl sulfonate (PVS) and poly(diallyldimethylammonium chloride) (PDADMAC) which induce an improved ionic flux of Li^+ to and from the electrode and improve conductivities and electrolyte wettability and reduce shrinking at 134 °C [32].

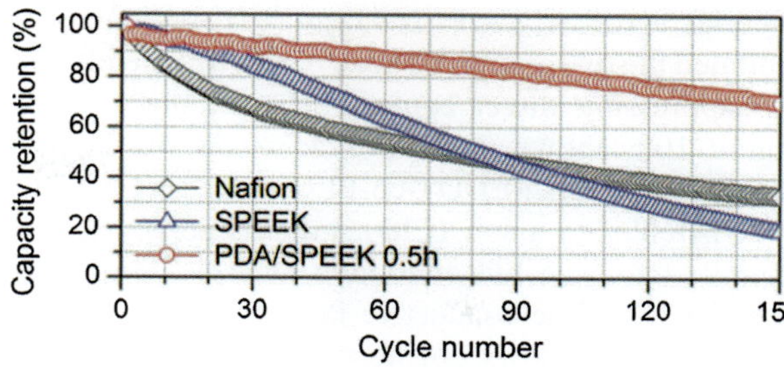

Fig. 10.5 Cycle performance of vanadium redox flow batteries with Nafion 117, SPEEK, and polydopamine (PDA)/SPEEK-modified membranes. (Reproduced from Xi et al. [86])

Flow battery separators. Separators are also crucial in flow batteries which utilize liquid anodes (anolytes) and cathodes (catholytes) flown over a selective membrane which would only allow the transfer of a single cation. A commercial flow battery is based on vanadium solutions. The challenge in this field is that in addition to the diffusion of the desired ion, there also occurs the diffusion of undesired ions which reduces the efficiency and cycle life of the battery. Xi group reported the polydopamine (PDA) coating of sulfonated poly(ether ether ketone) (SPEEK) separators which improves stability in highly oxidative vanadium electrolytes, higher mechanical strength, as well as reduced vanadium ion permeability [86]. Figure 10.5 displays the improved cycle life of a vanadium battery with a PDA/SPEEK separator membrane. Yuan group reported improved selectivity of a poly(ether sulfone) (PES) membrane separator upon spraying with zeolite flakes [98]. Zeolites are ion-conductive crystalline aluminosilicates with pore sizes between 0.3 and 1 nm which may function well in separator applications. Gong group [28] and Cho group [12] report the coating of urushi fibers with the charged polymers chitosan and poly(diallyldimethylammonium chloride) (PDADMAC). Urushi is a natural polymer with high mechanical robustness, and the polyelectrolytes impart ionic selectivity. The result is a promising alternative to traditional separator membranes for redox flow batteries

10.2.4 Cathodes

Simple polymeric coatings have been used to improve the operation of cathodes for lithium-ion, advanced lithium-ion, and post-lithium-ion batteries. Their use has two main targets. (1) First is the stable interface of advanced high-voltage cathodes with electrolytes and reducing electrolyte denaturation at that interface. (2) Second is the reduction of soluble species loss during cycle life such as soluble metal ions or polysulfides.

Lithium-ion. Poly(3,4-ethylenedioxythiophene)-polystyrene sulfonate (PEDOT: PSS) has been used to coat $LiFePO_4$ active material particles. 4% dispersion of this conductive polymer to the active mass of $LiFePO_4$ leads to a 30% improvement in capacity (160 mAh/g) [21]. Similarly, a coating of lithium cobalt oxide ($LiCoO_2$) active material with cross-linked PAN outperforms popular alumina (Al_2O_3) coatings in cycle life experiments [95]. Lee group reported the coating of lithium nickel cobalt manganese oxide ($LiNi_{0.6}Co_{0.2}Mn_{0.2}O_2$) (NCM622) cathode active material with Al_2O_3/PEDOT-co-PEG polymer [48]. This protective layer inhibited the dissolution of Ni, Co, and Mn metals and improved thermal stability by suppressing direct contact between the electrolyte and delithiated NCM. Figure 10.6 shows the improved cycle life at 55 °C and increased thermal stability of Al_2O_3/PEDOT-co-PEG-coated NCM622.

A perchlorate anion (ClO_4^-)-doped polypyrrole (Ppy) coating is reported to improve the cycle life of sodium manages ferrocene [$Na_{1+x}MnFe(CN)_6$] sodium cathodes [52]. The conductive Ppy coating acts as a barrier to loss of Mn in the electrolyte and allows operation of the electrode at very high rates (46C or full discharge in 78 s). 200 cycles are reported at 90 mAh/g and a rate of C/10.

Advanced lithium-ion. Lithium- and nickel-rich cathode materials hold the promise of improving the capacity of cathodes from a current maximum of ~180 to 250 mAh/g. Poly(3,4-ethylenedioxythiophene)-polystyrene sulfonate (PEDOT:PSS) polymer was used to coat lithium-rich-layered oxides such as $Li_{1.17}Mn_{0.56}Co_{0.09}Ni_{0.175}O_2$ [46]. While the precise mechanism for capacity fade in lithium-rich materials is still a matter of intense debate, polymer coatings have been shown to suppress the growth of blocking layers over cycling. Wu group has also used PEDOT:PSS surface particle coatings to improve the cycle life of lithium-rich $Li_{1.2}Ni_{0.2}Mn_{0.6}O_2$ [85]. Another class of cheap advanced lithium-ion cathode is vanadium compounds which can intercalate multiple lithiums per unit structure. Lithium vanadium phosphate [$Li_3V_2(PO_4)$] is gathering significant interest; however it is limited by poor electronic conductivity. Coating of active material particles with PEDOT:PSS improves the electronic

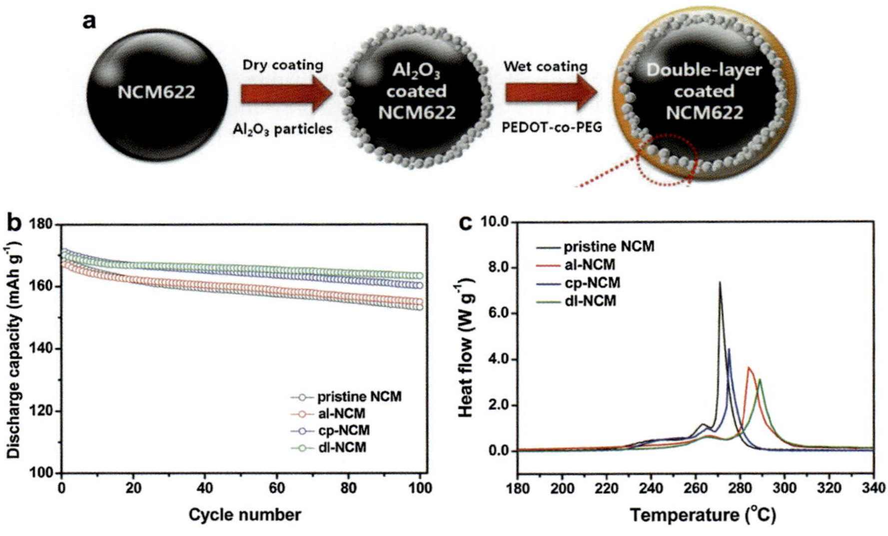

Fig. 10.6 (**a**) Schematic of the double-layer coating of $LiNi_{0.6}Co_{0.2}Mn_{0.2}O_2$ material with Al_2O_3 nanoparticles and PEDOT-co-PEG coating. (**b**) Discharge capacities of lithium-ion cells assembled with pristine NCM and surface-modified NCM electrodes. (**c**) DSC thermograms of delithiated NCM materials charged to 4.3 V after 100 repeated cycles. (Reproduced from Lee et al. [48])

conductivity of the material allowing stable cycling at a fast rate of 3C [93]. 500 cycles offered 120 mAh/g or 0.3 mAh/cm^2.

Post-lithium-ion. A sulfur cathode is one of the most promising post-lithium-ion cathodes due to its low cost and high capacity (5× higher than currently used NCM or NCA cathodes). A sulfur cathode has two main challenges. (1) Sulfur is an insulator and requires large amounts of conductive additives which makes it very difficult to assemble into thick electrodes. (2) Discharge intermediates called polysulfides dissolve and are lost in the electrolyte reducing cycle life. Both of these challenges can be mitigated by utilizing polymeric coatings. In an attempt to achieve high sulfur loadings, a conductive and robust biopolymer network built from guar gum (GG) and xanthan gum (XG) binder which contain large ratios of oxygen-containing functional groups can trap polysulfides [57]. An impressive 800 mAh/g or 9.5 mAh/cm^2 is obtained over 60 stable cycles in a lithium-sulfur cell at a rate of C/6. Cycle life is limited by the effect of lithium metal anode when large capacities such as these are extracted. Lim group reported a copper-anchored cross-linked PEO polymer-coated sulfur electrode which improves the cycle life [56]. High rate cycling of 100 cycles at 2C is reported for this sulfur cathode. Conductive PEDOT:PSS coatings have been reported by Lee group to improve sulfur cathodes [45]. In this example, commercial sulfur is coated by the polymer in water/ethanol mixtures, followed by drying. A similar approach was taken by Li group who doctor bladed a Sigma Aldrich suspension of 1.3% aqueous PEDOT: PSS onto a precast pure sulfur cathode [53]. Improvements are observed over untreated sulfur in both cases. A unique approach to a sulfur cathode was taken by Zhou group who used vanadium tetrasulfide (VS_4) as the source of sulfur. VS_4 microspheres were sequentially coated by PEDOT/PPY/PANI [107]. This conductive coating activated more of the available capacity and inhibited the loss of soluble polysulfides. Another sequential coating of sulfur particles has been reported by Yan group who coats sulfur cores with PEDOT, then with MnO_2 [92]. This approach provides improved electrolyte penetration into the cathode and 200 stable cycles at 800 mAh/g at C/5. Figure 10.7 shows the schematic of the coating as well as the cycle life.

A poly(3,4-ethylenedioxythiophene)-polystyrene sulfonate (PEDOT:PSS)/carbon-based multifunctional polysulfide blocking layer was electrosprayed on a sulfur electrode. 1000 cycles are reported with an average capacity of 500 mAh/g or 1 mAh/cm^2 [66]. A graphene/PANI/sulfur-layered composite has been investigated as a sulfur cathode by Moon group [62]. The layered structure is held together by cross-linking the PANI layer. 500 cycles are reported with a capacity of 900 mAh/g at a rate of 1C. A similar graphene/PANI approach was used to build selenium core-shell fibers which can cycle for 200 cycles with a near theoretical capacity of 600 mAh/g or 1.8 mAh/cm^2 [100].

10.3 Layer-by-Layer (LBL) Membranes

While simple polymeric coatings often improve the performance of battery materials, there is a dedicated method used for adhering polymers to surfaces which was first reported by Decher [14]. This layer-by-layer (LBL) method utilizes charged

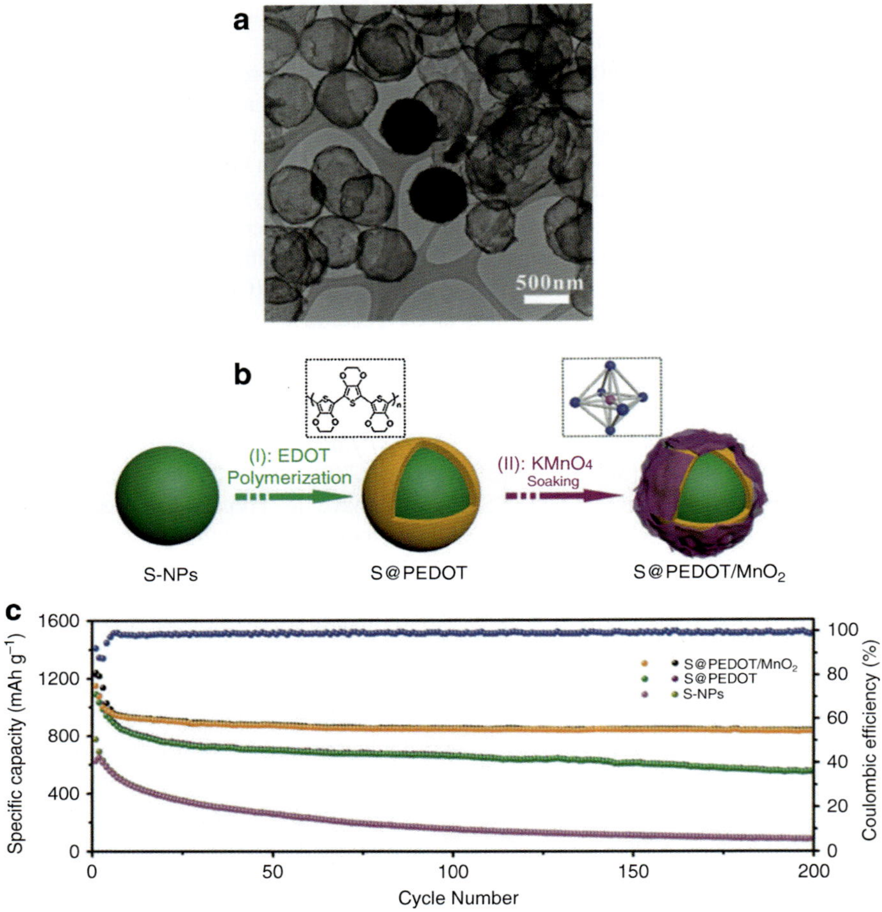

Fig. 10.7 (**a**) TEM characterization of S-PEDOT/MnO_2. (**a**) Bright-field (BF) image at low magnification. (**b**) Schematic illustration of the synthesis of the S-PEDOT/MnO_2 composite. (**c**) The cycling performances of S-PEDOT/MnO_2 electrodes at a current density of 0.2C. (Reproduced from Yan et al. [92])

moieties on polymer backbones to form ionic pairs or cross-links with substrates or additional polymeric layers. By this method, a polymer-polymer link (also called an "intrinsic" site) forms when the original counter ion of the charged polymer is exchanged with an oppositely charged moiety on a counter-polymer. This type of interaction is responsible for keeping the membrane together. Diffusion through the membrane has been modeled to correspond to hops between "extrinsic" sites, which correspond to original polymer-counter ion sites which are not unpaired by the counter-polymer [22, 23, 26, 71]. The ratio of extrinsic compensation in an LBL membrane can be fine-tuned by the ionic strength of the assembly solutions. A higher ionic strength corresponds to more extrinsic sites and a more permeable membrane. Permeability also hinges on the thermodynamic propensity of the original counter ion to partner exchange with an oppositely charged moiety on the counter-polymer. This depends on the negative base-10 logarithm of the acid dissociation constant (pKa) of the charged moieties on the polymer backbones and may also be used to fine-tune the permeability of the membrane (a strong vs. a weak polyacid, etc.). LBL membranes have also been reported to assemble based on H-bonds, even in the absence of an electrostatic drive. In this instance, tuning of membrane permeability may be achieved by changing the assembly solution negative logarithmic potential of hydrogen (pH) which varies the degree of protonation on functionalities of interest.

Experimentally, the LBL method requires the solvation (or suspension) of polymers in a compatible solvent (usually water is used) and the alternation of the substrate between solutions of polymer building blocks with rinses in between. The advantage over simple template polymerization or spin coating of single polymers is simplicity, access to complex substrates, and the availability of a wide variety of building blocks which increases the scientists' ability to innovate. The LBL method can assemble uniform, nearly pinhole-free films with exceptional mechanical properties such as robustness [80] or self-healing [108]. Much needed fundamental work regarding film assembly and permeability to ions has been continuously published by the groups of Schlenoff and Decher [15]. The groups of Lvov, Mohwald, and Caruso have demonstrated the application of layer-by-layer capsules for biomedical applications such as controlled drug delivery [1, 7, 8, 77]. As far as energy applications, the early works reported from the groups of Hammond and Rubner have led the way to modern applications in a battery such as those discussed here [10, 16, 47, 76]. The evolution of LBL films to mitigate the most urgent challenges of batteries truly stands on the shoulders of these early pioneers in the field.

10.3.1 Polymer/Polymer Multilayered Films

The use of LBL for battery application is a novel field which has already made a recognizable impact. Conductive polymers such as Nafion, PEDOT:PSS, PANI, and PEO have all been used as layers in functional membranes. For example, a lithium-conductive PEO polymer was considered to assemble LBL films by H-bonding. Tung et al. demonstrated the extreme temperature stability of a composite made from aramid nanofibers (Kevlar) and PEO [80]. Films can be grown on a flat and stiff substrate and then chemically delaminated and used as thin, strong battery separators with ideal Li^+ conductivities and electrolyte wettability. The aramid nanofiber/PEO separator is stable to temperatures in excess of 500 °C which is the highest performance reported for a polymeric separator (Fig. 10.8). Kim et al. have used a PEO/poly(acrylic acid) (PAA) LBL membrane to protect a sulfur cathode cast on the aluminum current collector [41]. Five cycles improved the cathode wettability and cyclability of the cathode in a lithium-sulfur battery. Bucur et al. reported that a PEO/PAA membrane based on H-bonding can improve the cycle life of sulfur cathodes [3]. 500 cycles were reported with capacities of 700 mAh/g at a fast rate of 2C.

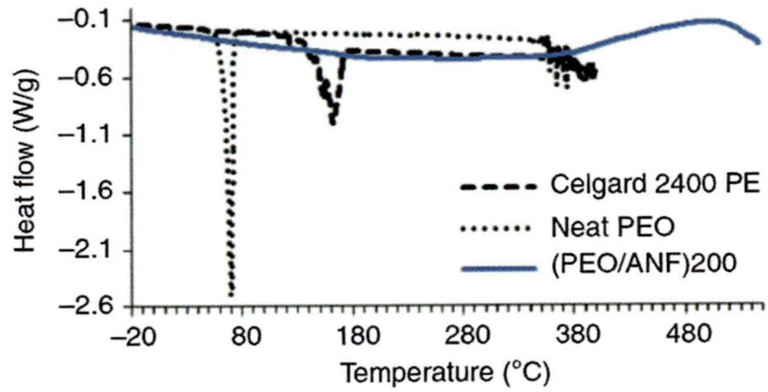

Fig. 10.8 DSC and analysis curves for (PEO/aramid fiber)$_{200}$ separator (blue solid), neat PEO (black dotted), and Celgard$_{2400}$ PE (black dashed). (Reproduced from Tung et al. [80])

Polyaniline (PANI) is also a polymer of interest in the battery community due to its inherent electronic conductivity. Like other conjugated polymers, it does not dissolve well in solvents but forms fine suspensions. However, Jeon group was able to build ultrathin (6 μm) LBL polymeric cathodes by spraying PANI and PAA on a current collector [36]. PANI can carry a charge if the pH is carefully controlled and can pair with negatively charged polymers such as PAA. A specific capacity of 100 mAh/g was obtained from this cathode with stable cycling over 1000 cycles. Christinelli et al. have reported that an LBL assembled capacitor can deliver 3000 cycles with 100 F/g capacities [13]. The electrode was assembled from 60 bilayers of poly(*o*-methoxyaniline) (POMA)/poly(3-thiophene acetic acid) (PTAA). The electrochemical performance of graphite anodes was demonstrated to improve when coated with a PAA/PEG multilayer [50]. The self-assembled film improves the stability of the SEI and improves the reversible capacity of artificial, natural, and silicon graphite. Wu et al. reported that a PEI/polystyrene sulfonate (PSS) coating of sulfur/graphite composites improves the polysulfide retention and increases cycle life [84]. 200 cycles are reported at 900 mAh/g or 1.71 mAh/cm^2.

The poly(3,4-ethylenedioxythiophene)-polystyrene sulfonate (PEDOT:PSS) is a conductive polymer which suspends well in water due to the negatively charged PSS component. Bucur group reported that a PEDOT:PSS/poly(diallyldimethylammonium chloride) (PDADMAC) multilayer membrane can improve the cycle life of sulfur cathodes [3] as well as improve the stripping efficiency of lithium metal anodes. There was no evidence of dendritic growth of lithium [2]. Another negatively charged conductive polymer is Nafion. Osada group showed that a Nafion/PDAD LBL membrane can also improve the cycle life of a sulfur cathode due to the blocking of polysulfide diffusion. High capacities of 850 mAh/g or 4.25 mAh/cm^2 are reported over 30 cycles. An interesting report was submitted by Wang group who sprayed an LBL coating on a cast sulfur cathode. The membrane was composed of Nafion and polyvinylpyrrolidone (PVP) which afforded 100 stable cycles with a capacity of 1100 mAh/g or 1.5 mAh/cm^2. Austing group reported the assembly of a Nafion/polyethylenimine (PEI) multilayer separator for vanadium/air redox flow batteries [29]. The LBL separator provided higher capacities due to a 21× higher selectivity for protons as well as an increase in charge/discharge efficiency. Yoo group also assembled a membrane for a vanadium redox flow battery from a Nafion/sulfonated poly(phenylene oxide) (SPPO) multilayer [97]. Improved proton conductivity was reported.

A common LBL building block is poly(diallyldimethylammonium chloride) (PDADMAC) which is a strong polyelectrolyte with a positive charge. A series of recent reports have been used to form protecting thin multilayers with polyanions such as PSS or PAA. Even in the absence of conductive components, the conformity and thin nature of LBL coatings enable their use in the battery field. Qu group reported the LBL assisted the growth of molybdenum disulfide (MoS_2) for battery anodes, which are layered along with PDAD/PSS on carbon fibers [70]. 200 stable cycles are reported with an electrode capacity of 1000 mAh/g or 1.13 mAh/cm^2 (Fig. 10.9). Li group used PDAD/PSS to coat graphite-silicon and pure silicon anodes [51]. Improved capacity retention is reported for more than 100 cycles with stable capacities of 450 mAh/g and 2750 mAh/g, respectively, which correspond to 2.25 mAh/cm^2 and 13.75 mAh/cm^2. Coated materials have the unique feature of stability on a fast charge with minimal differences between 0.1C and 10C. This is the most important benefit for anodes. Currently, commercial graphite anodes can only be operated at low rates of C/2 for long-term cycling. Wang group has also used PDAD/PSS to assemble a carbon fiber/iron oxide (Fe_3O_4) high-capacity anode [83]. 150 stable cycles with a high capacity of 1250 mAh/g were obtained with large improvements over pristine Fe_3O_4. A vanadium flow battery was enabled by a PDAD/PAA-coated SPEEK membrane. Improved Coulombic efficiency is reported by Zhao group due to the superior selectivity of the coated SPEEK membrane [104].

An interesting concept involving liquid electrolytes containing dissolved polyelectrolytes has been reported by Porcarelli and Buss groups. These approaches improve the thermal stability of liquid electrolytes as well as the conductivity and voltage stability. In the case of Porcarelli group, a new family of single-ion conducting block copolymer polyelectrolytes has been obtained via reversible addition-fragmentation chain transfer polymerization. The obtained charged polymer is poly(lithium 1-[3-(methacryloyloxy)-propylsulfonyl]-1-(trifluoromethylsulfonyl)imide) [69]. Compatibility with lithium metal is claimed in this report. Similarly, Buss et al. proposed lithium-neutralized polyanions in solution as a strategy to attain high conductivities and transfer numbers for Li^+. The polyelectrolyte used is poly(allyl glycidyl ether) (PAGELS) [5]. A new class of charged polymer electrolytes was also reported by Obadia et al. Poly(1,2,3-triazolium) polycationic ionic liquids have been developed based on simple click chemistry [67].

10.3.2 Polymer/Carbon Multilayered Films

Several reports couple a polymer with conductive carbon such as graphene oxide to assemble LBL coatings with enhanced electronic conductivity. Jeon et al. coupled conductive PANI (+) with graphene oxide (GO) (−) to build LBL films on cotton fabric support [37]. A thin film battery is assembled with this PANI/GO

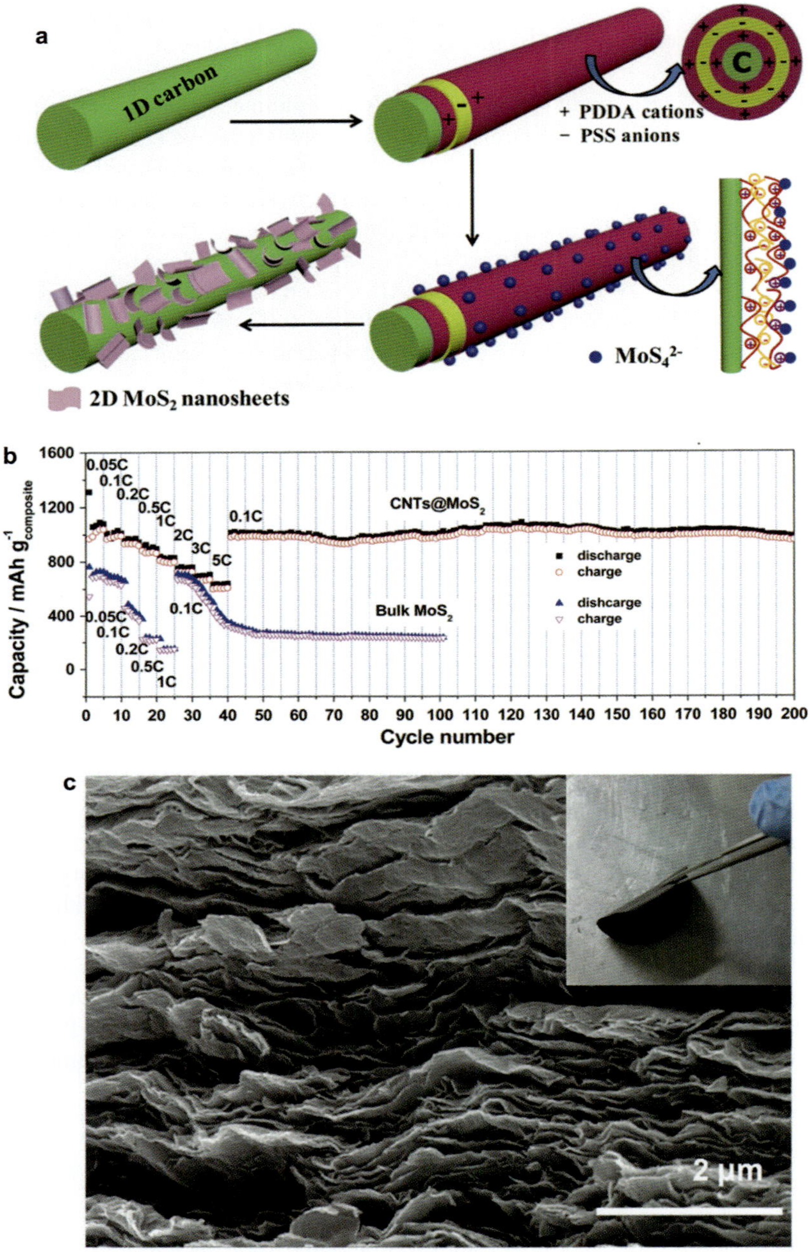

Fig. 10.9 (**a**) Schematic illustration of the formation of 2D MoS_2 nanosheets on 1D carbon. (**b**) Capacity evolution of CNTs – MoS_2 and bulk MoS_2 during repeated cycling at varying current rates and then at a constant rate of 0.1C. (Reproduced from Qu et al. [70]). (**c**) LBL stabilized graphene nanoplatelet electrode. (Reproduced from Kim et al. [40])

cathode. A 0.27 μm cathode operated with no capacity fade for 1000 cycles at high rates of 35 A/g. Impressively, this cathode offers ~200 mAh/g capacity at 1 A/g rates at an average voltage of 3 V. PDAD is another positively charged polymer which was paired with graphene oxide for lithium-sulfur batteries. Lee group assembled PDAD/GO/sulfur composites which offer improved capacity activation and retention [49]. Transparent cathodes were assembled from sprayed alternate layers of single-walled carbon nanotubes (SWNT), polystyrene sulfonate (PSS)/polyvinyl alcohol (PVA)/vanadium (V) oxide (V_2O_5), [(PSS+SWNT)/(PVA+V_2O_5)] (300 bilayers), and PEO+LiTFSI/PAA+V_2O_5 (100 bilayers) [27]. 3 μm electrodes can be deposited with capacities of 40 $\mu Ah/cm^2$. 100 cycles are reported with 5 $\mu Ah/cm^2$. Kim group stabilized graphene nanoplatelets with a water/PSS solution and layered them with PDAD/polyvinyl alcohol (PVA) on a substrate (Fig. 10.9) [40]. The LBL graphene containing material can be used as high-capacity anodes for batteries. A (PANI+sulfur)/(PSS+CNT) layer-by-layer sulfur cathode is demonstrated by Yan group [91]. 600 cycles with an average capacity of 1000 mAh/g and 1.85 mAh/cm^2 are reported. Luo group also stabilized a sulfur cathode with a trifunctional interlayer assembled in an LBL fashion from MWCNT and PEG [58]. This interlayer is placed in between a pure sulfur cathode and a Celgard separator. High capacities of 3 mAh/cm^2 are obtained over 200 cycles. Silicon anodes were improved by the LBL architecture of bamboo-derived ultrafine silicon particles with graphene oxide and PDAD [82].

10.3.3 Polymer/Inorganic Multilayered Films

LBL assembly has been successfully used to build thin and conformal coatings from polymer/inorganic building blocks. ZrO_2 was paired with poly(acrylic acid) (PAA). The H-bonding between zirconium (Zr) and metal center and the COO^- functionality of PAA ensure film assembly. Xu group demonstrated the PAA/ZrO_2 LBL film enhancing a Celgard polyethylene (PE) separator [87]. The enhanced separator shows higher transference numbers for Li^+ and improved stabilization of the SEI on the anode. The result is higher rates of operation and improved cycle life. Interestingly, zirconia (ZrO_2) has also been used by Chi group to build a ZnO_2/silsesquioxane (POSS) multilayered assembly aided by polyethylenimine PEI and 1-(3-dimethylaminopropyl)-3-ethylcarbodiimide hydrochloride (EDC)/*N*-hydroxysuccinimide (NHS) [11]. Improvements in cycle life are observed. A pinhole-free multilayer membrane assembled from branched PEI and commercial inorganic clay flakes such as Cloisite or montmorillonite was reported by Bucur group to reduce the loss of polysulfides in high-loading sulfur cathodes [4]. Near theoretical capacity for the sulfur cathode is obtained (~1500 mAh/g) at high loadings of 4 mAh/cm^2 (Fig. 10.10). The cycle life in this battery was shown to be limited by electrolyte decomposition at the lithium anode rather than by the loss of polysulfides at the cathode. The paper also describes a similar clay flake containing multilayer used to protect lithium metal anodes.

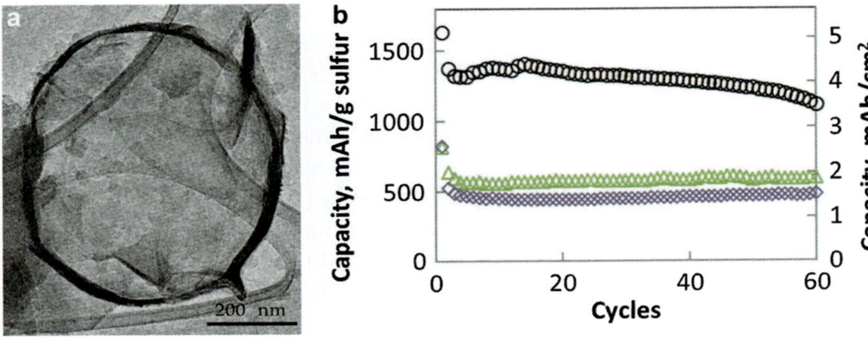

Fig. 10.10 (**a**) TEM tomography still for PEI/clay flake LBL film-coated sulfur particle. (**b**) Cycle life and energy density metrics for a cathode. (Reproduced from Bucur et al. [4])

10.3.4 Active Material/Carbon Multilayered Films

The layer-by-layer method can even be used to couple layers of oppositely charged carbon such as graphene oxide sheets and anode/cathode active material such as silicon, cobalt (II,III) oxide (Co_3O_4) or lithium iron phosphate ($LiFePO_4$), vanadium oxide (VO_x), and lithium trivanadate (LiV_3O_8). Graphene is usually used as obtained from the modified Hummers method. Graphene oxide typically has an overall negative charge. This approach allows for very high rates of operation and may open the door to commercial batteries which charge in under 5 min and have >10,000 cycle life. In addition, incorporation of graphene into electrodes also improves heat dissipation. Huawei has recently reported commercial batteries for cell phones which have long cycle life while operating at temperatures as high as 60 °C due to the incorporation of graphene.

Anodes. Kim group built an MWCNT/SiO_2 LBL structure which is used in a clip-like configuration to protect lithium metal when used with a sulfur cathode [44]. 250 stable cycles are afforded by the anode protection with an impressive capacity of 6.2 mAh/cm^2. This is an important report because the operation of sulfur cathodes hinge on high capacities of stable lithium utilization. Zhang group also assembled a freestanding Si/rGO LBL anode which can operate even if bent [102]. 150 cycles are reported with 600 mAh/g. LBL graphene/Co_3O_4 nanosheet hybrids have been reported by Yang group as stable anode materials with high capacities of 1000 mAh/g [96]. 50 stable cycles are reported. Dou group also demonstrated a layer-by-layer Co_3O_4/graphene composite which can cycle 2000× but at a low areal capacity of 0.9 mAh/cm^2 [20]. However, high rates of operation are shown for this thin film anode of >2 A/g. Another prospect for a high-capacity anode is Fe_3O_4. Tan group reported the Fe_3O_4 nanospheres/MWCNT LBL anode [79]. A stable capacity of 475 mAh/g is obtained for 300 stable cycles at a very high rate of 3.2 A/g (6.75C rate). The areal loadings were low, 0.475 mAh/cm^2.

Cathodes. Chen et al. assembled a hierarchical VO_x/graphene cathode sheet in an aqueous solution [9]. Capacities of 200 mAh/g (3.5 V) were stable for 500 cycles. Mo et al. also reported a vanadium cathode [61]. LiV_3O_8/graphene oxide LBL assembled cathodes were obtained by PDAD-treated LiV_3O_8 which results in a positive charge and allows LBL pairing with graphene oxide. Capacities of 225 mAh/g (2.5 V) (0.5 mAh/cm^2) can be maintained stable over 300 cycles at rates of 1.3C (0.3 A/g). Higher rates of 40C (4 A/g) and 115C (8 A/g) still offer stable capacities of 100 mAh/g and 70 mAh/g, respectively. This performance is impressive because commercial graphite anodes cannot charge faster than C/2 for extended cycle life. Mo et al. also reported the extremely high rate of operation of LBL assembled 3D composites of $LiFePO_4$/graphene oxide, by similar techniques [60]. 150 mAh/g (0.81 mAh/cm^2) is shown at 1C (a 25% improvement over bulk $LiFePO_4$) for 100 stable cycles. However, the LBL composite can be cycled at 160C and still deliver 50 mAh/g. Yan et al. assembled an LBL Na_2FePO_4/CNT cathode for sodium batteries. 400 cycles are reported at a rate of 0.4C with capacities of 100 mAh/g [90]. Another sodium cathode was reported by Xu group where sodium vanadium phosphate [$Na_3V_2(PO_4)_3$] was assembled by the LBL method with graphene oxide. Very high cycle life is reported [88]. 15,000 stable cycles at a fast rate of 50C (full discharge in 72 s) with average capacities of 70 mAh/g at a voltage of 3.25 V (Fig. 10.11) are reported. The areal capacity loading was 0.13 mAh/cm^2. Capacities as high as 120 mAh/g can be obtained at rates of 1C. Impressive cycle life is also demonstrated at high temperatures of 60 °C. These are impressive findings due to the low cost of sodium which may open the door to cheap sodium-ion batteries which can fully charge in a less than 2 min. A 15,000 cycle life would be ideal for commercial vehicles. The LBL approach has also been used by Gupta group to build capacitors from PEI-graphene coupled with MWCNT [31]. An impressive capacity of 75 Wh/kg (similar to nickel-metal hydride, NiMH) and 450 F/g is reported for this capacitor.

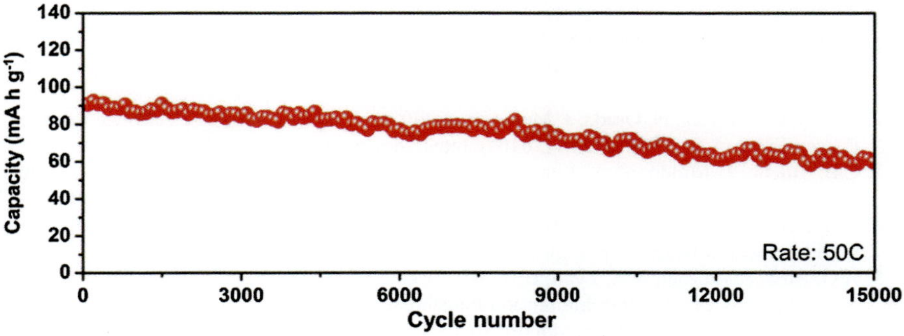

Fig. 10.11 Ultra-long cycling stability of the NVP/rGO for 15,000 cycles at a high rate of 50C. (Reproduced from Xu et al. [88])

10.4 Conclusion

Polymers enable the simple deposition of thin and stable coatings on anodes, separators, and cathode materials which improve the electrochemical performance of the battery. Careful selection of polymers can solve many problems associated with promising new batteries, such as inhibiting dissolution of soluble electrode species and stabilizing electrolyte/electrode interfaces during operation. The end result is a higher energy density and longer cycle life for the battery. For example, by reducing the direct contact of high-voltage cathode materials with the electrolyte, less electrolyte is decomposed, and cycle life is improved. Similarly, coatings on graphite improve the rate of operation of this commercial anode material. A special class of polymeric coatings is the layer-by-layer method introduced by Decher [14]. This approach minimizes the amount of material adsorbed and assembles conformal and pinhole-free coatings which improve conductivity in the battery. It also gives the battery chemist a rich toolset with many new building blocks which can be utilized to assemble LBL films. PANI, Nafion, PEDOT:PSS, graphene oxide, PDAD, and others are all charged polymers or solid materials which have been demonstrated to improve the performance of batteries. The most attractive feature of this method is its addictive nature which can be scaled up to accommodate any battery requirement. The most substantial benefit of LBL assembled electrodes has been the large increase in the rate of operation and cycle life which are key indicators in the evolution of battery technology. In order for widespread acceptance, it will be important for the field of LBL multilayers in the battery space to prove that high capacity loadings in excess of 3 or 4 mAh/cm^2 are possible. In addition, LBL may solve many of the challenges encountered in post-lithium-ion fields such as solid-state and multivalent (magnesium) batteries which are currently plagued by low rates of operation.

References

1. H. Ai, M. Fang, S.A. Jones, Y.M. Lvov, Electrostatic layer-by-layer nanoassembly on biological microtemplates: platelets. Biomacromolecules **3**, 560–564 (2002). https://doi.org/10.1021/bm015659r
2. C.B. Bucur, A. Lita, N. Osada, J. Muldoon, A soft, multilayered lithium–electrolyte interface. Energy Environ. Sci. **9**, 112–116 (2016a). https://doi.org/10.1039/C5EE03056K
3. C.B. Bucur, J. Muldoon, A. Lita, A layer-by-layer supramolecular structure for a sulfur cathode. Energy Environ. Sci. **9**, 992–998 (2016b). https://doi.org/10.1039/C5EE02367J
4. C.B. Bucur, M. Jones, M. Kopylov, et al., Inorganic–organic layer by layer hybrid membranes for lithium-sulfur batteries. Energy Environ. Sci. (2017). https://doi.org/10.1039/C7EE00398F
5. H.G. Buss, S.Y. Chan, N.A. Lynd, B.D. McCloskey, Nonaqueous polyelectrolyte solutions as liquid electrolytes with high lithium ion transference number and conductivity. ACS Energy Lett. **2**, 481–487 (2017). https://doi.org/10.1021/acsenergylett.6b00724
6. Z. Cao, H. Meng, P. Dou, et al., Effects of solid polymer electrolyte coating on the composition and morphology of the solid electrolyte interphase on Sn anodes. J. Solid State Electrochem. **21**, 955–966 (2017). https://doi.org/10.1007/s10008-016-3440-z

7. F. Caruso, R.A. Caruso, H. Möhwald, Nanoengineering of inorganic and hybrid hollow spheres by colloidal templating. Science **282**, 1111–1114 (1998a). https://doi.org/10.1126/science.282.5391.1111
8. F. Caruso, H. Lichtenfeld, M. Giersig, H. Möhwald, Electrostatic self-assembly of silica nanoparticle−polyelectrolyte multilayers on polystyrene latex particles. J. Am. Chem. Soc. **120**, 8523–8524 (1998b). https://doi.org/10.1021/ja9815024
9. J. Chen, W. Li, J. Jiang, et al., Facile and creative design of hierarchical vanadium oxides@graphene nanosheet patterns. RSC Adv. **6**, 13323–13327 (2016). https://doi.org/10.1039/C5RA25521J
10. J.H. Cheung, W.B. Stockton, M.F. Rubner, Molecular-level processing of conjugated polymers. 3. Layer-by-layer manipulation of polyaniline via electrostatic interactions. Macromolecules **30**, 2712–2716 (1997). https://doi.org/10.1021/ma970047d
11. M. Chi, L. Shi, Z. Wang, et al., Excellent rate capability and cycle life of Li metal batteries with ZrO_2/POSS multilayer-assembled PE separators. Nano Energy **28**, 1–11 (2016). https://doi.org/10.1016/j.nanoen.2016.07.037
12. E. Cho, J. Won, Novel composite membrane coated with a poly(diallyldimethylammonium chloride)/urushi semi-interpenetrating polymer network for a non-aqueous redox flow battery application. J. Power Sources **335**, 12–19 (2016). https://doi.org/10.1016/j.jpowsour.2016.10.020
13. W.A. Christinelli, R. Gonçalves, E.C. Pereira, Optimization of electrochemical capacitor stability of poly(*o*-methoxyaniline)-poly(3-thiophene acetic acid) self-assembled films. Electrochim. Acta **196**, 741–748 (2016). https://doi.org/10.1016/j.electacta.2016.02.187
14. G. Decher, Fuzzy nanoassemblies: toward layered polymeric multicomposites. Science **277**, 1232–1237 (1997). https://doi.org/10.1126/science.277.5330.1232
15. G. Decher, J. B. Schlenoff (eds.), *Multilayer Thin Films: Sequential Assembly of Nanocomposite Materials*, 2nd edn. (Wiley-VCH, Weinheim, 2012)
16. D. DeLongchamp, P.T. Hammond, Layer-by-layer assembly of PEDOT/polyaniline electrochromic devices. Adv. Mater. **13**, 1455–1459 (2001). https://doi.org/10.1002/1521-4095(200110)13:19<1455::AID-ADMA1455>3.0.CO;2-7
17. G. Ding, B. Qin, Z. Liu, et al., A polyborate coated cellulose composite separator for high-performance lithium-ion batteries. J. Electrochem. Soc. **162**, A834–A838 (2015). https://doi.org/10.1149/2.0261506jes
18. P.T. Dirlam, R.S. Glass, K. Char, J. Pyun, The use of polymers in Li-S batteries: a review. J. Polym. Sci. A Polym. Chem. **55**, 1635–1668 (2017). https://doi.org/10.1002/pola.28551
19. P. Dou, Z. Cao, J. Zheng, et al., Solid polymer electrolyte coating three-dimensional Sn/Ni bimetallic nanotube arrays for high-performance lithium-ion battery anodes. J. Alloys Compd. **685**, 690–698 (2016a). https://doi.org/10.1016/j.jallcom.2016.05.218
20. Y. Dou, J. Xu, B. Ruan, et al., Atomic layer-by-layer Co_3O_4/graphene composite for high-performance lithium-ion batteries. Adv. Energy Mater. (2016b). https://doi.org/10.1002/aenm.201501835
21. S.N. Eliseeva, O.V. Levin, E.G. Tolstopyatova, et al., Effect of addition of a conducting polymer on the properties of the $LiFePO_4$-based cathode material for lithium-ion batteries. Russ. J. Appl. Chem. **88**, 1146–1149 (2015). https://doi.org/10.1134/S1070427215070071
22. T.R. Farhat, J.B. Schlenoff, Ion transport and equilibria in polyelectrolyte multilayers. Langmuir **17**, 1184–1192 (2001). https://doi.org/10.1021/la001298+
23. T.R. Farhat, J.B. Schlenoff, Doping-controlled ion diffusion in polyelectrolyte multilayers: mass transport in reluctant exchangers. J. Am. Chem. Soc. **125**, 4627–4636 (2003). https://doi.org/10.1021/ja021448y
24. H. Gao, B. Guo, J. Song, et al., A composite gel–polymer/glass–fiber electrolyte for sodium-ion batteries. Adv. Energy Mater. **5** (2015). https://doi.org/10.1002/aenm.201402235
25. M. Genovese, K. Lian, Polyoxometalate modified inorganic-organic nanocomposite materials for energy storage applications: a review. Curr. Opin. Solid State Mater. Sci. **19**, 126–137 (2015). https://doi.org/10.1016/j.cossms.2014.12.002

26. R.A. Ghostine, J.B. Schlenoff, Ion diffusion coefficients through polyelectrolyte multilayers: temperature and charge dependence. Langmuir **27**, 8241–8247 (2011). https://doi.org/10.1021/la2015258
27. F.S. Gittleson, D. Hwang, W.-H. Ryu, et al., Ultrathin nanotube/nanowire electrodes by spin–spray layer-by-layer assembly: a concept for transparent energy storage. ACS Nano **9**, 10005–10017 (2015). https://doi.org/10.1021/acsnano.5b03578
28. S.-J. Gong, D. Kim, E. Cho, et al., A chitosan/urushi anion exchange membrane for a non–aqueous redox flow battery. ChemistrySelect **2**, 1843–1849 (2017). https://doi.org/10.1002/slct.201601772
29. J. Grosse Austing, C. Nunes Kirchner, L. Komsiyska, G. Wittstock, Layer-by-layer modification of Nafion membranes for increased lifetime and efficiency of vanadium/air redox flow batteries. J. Membr. Sci. **510**, 259–269 (2016). https://doi.org/10.1016/j.memsci.2016.03.005
30. K. Grygiel, J.-S. Lee, K. Sakaushi, et al., Thiazolium poly(ionic liquid)s: synthesis and application as binder for lithium-ion batteries. ACS Macro Lett. **4**, 1312–1316 (2015). https://doi.org/10.1021/acsmacrolett.5b00655
31. S. Gupta, E. Heintzman, C. Price, Electrostatic layer-by-layer self-assembled graphene/multi-walled carbon nanotubes hybrid multilayers as efficient "all carbon" supercapacitors. J. Nanosci. Nanotechnol. **16**, 4771–4782 (2016)
32. M. Han, D.-W. Kim, Y.-C. Kim, Charged polymer-coated separators by atmospheric plasma-induced grafting for lithium-ion batteries. ACS Appl. Mater. Interfaces **8**, 26073–26081 (2016). https://doi.org/10.1021/acsami.6b08781
33. Y. Hishinuma, T. Ogihara, Preparation of silica glass films on the surface of polypropylene microporous membrane separators by dip coating with polysilazane and their application in lithium-ion batteries. J. Ceram. Soc. Jpn. **124**, 480–483 (2016). https://doi.org/10.2109/jcersj2.15306
34. Y. Hu, X. Sun, Flexible rechargeable lithium-ion batteries: advances and challenges in materials and process technologies. J. Mater. Chem. A **2**, 10712–10738 (2014). https://doi.org/10.1039/C4TA00716F
35. H.J. Hwang, W.S. Chi, O. Kwon, et al., Selective ion transporting polymerized ionic liquid membrane separator for enhancing cycle stability and durability in secondary zinc–air battery systems. ACS Appl. Mater. Interfaces **8**, 26298–26308 (2016). https://doi.org/10.1021/acsami.6b07841
36. J.-W. Jeon, S.R. Kwon, F. Li, J.L. Lutkenhaus, Spray-on polyaniline/poly(acrylic acid) electrodes with enhanced electrochemical stability. ACS Appl. Mater. Interfaces **7**, 24150–24158 (2015a). https://doi.org/10.1021/acsami.5b07459
37. J.-W. Jeon, S.R. Kwon, J.L. Lutkenhaus, Polyaniline nanofiber/electrochemically reduced graphene oxide layer-by-layer electrodes for electrochemical energy storage. J. Mater. Chem. A **3**, 3757–3767 (2015b). https://doi.org/10.1039/C4TA04697H
38. B. Jiang, Y. He, B. Li, et al., Polymer-templated formation of polydopamine-coated SnO_2 nanocrystals: anodes for cyclable lithium-ion batteries. Angew. Chem. Int. Ed. **56**, 1869–1872 (2017). https://doi.org/10.1002/anie.201611160
39. R. Jin, Q. Wang, H. Li, et al., Polypyrrole layer coated MnO_x/Fe_2O_3 nanotubes with an enhanced electrochemical performance for lithium ion batteries. Appl. Surf. Sci. **403**, 62–70 (2017). https://doi.org/10.1016/j.apsusc.2017.01.120
40. M. Kim, W.J. Hyun, S.C. Mun, O.O. Park, Preparation of functional composite materials based on chemically derived graphene using the solution process. J. Phys. Conf. Ser. **602**, 012027 (2015). https://doi.org/10.1088/1742-6596/602/1/012027
41. E.T. Kim, J. Park, C. Kim, et al., Conformal polymeric multilayer coatings on sulfur cathodes via the layer-by-layer deposition for high capacity retention in Li–S batteries. ACS Macro Lett. **5**, 471–475 (2016a). https://doi.org/10.1021/acsmacrolett.6b00144
42. J.H. Kim, J. Seo, J. Choi, et al., Synergistic ultrathin functional polymer-coated carbon nanotube interlayer for high-performance lithium-sulfur batteries. ACS Appl. Mater. Interfaces **8**, 20092–20099 (2016b). https://doi.org/10.1021/acsami.6b06190

43. J.-H. Kim, H.-S. Woo, W.K. Kim, et al., Improved cycling performance of lithium–oxygen cells by use of a lithium electrode protected with conductive polymer and aluminum fluoride. ACS Appl. Mater. Interfaces **8**, 32300–32306 (2016c). https://doi.org/10.1021/acsami.6b10419
44. M.S. Kim, L. Ma, S. Choudhury, et al., Fabricating multifunctional nanoparticle membranes by a fast layer-by-layer Langmuir–Blodgett process: application in lithium-sulfur batteries. J. Mater. Chem. A **4**, 14709–14719 (2016d). https://doi.org/10.1039/C6TA06018H
45. J. Lee, W. Choi, Surface modification of sulfur cathodes with PEDOT: PSS conducting polymer in lithium-sulfur batteries. J. Electrochem. Soc. **162**, A935–A939 (2015). https://doi.org/10.1149/2.0651506jes
46. J. Lee, W. Choi, Surface modification of over-lithiated layered oxides with PEDOT: PSS conducting polymer in lithium-ion batteries. J. Electrochem. Soc. **162**(4), A743–A748 (2015). https://www.researchgate.net/publication/272377276_Surface_Modification_of_Over-Lithiated_Layered_Oxides_with_PEDOTPSS_Conducting_Polymer_in_Lithium-Ion_Batteries. Accessed 4 Apr 2017
47. S.W. Lee, B.-S. Kim, S. Chen, et al., Layer-by-layer assembly of all carbon nanotube ultrathin films for electrochemical applications. J. Am. Chem. Soc. **131**, 671–679 (2009). https://doi.org/10.1021/ja807059k
48. Y.-S. Lee, W.-K. Shin, A.G. Kannan, et al., Improvement of the cycling performance and thermal stability of lithium-ion cells by double-layer coating of cathode materials with Al_2O_3 nanoparticles and conductive polymer. ACS Appl. Mater. Interfaces **7**, 13944–13951 (2015). https://doi.org/10.1021/acsami.5b02690
49. H.-Y. Lee, Y. Jung, S. Kim, Conducting polymer coated graphene oxide electrode for rechargeable lithium-sulfur batteries. J. Nanosci. Nanotechnol. **16**, 2692–2695 (2016). https://doi.org/10.1166/jnn.2016.11061
50. F.-S. Li, Y.-S. Wu, J. Chou, et al., A mechanically robust and highly ion-conductive polymer-blend coating for high-power and long-life lithium-ion battery anodes. Adv. Mater. **27**, 130–137 (2015a). https://doi.org/10.1002/adma.201403880
51. F.-S. Li, Y.-S. Wu, J. Chou, N.-L. Wu, A dimensionally stable and fast-discharging graphite–silicon composite Li-ion battery anode enabled by electrostatically self-assembled multifunctional polymer-blend coating. Chem. Commun. **51**, 8429–8431 (2015b). https://doi.org/10.1039/C4CC09825K
52. W.-J. Li, S.-L. Chou, J.-Z. Wang, et al., Multifunctional conducting polymer coated $Na_{1+x}MnFe(CN)_6$ cathode for sodium-ion batteries with superior performance via a facile and one-step chemistry approach. Nano Energy **13**, 200–207 (2015c). https://doi.org/10.1016/j.nanoen.2015.02.019
53. Y. Li, L. Yuan, Z. Li, et al., Improving the electrochemical performance of a lithium-sulfur battery with a conductive polymer-coated sulfur cathode. RSC Adv. **5**, 44160–44164 (2015d). https://doi.org/10.1039/C5RA05481H
54. Y. Li, B. Xu, H. Xu, et al., Hybrid polymer/garnet electrolyte with a small interfacial resistance for lithium-ion batteries. Angew. Chem. Int. Ed. **56**, 753–756 (2017). https://doi.org/10.1002/anie.201608924
55. J.-Y. Liao, S.-M. Oh, A. Manthiram, Core/double-shell type gradient Ni-rich $LiNi_{0.76}Co_{0.10}Mn_{0.14}O_2$ with high capacity and long cycle life for lithium-ion batteries. ACS Appl. Mater. Interfaces **8**, 24543–24549 (2016). https://doi.org/10.1021/acsami.6b06172
56. S. Lim, R. Lilly Thankamony, T. Yim, et al., Surface modification of sulfur electrodes by chemically anchored cross-linked polymer coating for lithium-sulfur batteries. ACS Appl. Mater. Interfaces **7**, 1401–1405 (2015). https://doi.org/10.1021/am508528p
57. J. Liu, D.G.D. Galpaya, L. Yan, et al., Exploiting a robust biopolymer network binder for an ultrahigh-areal-capacity Li–S battery. Energy Environ. Sci. (2016). https://doi.org/10.1039/C6EE03033E
58. L. Luo, S.-H. Chung, A. Manthiram, A trifunctional multi-walled carbon nanotubes/polyethylene glycol (MWCNT/PEG)-coated separator through a layer-by-layer coating strategy for high-energy Li–S batteries. J. Mater. Chem. A **4**, 16805–16811 (2016). https://doi.org/10.1039/C6TA07709A

59. J. Luo, R.-C. Lee, J.-T. Jin, et al., A dual-functional polymer coating on a lithium anode for suppressing dendrite growth and polysulfide shuttling in Li–S batteries. Chem. Commun. **53**, 963–966 (2017). https://doi.org/10.1039/C6CC09248A
60. R. Mo, S.O. Tung, Z. Lei, et al., Pushing the limits: 3D layer-by-layer-assembled composites for cathodes with 160 C discharge rates. ACS Nano **9**, 5009–5017 (2015a). https://doi.org/10.1021/nn507186k
61. R. Mo, F. Zhang, Y. Du, et al., Sandwich nanoarchitecture of LiV_3O_8/graphene multilayer nanomembranes via layer-by-layer self-assembly for long-cycle-life lithium-ion battery cathodes. J. Mater. Chem. A **3**, 13717–13723 (2015b). https://doi.org/10.1039/C5TA02562A
62. S. Moon, Y.H. Jung, D.K. Kim, Enhanced the electrochemical performance of a crosslinked polyaniline-coated graphene oxide-sulfur composite for rechargeable lithium-sulfur batteries. J. Power Sources **294**, 386–392 (2015). https://doi.org/10.1016/j.jpowsour.2015.06.011
63. J. Muldoon, C.B. Bucur, N. Boaretto, et al., Polymers: opening doors to future batteries. Polym. Rev. **55**, 208–246 (2015). https://doi.org/10.1080/15583724.2015.1011966
64. W. Na, A.S. Lee, J.H. Lee, et al., Lithium dendrite suppression with UV-curable polysilsesquioxane separator binders. ACS Appl. Mater. Interfaces **8**, 12852–12858 (2016). https://doi.org/10.1021/acsami.6b02735
65. W. Ni, D. Yang, J. Cheng, et al., Gel-type polymer separator with higher thermal stability and effective overcharge protection of 4.2 V for secondary lithium-ion batteries. RSC Adv. **6**, 52966–52973 (2016). https://doi.org/10.1039/C6RA11638H
66. S. Niu, W. Lv, G. Zhou, et al., Electrostatic-spraying an ultrathin, multifunctional and compact coating onto a cathode for a long-life and high-rate lithium-sulfur battery. Nano Energy **30**, 138–145 (2016). https://doi.org/10.1016/j.nanoen.2016.09.044
67. M.M. Obadia, E. Drockenmuller, Poly(1,2,3-triazolium)s: a new class of functional polymer electrolytes. Chem. Commun. **52**, 2433–2450 (2016). https://doi.org/10.1039/C5CC09861K
68. L. Pan, H. Wang, C. Wu, et al., Tannic-acid-coated polypropylene membrane as a separator for lithium-ion batteries. ACS Appl. Mater. Interfaces **7**, 16003–16010 (2015). https://doi.org/10.1021/acsami.5b04245
69. L. Porcarelli, A.S. Shaplov, M. Salsamendi, et al., Single-ion block copoly(ionic liquid)s as electrolytes for all-solid-state lithium batteries. ACS Appl. Mater. Interfaces **8**, 10350–10359 (2016). https://doi.org/10.1021/acsami.6b01973
70. Q. Qu, F. Qian, S. Yang, et al., Layer-by-layer polyelectrolyte assisted growth of 2D ultrathin MoS_2 nanosheets on various 1D carbons for superior Li-storage. ACS Appl. Mater. Interfaces **8**, 1398–1405 (2016). https://doi.org/10.1021/acsami.5b10497
71. J.B. Schlenoff, H. Ly, M. Li, Charge and mass balance in polyelectrolyte multilayers. J. Am. Chem. Soc. **120**, 7626–7634 (1998). https://doi.org/10.1021/ja980350+
72. S.M. Seidel, S. Jeschke, P. Vettikuzha, H.-D. Wiemhöfer, PVDF-HFP/ether-modified polysiloxane membranes obtained via airbrush spraying as active separators for application in lithium-ion batteries. Chem. Commun. **51**, 12048–12051 (2015). https://doi.org/10.1039/C5CC04424C
73. C. Shi, J. Dai, S. Huang, et al., A simple method to prepare a polydopamine modified core-shell structure composite separator for application in high-safety lithium-ion batteries. J. Membr. Sci. **518**, 168–177 (2016). https://doi.org/10.1016/j.memsci.2016.06.046
74. W.-K. Shin, A.G. Kannan, D.-W. Kim, Effective suppression of dendritic lithium growth using an ultrathin coating of nitrogen and sulfur codoped graphene nanosheets on polymer separator for lithium metal batteries. ACS Appl. Mater. Interfaces **7**, 23700–23707 (2015a). https://doi.org/10.1021/acsami.5b07730
75. W.-K. Shin, J.-H. Yoo, D.-W. Kim, Surface-modified separators prepared with conductive polymer and aluminum fluoride for lithium-ion batteries. J. Power Sources **279**, 737–744 (2015b). https://doi.org/10.1016/j.jpowsour.2015.01.047
76. W.B. Stockton, M.F. Rubner, Molecular-level processing of conjugated polymers. 4. Layer-by-layer manipulation of polyaniline via hydrogen-bonding interactions. Macromolecules **30**, 2717–2725 (1997). https://doi.org/10.1021/ma9700486

77. G.B. Sukhorukov, E. Donath, H. Lichtenfeld, et al., Layer-by-layer self-assembly of polyelectrolytes on colloidal particles. Colloids Surf. A Physicochem. Eng. Asp. **137**, 253–266 (1998). https://doi.org/10.1016/S0927-7757(98)00213-1
78. Y. Sun, J. Lopez, H.-W. Lee, et al., A stretchable graphitic carbon/Si anode enabled by conformal coating of a self-healing elastic polymer. Adv. Mater. **28**, 2455–2461 (2016). https://doi.org/10.1002/adma.201504723
79. H. Tan, K. Huang, Y. Bao, et al., Rationally designed layer-by-layer structure of Fe_3O_4 nanospheres@MWCNTs/graphene as an electrode for lithium-ion batteries with enhanced electrochemical performance. J. Alloys Compd. **699**, 812–817 (2017). https://doi.org/10.1016/j.jallcom.2016.12.393
80. S.-O. Tung, S. Ho, M. Yang, et al., A dendrite-suppressing composite ion conductor from aramid nanofibres. Nat. Commun. **6**, 6152 (2015). https://doi.org/10.1038/ncomms7152
81. H. Wang, L. Pan, C. Wu, et al., Pyrogallic acid coated polypropylene membranes as separators for lithium-ion batteries. J. Mater. Chem. A **3**, 20535–20540 (2015a). https://doi.org/10.1039/C5TA06381G
82. L. Wang, B. Gao, C. Peng, et al., Bamboo leaf derived ultrafine Si nanoparticles and Si/C nanocomposites for high-performance Li-ion battery anodes. Nanoscale **7**, 13840–13847 (2015b). https://doi.org/10.1039/C5NR02578H
83. Y. Wang, Q. Qu, Y. Han, et al., Robust 3D nanowebs assembled from interconnected and sandwich-like C@Fe_3O_4@C coaxial nanocables for enhanced Li-ion storage. J. Mater. Chem. A **4**, 10314–10320 (2016). https://doi.org/10.1039/C6TA03118H
84. F. Wu, J. Li, Y. Su, et al., Layer-by-layer assembled architecture of polyelectrolyte multilayers and graphene sheets on hollow carbon spheres/sulfur composite for high-performance lithium-sulfur batteries. Nano Lett. **16**, 5488–5494 (2016a). https://doi.org/10.1021/acs.nanolett.6b01981
85. F. Wu, J. Liu, L. Li, et al., Surface modification of Li-rich cathode materials for lithium-ion batteries with a PEDOT: PSS conducting polymer. ACS Appl. Mater. Interfaces **8**, 23095–23104 (2016b). https://doi.org/10.1021/acsami.6b07431
86. J. Xi, W. Dai, L. Yu, Polydopamine coated SPEEK membrane for a vanadium redox flow battery. RSC Adv. **5**, 33400–33406 (2015). https://doi.org/10.1039/C5RA01486G
87. W. Xu, Z. Wang, L. Shi, et al., Layer-by-layer deposition of organic-inorganic hybrid multilayer on microporous polyethylene separator to enhance the electrochemical performance of lithium-ion battery. ACS Appl. Mater. Interfaces **7**, 20678–20686 (2015). https://doi.org/10.1021/acsami.5b05457
88. Y. Xu, Q. Wei, C. Xu, et al., Layer-by-layer $Na_3V_2(PO_4)_3$ embedded in reduced graphene oxide as superior rate and ultralong-life sodium-ion battery cathode. Adv. Energy Mater. **6** (2016). https://doi.org/10.1002/aenm.201600389
89. R. Xu, X. Huang, X. Lin, et al., The functional aqueous slurry coated separator using polyvinylidene fluoride powder particles for lithium-ion batteries. J. Electroanal. Chem. **786**, 77–85 (2017). https://doi.org/10.1016/j.jelechem.2017.01.016
90. J. Yan, X. Liu, B. Li, Nano-assembled Na_2FePO_4F/carbon nanotube multi-layered cathodes for Na-ion batteries. Electrochem. Commun. **56**, 46–50 (2015a). https://doi.org/10.1016/j.elecom.2015.04.009
91. J. Yan, X. Liu, M. Yao, et al., Long-life, high-efficiency lithium–sulfur battery from a nanoassembled cathode. Chem. Mater. **27**, 5080–5087 (2015b). https://doi.org/10.1021/acs.chemmater.5b01780
92. M. Yan, Y. Zhang, Y. Li, et al., Manganese dioxide nanosheet functionalized sulfur@PEDOT core-shell nanospheres for advanced lithium-sulfur batteries. J. Mater. Chem. A **4**, 9403–9412 (2016). https://doi.org/10.1039/C6TA03211G
93. H. Yan, G. Zhang, Y. Li, Synthesis and characterization of advanced $Li_3V_2(PO_4)_3$ nanocrystals@conducting polymer PEDOT for high energy lithium-ion batteries. Appl. Surf. Sci. **393**, 30–36 (2017). https://doi.org/10.1016/j.apsusc.2016.09.156
94. C. Yang, H. Wei, L. Guan, et al., Polymer nanocomposites for energy storage, energy saving, and anticorrosion. J. Mater. Chem. A **3**, 14929–14941 (2015a). https://doi.org/10.1039/C5TA02707A

95. X. Yang, L. Shen, B. Wu, et al., Improvement of the cycling performance of $LiCoO_2$ with the assistance of cross-linked PAN for lithium-ion batteries. J. Alloys Compd. **639**, 458–464 (2015b). https://doi.org/10.1016/j.jallcom.2015.03.153
96. Q. Yang, J. Wu, K. Huang, et al., Layer-by-layer self-assembly of graphene-like Co_3O_4 nanosheet/graphene hybrids: towards high-performance anode materials for lithium-ion batteries. J. Alloys Compd. **667**, 29–35 (2016). https://doi.org/10.1016/j.jallcom.2016.01.136
97. H.Y. Yoo, A. Heo, C.G. Cho, Crosslinkable layer-by-layer assembled sulfonated poly(phenylene oxide) membrane based on Nafion for vanadium redox flow battery. J. Nanosci. Nanotechnol. **16**, 10515–10519 (2016). https://doi.org/10.1166/jnn.2016.13186
98. Z. Yuan, X. Zhu, M. Li, et al., A highly ion-selective zeolite flake layer on porous membranes for flow battery applications. Angew. Chem. **128**, 3110–3114 (2016). https://doi.org/10.1002/ange.201510849
99. Y. Zhai, K. Xiao, J. Yu, B. Ding, Closely packed x-poly(ethylene glycol diacrylate) coated polyetherimide/poly(vinylidene fluoride) fiber separators for lithium-ion batteries with enhanced thermostability and improved electrolyte wettability. J. Power Sources **325**, 292–300 (2016). https://doi.org/10.1016/j.jpowsour.2016.06.050
100. J. Zhang, Y. Xu, L. Fan, et al., Graphene–encapsulated selenium/polyaniline core-shell nanowires with an enhanced electrochemical performance for Li–Se batteries. Nano Energy **13**, 592–600 (2015a). https://doi.org/10.1016/j.nanoen.2015.03.028
101. Z. Zhang, Z. Zhang, J. Li, Y. Lai, Polydopamine-coated separator for high-performance lithium-sulfur batteries. J. Solid State Electrochem. **19**, 1709–1715 (2015b). https://doi.org/10.1007/s10008-015-2797-8
102. H. Zhang, S. Jing, Y. Hu, et al., A flexible freestanding Si/rGO hybrid film anode for stable Li-ion batteries. J. Power Sources **307**, 214–219 (2016). https://doi.org/10.1016/j.jpowsour.2015.12.107
103. Y. Zhao, N.B. Mercier, H.R. Byon, An aqueous lithium–iodine battery with solid polymer electrolyte-coated metallic lithium anode. ChemPlusChem **80**, 344–348 (2015). https://doi.org/10.1002/cplu.201402038
104. Y. Zhao, Z. Yuan, W. Lu, et al., The porous membrane with tunable performance for vanadium flow battery: the effect of charge. J. Power Sources **342**, 327–334 (2017). https://doi.org/10.1016/j.jpowsour.2016.12.058
105. G. Zheng, C. Wang, A. Pei, et al., High-performance lithium metal negative electrode with a soft and flowable polymer coating. ACS Energy Lett. **1**, 1247–1255 (2016). https://doi.org/10.1021/acsenergylett.6b00456
106. G. Zhou, L. Li, D.-W. Wang, et al., A flexible sulfur-graphene-polypropylene separator integrated electrode for advanced Li–S batteries. Adv. Mater. **27**, 641–647 (2015). https://doi.org/10.1002/adma.201404210
107. Y. Zhou, Y. Li, J. Yang, et al., Conductive polymer-coated VS4 submicrospheres as advanced electrode materials in lithium-ion batteries. ACS Appl. Mater. Interfaces **8**, 18797–18805 (2016). https://doi.org/10.1021/acsami.6b04444
108. Y. Zhu, C. Yao, J. Ren, et al., Graphene improved electrochemical property in self-healing multilayer polyelectrolyte film. Colloids Surf. A Physicochem. Eng. Asp. **465**, 26–31 (2015). https://doi.org/10.1016/j.colsurfa.2014.10.035

Functionalized Ionic Liquid-Based Electrolytes for Li-Ion Batteries

11

Aarti Tiwari, Tharamani C. Nagaiah, Debaprasad Mandal, and Santosh N. Chavan

Contents

Abstract

Ionic liquids (ILs) are molten salts with an immense potential as an electrolyte in battery which fulfills the two most demanding aspects, negligible volatility, and non-flammability along with low melting temperature affecting their wide range applicability. Along with these aspects, ILs need suitable functionalization of the chosen cationic and/or anionic component for high stability within the working potential of the battery and suitable ionic conductivity. The chemistry of the substituent groups and the interaction between this cation and anion of an ionic liquid are the key to fine-tune the electrolyte. The present chapter deals with the functionalization of alkoxy (R–O–R) and siloxy (Si–O–Si) groups as substituents over the imidazolium cation. This modified cation was associated with an electronically diffused anion, N, N- bis(trifluoromethane)sulfonimide (TFSI, $CF_3SO_2)_2NH$) to control the extent of cluster formation prevalent in the presence

A part of the work was contributed to the thesis of Dr. S. N. Chavan. The authors further acknowledge that there is no financial relationship with the editors or publisher and have contributed original work in this chapter.

A. Tiwari · T. C. Nagaiah (✉) · D. Mandal (✉) · S. N. Chavan
Department of Chemistry, Indian Institute of Technology Ropar, Rupnagar, Punjab, India
e-mail: aarti.tiwari@iitrpr.ac.in; tharamani@iitrpr.ac.in; dmandal@iitrpr.ac.in; santosh.chavan@iitrpr.ac.in

Q. Zhen et al. (eds.), *Nanostructured Materials for Next-Generation Energy Storage and Conversion*, https://doi.org/10.1007/978-3-662-58675-4_11

of high concentration of lithium-ion (Li^+, 1 mol/kg solvent). This cluster formation regulates the viscosity which in turn controls the obtainable ionic conductivity essential for good charge-discharge characteristics. Another issue of high-energy electrodes reacting slowly with the ionic liquid electrolyte and requiring a long time for a stable solid electrolyte interface formation was addressed by using an organic additive, namely, propylene carbonate (PC, $C_4H_6O_3$). The concept of a functionalized electrolyte mixture comprising of an ionic liquid, organic additive, and Li-ion source is pursued in depth, and its impact over the various parameters of the electrolyte was analyzed. The probable impact of these ionic liquid mixtures in the battery electrolyte was assessed both physical and electrochemical characterization. The functional applicability was finally tested by performing charge-discharge cycling in a compiled battery consisting of a graphite (C) anode and lithium cobalt(III) oxide ($LiCoO_2$) cathode separated by the ionic liquid mixture soaked separator.

11.1 Introduction

The sustainability of energy supply is an emerging key aspect, which until now has been dominated by the fossil fuel sector. The rapid overutilization of these non-renewable resources has endangered the globalization and eventually hampers the natural habitability owing to pollution and bloated-up greenhouse emissions. This drove the scientific community to look for an alternative renewable energy supply for which Lithium-ion (Li-ion, Li^+) batteries are an imperative contender [3, 45]. The applicability of any energy conversion or storage device is highly dependent upon the available specific power against high-energy density. Considering this balance between the obtainable power and energy, batteries lead the forefront as the alternatives like fuel cells have a high-energy density but are limited by their power, whereas the case is vice versa for supercapacitors.

The history of Li-ion battery dates back to the 1900s with the advent of high-energy dense but unsafe lithium batteries; following these a relatively low-energy dense but safer Li-ion batteries were developed by utilizing the phenomena of intercalation [51] of Li-ions into layered graphite anodes having better safety features compared to the former. The efforts of Yoshino et al. [52] in the year 1987 were patented describing the use of carbon-based (C) anode along with lithium cobalt(III) oxide ($LiCoO_2$) cathode in building an initial prototype whose designing was employed in manufacturing early Li-ion batteries. This led to the eventual commercialization of Li-ion batteries by Sony [3, 22], and the untiring efforts of Goodenough, Yazami, and Yoshino were awarded the Institute of Electrical and Electronics Engineers (IEEE) Medal in 2012 for their unmatchable pioneering innovation. Since then, the battery has revolutionized the energy sector especially in the sphere of portable electronics and at present is swiftly moving toward heavy load applications like in power grids, hybrid electric vehicles, and the like [45].

The rechargeable Li-ion batteries have a unique capability of holding and subsequently delivering a large amount of charge per unit of Li-ions involved. This phenomenon could be described by considering a battery system composed of a graphite (C) anode and a lithium cobalt(III) oxide ($LiCoO_2$) cathode, for instance, both separated and yet ionically connected by an electrolyte which till date are predominantly organic electrolyte mixtures. The reactions involved in these electrodes during charging are:

$$\textbf{Anode}: C_n + xLi^+ + xe^- \rightarrow C_nLi_x$$

$$\textbf{Cathode}: LiCoO_2 \rightarrow Li_{1-x}CoO_2 + xLi^+ + xe^-$$

The reactions involved during discharging are:

$$\textbf{Anode}: C_nLi_x \rightarrow C_n + xLi^+ + xe^-$$

$$\textbf{Cathode}: Li_{1-x}CoO_2 + xLi^+ + xe^- \rightarrow LiCoO_2$$

Overall reaction can be represented as follows wherein the forward reaction depicts charging and the reverse denotes discharging:

$$LiCoO_2 + C_n \rightleftharpoons Li_{1-x}CoO_2 + C_nLi_x$$

Simply, the energy is stored in the anode upon providing external electron flow forcing electronic movement from the cathode to the anode wherein the Li-ions tend to intercalate between the layered graphite anodes. This overall process of storing the electrical energy chemically in the cathode in the form of chemical potentials is termed as charging. The opposite occurs when an external load is powered by this charged battery wherein the electrons move externally from anode to cathode, while the Li-ions simultaneously intercalate the cathode by shuttling via the electrolyte. Considering these electrode chemical potentials which invariably depend upon the electrode material employed the theoretical voltage of the battery cell which can be obtained by estimating the total Gibbs free energy using:

$$\Delta E = -\Delta G/nF$$

where, the cell voltage, E (V), is directly proportional to the change in Gibbs free energy, G (kJ/mol), and the number of electrons involved, n, along with the Faraday constant, F (96485.3 C mol^{-1}). This cell voltage obtainable from Li-ion battery is 4.2 V which is very high compared to any other battery currently commercialized in the market. This is the reason wherein the number of cells required to power suitable electronics has reduced favoring the compact design of gadgets and is the sole reason of wide verse application in a host of portable electronics.

The wide-scale applicability of Li-ion battery derives further research toward betterment by improving the various battery components, namely, anode, cathode, and electrolyte. The evolution of better electrodes strives to improve the inherent energy as well as power density, but these high-energy electrodes have a major disadvantage of instability in terms of their interaction with the battery electrolyte upon charging and discharging. Therefore, improvement toward high-energy electrodes demands simultaneous research for better electrolytes both in terms of their functioning and safety considerations. The electrolyte has a vital role to play in the overall functioning of the Li-ion battery and lies at the heart of a smoothly operating battery [22]. Looking more closely at the electrolyte, it essentially is an electronic insulator not allowing any electrons resulting from the oxidation of Li-ions at the anode during discharging to pass through which would otherwise lead to short-circuiting and eventually blasting of the battery. Rather, it is required to be an excellent ionic conductor which would facilitate the movement of these resulting Li-ions at the anode to the cathode through the separator preferably at a pace similar to that of the electrons reaching the cathode through the external circuit.

In this pursuit, the electrolyte is expected to have a high conductivity in the range of 1–10 mS/cm [16] and a low viscosity for efficient battery cycling. The electrolyte thereby plays an important role in the overall functioning of battery acting as the reservoir of Li-ions which subsequently undergoes intercalation in the cathode during discharging whereas in the anode during the charging process from the electrolyte. So, overall the electrolyte facilitates the diffusion of Li-ions under the influence of electric charge and is therefore expected to possess a very high ionic conductivity so that it affords minimum resistance toward diffusion. Besides this basic function of shuttling Li-ions effectively between the two electrodes, the battery electrolyte should also be electrochemically stable within the working potential range of the battery. A measure of this safety is usually reflected by a wide potential window termed as electrochemical stability window (ESW) in which the electrolyte does not react with the electrode material which may otherwise lead to its degradation. Collectively, the desirable aspects of a successful Li-ion battery electrolyte include:

- High conductivity
- High transference number
- Wide thermal stability
- Wide electrochemical stability window (ESW)
- Unreactive with active electrode material
- Passivation ability
- Nonvolatile and non-flammable
- Low environmental impact
- Cost-effective

The common commercialized Li-ion battery uses organic solvents like alkyl carbonates as an electrolyte [14] which affords high conductivity for the Li-ions but suffers from the inherent safety issues. The major drawback of the organic solvents as electrolytes is their high volatility and flammability; besides they are also limited by their ESW due to their instability/decomposition at the working potentials of the battery. These organic electrolytes cannot be employed in case of the high-energy electrodes as they have a low flash point which ultimately hampers obtainable output in terms of energy and power density for heavy load application.

Developing a safe electrolyte requires either stabilization of the solid electrolyte interphase (SEI) layer by introducing additives and/or enlarging the electrochemical stability window by widening the highest occupied molecular orbital-lowest unoccupied molecular orbital (HOMO-LUMO) gap for the designed electrolyte which subsequently prevents dendrite formation over the anode surface. Considering the battery stability in terms of electronic energies of the anode, cathode, and electrolyte requires the electrolyte stability window to lie between the energy of the anode (reductant; constituting the LUMO) and that of the cathode (oxidant; constituting the HOMO) [21]. If the anode has a higher energy than the LUMO of the electrolyte, it will increasingly reduce the electrolyte unless a stable insulating barrier separates the two. Similarly, for the cathode, if the energy of the cathode lies below the HOMO of the electrolyte, then the electrolyte is increasingly prone to oxidative degradation unless a passivation layer forms atop the cathode barring electronic interaction with electrolyte. So essentially SEI layer formation imparts kinetic stability to the battery against high operating potential. Therefore, designing a stable electrolyte requires:

- Appropriate matching between the electrolyte HOMO-LUMO levels with the cathode and anode, respectively
- Rapid formation of a passivating SEI, which would self-heal upon cracking during charge-discharge cycling, thereby preventing dendrite formation
- Overall the formed SEI must allow a fast transfer of Li-ions across the electrode-electrolyte interface for effective cycling

Other major challenges associated with organic electrolytes are the instability of the formed SEI passivating the reactive electrode surface which tends to intercalate within the graphite anode and exfoliates with successive cycling [55]. This exfoliation in coherence with the dendrite formation at the lithium (Li) anode poses grave concern both toward safety and achievable specific capacity from the battery. These shortcomings of the organic electrolytes causing serious functioning issues demand further research toward alternative electrolytes. The suitable replacement options available include employing fluidic salts or room temperature ionic liquids (ILs) and the use of solid electrolytes like polymer-based electrolytes and inorganic electrolytes [5, 6, 17, 43]. Application of the solid electrolytes is an ideal alternative considering the safety features and reduced impact of dendrite formation, but they are potentially challenged by their greatly reduced ionic conductivities. The conductivity obtainable by solid electrolytes is of the order of 10^{-5}–10^{-8} S/cm compared to 1–20 mS/cm for organic electrolytes which is the result of high interfacial resistance between the solid-solid interface of electrode and electrolyte [9]. Besides, the kinetics of intercalation-de-intercalation is also suppressed greatly posing a grave challenge for solid electrolytes. On the other hand ILs also tend to have a slightly reduced ionic conductivity owing to their viscous nature, but the trade-off between slight conductivity losses to improved safety is a viable alternative.

Thereby, ionic liquids (ILs) come as handy electrolytes allowing a great degree of tunability and stability over a wide temperature range. Ionic liquids have already found diverse applicability in the fields ranging from solvents in green synthesis, as lubricants in machining, gas storage and modification, waste recycling and as a processing agent in different industries [20]. The various attributes of ionic liquids which attract wide-sphere application are [2]:

- Nonvolatile
- Non-flammable
- High thermal stability
- Electrochemical stability
- Improved solvation
- Ionic conductivity

These features enable ILs to be a favorable alternative as electrolytes in the battery. The most important aspect of ILs is their enormous tuning ability in both the cation and the anion exhibiting weak interionic interactions resulting in low crystallization temperature and therefore wide liquid range along with low vapor pressure. Due to the ionic nature, ILs easily solvate Li-ions and subsequently carry them to respective electrodes under electric field, and the extent of this shuttling can be fine-tuned by tailoring the viscosity of the ILs depending on its constituents [41]. While designing the battery electrolyte, another major aspect is the electrochemical stability within the potential window of battery functioning which is hard to achieve in case of organic electrolytes as they are prone to degradation either at the anodic or at the cathodic limit ultimately restraining the obtainable voltage output from the battery during safe operation. Most ILs are stable in the wide potential range and thereby counteract this utmost major functional as well as safety limitation posed in case of organic electrolytes. Taking these features into consideration, there is a need to move toward IL-based electrolytes in the battery.

The most common and simple IL in present times [with common chemical formulae] is derived from the quaternary ammonium salts [$(CH_3)_4N(Cl)$] [24] and cyclic amines like pyrrolidinium [$C_5H_{11}BF_3N$], piperidinium [$C_6H_{13}BF_3N$], and aromatic amines like pyridinium [$C_5H_6NClCrO_3$] and imidazolium [$C_6H_{10}Cr_2N_4O_7$] [15, 18, 36, 49]. Likewise, other cations are sulfonium [$C_6H_{14}ClNO_2S$], phosphonium [$C_{16}H_34BrO_2P$], guanidinium [CH_6BrN_3], and the like [31, 48]. These cations constitute the core of the IL, however, functionalizing these cores with a variety of substituents, and design them as functional cations depending upon the desired application [20, 40, 50]. Similarly, the anions range from small and charge dense inorganic anions like halide $[F]^-$, $[Cl]^-$, $[Br]^-$, $[I]^-$, and acetato $[OAc^-]$ to delocalized inorganic anions tetrafluoroborate $[BF_4]^-$, hexafluorophosphate $[PF_6]^-$, hexafluoroarsenate(V) $[AsF_6]^-$, or amide $[N(CN)_2]^-$ to charge-delocalized organic anions like perfluorobutane sulfonate $[C_4F_9SO_3]^-$, trifluoroacetate $[CF_3CO_2]^-$, triflate $[CF_3SO_3]^-$, amides ($[N(CF_3SO_2)_2]^-$ and $[CF_3CONCF_3SO_2]^-$), and methide $[C(CF_3SO_2)_3]^-$ [18, 34, 42].

Among these large number of choices for cationic core like imidazolium, quaternary ammonium, phosphonium, piperidinium, pyrrolidinium, guanidinium, and phosphonium which mainly control various physicochemical and electrochemical properties of an IL, imidazolium-based ILs have found wide-sphere applications

[2, 21] especially in the field of battery electrolytes because of its relative ease of synthesis with different substituents, relatively low viscosity, high ionic conductivity, low melting point, resistive toward moisture and temperature, and a favorable electrochemical window which can be fine-tuned by the substituents. These imidazolium cation-based ILs can act as a good electrolyte for transporting the Li-ion in between the two electrodes only when it is teamed up with a suitable anion. The choice of anions ranges from small and charge dense inorganic anions like halides to charge-delocalized organic anions like amides $[N(CF_3SO_2)_2]^-$ or dicyandiamide $[N(CN)_2]^-$. However, a weak interaction between these two components of the ionic liquid is usually preferred by combining a large cation with a highly delocalized anion resulting in a small charge to radius ratio [41]. This confers low lattice energies resulting in weak electrostatic bonding yielding low crystallization tendency for obtaining low melting point and low viscosity in turn high ionic conductivity, which are two integral requirements of a battery electrolyte [23]. In this regard, N, N- bis(trifluoromethane)sulfonimide (TFSI, $[(CF_3SO_2)_2N]^-$) anion is a good choice as it possesses good ionic conductivity, thermal and electrochemical stability, and moisture resistive and also induces the formation of stable SEI layer to achieve long and safe battery cycling [4, 32]. Therefore, combining an imidazolium cation with delocalized TFSI anion would be a suitable choice for alternative IL electrolyte for the battery. Since the electrolyte is the reservoir of Li-ions in a battery, the addition of a Li-salt would provide the complete configuration of the electrolyte, but since it is preferred to use the salt with the similar anion to maximize the amount of soluble Li-ion, the choice of salt is lithium bis(trifluoromethanesulfonyl)imide (LiTFSI, $LiC_2F_6NO_4S_2$) in the present scenario. Besides, the usage of ILs also counteracts the problem of solvent co-intercalation in the electrode as the Li-ions are solvated by the anions and they lack the co-intercalation tendency at potentials prevalent at the anode.

The use of IL electrolyte comes handy as they perform the shuttling of Li-ions at the positive electrode (cathode) without any problem due to their high oxidative stability even against highly oxidizing materials like lithium manganese(III) phosphate ($LiMnPO_4$) [2]. But by using a simple imidazolium-based cation, it will be difficult to withstand the high negative potentials required for Li-ion deposition; however if somehow a stable passivating layer forms atop the anode of the battery quickly, this provides kinetic metastability to the cation of the IL electrolyte. To avoid this unlikely reaction, two significant alternatives would be appropriately functionalizing the imidazolium core [38] and the addition of small amount of organic solvent as an additive to aggravate the SEI layer formation over the high-energy anode [21, 32]. Similar effort toward functionalization of the imidazolium core was attempted by Ferrari group [15] wherein they have reported that introducing ether side chains to the central imidazolium cation brings down the crystallization temperature of the IL because of the flexibility in the ether chain and a weak interaction between the oxygen groups of the ether and the large TFSI anion. Besides it also exhibited decreased viscosity due to the presence of labile ether functionalization resulting in the ionic conductivity of 2.9 mS/cm at 20 °C. Higher viscosity resulting from the large ions usually having a low charge to radius ratio can be reduced by mixing with the organic solvent like ethylene carbonate (EC, $C_3H_4O_3$), propylene carbonate (PC, $C_4H_6O_3$), etc. These organic additives being more reactive compared to the ILs tend to react predominantly over the high-energy anode forming passivating surface films essentially stabilizing the battery system [29]. These electrolyte additives facilitate the battery performance by aggravating the SEI layer formed over the high-energy electrode, stabilize the formed SEI to minimize its cracking by gas evolution due to decomposition of the electrolyte and/or dissolution with cycling, improves the physicochemical properties of the ILs in terms of their obtainable conductivity, wettability, safety and likewise. Further, the addition of polymerizable additives is also popular since they form a mesh-like structure over the reactive electrodes, thereby preventing deterioration. Important examples are propylene carbonate (PC, $C_4H_6O_3$), methyl propargyl sulfone (PMS)[$HC \equiv CCH_2SO_2CH_3$, 3-methanesulfonylprop-1-yne], and the like which upon reduction tend to form good passivating moieties allowing facile Li-ion diffusion and restricted detrimental reaction [14]. So essentially, a functionalized IL mixture with a suitable stabilizing additive is an appropriate methodology for designing a composite battery electrolyte [49].

11.2 Functionalized Ionic Liquid as Battery Electrolyte

The challenge of modeling an electrolyte for successful application in Li-ion battery requires conceptual designing of the ionic liquid components to fine-tune the obtainable stability and ionic conductivity. The use of imidazolium-based cation functionalized at both the nitrogen is a suitable alternative although it requires fine-tuning in terms of the substituent groups and their respective chain lengths.

The present text details the effect of two different substituents, namely, alkoxy (R–O–R) and siloxy (Si–O–Si) groups over the imidazolium cation, teaming it up with a diffused anion, bis(trifluoromethylsulfonyl)imide [TFSI]. The trialkyl imidazolium anion can be substituted at the N_1 (leftmost) –C_2 (Center) or N_3^+ (rightmost) in the imidazolium ring, and a numbering/labeling scheme is read right to left. For example, an imidazolium (Im)-based bis(trifluoromethanesulfonyl)imide (TFSI) substitute at the N_1 (by CH_3), C_1 (with H), and N_2 (CH_2CH_3) would be classified as Im-1-H-2-TFSI. In a similar manner, ether substitutes are classified by carbon number between the oxygen atoms. For example, substitute at the N_1 (by CH_2CH_3), C_1 (with CH_3), and N_2 (with $CH_2CH_2OCH_2CH_3$) would be classified as Im-2o1-1-2-TFSI. The imidazolium ring was substituted with different length ethers yielding different ILs Im-1o2o2-H-2o1-TFSI (**1**), Im-1o2o2-H-1SioSi-TFSI (**2**), and Im-1o2-H-1SioSi-TFSI (**3**), respectively, and is represented in Scheme 11.1. The Im-1o2 group can be represented as $R_3 = CH_2CH_2OCH_2CH_3$ (at N_2^+). These unsymmetrical imidazolium-based ILs with alkoxy and siloxy groups were obtained from their N-substituted imidazolium iodides which were synthesized under microwave irradiation, a greener synthetic route for the bulk production of the ILs. For instance, Im-1o2o2-H-2o1-TFSI (**1**) were synthesized under microwave irradiation of N-alkoxy imidazole and 1-Iodo-3-methoxypropane in acetonitrile at a temperature

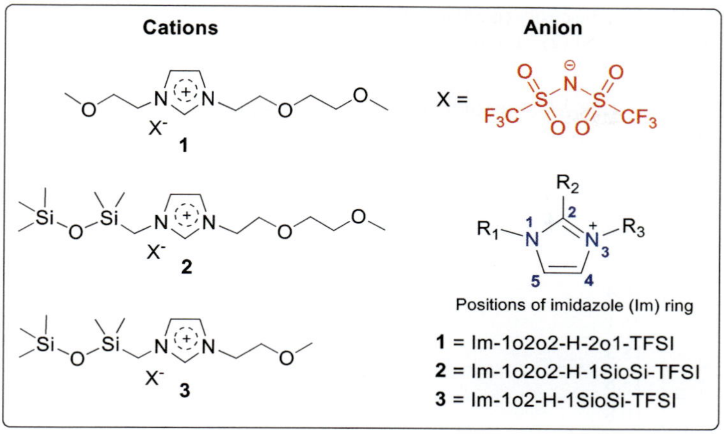

Scheme 11.1 Functionalized ionic liquids studied toward electrolyte application

of 150 °C for 40 min. The obtained IL [Im-1o2o2-H-2o1]I was subsequently anion exchanged with LiTFSI in acetone to give ether-ether IL (**1**). The other ILs followed the same synthetic protocol [7].

Viscosity is an important parameter considering the electrolytic application of functionalized ionic liquid which is governed by the size, symmetry, and nature of interactions existing between the cation and anion of the IL [26]. The interactions like interionic interactions, van der Waals forces, and hydrogen bonding are responsible for the inherent viscosity of the ILs or any system, for instance [19]. This parameter is highly dependent upon the working temperature and is well explained by the Arrhenius relation excluding the behavior wherein the IL approaches the melting point. Generally, these molten salts (ILs) are quite viscous at reduced temperatures due to the high interionic interactions.

The low viscosity of electrolyte will have higher ionic conductivity, and so better Li-ions transport through the electrolyte during the charge-discharge process. The viscosity of these imidazolium-based ILs depends upon the nature of substituents on imidazolium core and the interactions between the cation and anion. In the present scenario, all three ILs possess imidazolium cation having alkoxy and siloxy side chains on both nitrogen and TFSI as the anion and their differential interaction is reflected on the observed viscosity and in turn conductivity. Therefore, the substituents were designed to possess two CH_2 spacers to minimize the effect of electron-withdrawing by an alkoxy group. While the siloxy group was introduced to study its electron donating effect whereby the positive charge of the core is partially balanced and decreases its interaction with the anion. Screening of viscosities of the three designed ILs Im-1o2o2-H-2o1-TFSI (**1**), Im-1o2o2-H-1SioSi-TFSI (**2**), and Im-1o2-H-1SioSi-TFSI (**3**) is detailed in Table 11.1 for a wide temperature range (10–100 °C). The viscosities of both Im-1o2o2-H-2o1-TFSI (**1**) and Im-1o2o2-H-1SioSi-TFSI (**2**) ILs were found to be quite close, whereas that for the Im-1o2-H-1SioSi-TFSI (**3**) IL was on the higher

Table 11.1 Dynamic viscosity (η/mPa·s) at varying temperatures for Im-1o2o2-H-2o1-TFSI (**1**), Im-1o2o2-H-1SioSi-TFSI (**2**), and Im-1o2-H-1SioSi-TFSI (**3**) ILs. (Ref. [7], with permission from RSC)

	Dynamic viscosity (η/mPa.s)		
Temp. (°C)	Im-1o2o2-H-2o1-TFSI	Im-1o2o2-H-1SioSi-TFSI	Im-1o2-H-1SioSi-TFSI
10	174.1	174.2	194.7
20	94.0	96.9	106.5
25	71.9	74.7	81.9
30	56.3	59.2	64.3
40	36.4	39.2	41.3
50	25.1	27.3	28.6
60	18.3	20.0	20.8
70	13.8	15.2	15.7
80	10.8	11.9	12.2
90	8.7	9.5	9.8
100	7.1	7.6	7.9

end. This could be due to the high molecular mass of the Im-1o2-H-1SioSi-TFSI (**3**) IL without the additional benefit from molecular flexion from single ether substituent versus the two ether groups. Hence, IL (**3**) has not considered for further study as a battery electrolyte.

The basic property of redox stability especially in the working potential range of the battery is well described by voltammetric studies for both the designed ether-ether (**1**) and ether-siloxane and (**2**) ILs. In addition, the process of rapid SEI layer formation over the reactive anode was addressed by incorporating an additive, i.e., propylene carbonate (PC) in the IL electrolyte containing lithium bis9trifluoromethanesulfonimidate (LiTFSI) as the source of Li-ion. This stable passivating film protects further degradation of both the anode and electrolyte while allowing diffusion of only Li-ions for the continuous charge-discharge process. Electrochemical stability of the IL (**1** and **2**) was studied by linear sweep voltammetric (LSV) study in the potential range between 3.5 to −5.0 V. The LSV for the two variants of electrolytes being studied, viz., ether-ether (**1**) and ether-siloxane (**2**) as neat, IL with PC (48:52) and IL + PC + Li (37:41:22 *c:c:c*), is shown in Fig. 11.1. The electrochemical stability window (ESW) is the intersection point of the mean current with respect to (*w.r.t.*) voltage for both the oxidative and reductive sections giving anodic and cathodic limits derived from the LSVs.

The potential range between the anodic and the cathodic limits signifies the ESW and is 5.0 V for neat ether-ether IL, whereas it is 5.5 V for ether-siloxane IL, respectively. The high potential stability window for neat ether-siloxane could be attributed to the presence of *d*-orbitals in Si present as the cationic substituent. However, upon addition of the PC, the ESW in ether-siloxane increased slightly to 5.7 V, whereas that for an ether-ether slight decrease to 4.8 V was observed. However, upon addition of LiTFSI into these PC and IL mixture, a noticeable increase in the ESW for ether-ether IL from 5.0 to 5.9 V was observed, whereas

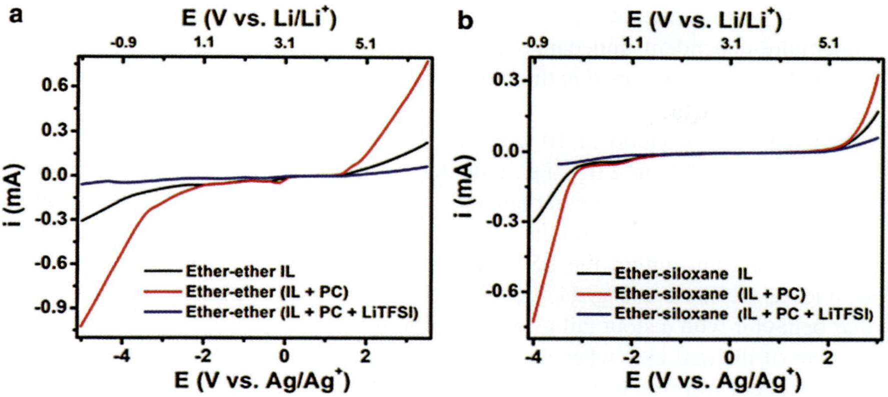

Fig. 11.1 Linear sweep voltammograms of both (**a**) ether-ether IL and (**b**) ether-siloxane IL-based electrolyte as neat IL, (IL + PC) and (IL + PC + LiTFSI) mixtures at a sweep rate of 50 mV/s and 25 °C. WE, glassy carbon (GC); CE, graphite rod; RE, Ag/Ag+/PC. (Reproduced from Ref. [8] with permission from the PCCP Owner Societies)

Table 11.2 Electrochemical stability window (ESW) at various temperatures

	Electrochemical stability window (V)				
Electrolytes	10 °C	25 °C	40 °C	50 °C	60 °C
Ether-ether IL	4.7	5.0	5.0	4.9	4.7
Ether-ether IL + PC	4.8	4.8	4.6	4.6	4.7
Ether-ether IL + PC + LiTFSI	5.6	5.9	5.7	5.6	5.4

ESW decreased drastically from 5.5 to 4.2 V for ether-siloxane IL. These striking observations could be due to the aggregation being more prevalent for ether-siloxane IL which upon Li-salt addition forms aggregates with Li-ion while exposing the organic cation making it more susceptible to potential-dependent degradation. These observations were further validated from electrochemical conductivity analysis. Considering the profound electrochemical stability of ether-ether IL, temperature-dependent ESW analysis was studied as given in Table 11.2. Upon increasing the temperature from 10 °C to 50 °C, the window remained nearly 5 V for neat with a slight decrease at 60 °C, whereas upon addition of PC, the ESW decreased compared to the neat IL across all temperatures; however, upon further addition of LiTFSI, the stability window increased for all temperatures and decreased slightly at elevated temperatures. Further, electrochemical cyclability test for the solvent mixture having ether-ether IL + PC + LiTFSI showed a stable behavior for 50 cycles.

Apart from the potentiodynamic analysis, another important criterion for an electrolyte to furnish in the battery is the transportation of Li-ions across the electrodes during the charge-discharge process. This depends on the intercalation-de-intercalation phenomena, and the speed of Li-ions shuttling between the electrodes; faster the Li-ions shuttling better is the battery response. In order to assess this feature, the conductivity of the two functionalized ILs was studied by temperature-dependent impedance spectroscopic under argon atmosphere. Impedance analysis was performed in the frequency range of 0.1 Hz to 100 kHz stepped-up logarithmically using a suitable direct current (DC) potential over an alternating current (AC) perturbation of 10 mV. The potential-dependent impedance analysis ranging from −3.0 V to 2.0 V, i.e., potentials within the ESW, was studied initially for both the ether-ether and ether-siloxane IL electrolyte comprising of the additive PC and LiTFSI which forms the mixture IL + PC + LiTFSI. As expected, this potential variation within the ESW did not have any effect upon the impedance behavior as depicted in Fig. 11.2a, b. The resultant Nyquist plot showed a semicircular behavior with a stout tail giving solution resistance (R_s) as the high-frequency intercept of the real axis whereas polarization resistance (R_p) as the low-frequency intercept and a Warburg diffusion feature at low frequency in the presence of LiTFSI owing to the diffusion of the Li-ion. Since the potential in the entire range within ESW had no role in the impedance behavior, further studies at a fixed potential of −2 V were assessed by varying the temperatures from 10 °C to 60 °C for the (IL + PC + LiTFSI) electrolyte mixtures of both the ILs. The features observed remained the same across the temperature range with the only striking observation being a reduction in the diameter of the semicircle with increasing temperature for both the ILs but with varying extent. This observation was materialized by computing the charge transfer resistance ($R_{ct,\ \Omega}$) which is obtained from the difference between the R_p and R_s and is the diameter of the observed semicircle in the Nyquist plot.

The R_{ct} in conjunction with the cell constant of the electrochemical cell used yielded the conductivity offered by the electrolyte and is given in Table 11.3. As evident from Fig. 11.2c, d, the conductivity of ether-ether IL is almost twice that of the ether-siloxane IL at room temperature and tends to increase significantly upon addition of the organic additive (PC) to almost three times for both the ILs, but the conductivity decreased drastically upon addition of LiTFSI salt (1 mol/kg). The conductivity of the ether-ether IL in the IL + PC + LiTFSI mixture as a battery electrolyte was found to be 2.2 mS/cm whereas merely 0.5 mS/cm for ether-siloxane-based electrolyte compared to their neat IL being 4.6 and 2.5 mS/cm, respectively. These observations are attributed to the solvation of LiTFSI forming clusters leading to slower Li-ion diffusion across the electrodes. However, as expected the conductivity increased with increase in temperature for the different compositions due to thermal convection which decreases the viscosity and enhances the Li-ion shuttling across the electrodes (detailed in an upcoming section). The detailed impedance analysis suggested that potential is insignificant toward the impedance irrespective of the electrolyte system within the ESW. Furthermore, the ether-ether IL provides better conductivity toward Li-ions even in a very high LiTFSI concentration of 1 mol/kg solvent.

The electrochemical conductivity measurements were also supported by viscosity analysis since the viscosity is a key element for the conductivity and is inversely proportional to the conductivity of the various electrolytes (Fig. 11.2e–f). Viscosity studies in the temperature ranging from 10 °C to 100 °C at a heating rate of 5 °C/min

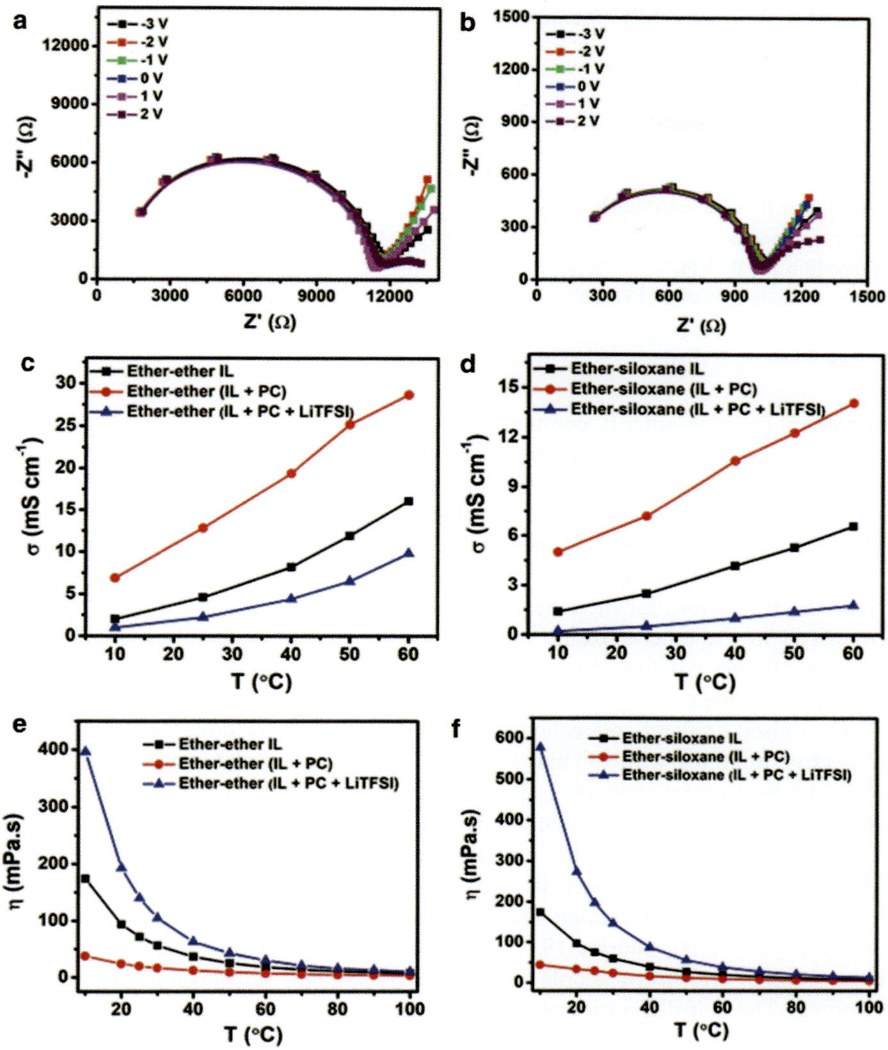

Fig. 11.2 Impedance response of both (**a**) ether-ether IL + PC + LiTFSI and (**b**) ether-siloxane IL + PC + LiTFSI electrolyte at 25 °C. Temperature-dependent conductivity (**c** and **d**) derived from impedance measurement and viscosity (**e** and **f**) for ether-ether and ether-siloxane ionic liquid-based electrolyte. Working electrode (WE), glassy carbon (GC); counter electrode (CE), graphite rod; reference electrode (RE), $Ag/Ag^{+}/PC$. (Reproduced from Ref. [8] with permission from the PCCP Owner Societies)

under inert atmosphere are detailed in Table 11.4 wherein both the neat ILs possess nearly the same viscosity which decreases upon increasing the temperature with ether-ether IL being slightly less viscous. When these ILs are in combination with the organic additive such as PC, the viscosity decreases drastically which is quite

Table 11.3 The conductivity of various electrolytes extracted from impedance analysis at different temperatures

	Conductivity (mS/cm)				
Electrolytes	10 °C	25 °C	40 °C	50 °C	60 °C
Ether-ether IL[a]	2.0	4.6	8.2	11.9	16.1
Ether-ether IL + PC[a]	6.9	12.8	19.4	25.2	28.7
Ether-ether IL + PC + LiTFSI[a]	1.0	2.2	4.4	6.5	9.8
Ether-siloxane IL[b]	1.4	2.5	4.2	5.3	6.6
Ether-siloxane IL + PC[b]	5.0	7.2	10.6	12.3	14.1
Ether-siloxane IL + PC + LiTFSI[b]	0.2	0.5	1.0	1.4	1.8

[a]Cell constant = 25.1 cm^{-1}
[b]Cell constant = 7.07 cm^{-1}

prominent for ether-ether IL compared to ether-siloxane IL. Finally, upon addition of LiTFSI, the viscosity of the mixture increased manifolds, i.e., 197 mPa.s for ether-siloxane IL + PC + LiTFSI whereas only 140 mPa.s for ether-ether IL + PC + LiTFSI compared to the viscosity of neat ILs which were 74 and 71 mPa.s, respectively at 25 °C. However, as thermal convection sets in with increasing temperature, the viscosity drops down significantly. Considering the viscosity of these two electrolytes, the IL mixture for the ether-ether variant exhibits significantly less viscosity compared to the ether-siloxane electrolyte mixture and is therefore expected to possess a higher conductivity which matches with the impedance-derived conductivity measurement.

The pulsed gradient spin echo (PGSE) NMR analysis for both the ILs and their respective mixtures with PC and LiTFSI (1 mol/kg electrolyte) was measured at 25 °C under an inert atmosphere to determine the self-diffusion coefficients of the various components. Calibration of the self-diffusion coefficient was standardized against the standard 2.27×10^{-9} m^2 s^{-1} value [27] for water (H_2O) at 25 °C. The determination of self-diffusion coefficients of the various electrolyte components, namely, the imidazolium cation and Li-ion and TFSI anion, was made by analyzing the hydrogen-1 (^1H), fluorine-19 (^{19}F), and lithium-7 (^7Li) NMR and by the careful elimination of error from sample motion by stagnant NMR analysis without sample spinning. The self-diffusion coefficients were determined by computing the echo signal attenuation (E), as follows [47]:

$$\ln(E) = \ln\left\{\frac{S}{S_{g-o}}\right\} = -\gamma^2 g^2 D\delta^2 \left\{\Delta - \frac{\delta}{3}\right\}$$

where D is the diffusion coefficient (m^2/s), S is the spin echo signal intensity (T), δ is the duration of the field gradient with magnitude g *(T/m)*, γ is the gyromagnetic ratio (rad T^{-1} s^{-1}), and Δ is the interval between the two gradient pulses (ms). The self-diffusion coefficients for imidazolium cation was 4.16×10^{-11} m^2 s^{-1} for ether-ether IL whereas only 3.76×10^{-11} m^2 s^{-1} for ether-siloxane IL. Similarly, the TFSI anion (by ^{19}F NMR) was 4.52×10^{-11} for the former and 4.49×10^{-11} m^2 s^{-1} for the latter IL. The ability of the designed ILs to conduct Li-ion was further hinted by

Table 11.4 Dynamic viscosity (η/mPa·s) at varying temperatures for ether-ether and ether-siloxane ILs

	Neat IL		IL + PC		IL + PC + LiTFSI	
Temp. (°C)	Ether-ether	Ether-siloxane	Ether-ether	Ether-siloxane	Ether-ether	Ether-siloxane
10	174.1	174.3	37.5	43.6	396.0	577.3
20	94.0	97.0	23.8	33.2	192.7	272.9
25	72.0	74.7	19.5	28.7	140.0	196.8
30	56.4	59.2	16.3	23.2	105.1	146.1
40	36.5	39.2	11.8	16.1	62.91	87.2
50	25.2	27.3	8.9	12.0	42.6	55.9
60	18.3	20.0	6.9	9.2	30.0	38.4
70	13.9	15.2	5.4	7.3	20.6	27.8
80	10.8	11.9	4.4	5.9	15.5	20.9
90	8.7	9.6	3.6	4.9	12.9	15.8
100	7.1	7.7	3.1	4.2	9.67	12.5

the diffusion coefficient determination upon addition of LiTFSI to the ILs which visibly forms more viscous solution having lower fluidity. Under such circumstances, ether-ether IL + LiTFSI solution exhibited a greatly reduced self-diffusion coefficient of 0.32×10^{-11} m^2 s^{-1} for imidazolium cation (by ^{1}H NMR), 0.96×10^{-11} m^2 s^{-1} for TFSI (by ^{19}F NMR), and 0.08×10^{-11} m^2 s^{-1} for Li^+ (by ^{7}Li), whereas that for ether-siloxane IL + LiTFSI, it was 0.31×10^{-11} m^2 s^{-1}, 0.56×10^{-11} m^2 s^{-1}, and 0.11×10^{-11} m^2 s^{-1}, respectively. However, upon further addition of the additive (PC), i.e., in the solution of IL + PC + LiTFSI for ether-ether IL, the diffusion coefficient increased compared to the IL + TFSI solution giving 2.12×10^{-11} m^2 s^{-1} for imidazolium cation, 1.14×10^{-11} m^2 s^{-1} for TFSI, and 0.52×10^{-11} m^2 s^{-1} for Li^+, but for ether-siloxane it was only 1.98×10^{-11} m^2 s^{-1}, 1.89×10^{-11} m^2 s^{-1}, and 0.25×10^{-11} m^2 s^{-1}, respectively. These observations for self-diffusion coefficient measurements indicate that neat ether-ether IL exhibits higher ability of ionic diffusion or mobility compared to the ether-siloxane IL and upon addition of LiTFSI to these ILs the diffusion coefficient of imidazolium cation reduces by a factor of 13, whereas for TFSI anion it reduces almost four times for both the ILs. Thereby, the addition of very large amount of Li-salt (1 mol/kg) in the electrolyte causes highly reduced mobility of the Li-ions as indicated by their very low diffusion coefficients which are in coherence with the conductivity as well as viscosity measurements. On comparing the total diffusion (D_{sum}), the decrease on addition of LiTFSI is less pronounced, i.e., only six times for ether-ether IL, whereas it is around eight times for ether-siloxane IL.

These observations point toward a higher degree of aggregation prevalent in the ether-siloxane system forming highly solvated Li-ion clusters having a bigger effective Li^+ radius. Whereas for the ether-ether IL, the ether substituent provides more flexibility and the affinity of the ether moiety toward Li-ions which helps to collapse the formed $Li[TFSI]_{n+1}^{-n}$ clusters and discourage the formation of TFSI solvated Li^+ clusters. However, upon addition of PC, the diffusion of the overall electrolyte increased, but the relative increase was found to be the highest for Li-ions, thereby facilitating their movement during the charge-discharge process in battery. These diffusion coefficients are in-line with the conductivity measurements. The electrolyte system is composed of ether-ether IL + PC + LiTFSI possessing the flexible ether functionality with an affinity toward Li-ions and thereby improving the mobility as well as conductivity by incorporating more Li-ion at both of their side chains compared to the ether-siloxane IL-based electrolyte.

These findings suggest that the ether groups over the imidazolium core play an important role in preventing the self-aggregation of ions. The addition of a high concentration of LiTFSI (1 mol/kg solvent) increases the content of TFSI anion which forms a big solvation shell around the Li^+ ion decreasing its mobility and in turn conductivity, while the viscosity tends to shoot up. Another contributing factor from the formation of solvated $Li[\mathrm{TFSI}]_{n+1}^{n-}$ species is the ionic migration being governed by both the concentration as well as potential gradient owing to its overall negative charge. These propositions were testified further by IR studies, initially monitoring the variation of carbonyl (C=O) stretching frequency of PC (1786 cm^{-1}) which increased upon the addition of either of the ILs (1788.0 cm^{-1} for E-E IL and 1791.8 cm^{-1} for E-S IL) due to PC aggregate dispersion and again further decreased to 1786 cm^{-1} upon Li-ion addition in either of the ILs as expected due to reaggregation. The interaction between the cation and the TFSI anion was subsequently studied by monitoring the S-N-S stretching frequency which was 788.5 cm^{-1} for ether-ether whereas 789.0 cm^{-1} for ether-siloxane IL, respectively. This frequency decreased to 777.0 cm^{-1} for both the ILs upon inclusion of the additive (PC) which is indicative of a weak interaction between TFSI and imidazolium cation. Further addition of LiTFSI to the neat ether-ether IL increased the S-N-S stretching frequency from 788.5 cm^{-1} to 791.5 cm^{-1} which can be attributed to the solvation of Li-ions by TFSI. The effect is more pronounced in the case of ether-siloxane (789.0 cm^{-1} to 794.0 cm^{-1}) when more TFSI are involved in clusters formation as revealed from viscosity and conductivity studies.

The various studies for both the ether-ether (E-E) IL and ether-siloxane (E-S) IL mixture electrolytes in different aspects like potential stability, dynamic impedance-derived conductivity, viscosity, and current resistance (IR) measurements point toward E-E + PC + LiTFSI electrolyte to be the most promising out of the two as a potential battery electrolyte. Since the safety of battery even under thermal runaway is of utmost importance, therefore, the thermal stability of electrolyte was analyzed for both E-E and E-S IL, IL + LiTFSI, IL + PC, and IL + PC + LiTFSI by thermogravimetric analysis (TGA) and is shown in Fig. 11.3a–c. Both the neat IL and IL + LiTFSI undergo maximal weight loss (ca. 90%) between 350 °C and 530 °C, but when PC is the part of the electrolyte (i.e., in IL + PC and IL + PC + LiTFSI), weight loss occurred in two major steps, i.e., nearly 15% up to 245 °C which could be accounted for the additive, PC, and the major 75% from 350 °C to 530 °C due to the IL dissociation. An interesting observation is that the electrolyte is highly stable having a high thermal decomposition temperature (T_{start}) of 150 °C for E-E IL and 130 °C for E-S IL while being non-flammable catering to the safety concern. Applicability at the wider temperature range determines the

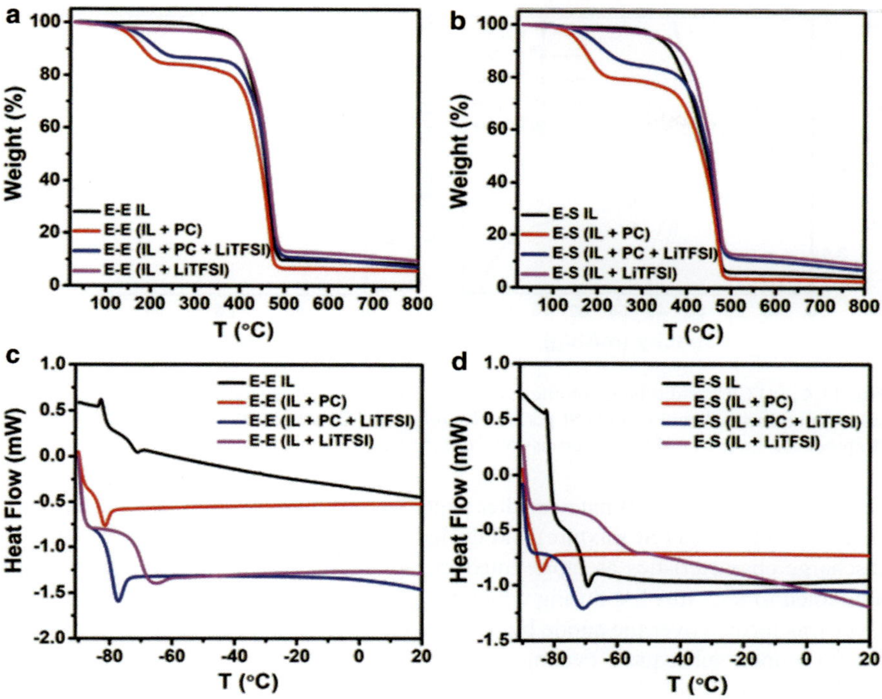

Fig. 11.3 (**a** and **c**) TGA and (**b** and **d**) differential scanning calorimetry (DSC) analysis of ether-ether IL and ether-siloxane IL along with their respective composites. (Reproduced from Ref. [8] with permission from the PCCP Owner Societies)

safety, but worldwide applicability also demands low crystallization temperature considering weather across the globe. This lower limit applicability of an electrolyte was effectively tested by performing differential scanning calorimetry (DSC) measurements in the temperature range of −90 to 20 °C at 5 °C/min. and is shown in Fig. 11.3b, d. The DSC studies did not reveal any crystallization; rather a glass transition (T_g) at −74 °C was observed for neat E-E IL and − 72 °C for E-S IL which remains nearly the same even after addition of LiTFSI but further reduces to −84 °C and − 84 °C when PC was added to E-E and E-S IL, respectively. In case of electrolyte mixture, viz., IL + PC + LiTFSI for E-E, T_g increased slightly to −81 °C and − 77 °C for E-S mixture but remained lower than the neat IL owing to its high flexibility. The maximization of the temperature range between the T_g and T_{start} (i.e., from −81 °C to 150 °C) allows the potential use of E-E IL mixture (with PC and LiTFSI) as an electrolyte in Li-ion batteries as it is functional in a wide temperature range and is also resistive toward thermal decomposition and runaway.

Following the detailed analysis of the E-E IL-based electrolyte features toward favorable application, the E-E IL was further assessed for battery performance analysis in a Swagelok cell as a proof of concept presented by the study. The compiled cell consisted of a graphite anode and lithium cobalt(III) oxide ($LiCoO_2$)

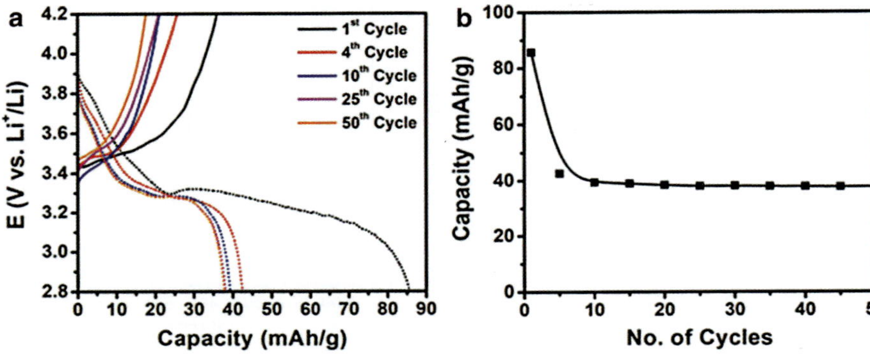

Fig. 11.4 (**a**) Charge-discharge profile and (**b**) cycling discharge capacity for ether-ether IL + PC solvent containing 1 mol of LiTFSI per kg electrolyte at 1C rate *w.r.t* the cathode for Li-ion cell. (Reproduced from Ref. [8] with permission from the PCCP Owner Societies)

cathode separated by a Whatman filter paper soaked with the E-E-based electrolyte (E-E IL + PC + LiTFSI mixture) and cycled galvanostatically to obtain the charge-discharge characteristics as represented in Fig. 11.4a. The first discharge capacity amounted to 85.7 mA h/g during which the stable solid electrolyte interphase (SEI) layer was formed over the anode by accelerated reaction with the propylene carbonate (PC) upon subsequent cycling and the capacity declines to 39.4 mA h/g after tenth cycle and finally became steady at 37.9 mA h/g after 50 cycles. This capacity variation with a number of cycles is shown in Fig. 11.4b which is fairly stable by the end of the 50th cycle and projects the final applicability of the designed appropriately functionalized IL system in Li-ion batteries.

This charge-discharge analysis clearly projects the formation of a stable SEI layer following the first few cycles which has subsequently yielded a stable cycling performance. The formation of this stable SEI layer is a key aspect for high-energy electrodes to constantly attain reproducible and long charge-discharge cycling. Thereby, understanding the composition and mechanism of its formation is of utmost necessity to achieve a stable and long-lived battery cycling. Various spectroscopic studies toward analyzing the SEI composition [54] have suggested that the SEI is essentially composed of the remnants from the active reaction between the reactive electrode and the electrolyte and/or additive by large forming carbonates (lithium carbonate, lithium alkyl carbonate, etc.), lithium oxides (Li_2O), alkoxides ($LiOCH_3$), and fluorides (LiF) along with salts [11].

11.3 C2-Substitution on Imidazolium Ionic Liquids Toward Improved Stability and Performance

Presently, the portable energy market is dominated by Li-ion batteries; however their energy density is insufficient to support highly demanding applications like electric vehicles and smart grid power supply [44]. Besides limited specific capacity, another appealing challenge associated with these future high-energy batteries is their safety while providing high voltages especially because of electrolyte limitations [49]. An electrolyte which can sustain a wide range of potential application without hampering the battery performance is urgently required [21]. This section details an approach to modify the ILs toward application in high-energy batteries research. Majority of the above discussion details the role of functionalizing **1** and **3** positions of the imidazolium cation core. Varying the functional groups attached at **1** and **3** positions significantly affected the physicochemical properties of the ILs and their applicability as an electrolyte in Li-ion batteries. However, another important aspect was the acidic proton at the C-2 position of these imidazolium-derived ILs. The C-2 proton can be easily abstracted, and the deprotonated imidazolium core forms carbene (R = C:) which is highly reactive especially in contact with Li-metal anode undergoing side reactions [25]. These, in turn, cause detrimental effect over the battery cell electrode hampering its longevity. Substitution at the C-2 position will prevent generation of reactive carbenes. However, the replacement with an electron-withdrawing group leads to charge localization over the cation which further forms a strong cation-anion pair and hence increases the viscosity by several folds [30]. On the other hand, attachment of an electron-donating substituent causes a decreased charge which is delocalized over the cation, thereby weakening the ion pairing, and hence the increase in viscosity is comparatively lesser than the electron-withdrawing substituent [46]. Although C-2 substitution generally leads to a better thermal stability along with an enhanced ESW, the impact on viscosity is substantial. This fact was further detailed by Seki group [46] in their attempt to design the IL toward the improvement of both safety and performance by introducing an electron-donating substituent at the C-2 position.

It was observed that C-2 substituted 1,3-dialkyl imidazolium [similar to $C_8H_{14}N_2O_2$] IL shows a good ionic conductivity and most importantly it retained discharge characteristics even after 50 charge-discharge cycles as opposed to the unsubstituted 1,3-dialkyl imidazolium cation. Therefore it is evident that the substitution of C-2 position by a stable and nonreactive substituent further improves the overall ESW while retaining the charge-discharge profile to yield a stable cell capacity.

Interestingly, the effect of C-2 substitution on the solvation properties of IL and their molecular dynamics were explored simultaneously from the perspective of the application as green electrolytes. The acidic C-2 hydrogen in unsubstituted variants is considered to provide hydrogen bonding interactions which dominate the ion solvation and cation-anion pairing, eventually affecting their melting point and viscosity. Detailed studies by IR spectroscopy and dynamics from time-resolved Fourier-transform infrared (FTIR) analysis suggest a negligible effect on the degree of interaction exhibited by 2-methyl-imidazolium core having the same anion [10]. Subsequently, the increase in the ordering of C-2 substituted imidazolium IL upon favorable cation-anion interaction was also proved by Nuclear Overhauser effect (NOE) NMR spectroscopy [1]. These experimental observations were then explored by Hunt [28] using ab initio quantum-chemical calculations to study the gas-phase ion pairs for 1-butyl-3-methyl-imidazolium (unsubstituted, $C_9H_{15}N_2$)

versus 1-butyl-2,3-dimethyl-imidazolium (C-2 substituted) chloride [$C_9H_{17}ClN_2$] IL in an attempt to explain the increased melting point (mp) and viscosity for the latter. Although a reduced mp and decreased viscosity would be expected considering the loss of hydrogen bonding upon substitution, the observation is vice versa due to predominant entropy loss and leads to a greater ordering. The similar observation has also been supported experimentally having contribution from increased rotational hindrance from the butyl side chain at 1-position increasing both mp and viscosity. These effects were subsequently revisited by a host of consequent reports employing temperature-dependent FTIR studies [35], molecular dynamics simulations [53], large angle X-ray scattering (LAXS) [33], calorimetry along with Raman spectroscopy [13], combination of FTIR, Raman spectroscopy with NMR spectroscopy [39], rotational dynamic studies from NMR spectroscopy [12], and the like.

Following these reports, the applicability of C-2 substituted ILs was explored as an electrolyte in Li-ion batteries. A study by Liao group [37] detailed the impact of varying the substituent as methyl, cyano, and ether moiety (-$CH_2OCH_2CH_2CH_2CH_3$) which was correlated with the alteration in their lowest unoccupied molecular orbital (LUMO) energy level. The cyano (-CN) group possesses a strong electron-withdrawing ability, thereby lowering LUMO which facilitates reduction ability of the imidazolium cation core decreasing the electrochemical stability window (ESW). However, ether moiety at C-2 along with a suitable substituent at 3 position raises the LUMO, enhanced its reduction stability while exhibiting lowest solid electrolyte interfacial resistance (R_{SEI}), and also allowed for a high Li-ion transference number. In the following year, Jin group [30] reported an entire family of 30 ether-functionalized ILs over a trialkyl imidazolium cation core with either one or two ether groups paired with bis(trifluoromethylsulfonyl) imide ($LiC_2F_6NO_4S_2$, $TFSI^-$) anion. A structure-physicochemical activity relationship was drawn based on their extensive characterizations which suggested that alkylation at C-2 position enhanced the cathodic stability, but lead to an increase in the size of cation allowing for an increase in viscosity. However, the incorporation of an ether substituent at the 1- and/or 3- positions, compensates the effect of C-2 substitution and thereby comparatively reduces the viscosity and mp. An important observation presented in this study was that incorporation of an ether substituent on the cation core having alkyl group at the C-2 position did not impact both the thermal and electrochemical stability.

Insights obtained from these studies could further be improvised upon combining the favorable outcome of our detailed analysis wherein Im-1o2o2-H-2o1-TFSI (**1**) outperformed the siloxy-substituted imidazolium IL. This can be done by introducing a methyl substituent at the C-2 position which could possibly enhance the stability of the IL. However, to rationalize the substitutional effect, appropriately two variants were prepared, namely, an unsymmetrical Im-1o2o2-1-2o1-TFSI (**4**) IL and another symmetrical Im-1o2o2-1-2o2o1-TFSI (**5**) IL *w.r.t.* the ether substitution as shown in Scheme 11.2 and compared with IL (**1**).

The impact of C2 methyl on viscosity is expected to increase in these two ILs, which was studied by temperature-dependent (10–100 °C) dynamic viscosity

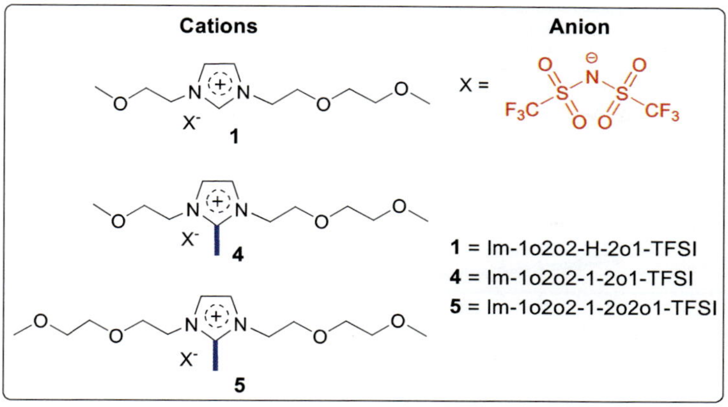

Scheme 11.2 C-2 functionalized ionic liquids studied toward electrolyte application

Table 11.5 Dynamic viscosity (η/mPa·s) at varying temperatures for Im-1o2o2-H-2o1-TFSI IL (**1**), Im-1o2o2-1-2o1-TFSI IL (**4**), and Im-1o2o2-1-2o2o1-TFSI IL (**5**) along with their blended mixtures with PC and both PC + LiTFSI

	Dynamic Viscosity (η/mPa.s)					
Temp. (°C)	IL (**1**)	(**1**) + PC + LiTSI	IL (**4**)	(**4**) + PC + LiTSI	IL (**5**)	(**5**) + PC + LiTSI
25	71.9	140.0	98.7	183.0	135.0	308.8
30	56.3	105.1	79.2	135.2	109.0	218.1
40	36.4	62.9	56.6	74.6	67.5	126.1
60	18.3	30.0	27.3	37.8	31.8	51.7
80	10.8	15.5	14.6	20.5	18.1	29.6
100	7.1	9.6	8.6	12.1	11.6	17.7

measurements under an inert atmosphere. The analysis further extended to study the variation upon blending with PC and also PC + LiTFSI and detailed in Table 11.5 and depicted in Fig. 11.5a, b.

The viscosity for neat ILs at 25 °C shows that unsubstituted IL (**1**) possesses the least viscosity among the three ILs tested. This is in accordance with the previous discussion as the C-2 substitution is known to increase the viscosity of ILs. Another important observation is that the symmetrical IL (**5**) with the higher molecular weight has a viscosity of 135 mPa.s at 25 °C, whereas the unsymmetrical IL (**4**) is ca. 1.4 times less viscous (98.7 mPa.s). This is attributed to the effect on tight ion pairing besides the molecular weight, where the unsymmetrical variant is imparting more flexibility and therefore lower extent of ordering. Further, the viscosity decreased considerably for all these ILs when blended with PC due to the reduced interionic interaction allowing better mobility. However, upon addition of (1 mol/kg) LiTFSI, an increase in viscosity by a factor of two was observed in ILs + PC + LiTFSI. This could be explained on the basis of ion aggregation limiting the overall ionic

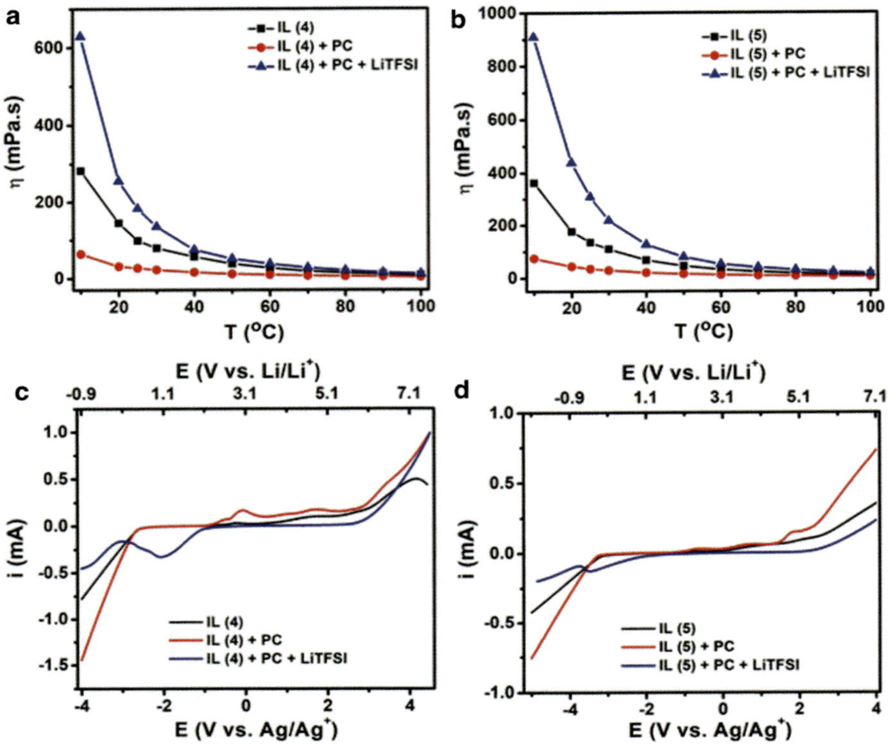

Fig. 11.5 Temperature-dependent viscosity and linear sweep voltammograms at a sweep rate of 50 mV/s and 25 °C for (**a** and **c**) Im-1o2o2-1-2o1-TFSI (**4**) and (**b** and **d**) Im-1o2o2-1-2o2o1-TFSI (**5**) as neat IL, (IL + PC) and (IL + PC + LiTFSI) mixtures WE, glassy carbon (GC); CE, graphite rod; RE, $Ag/Ag^+/PC$

mobility. Further insight on ion aggregation can be possible by the study of self-diffusion of different ions by pulsed field gradient (PFG) NMR and different in situ spectroscopy measurements. The self-diffusion coefficient of Li-ion in IL + PC + LiTFSI was found to be 5.82×10^{-11} m^2 s^{-1} for IL (**4**) two times higher than for IL (**5**) and ten times better than IL (**1**). The conductivity of ILs **4** and **5** was subsequently estimated by EIS studies similar to the previous set of ILs (**1**, **2**, and **3**). The unsymmetrical IL **4** exhibited a strikingly high conductivity of 13.5 mS cm^{-1} at 25 °C which is ca. three times higher than IL **1** (4.6 mS cm^{-1}) and 1.2 times higher than **5** (10.5 mS cm^{-1}). However, the addition of PC to these ILs resulted in a drastic increase in conductivity to ca. 24 mS cm^{-1} for IL **5** and 31 for **4**; however it was only 12.8 mS cm^{-1} for IL **1**. Further, the IL + PC + LiTFSI mixture followed the trend **4** (8.7 mS cm^{-1}) > **5** (5.3 mS cm^{-1}) > **1** (2.2 mS cm^{-1}). This trend of conductivity was found to support the viscosity and self-diffusion studies, and it is attributed that the effect of C-2 substitution on these imidazolium ILs is compensated by introducing ether functionality at the 1 and 3 positions of trialkyl imidazolium cation core.

Although C-2 substitution leads to increased viscosity whereas conductivity is enhanced but the primary aim is the electrochemical stability enhancement, especially the reduction stability. The ESW was studied by linear sweep voltammetric measurements. Figure 11.5c, d depicts this response for all the three ILs, and their respective blends in the potential range 4 V to −5 V vs. $Ag/Ag^{+}/PC$ at a scan rate of 50 mV s^{-1}. Similar to the previous study [8] detailed in Sect. 2, the ESW for neat IL (**1**), IL (**4**), and IL (**5**) along with its respective blends initially with PC only and subsequently with PC + LiTFSI is detailed in Table 11.6.

Strikingly, the ESW at 25 °C was enhanced by 900 mV for neat IL (**5**) and by 700 mV for IL (**4**) as compared to IL (**1**) which gives an experimental evidence for the increased stability upon C-2 substitution. This provides a physical correlation of the effect of substituting the acidic C-2 proton signifying that it otherwise involves in various side reactions hampering the cell's output and cycling stability. Addition of PC to IL (**1**) led to decrease in ESW by 200 mV, but contrastingly, it was found to increase drastically (by 500 mV) for IL (**4**) and IL (**5**). Ideally, PC has a narrower ESW which caused the decrease for IL (**1**), but on C-2 methyl substitution, there is a possibility of enhanced interaction between PC and these substituents which eventually increased the ordering and hence ESW. The electrolyte blend resulting from a mixture of both PC and LiTFSI, on the other hand, further increases the ESW for all the three ILs which can be accounted for by the enhanced ion interaction due to Li-ion solvation.

Subsequently, the thermal stability of these ILs was compared to account for C-2 substitution by TGA and DSC as in Fig. 11.6. TGA analysis under an inert atmosphere at a heating rate of 10 °C min^{-1} in the range 30–800 °C yielded only one major step for neat ILs corresponding to a weight loss of ca. 90% in the temperature range between 360 °C and 490 °C (Fig. 11.6a, b). The highest stability was displayed by the unsymmetrical IL (**4**) having a T_{start} of 379.1 °C which is comparable with IL (**5**) and 70 °C higher than the C-2 unsubstituted IL (**1**). For PC + ILs, a two-step weight loss behavior was observed wherein the first one around 110 °C to 250 °C accounted for 15–20% weight loss and a T_{start} between 360 °C and

Table 11.6 Electrochemical stability window (ESW) for Im-1o2o2-H-2o1-TFSI IL (**1**), Im-1o2o2-1-2o1-TFSI IL (**4**), and Im-1o2o2-1-2o2o1-TFSI IL (**5**) along with their blended mixtures with PC and LiTFSI at various temperatures

	Electrochemical stability window (V)		
Electrolytes	25 °C	40 °C	60 °C
Im-1o2o2-H-2o1-TFSI (**1**)	5.0	5.0	4.7
Im-1o2o2-H-2o1-TFSI (**1**) + PC	4.8	4.6	4.7
Im-1o2o2-H-2o1-TFSI (**1**) + PC + LiTFSI	5.9	5.7	5.4
Im-1o2o2-1-2o1-TFSI (**4**)	5.7	5.6	5.5
Im-1o2o2-1-2o1-TFSI (**4**) + PC	6.2	6.1	5.8
Im-1o2o2-1-2o1-TFSI (**4**) + PC + LiTFSI	6.8	6.6	6.3
Im-1o2o2-1-2o2o1-TFSI (**5**)	5.9	5.6	5.2
Im-1o2o2-1-2o2o1-TFSI (**5**) + PC	6.4	6.1	5.9
Im-1o2o2-1-2o2o1-TFSI (**5**) + PC + LiTFSI	7.0	6.8	6.5

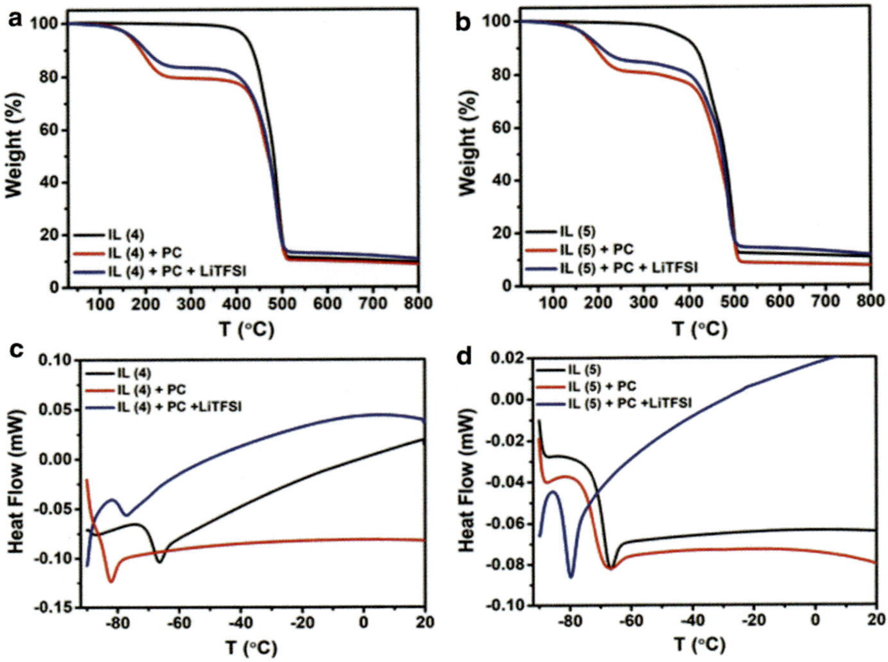

Fig. 11.6 Thermogravimetry analysis (TGA) and differential scanning calorimetry (DSC) analysis of (**a** and **c**) Im-1o2o2-1-2o1-TFSI (**4**) and (**b** and **d**) Im-1o2o2-1-2o2o1-TFSI (**5**) as neat IL, (IL + PC) and (IL + PC + LiTFSI) mixtures

380 °C for second step. IL + PC + LiTFSI also shows weight loss in two steps similar to IL + PC which could be accounted for the loss of organic PC component, but the stability was enhanced upon LiTFSI inclusion increasing the T_{start} for the first step by 20–40 °C, whereas the second step temperature remains unaffected. Additionally, DSC measurements (Fig. 11.6c, d) in the temperature range − 90 to 20 °C (at 5 °C min^{-1}) revealed the lowest glass transition temperature (T_g) for unsubstituted IL (**1**) which was comparable to the unsymmetrical IL (**4**) while being lowest for the symmetrical IL (**5**) [i.e., −74.5 °C > −70.6 °C > −68.2 °C]. This lower working temperature limit was enhanced upon PC addition which further decreased in IL + PC + LiTFSI; however the trend remained the same.

11.4 Conclusion

The safety aspect of the battery technology is of utmost importance irrespective of the obtainable capacity especially when it is composed of a highly flammable component as in the case of Li-ion batteries. The most widely employed electrolyte system comprises of the organic electrolyte mixture which provides high conductivity at the cost of safety owing to their flammable nature which has proven disastrous in the past under stressed conditions causing thermal runaways. The electrolytes composed of the highly stable ionic liquids with a wide liquid range, high thermal stability, and substantial conductivity would be a reasonable alternative to the present thermally unstable electrolytes in practice. The elaborate study of two newly designed imidazolium ILs functionalized with ether-ether and ether-siloxane was analyzed toward each aspect of an electrolyte ranging from the potential stability, conductivity, viscosity, thermal stability, and electrolyte range to battery testing. The electrolyte system was also modified with 55 mol % of the organic solvent, viz., propylene carbonate as an additive to accelerate the SEI layer formation and stabilization along with 1 mol LiTFSI/kg solvent as the source of Li-ion. Preliminary testing of both the IL systems suggested E-E-based electrolyte possesses high stability both thermal and electrochemical with heightened conductivity toward Li-ions even in the presence of high Li-ion content. High LiTFSI concentration causes severe issues by aggregated cluster formation in the form of $Li[\mathrm{TFSI}]_{n+1}^{n-}$ species which hamper the mobility of Li-ions during the charge-discharge process. The ether-ether substituent on the imidazolium core of the cation bestows side chain flexibility reducing the ion symmetry resulting into a diffused cation which upon teaming up with a large diffused anion, namely, TFSI, interacts electrostatically overcoming the tendency of cluster formation. Besides the ether functionality aids in holding more Li-ions for its shuttling safely during the charge-discharge process proving to be an appropriate and potential electrolyte in Li-ion batteries. Further, the stability issue resulting in limited cycling was addressed by eliminating the acidic C-2 proton by methyl. This C-2 methyl imidazolium IL with ether functionalities unsymmetrically substituted at 1 and 3 positions provided a suitable combination to be explored further for a probable Li-ion battery electrolyte application. Functionalized ionic liquids therefore hold a unique potential of wide fluidity and stability which lends a new direction toward fine-tuning its cationic and anionic components to achieve a safe electrolyte for Li-ion batteries.

Acknowledgments This research is supported by the Department of Atomic Energy (DAE), India (2013/37C/57/BRNS). Dr. Tharamani C. Nagaiah thanks the Department of Science and Technology (DST) for the Ramanujan Fellowship (SR/S2/RJN-26/2012). Aarti Tiwari thanks IIT Ropar for Fellowship.

The authors thank the editors in allowing us to extend our previously published work [7] with new data from [8] based on our research and other cited works in the field.

References

1. M. Andrea, R. Greta, G. Matteo, R. Enzio, F. Giovanni, R. Guido, M. Valentina, The Local structure of ionic liquids: Cation–cation NOE interactions and internuclear distances in neat [BMIM][BF4] and [BDMIM][BF4]. Angew. Chem. Int. Ed. **45**(7), 1123–1126 (2006). https://doi.org/10.1002/anie.200503745
2. M. Armand, F. Endres, D.R. MacFarlane, H. Ohno, B. Scrosati, Ionic-liquid materials for the electrochemical challenges of the future. Nat. Mater. **8**(8), 621–629 (2009)
3. M. Armand, J.M. Tarascon, Building better batteries. Nature **451**(7179), 652–657 (2008)

4. M.B. Armand, F.E.K.C.E. Moursli, Bis perhalogenoacyl -or sulfonyl-imides of alkali metals, their solid solutions with plastic materials and their use to the constitution of conductor elements for electrochemical generators: Google Patents (1985)
5. D. Baril, C. Michot, M. Armand, Electrochemistry of liquids vs. solids: Polymer electrolytes. Solid State Ionics **94**(1), 35–47 (1997). https://doi.org/10.1016/S0167-2738(96)00614-5
6. F. Capuano, F. Croce, B. Scrosati, Composite polymer electrolytes. J. Electrochem. Soc. **138**(7), 1918–1922 (1991). https://doi.org/10.1149/1.2085900
7. S.N. Chavan, D. Mandal, The combined effect of ether and siloxane substituents on imidazolium ionic liquids. RSC Adv. **5**(80), 64821–64831 (2015). https://doi.org/10.1039/C5RA07466E
8. S.N. Chavan, A. Tiwari, T.C. Nagaiah, D. Mandal, Ether and siloxane functionalized ionic liquids and their mixtures as an electrolyte for lithium-ion batteries. Phys. Chem. Chem. Phys. **18**(24), 16116–16126 (2016). https://doi.org/10.1039/C6CP01519K
9. F. Croce, G.B. Appetecchi, L. Persi, B. Scrosati, Nanocomposite polymer electrolytes for lithium batteries. Nature **394**(6692), 456–458 (1998)
10. K. Dahl, G.M. Sando, D.M. Fox, T.E. Sutto, J.C. Owrutsky, Vibrational spectroscopy and dynamics of small anions in ionic liquid solutions. J. Chem. Phys. **123**(8), 084504 (2005). https://doi.org/10.1063/1.2000229
11. Y. Ein-Eli, S.F. McDevitt, D. Aurbach, B. Markovsky, A. Schechter, Methyl propyl carbonate: A promising single solvent for Li-Ion battery electrolytes. J. Electrochem. Soc. **144**(7), L180–L184 (1997). https://doi.org/10.1149/1.1837792
12. T. Endo, M. Imanari, H. Seki, K. Nishikawa, Effects of methylation at position 2 of cation ring on rotational dynamics of imidazolium-based ionic liquids investigated by NMR Spectroscopy: [C4mim]Br vs [C4C1mim]Br. Chem. A Eur. J. **115**(14), 2999–3005 (2011). https://doi.org/10.1021/jp200635h
13. T. Endo, T. Kato, K. Nishikawa, Effects of methylation at the 2 position of the cation ring on phase behaviors and conformational structures of imidazolium-based ionic liquids. J. Phys. Chem. B **114**(28), 9201–9208 (2010). https://doi.org/10.1021/jp104123v
14. V. Etacheri, R. Marom, R. Elazari, G. Salitra, D. Aurbach, Challenges in the development of advanced Li-ion batteries: a review. Energy Environ. Sci. **4**(9), 3243–3262 (2011). https://doi.org/10.1039/C1EE01598B
15. S. Ferrari, E. Quartarone, C. Tomasi, D. Ravelli, S. Protti, M. Fagnoni, P. Mustarelli, Alkoxy substituted imidazolium-based ionic liquids as electrolytes for lithium batteries. J. Power Sources **235**, 142–147 (2013). https://doi.org/10.1016/j.jpowsour.2013.01.149
16. T. Frömling, M. Kunze, M. Schönhoff, J. Sundermeyer, B. Roling, Enhanced lithium transference numbers in ionic liquid electrolytes. J. Phys. Chem. B **112**(41), 12985–12990 (2008). https://doi.org/10.1021/jp804097j
17. Z. Gadjourova, Y.G. Andreev, D.P. Tunstall, P.G. Bruce, Ionic conductivity in crystalline polymer electrolytes. Nature **412**(6846), 520–523 (2001)
18. M. Galiński, A. Lewandowski, I. Stępniak, Ionic liquids as electrolytes. Electrochim. Acta **51** (26), 5567–5580 (2006). https://doi.org/10.1016/j.electacta.2006.03.016
19. B. Garcia, S. Lavallée, G. Perron, C. Michot, M. Armand, Room temperature molten salts as lithium battery electrolyte. Electrochim. Acta **49**(26), 4583–4588 (2004). https://doi.org/10.1016/j.electacta.2004.04.041
20. R. Giernoth, Task-specific Ionic Liquids. Angew. Chem. Int. Ed. **49**(16), 2834–2839 (2010). https://doi.org/10.1002/anie.200905981
21. J.B. Goodenough, Y. Kim, Challenges for rechargeable Li Batteries. Chem. Mater. **22**(3), 587–603 (2010). https://doi.org/10.1021/cm901452z
22. J.B. Goodenough, K.-S. Park, The Li-Ion rechargeable battery: A perspective. J. Am. Chem. Soc. **135**(4), 1167–1176 (2013). https://doi.org/10.1021/ja3091438
23. A. Guerfi, M. Dontigny, P. Charest, M. Petitclerc, M. Lagacé, A. Vijh, K. Zaghib, Improved electrolytes for Li-ion batteries: Mixtures of ionic liquid and an organic electrolyte with

enhanced safety and electrochemical performance. J. Power Sources **195**(3), 845–852 (2010). https://doi.org/10.1016/j.jpowsour.2009.08.056
24. H.-B. Han, K. Liu, S.-W. Feng, S.-S. Zhou, W.-F. Feng, J. Nie, et al., Ionic liquid electrolytes based on multi-methoxyethyl substituted ammoniums and perfluorinated sulfonimides: Preparation, characterization, and properties. Electrochim. Acta **55**(23), 7134–7144 (2010). https://doi.org/10.1016/j.electacta.2010.06.063
25. S.T. Handy, M. Okello, The 2-Position of Imidazolium Ionic Liquids: Substitution and Exchange. J. Org. Chem. **70**(5), 1915–1918 (2005). https://doi.org/10.1021/jo0480850
26. P. Hapiot, C. Lagrost, Electrochemical reactivity in room-temperature ionic liquids. Chem. Rev. **108**(7), 2238–2264 (2008). https://doi.org/10.1021/cr0680686
27. K. Hayamizu, Y. Aihara, S. Arai, W.S. Price, Self-diffusion coefficients of lithium, anion, polymer, and solvent in polymer gel electrolytes measured using 7Li, 19F, and 1H pulsed-gradient spin-echo NMR. Electrochim. Acta **45**(8), 1313–1319 (2000). https://doi.org/10.1016/S0013-4686(99)00338-2
28. P.A. Hunt, Why Does a Reduction in Hydrogen Bonding Lead to an Increase in Viscosity for the 1-Butyl-2,3-dimethyl-imidazolium-Based Ionic Liquids? J. Phys. Chem. B **111**(18), 4844–4853 (2007). https://doi.org/10.1021/jp067182p
29. M. Ishikawa, T. Sugimoto, M. Kikuta, E. Ishiko, M. Kono, Pure ionic liquid electrolytes compatible with a graphitized carbon negative electrode in rechargeable lithium-ion batteries. J. Power Sources **162**(1), 658–662 (2006). https://doi.org/10.1016/j.jpowsour.2006.02.077
30. Y. Jin, S. Fang, M. Chai, L. Yang, S.-I. Hirano, Ether-Functionalized Trialkylimidazolium Ionic Liquids: Synthesis, Characterization, and Properties. Ind. Eng. Chem. Res. **51**(34), 11011–11020 (2012). https://doi.org/10.1021/ie300849u
31. Y. Jin, S. Fang, L. Yang, S.-I. Hirano, K. Tachibana, Functionalized ionic liquids based on guanidinium cations with two ether groups as new electrolytes for a lithium battery. J. Power Sources **196**(24), 10658–10666 (2011). https://doi.org/10.1016/j.jpowsour.2011.08.008
32. J. Kalhoff, D. Bresser, M. Bolloli, F. Alloin, J.-Y. Sanchez, S. Passerini, Enabling LiTFSI-based electrolytes for Safer Lithium-Ion Batteries by Using Linear Fluorinated Carbonates as (Co) Solvent. ChemSusChem **7**(10), 2939–2946 (2014). https://doi.org/10.1002/cssc.201402502
33. F. Kenta, M. Takushi, T. Toshiyuki, Y. Toshio, U. Yasuhiro, I. Shin-ich, Effect of Methylation at the C2 Position of Imidazolium on the Structure of Ionic Liquids Revealed by Large Angle X-ray Scattering Experiments and MD Simulations. Chem. Lett. **38**(4), 340–341 (2009). https://doi.org/10.1246/cl.2009.340
34. V.R. Koch, C. Nanjundiah, G.B. Appetecchi, B. Scrosati, The interfacial stability of Li with Two New Solvent-Free Ionic Liquids: 1,2-Dimethyl-3-propylimidazolium Imide and Methide. J. Electrochem. Soc. **142**(7), L116–L118 (1995). https://doi.org/10.1149/1.2044332
35. F. Koichi, W. Alexander, L. Ralf, Strong, localized, and directional hydrogen bonds fluidize ionic liquids. Angew. Chem. Int. Ed. **47**(45), 8731–8734 (2008). https://doi.org/10.1002/anie.200803446
36. A. Lewandowski, A. Świderska-Mocek, Ionic liquids as electrolytes for Li-ion batteries – An overview of electrochemical studies. J. Power Sources **194**(2), 601–609 (2009). https://doi.org/10.1016/j.jpowsour.2009.06.089
37. C. Liao, N. Shao, K.S. Han, X.-G. Sun, D.-E. Jiang, E.W. Hagaman, S. Dai, Physicochemical properties of imidazolium-derived ionic liquids with different C-2 substitutions. Phys. Chem. Chem. Phys. **13**(48), 21503–21510 (2011). https://doi.org/10.1039/C1CP22375E
38. D.R. MacFarlane, N. Tachikawa, M. Forsyth, J.M. Pringle, P.C. Howlett, G.D. Elliott, et al., Energy applications of ionic liquids. Energy Environ. Sci. **7**(1), 232–250 (2014). https://doi.org/10.1039/C3EE42099J
39. K. Noack, P.S. Schulz, N. Paape, J. Kiefer, P. Wasserscheid, A. Leipertz, The role of the C2 position in interionic interactions of imidazolium-based ionic liquids: a vibrational and NMR spectroscopic study. Phys. Chem. Chem. Phys. **12**(42), 14153–14161 (2010). https://doi.org/10.1039/C0CP00486C

40. S. Pandian, S.G. Raju, K.S. Hariharan, S.M. Kolake, D.-H. Park, M.-J. Lee, Functionalized ionic liquids as electrolytes for lithium-ion batteries. J. Power Sources **286**, 204–209 (2015). https://doi.org/10.1016/j.jpowsour.2015.03.130
41. M. Park, X. Zhang, M. Chung, G.B. Less, A.M. Sastry, A review of conduction phenomena in Li-ion batteries. J. Power Sources **195**(24), 7904–7929 (2010). https://doi.org/10.1016/j.jpowsour.2010.06.060
42. J. Reiter, E. Paillard, L. Grande, M. Winter, S. Passerini, Physicochemical properties of N-methoxyethyl-N-methyl pyrrolidinium ionic liquids with perfluorinated anions. Electrochim. Acta **91**, 101–107 (2013). https://doi.org/10.1016/j.electacta.2012.12.086
43. A.D. Robertson, A.R. West, A.G. Ritchie, Review of crystalline lithium-ion conductors suitable for high-temperature battery applications. Solid State Ionics **104**(1), 1–11 (1997). https://doi.org/10.1016/S0167-2738(97)00429-3
44. B. Scrosati, J. Garche, Lithium batteries: Status, prospects, and future. J. Power Sources **195**(9), 2419–2430 (2010). https://doi.org/10.1016/j.jpowsour.2009.11.048
45. B. Scrosati, J. Hassoun, Y.-K. Sun, Lithium-ion batteries. A look into the future. Energy Environ. Sci. **4**(9), 3287–3295 (2011). https://doi.org/10.1039/C1EE01388B
46. S. Seki, Y. Kobayashi, H. Miyashiro, Y. Ohno, A. Usami, Y. Mita, et al., Lithium secondary batteries using modified-Imidazolium Room-temperature ionic liquid. J. Phys. Chem. B **110**(21), 10228–10230 (2006). https://doi.org/10.1021/jp0620872
47. E.O. Stejskal, Use of spin echoes in a pulsed magnetic-field gradient to study anisotropic, restricted diffusion and flow. J. Chem. Phys. **43**(10), 3597–3603 (1965). https://doi.org/10.1063/1.1696526
48. K. Tsunashima, M. Sugiya, Physical and electrochemical properties of low-viscosity phosphonium ionic liquids as potential electrolytes. Electrochem. Commun. **9**(9), 2353–2358 (2007). https://doi.org/10.1016/j.elecom.2007.07.003
49. T. Vogl, S. Menne, A. Balducci, Mixtures of protic ionic liquids and propylene carbonate as advanced electrolytes for lithium-ion batteries. Phys. Chem. Chem. Phys. **16**(45), 25014–25023 (2014). https://doi.org/10.1039/C4CP03830D
50. T. Welton, Room-temperature ionic liquids. Solvents for synthesis and catalysis. Chem. Rev. **99**(8), 2071–2084 (1999). https://doi.org/10.1021/cr980032t
51. M.S. Whittingham, Electrical energy storage and intercalation chemistry. Science **192**(4244), 1126–1127 (1976)
52. A. Yoshino, K. Sanechika, T. Nakajima, Secondary battery: Google Patents (1987)
53. S. Zahn, G. Bruns, J. Thar, B. Kirchner, What keeps ionic liquids in flow? Phys. Chem. Chem. Phys. **10**(46), 6921–6924 (2008). https://doi.org/10.1039/B814962N
54. S.S. Zhang, A review on electrolyte additives for lithium-ion batteries. J. Power Sources **162**(2), 1379–1394 (2006). https://doi.org/10.1016/j.jpowsour.2006.07.074
55. H. Zhou, Y. Wang, H. Li, P. He, The Development of a new type of rechargeable batteries based on hybrid electrolytes. ChemSusChem **3**(9), 1009–1019 (2010). https://doi.org/10.1002/cssc.201000123

Multi-physics Simulation of Charge-Transfer Reaction and Mass Transport in Lithium-Ion Battery Cathode

12

Michihisa Koyama and Baber Javed

Contents

Abstract

The practical utilization of computer simulations is expected in many disciplines as results of advancements of computers, computer science, and simulation methods in the last decades. Perceptions of simulation technologies in the past would be that it rarely contributes to the practical battery developments; however, their use is expected to be inevitable in the battery research and development in the future. In this chapter, authors will introduce the multi-physics simulations of battery cathodes. An approach to modeling the complex microstructure of

Author Contribution: The manuscript was written by MK. Part of simulation data is from BJ.

M. Koyama (✉)
National Institute for Materials Science, Global Research Center for Environment and Energy Based on Nanomaterials Science, Tsukuba, Ibaraki, Japan

Center for Energy and Environmental Science, Shinshu University, Nagano, Japan

Graduate School of Engineering, Hiroshima University, Higashi-Hiroshima, Hiroshima, Japan
e-mail: Michihisa@nims.go.jp; michihisakoyama3@gmail.com

B. Javed
National Institute for Materials Science, Global Research Center for Environment and Energy Based on Nanomaterials Science, Tsukuba, Ibaraki, Japan

Q. Zhen et al. (eds.), *Nanostructured Materials for Next-Generation Energy Storage and Conversion*, https://doi.org/10.1007/978-3-662-58675-4_12

composite cathode is first introduced, followed by the discussion of the physical properties to be used in the simulation. The importance of interfacial resistance in the composite cathode is discussed on the basis of simulation results for lithium-ion battery cathode. Preliminary application of the approach to the solid-state battery system is also introduced. Finally, future directions are introduced.

12.1 Introduction

The rapid development of highly functional cathode has been one of the central issues in lithium-ion battery (LIB) research; thus intensive studies have been conducted after its emergence into commercial market decades ago. Approaches for improving the cathode performance can be classified into three types as schematically illustrated in Fig. 12.1: the development of novel functional materials, manipulating interface for faster electrochemistry, and controlling microstructure for better multi-physics of electrochemical reaction and mass transport. The examples of the former two can be found much in literature and other chapters of this book; therefore, this chapter will focus on the final approach. In general, to increase the storage density of LIB, thicker cathode with a higher portion of active material particles is favored, which usually results in lower output power density limited by mass transfer processes involved in electrode reaction. Thus, it is demanded to optimize microstructure for cathode with both higher storage density and output power density at the same time. However, the complexity of the cathode microstructure has been a big obstacle for analyzing elementary processes. Therefore, traditional researches aiming at optimization of microstructure have been researches investigating electrode characteristics as a function of conditions of the

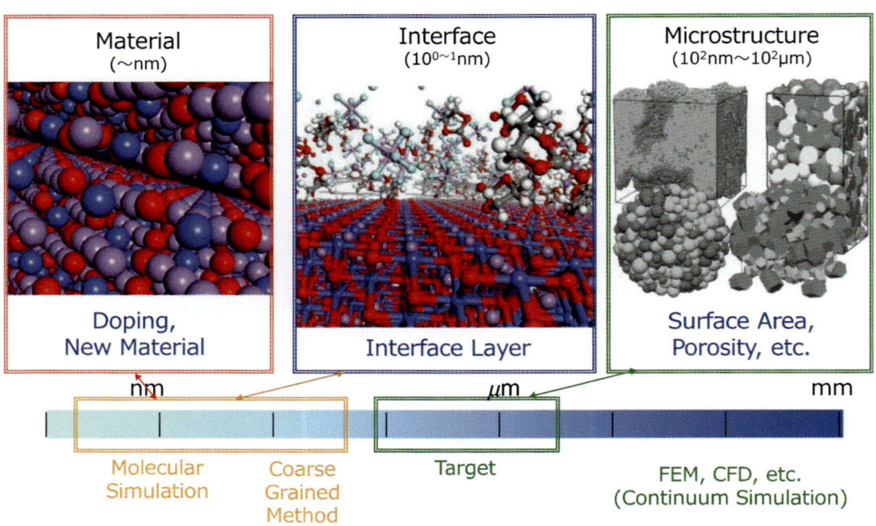

Fig. 12.1 Approaches for improving lithium-ion battery (LIB) cathode performance

fabrication process as parameters. Though such approach is important to accumulate tacit process knowledge and useful for establishing fabrication recipes of the better electrode, breakthroughs are awaited in order to obtain mechanistic and essential understandings, which will be inevitable to design electrode, for example, free from failures with microstructure origin.

LIB cathode typically consists of active material, a conductive additive, binder, and electrolyte. Those components interact with each other, forming a complicated structure field for lithium-ion transfer, charge-transfer, and electron transfer. To best design the cathode microstructure, understanding the rate-limiting process is important. In fact, one can see that single-particle cathode, which is free from the complexity of the microstructure, shows an excellent performance [1–7]; e.g., ca. 75% of capacity is retained in the single-particle cathode of lithium cobalt oxide (Li_xCoO_2) even at a high discharge rate of 300 C [1]. A large difference with actual operation in automobile application is not only due to the limitations of heat dissipations for securing the safety but also due to the different rate-limiting process in the composite electrode system. The effort to quantify the relation between the complex microstructure of battery electrodes and the charge/discharge characteristics is driven by the utilization of focused-ion-beam scanning electron microscopy (FIB-SEM) and nano/micro-X-ray computed tomography (CT), which can analyze the actual electrode microstructure three-dimensionally followed by computer reconstructions [8–15]. When we apply those technologies to the composite cathode systems, we can easily differentiate the active materials and other phases; however, it is difficult to differentiate conductive additive, binder, and void filled by the liquid electrolyte. This makes virtual modeling approaches [16–22] alternatively useful and effective to understand the multi-physics in composite electrode. We can find pioneering studies applying virtual modeling to LIB. In the past several years, many studies have addressed the issue of the influence of the composite cathode microstructure. Goldin et al. have carefully discussed the influence of microstructure on the discharge characteristics of LIBs based on the electrode structures represented as orderly packed spheres [20]. Latz and Zausch have performed a multi-scale simulation of LIBs using a three-dimensional structure model based on the homogenization method [21]. However, most of those explicitly consider active materials only, and the gap between such a simplified structure and the actual composite structure is needed to be bridged if we target the simulation-driven microstructure optimization. In this chapter, we will introduce an example of such efforts by Kikukawa et al. [16].

12.2 Modeling Composite Electrode Structure for Multi-physics Simulation of LIB Cathode

In order to investigate the multi-physics within the composite electrode, it is necessary to explicitly model the electrode structure comprising of active material, conductive additive, binder, and electrolyte distinctly. In addition, the existence of an interface between particles should be noted as a difference between the single particle and composite cathodes. Kikukawa et al. have modeled the composite cathode as schematically illustrated in Fig. 12.2. First, three distinct phases of active materials, conductive additive, and binder are randomly packed into the simulation cell by using a porous structure simulator [19]. Note that the overlap of each entity is permitted; thus two-dimensional interfaces are formed. The open void will be assumed to be filled with liquid electrolyte in the later finite element simulation. Here, it is noted that the boundary between active materials is treated explicitly by inserting a circular disc to discuss its influence on the discharge characteristics. Then, an in-house code is used to generate the two-dimensional image along the thickness direction by defining the appropriate unit step. At each unit step, the 2D slice and view images are saved and are further used to generate a three-dimensional voxel model by importing them in a three-dimensional image data processing software, AVIZO [23]. Finally, meshes for finite element simulation are generated from the voxel model by using a function available in multi-physics simulation package, COMSOL Multiphysics® [24]. This process is compatible with the process of the three-dimensional reconstruction based on FIB-SEM, while it should be noted that the clear identification of interface between particles is a remaining big challenge.

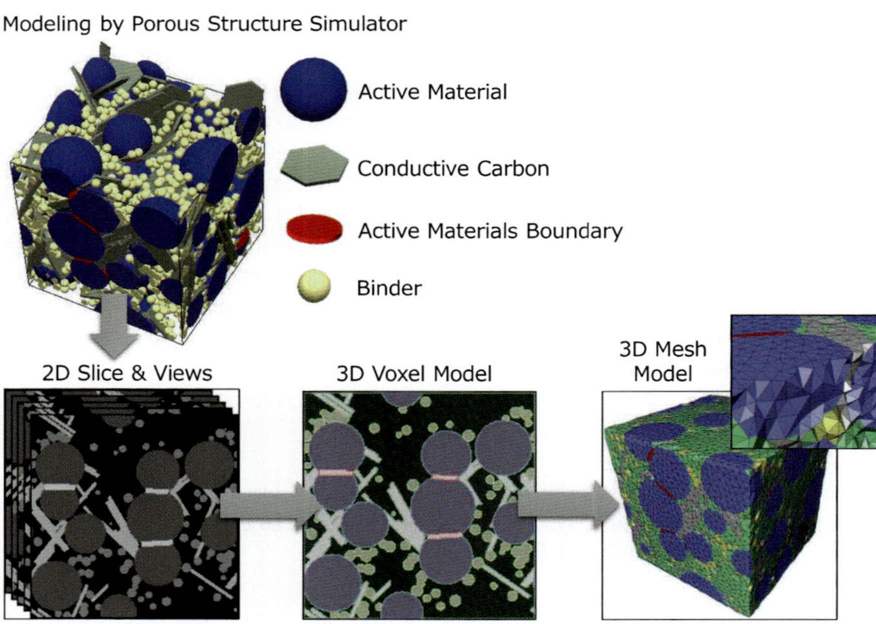

Fig. 12.2 The modeling process for the three-dimensional electrode structure. (Reprinted from Kikukawa et al. [16], Copyright 2018, with permission from Elsevier)

12.3 Multi-physics Simulation of Liquid Electrolyte LIB Cathode

Based on the constructed composite cathode structure model, multi-physics of charge-transfer reaction and mass transfer was simulated by finite element method. The following elementary physics are considered for each component phase of the composite cathode, which is also illustrated in Fig. 12.3a.

- Electron conduction: Active material, a conductive additive, current collector, and the interface between those components.
- Lithium diffusion: Active material, and the interface between active materials.
- Lithium-ion conduction: Electrolyte.
- Charge-transfer reaction: Interface between active material and electrolyte.

Note that binder is treated as inert phase just acting as inhibition phase; thus no physics are applied. In addition, it is noted that the interface between active materials is modeled as a separate phase; therefore, the parameters different from bulk active

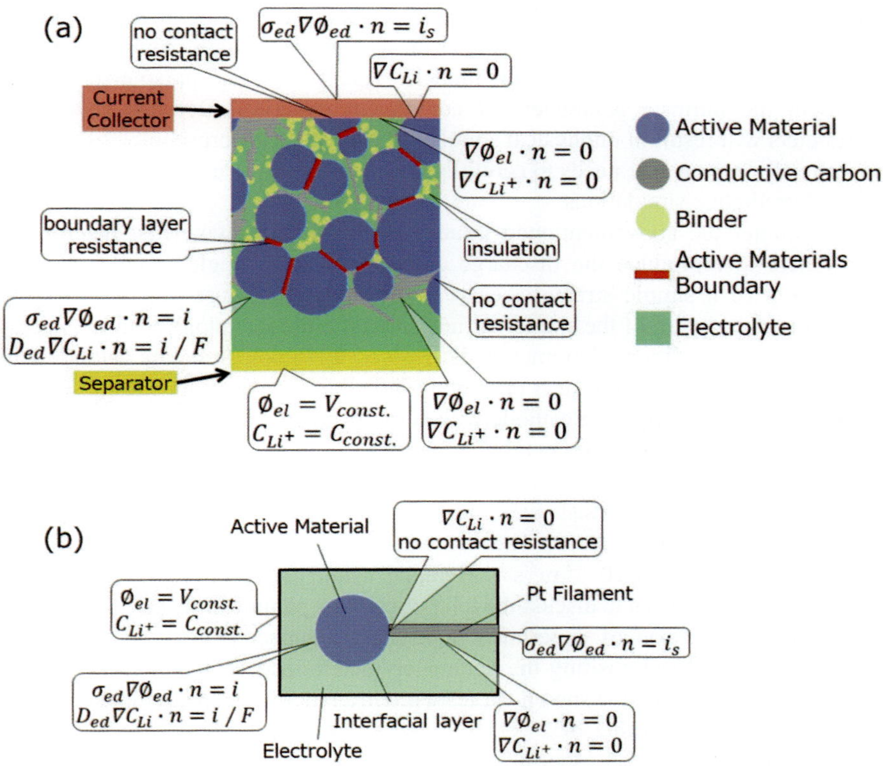

Fig. 12.3 Schematic cross-sectional images of models used for the (**a**) composite cathode and (**b**) single-particle cathode. (Reprinted from Kikukawa et al. [16], Copyright 2018, with permission from Elsevier)

material phase can be set. Butler-Volmer equation was adopted to represent the charge-transfer reaction at the electrode/electrolyte interface. In addition, film resistance is set referring to the preceding literature [25]. Further details are available in the original manuscript [16].

Li_xCoO_2 and carbon were assumed as an active material and a conductive additive, respectively. 1 mol/dm^3 of lithium perchlorate ($LiClO_4$) solvated in propylene carbonate ($C_4H_6O_3$), and ethylene carbonate ($C_3H_4O_3$) mixture with the composition of 1:1 v/v was assumed as electrolyte [4].

Another difficulty in the multi-physics simulation of the composite cathode is the reliability of the physical properties. Conventionally, complex microstructure and physics are simplified in many of preceding studies [14, 17–22, 26]. Of course, we must admit that simulation results from such simplified models are very useful for the interpretations of observable charge/discharge characteristics. However, we will notice that different parameters are set for, for example, the same physical properties if we closely examine those preceding studies. This will be associated with the parameter fitting process in that simulation, typically conducted in order to reproduce the experimentally observed characteristics. In the parameter fitting process, parameters such as physical property values are changed usually; however, parameters determined in such a way result in the values into which many differences between experiment and simulation models. Therefore, those parameters do not represent the intrinsic parameters of component materials. Because unphysical parameters will result in unphysical simulation even if the macroscopic observation superficially agrees, one needs to carefully consider this issue in order to transferrable and predictive simulations.

To counter the abovementioned issue, Kikukawa et al. have first determined parameters to reproduce the discharge characteristics of single-particle cathode, comprising of a simple structure, easy to be modeled. Figure 12.3b shows the schematic description of the single-particle simulation model along with the necessary boundary conditions. Parameters determined for simple structure cathode will represent more intrinsic physical properties and should be used for the subsequent simulations of the composite cathode with complex microstructure. Kikukawa et al. have carefully examined parameters of the diffusion coefficient of lithium (Li) in active material, the conductivity of the active material, the rate constant for exchange current density, and film resistance of an active material to reproduce experimental observations by Dokko et al. [4]. They have determined lithium diffusion coefficient in active material as 3×10^{-13} m^2 s^{-1}. Here, we would introduce another example of parameter determination to discuss the reliability of the parameter values. Figure 12.4 shows the comparison of experimental discharge characteristics and simulation results obtained after revisiting the parameters determined by Kikukawa et al. for more precision. Among the shown curves with different diffusion coefficient values, one can see that the value of 2×10^{-13} m^2 s^{-1} shows the best agreement with experimental observations especially up to the current of 20 nA, which is actually better than the curves by Kikukawa et al. in terms of both the initial voltage and capacity retention ratio. This means that we should admit such range of determined

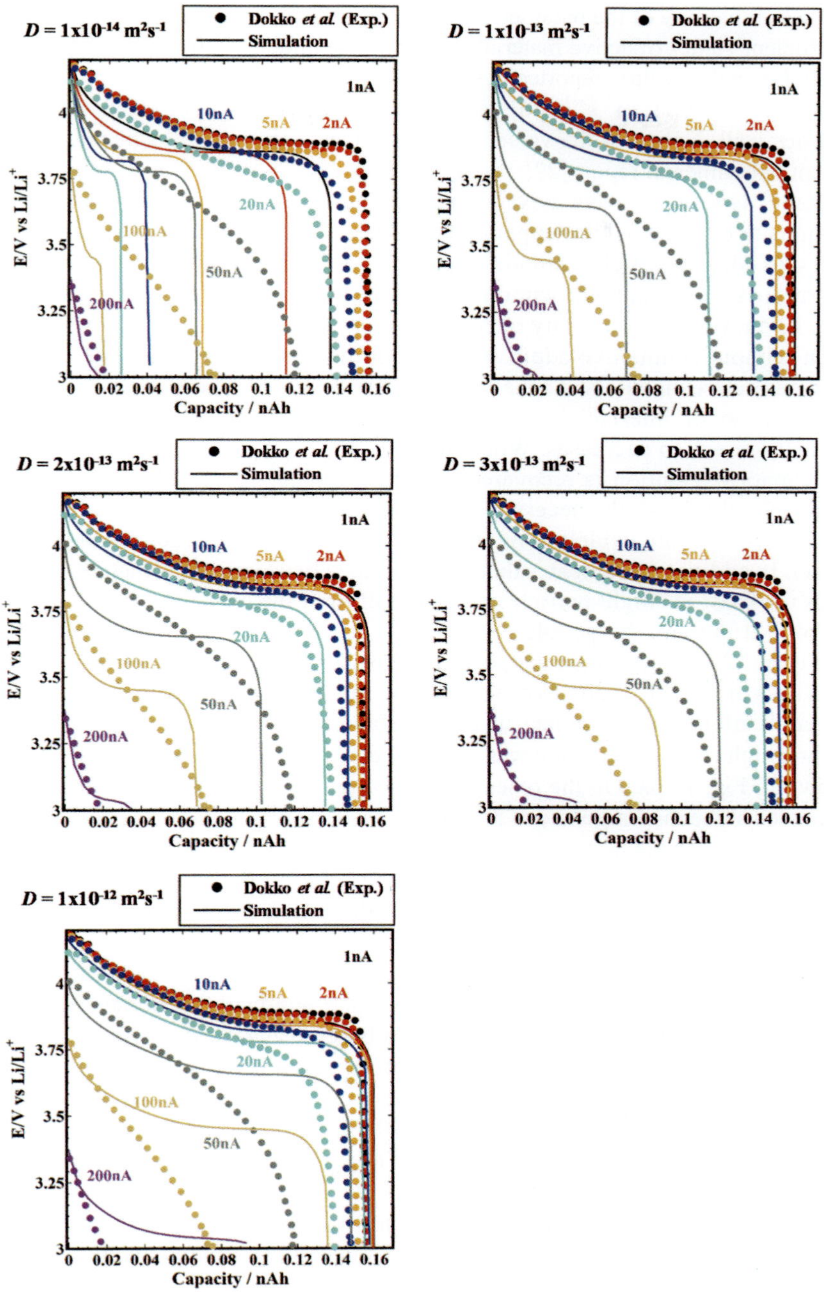

Fig. 12.4 Discharge characteristics of single-particle cathode: solid lines are simulation results [28]; filled circles are from the experiment by Dokko et al. [4]

value as reliability of the parameters; however, we can understand that the diffusion coefficient values of active materials with the magnitude of 10^{-13} m^2 s^{-1}, which is much larger than values reported in simulation researches such as 3.7×10^{-16} m^2 s^{-1} [20] and 1×10^{-14} m^2 s^{-1} [26], can reproduce the single-particle cathode characteristics well. Note that the values of 10^{-13} m^2 s^{-1} are in good agreement with a recent experimental report [27].

Kikukawa et al. have next simulated the discharge characteristics of the composite cathode using the parameters determined to reproduce the characteristics of single-particle cathode. Initially, they have simulated the discharge characteristics of cathode without the conductive additive at 10 C. As seen in Fig. 12.5a, very high capacity retention was obtained at the discharge rate of 10 C even without conductive additive if they do not set any interparticle resistance between active materials. As the relative interparticle resistance, which is defined as the ratio of interfacial resistivity against that of the bulk, increases, the capacity retention decreases. When the carbon is added as a conductive additive, the capacity retention is recovered as shown in Fig. 12.5b. This observed tendency clearly shows the necessity of conductive additive to retain the capacity at high discharge rate, which have provided an alternative path for the electrons to travel. In addition, it indicates that the resistance between active materials, which does not exist in single-particle cathode, will limit the capacity of the composite cathode at high discharge rate.

Figure 12.6 shows the discharge characteristics of the composite cathode with the thickness of 30 and 60 μm. Note that relative interparticle resistivity between active materials of 10^4 was assumed in the simulation. When the cathode thickness is 30 μm, high capacity retention was achieved even at a high discharge rate of 10 C as shown in Fig. 12.6a. On the other hand, 60% and 30% of capacity retention were estimated at the discharge rate of 5 C and 10 C, respectively, from the simulation for 60-μm-thick cathode as seen from Fig. 12.6b.

To see the reason more closely, concentration profile in composite cathode visualized in Fig. 12.7 that shows the lithium-ion concentration within the electrolyte phase (Fig. 12.7a, b). Lithium concentration is homogeneous at the early stage of discharge in both 30- and 60-μm-thick cathodes. Lithium-ion concentration near the current collector side gradually becomes scarce in 30-μm-thick composite cathode as the discharge reaches intermediate and final stages (Fig. 12.7a). On the contrary, lithium-ion concentration within 60-μm-thick composite cathode becomes very scarce even at the intermediate stage of discharge (Fig. 12.7b). This clearly shows the typical lithium-ion depletion in the electrolyte phase of composite cathode. Figure 12.7c, d show the lithium concentration within the active material phase of the composite cathode at the final stage of the discharge. Lithium-ion is homogeneously inserted into active materials in the 30-μm-thick composite cathode; thus, only a limited number of active materials with slightly lower concentration are noticed near the current collector side at high discharge capacity of 10 C (Fig. 12.7c). From Fig. 12.7d, homogeneous insertion was achieved at the discharge rates of 0.2 and 1 C in 60-μm-thick

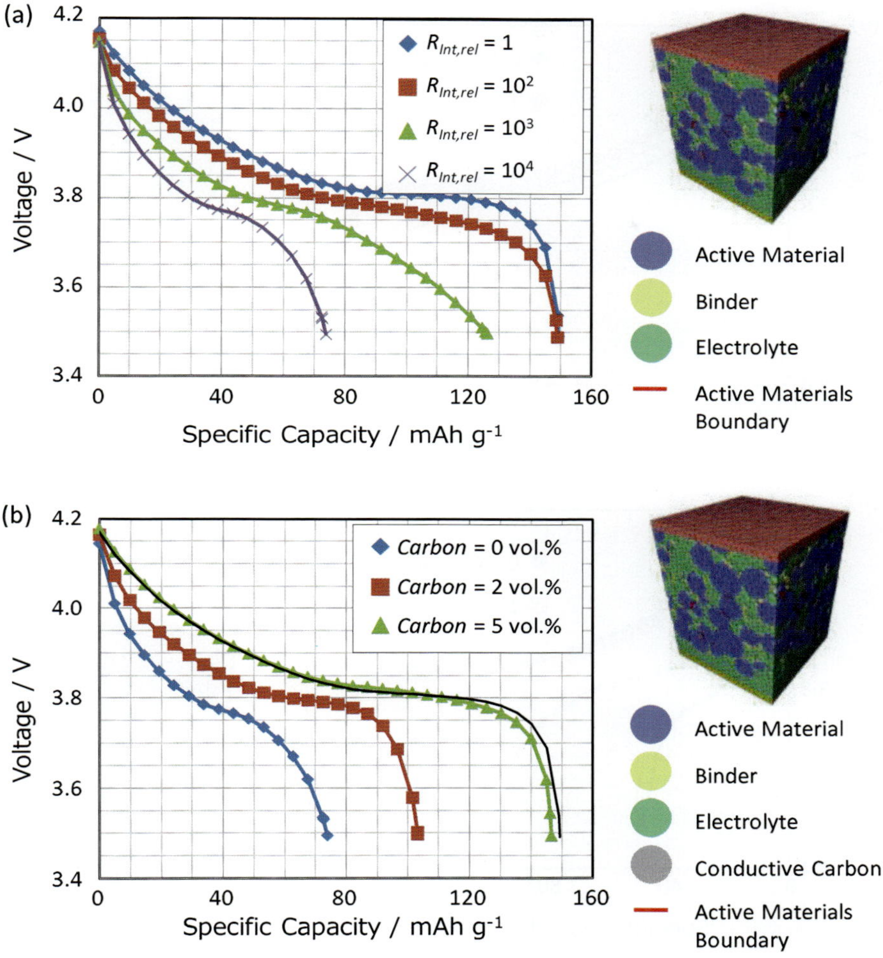

Fig. 12.5 Discharge curves for (**a**) different relative interparticle resistivity between active materials and (**b**) different carbon content with the relative interparticle resistivity of 10^4. (Reprinted from Kikukawa et al. [16], Copyright 2018, with permission from Elsevier)

composite cathode, and insufficient insertion near the current collector side was noticed at 5 C. One can clearly see that the insertion only occurs near the separator side at 10 C, corresponding to the lithium-ion depletion in the electrolyte phase shown in Fig. 12.6d. We should note that those results qualitatively reproduce the experimental observations; therefore, the next step is the validation against experimental observations, which will be a starting point toward the computer-aided design of composite cathode microstructure.

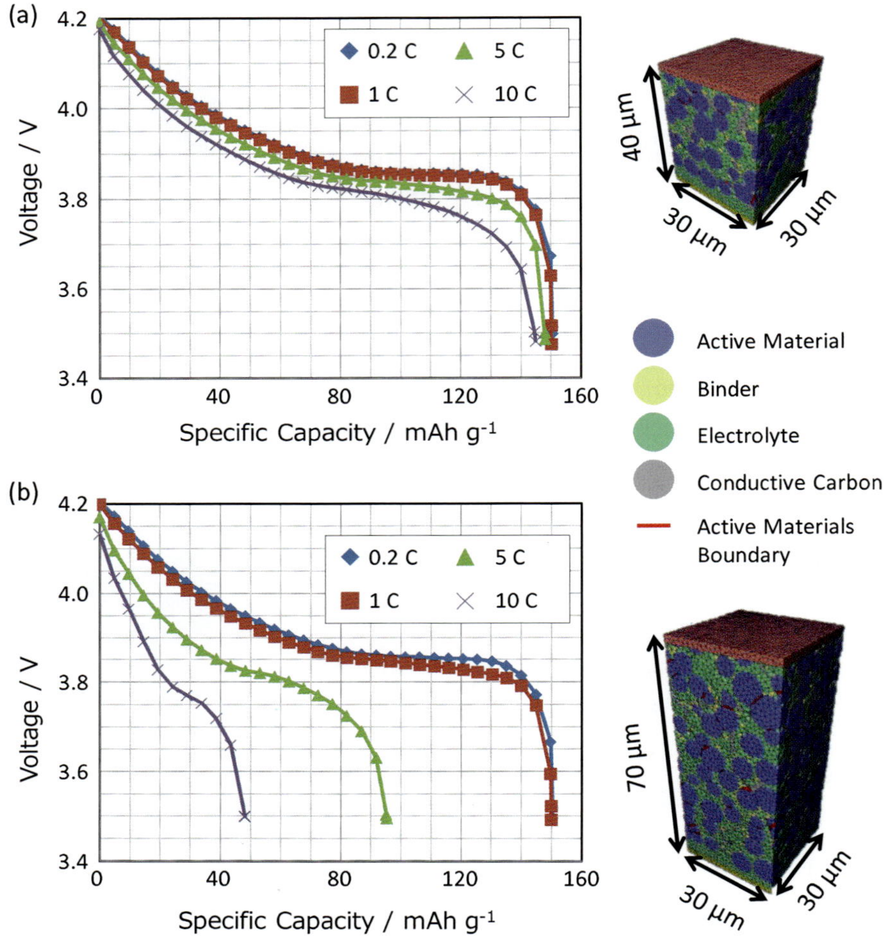

Fig. 12.6 Discharge curves for cathodes with a thickness of (**a**) 30 μm and (**b**) 60 μm. (Reprinted from Kikukawa et al. [16], Copyright 2018, with permission from Elsevier)

12.4 Multi-physics Simulation of Solid-State Battery Cathode

As a next-generation battery of LIB, all-solid-state batteries (SSB) have received much attention, due to its high energy density, non-flammability, and large electrochemical window compared to the conventional liquid electrolytes. As results of the development of electrolytes with high conductivity comparable to that of the liquid electrolyte, many efforts are devoted to achieving the characteristics satisfying the specifications required for an automobile or stationary applications. Following the approach for liquid electrolyte LIB described in the preceding section, the existence

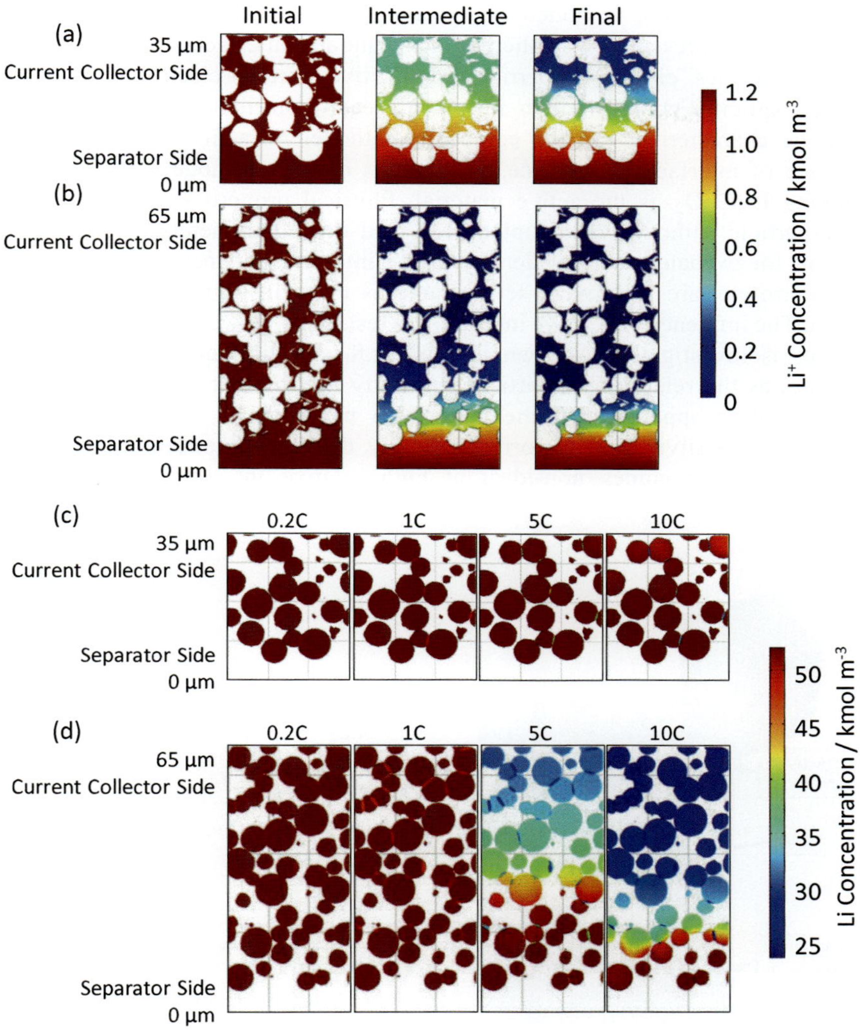

Fig. 12.7 Lithium-ion concentration in the electrolyte phase for cathodes with a thickness of (**a**) 30 μm and (**b**) 60 μm, and lithium concentration in active materials at the final phase of discharging for cathodes with a thickness of (**c**) 30 μm and (**d**) 60 μm. (Reprinted from Kikukawa et al. [16], Copyright 2018, with permission from Elsevier)

of various interfaces determines the charge/discharge characteristics of the practical composite cathode, the simulations considering interfacial resistances will help in the design of composite cathode for SSB.

As a first step, we have conducted preliminary simulations of SSB composite cathode considering interfacial resistances as schematically illustrated in Fig. 12.8.

The composite cathode is modeled by packing active material and solid electrolyte particles represented by spheres. In the model, interfaces between active material particles, electrolyte particles, and active material-electrolyte particles can be explicitly considered, to investigate each contribution to the charge/discharge characteristics of the composite cathode. As a first step to see the influence of interfacial resistance, simulations of SSB cathode are conducted assuming Li_xCoO_2 as the active material. Physical properties determined for single-particle cathode was adopted [16], and other parameters, such as rate constant for exchange current density at the interface between active material and electrolyte, are tentatively set as same as those in liquid electrolyte LIB system. The influence of relative interparticle resistivity between solid electrolyte particles is investigated. As seen in Fig. 12.8, capacity retention drastically decreases as the relative interparticle resistivity increases at the discharge rate of 10 C. It is apparent that the battery no more works when the relative interparticle resistivity is 10^3. Correspondingly, the lithium concentration within active material becomes non-ideal at high relative interparticle resistivity.

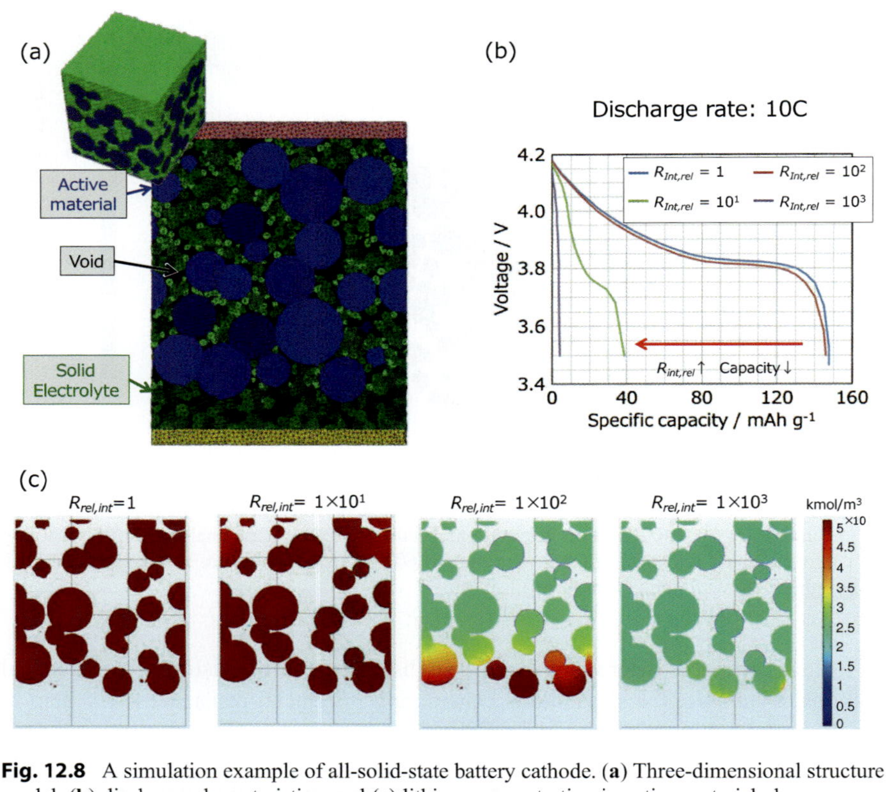

Fig. 12.8 A simulation example of all-solid-state battery cathode. (**a**) Three-dimensional structure model, (**b**) discharge characteristics, and (**c**) lithium concentration in active material phase

Presently, authors are conducting simulations on the SSB cathode intensively, to reproduce experimental observations and then to obtain insights and guiding principles for higher and robust performance [29, 30].

12.5 Summary and Future Directions

In this chapter, authors have introduced multi-physics simulations of LIB cathode. By three-dimensionally modeling the composite cathode microstructure and considering the interparticle resistance, discharge characteristics of the composite cathode were reasonably simulated adopting the bulk properties of component materials. This clearly verifies the importance to consider the interfacial resistances in the numerical simulations of composite cathode. The same approach was applied to the SSB cathode. While the immaturity of the simulation must be admitted, the importance of considering interfacial resistances was confirmed.

Future directions of the research can be threefold. One is the validation of the simulation results against experimental observations, which is ongoing for the solid-state battery (SSB) cathode. The second is to incorporate the microstructure of actual cathode into the simulation. This requires the advancement of reconstructions from FIB-SEM or X-ray CT observations. The final one is the multi-scale simulation. As results of the advancements of computers and theoretical methods, many atomic-level insights on the interfaces are reported. Also, we can find considerable numbers of experimental reports on the atomically well-defined model interface systems. Incorporating such information into the multi-physics simulations will be an important step toward rationally designing best composite cathode microstructure, satisfying the specification requirements as the battery device.

Acknowledgments The part of the research is supported by JSPS KAKENHI Grant Number JP25709007 and MEXT Program for Integrated Materials Development. The authors wish to thank Dr. Kazunori Takada of National Institute for Materials Science and Prof. Kiyoshi Kanamura of Tokyo Metropolitan University for fruitful discussion.

References

1. K. Dokko, M. Mohamedi, Y. Fujita, T. Itoh, M. Nishizawa, M. Umeda, I. Uchida, Kinetic characterization of single particles of $LiCoO_2$ by AC impedance and potential step methods. J. Electrochem. Soc. **148**(5), A422–A426 (2001)
2. K. Dokko, M. Mohamedi, M. Umeda, I. Uchida, Kinetic study of Li-ion extraction and insertion at $LiMn_2O_4$ single particle electrodes using potential step and impedance methods. J. Electrochem. Soc. **150**(4), A425–A429 (2003)
3. A. Palencsár, D.A. Scherson, Electrochemical and in situ optical characterization of single micrometer-size particles of spherical nickel oxide in alkaline aqueous electrolytes. Electrochem. Solid-State Lett. **6**(4), E1–E4 (2003)
4. K. Dokko, N. Nakata, K. Kanamura, High rate discharge capability of single particle electrode of $LiCoO_2$. J. Power Sources **189**(1), 783–785 (2009)

5. H. Munakata, B. Takemura, T. Saito, K. Kanamura, Evaluation of real performance of $LiFePO_4$ by using single particle technique. J. Power Sources **217**, 444–448 (2012)
6. Y.-H. Huang, F.-M. Wang, T.-T. Huang, J.-M. Chen, B.-J. Hwang, J. Rick, Micro-electrode linked cyclic voltammetry study reveals ultra-fast discharge and high ionic transfer behavior of $LiFePO_4$. Int. J. Electrochem. Sci. **7**, 1205–1213 (2012)
7. T. Li, B. Song, L. Lu, K. Zeng, Voltage induced electrochemical reactions in the single lithium-rich layer-oxide nanoparticles. Phys. Chem. Chem. Phys. **17**(15), 10257–10264 (2015)
8. P.R. Shearing, L.E. Howard, P.S. Jorgensen, N.P. Brandon, S.J. Harris, Characterization of the 3-dimensional microstructure of a graphite negative electrode from a Li-ion battery. Electrochem. Commun. **12**(3), 374–377 (2010)
9. J.R. Wilson, J.S. Cronin, S.A. Barnett, S.J. Harris, Measurement of three-dimensional microstructure in a $LiCoO_2$ positive electrode. J. Power Sources **196**(7), 3443–3447 (2011)
10. M. Ender, J. Joos, T. Carraro, E. Ivers-Tiffe, Three-dimensional reconstruction of a composite cathode for lithium-ion cells. Electrochem. Commun. **13**(2), 166–168 (2011)
11. A.H. Wiedemann, G.M. Goldin, S.A. Barnett, H. Zhu, R.J. Kee, Effects of three-dimensional cathode microstructure on the performance of lithium-ion battery cathodes. Electrochim. Acta **88**, 580–588 (2013)
12. T. Hutzenlaub, S. Thiele, N. Paust, R. Spotnitz, R. Zengerle, C. Walchshofer, Three-dimensional electrochemical Li-ion battery modelling featuring a focused ion-beam/scanning electron microscopy based three-phase reconstruction of a $LiCoO_2$ cathode. Electrochim. Acta **115**, 131–139 (2014)
13. B. Yan, C. Lim, L. Yin, L. Zhu, Simulation of heat generation in a reconstructed $LiCoO_2$ cathode during galvanostatic discharge. Electrochim. Acta **100**, 171–179 (2013)
14. B. Yan, C. Lim, L. Yin, L. Zhu, Three dimensional simulation of Galvanostatic discharge of $LiCoO_2$ cathode based on X-ray Nano-CT images. J. Electrochem. Soc. **159**(10), A1604–A1614 (2012)
15. P. Pietsch, V. Wood, X-ray tomography for Lithium ion battery research: A practical guide. Annu. Rev. Mater. Res. **47**, 451–479 (2017)
16. H. Kikukawa, K. Honkura, M. Koyama, Influence of inter-particle resistance between active materials on the discharge characteristics of the positive electrode of lithium ion batteries. Electrochim. Acta **278**, 385–395 (2018)
17. M. Koyama, H. Tsuboi, N. Hatakeyama, A. Endou, H. Takaba, M. Kubo, C.A. Del Carpio, A. Miyamoto, Development of three-dimensional porous structure simulator for optimizing microstructure of SOFC anode. ECS Trans. **7**(1), 2057–2064 (2007)
18. C.W. Wang, A.M. Sastry, Mesoscale modeling of a Li-ion polymer cell. J. Electrochem. Soc. **154**(11), A1035–A1047 (2007)
19. M. Koyama, K. Ogiya, T. Hattori, H. Fukunaga, A. Suzuki, R. Sahnoun, H. Tsuboi, N. Hatakeyama, A. Endou, H. Takaba, M. Kubo, C.A. Del Carpio, A. Miyamoto, Development of three-dimensional porous structure simulator $POCO_2$ for simulations of irregular porous materials. J. Comput. Chem. Jpn. **7**(2), 55–62 (2008)
20. G.M. Goldin, A.M. Colclasure, A.H. Widemann, R.J. Kee, Three-dimensional particle-resolved models of Li-ion batteries to assist the evaluation of empirical parameters in one-dimensional models. Electrochim. Acta **64**, 118–129 (2012)
21. A. Latz, J. Zausch, Multiscale modeling of lithium ion batteries: thermal aspects. Beilstein J. Nanotechnol. **6**, 987–1007 (2015)
22. R.E. García, Y.-M. Chiang, Spatially resolved modeling of microstructurally complex battery architectures. J. Electrochem. Soc. **154**(9), A856–A864 (2007)
23. https://www.fei.com/software/amira-avizo/. Accessed 8 Oct 2018
24. https://www.comsol.com/. Accessed 8 Oct 2018
25. M. Doyle, J. Newman, A.S. Gozdz, C.N. Schmutz, J.-M. Tarascon, Comparison of modeling predictions with experimental data from plastic Lithium ion cells. J. Electrochem. Soc. **143**(6), 1890–1903 (1996)

26. D.K. Karthikeyan, G. Sikha, R.E. White, Thermodynamic model development for lithium intercalation electrodes. J. Power Sources **185**(2), 1398–1407 (2008)
27. K. Kanamura, Y. Yamada, K. Annaka, N. Nakata, H. Munakata, Electrochemical evaluation of active materials for Lithium ion batteries by one (single) particle measurement. Electrochemistry **84**(10), 759–765 (2016)
28. M. Koyama, B. Javed, G. Inoue, Chapter 14, section 1, Chikuden Debaisu Niokeru Keisan Gijutsu No Katsuyo (Utilization of simulation technologies in storage devices), in *Ritiumu Ion Denchi Niokeru Kouyouryouka, Koudenatsuka Gijutsu To Anzen Taisaku* (Technologies for high capacity and high voltage and safety management in lithium-ion battery). (Technical Information Institute Co. Ltd., 2018 in press). (in Japanese) Published in Oct 2018. ISBN 978-4-86104-728-2
29. B. Javed, M. Koyama, G. Inoue, Three-dimensional model based simulation of the solid state batteries, ECS meeting abstracts, MA2018-02, 530, 2018
30. B. Javed, M. Koyama, Multi-physics Simulation of Solid-State Batteries with Active Material Coating, submitted

Operating Voltage of Li-Ion Batteries on the Basis of Phase Diagram and Thermodynamics

13

Dajian Li, Weibin Zhang, and Song-Mao Liang

Contents

Author Contribution: Dajian Li formulated structure of the manuscript and wrote Li-Sn, Li-Si-Sn, and combined battery part. Weibin Zhan and Song-Mao Liang wrote Li-Mn-O part and Li-Si part, respectively. The authors worked together for writing the introduction part as well as reviewing the total manuscript.

D. Li (✉)
Institute for Applied Materials – Applied Materials Physics (IAM-AWP), Karlsruhe Institute of Technology (KIT), Eggenstein-Leopoldshafen, Germany
e-mail: dajian.li@kit.edu; lidajian@gmail.com

W. Zhang
Key Laboratory for Liquid-Solid Structural Evolution and Processing of Materials (Ministry of Education), Shandong University, Jinan, People's Republic of China

S.-M. Liang
University of Wisconsin Madison, Madison, WI, USA
e-mail: sliang73@wisc.edu; songmao.liang@gmail.com

Q. Zhen et al. (eds.), *Nanostructured Materials for Next-Generation Energy Storage and Conversion*, https://doi.org/10.1007/978-3-662-58675-4_13

Abstract
The behavior of Li-ion batteries is explained on the basis of phase diagram and thermodynamics. Calculation of phase diagram (Calphad) technique is introduced and applied for obtaining reliable thermodynamic models of corresponding material systems. Si and Sn as well as mixed conductor Si-Sn alloys are selected as example anode materials. $LiMn_2O_4$ spinel cathode is used to explain behavior of the cathode side. Finally, operating voltages of an imaginary battery series composed of Si-Sn mixed conductor anodes and $LiMn_2O_4$ spinel cathodes are shown using calculated results obtained with established thermodynamic models.

13.1 Introduction

Lithium-ion batteries (LIBs) are prerequisites for designing advanced electrode materials. The application of a battery is actually conversing chemical energy into electricity to provide energy. Complete mass transfer for LIBs during battery application includes electrons transfer via out circuit and Li^+ transport inside the battery cell. As a result, Li atoms move from anode side to cathode side, and the cell provides electricity to the environment during discharge process. The driving force for such a process is actually different Li chemical potentials (μLi) at anode and cathode sides. Chemical potential of Li mainly depends on electrode composition, temperature, and crystal structure, in short, the state of electrodes. This critical information for LIBs can be describe using phase diagrams, which are also called as state diagrams. Therefore, phase diagrams, that often provide quantitative relationship of the composition and temperature, are useful thinking tools as suggested in [1] to help understanding the fundamental electrochemical properties of electrodes. The phase diagrams can be under equilibrium condition or metastable state with some phases being suspended. With given composition and other conditions in phase diagram, operating voltage as well as other electrochemical behaviors of Li-ion batteries can be predicated, especially using thermodynamic descriptions constructed using Calphad approach.

The Calphad approach, shorted from Calculation of phase diagrams, together with computational thermodynamics has reached majority providing a powerful tool for new materials exploration. This approach includes two sections. First is development of self-consistent thermodynamic databases. The second part is calculation of different types of phase diagrams and property diagrams for a wide field of applications. Originally, Calphad only refers to the first part and the computational thermodynamics for the second, but now researchers gradually use Calphad as an integrated approach for both database development and following calculations for applications.

Relating to the application of Li-ion batteries, the chemical potential phase diagrams provide more information on the electrochemical properties of the system. The chemical potential of Li from the Calphad calculation is related to the electromotive force (emf), E, by the Nernst equation:

$$\mu_{Li} - \mu_{Li}^{o,ref} = -nFE \tag{13.1}$$

where F is Faraday's constant and n the number of electrons, $n = 1$ for Li. This voltage is also denoted as equilibrium open-circuit voltage (OCV).

Successful applications of Calphad approach relay on reliability of multicomponent databases, for which subsystems must be previously assessed. The core of a thermodynamic database for Calphad approach is a set of thermodynamic description of Gibbs energy of all phases, which include specific models and values of adjustable parameters. In order to develop a reliable thermodynamic description of an alloy system, all available experimental and theoretical data have to be collected and evaluated thoroughly. Then, one must select proper models for all the phases in the system. At last, parameters of corresponding Gibbs energy functions are optimized based on all weighted experimental and theoretical data using optimization software packages, often through least square method. Currently, commercial software packages are Thermo-calc, Pandat, FactSage, and MatData, and two open-source software packages are OpenCalphad and Pycalphad.

It must be stressed a reliable thermodynamic description reproduces not only correlating phase boundaries but also the correct Gibbs energy values under different conditions. Precise reproducibility of phase diagram data, e.g., phase boundaries, is often considered as the most important judging factor for the qualities of a database. But as described previously, the Calphad approach detect the minimum overall Gibbs energy by distributing the components into different phases, whose Gibbs energy has been given as functions of temperature, compositions, etc. Therefore, incorrect Gibbs energy functions for individual phases may also lead to similar phase relations to the reality when specific relation is fulfilled. However, this kind of unrealistic description cannot provide useful thermochemical information such as chemical potential from which OCV is calculated. A schematic example is given in Fig. 13.1, where blue and red curves show the "correct" and "wrong" Gibbs energy curves of α and β phases, respectively. The two vertical lines in Fig. 13.1 show the phase boundaries of α and β phases in equilibrium with each other, which are points of contact of corresponding common tangents with Gibbs energy curves. Clearly, the two sets of Gibbs energy descriptions lead to the same phase boundaries on the phase diagram. However, the corresponding chemical potential of element a and b is far away from each other, as shown in Fig. 13.1 as the end points of the common tangents at pure a or b compositions. According to Nernst equation (13.1), they will make different predications for the OCV when a or b is the charge carrier.

A multicomponent database is constructed by combining fully accessed self-consistent thermodynamic descriptions of all binary, ternary, and quaternary subsystems. Systems with more than four elements are essentially based on certain extrapolation schemes. With well-developed thermodynamic database, a series of the phase diagrams and property diagrams are able to be calculated. In addition to the typical composition and temperature relationship phase diagram, the chemical potential phase diagrams and property diagrams, such as OCV and Gibbs energy of formation, are of great interests for the Li-ion battery application. A hypothetical Li-ion battery with Si-Sn mixed conductor anode and Li-Mn-O spinel cathode will be given as an example in the following text.

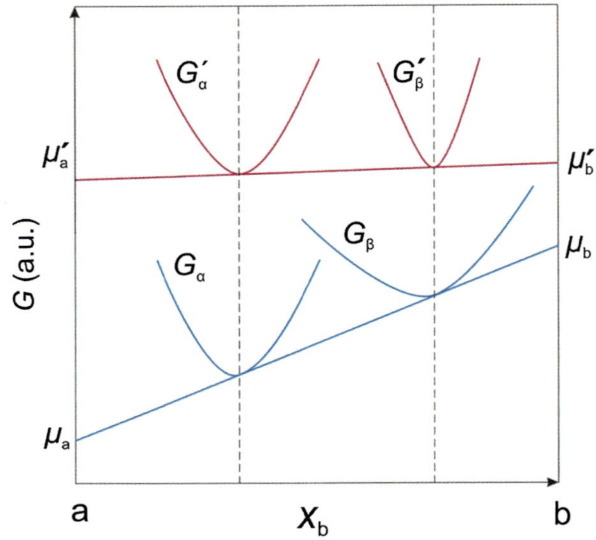

Fig. 13.1 Schematic example for different Gibbs energy description sets representing the same phase boundaries

13.2 Alloying Anode

13.2.1 Li-Si System

Figure 13.2 shows calculated phase diagram of Li-Si system with recently developed thermodynamic description [2]. The equilibrium-state phase diagram is given as the solid line. Five Li-Si intermediate compounds are stable at ambient temperature, and one compound, $Li_{4.11}Si$ phase, is stable between 500 °C and 620 °C. The metastable liquidus line of $Li_{15}Si_4$ and the metastable extension of the liquidus line of (Si) are superimposed pink dash-dot lines. The metastable congruent melting point of $Li_{15}Si_4$ is at 649 °C.

Figure 13.3 represents the Calphad calculated chemical potential phase diagram [2] in good agreement with the emf μLi-data of the solid-state two-phase equilibria. A specific example of the OCV curves at 415 °C is shown in Fig. 13.4a by comparing the Calphad calculation [2] with the experimental data [3].

Compared to the equilibrium state at elevated temperature, the Li lithiation and delithiation at room temperature under metastable state is of more interest for exploring the properties of Si as anode in Li-ion batteries, as the charging and discharging process of the battery is often in non-equilibrium state. A method for OCV curves involving the metastable $Li_{15}Si_4$ phase and amorphous Si(Li) phase was developed based on Calphad approach [2]. The Gibbs energy of the Li-Si amorphous alloys is described by the extrapolation of Gibbs energy of the undercooled liquid phase. Neglecting the kinetic factors, the Calphad calculated metastable OCV curves are surprisingly in qualitatively good agreement with the experimental data as shown in Fig. 13.4b.

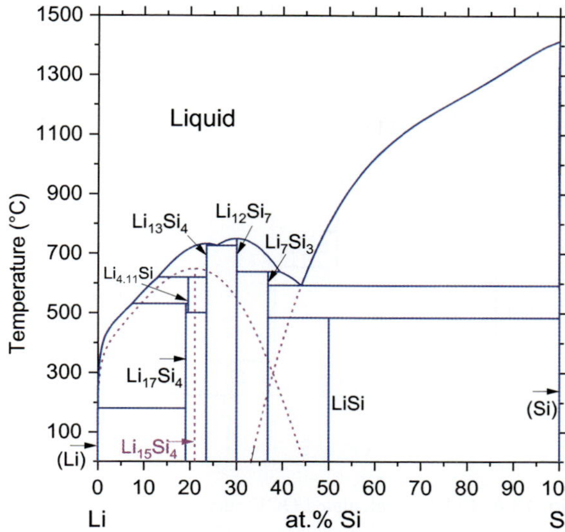

Fig. 13.2 Calphad calculated phase diagram of Li-Si system under equilibrium condition (blue solid lines) and metastable state (pink dashed lines)

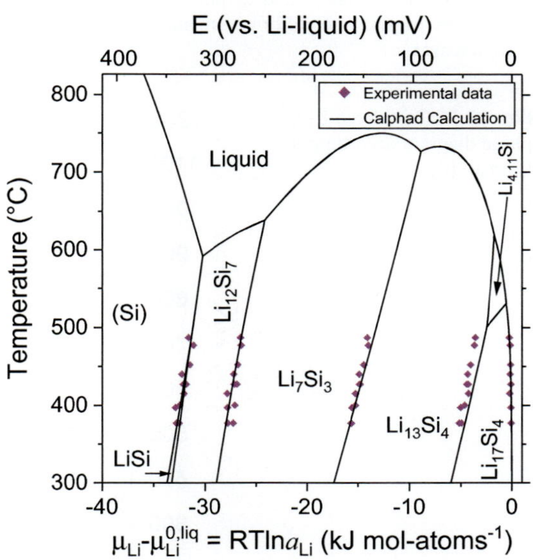

Fig. 13.3 The calculated Li-Si chemical potential phase diagram compared with experimental data. The emf scale (top) displays the voltage in an electrochemical cell between pure liquid Li and Li-Si alloy. The source of the experimental data refers to [2]

13.2.2 Li-Sn System

Figure 13.5 shows calculated phase diagram of Li-Sn system using thermodynamic description obtained by Calphad approach [6]. The Li-Sn phase diagram is characterized by a series of intermetallic compounds and peritectic reactions. Seven Li-Sn intermetallic phases are stable at ambient temperature, in which three are congruent melting compounds and four are formed through peritectic reactions during cooling process.

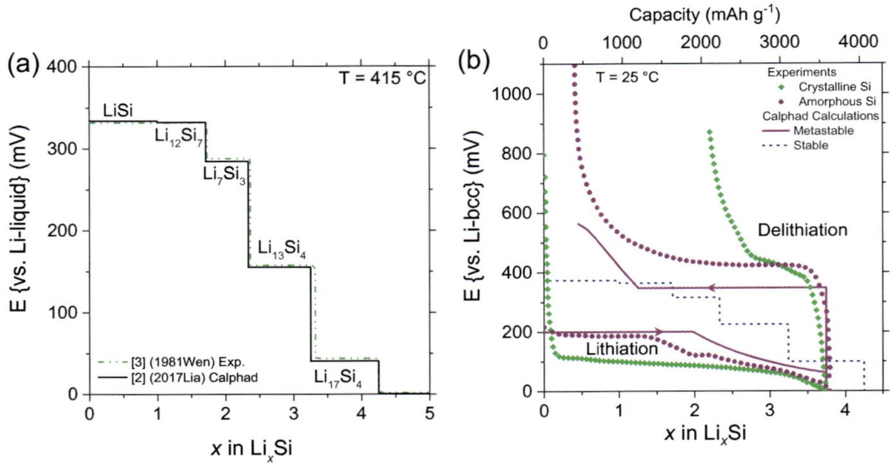

Fig. 13.4 Calphad calculated OCV curves in the Li-Si system compared with experimental data. (**a**) Equilibrium of stable structures at 668 K compared with experimental results [3]. (**b**) Metastable simulation using models of amorphous $\alpha - Li_xSi$ alloy, metastable $Li_{15}Si_4$ phase, and crystalline (Si) during first lithiation and delithiation cycle compared with experimental data [4, 5]; present calculation of stable equilibrium relation at 298 K (dash dot line) is shown for the sake of completeness

Fig. 13.5 Calculated Li-Sn phase diagram according to Li et al. [6]

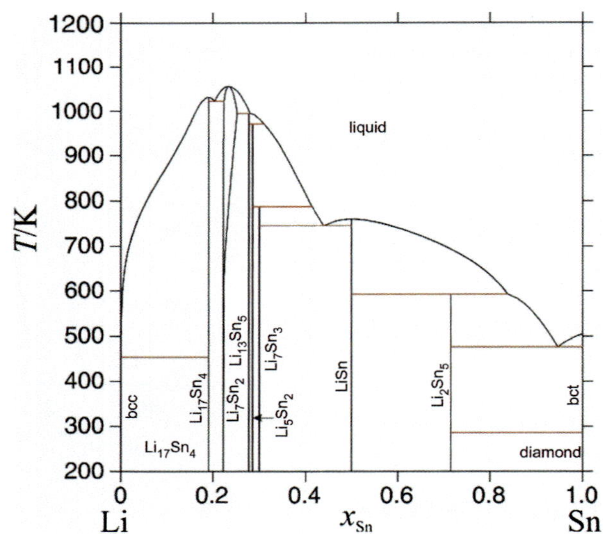

Figure 13.6 shows calculated OCV curves at different temperatures in comparison with experimental data (415 °C [7], 550 °C [9]) and ab initio [8] results at room temperature. A good agreement between the Calphad calculation and the experimental data can be observed. Especially for room temperature, the ab initio

calculation [8] was not used to adjust the Calphad parameters but still shows excellent repeatability between the two groups of data. This is a strong validation for both methods.

Because the reported emf data at various temperatures, the Li-Sn system is a good example to demonstrate importance of well-described thermochemical data to usability of a thermodynamic database. The calculated Gibbs energies of Li-Sn liquid and solid alloys using Calphad model are shown in Figs. 13.7 and 13.8, respectively. Good representability of Gibbs energy covering wide composition and

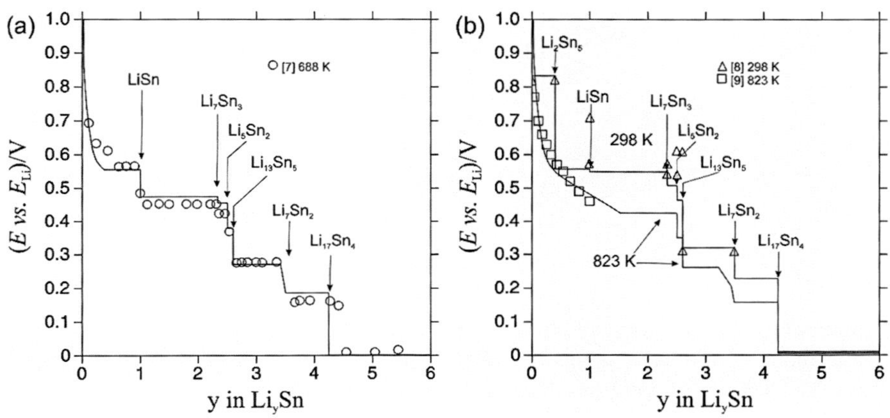

Fig. 13.6 Calculated open circle voltage of Li-Sn alloy against pure Li in comparison of experimental data: (**a**) Wen and Huggins [7] at 688 K; (**b**) Courtney et al. [8] at 298 K and Morachevskii et al. [9] at 823 K

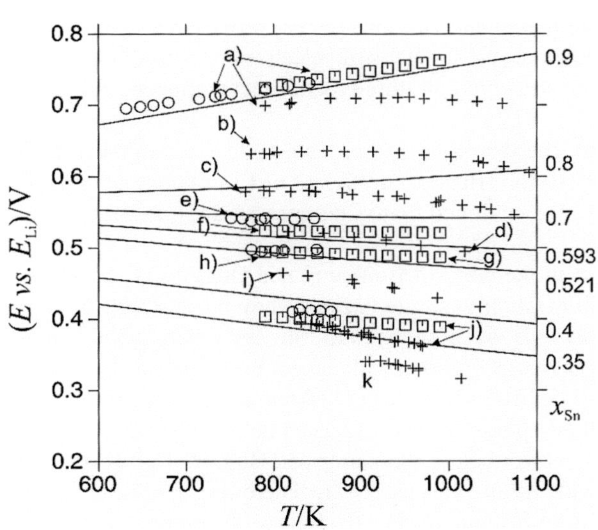

Fig. 13.7 Calculated emf for the liquid phase at various compositions in comparison with experimental data. The alloy compositions used for the compositions are given on the right side. Symbols: circle [7], cross [10], square [11]. Sample compositions are given as x_{Sn}; (a) 0.9

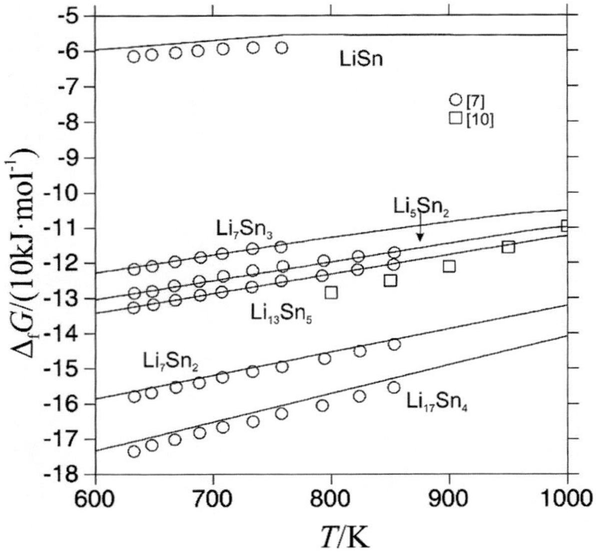

Fig. 13.8 Calculated Gibbs energies of formation for the Li-Sn alloys in comparison with experimental data from Wen and Huggins [7] and Foster et al. [10]. Reference state: pure liquid components. Values are normalized for 1 molar of Sn

temperature range ensures reliability of predicated emf values using the thermodynamic model. Moreover, precision of the binary system descriptions can strongly influence usability of combined thermodynamic models for higher order systems, as the following example of the Li-Si-Sn system.

13.2.3 Si-Sn System as a Mixed Conductor Matrix Electrodes

In order to overcome the problem of volume expansion related to metallic anodes, Boukamp and Huggins [1, 12] proposed "mixed conductor" which is composed by reactant and another solid electronically conducting matrix. In order to obtain a good mixed conductor, three requirements must be fulfilled:

- The matrix compound is a good electronic conductor.
- High diffusion coefficient of Li in both reactant and matrix compound.
- Reactant and matrix compound does not react with each other.

In the work of Boukamp and Huggins [1, 12], Si and Sn mixture was chosen as illustrating example because they meet all the three requirements. The first two is obvious as both Si and Sn are solely anode candidates for Li-ion batteries. Figure 13.9 shows calculated Sn-Si phase diagram using parameters reported in COST 507 database [13], where no Sn-Si compound, even obvious solubility in the liquid phase, can be observed below 500 °C.

Nevertheless, Boukamp and Huggins [1, 12] started with a mixture of Si – $Li_{2.6}Sn$ with Si/Sn ratio as 0.671. Therefore, Li-Si-Sn dataset combined using Li-Si [2], Li-Sn [6], and Si-Sn [13] binary description was applied for necessary calculation.

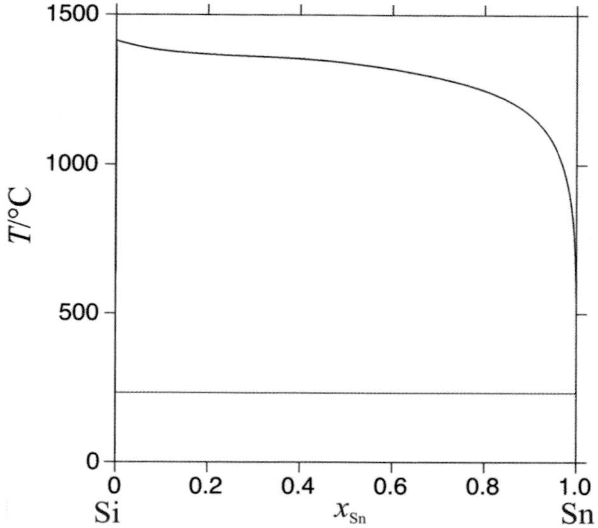

Fig. 13.9 Calculated Si-Sn phase diagram following COST 507 database [13]

Figure 13.10 shows calculated Li-Si-Sn isothermal section at 415 °C as the experimental condition of Boukamp and Huggins [1, 12]. According to Fig. 13.10, their starting composition locates at the edge of Si + LiSi + $Li_{13}Sn_5$ ternary phase field without appearance of liquid.

Figure 13.11a shows calculated emf curve by lithithing a Si-Sn alloy with nSi: nSn = 0.671 at 415 °C. One can see many voltage plateaus corresponding to different ternary phase regions on the Li-Si-Sn isothermal section. The emf profile measured by Boukamp and Huggins [1, 12] is shown in Fig. 13.11b, starting with a two-phase mixture Si + $Li_{13}Sn_5$, ending up with $Li_{13}Sn_5$ + Li_7Sn_2 + Li_7Sn_3. A good agreement between calculation and experimental results can be observed. The entire calculated curve locates between measured charge and discharge profiles. The calculated vertical sections, which correspond to phase boundaries between different three-phase regions, also agree well with the experimental results. Because the calculation and experimental results are independent from each other, the good agreement between the measurement and calculation is a strong evidence that the current combined thermodynamic description can reliably predict corresponding battery behaviors.

Although high temperatures are more suitable for laboratory investigations due to fast kinetic, most battery applications are at room temperature. Figure 13.12 shows calculated Li-Sn-Sn isothermal section at room temperature, which consists of 12 three-phase regions. In each three-phase region, the chemical potential of Li is a constant, corresponding to a fixed emf potential for the battery application. Figure 13.13 shows calculated OCV curves when adding Li into 1 molar Si-Sn mixture at room temperature under equilibrium state. Besides the two initial conditions starting with pure Si or Sn, each OCV curve contains 12 voltage plateaus, corresponding to the 12 three-phase regions shown in Fig. 13.12. Lithiation process of a Si-Sn mixed conductor is given in Table 13.1.

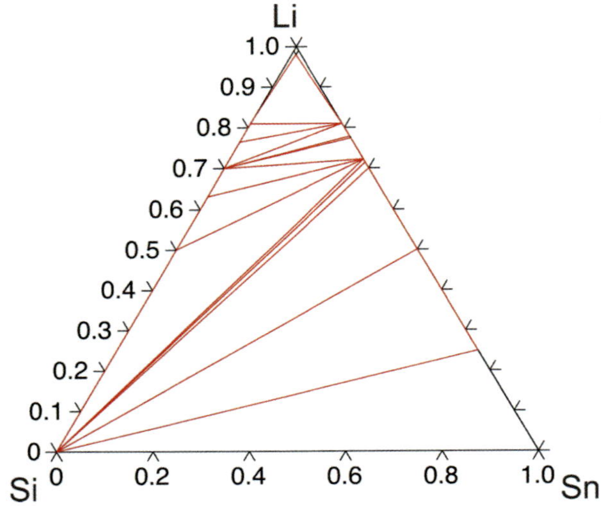

Fig. 13.10 Li-Si-Sn isothermsection at 415 °C

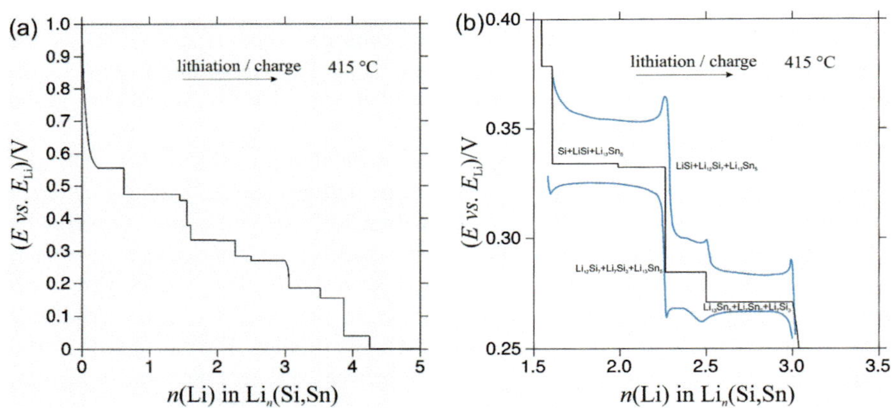

Fig. 13.11 Calculated emf profile for lithiation/delithiation with 1 mol Si-Sn mixed anode alloy, in which nSi:nSn = 0.671 at 415 °C. (**a**) complete emf profile; (**b**) compare this diagram from [12] with calculation

The maximum capability is not changed with initial composition because the Li-richest compounds are $Li_{17}Si_4$ and $Li_{17}Sn_4$ for Si and Sn, respectively. With different initial Si-Sn compositions, the voltage plateaus have different capacities. Naturally, even distribution of capacity to different voltage plateaus, which correspond to different reactions, plus alternative reaction with Si and Sn can avoid accumulation of internal forces in the electrode materials and brake of the active materials. As shown in Fig. 13.13, when Si and Sn amount is close to each other ($x = 0.4$ and 0.6), the overall OCV curve looks smoother than other compositions. If we take into consideration that the molar mass of Si and Sn are 28.09 and

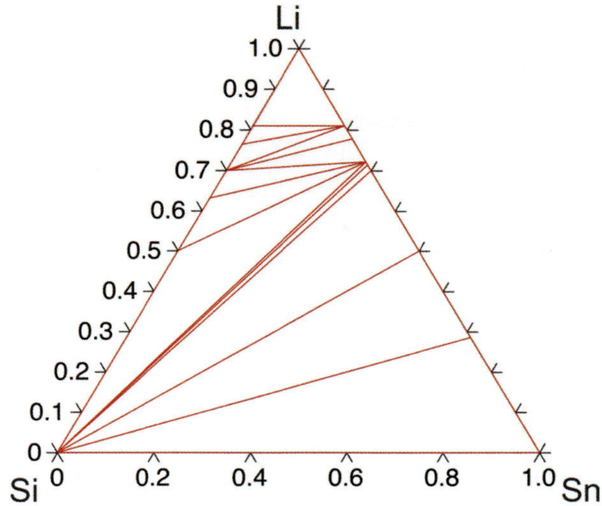

Fig. 13.12 Li-Sn-Sn isothermal section at RT

118.71 g mol^{-1}, respectively, higher Si content is always preferred for its higher mass capacity. Additionally, as the overall battery potential is the difference of electric potential of Li between anode side and cathode side, this is essential for the battery performance design and evaluation.

13.3 A Li-Transition-Metal Oxide Cathode: Li-Mn-O System

Li-Mn-O spinel is a series of attractive cathode materials for rechargeable lithium-ion batteries. However, a wide range of solid solution can be directly sintered within the Li-Mn-O spinel compound composition range. Moreover, the battery-related properties of sintered Li-Mn-O spinels can differ from each other even with the same composition when sintering process has been changed. This makes it difficult to understand its intrinsic properties and evaluate relating battery performance. To overcome this difficulty, a high-throughput computational framework based on the Calphad approach was developed [14] to systematically describe the intrinsic properties of the Li-Mn-O spinel compounds under the "sintered" and "battery" states, indicating the cathode material states just sintered and during battery operations, respectively. Calculated room temperature phase diagrams of the Li-Mn-O system for "sintered state" and "battery state" are shown in Fig. 13.14a, b, respectively.

One critical cathode material property, which is essential for the safety of lithium-ion battery, is stability, which means (1) there is stable temperature/composition range of the electrode and (2) cathode material may release oxygen under battery operation when it reaches certain composition/temperature. Enthalpy of formation represents thermodynamic stability of the electrode materials. Figure 13.15 shows calculated enthalpies of formation per mole atom for spinel based on the "sintered state" within $LiMn_2O_4$ – $Li_4Mn_5O_{12}$ – $Li_2Mn_4O_9$ composition triangle

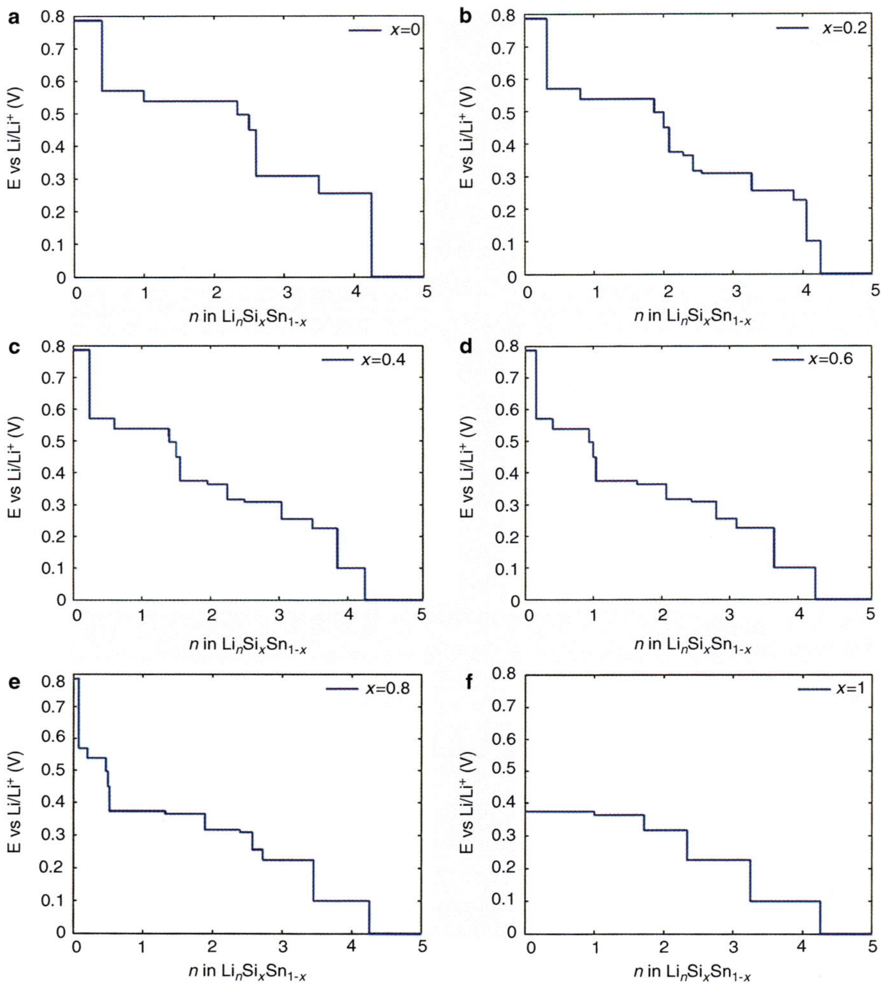

Fig. 13.13 Calculated emf profile for lithiation/delithiation with 1 mol Si_xSn_{1-x} mixed anode alloy at room temperature

and "battery state" within $LiMn_2O_4 - Li_4Mn_5O_{12} - \lambda - MnO_2$ composition triangle. Enthalpy of formation under "sintered state" becomes more negative from $Li_2Mn_4O_9$ compound to $LiMn_2O_4 - Li_4Mn_5O_{12}$ edge. Therefore, $Li_{1+y}Mn_{2-y}O_4$ $(0 \leq y \leq 0.33)$ is more stable than the $Li_2Mn_4O_9$ stoichiometric compound. The calculated enthalpies of formation of the $Li_{1+y}Mn_{2-y}O_4$ $(0 \leq y \leq 0.33)$ spinels increase slightly with decreasing lithium content, indicating a slightly decreased thermodynamic stability. An increment of enthalpies of formation of $Li_{1-w}Mn_2O_4$ $(0 \leq w \leq 1)$ spinels under the "battery state" leads to the decreasing phase stability during the delithiation (charge of battery) process.

Table 13.1 Lithiation process of a Si-Sn mixed conductor; PB stands for phase boundary/two-phase region

Stat/reaction	Phase components	Emf vs. Li/Li$^+$ at RT (V)
Initial state	Si + Sn	
Li + Sn → Li_2Sn_5	Si + Sn + Li_2Sn_5	0.79
PB	Si + Li_2Sn_5	
Li + Li_2Sn_5 → LiSn	Si + LiSn + Li_2Sn_5	0.57
PB	Si + LiSn	
Li + LiSn → Li_7Sn_3	Si + LiSn + Li_7Sn_3	0.54
PB	Si + Li_7Sn_3	
Li + Li_7Sn_3 → Li_5Sn_2	Si + Li_7Sn_3 + Li_5Sn_2	0.50
PB	Si + Li_5Sn_2	
Li + Li_5Sn_2 → $Li_{13}Sn_5$	Si + Li_5Sn_2 + $Li_{13}Sn_5$	0.45
PB	Si + $Li_{13}Sn_5$	
Li + Si → LiSi	Si + LiSi + $Li_{13}Sn_5$	0.37
PB	LiSi + $Li_{13}Sn_5$	
Li + LiSi → $Li_{12}Si_7$	LiSi + $Li_{12}Si_7$ + $Li_{13}Sn_5$	0.36
PB	$Li_{12}Si_7$ + $Li_{13}Sn_5$	
Li + $Li_{12}Si_7$ → Li_7Si_3	$Li_{12}Si_7$ + Li_7Si_3 + $Li_{13}Sn_5$	0.32
PB	Li_7Si_3 + $Li_{13}Sn_5$	
Li + $Li_{13}Sn_5$ → Li_7Sn_2	Li_7Si_3 + $Li_{13}Sn_5$ + Li_7Sn_2	0.31
PB	Li_7Si_3 + Li_7Sn_2	
Li + Li_7Sn_2 → $Li_{17}Sn_4$	Li_7Si_3 + Li_7Sn_2 + $Li_{17}Sn_4$	0.26
PB	Li_7Si_3 + $Li_{17}Sn_4$	
Li + Li_7Si_3 → $Li_{13}Si_4$	Li_7Si_3 + $Li_{13}Si_4$ + $Li_{17}Sn_4$	0.22
PB	$Li_{13}Si_4$ + $Li_{17}Sn_4$	
Li + $Li_{13}Si_4$ → $Li_{17}Si_4$	$Li_{13}Si_4$ + $Li_{17}Si_4$ + $Li_{17}Sn_4$	0.10
PB, maximum Li content	$Li_{17}Si_4$ + $Li_{17}Sn_4$	Drop to 0

Figure 13.16a, b shows calculated voltage-composition profiles of Li/Li_x-Mn_2O_4($0 \leq x \leq 2$) and Li/$Li_xMn_{1.85}O_4$($0 \leq x \leq 2$) at room temperature along with the literature data [15, 16]. $LiMn_2O_4$ as electrode material provides the possibility of both Li extraction and insertion. In the spinel single-phase region ($0 \leq x \leq 1$), Li-ions on the 8a sites undergo intercalation/deintercalation around 4 V. Once the phase transition from the spinel $LiMn_2O_4$ to tetragonal $t - LiMnO_2$ because of the Jahn-Teller distortion of Mn^{3+} ion, the voltage drops suddenly to 2.95 V at $x = 1$ and remains at this value in the two-phase region ($1 \leq x \leq 2$). Compared to $LiMn_2O_4$, $Li_{1.15}Mn_{1.85}O_4$ displays more complicated electrochemical behavior during cycling. In the spinel single-phase region ($0.60 \leq x \leq 1.15$), the voltage plateau is around 4 V. When the composition is $1.15 \leq x \leq 1.375$, the voltage plateau is around 3.1 V in the spinel + Li_2MnO_3 two-phase region. Li_2MnO_3 particles can facilitate Li-ion transport in the cathode material and benefit the cyclability of the Li-ion battery [17]. Further discharge to $1.375 \leq x \leq 2$, a flat 2.95 V plateau can be observed in the spinel+Li_2MnO_3 + $t - LiMnO_2$ three-phase region. In this region, the formation of the t-$LiMnO_2$ via the Jahn-Teller distortion leads to

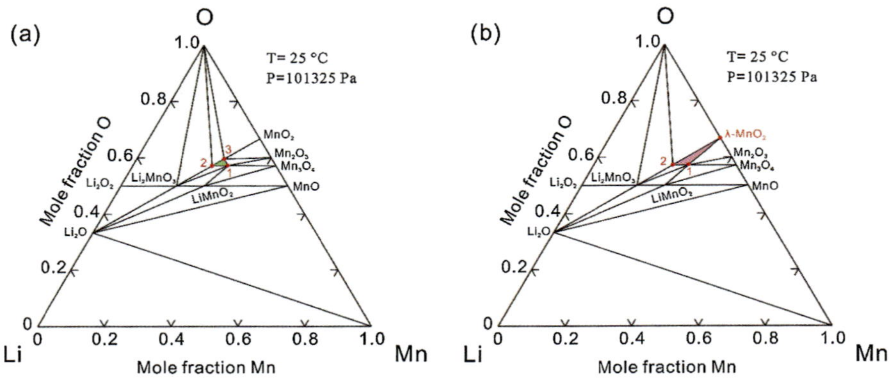

Fig. 13.14 Calculated phase diagrams of the Li-Mn-O system at the room temperature for (**a**) "sintered state" and (**b**) "battery state" (metastable with MnO_2 phase being suspended). The shaded regions of the phase diagram represent the spinel single phase. 1, 2, and 3 represent the stoichiometric spinel compounds $LiMn_2O_4$, $Li_4Mn_5O_{12}$, and $Li_2Mn_4O_9$, respectively

the capacity loss. Therefore, the spinel + Li_2MnO_3 two-phase region can act as a buffer region, which can prohibit the Jahn-Teller distortion during battery operation under non-equilibrium state. Figure 13.16c shows calculated cell voltage relating to composition within the $t - LiMnO_2 - Li_2MnO_3 - \lambda - MnO_2$ triangle. Obviously, different Li activities in different phase regions lead to different voltage plateaus. The spinel single phase is related to the upper 4 V voltage plateau. Around 3.1 V voltage plateau is due to coexistence of the spinel + Li_2MnO_3 phases. The voltage within the three-phase spinel+ $Li_2MnO_3 + t - LiMnO_2$ region is 2.95 V flat plateau.

13.4 Matching of the Anodes and Cathodes, Si/Sn Anodes Against $LiMn_2O_4$ Cathodes

With knowledge of both anode and cathode materials, output voltage of a complete battery can be calculated as the difference between Li chemical potentials at the anode and cathode sides. As discussed previously, a 0.6Si–0.4Sn mixture may provide a good battery performance. Therefore, in the following imaginary batteries, Si to Si ratio is always kept to be 3:2 for the anode side. A schematic drawing of such an electrochemical cell is given in Fig. 13.17 for its initial state. By controlling initial Li content (x value) in anode material, and amount of active electrode materials (n_A.and n_C in Fig. 13.17 for anode and cathode materials, respectively), different cell behaviors can be predicated.

13.4.1 The Same Amount of Charge Capacity

To achieve the maximum cell capacity with minimized amount of active material, clearly anode side should initially contain no Li, i.e., $x = 0$ to have maximum charge capacity. Meanwhile, the charge capacity must match for the anode and cathode side. Because the highest Li contents are $Li_{17}Si_4$ and $Li_{17}Sn_4$ for Si and

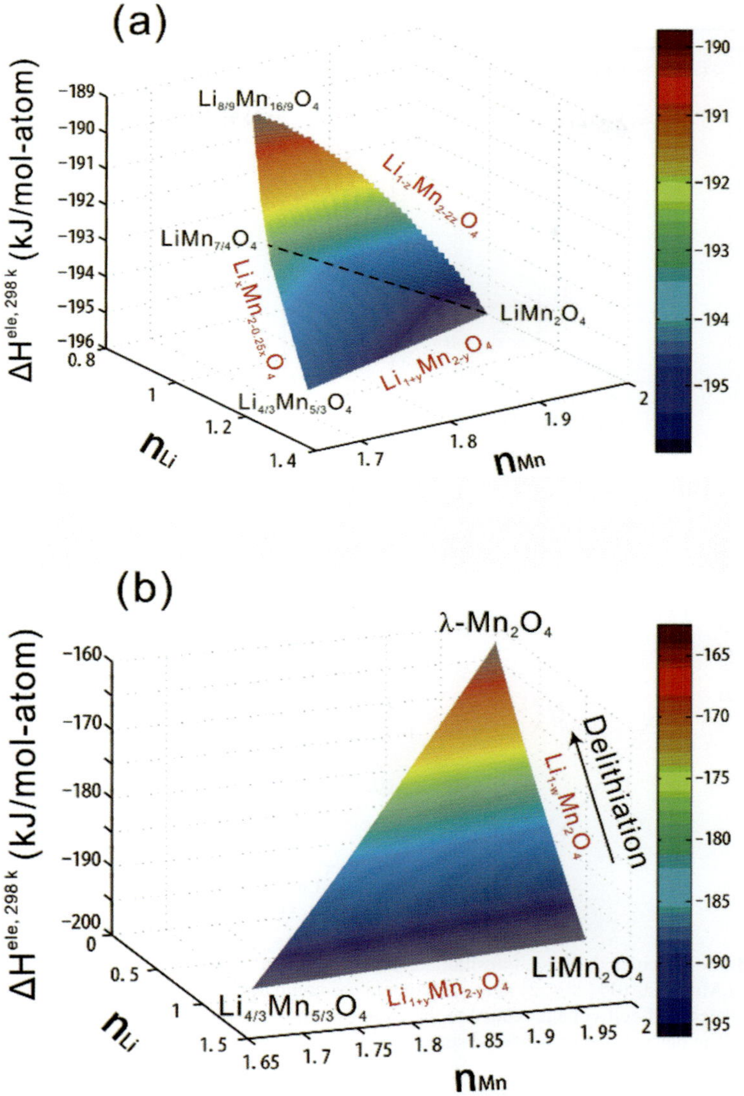

Fig. 13.15 Calculated enthalpies of formation per mole atom for spinel based on (**a**) "sintered state" within $LiMn_2O_4$–$Li_4Mn_5O_{12}$–$Li_2Mn_4O_9$ composition triangle and (**b**) "battery state" within $LiMn_2O_4$–$Li_4Mn_5O_{12}$–λ–MnO_2 composition triangle

Sn, respectively, the maximum Li content will be 4.25 n_A.for anode side. Theoretically, all Li (n_C) can be extracted from $LiMn_2O_4$ spinel material. Therefore, when $4.25n_A = n_C$, all Li can move between anode and cathode materials using 100% charge capacity of both sides.

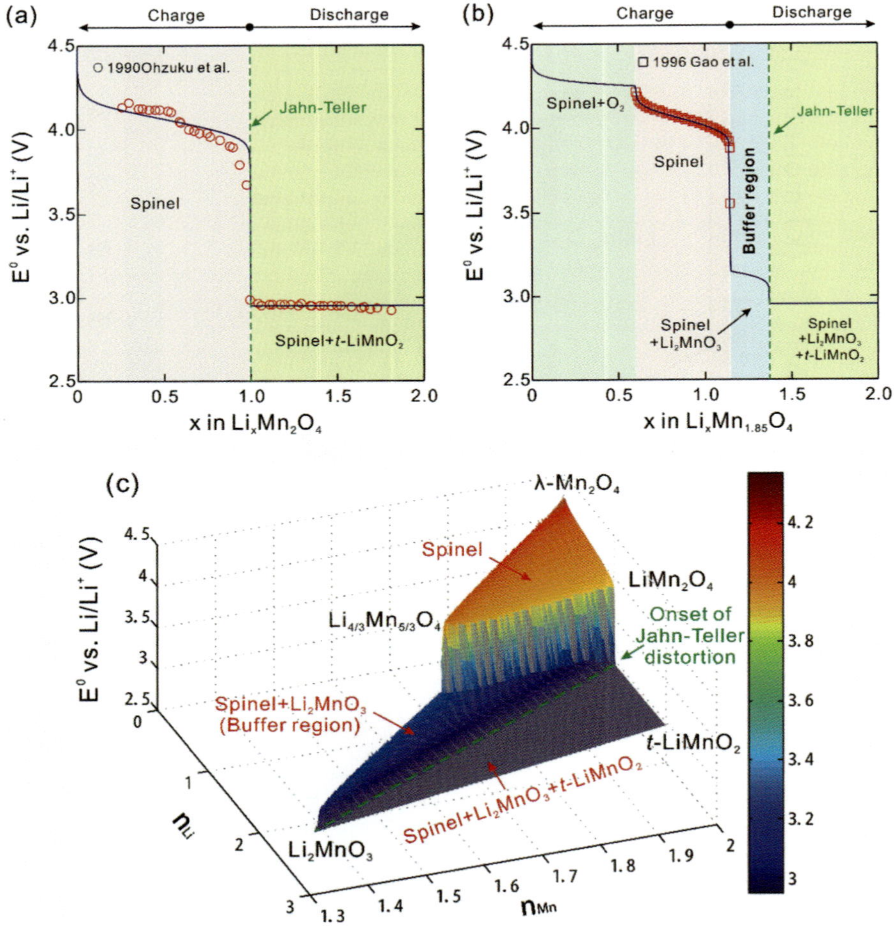

Fig. 13.16 Calculated phase transition and cell voltages of (**a**) Li/$Li_xMn_2O_4$ ($0 \le x \le 2$), (**b**) Li/$Li_xMn_{1.85}O_4$ ($0 \le x \le 2$) at room temperature compared to the literature data [15, 16], and (**c**) calculated half-cell voltage related to the composition within the t-$LiMnO_2$–Li_2MnO_3–λ-MnO_2 triangle. The dashed line presents the starting composition of Jahn-Teller distortion

Calculated emf curves using a cell composed by 1 mol 0.6Si–0.4 Sn anode and 4.25 mol $LiMn_2O_4$ cathode material are shown in Fig. 13.18 for (a) electrodes and (b) complete cell. One can see that the OCV increases with charge and decreases with discharge process. The calculated OCV can be directly used for the battery management system.

13.4.2 Distributed Charge Capacity Between Si and Sn Reaction

In reality, a battery cannot reach its theoretical capacity. Li can be consumed by side reactions with electrolyte or formation of SEI. Electrodes can also degenerate during

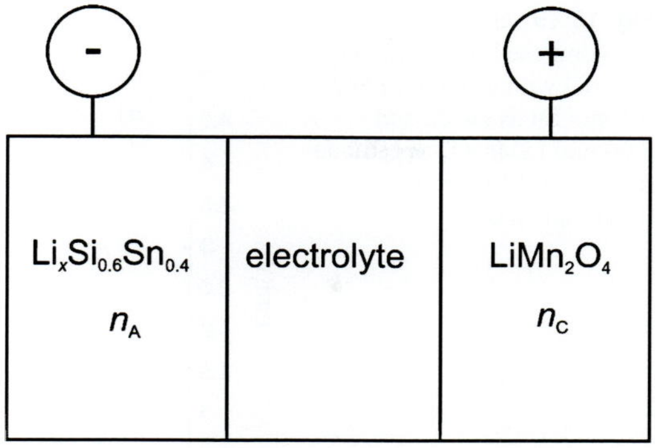

Fig. 13.17 Schematic drawing of electrochemical cell with 0.6 Si–0.4Sn and Li

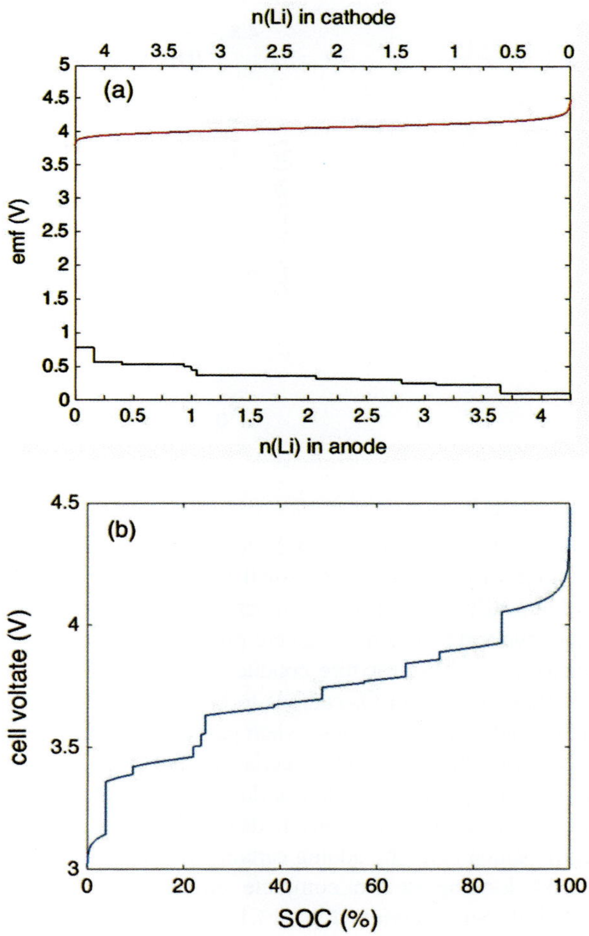

Fig. 13.18 emf curve of a cell composed by 1 mol 0.6Si–0.4 Sn anode and 4.25 mol $LiMn_2O_4$ cathode material. (**a**) Electrodes; (**b**) complete cell

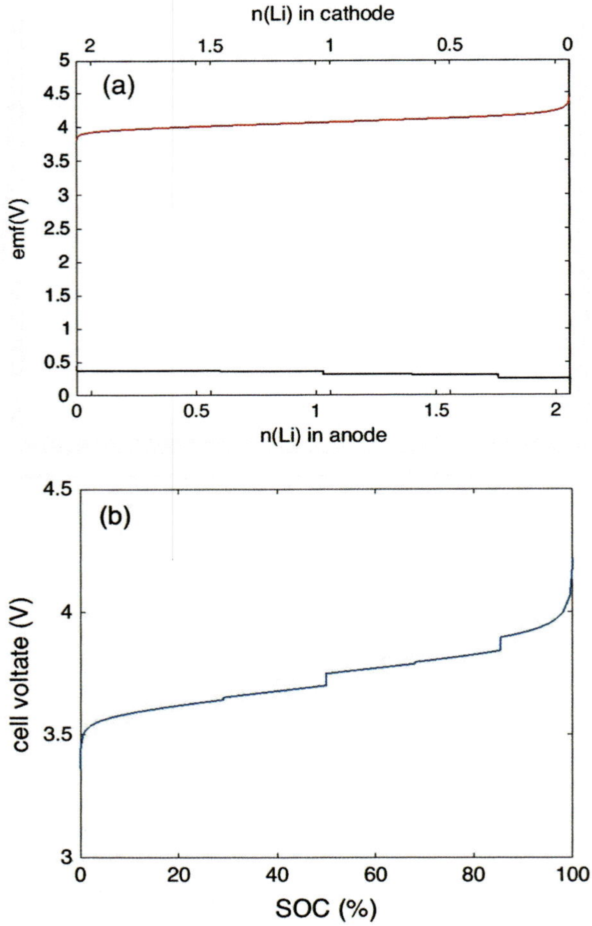

Fig. 13.19 emf curve of a cell composed by Si-$Li_{13}Sn_5$ material with 0.6 mol Si and 0.4 mol Sn as anode and 2.06 mol $LiMn_2O_4$ as cathode material. (a) Electrodes; (b) complete cell

cycling. At the current state of Li-ion battery development, charge capacity is limited by cathode material due to its small theoretical capacity compared to anode materials. Additionally, cathode price is much higher than that of anode materials. From the other side, cyclability is a more severe problem than charge capacity for alloy-type anodes, such as the Si-Sn mixture conductor. Therefore, extra Si-Sn anode material may be matched with $LiMn_2O_4$ cathode to be able to use the full capacity of cathode material and avoid cell capacity loss when part of the anode material is degenerated. Additionally, composition of anode material will not cross all three-phase regions, which means, less voltage plateaus will appear during charge/discharge process. If only part of the total possible compositions at the anode side will be used, it is possible to select the "active" composition range by adding certain amount of Li to the initial anode material.

By looking at the complete emf curve during lithiation/delithiation of the 0.6Si–0.4Sn material (Fig. 13.13), we can find that the first several plateaus corresponding to Li-Sn reactions show relatively big voltage jump at phase boundaries and could be avoided.

Composition range from Si-$Li_{13}Sn_5$ to Li_7Si_3 - $Li_{17}Sn_4$ is selected as an example because of the containing reactions during lithiation/delithiation process. From Table 13.1, one can see this corresponds to an upper voltage limit between 0.37 V and 0.45 V and a lower voltage between 0.26 V and 0.22 V and includes three reaction steps between Li and Si followed by two reaction steps between Li and Sn. For 1 mol selected 0.6Si–0.4Sn material, the limit amount of Li is calculated to be 1.04 mol and 3.1 mol, with difference of 2.06 mol. This means, the starting Si-$Li_{13}Sn_5$ material with 0.6 mol Si and 0.4 mol Sn should match with 2.06 mol $LiMn_2O_4$. The calculated emf for the anode and cathode material with different SOC is given in Fig. 13.19.

Comparing the emf profiles given in Figs. 13.18 and 13.19, one can see the OCV of entire cell is smoother and flatter with partially lithiation/delithiation process of Si-Sn anode material, which is beneficial for battery management system. Certainly, more variations such as Li-rich spinel can be used for simulating cell performance, but they will be skipped from the current text.

13.5 Conclusion

Operating voltage of lithium-ion batteries was explained from the viewpoint of thermodynamics and phase diagrams. Importance of calculated phase diagram obtained by Calphad approach was expressed using a series of imaginary LIBs with Si-Sn mixed conductor anode and Li-Mn-O spinel cathode as examples. The explanation was given in a sequence of Si/Sn unitary alloy anode, Si-Sn mix conductor anodes, Li-Mn-O spinel cathodes, and combination of anode and cathodes. Metastable conditions were also discussed for the Si anode and Li-Mn-O spinel cathodes.

Acknowledgments Dajian Li appreciates financial support from the German Research Foundation DFG project (LI 2839/1-1). Weibin Zhang was supported by Qilu Young Scholar Program in Shandong University. Song-Mao Liang acknowledges DFG Priority Programme "WendeLiB SPP 1473" under grant no. Schm 588/37.

References

1. R.A. Huggins, *Advanced Batteries Materials Science Aspects* (Springer, New York, 2009)
2. S.-M. Liang et al., Thermodynamics of Li-Si and Li-Si-H phase diagrams applied to hydrogen absorption and Li-ion batteries. Intermetallics **81**, 32–46 (2017)
3. C.J. Wen, R.A. Huggins, Chemical diffusion in intermediate phases in the lithium-silicon system. J. Solid State Chem. **37**(3), 271–278 (1981)
4. M.N. Obrovac, L. Christensen, Structural changes in silicon anodes during lithium insertion/extraction. Electrochem. Solid-State Lett. **7**(5), A93–A96 (2004)
5. T.D. Hatchard, J.R. Dahn, In situ XRD and electrochemical study of the reaction of lithium with amorphous silicon. J. Electrochem. Soc. **151**(6), A838–A842 (2004)

6. D. Li et al., Thermodynamic assessment and experimental investigation of the Li-Sn system. Calphad: Comput. Coupling Phase Diagrams Thermochem. **47**, 181–195 (2014)
7. C.J. Wen, R.A. Huggins, Thermodynamic study of the lithium-tin system. J. Electrochem. Soc. **128**(6), 1181–1187 (1981)
8. I.A. Courtney et al., Ab initio calculation of the lithium-tin voltage profile. Phys. Rev. B **58**(23), 15583–15588 (1998)
9. A.G. Morachevskii et al., Thermodynamic properties of molten Lithium-Tin alloys. Sov. Electrochem., **8**, 1578–1580 (1972)
10. M.S. Foster, C.E. Crouthamel, S.E. Wood, Thermodynamics of binary alloys. II. The lithium-tin system1. J. Phys. Chem. **70**(10), 3042–3045 (1966)
11. W. Gasior, Z. Moser, Thermodynamic properties of Li-Sn (lithium-tin) liquid solutions. Arch. Metall. Mater. **44**(1), 83–92 (1999)
12. B.A. Boukamp, G.C. Lesh, R.A. Huggins, All-solid lithium electrodes with mixed-conductor matrix. J. Electrochem. Soc. **128**(4), 725–729 (1981)
13. I. Ansara, E. Gemeinschaften, *Definition of Thermochemical and Thermophysical Properties to Provide a Database for the Development of New Light Alloys: COST 507. Vol 2. Thermochemical Database for Light Metal Alloys* (Office for Official Publications of the European Communities, Luxembourg, 1998)
14. W. Zhang et al., High-throughput description of infinite composition-structure-property-performance relationships of lithium-manganese oxide spinel cathodes. Chem. Mater. **30**(7), 2287–2298 (2018)
15. T. Ohzuku, M. Kitagawa, T. Hirai, Electrochemistry of manganese-dioxide in lithium nonaqueous cell. III. X-ray diffractional study on the reduction of spinel-related manganese dioxide. J. Electrochem. Soc. **137**(3), 769–775 (1990)
16. Y. Gao, J.N. Reimers, J.R. Dahn, Changes in the voltage profile of Li/Li1+xMn2-xO4 cells as a function of x. Phys. Rev. B **54**(6), 3878–3883 (1996)
17. M.M. Thackeray et al., Li2MnO_3-stabilized $LiMO_2$ (M = Mn, Ni, Co) electrodes for lithium-ion batteries. J. Mater. Chem. **17**(30), 3112–3125 (2007)

Lithium-Ion Batteries (LIB): A Postface

Qiang Zhen

qzhen@staff.shu.edu.cn
Research Center of Nano Science and Technology, Shanghai University Shanghai, China.

This postface is dedicated to the Professor Peter J Derrick

The accelerated usage of consumer electronics and the realization of all-electric or hybrid vehicles and smart grid have enabled great strides to be made in lithium-ion battery chemistry to meet anticipated demand for portable power.

There is a need to develop Li-ion technology which is resource-efficient and economically feasible through point-of-source recycling to preserve the future supply of LIB component materials. Unfortunately, there is no simple way to recycle, and therefore great efforts should be made in battery redesign, battery monitoring and development of novel material separation technology on several scales, and modular design for recycling to meet the demand for future consumer electrics and automobiles.

Further advances in electrolyte chemistry are required to eliminate dendrite formation and increase the depth of discharge (DoD) and state of charge (SoC) profiles. Electrolyte additives have increased electrochemical performance by a variety of distinct methods, such as developing new electrode materials and new morphology and framework of the electrode materials, modifying the surface area of materials and exploring a new type of electrolytes. More improvement is required in four general areas:

Electrode materials: these are prone to electrode dendrite formation and dissolution in aqueous electrolyte. Incorporation of alloying intrinsic and composite defects may alleviate these drawbacks and extend operational lifespan of LIBs through surface modification conditioning.

Undesired reactions: the side reactions between the electrode material and the electrolyte create imperfections, dendrite formation, and lower SoC and operational lifetime. Careful dimensional crafting of the electrode, using binary or

Q. Zhen et al. (eds.), *Nanostructured Materials for Next-Generation Energy Storage and Conversion*, https://doi.org/10.1007/978-3-662-58675-4

ternary spinel-like (or reverse-spinel) geometries, can minimize or eliminate the side reactions.

Limitations of kinetics and thermodynamics: While delta G defines spontaneity, the rate of many reactions is limited by operational temperatures. In addition, understanding in the discharge-charge reaction mechanism of electrode materials is limited. Therefore, electrode chemistry was developed using more combinatorial chemistry than DFT and hypothesis-driven argument. A better understanding in the reaction rate and mechanism will allow us to formulate novel and more synergic materials, for example, tailored metal-organic framework for favorable surface kinetics.

Development of aqueous electrolyte: While as-solid-state electrolyte batteries are emerging, development in the aqueous electrolyte is still an active area of research with respect to cost and safety. Examples include metal dissolution and dendrite formation in the aqueous electrolyte. Electrolysis of trace water generates hydrogen and oxygen, which affect the chemical stability of activity materials and the safety of the cell. Restricted voltage (~1.23 V) of the aqueous battery can also be ascribed to its aqueous electrolyte. Co-intercalation of proton and generation of surface defect sites and transport properties of aqueous electrolyte impact ion diffusivity, conductivities, electrolyte viscosity, battery kinetics, and operational lifespan. The use of solid-state and "water-in-salt" electrolyte and electrolyte additives has improved the electrochemical stability of LIBs, but more work is required.

In addition to consumer supply, lithium-ion batteries are the preferred option for the emerging sustainable solar- and wind-related power supply systems, with great storage potential to improve energy sustainability and substantial reductions in carbon emissions (0.05 kg CO_2/kWh). Within this context, developing countries must develop roadmaps to migrate from a central grid to a decentralized grid system as their infrastructure costs are not as high as first-world countries.

The common electrode materials in LIB are still carbon based, owing to their low operational cost, high abundance, and large working voltage, appropriate for electrochemical characterization of electrodes used in aluminum ion batteries and dual-ion batteries, respectively. These batteries exhibit low Columbic efficiencies (CE) due to instability of the intercalation voltage of electrolyte from cathode, resulting in the generation of an unstable solid electrolyte interphase (SEI) formed at the anode when using high specific capacity and low reaction potential alloying type anode (Sn). The CE can be increased by flowing LIB solutions through electrode (dimensional engineering to nanoscale) and electrolyte (supplementation with additives (stable SEI)) modification. However, engineering an inert, stable graphitic cathode with fast kinetics in the suitable aqueous electrolyte is non-trivial as outlined earlier. Current specific capacities of cathode materials are lower than 124 mA hg^{-1}, insufficient to meet anticipated demand.

Finally, the electrolyte dependence nature of dual-ion batteries suggests that electrolyte with higher salt concentration is required. An optimal ratio of anode and cathode materials has not been identified using ab initio calculations and the state-of-the-art density functional theory. Molecular dynamics (MD) is able to predict or totally account for anion intercalation behavior of graphite or its interfacial effect between the electrolyte and graphite. Anion intercalation system is electrochemically probed to investigate material bonding charge transfer, electrode polarization, and bulk many-body effects. These intrinsic properties determined theoretically are then compared and contrasted with equivalent *in situ* experimental data to determine electrocatalyticity, electrode robustness, and reaction kinetics of the cell.

Incorporating multi-valent anion intercalation graphite cathode and introducing other layered materials (e.g., molybdenum carbide) other than graphite promote anion intercalation and positive charge-accepting cathode materials such as tetrathiafulvalene as novel cathode materials with higher specific capacity and lower reaction voltage, higher stability, and longer operational life than lithium iron phosphate and lithium cobalt oxide batteries. The future strategies were outlined in the earlier chapters.

We would also like to extend our deep appreciation to our institution, colleagues, and students as well as the editorial staff at Springer who have strived to deliver the best possible scholarly product. As always, errors and omission are the responsibility of the editors, for which we as preface authors ask forgiveness. We would also like to acknowledge the Robert Welch Foundation (AC-0006) Award for assisting part of the research.

Index

Q. Zhen et al. (eds.), *Nanostructured Materials for Next-Generation Energy Storage and Conversion*, https://doi.org/10.1007/978-3-662-58675-4

M

N

O

P

R

S

9783662586730

Printed in Japan

J0037290003